T0183937

Lecture Notes in Computer Science 9214

Commenced Publication in 1973
Founding and Former Series Editors:
Gerhard Goos, Juris Hartmanis, and Jan van Leeuwen

More information about this series at http://www.springer.com/series/7407

Frank Dehne · Jörg-Rüdiger Sack
Ulrike Stege (Eds.)

Algorithms and Data Structures

14th International Symposium, WADS 2015
Victoria, BC, Canada, August 5–7, 2015
Proceedings

Springer

Editors
Frank Dehne
Carleton University
Ottawa
Canada

Jörg-Rüdiger Sack
Carleton University
Ottawa
Canada

Ulrike Stege
University of Victoria
Victoria
Canada

ISSN 0302-9743   ISSN 1611-3349 (electronic)
Lecture Notes in Computer Science
ISBN 978-3-319-21839-7   ISBN 978-3-319-21840-3   (eBook)
DOI 10.1007/978-3-319-21840-3

Library of Congress Control Number: 2015944438

LNCS Sublibrary: SL1 – Theoretical Computer Science and General Issues

Springer Cham Heidelberg New York Dordrecht London

Springer International Publishing AG Switzerland is part of Springer Science+Business Media
(www.springer.com)

Preface

This volume contains the papers presented at WADS 2015—Algorithms and Data Structures Symposium—which was held during August 4–6, 2015, in Victoria, BC. WADS alternates with the Scandinavian Workshop on Algorithms Theory (SWAT), continuing the tradition of SWAT and WADS starting with SWAT 1988 and WADS 1989.

In response to the call for papers, 148 papers were submitted. From these submissions, the Program Committee selected 51 papers for presentation at WADS 2015. In addition, invited lectures were given by the following distinguished researchers: Bernard Chazelle (Princeton), Cyrus Shahabi (USC), and Bodo Manthey (University of Twente).

On behalf of the Program Committee, we would like to express our appreciation to the invited speakers, reviewers, and all authors who submitted papers. We would also like to thank the WADS 2015 sponsors: SAP Inc., Semaphore Solutions Inc., Barrodale Computing Services Ltd., Semaphore Solutions Inc., the Pacific Institute for the Mathematical Sciences, and the University of Victoria.

June 2015

Frank Dehne
Jörg-Rüdiger Sack
Ulrike Stege

Organization

Conference Chair and Local Arrangements Chair

Ulrike Stege   University of Victoria, Canada

Program Committee Co-chairs

Frank Dehne   Carleton University, Canada
Jörg-Rüdiger Sack   Carleton University, Canada
Ulrike Stege   University of Victoria, Canada

Committee Members

Faisal Abu-Khzam   Lebanese American University, Lebanon
Evripidis Bampis   LIP 6, Paris, France
Danny Chen   Notre Dame University, USA
Jianer Chen   Texas A&M University, USA
Alfredo Cuzzocrea   University of Calabria, Italy
Mark de Berg   University of Technology Eindhoven,
     The Netherlands
Faith Ellen   University of Toronto, Canada
Will Evans   University of British Columbia, Canada
Rudolf Fleischer   GUtech, Oman
Stefan Funke   University of Stuttgart, Germany
Michael Goodrich   UC Irvine, USA
Valentine Kabanets   Simon Fraser University, Canada
Rolf Klein   University of Bonn, Germany
Antonina Kolokova   Memorial University, Canada
Mike Langston   University of Tennessee, USA
Ulrich Meyer   University of Paderborn, Germany
Matthias   Martin Luther University Halle-Wittenberg, Germany
     Mueller-Hannemann
Gonzalo Navarro   University of Chile, Chile
Rolf Niedermeier   TU Berlin, Germany
Kirk Pruhs   University of Pittsburgh, USA
Roberto Solis-Oba   Western University, Canada
Venkatesh Srinivasan   University of Victoria, USA
Sergei Vassilvitski   Stanford University, USA
László A. Végh   London School of Economics, UK

Carola Wenk   Tulane University, USA
Peter Widmayer   ETH Zuerich, Switzerland
Gerhard Woeginger   TU Eindhoven, The Netherlands
Ke Yi   Hong Kong University of Science and Technology,
                           Hong Kong
Norbert Zeh   Dalhousie University, Canada

Proceedings Editors

Frank Dehne   Carleton University, Canada
Jörg-Rüdiger Sack   Carleton University, Canada
Ulrike Stege   University of Victoria, Canada

Contents

Contact Graphs of Circular Arcs

Md. Jawaherul Alam[1]([✉]), David Eppstein[2], Michael Kaufmann[3],
Stephen G. Kobourov[1], Sergey Pupyrev[1,4], André Schulz[5],
and Torsten Ueckerdt[6]

[1] Department of Computer Science, University of Arizona, Tucson, USA
mjalam@email.arizona.edu
[2] Computer Science Department, University of California, Irvine, USA
[3] Wilhelm-Schickard-Institut Für Informatik, Universität Tübingen,
Tübingen, Germany
[4] Institute of Mathematics and Computer Science,
Ural Federal University, Yekaterinburg, Russia
[5] Institut Math. Logik und Grundlagenforschung,
Universität Münster, Münster, Germany
[6] Department of Mathematics, Karlsruhe Institute of Technology,
Karlsruhe, Germany

Abstract. We study representations of graphs by *contacts of circular arcs*,
CCA-representations for short, where the vertices are interior-disjoint circular arcs in the plane and each edge is realized by an endpoint of one arc
touching the interior of another. A graph is $(2, k)$-*sparse* if every s-vertex
subgraph has at most $2s - k$ edges, and $(2, k)$-*tight* if in addition it has
exactly $2n - k$ edges, where n is the number of vertices. Every graph with a
CCA-representation is planar and $(2, 0)$-sparse, and it follows from known
results that for $k \geq 3$ every $(2, k)$-sparse graph has a CCA-representation.
Hence the question of CCA-representability is open for $(2, k)$-sparse graphs
with $0 \leq k \leq 2$. We partially answer this question by computing CCA-
representations for several subclasses of planar $(2, 0)$-sparse graphs. Next,
we study CCA-representations in which each arc has an empty convex hull.
We show that every plane graph of maximum degree 4 has such a representation, but that finding such a representation for a plane $(2, 0)$-tight graph
with maximum degree 5 is NP-complete. Finally, we describe a simple algorithm for representing plane $(2, 0)$-sparse graphs with *wedges*, where each
vertex is represented with a sequence of two circular arcs (straight-line
segments).

1 Introduction

In a *contact representation* of a planar graph, the vertices are represented by
non-overlapping geometric objects such as circles, polygons, or line segments and
the edges are realized by a prespecified type of contact between these objects.
Contact graphs of circles, made famous by the Koebe–Andreev–Thurston circle
packing theorem [16], have a large number of applications in graph drawing (see
e.g. [2] for many references) and this success has motivated the study of many

© Springer International Publishing Switzerland 2015
F. Dehne et al. (Eds.): WADS 2015, LNCS 9214, pp. 1–13, 2015.
DOI: 10.1007/978-3-319-21840-3_1

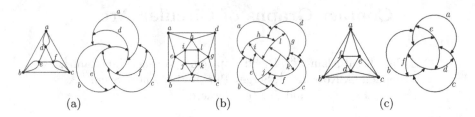

Fig. 1. CCA representations of a multigraph (a) and two simple graphs (b)–(c)

other contact representations [9,12]. The special cases of contact representations with curves and line segments are of particular interest [5,8,15]. We consider a novel type of contact representation where a vertex is represented by a circular arc, and an edge corresponds to an endpoint of one arc touching an interior point of another (Fig. 1). Tangencies between interior points of arcs do not count as contacts as this would trivialize the problem. These representations are a generalization of the contacts of straight-line segments (which are circular arcs with infinite radius) and we call them *contacts of circular arcs* or *CCA-representations* for short.

Every k-vertex induced subgraph of a contact graph of curves in the plane has at most $2k$ edges, because every edge uses up one of the curve endpoints. This motivates us to study classes of sparse graphs defined by limits on the numbers of edges in their subgraphs. A graph $G = (V, E)$ is said to be (p, k)-*sparse* [17] if for every $W \subseteq V$ we have $|E[W]| \leq \max\{p|W| - k, |W| - 1\}$; it is (p, k)-*tight* if in addition $|E| = p|V| - k$. For example, $(1, 1)$-sparse graphs are exactly forests, while $(1, 1)$-tight graphs are trees, and the observation above can be rephrased as stating that all graphs representable by circular arcs are $(2, 0)$-sparse. This definition makes sense only for $k < 2p$: for larger k, each two vertices would induce no edges and the graph would be empty. However, we may extend the definition by restricting $|W|$ to be larger than two. Thus, we define a graph to be $(2, 4)$-sparse if every s-vertex subgraph with $s \geq 3$ has at most $2s - 4$ edges, and $(2, 4)$-tight if in addition it has exactly $2|V| - 4$ edges. A planar graph is $(2, 4)$-sparse if and only if it is triangle-free and $(2, 4)$-tight if and only if it is a maximal bipartite planar graph. The same idea can be extended to larger k by restricting $|W|$ to be even greater, but in the remainder we consider only planar $(2, k)$-sparse and planar $(2, k)$-tight graphs for $k \in \{0, 1, 2, 3, 4\}$.

A graph admits a curve contact representation if and only if it is planar (because the curves do not cross) and $(2, 0)$-sparse (each subset of s curves has at most $2s$ contacts) [15]. On the other hand, a planar graph has a contact representation with line segments if and only if it is $(2, 3)$-sparse [1]. Hence, natural questions arise: What are the simplest curves that can represent all planar $(2, 0)$-sparse graphs? Perhaps the simplest non-straight curves are circular arcs, so how powerful are circular arcs in terms of contact representations? In particular, does every planar $(2, k)$-sparse graph have a CCA-representation for $k \in \{0, 1, 2\}$? We partially answer these questions by computing circular-arc contact representation for several subclasses of planar $(2, 0)$-sparse graphs, and by finding a $(2,0)$-sparse plane multigraph that does not have such a representation.

As another contribution, we resolve an open problem by de Fraysseix and de Mendez [8]. They proved that any contact representation with curves is homeomorphic to one with polylines composed of three segments, and asked if two segments per polyline is sufficient. We affirmatively answer the question.

Preliminaries. In order to state our results, we need some structural information about sparse planar graphs. The proofs of the following two auxiliary lemmas are in the full version of the paper [3].

Lemma 1 (Augmentations)

- *For every integer $k \in \{0,1,2,3\}$, every plane $(2,k)$-sparse graph is a spanning subgraph of some plane $(2,k)$-tight graph.*
- *A $(2,4)$-sparse graph forms a subgraph of a $(2,4)$-tight graph if and only if it is bipartite. In particular, the 5-cycle is $(2,4)$-sparse but not a subgraph of a $(2,4)$-tight graph.*

Lemma 2 (Plane Duals)

- *For every $k \in \{0,1\}$ and every integer $\ell \in \mathbb{Z}$, there is a plane $(2,k)$-tight graph whose dual is not $(2,\ell)$-sparse.*
- *For every $k \in \{2,3,4\}$, every plane $(2,k)$-tight graph has a $(2,4-k)$-tight dual.*

In particular, duality is an involution on the plane $(2,2)$-tight graphs, so every plane $(2,2)$-tight graph is the dual of another plane $(2,2)$-tight graph. However, for $k \in \{3,4\}$, the duals of plane $(2,k)$-tight graphs form a proper subclass of all plane $(2,4-k)$-tight graphs. In fact, we prove that the dual of every plane $(2,3)$-tight graph is a co-Laman graph (a graph where $|E| = 2|V|-1$ and $E[W] \leq 2|W|-2$ for all $W \subsetneq V$; see Fekete et al. [11]). The duals of the $(2,4)$-tight graph are exactly the 4-regular plane graphs.

New Results. Our main results are:

Theorem 1 (CCA-Representations)

- *Every plane $(2,2)$-sparse graph admits a CCA-representation.*
- *Every plane co-Laman multigraph admits a CCA-representation.*
- *Every plane graph with maximum degree 4 admits a CCA-representation.*
- *There is a plane $(2,0)$-tight multigraph with no CCA-representation.*

The theorem above directly implies the following corollary.

Corollary 1. *For every $k \in \{0,1,2,3,4\}$ and every plane $(2,k)$-tight graph G, the plane dual G^* of G has a contact representation with circular arcs, whenever G^* is $(2,4-k)$-tight.*

We use two different approaches to construct CCA representations. The first approach is a constructive one, and it can be used for plane $(2,2)$-tight graphs and plane co-Laman graphs. We find a special construction sequence for each graph in these two classes (similar to the Henneberg moves [14] for $(2,3)$-tight graphs, also see [11,21,23]), and we show that this construction sequence can be modified into a construction sequence for a CCA-representation. The second approach is structural. For some planar $(2,0)$-sparse graphs, in particular for all graphs of maximum degree 4, we can obtain a stronger form of contact representation where the convex hull of each arc is empty. For these graphs we define the notion of a *good 2-orientation*, and use a circle packing construction to find this stronger CCA-representation from the good orientation. However, we show

Theorem 2. *Testing whether a planar $(2,0)$-tight graph has a contact representation where the convex hull of each arc is empty is NP-complete, even for graphs of maximum degree 5.*

Finally we consider contact representation with *wedges* (that is, polyline segments with at most one bend). A wedge can be viewed as a sequence of two circular arcs (straight-line segments). It is not difficult to prove that every planar $(2,0)$-sparse graph has a contact representation with polylines composed of three segments [8]. On the other hand, as pointed out earlier, one segment per polyline is not sufficient. This raises a question (asked in [8]) whether every planar $(2,0)$-sparse graph has a contact representation with polylines composed of two segments. We resolve the question by showing that every plane $(2,0)$-sparse graph has a contact representation with wedges.

Theorem 3. *Every plane $(2,0)$-sparse graph has a contact representation where each vertex is represented by a wedge.*

2 Contact Representations from Henneberg Moves

Here we prove the existence of CCA-representations for $(2,2)$-sparse and co-Laman graphs, the first two cases of Theorem 1. We defer the degree-4 and $(2,0)$-tight cases to Section 3.

We begin by describing a set of moves which can be applied to a plane $(2,k)$-tight graph, in order to obtain a larger plane $(2,k)$-tight graph (with more vertices), where $k \in \{0,1,2,3,4\}$ depends on the type of move. Afterwards we show that certain subsets of these moves can be used to generate all plane $(2,k)$-tight graphs of a certain class of graphs, starting from one concrete base graph. All but one of these moves are well-known and have already successfully been used for this purpose; see Fig. 2.

Definition 1 (Moves). *Let $G = (V, E)$ be a plane $(2,k)$-graph for some $k \in \{0,1,2,3,4\}$.*

The Henneberg 1 move H_1. *For a face f of G and two distinct vertices u, v on f, introduce a new vertex x inside f and add edges from x to u and v.*

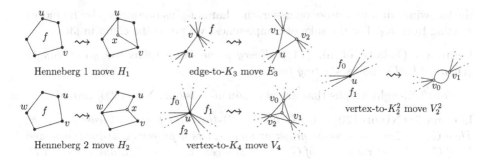

Fig. 2. The moves H_1, H_2, E_3, V_4 and V_2^2

The Henneberg 2 move H_2. *For a face f of G and an edge $e = (u,v)$ on f and a third vertex $w \neq u,v$ on f, introduce a new vertex x inside f, add edges from x to u, v and w, and remove the edge e.*

The edge-to-K_3 move E_3. *For an edge $e = (u,v)$ of G and a face f incident to v, replace v by two vertices v_1, v_2 connected by an edge (v_1,v_2), and add edges from v_1 (v_2) to each neighbor of v that lies clockwise (counterclockwise) between f and e (included) around v.*

The vertex-to-K_4 move V_4. *For a vertex u of G and three (not necessarily distinct) faces f_0, f_1, f_2 incident to u, appearing in that clockwise order around u, replace u by a plane K_4 with outer vertices v_0, v_1, v_2, and add edges from v_i to every neighbor of u that lies clockwise between f_i and f_{i+1} around u, $i = 0,1,2$, where indices are taken modulo 3.*

The vertex-to-K_2^2 move V_2^2. *For a vertex u of G and two (not necessarily distinct) faces f_0, f_1 at u, replace u by two vertices v_1, v_2 connected by two parallel edges, and add edges from v_i to every neighbor of u that lies clockwise between f_i and f_{i+1} around u, $i = 0,1$, where indices are taken modulo 2.*

Henneberg moves H_1 and H_2 were introduced by Henneberg [14], moves E_3 and V_2^2 were defined by Whiteley [23], E_3 also appears in [11] under the name *vertex-splitting*, and the move V_4 was introduced by Nixon and Owen [21].

Part (i) of Lemma 3 is due to Henneberg [14], see also Haas et al. [13].

Lemma 3. *Each of the following holds.*

(i) *All plane $(2,3)$-tight graphs can be generated by H_1 and H_2 moves starting from a triangle [13, 14].*

(ii) *All duals of plane $(2,3)$-tight graphs can be generated by E_3 and V_2^2 moves starting from three parallel edges.*

(iii) *All plane $(2,2)$-tight graphs can be generated by E_3 and V_4 moves starting from an isolated vertex.*

In order to prove (iii) we need one more concept from the literature. A *Laman-plus-one* graph is a simple graph G with an edge $e = uv$, so that $\deg(u) \geq 2$ and $\deg(v) \geq 2$, and $G - e$ is a $(2,3)$-tight (Laman) graph. Laman-plus-one graphs form a proper subclass of $(2,2)$-tight graphs. Fekete et al. [11] claimed the following without a proof on generating Laman-plus-one graphs by E_3 moves starting from K_4. For the sake of completeness we prove the claim in [3].

Lemma 4 (Fekete et al. [11]). *Every plane Laman-plus-one graph can be generated by E_3 moves starting from K_4.*

For $(2,2)$-tight graphs that are not Laman-plus-one, Nixon [20] proved:

Lemma 5 (Nixon [20]). *Let G be a $(2,2)$-tight graph with at least one edge. Then G is a Laman-plus-one graph or there exists a proper $(2,2)$-tight subgraph H of G such that no vertex of $G - H$ is adjacent to more than one vertex in H.*

For a subgraph H of a graph G, let $V(H)$ and $E(H)$ be the vertex set and the edge set of H. Denote by $G - H$ the subgraph of G induced by the vertices $V(G) \setminus V(H)$. For two subgraphs H_1 and H_2 of G, let $H_1 \cup H_2$ be the subgraph $H = (V(H_1) \cup V(H_2), E(H_1) \cup E(H_2))$. Let $E(H_1, H_2)$ be the set of edges between the vertices of H_1 and H_2. The following Lemma is proved in the full paper [3].

Lemma 6. *Let G be a $(2,k)$-tight graph and let H be a proper $(2,k)$-tight subgraph of G, for $k > 0$. Let C and D be a partition of the vertices of $G - H$ such that there is no edge between vertices in C and vertices in D. Then both the graphs induced by the vertices of $H \cup C$ and $H \cup D$ are also $(2,k)$-tight.*

We are now ready to prove Lemma 3.

Proof. [Proof of Lemma 3] Part (i), that every plane Laman graph can be generated by H_1 and H_2 moves starting from a triangle, is already known [13,14].

Now, if G^* is the plane dual of a plane Laman graph G, then we follow the construction sequence of G with H_1 and H_2 moves in the dual and observe that this gives a construction sequence of G^* with V_2^2 and E_3 moves, starting with three parallel edges. This proves (ii).

Let G be a plane $(2,2)$-tight graph. We prove by induction on $|V(G)|$, that G can be generated by E_3 and V_4 moves, starting with a vertex. If $|V(G)| = 1$, this clearly holds. Assume that $|V(G)| \geq 2$. If G is Laman-plus-one, it can be obtained from a single vertex by a single V_4 move, followed by a number of E_3 moves, by Lemma 4, and the claim follows. Otherwise, by Lemma 5, there exists a proper $(2,2)$-tight subgraph H of G such that no vertex of $G - H$ is adjacent to more than one vertex in H. Furthermore, since G is $(2,2)$-tight, H is connected. Since H is a proper subgraph of G, assume without loss of generality that the outer face of H is not vertex-empty in G (otherwise at least one internal face is not vertex-empty and a similar reasoning holds). Let H' be the subgraph of G consisting of H and all the vertices inside the outer boundary of H. By Lemma 6, H' is also a planar $(2,2)$-tight graph, which is a proper subgraph of G. Thus by the induction hypothesis, H' can be constructed from a single vertex by E_3 and V_4 moves. Let Π_1 denote this sequence of these two moves. The graph G' obtained from G by merging H' into a single vertex is simple and planar $(2,2)$-tight. By the induction hypothesis, G' can be obtained from a single vertex by a sequence Π_2 of E_3 and V_4 moves. The sequence Π_2 followed by the sequence Π_1 generates G from a vertex. $\square$

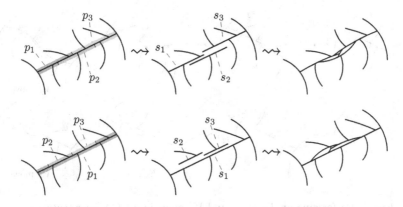

Fig. 3. Performing a V_4 move in a CCA-representation

Lemma 7. *Let G be a plane graph with a CCA-representation and G' be a plane graph obtained from G by a V_4, E_3 or V_2^2 move. Then G' admits a CCA-representation as well.*

Proof. Let G' be obtained from G by a V_4, E_3 or V_2^2 move. In each case, we show how to locally modify a given CCA-representation of G into a CCA-representation of G'. All the cases are similar, so we restrict ourselves to a careful description of the first case only, and provide figures illustrating the remaining cases.

Let v be the vertex in G and $\{v_1, v_2, v_3, v_4\}$ be the four vertices in G' that replace v. Let $S_i = N_{G'}(v_i) \setminus \{v_1, v_2, v_3, v_4\}$, $i = 1, 2, 3, 4$. By definition, we have $S_i \cap S_j = \emptyset$ for $i \neq j$, $S_4 = \emptyset$, $S_1 \cup S_2 \cup S_3 = N_G(v)$ and each of S_1, S_2, S_3 forms a subset of $N_G(v)$ that appears consecutively in the clockwise order around v in G. Assume without loss of generality that the circular arc c_v for v in the given CCA-representation of G is a straight segment. The boundary of c_v can be partitioned into three consecutive pieces p_1, p_2, p_3, so that p_i contains exactly the contacts corresponding to vertices in S_i, $i = 1, 2, 3$; see Fig. 3.

From the pieces p_1, p_2, p_3, we define straight segments s_1, s_2, s_3 parallel to c_v so that each s_i intersects exactly the circular arcs for vertices in S_i, $i = 1, 2, 3$. Then, each s_i is "curved" into a circular arc, so that s_1, s_2, s_3 form a triangle with one free endpoint on the inside. We add a fourth circular arc for v_4 in the triangle, containing the free endpoint in its interior and touching the other two circular arcs with its two endpoints; see Fig. 3.

The cases for E_3 move or V_2^2 are similar; the only difference is that we define only two sets S_1, S_2 (with $S_1 \cap S_2 = \{u\}$ for an E_3 move) and consequently only two pieces p_1, p_2 and two straight segments s_1, s_2; see Fig. 4.   □

Finally, we prove the main theorem of this section.

Proof. [Proof of Theorem 1, Cases 1 and 2] Let G be a plane graph. We show that G admits a CCA-representation, provided it is $(2, 2)$-sparse or a co-Laman graph.

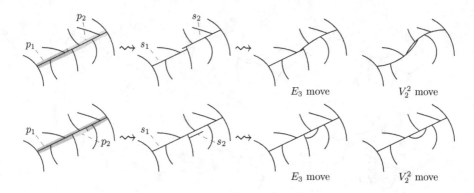

Fig. 4. Performing an E_3 move and V_2^2 move in a CCA-representation

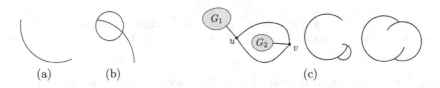

(a)   (b)   (c)

Fig. 5. The base case G_0 for (a) $(2,2)$-tight graphs and (b) co-Laman graphs. (c) A plane $(2,0)$-tight graph that does not admit a CCA-representation.

Case 1, G is $(2,2)$-tight. By Lemma 3, G can be generated by E_3 and V_4 moves, starting from a graph G_0 with a single vertex. Since G_0 admits a CCA-representation (Fig. 5(a)), by Lemma 7, G also has a CCA-representation.

Case 2, G is co-Laman. By Lemma 3, G can be generated by E_3 and V_2^2 moves, starting from a graph G_0 with two vertices and three edges. Since G_0 has a CCA-representation (Fig. 5(b)), by Lemma 7, so does G.   □

3   Good Orientations and One-Sided Representations

Next we consider the third case of Theorem 1, graphs of maximum degree four.

An *orientation* of a graph G is a directed graph whose underlying undirected graph is G. We call it a *2-orientation* if every vertex has out-degree exactly 2, and a *2^--orientation* if every vertex has out-degree at most 2. An orientation of G is called *good* if, for every vertex v of G, all the outgoing edges (equivalently incoming edges) incident to v are consecutive in the circular ordering of the edges around v. A CCA-representation is called *one-sided* if, for each arc a, the endpoints of other arcs that touch a all do so on one side of a. This analogous to the concept of one-sided segment contact representations [10,15]. A CCA-representation is *interior-disjoint* if each arc has nonzero curvature and the interior of the convex hull of each arc is disjoint from all the other arcs.

Lemma 8. *A simple plane graph with a good 2^--orientation has an interior-disjoint CCA-representation.*

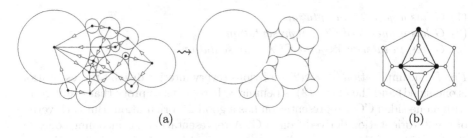

(a)   (b)

Fig. 6. (a) From a contact representation with disks and a good 2^--orientation to a CCA-representation. (b) A planar Laman graph with the minimum degree 3, that has no good 2-orientation.

Proof. Consider a plane graph G with a good 2^--orientation. As with any plane graph, G has a contact representation with disks [16]. For each vertex v of G with out-deg$(v) = 2$, the two outgoing edges of v define two points, p and p', on the circle $C(v)$ representing v. If out-deg$(v) = 1$, the outgoing edge defines p and we choose $p' \in C(v)$ very close to it. If out-deg$(v) = 0$ we choose p and p' distinct from all contacts of $C(v)$ and close to each other. In both cases the two points, p and p', split $C(v)$ into two circular arcs. Since the 2^--orientation is good, one of these two arcs contains none of the contacts of $C(v)$ with other disks. We represent each vertex v by the other circular arc defined by $C(v), p$ and p', which contains all the contacts of $C(v)$; see Fig. 6(a). □

Lemma 9. *Every plane graph G with maximum degree 4 has a good 2^--orientation.*

Proof. First note that vertices of degree strictly less than 4 are harmless, as long as they have at most two outgoing edges: their outgoing edges (if any) cannot be non-consecutive. In order to find a 2^--orientation of G in which every vertex of degree 4 has consecutive outgoing edges, we define a number of walks in G. Start with any edge and define a walk so that, when entering some vertex v of degree 4 via edge e the walk always continues with the edge e' that lies opposite of e at v. At a degree-3 vertex that has not already been made part of one of these walks, continue the walk with an arbitrary incident edge, and otherwise stop. Orienting every edge in this walk consistently and starting another iteration with any so-far unoriented edge (if any exists), eventually gives a good 2^--orientation of G. □

Not every planar $(2,0)$-sparse graph has a good orientation; a counterexample is easy to construct by adding sufficiently many degree-2 vertices. Moreover, there is a counterexample with minimum degree 3; see Fig. 6(b). Indeed, the bold subgraph (induced by the black vertices) has five edges and four vertices. Thus, in any 2^--orientation, at least one black vertex has two bold outgoing edges. At this vertex, all light edges are incoming, breaking up its outgoing edges.

Lemma 10. *For a simple plane graph G, the following statements are equivalent.*

(1) *G has a good 2^--orientation*
(2) *G has a one-sided CCA-representation*
(3) *G has an interior-disjoint CCA-representation*

Proof. The implication (3) $\Rightarrow$ (2) is obvious (every interior-disjoint representation is one-sided) and the (1) $\Rightarrow$ (3) is Lemma 8. It remains to prove that every graph with a one-sided CCA-representation has a good 2^--orientation. But each vertex of the 2^--orientation derived from a CCA-representation has incoming edges on the two sides of the corresponding arc separated by outgoing edges at the two arc endpoints; in a one-sided representation, one set of incoming edges is empty and cannot separate the outgoing edges.   □

Proof. [Proof of Theorem 1, Cases 3, 4] Let G be a plane graph with the maximum degree 4. By Lemma 9, G admits a good 2^--orientation; and hence by Lemma 8 G admits a CCA-representation. This completes Case 3.

A plane $(2,0)$-tight multigraph with no CCA-representation is shown in Fig. 5(c). It has two vertices u, v joined by two parallel edges e, e', and two plane $(2,0)$-tight subgraphs G_1 and G_2. G_1 lies in the unbounded region and G_2 lies in the bounded region defined by e, e', and u and v are connected by an edge to a vertex in G_1 and G_2, respectively. G is plane and $(2,0)$-tight and admits no CCA-representation since two touching circular arcs have their free ends either both in the bounded or both in the unbounded region defined by the closed created curve (Fig. 5(c)). Note that whether every *planar* $(2,0)$-tight multigraph has a plane embedding that has a CCA-representation is an open question.   □

As noted in the introduction, Case 3 of the proof of Theorem 1 always constructs an interior-disjoint CCA-representation for graphs of maximum degree 4. We now show that finding such representations without the degree constraint is hard. We only provide a sketch here; the details are in [3].

Proof Sketch. [Theorem 2] It is sufficient to prove that finding a good representation of a $(2,0)$-tight graph is NP-complete. We first prove this for plane multigraphs. We reduce from the known NP-complete problem Positive Planar 1-in-3SAT [18]. Our reduction uses *wire, splitter,* and *clause* gadgets (Fig. 7) based on the fact that, for a good 2-orientation at a degree-4 vertex,

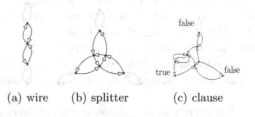

(a) wire   (b) splitter   (c) clause

Fig. 7. The gadgets used in the reduction; gray edges show how adjacent gadgets are connected

each incoming edge is opposite to an outgoing edge and vice versa. Each variable of the 3SAT formula is a wire gadget, that is closed to a circle with doubled edges. There are two good 2-orientations of this circle, encoding the truth value of the variable. A splitter gadget with a short piece of wire propagate this signal to the clause gadgets, representing clauses. Due to the degree-4 vertices, there is only one good 2-orientation for the splitter that extends a "wire signal". The degree-3 vertex in the clause gadget verifies that exactly two of the attached wires carry a false signal. Finally we convert this multigraph to a $(2, 0)$-tight simple planar graph, keeping the existence of a good 2-orientation; see [3] for details.   □

4   Contact Representations with Wedges

A *wedge* is a polyline segment with at most one bend (thus two circular arcs). We show that plane $(2, 0)$-sparse graphs have a contact representations with wedges.

Theorem 3. *Every plane $(2, 0)$-sparse graph has a contact representation where each vertex is represented by a wedge.*

Proof. A plane $(2, 0)$-sparse graph G has a 2^--orientation [4,6,19]. Consider a straight-line drawing of G. For each vertex v, the wedge for v is the union of the line segments representing the outgoing edges from v. Here all the contacts representing the incoming edges for a vertex is at the bend-point of the wedge, but a small perturbation of the representation is sufficient to get rid of this degeneracy; see Fig. 8. Indeed, one

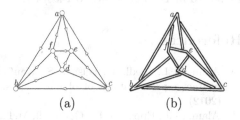

Fig. 8. (a) A straight-line drawing and a 2-orientation of a $(2, 0)$-sparse graph G, (b) a contact representation of G with wedges.

can slide the endpoint of every wedge a bit along the wedge it touches, ensuring that the endpoints of wedges with smaller incoming angle slide a bit further.   □

5   Conclusion and Open Questions

We presented new results about contact representations of graphs with circular arcs. Although every graph with such a contact representation is planar and $(2, 0)$-sparse, we provided a $(2, 0)$-tight plane multigraph that does not admit such a representation. On the other hand, we identified several subclasses of plane $(2, 0)$-sparse graphs that have CCA representations. The natural question remains open: does every simple planar $(2, 0)$-sparse graph have a circular-arc contact representation, if we allow changing the embedding?

A circular-arc contact representation Γ for a $(2, 0)$-tight graph G defines a 3-regular *skeleton graph* [22], where the points of contact are vertices and arcs between contacts are edges (Fig. 1). Each vertex of G with degree $d \geq 3$ corresponds to a path of $d - 2$ vertices in Γ along one circular arc. Thus one way to find a circular-arc contact representation for a plane $(2, 0)$-tight graph G is to find a 3-regular graph by splitting each vertex of degree $d > 3$ into $(d - 2)$ degree-3 vertices (each choice of splitting corresponds to a different 2^--orientation), and align the path associated with each vertex of G into a circular arc. Given a 3-regular graph with a path-cover, to find a representation with each path aligned as a circular arc is related to the stretchability question [7], which is still open.

We also showed that every plane $(2, 0)$-sparse graph has a contact representation with polyline segments with a single bend (wedges). In this context, several questions seem interesting: does every $(2, 0)$-sparse graph admit a contact representation with *equilateral wedges* (i.e., wedges with equal-length segments)? Can we bound the smallest angle at the corner of the wedges (to say, $45°$)?

Acknowledgments. This work was initiated at the Dagstuhl workshop on Drawing Graphs and Maps with Curves. We thank Éric Fusy for helpful comments. M. J. Alam and S. G. Kobourov are supported by NSF grant CCF-1115971. D. Eppstein is supported by in part NSF grant 1228639 and ONR grant N00014-08-1-1015. A. Schulz is supported by DFG grant SCHU 2458/4-1.

References

1. Alam, M.J., Biedl, T., Felsner, S., Kaufmann, M., Kobourov, S.G.: Proportional contact representations of planar graphs. J. Graph Algor. Appl. **16**(3), 701–728 (2012)
2. Alam, M.J., Eppstein, D., Goodrich, M.T., Kobourov, S.G., Pupyrev, S.: Balanced circle packings for planar graphs. In: Duncan, C., Symvonis, A. (eds.) GD 2014. LNCS, vol. 8871, pp. 125–136. Springer, Heidelberg (2014)
3. Alam, M.J., Eppstein, D., Kaufmann, M., Kobourov, S.G., Pupyrev, S., Schulz, A., Ueckerdt, T.: Contact representations of sparse planar graphs. CoRR, abs/1501.00318 (2015)
4. Bernardi, O., Fusy, É.: A bijection for triangulations, quadrangulations, pentagulations, etc. J. Combinatorial Th., Ser. A **119**(1), 218–244 (2012)
5. de Castro, N., Cobos, F., Dana, J., Márquez, A., Noy, M.: Triangle-free planar graphs and segment intersection graphs. J. Graph Algor. Appl. **6**(1), 7–26 (2002)
6. de Fraysseix, H., Ossona de Mendez, P.: On topological aspects of orientations. Discrete Math. **229**(1–3), 57–72 (2001)
7. de Fraysseix, H., Ossona de Mendez, P.: Barycentric systems and stretchability. Discrete Applied Math. **155**(9), 1079–1095 (2007)
8. de Fraysseix, H., Ossona de Mendez, P.: Representations by contact and intersection of segments. Algorithmica **47**(4), 453–463 (2007)
9. Duncan, C.A., Gansner, E.R., Hu, Y., Kaufmann, M., Kobourov, S.G.: Optimal polygonal representation of planar graphs. Algorithmica **63**(3), 672–691 (2012)
10. Eppstein, D., Mumford, E., Speckmann, B., Verbeek, K.: Area-universal and constrained rectangular layouts. SIAM J. Comput. **41**(3), 537–564 (2012)
11. Fekete, Z., Jordán, T., Whiteley, W.: An inductive construction for plane laman graphs via vertex splitting. In: Albers, S., Radzik, T. (eds.) ESA 2004. LNCS, vol. 3221, pp. 299–310. Springer, Heidelberg (2004)
12. Gonçalves, D., Lévêque, B., Pinlou, A.: Triangle contact representations and duality. Discrete Comput. Geom. **48**(1), 239–254 (2012)
13. Haas, R., Orden, D., Rote, G., Santos, F., Servatius, B., Servatius, H., Souvaine, D.L., Streinu, I., Whiteley, W.: Planar minimally rigid graphs and pseudo-triangulations. Comput. Geom. Th. Appl. **31**(1–2), 31–61 (2005)

14. Henneberg, L.: Die graphische Statik der starren Systeme. BG Teubner (1911)
15. Hliněný, P.: Classes and recognition of curve contact graphs. J. Combinatorial Th., Ser. B **74**(1), 87–103 (1998)
16. Koebe, P.: Kontaktprobleme der konformen Abbildung. Ber. Sächs. Akad., Math.-Phys. Klasse **88**, 141–164 (1936)
17. Lee, A., Streinu, I.: Pebble game algorithms and sparse graphs. Discrete Math. **308**(8), 1425–1437 (2008)
18. Mulzer, W., Rote, G.: Minimum-weight triangulation is NP-hard. J. ACM 55(2) (2008)
19. Nash-Williams, C.S.J.A.: Edge-disjoint spanning trees of finite graphs. J. London Math. Soc. **36**, 445–450 (1961)
20. Nixon, A.: Rigidity on Surfaces. PhD thesis, Lancaster University (2011)
21. Nixon, A., Owen, J.: An inductive construction of (2, 1)-tight graphs (2011) (unpublished preprint)
22. Schulz, A.: Drawing graphs with few arcs. In: Brandstädt, A., Jansen, K., Reischuk, R. (eds.) WG 2013. LNCS, vol. 8165, pp. 406–417. Springer, Heidelberg (2013)
23. Whiteley, W.: Some matroids from discrete applied geometry. In: Matroid Theory. Contemp. Math., vol. 197, pp. 171–311. Amer. Math. Soc. (1996)

Contact Representations of Graphs in 3D

Jawaherul Alam[1]([✉]), William Evans[2], Stephen Kobourov[1], Sergey Pupyrev[1,3],
Jackson Toeniskoetter[1], and Torsten Ueckerdt[4]

[1] Department of Computer Science, University of Arizona, Tucson, USA
mjalam@email.arizona.edu
[2] Department of Computer Science, University of British Columbia,
Vancouver, Canada
[3] Institute of Mathematics and Computer Science, Ural Federal University,
Yekaterinburg, Russia
[4] Department of Mathematics, Karlsruhe Institute of Technology,
Karlsruhe, Germany

Abstract. We study contact representations of non-planar graphs in
which vertices are represented by axis-aligned polyhedra in 3D and edges
are realized by non-zero area common boundaries between corresponding
polyhedra. We present a liner-time algorithm constructing a representation of a 3-connected planar graph, its dual, and the vertex-face incidence
graph with 3D boxes. We then investigate contact representations of 1-
planar graphs. We first prove that optimal 1-planar graphs without separating 4-cycles admit a contact representation with 3D boxes. However,
since not every optimal 1-planar graph can be represented in this way, we
also consider contact representations with the next simplest axis-aligned
3D object, L-shaped polyhedra. We provide a quadratic-time algorithm
for representing optimal 1-planar graphs with L-shapes.

1 Introduction

Graphs are often used to describe relationships between objects, and graph
embedding techniques allow us to visualize such relationships. There are com-
pelling theoretical and practical reasons to study *contact representations* of
graphs, where vertices are interior-disjoint geometric objects and edges correspond to pairs of objects touching in some specified fashion. In practice, 2D
contact representations with rectangles, circles, and polygons of low complexity
are intuitive, as they provide the viewer with the familiar metaphor of geograph-
ical maps. Such representations are preferred in some contexts over the standard
node-link representations for displaying relational information [9].

A large body of work considers representing graphs by contacts of simple
curves or polygons in 2D. Graphs that can be represented in this way are planar
and Koebe's 1936 theorem established that *all* planar graphs can be represented
by touching disks [18]. Every planar graph also has a contact representation with
triangles [15]. Curves, line-segments, and *L*-shapes have also been used [14,17].
In particular, it is known that all planar bipartite graphs can be represented
by contacts of axis-aligned segments [10]. For non-planar graphs such contact

© Springer International Publishing Switzerland 2015
F. Dehne et al. (Eds.): WADS 2015, LNCS 9214, pp. 14–27, 2015.
DOI: 10.1007/978-3-319-21840-3_2

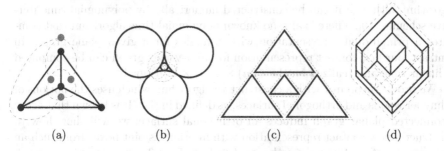

Fig. 1. (a) A plane graph K_4 and its dual; primal-dual contact representations of the graph with (b) circles and (c) triangles. (d) The primal-dual box-contact representation of K_4 with dual vertices shown dashed. The outer box (shell) contains all other boxes.

representations in 2D are impossible. In a natural generalization for non-planar graphs, vertices can be represented with 3D-polyhedra. For example, representations of complete graphs and complete bipartite graphs using spheres and cylinders have been considered [5,16]. Overall, very little is known about such contact representations of non-planar graphs.

As a first step towards representing non-planar graphs, we consider *primal-dual* contact representations, in which a plane graph (a planar graph with a fixed planar embedding), its dual graph, and the face-vertex incidence graph are all represented simultaneously. More formally, in such a representation vertices and faces are represented by some geometric objects so that:

(i) the objects for the vertices are interior-disjoint and induce a contact representation for the primal graph;

(ii) the objects for the faces are interior-disjoint except for the object for the outer face, which contains all the objects for the internal faces, and together they induce a contact representation of the dual graph;

(iii) the objects for a vertex v and a face f intersect if and only if v and f are incident.

Primal-dual representations of plane graphs have been studied in 2D. Every 3-connected plane graph has a primal-dual representation with circles [2] and triangles [15]; see Fig. 1(a)–(c). Our first result in this paper is an analogous primal-dual representation using axis-aligned 3D boxes. While it is known that every planar graph has a contact representation with 3D boxes [7,12,23], Theorem 1 strengthens the result; see Fig. 1d.

Theorem 1. *Every 3-connected plane graph $G = (V, E)$ admits a proper primal-dual box-contact representation in 3D and it can be computed in $\mathcal{O}(|V|)$ time.*

Before proving this theorem we point out two important differences between our result for box-contact representation and the earlier primal-dual representations for circles and triangles [2,15]. First, the existing constructions induce *non-proper* (point) contacts, while our contacts are always *proper*, that is, have non-zero areas. Second, for a given 3-connected plane graph, it is not always possible to find a primal-dual representation with circles by a polynomial-time algorithm, although it can be constructed numerically by polynomial-time iterative schemes [19]. There is also no known polynomial-time algorithm that computes a primal-dual representation with triangles for a given plane graph. In contrast, our box-contact representation for an n-vertex graph can be computed in linear time and realized on the $\mathcal{O}(n) \times \mathcal{O}(n) \times \mathcal{O}(n)$ grid.

We prove Theorem 1 with a constructive algorithm, which uses the notions of Schnyder woods and orthogonal surfaces, as defined in [13]. It is known that every 3-connected planar graph induces an orthogonal surface; we will show how to construct a new contact representation with interior-disjoint boxes from such an orthogonal surface. Since the orthogonal surfaces for a 3-connected planar graph and its dual coincide *topologically*, we show how to *geometrically* realize the primal and the dual box-contact representations so that they fit together to realize all the desired contacts. The construction idea is inspired by recent box-contact representation algorithms for maximal planar graphs [7]. Note, however, that we generalize one such algorithm to handle 3-connected planar graphs (rather than maximal-planar graphs) and show how to combine the primal and dual representations. Our method relies on a correspondence between Schnyder woods and generalized canonical orders for 3-connected planar graphs. Although the correspondence has been claimed in [3], the earlier proof appears to be incomplete. We provide a complete proof of the claim in the full version of the paper [1].

The representation in Theorem 1 immediately gives box-contact representations for a special class of non-planar graphs that are formed by the union of a planar graph, its dual, and the vertex-face incidence graph. The graphs were called *prime* by Ringel [20], who studied them in the context of simultaneously coloring a planar graph and its dual, and are defined as follows. A simple graph $G = (V, E)$ is said to be 1-*planar* if it can be drawn on the plane so that each of its edges crosses at most one other edge. A 1-planar graph has at most $4|V| - 8$ edges and it is *optimal* if it has exactly $4|V| - 8$ edges [11], that is, it is the densest 1-planar graph on the vertex set. An optimal 1-planar graph is called *prime* if it has no separating 4-cycles, that is, cycles of length 4 whose removal disconnects the graph. These optimal 1-planar graphs are exactly the ones that are 5-connected; alternatively, these graphs can be obtained as the union of a 3-connected simple plane graph, its dual and its vertex-face-incidence graph [21].

As in earlier primal-dual contact representations, it is not possible to have all vertex-objects interior disjoint. Specifically, one vertex-object (be it triangle, circle, or box) contains all the others. We call this special box the *shell* and such a representation a *shelled* box-contact representation. Here all the vertices are represented by 3D boxes, except for one vertex, which is a shell, and the interiors of all boxes and the exterior of the shell are disjoint. Note that a similar shell is required in circle-contact and triangle-contact representations; see Fig. 1. The following is a direct corollary of Theorem 1.

Corollary 1. *Every prime 1-planar graph G has a shelled box-contact representation in 3D and it can be computed in linear time.*

One may wonder whether every 1-planar graph admits a box-contact representation in 3D, but it is easy to see that there are 1-planar graphs, even as simple as K_5, that do not admit a box-contact representation. Furthermore, there exist optimal 1-planar graphs (which contain separating 4-cycles) that have neither a box-contact representation nor a shelled box-contact representation; see the full paper [1].

Fig. 2. An *L*-shaped polyhedron

Therefore, we consider representations with the next simplest axis-aligned object in 3D, an *L*-shaped polyhedron or simply an $\mathcal{L}$, which is an axis-aligned box minus the intersection of two axis-aligned half-spaces; see Fig. 2. An $\mathcal{L}$ can also be considered the union of two 3D boxes. Note that the union of two axis-aligned boxes does not always form an $\mathcal{L}$ (e.g., it could form a T-shape); an $\mathcal{L}$ is the simplest of all such polyhedra. We provide a quadratic-time algorithm for representing every optimal 1-planar graph with $\mathcal{L}$'s (note that a 3D box is simply a degenerate $\mathcal{L}$).

Theorem 2. *Every embedded optimal 1-planar graph $G = (V, E)$ has a proper $\mathcal{L}$-contact representation in 3D and it can be computed in $\mathcal{O}(|V|^2)$ time.*

Our algorithm is similar to a recursive procedure used for constructing box-contact representations of planar graphs in [12,23]. The basic idea is to find separating 4-cycles and represent the inner and the outer parts of the graph induced by the cycles separately. Then these parts are combined to produce the final representation. Since the separating 4-cycles can be nested inside each other, the running time of our algorithm is dominated by the time required to find separating 4-cycles and their nested structure. Unlike the early algorithms for box-contact representations of planar graphs [12,23], our algorithms produce proper contacts between the 3D objects (boxes and $\mathcal{L}$'s).

2   Primal-Dual Contact Representations

In this section we prove Theorem 1. Specifically, we describe a linear-time algorithm that computes a box-contact representation for the primal graph and the dual graph separately and then fits them together to also realize the face-vertex incidence graph. We first require some concepts about Schnyder woods and ordered path partitions.

Let G be a 3-connected plane graph with a specified pair of vertices $\{v_1, v_2\}$ and a third vertex $v_3 \notin \{v_1, v_2\}$, such that v_1, v_2, v_3 are all on the outer face in that counterclockwise order. Add the edge (v_1, v_2) to the outer face of G (if it does not already contain it) such that v_3 remains on the outerface and call the augmented graph G'. Let $\Pi = (V_1, V_2, \ldots, V_L)$ be a partition of the vertices of G such that each V_i induces a path in G; Π is an *ordered path partition* [3] of G if the following conditions hold:

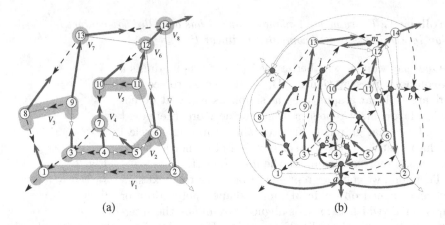

(a)   (b)

Fig. 3. (a) An ordered path partition and its corresponding Schnyder wood for a 3-connected graph G. (b) The Schnyder woods for the primal and the dual of G. The thick solid red, dotted blue and thin solid green edges represent the three trees in the Schnyder wood.

(1) V_1 contains the vertices on the counterclockwise path from v_1 to v_2 on the outer cycle; $V_L = \{v_3\}$;

(2) for $1 \leq k \leq L$, the subgraph G_k of G' induced by the vertices in $V_1 \cup \ldots \cup V_k$ is 2-connected and internally 3-connected (that is, removing two internal vertices of G_k does not disconnect it); hence the outer cycle C_k of G_k is a simple cycle containing the edge (v_1, v_2);

(3) for $2 \leq k \leq L$, each vertex on C_{k-1} has at most one neighbor in V_k.

The pair (v_1, v_2) forms the *base-pair* for Π and v_3 is the *head vertex* of Π. For an ordered path partition $\Pi = (V_1, V_2, \ldots, V_L)$ of G, a vertex v of G has *label* k if $v \in V_k$. The *predecessors* of v are the neighbors of v with equal or smaller labels; the *successors* of v are the neighbors of v with equal or larger labels; see Fig. 3a.

Again consider the three specified vertices v_1, v_2, v_3 in that counterclockwise order on the outer face of G. For $i \in \{1, 2, 3\}$, add a half-edge from v_i reaching into the outer face. A *Schnyder wood* [6] is an orientation and a coloring of the edges of G (including the added half-edges) with the colors $1, 2, 3$ satisfying the following conditions:

(1) every edge e is oriented in either one (*uni-directional*) or two opposite directions (*bi-directional*). The edges are colored so that if e is bi-directional, the two directions (half-edges) have distinct colors;

(2) the half-edge at v_i is directed outwards and colored i;

(3) each vertex v has out-degree exactly one in each color, and the counterclockwise order of edges incident to v is: outgoing in color 1, incoming in color 2, outgoing in color 3, incoming in color 1, outgoing in color 2, incoming in color 3;

(4) there is no interior face whose boundary is a directed cycle in one color.

These conditions imply that for $i \in \{1, 2, 3\}$, the edges with color i induce a tree $\mathcal{T}_i$ rooted at v_i, where all edges of $\mathcal{T}_i$ are directed towards the root. Denote by $\mathcal{T}_i^{-1}$ the tree with all the edges of $\mathcal{T}_i$ reversed, and the Schnyder wood by $(\mathcal{T}_1, \mathcal{T}_2, \mathcal{T}_3)$. Every 3-connected plane graph has a Schnyder wood [4,13]. From a Schnyder wood of a 3-connected plane graph G, one can construct a *dual Schnyder wood* (the Schnyder wood for the dual of G). Consider the dual graph G^* of G in which the vertex for the outer face of G has been split into three vertices forming a triangle. These three vertices represent the three regions between pairs of half edges from the outer vertices of G. Then a Schnyder wood for G^* is formed by orienting and coloring the edges so that between an edge e in G and its dual e^* in G^*, all three colors $1, 2, 3$ have been used. In particular, if e is uni-directional in color i, $i \in \{1, 2, 3\}$, then e^* is bi-directional in colors $i - 1$, $i + 1$ and vice versa; see Fig. 3b; also see [6].

It is known that an ordered path partition of G defines a Schnyder wood on G, where the three outgoing edges for each vertex are to its (1) leftmost predecessor, (2) rightmost predecessor, and (3) highest-labeled successor [4,13]. We call an ordered path partition and the corresponding Schnyder wood computed this way to be *compatible* with each other. Badent et al. [3] argue that the converse can also be done, that is, given a Schnyder wood on G, one can compute an ordered path partition, compatible with the Schnyder wood (and hence, there is a one-to-one correspondence between the concepts). However, the algorithm in [3] for converting a Schnyder wood to a compatible ordered path partition is incomplete, that is, the computed ordered path partition is not always compatible with the Schnyder wood. In the full version of the paper [1] we show such an example and provide a correction of the algorithm. Hence, we have:

Lemma 1. *Let $(\mathcal{T}_1, \mathcal{T}_2, \mathcal{T}_3)$ be a Schnyder wood of a 3-connected plane graph G with three specified vertices v_1, v_2, v_3 in that counterclockwise order on the outer face. Then for $i \in \{1, 2, 3\}$, one can compute in linear time an ordered path partition Π_i compatible with $(\mathcal{T}_1, \mathcal{T}_2, \mathcal{T}_3)$ such that Π_i has (v_{i-1}, v_{i+1}) as the base-pair and v_i as the head. Furthermore Π_i is consistent with the partial order defined by $\mathcal{T}_{i-1}^{-1} \cup \mathcal{T}_{i+1}^{-1} \cup \mathcal{T}_i$.*

We denote a connected region in a plane embedding of a graph by a *face*, and a side of a 3D shape by a *facet*. For a 3D box R, call the facet with highest (lowest) x-coordinate as the x^+-facet (x^--facet) of R. The y^+-facet, y^--facet, z^+-facet and z^--facet of R are defined similarly. For convenience, we denote the x^+-, x^--, y^+-, y^--, z^+- and z^--facets of R as the *right, left, front, back, top* and *bottom* facets of R, respectively. We now sketch a proof for Theorem 1; see [1] for a complete version.

Theorem 1. *Every 3-connected plane graph $G = (V, E)$ admits a proper primal-dual box-contact representation in 3D and it can be computed in $\mathcal{O}(|V|)$ time.*

Proof sketch. Our algorithm consists of the following steps. Let v_1, v_2 and v_3 be three vertices on the outer face of G in the counterclockwise order. First, we create a Schnyder wood $(\mathcal{T}_1, \mathcal{T}_2, \mathcal{T}_3)$ such that for $i \in \{1, 2, 3\}$, $\mathcal{T}_i$ is rooted at

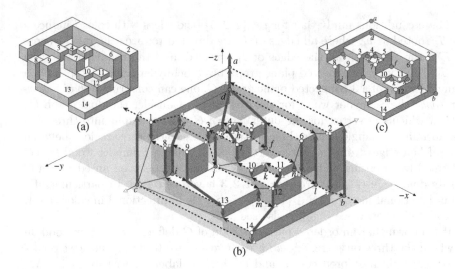

Fig. 4. Box-contact representation (a) for the graph in Fig. 3 with its primal-dual Schnyder wood (b) and the associated orthogonal surface (c). The thick solid red, dotted blue and thin solid green edges represent the three trees in the Schnyder wood.

v_i. Then using Lemma 1, we compute three ordered path partitions compatible with $(\mathcal{T}_1, \mathcal{T}_2, \mathcal{T}_3)$. Next the ordered path partitions are used to calculate the coordinates of 3D boxes that form a contact representation for the primal graph G; a number of local modifications is performed to obtain proper contacts. Finally, the same steps are applied, starting with the dual Schnyder wood of $(\mathcal{T}_1, \mathcal{T}_2, \mathcal{T}_3)$, to construct the representation of the dual graph G^*. These two representations induce the same orthogonal surfaces [13]; hence, they can be combined together to form a primal-dual box-contact representation.

Note that a similar idea is used in [7] to compute a box-contact representation for a maximal planar graph. We strengthen the result by (1) generalizing the method to 3-connected planar graphs and (2) computing an ordered path partition compatible with a Schnyder wood. The latter guarantees the fit between the primal and the dual.

We sketch the steps for computing the primal representation from a Schnyder wood $(\mathcal{T}_1, \mathcal{T}_2, \mathcal{T}_3)$; the computation for the dual representation is analogous. By Lemma 1, for $i \in \{1, 2, 3\}$, one can compute a compatible ordered path partition with the base-pair (v_{i-1}, v_{i+1}) and head v_i, consistent with the partial order defined by $\mathcal{T}_{i-1}^{-1} \cup \mathcal{T}_{i+1}^{-1} \cup \mathcal{T}_i$. Denote by $<_X$, $<_Y$ and $<_Z$ the three ordered path partitions compatible with $(\mathcal{T}_1, \mathcal{T}_2, \mathcal{T}_3)$, that are consistent with $\mathcal{T}_3^{-1} \cup \mathcal{T}_2^{-1} \cup \mathcal{T}_1$, $\mathcal{T}_1^{-1} \cup \mathcal{T}_3^{-1} \cup \mathcal{T}_2$, and $\mathcal{T}_2^{-1} \cup \mathcal{T}_1^{-1} \cup \mathcal{T}_3$, respectively. For a vertex u, let $x_M(u)$, $y_M(u)$, and $z_M(u)$ be the labels of u in the ordered path partitions $<_X$, $<_Y$, and $<_Z$, respectively. Define $x_m(u) = x_M(b)$, $y_m(u) = y_M(g)$ and $z_m(u) = z_M(r)$, where b, g and r are the parents of u in $\mathcal{T}_1$, $\mathcal{T}_2$ and $\mathcal{T}_3$, respectively, when the parents are defined. For each special vertex v_i, $i \in \{1, 2, 3\}$, the parent is not defined in T_i. Assign $x_m(v_1) = 0$, $y_m(v_2) = 0$ and $z_m(v_3) = 0$. For each vertex u, define a box $R(u)$ as $[x_M(u), x_m(u)] \times [y_M(u), y_m(u)] \times [z_M(u), z_m(u)]$.

The boxes defined above yield a box-contact representation for G. Similarly, a representation for the dual graph G^* is computed. These representations can be combined together; see [1] for details. Finally, the three boxes for the three outer vertices of G^* are replaced by a single shell-box, which forms the boundary of the entire representation.

The algorithm runs in $\mathcal{O}(|V|)$ time since computing the primal and the dual Schnyder woods [13], computing ordered path partitions from Schnyder woods (Lemma 1), and the computation of the coordinates all can be accomplished in linear time.   □

3 L-Contact Representation of Optimal 1-Planar Graphs

In this section we prove Corollary 1 and Theorem 2. Throughout, let G be an optimal 1-planar graph with a fixed 1-planar embedding. An edge is *crossing* if it crosses another edge, and *non-crossing* otherwise. A cycle in a connected graph is *separating* if removing it disconnects the graph. We list some properties of optimal 1-planar graphs.

Lemma 2 (Brinkmann et al. [8], Suzuki [22])

- *The subgraph of an embedded optimal 1-planar graph G induced by the non-crossing edges is a plane quadrangulation Q with bipartition classes W, B.*
- *The induced subgraphs $G_W = G[W]$ and $G_B = G[B]$ on white and black vertices, respectively, are planar and dual to each other.*
- *G_B and G_W are 3-connected if and only if Q has no separating 4-cycles.*
- *There exists a simple optimal 1-planar graph with quadrangulation Q if and only if Q is 3-connected.*

An optimal 1-planar graph is *prime* if its quadrangulation has no separating 4-cycle.

Corollary 1. *Every prime 1-planar graph G has a shelled box-contact representation in 3D and it can be computed in linear time.*

Proof. Let Q be the quadrangulation of G and let B, W be the bipartition classes of Q. By Lemma 2, $G_B = G[B]$ and $G_W = G[W]$ are 3-connected planar and dual to each other. By Theorem 1, a primal-dual box-contact representation Γ of G_B can be computed in linear time. We claim that Γ, with the outer face of G_B as bounding box, is a contact representation of G. Indeed, the edges of G are partitioned into G_B, G_W, Q. Each edge in G_B is realized by contact of two "primal" boxes, each edge in G_W by contact of "dual" boxes, and each edge in Q by contact of a primal and a dual box.   □

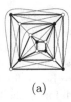   

(a)   (b)   (c)

Fig. 5. (a) An embedded optimal 1-planar graph, its quadrangulation Q (bold) and the partition into white and black vertices. (b) The graph G_{out} produced by removing the interior of separating 4-cycle C. (c) The graph $G_{in}(C)$ comprised of the separating 4-cycle and its interior.

Next, assume that G is any (not necessarily prime) optimal 1-planar graph. To find an $\mathcal{L}$-representation for G, we find all separating 4-cycles in G, replace their interiors by a pair of crossing edges and construct an $\mathcal{L}$-representation Γ_{out} of the obtained prime 1-planar graph G_{out} from a shelled box-contact representation given by Corollary 1. We ensure that Γ_{out} has some "available space" where we place the $\mathcal{L}$-representations for the removed subgraph in each separating 4-cycle, which we construct recursively. We remark that similar procedures were used, e.g., for maximal planar graphs and their separating triangles [12,23]. A separating 4-cycle is *maximal* if its interior is inclusion-wise maximal. A 1-planar graph with at least 5 vertices is *almost-optimal* if its non-crossing edges induce a quadrangulation Q and inside each face of Q, other than the outer face, there is a pair of crossing edges.

Algorithm. L-*Contact*(optimal 1-planar graph G)
1.   Find all separating 4-cycles in the quadrangulation Q of G
2.   **if** some inner vertex w of Q is adjacent to two outer vertices of Q
3.   **then** $\mathcal{C} =$ the two 4-cycles containing w and 3 outer vertices of Q. (**Case 1**)
        else $\mathcal{C} =$ set of all maximal separating 4-cycles in Q. (**Case 2**)
4.   Take the optimal 1-planar (multi)graph G_{out} obtained from G by replacing for each 4-cycle $C \in \mathcal{C}$ all vertices strictly inside C by a pair of crossing edges; see Fig. 5b.
5.   Compute an $\mathcal{L}$-representation of G_{out} with "some space" at each 4-cycle $C \in \mathcal{C}$. In Case 2, this is based on the box-contact representation of G_{out} in Corollary 1.
6.   For each $C \in \mathcal{C}$, take the almost-optimal 1-planar subgraph $G_{in}(C)$ induced by C and all vertices inside C; see Fig. 5c. Recursively compute an $\mathcal{L}$-representation of $G_{in}(C)$ and insert into the corresponding "space" in the $\mathcal{L}$-representation of G_{out}.

Let us formalize the idea of "available space" mentioned in steps 5 and 6. Let Γ be any $\mathcal{L}$-representation of some graph G and C be a 4-cycle in G. A *frame for C* is a 3-dimensional axis-aligned box F together with an injective mapping of $V(C)$ onto the facets of F such that the two facets without a preimage are adjacent. Every frame has one of two possible types. If two opposite vertices of C are mapped onto two opposite facets of F, then F has type $(\perp-\|)$; otherwise,

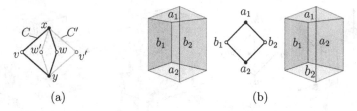

(a)   (b)

Fig. 6. (a) Illustration for Lemma 3. (b) A frame of type $(\bot - \|)$ (left) and of type $(\bot - \bot)$ (right).

F has type $(\bot - \bot)$; see Fig. 6b. Finally, for an almost-optimal 1-planar graph G with corresponding quadrangulation Q and outer face C, and a given frame F for C, we say that an $\mathcal{L}$-representation Γ of G *fits into* F if replacing the boxes or $\mathcal{L}$'s for the vertices in C by the corresponding facets of F yields a proper contact representation of $G - E(G[C])$ that is strictly contained in F.

Before we prove Theorem 2, we need one last lemma addressing the structure of maximal separating 4-cycles in almost-optimal 1-planar graphs.

Lemma 3. *Let G be an almost-optimal 1-planar graph with corresponding quadrangulation Q. Then all maximal separating 4-cycles of Q are interior-disjoint, unless two inner vertices w and w' of Q are adjacent to two outer vertices of Q.*

Proof. When two maximal separating 4-cycles C and C' are not interior-disjoint, then some vertex from C lies strictly inside C' and some vertex from C' lies strictly inside C. It follows that $V(C) \cap V(C')$ is a pair x, y of two vertices from the same bipartition class of Q, say $x, y \in B$, and that some $v \in V(C)$ lies strictly outside C' and some $v' \in V(C')$ lies strictly outside C. We have $v, v' \in W$ and that $C^* = (x, v, y, v')$ is a 4-cycle whose interior strictly contains C and C'. By the maximality of C and C', C^* is not separating. Since the vertices $w \in V(C) \setminus V(C^*)$ and $w' \in V(C') \setminus V(C^*)$ lie strictly inside C^*, C^* is the outer cycle of Q and w, w' are the desired vertices.   □

Theorem 2. *Every embedded optimal 1-planar graph $G = (V, E)$ has a proper $\mathcal{L}$-contact representation in 3D and it can be computed in $\mathcal{O}(|V|^2)$ time.*

Proof. Let Q be the quadrangulation of G with outer cycle C_{out}. Following algorithm **L-Contact**, we distinguish two cases. If (**Case 1**) some inner vertex w of Q has two neighbors on C_{out} we let $\mathcal{C}$ be the set of the two 4-cycles in Q that consist of w and 3 vertices of C_{out}. Otherwise (**Case 2**), let $\mathcal{C}$ be the set of all maximal separating 4-cycles in Q. By Lemma 3 the cycles in $\mathcal{C}$ are interior-disjoint. As in step 4 we define G_{out} to be the optimal 1-planar (multi)graph obtained from G by replacing for each $C \in \mathcal{C}$ all vertices strictly inside C by a pair of crossing edges. Note that in Case 1 the quadrangulation corresponding to G_{out} is $K_{2,3}$ with inner vertex w. We proceed by proving the following lemma, which corresponds to step 5 in the algorithm.

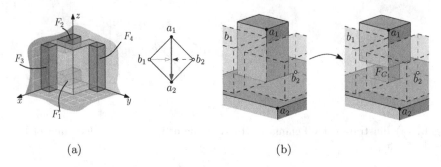

Fig. 7. Illustration for Lemma 4: (a) Case 1 construction, (b) Creating a frame F_C in Case 2 for an inner facial cycle $C = (a_1, b_1, a_2, b_2)$ of Q_H by releasing the contact between a_1 and a_2

Lemma 4. *Let H be an almost-optimal 1-planar (multi)graph whose corresponding quadrangulation Q_H is either $K_{2,3}$ or has no separating 4-cycles. Let C be a set of facial 4-cycles of Q_H, different from its outer cycle C_o, and H' be the graph obtained from H by removing the crossing edges in each $C \in C$. Then for any given frame F for C_o, one can compute an $\mathcal{L}$-representation Γ of H' fitting into F so that there is a frame $F_C \subseteq F$ for every $C \in C$ that is interior-disjoint from all boxes and $\mathcal{L}$'s in Γ.*

Proof. **Case 1, $Q_H = K_{2,3}$.** Let w be the inner vertex of H. Without loss of generality let $F = [0,5] \times [0,5] \times [0,4]$ and let $V(C_o)$ be mapped onto the top, back left, bottom and back right facets of F. Define the $\mathcal{L}$ for w to be the union of $[0,3] \times [2,3] \times [0,4]$ and $[2,3] \times [0,3] \times [0,4]$. Define four boxes $F_1 = [0,2] \times [0,1] \times [0,1]$, $F_2 = [0,2] \times [0,1] \times [3,4]$, $F_3 = [3,4] \times [0,1] \times [0,4]$ and $F_4 = [0,1] \times [3,4] \times [0,4]$, each completely contained in F and disjoint from the $\mathcal{L}$ for w; see Fig. 7a. Each F_i, $i \in \{1,2,3,4\}$ is a frame for a 4-tuple containing w and three vertices of C_o. Thus independent of the type of F and the neighbors of w in Q_H, we find a frame for the inner faces of Q_H.

Case 2, $Q_H \neq K_{2,3}$. Let B and W be the bipartition classes of Q_H and $C_o = (v_1, w_1, v_2, w_2)$ with $v_i \in B$ and $w_i \in W$, $i = 1,2$. Without loss of generality v_1, v_2, w_1 are mapped onto the back left, back right and top facets of F, respectively, and w_2 is mapped onto the bottom facet if (**Case 2.1**) F has type ($\perp$−$\|$) and onto the front left facet if (**Case 2.2**) F has type ($\perp$−$\perp$). Let H^* be the graph obtained from H by inserting a pair of crossing edges in C_o, leaving v_1, w_2 and v_2 on the unbounded region. By assumption, H^* is a prime 1-planar graph and thus by Lemma 2 $H_B^* = H^*[B]$ and $H_W^* = H^*[W]$ are planar 3-connected and dual to each other. We choose v_3 to be the clockwise next vertex after v_2 on the outer face of H_B^* and compute (using Corollary 1) a shelled box-contact representation Γ^* of H^*, in which w_2 is represented as the bounding box $F^* = [0,n]^3$, $n \in \mathbb{N}$, and v_1, v_2, w_1 as $[0,n] \times [0,1] \times [0,n]$, $[0,1] \times [0,n] \times [1,n] \times [1,n] \times [n-1,n]$, i.e., these boxes constitute the back left, back right and top facets of F^*, respectively.

Next we show how to create a frame for each facial 4-cycle $C \in C$. Let a_1, b_1, a_2, b_2 be the vertices of C in cyclic order. Assume without loss of generality

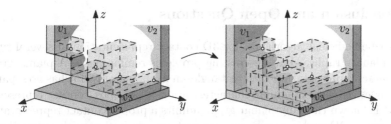

Fig. 8. Modifying Γ' when F has type $(\perp-\|)$ (Case 2.1) to find a representation fitting F

that $a_1, a_2 \in W$ and $b_1, b_2 \in B$. Thus (a_1, a_2) and (b_1, b_2) are crossing edges of H_W^* and H_B^*, respectively. In the Schnyder wood of H_W^* underlying Corollary 1 exactly one of (a_1, a_2), (b_1, b_2) is uni-directed, say (a_1, a_2) is uni-directed in tree $\mathcal{T}_1$. Then there is a point in $\mathbb{R}^3$ in common with all four boxes in Γ^* corresponding to vertices of C. Moreover, by construction boxes b_1, a_2, b_2 touch box a_1 with their y^+, z^+, y^- facets, respectively; see Fig. 7b. Now we can increase the lower z-coordinate of the box a_1 by some $\varepsilon > 0$ so that a_1 and a_2 lose contact and between these two boxes a cubic frame F_C with side length ε is created; see again Fig. 7b. Note that the z^- facet of a_1 makes contact only with a_2 and hence if ε is small enough all other contacts in Γ^* are maintained. We apply this operation to each $C \in \mathcal{C}$ and obtain a shelled box-representation Γ' of H'.

Finally, we show how to modify Γ' to obtain an $\mathcal{L}$-representation of H' fitting the given frame F. If (**Case 2.1**) F has type $(\perp-\|)$, we define a new box for w_2 to be $[0, n+1] \times [0, n] \times [-1, 0]$. For each white neighbor of w_2 we union the corresponding box with another box that is contained in $[n, n+1] \times [0, n] \times [0, n]$ with bottom facet at $z = 0$ so that the result is an $\mathcal{L}$-shape. For each black neighbor of w_2 we set the lower z-coordinate of the corresponding box to 0; see Fig. 8. (This requires the proper contacts for outer edges of G_B, except for (v_1, v_2), to be parallel to the xz-plane, which we can easily guarantee.) We then apply an affine transformation mapping $[1, n+1] \times [1, n] \times [0, n-1]$ onto F. If (**Case 2.2**) F has type $(\perp-\perp)$, define a new box for w_2 to be $[0, n] \times [n, n+1] \times [0, n]$ and apply an affine transformation mapping $[1, n] \times [1, n] \times [0, n-1]$ to F. In both cases we have an $\mathcal{L}$-representation of H' fitting F.   □

By the lemma above we can compute an $\mathcal{L}$-representation Γ_{out} of G_{out} fitting any given frame F_{out} for C_{out} in $\mathcal{O}(|V(G_{out})|)$ time. Moreover, Γ_{out} has a set of disjoint frames $\{F_C \mid C \in \mathcal{C}\}$. Following step 6 of algorithm **L-Contact**, for each $C \in \mathcal{C}$, let $G_{in}(C)$ be the almost-optimal 1-planar graph given by all vertices and edges of G on and strictly inside C. Recursively applying the lemma we can compute an $\mathcal{L}$-representation Γ_C of $G_{in}(C)$ fitting the frame F_C for C in Γ_{out}. Clearly, $\Gamma = \Gamma_{out} \cup \bigcup_{C \in \mathcal{C}} \Gamma_C$ is an $\mathcal{L}$-representation of G fitting F_{out}. We pick a frame F_{out} of arbitrary type for C_{out} to complete the construction. Although computing an $\mathcal{L}$-representation of G_{out} takes $\mathcal{O}(|V(G_{out})|)$ time, recursive computation and affine transformations on the $\mathcal{L}$'s for the vertices in $G_{in}(C)$ for each $C \in \mathcal{C}$ require $\mathcal{O}(|V|^2)$ time.   □

4   Conclusion and Open Questions

We described efficient algorithms for 3D contact representation of several types on non-planar graphs. Many interesting problems remain open. A planar graph has a contact representation with rectangles in 2D if and only if it has no separating triangles. Is there a similar characterization for 3D box-contact representations? It is known that any planar graph admits a proper contact representation with boxes in 3D and a non-proper contact representation with cubes (boxes with equal sides). Does every planar graph admit a proper contact representation with cubes? Representing graphs with contacts of constant-complexity 3D shapes, such as $\mathcal{L}$'s, is open for many graph classes with a linear number of edges, such as 1-planar, quasi-planar and other nearly planar graphs.

Acknowledgments. Work on this problem began at the 9th Bertinoro Workshop on Graph Drawing. J. Alam and S. Kobourov are supported by NSF grant CCF-1115971. W. Evans is supported by NSERC. We thank M. Bekos, T. Biedl, F. Brandenburg, M. Kaufmann, G. Liotta for useful discussions.

References

1. Alam, M.J., Evans, W.S., Kobourov, S.G., Pupyrev, S., Toeniskoetter, J., Ueckerdt, T.: Contact representations of graphs in 3D. CoRR abs/1501.00304 (2015)
2. Andreev, E.: On convex polyhedra in Lobachevskii spaces. Mat. Sb. **123**(3), 445–478 (1970)
3. Badent, M., Brandes, U., Cornelsen, S.: More canonical ordering. Journal of Graph Algorithms and Applications **15**(1), 97–126 (2011)
4. Bernardi, O., Fusy, E.: Schnyder decompositions for regular plane graphs and application to drawing. Algorithmica **62**(3–4), 1159–1197 (2012)
5. Bezdek, A.: On the number of mutually touching cylinders. Combinatorial and Computational Geometry **52**, 121–127 (2005)
6. Bonichon, N., Felsner, S., Mosbah, M.: Convex drawings of 3-connected plane graphs. Algorithmica **47**(4), 399–420 (2007)
7. Bremner, D., et al.: On representing graphs by touching cuboids. In: Didimo, W., Patrignani, M. (eds.) GD 2012. LNCS, vol. 7704, pp. 187–198. Springer, Heidelberg (2013)
8. Brinkmann, G., Greenberg, S., Greenhill, C., McKay, B., Thomas, R., Wollan, P.: Generation of simple quadrangulations of the sphere. Discrete Math. **305**(1–3), 33–54 (2005)
9. Buchsbaum, A.L., Gansner, E.R., Procopiuc, C.M., Venkatasubramanian, S.: Rectangular layouts and contact graphs. ACM Transactions on Algorithms **4**(1) (2008)
10. Czyzowicz, J., Kranakis, E., Urrutia, J.: A simple proof of the representation of bipartite planar graphs as the contact graphs of orthogonal straight line segments. Information Processing Letters **66**(3), 125–126 (1998)
11. Fabrici, I., Madaras, T.: The structure of 1-planar graphs. Discrete Mathematics **307**(7–8), 854–865 (2007)
12. Felsner, S., Francis, M.C.: Contact representations of planar graphs with cubes. In: Hurtado, F., van Kreveld, M.J. (eds.) SOCG, pp. 315–320. ACM (2011)

13. Felsner, S., Zickfeld, F.: Schnyder woods and orthogonal surfaces. Discrete & Computational Geometry **40**(1), 103–126 (2008)
14. de Fraysseix, H., de Mendez, P.O.: Representations by contact and intersection of segments. Algorithmica **47**(4), 453–463 (2007)
15. Gonçalves, D., Lévêque, B., Pinlou, A.: Triangle contact representations and duality. Discrete & Computational Geometry **48**(1), 239–254 (2012)
16. Hliněný, P., Kratochvíl, J.: Representing graphs by disks and balls (a survey of recognition-complexity results). Discrete Mathematics **229**(1–3), 101–124 (2001)
17. Kobourov, S.G., Ueckerdt, T., Verbeek, K.: Combinatorial and geometric properties of planar Laman graphs. In: Khanna, S. (ed.) SODA, pp. 1668–1678. SIAM (2013)
18. Koebe, P.: Kontaktprobleme der konformen Abbildung. Berichte über die Verhandlungen der Sächsischen Akad. der Wissen. zu Leipzig. Math.-Phys. Klasse **88**, 141–164 (1936)
19. Mohar, B.: Circle packings of maps in polynomial time. European Journal of Combinatorics **18**(7), 785–805 (1997)
20. Ringel, G.: Ein Sechsfarbenproblem auf der Kugel. Abhandlungen aus dem Mathematischen Seminar der Universitat Hamburg **29**(1–2), 107–117 (1965)
21. Schumacher, H.: Zur struktur 1-planarer graphen. Math. Nachrichten **125**, 291–300 (1986)
22. Suzuki, Y.: Re-embeddings of maximum 1-planar graphs. SIAM Journal on Discrete Mathematics **24**(4), 1527–1540 (2010)
23. Thomassen, C.: Interval representations of planar graphs. Journal of Combinatorial Theory Series B **40**(1), 9–20 (1988)

Minimizing the Aggregate Movements
for Interval Coverage

Aaron M. Andrews and Haitao Wang$^{(\boxtimes)}$

Department of Computer Science, Utah State University, Logan, UT 84322, USA
aaron.andrews@aggiemail.usu.edu, haitao.wang@usu.edu

Abstract. We consider an interval coverage problem. Given n intervals of the same length on a line L and a line segment B on L, we wish to move the intervals along L such that every point of B is covered by at least one interval and the sum of the moving distances of all intervals is minimized. As a basic geometry problem, it also has applications in mobile sensor barrier coverage. The previous work solved the problem in $O(n^2)$ time. In this paper, we present an $O(n \log n)$ time algorithm.

1 Introduction

We consider an interval coverage problem, which has applications in barrier coverage of mobile sensors. For convenience, we will introduce and discuss the problem from the barrier coverage point of view. Given a set of n points $S = \{s_1, s_2, \ldots, s_n\}$ on a line L, say, the x-axis, each point s_i represents a sensor. Let x_i be the coordinate of s_i on L for each $1 \leq i \leq n$. For any two coordinates x and x' with $x \leq x'$, we use $[x, x']$ to denote the interval of L between x and x'. The sensors of S have the same *covering range*, denoted by z, such that for each $1 \leq i \leq n$, sensor s_i *covers* the interval $[x_i - z, x_i + z]$. Let B be a line segment of L and we call B a "barrier". We assume that the length of B is at most $2z \cdot n$ since otherwise B could not be fully covered by these sensors. The problem is to move all sensors along L such that each point of B is covered by at least one sensor of S and the sum of the moving distances of all sensors is minimized. Note that although sensors are initially on L, they may not be on B. We call this problem the *min-sum barrier coverage*, denoted by MSBC.

The problem MSBC has been studied before and Czyzowicz et al. [6] gave an $O(n^2)$ time algorithm. In this paper, we present an $O(n \log n)$ time algorithm and we also show an $\Omega(n \log n)$ time lower bound for this problem.

Related Work. A number of related problems have been studied in the literature. If sensors have different ranges, Czyzowicz et al. [7] proved that the problem MSBC is NP-hard. In the *min-max* version of MSBC, the objective is to minimize the maximum movement of all sensors. If the sensors have the same range, Czyzowicz et al. [6] gave an $O(n^2)$ time algorithm, and later Chen et al. presented

---

This research was supported in part by NSF under Grant CCF-1317143.

F. Dehne et al. (Eds.): WADS 2015, LNCS 9214, pp. 28–39, 2015.
DOI: 10.1007/978-3-319-21840-3_3

an $O(n \log n)$ time solution [4]. If sensors have different ranges, Chen *et al.* [4] gave an $O(n^2 \log n)$ time algorithm.

Mehrandish *et al.* [9,10] considered another variant, where the goal is to move the minimum number of sensors to form a barrier coverage. They [9,10] proved the problem is NP-hard if sensors have different ranges and gave polynomial time algorithms otherwise. In addition, Li *et al.* [8] considered setting an energy for each sensor to form a coverage such that the cost of all sensors is minimized. There [8], the sensors are not allowed to move, and the more energy a sensor has, the larger the covering range of the sensor and the larger the cost of the sensor. Another problem variation is considered in [2], where the goal is to maximize the barrier coverage lifetime subject to the limited battery powers.

Bhattacharya *et al.* [3] studied a two-dimensional barrier coverage in which the barrier is a circle and the sensors, initially located inside the circle, are moved to the circle to minimize the sensor movements; the ranges of the sensors are not explicitly specified but the destinations of the sensors are required to form a regular n-gon on the circle. Algorithms for both min-sum and min-max versions were given in [3] and subsequent improvements were made in [5].

Outline. In Section 2, we introduce some notations. If the covering intervals of all sensors intersect B, we call it the *containing case*. If the sensors whose covering intervals do not intersect B are all in one side of B, it is the *one-sided case*. Otherwise, it is the *general case*. We solve the three cases respectively in Sections 3, 4, and 5. Based on the $O(n^2)$ time algorithm in [6], we solve the containing case in $O(n \log n)$ time by using a more efficient implementation. To solve the one-sided case, we use our containing case algorithm as an initial step and apply a sequence of so-called "reverse operations", which is based on a number of interesting observations on the structure of the optimal solution. For solving the general case, we generalize the techniques for the one-sided case. In addition, we prove the $\Omega(n \log n)$ time lower bound by a reduction from sorting.

Due to the space limit, proofs and many details are omitted but can be found in the full version of the paper [1].

2 Preliminaries

A line segment of L is also an interval and vice versa. Let β denote the length of B. Without loss of generality, we assume the barrier B is the interval $[0, \beta]$. For short, sensor covering intervals are called *sc-intervals*.

We assume the sensors of S are already sorted, i.e., $x_1 \leq x_2 \leq \cdots \leq x_n$. For each sensor s_i, we use $I(s_i)$ to denote its covering interval. Recall that z is the covering range of each sensor and the length of each sc-interval is $2z$. We assume $2z < \beta$ since otherwise the solution would be trivial. A crucial observation given in [7] is the following *order preserving property*: there always exists an optimal solution where the order of the sensors is the same as that in the input. Note that this property does not hold if sensors have different ranges.

Sensors will be moved during the algorithm. For any sensor s_i, suppose its location at some moment is y_i; the value $x_i - y_i$ is called the *displacement* of s_i.

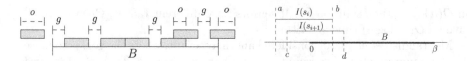

Fig. 1. Illustrating gaps (denoted by g) and overlaps (denoted by o)

Fig. 2. $I(s_i) \cap I(s_{i+1})$ contains 0 in its interior

Hence, if the displacement of s_i is positive (resp., negative), then it is to the left (resp., right) of its original location in the input.

As in [7], we define two concepts: *gaps* and *overlaps*. A *gap* refers to a maximal sub-segment of B such that each point of the sub-segment is not covered by any sensors (e.g., see Fig. 1). Each endpoint of any gap is an endpoint of either an sc-interval or B. Specifically, consider two adjacent sensors s_i and s_{i+1} such that $x_i+z < x_{i+1}-z$. If $0 \le x_i+z$ and $x_{i+1}-z \le \beta$, then the interval $[x_i+z, x_{i+1}-z]$ is on B and defines a gap, and s_i and s_{i+1} are called the left and right *generators* of the gap, respectively. If $x_i + z < 0 < x_{i+1} - z \le \beta$, then $[0, x_{i+1} - z]$ is a gap and s_{i+1} is the only *generator* of the gap. Similarly, if $0 \le x_i+z < \beta < x_{i+1}-z$, then $[x_i + z, \beta]$ is a gap and s_i is the only *generator*.

Consider two adjacent sensors s_i and s_{i+1}. The intersection $I(s_i) \cap I(s_{i+1}) \cap B$ defines an overlap if it is not empty (e.g., see Fig. 1), and we call s_i and s_{i+1} the left and right *generators* of the overlap, respectively. Consider any sensor s_i. If $I(s_i)$ is not completely on B, then the sub-interval of $I(s_i)$ that is not on B defines an overlap and s_i is its only generator (e.g., see Fig. 1). A subtle situation appears when $I(s_i) \cap I(s_{i+1})$ contains an endpoint of B in its interior. Refer to Fig. 2 as an example, where 0 is in the interior of $I(s_i) \cap I(s_{i+1})$ with $I(s_i) = [a, b]$ and $I(s_{i+1}) = [c, d]$. According to our definition, s_i and s_{i+1} together define an overlap $[0, b]$; s_i itself defines an overlap $[a, 0]$; s_{i+1} itself defines an overlap $[c, 0]$. However, to make the discussions easier, we consider the union of $[c, 0]$ and $[0, b]$ as a single overlap $[c, b]$ defined by s_i and s_{i+1} together, but s_i still itself defines the overlap $[a, 0]$. Symmetrically, if $I(s_i) \cap I(s_{i+1})$ contains β in its interior, then we consider $I(s_i) \cap I(s_{i+1})$ as a single overlap defined by s_i and s_{i+1}, and s_{i+1} itself defines an overlap that is the portion of $I(s_{i+1})$ outside B.

We should point out that according to our above definition, if an overlap has two generators, then these two generators must be two adjacent sensors (e.g., s_i and s_{i+1} for some i). In other words, even if the sc-intervals of two non-adjacent sensors (e.g., s_i and s_{i+2}) intersect, their intersection does not define any overlap.

For any gap or overlap a, we use $|a|$ to denote its length; if a has only one generator s_i, then both the left and the right generators of a refer to s_i.

The total number of overlaps and gaps is $O(n)$. To solve MSBC, the goal is to move sensors to cover all gaps by eliminating overlaps. We say a gap/overlap go_1 is to the *left* (resp., *right*) of another gap/overlap go_2 if the left generator of go_1 is to the left (resp., right) of the left generator of go_2 (in the case of Fig. 2, where overlaps $[c, b]$ and $[a, 0]$ have the same left generator s_i, $[a, 0]$ is considered to the left of $[c, b]$). For any i and j with $i \le j$, let $S(i, j) = \{s_i, s_{i+1}, \ldots, s_j\}$.

3   The Containing Case

The high-level scheme of our algorithm for the containing case is the same as the $O(n^2)$ time algorithm in [7], but with an $O(n \log n)$ time implementation using efficient data structures. Below, we sketch the algorithmic scheme and our improvement (see the full paper [1] for details), and this will help us explain our algorithms in Sections 4 and 5. The algorithm "greedily" covers all gaps from left to right one by one, by eliminating overlaps. Suppose the first $i - 1$ gaps have just been covered completely and the algorithm is about to cover gap g_i.

Let o_i^r (resp., o_i^l) be the closest overlap to the right (resp., left) of g_i. We will cover g_i using either o_i^r or o_i^l, depending on their costs $C(o_i^r)$ and $C(o_i^l)$, defined as follows. Let $S_r(g_i)$ be the set of sensors between the right generator of g_i and the left generator of o_i^r. Define $C(o_i^r)$ to be $|S_r(g_i)|$. The intuition of this definition is that suppose we shift all sensors of $S_r(g_i)$ to the left for an infinitesimal distance $\epsilon > 0$ (such that g_i becomes ϵ shorter), then the sum of the moving distances of all sensors of $S_r(g_i)$ is $\epsilon \cdot C(o_i^r)$. As will be clear later, the current displacement of each sensor in $S_r(g_i)$ may be positive but cannot be negative. For $C(o_i^l)$, it is defined differently. Let $S_l(g_i)$ be the set of sensors between the left generator of g_i and the right generator of o_i^l, and let $S_l'(g_i)$ be the subset of sensors of $S_l(g_i)$ whose displacements are positive. If we shift all sensors in $S_l(g_i)$ to the right for an infinitesimal distance $\epsilon > 0$, although the sum of the moving distances of all sensors of $S_l(g_i)$ is $\epsilon \cdot |S_l(g_i)|$, the total moving distance contributed to the sum of the moving distances of all sensors of S is actually $\epsilon \cdot (|S_l(g_i)| - 2 \cdot |S_l'(g_i)|)$ because the sensors of $S_l'(g_i)$ are moved towards their original locations. Hence, the cost $C(o_i^l)$ is defined to be $|S_l(g_i)| - 2 \cdot |S_l'(g_i)|$. Note that the sensors in $S_r(g_i)$ or $S_l(g_i)$ are consecutive in their index order.

If $C(o_i^r) < C(o_i^l)$, we move each sensor in $S_r(g_i)$ leftwards by $\min\{|o_i^r|, |g_i|\}$, and we call this a *left-shift process*. Note that if there is any gap g_j between two sensors in $S_r(g_i)$, then the above shift process will move g_j leftwards as well, but the size and the generators of g_j do not change, and thus we can still use g_j without causing any problems. If $|g_i| \leq |o_i^r|$, then after the left-shift process g_i is covered completely and we proceed on the next gap g_{i+1}. Otherwise, o_i^r is eliminated and g_i is only partially covered. We proceed on the remaining g_i.

If $C(o_i^r) \geq C(o_i^l)$, we move each sensor in $S_l(g_i)$ rightwards by distance $\min\{|o_i^l|, |g_i|, \alpha\}$, where α is the smallest displacement of the sensors in $S_l'(g_i)$, and we call this a *right-shift process*. After the process, if g_i is only partially covered, we proceed on the remaining g_i; otherwise we proceed on g_{i+1}.

The algorithm finishes after all gaps are covered. It can be shown that there are $O(n)$ shift processes in total. The algorithm in [7] implements each shift process in $O(n)$ time, and thus the overall time is $O(n^2)$. Instead, we implement each shift process in $O(\log n)$ amortized time. Specifically, we design an *overlap tree* T_o, a *position tree* T_p, and a *left-shift tree* T_l. We store gaps and overlaps by their generators. The tree T_o maintains all overlaps such that o_i^r and o_i^l can be determined in $O(\log n)$ time (after having o_i^r and o_i^l, $|S_r(g_i)|$ and $|S_l(g_i)|$ are known immediately). The tree T_p implicitly maintains the positions of each sensor such that each of the following *shift operation* can be done in $O(\log n)$

time: given j and k with $j \leq k$ and a distance δ, move all sensors of $S(j,k)$ by δ. The main difficulty of our approach is to determine the values $|S_l'(g_i)|$ and α in $O(\log n)$ time, which is done by the left-shift tree T_l. The details are omitted.

Only the position tree T_p is used in Sections 4 and 5 and we briefly explain it here. T_p is a complete binary tree of n leaves and $O(\log n)$ height. The leaves from left to right correspond to the sensors in their index order. For each $1 \leq j \leq n$, leaf j (i.e., the j-th leaf from the left) stores the original location x_j of sensor s_j. Each node of T_p (either an internal node or a leaf) is associated with a *shift* value, which is zero initially. At any moment during the algorithm, the actual location of each sensor s_j is x_j plus the sum of the shift values of the nodes in the path from the root to leaf j, which can be obtained in $O(\log n)$ time.

Consider a right-shift process that moves sensors in $S(j,k)$ for $j \leq k$ rightwards by a distance δ. We first find a set V_{jk} of $O(\log n)$ nodes of T_p such that the leaves of the subtrees of all these nodes correspond to exactly the sensors in S_{jk}, which can be done in $O(\log n)$ time by a standard approach. Then, for each node in V_{jk}, we increase its shift value by δ. This finishes the right-shift process. Similarly, each left-shift process can also be done in $O(\log n)$ time.

4 The One-Sided Case

Without loss of generality, we assume that the sensors whose covering intervals do not intersect B are all to the right side of B, i.e., $0 \leq x_1 + z$ holds. We assume at least one sc-interval does not intersect B since otherwise it would become the containing case. Note that this implies $\beta < x_n - z$.

We use *configuration* to refer to a specification of where each sensor is located. For example, in the input configuration, each sensor s_i is located at x_i.

A sequence of consecutive sensors $s_i, s_{i+1}, \ldots s_j$ are said to be in *attached positions* if for each $i \leq k \leq j-1$, the right endpoint of the covering interval $I(s_k)$ of s_k is at the same position as the left endpoint of $I(s_{k+1})$.

The following lemma solves a special case where no sc-interval intersects B.

Lemma 1. *If $\beta < x_1 - z$, we can find an optimal solution in $O(n)$ time.*

In the following, we assume $\beta \geq x_1 - z$, i.e., $I(s_1)$ intersects B. Let m be the largest index such that $I(s_m)$ intersects B. Note that $m < n$ due to $\beta < x_n - z$. To simplify the notation, let $S_I = S(1,m)$ and $S_R = S(m+1,n)$.

Our containing case algorithm is not applicable here and one can easily verify that the cost functions we used in the containing case do not work for the sensors in S_R. More specifically, suppose we want to move a sensor s_i in S_R leftwards to cover a gap; there will be an "additive" cost $x_i - z - \beta$, i.e., $I(s_i)$ has to move leftwards by that distance before it touches B. Recall that the cost we defined on overlaps in the containing case is a "multiplicative" cost, and the above additive cost is not consistent with the multiplicative cost. To overcome this difficulty, we have to use a different approach to solve the one-sided case.

Our main idea is to somehow reduce the one-sided case to the containing case so that we can use our containing case algorithm. Let D_{opt} be any optimal solution for our problem. By slightly abusing notation, depending on the context, a "solution" may either refer to the configuration of the solution or the sum of moving distances of all sensors in the solution. If no sensor of S_R is moved in D_{opt}, then we can compute D_{opt} by running our containing case algorithm on the sensors in S_I. Otherwise, let m^* be the largest index such that sensor $s_{m^*} \in S_R$ is moved in D_{opt}. If we know m^*, then we can easily compute D_{opt} in $O(n \log n)$ time as follows. First, we "manually" move all sensors in $S(m+1, m^*)$ leftwards to $\beta + z$ such that the left endpoints of their covering intervals are at β. Then, we apply our containing case algorithm on all sensors in S_{1m^*}, which now all have their covering intervals intersecting B (which is an instance of the containing case), and let $D(m^*)$ be the solution obtained above. Based on the order preserving property, we can show that $D(m^*)$ is D_{opt}.

By the above discussion, one main task of our algorithm is to determine m^*.

For each j with $m < j \le n$, let $D_s(j) = \sum_{i=m+1}^{j}(x_i - z - \beta)$, i.e., the sum of the moving distances for "manually" moving all sensors in $S(m+1, j)$ leftwards to $\beta + z$, and we use F_j to denote the configuration after the above manual movement and we let F_j contain only the sensors in $S(1, j)$ (i.e., sensors in $S(j+1, n)$ do not exist in F_j). Let $D_s(m) = 0$ and F_m be the input configuration but containing only sensors in $S(1, m)$. For each $m \le j \le n$, suppose we apply our containing case algorithm on F_j and denote by $D_c(j)$ the solution (in the case where $\beta > 2zj$, we let $D_c(j) = +\infty$), and further, let $D(j) = D_s(j) + D_c(j)$.

The above discussion leads to the following lemma.

Lemma 2. $D_{opt} = \min_{m \le j \le n} D(j)$ *and* $m^* = \arg\min_{m \le j \le n} D(j)$.

Our algorithm will compute $D(j)$ for all $j = m, m+1, \ldots, n$. Recall that $D(j) = D_s(j) + D_c(j)$. Since it is easy to compute all $D_s(j)$'s in $O(n)$ time, we focus on computing $D_c(j)$'s. The main idea is the following. Suppose we already have the solution $D_c(j-1)$, which can be considered as being obtained by our containing case algorithm. To compute $D_c(j)$, since we have an additional overlap defined by s_j at $\beta + z$, i.e., the sc-interval $I(s_j)$, we modify $D_c(j-1)$ by "reversing" some shift processes that have been performed in the containing algorithm when computing $D_c(j-1)$, i.e., using $I(s_j)$ to cover some gaps that were covered by other overlaps in $D_c(j-1)$. The details are given below.

We first compute $D_c(m)$ on the configuration F_m. If $2zm < \beta$, then $D_c(j) = +\infty$ for each $m \le j < \lceil \frac{\beta}{2z} \rceil$; in this case, we can start from computing $D_c(\lceil \frac{\beta}{2z} \rceil)$ and use the similar idea as the following algorithm. To make it more general, we assume $m \ge \lceil \frac{\beta}{2z} \rceil$, and thus $D_c(m) < +\infty$.

Consider our containing case algorithm for computing $D_c(m)$. Recall that our containing case algorithm consists of shift processes and each shift process covers a gap using an overlap. Let $p_1, p_2, \ldots, p_q$ be the shift processes performed in the algorithm in the *inverse* order of time (e.g., p_1 is the last process), where q is the total number of processes in the algorithm. For each $1 \le i \le q$, let g_i be the gap covered in the process p_i by using/eliminating an overlap, denoted by o_i. Note that each gap/overlap above may not be an original gap/overlap in the input configuration but only a subset of an original gap/overlap. It holds that $|o_i| = |g_i|$ for each $1 \leq i \leq q$. We call $G = \{g_1, g_2, \ldots, g_q\}$ the *gap list* of $D_c(m)$. For each i, we use $C(o_i)$ to denote the cost of o_i when the algorithm uses o_i to cover g_i in the process p_i. Note that the above *process information* can be explicitly stored during our containing case algorithm without affecting the overall running time asymptotically. We will use these information later. Note that according to our algorithm the gaps in G are sorted from right to left.

Next, we compute $D_c(m + 1)$, by modifying the configuration $D_c(m)$. Comparing with F_m, the configuration F_{m+1} has an additional overlap defined by s_{m+1} at $\beta + z$, and we use $o(s_{m+1})$ to denote it. We have the following lemma.

Lemma 3. $D_c(m + 1) = D_c(m)$ *holds if one of the following happens: (1) the coordinate of the right endpoint of $I(s_m)$ is strictly larger than β; (2) o_1 is to the right of g_1; (3) o_1 is to the left of g_1 and the cost $C(o_1)$ is not greater than the number of sensors between g_1 and s_{m+1}.*

To compute $D_c(m+1)$, we first check whether one of the three cases in Lemma 3 happens, which can be done in constant time by the above process information stored when computing $D_c(m)$. If any of the three cases happens, we are done for computing $D_c(m + 1)$. Below, we assume none of the cases happens.

Let $C(s_{m+1}, g_1)$ be the number of sensors between g_1 and s_{m+1}, which would be the cost of the overlap $o(s_{m+1})$ if it were there right before we cover g_1. Note that since we know the generators of g_1, $C(s_{m+1}, g_1)$ can be computed in constant time (e.g., if g_1 has two generators, $C(s_{m+1}, g_1) = m + 1 - a + 1$, where a is the index of the right generator of g_1). Define $R(g_1)$ to be $C(s_{m+1}, g_1) - C(o_1)$. We can consider $R(g_1)$ as the "unit revenue" (or savings) if we use $o(s_{m+1})$ to cover g_1 instead of using o_1. Note that $R(g_1) > 0$ otherwise the third case of Lemma 3 would happen. Hence, it is possible to obtain a better solution than $D_c(m)$ by using $o(s_{m+1})$ to cover g_1 instead of o_1. Note that $|g_1| \leq 2z$ and $|o(s_{m+1})| = 2z$.

If $|o(s_{m+1})| = |g_1|$, then we use $o(s_{m+1})$ to cover g_1. Specifically, we move all sensors in $S(a, m+1)$ leftwards by distance $|g_1|$, where a is the index of the right generator of g_1. The above essentially "restores" the overlap o_1 and covers g_1 by eliminating $o(s_{m+1})$. We refer to it as a *reverse operation* (i.e., it reverses the shift process that covers g_1 by using o_1 in the algorithm for computing $D_c(m)$). Due to $|o(s_{m+1})| = |g_1|$, after the reverse operation, g_1 is fully covered by $o(s_{m+1})$ and $o(s_{m+1})$ is eliminated. We can show that the current configuration is $D_c(m+1)$. Note that $D_c(m+1) = D_c(m) - R(g_1) \cdot |g_1|$. Again, o_1 is restored in $D_c(m+1)$. Finally, we remove g_1 from the list G.

If $|g_1| < |o(s_{m+1})|$, then we do a revere operation by using $o(s_{m+1})$ to cover g_1 and restore o_1, after which $o(s_{m+1})$ is not eliminated but becomes shorter. We remove g_1 from G and proceed on the next gap g_2.

In general, suppose we have covered gaps $g_1, g_2, \ldots, g_k$ by using $o(s_{m+1})$ and the overlap $o(s_{m+1})$ still partially remains (i.e., $\sum_{t=1}^{k} |g_i| < 2z$). The above gaps have all been removed from G. Let F' denote the current configuration. If G is now empty, then we are done with computing $D_c(m + 1)$, which is equal to $D_c(m) - \sum_{t=1}^{k} R(g_t) \cdot |g_t|$; otherwise, we consider gap g_{k+1}, as follows.

Similar to Lemma 3, we can show that F' is $D_c(m+1)$ if one of the following two cases happens: (1) o_{k+1} is to the right of g_{k+1}; (2) o_{k+1} is to the left of g_{k+1}

but $C(o_{k+1})$ is not greater than the number of sensors between g_{k+1} and s_{m+1}. If one of the two cases happens, then we are done with computing $D_c(m+1)$, which is equal to $D_c(m) - \sum_{t=1}^{k} R(g_t) \cdot |g_t|$. Otherwise, we do the following. Note that the length of $o(s_{m+1})$ in F' is $2z - \sum_{t=1}^{k} |g_t|$. Depending on whether $|o(s_{m+1})| \geq |g_{k+1}|$, there are two cases. As for g_1, we define $C(s_{m+1}, g_{k+1})$ as the number of sensors between g_{k+1} and s_{m+1}, and define $R(g_{k+1}) = C(o_{k+1}) - C(s_{m+1}, g_{k+1})$.

If $|o(s_{m+1})| \geq |g_{k+1}|$, then we do a reverse operation to cover g_{k+1} by using $o(s_{m+1})$. If $|o(s_{m+1})| = |g_{k+1}|$, $D_c(m+1)$ is obtained, which is equal to $D_c(m) - \sum_{t=1}^{k+1} R(g_t) \cdot |g_t|$; otherwise, we proceed on the next gap g_{k+2}. In either case, we remove g_{k+1} from G, and the reverse operation restores the overlap o_{k+1}.

If $|o(s_{m+1})| < |g_{k+1}|$, then $o(s_{m+1})$ is not long enough to cover g_{k+1}. We do a reverse operation to use $o(s_{m+1})$ to partially cover g_{k+1} of length $|o(s_{m+1})|$, and the remaining part of g_{k+1} is still covered by o_{k+1}. We are done with computing $D_c(m+1)$, which is equal to $D_c(m) - \sum_{t=1}^{k} R(g_t) \cdot |g_t| - R(g_{k+1}) \cdot |o(s_{m+1})|$. Since g_{k+1} still partially remains in $D_c(m+1)$, we do not remove g_{k+1} from G but change its size accordingly. Also, overlap o_{k+1} is partially restored in $D_c(m+1)$, with size $|o(s_{m+1})|$. The algorithm stops after $D_c(m+1)$ is obtained.

Next, we use the same approach to compute $D_c(m+2)$ by using the remaining gaps in G. Let G_m denote the remaining G. In order to correctly compute $D_c(m+2)$, one may wonder that we should use the corresponding gap list of $D_c(m+1)$ (i.e., the gap list of the containing case algorithm if we apply it on F_{m+1} to compute $D_c(m+1)$), which may not be the same as G_m. However, we can show that the result obtained using G_m is $D_c(m+2)$, and further, this can be generalized to the next solution until $D_c(n)$, i.e., we can use the same approach to compute $D_c(m+3), D_c(m+4), \ldots, D_c(n)$ by using the remaining gaps.

Our algorithm can be easily implemented in $O(n \log n)$ time to compute $D_c(i)$ for all $i = m, m+1, \ldots, n$. First, we compute $D_c(m)$ in $O(n \log n)$ time using our containing case algorithm. During the algorithm, we explicitly record the information of each shift process p_i, as discussed earlier. In fact, we only need to record all right-shift processes after the *last* left-shift process of the algorithm, and let G be the gap list for the above right-shift processes (i.e., for each gap g_i in G, o_i is to the left of g_i).

Next, we apply the reverse operations on G to compute solutions $D_c(j)$ for $m+1 \leq j \leq n$ one by one. To this end, we only need to use the position tree T_p (the other two trees are not necessary). Each reverse operation can be done in $O(\log n)$ time using T_p because the operation essentially moves a sequence of consecutive sensors leftwards by the same distance. If G becomes $\emptyset$ during the algorithm, then the current configuration is the solution we seek. The overall time for computing all solutions $D_c(j)$ for $m+1 \leq j \leq n$ is $O(K \cdot \log n)$, where K is the total number of reverse operations in the entire algorithm. Note that each reverse operation either covers completely a gap of G or eliminates an overlap $o(s_j)$ for $m+1 \leq j \leq n$. Therefore, $K \leq |G| + n - m = O(n)$.

In summary, we can compute the solutions $D_c(j)$ for all $m \leq j \leq n$ in $O(n \log n)$ time, and thus, the one-sided case is solved in $O(n \log n)$ time.

If there is more than one index $j \in [m, n]$ such that $D(j) = D_{opt}$, then we let m^* refer to the smallest such index. The following lemma, which will be useful in Section 5, shows a *unimodal property* of the values $D(j)$ for $j = m, m+1, \ldots, n$.

Lemma 4. *As j increases from m to n, the value $D(j)$ first strictly decreases until $D(m^*)$ and then strictly increases except that $D(m^*) = D(m^* + 1)$ may be possible. Formally, $D(j-1) > D(j)$ for any $m < j \leq m^*$; $D(m^*) \leq D(m^* + 1)$; $D(j-1) < D(j)$ for any $m^* + 2 < j \leq n$.*

5   The General Case

We assume there is at least one sensor whose covering interval intersects B. The case where this assumption does not hold can be solved using similar but simpler techniques (see the full paper for details).

Let s_l (resp., s_r) be the leftmost (resp., rightmost) sensor whose covering interval intersects B. We assume $1 < l$ and $r < n$, since otherwise it becomes the one-sided case. Let $S_L = S(1, l-1)$, $S_I = S(l, r)$, and $S_R = S(r+1, n)$.

Let $\lambda = \lceil \frac{\beta}{2z} \rceil$, i.e., the minimum number of sensors necessary to fully cover B. Consider any i with $1 \leq i \leq l$ and any j with $r \leq j \leq n$ such that $j - i + 1 \geq \lambda$. If $i \neq l$, define $D_s^L(i, j) = \sum_{t=i}^{l-1}(-z - x_t)$, i.e., the total sum of the moving distances for "manually" moving all sensors in $S(i, l-1)$ rightwards to $-z$ (such that the right endpoints of their covering intervals are all at 0); otherwise, $D_s^L(l, j) = 0$. Similarly, if $j \neq r$, define $D_s^R(i, j) = \sum_{t=r+1}^{j}(x_t - z - \beta)$; otherwise, $D_s^R(i, r) = 0$. Let $D_s(i, j) = D_s^L(i, j) + D_s^R(i, j)$. Let $F(i, j)$ denote the configuration after the above manual movements and including only sensors in $S(i, j)$. Hence, $F(i, j)$ is an instance of the containing case on sensors in $S(i, j)$. Let $D_c(i, j)$ be the solution obtained by applying our containing case algorithm on $F(i, j)$. Finally, let $D(i, j) = D_c(i, j) + D_s(i, j)$. For simplicity, for any i and j with $j - i + 1 < \lambda$, we let $D(i, j) = +\infty$, as $S(i, j)$ does not have enough sensors to fully cover B.

For each i with $1 \leq i \leq l$, define $f(i)$ to be the index in $[r, n]$ such that $D(i, f(i)) = \min_{r \leq j \leq n} D(i, j)$. Similarly, for each j with $r \leq j \leq n$, define $f(j)$ to be the index in $[1, l]$ such that $D(f(j), j) = \min_{1 \leq i \leq l} D(i, j)$.

Let D_{opt} denote the optimal solution. We have the following lemma.

Lemma 5. $D_{opt} = \min_{1 \leq i \leq l, r \leq j \leq n} D(i, j) = \min_{1 \leq i \leq l} D(i, f(i)) = \min_{r \leq j \leq n} D(f(j), j)$.

Let l^* and r^* be the indices with $1 \leq l^* \leq l$ and $r \leq r^* \leq n$ such that $D(l^*, r^*) = D_{opt}$. It is easy to see that $l^* = f(r^*)$ and $r^* = f(l^*)$.

To compute D_{opt}, if we know either l^* or r^*, then D_{opt} can be computed in additional $O(n \log n)$ time, as follows. Suppose l^* is known. We first "manually" move each sensor s_i for $l^* \leq i \leq l-1$ rightwards to $-z$ (this step is not necessary for the case $l^* = l$) and then apply our one-sided case algorithm on $S(l^*, n)$ (the obtained solution is D_{opt}). Hence, the key is to determine l^* or r^*.

Lemma 6. *If $|S_I| \geq \lambda$, then it holds that $f(i) = r^*$ for any $i \in [1, l]$ and $f(j) = l^*$ for any $j \in [r, n]$.*

By Lemma 6, if $|S_I| \geq \lambda$, then it holds that $f(1) = r^*$, which can be easily computed in $O(n \log n)$ time by applying our one-sided case algorithm on $S(1, n)$ after moving sensors in S_L rightwards to the position $-z$.

Below we assume $|S_I| < \lambda$. Note that $|S(l^*, r^*)| \geq \lambda$ always holds. Since both $|S(l^*, r^*)|$ and λ are integers, either $|S(l^*, r^*)| \geq \lambda + 1$ or $|S(l^*, r^*)| = \lambda$.

Lemma 7. *If $|S(l^*, r^*)| \geq \lambda + 1$, then $f(i) = r^*$ holds for any i with $1 \leq i < l^*$.*

By Lemma 7, if $|S(l^*, r^*)| \geq \lambda + 1$, then $f(1) = r^*$, which again can be computed in $O(n \log n)$ time.

It remains to handle the case where $|S(l^*, r^*)| = \lambda$. Due to $l^* \leq l$ and $r^* \geq r$, we have $\max\{1, r - \lambda + 1\} \leq l^* \leq \min\{l, n - \lambda + 1\}$. In the following, for simplicity of discussion, we assume $r - \lambda + 1 > 1$ and $l < n - \lambda + 1$ since the other cases can be solved similarly. Let $l' = r - \lambda + 1$. Thus, we have $l' \leq l^* \leq l$, and for any i with $i \geq 0$ and $r + i \leq n$, $|S(l' + i, r + i)| = \lambda$. Clearly, $D_{opt} = \min_{0 \leq i \leq l - l'} D(l' + i, r + i)$.

Let $l'' = l - l'$. In the following, we present an $O(n \log n)$ time algorithm that can compute $D(l' + i, r + i)$ for all $i = 0, 1, \ldots, l''$. Recall that $D(l' + i, r + i) = D_c(l' + i, r + i) + D_s(l' + i, r + i)$. We can easily compute $D_s(l' + i, r + i)$ for all $i = 0, 1, \ldots, l''$ in $O(n)$ time. Therefore, it is sufficient to compute the solutions $D_c(l' + i, r + i)$ for all $i = 0, 1, \ldots, l''$ in $O(n \log n)$ time, which is our focus below. To simplify the notation, we use $D_c(i)$ to represent $D_c(l' + i, r + i)$.

Below, unless otherwise stated, we assume all sensors in $S(1, l - 1)$ are at $-z$ and all sensors in $S(r + 1, n)$ are at $\beta + z$; sensors in $S(l, r)$ are in their original locations as input. In other words, we work on the configuration $F(1, n)$.

We first consider a special case where $\lambda = \frac{\beta}{2z}$, i.e., $\frac{\beta}{2z}$ is an integer. In this case, for each $0 \leq i \leq l''$, the configuration $D_c(i)$ has a special pattern: sensors in $S(l' + i, r + i)$ are in attached positions with $s_{l'+i}$ at z. Due to this property, we can easily compute D_{opt} in $O(n \log n)$ time and the algorithm is omitted.

In the following, we assume $\lambda \neq \frac{\beta}{2z}$, i.e., $\frac{\beta}{2z}$ is not an integer. This implies that there must be an overlap in any solution $D_c(i)$ for $0 \leq i \leq l''$.

Suppose we already have $D_c(0)$. Below, we compute $D_c(1)$ by modifying the configuration $D_c(0)$. The algorithm consists of two main steps. The first step is to compute $D_c(l', r + 1)$ by doing reverse operations on $D_c(l')$ (i.e., $D_c(l', r)$) with sensor s_{r+1} at $\beta + z$, in the same way as in our one-sided case. The second step is to compute $D_c(1)$ by modifying the configuration $D_c(l', r + 1)$, as follows.

Note that $D_c(1)$ is on the configuration $F(l' + 1, r + 1)$ with sensors in $S(l' + 1, r + 1)$ while $D_c(l', r + 1)$ is on $F(l', r + 1)$ with sensors in $S(l', r + 1)$. Hence, $s_{l'}$ is not used in $D_c(1)$ but may be used in $D_c(l', r + 1)$. If in $D_c(l', r + 1)$, $s_{l'}$ covers some portion of B that is not covered by any other sensor in $S(l' + 1, r + 1)$, then we should move sensors of $S(l' + 1, r' + 1)$ to cover the above portion and more specifically, that portion should be covered by eliminating some overlaps in $D_c(l', r + 1)$. The details are given below.

Consider the configuration $D_c(l', r + 1)$. If $s_{l'}$ is at $-z$, then $I(s_{l'}) \cap B = \emptyset$ and B is covered by sensors of $S(l' + 1, r + 1)$, implying that $D_c(1) = D_c(l', r + 1)$.

If $s_{l'}$ is not at $-z$, then let $g = I(s_{l'}) \cap B$. We can show that $s_{l'}$ is the only sensor that covers g in $D_c(l', r + 1)$. To obtain $D_c(l' + 1)$, we remove $s_{l'}$ and cover

g by eliminating overlaps of $D_c(l', r+1)$ from left to right until g is fully covered. Specifically, let $o_1, o_2, \ldots, o_k$ be the overlaps of $D_c(l', r+1)$ sorted from left to right. We move the sensors between g and o_1 leftwards by distance $\min\{|g|, |o_1|\}$. This movement can be done in $O(\log n)$ time by updating the position tree T_p. If $|g| \leq |o_1|$, then we are done. Otherwise, we consider the next overlap o_2. We continue this procedure until g is fully covered. Since $|S(l'+1, r+1)| = \lambda$, $\sum_{i=1}^{k} |o_i| \geq |g|$ holds, implying that g will eventually be fully covered. We can show that the obtained configuration is $D_c(1)$.

The above gives a way to compute $D_c(1)$ from $D_c(0)$. In general, for each $0 \leq i \leq l''$, if we know $D_c(i)$, we can use the same approach to compute $D_c(i+1)$.

We say a solution $D_c(i)$ for $i \in [0, l'']$ is *trivial* if the right endpoint of $I(s_{r+i})$ is strictly to the right of β. The algorithm for Lemma 8 is omitted.

Lemma 8. *Suppose k is the smallest index in $[0, l'']$ such that $D_c(k)$ is a trivial solution. We can compute $D_c(i)$ for all $i = k, k+1, \ldots, l''$ in $O(n \log n)$ time.*

In the following, we compute solutions $D_c(i)$ for all $i = 0, 1, \ldots, l''$ in $O(n \log n)$ time. Our algorithm will compute $D_c(i)$ in the order from 0 to l'' until either $D_c(l'')$ is obtained, or we find a trivial solution (and then we apply Lemma 8).

First, we compute $D_c(0)$ in $O(n \log n)$ time by applying our containing case algorithm on the configuration $F(l', r)$. As in our one-sided case algorithm, we also maintain the process information of the right-shift processes after the last left-shift process in the above algorithm. Let $P = \{p_1, p_2, \ldots, p_q\}$ be the above process list in the inverse time order (i.e., p_1 is the last process of the algorithm), where q is the number of these processes. Let $G = \{g_1, g_2, \ldots, g_q\}$ and $O = \{o_1, o_2, \ldots, o_q\}$ be the corresponding gap list and overlap list, i.e., for each $1 \leq i \leq q$, process p_i covers g_i by eliminating o_i. For each $1 \leq i \leq q$, we also maintain the cost $C(o_i)$ of the overlap o_i. As discussed in Section 4, the gaps of G are sorted from right to left while the overlaps of O are sorted from left to right. In addition, we maintain an extra overlap list $O' = \{o'_1, o'_2, \ldots, o'_h\}$, which are the overlaps in the configuration $D_c(0)$ sorted from left to right. The list O' will be used in the second main step for computing each $D_c(i)$. According to their definitions, all overlaps of O' are to the left of the overlaps of O.

To compute $D_c(1)$, the first main step is to compute $D_c(l', r+1)$ by doing the reverse operations on $D_c(0)$ with s_{r+1}, similar to the one-sided case. Let $o(s_{r+1})$ be the overlap $[\beta, \beta+2z]$ defined by s_{r+1} at $\beta+z$. In general, suppose during the reverse operations $g_1, g_2, \ldots, g_{t-1}$ are the gaps fully covered by $o(s_{r+1})$ and g_t is only partially covered by a length of d_t. Then, gaps $g_1, g_2, \ldots, g_{t-1}$ are removed from G, and g_t is still in G but its length is changed to its original length minus d_t. Correspondingly, the overlaps $o_1, o_2, \ldots, o_{t-1}$ are restored and o_t is partially restored with length d_t in $D_c(l', r+1)$. We append $o_1, o_2, \ldots, o_t$ at the end of O'. Since overlaps of O' are to the left of overlaps of O and overlaps of the two lists O and O' are both sorted from left to right, after the above "append" operation, the overlaps of the new list O' are still sorted from left to right.

The second main step is to compute $D_c(1)$ from $D_c(l', r+1)$, by eliminating overlaps of O' from left to right until $I(s_{l'}) \cap B$ is covered, as discussed earlier. For each overlap that is eliminated, we remove it from O'.

If $D_c(1)$ is a trivial solution, we are done. Otherwise, we continue to compute $D_c(2)$, again by first computing $D_c(l'+1, r+2)$. Let G_1 be the remaining gap list of G after $D_c(1)$ is computed. To compute $D_c(l'+1, r+2)$, we use G_1 to do the reverse operations on $D_c(1)$ with s_{r+2}. Although G_1 may not be the corresponding gap list for $D_c(1)$, we can show that the obtained result using G_1 is $D_c(2)$, and further, this can be generalized to $D_c(3), D_c(4), \ldots$ until $D_c(l'')$.

After obtaining $D_c(l'+1, r+2)$, we can use the same approach to compute $D_c(2)$ (i.e., cover $I(s_{l'+1}) \cap B$ by eliminating the overlaps of $D_c(l'+1, r+2)$ from left to right). We continue the same algorithm to compute $D_c(i)$ for $i = 3, 4, \ldots, l''$, until we find a trivial solution or $D_c(l'')$ is computed.

Lemma 9. *It takes $O(n \log n)$ time to compute $D_c(i)$ for $i = 0, 1, \ldots, l''$, until we find a trivial solution or $D_c(l'')$ is computed.*

As a summary, the general case is solvable in $O(n \log n)$ time. The $\Omega(n \log n)$ time lower bound is based on a reduction from sorting and is omitted.

References

1. Andrews, A., Wang, H.: Minimizing the aggregate movements for interval coverage (2014). arXiv:1412.2300
2. Bar-Noy, A., Rawitz, D., Terlecky, P.: Maximizing barrier coverage lifetime with mobile sensors. In: Bodlaender, H.L., Italiano, G.F. (eds.) ESA 2013. LNCS, vol. 8125, pp. 97–108. Springer, Heidelberg (2013)
3. Bhattacharya, B., Burmester, B., Hu, Y., Kranakis, E., Shi, Q., Wiese, A.: Optimal movement of mobile sensors for barrier coverage of a planar region. Theoretical Computer Science **410**(52), 5515–5528 (2009)
4. Chen, D., Gu, Y., Li, J., Wang, H.: Algorithms on minimizing the maximum sensor movement for barrier coverage of a linear domain. Discrete and Computational Geometry **50**, 374–408 (2013)
5. Chen, D., Tan, X., Wang, H., Wu, G.: Optimal point movement for covering circular regions. Algorithmica (2013), online First. doi:10.1007/s00453-013-9857-1
6. Czyzowicz, J., et al.: On minimizing the maximum sensor movement for barrier coverage of a line segment. In: Ruiz, P.M., Garcia-Luna-Aceves, J.J. (eds.) ADHOC-NOW 2009. LNCS, vol. 5793, pp. 194–212. Springer, Heidelberg (2009)
7. Czyzowicz, J., et al.: On minimizing the sum of sensor movements for barrier coverage of a line segment. In: Nikolaidis, I., Wu, K. (eds.) ADHOC-NOW 2010. LNCS, vol. 6288, pp. 29–42. Springer, Heidelberg (2010)
8. Li, M., Sun, X., Zhao, Y.: Minimum-cost linear coverage by sensors with adjustable ranges. In: Cheng, Y., Eun, D.Y., Qin, Z., Song, M., Xing, K. (eds.) WASA 2011. LNCS, vol. 6843, pp. 25–35. Springer, Heidelberg (2011)
9. Mehrandish, M.: On Routing, Backbone Formation and Barrier Coverage in Wireless Ad Doc and Sensor Networks. Ph.D. thesis, Concordia University, Montreal, Quebec, Canada (2011)
10. Mehrandish, M., Narayanan, L., Opatrny, J.: Minimizing the number of sensors moved on line barriers. In: Proc. of IEEE Wireless Communications and Networking Conference (WCNC), pp. 653–658 (2011)

Online Bin Packing with Advice of Small Size

Spyros Angelopoulos[1,2], Christoph Dürr[1,2], Shahin Kamali[3]([⊠]),
Marc Renault[1], and Adi Rosén[4]

[1] Sorbonne Universités, UPMC Univ Paris 06, UMR 7606, LIP6, 75005 Paris, France
[2] CNRS, UMR 7606, LIP6, 75005 Paris, France
[3] University of Waterloo, Waterloo, Canada
s3kamali@uwaterloo.ca
[4] CNRS and Université Paris Diderot, Paris, France

Abstract. In this paper, we study the advice complexity of the online bin packing problem. In this well-studied setting, the online algorithm is supplemented with some additional information concerning the input. We improve upon both known upper and lower bounds of online algorithms for this problem. On the positive side, we first provide a relatively simple algorithm that achieves a competitive ratio arbitrarily close to 1.5, using constant-size advice. Our result implies that 16 bits of advice suffice to obtain a competitive ratio better than any online algorithm without advice, thus improving the previously known bound of $O(\log(n))$ bits required to attain this performance. In addition, we introduce a more complex algorithm that still requires only constant-size advice, and which is below 1.5-competitive, namely has competitive ratio arbitrarily close to 1.47012. This is the currently best performance of any online bin packing algorithm with sublinear advice. On the negative side, we extend a construction due to Boyar *et al.* [10] so as to show that no online algorithm with sub-linear advice can be 7/6-competitive, which improves upon the known lower bound of 9/8.

1 Introduction

Bin packing is a fundamental optimization problem that has played an important role in the development of approximation and online algorithms. An instance of the problem is defined by a set of *items* of different sizes, and the objective is to place these items into a minimum number of bins. For convenience, it is often assumed that the bins have capacity 1 and items have sizes in the range $(0, 1]$. In the online setting, the input set is revealed in a sequential manner, and the online algorithm must make an irrevocable decision concerning the placement of an item without any knowledge about the forthcoming items. We follow the canonical framework of competitive analysis of online algorithms, in which the

S. Angelopoulos, C. Dürr, S. Kamali, M. Renault and A. Rosén—Research supported in part by project ANR-11-BS02-0015 "New Techniques in Online Computation–NeTOC".

S. Kamali—Research supported by the France Canada Research Fund.

© Springer International Publishing Switzerland 2015
F. Dehne et al. (Eds.): WADS 2015, LNCS 9214, pp. 40–53, 2015.
DOI: 10.1007/978-3-319-21840-3_4

performance of an algorithm $\mathbb{A}$ is determined by its competitive ratio, namely the maximum ratio between the cost of $\mathbb{A}$ (i.e., the number of bins opened by $\mathbb{A}$) and that of an optimal offline algorithm OPT for the same sequence. For the bin packing problem, in particular, we are interested in the *asymptotic* competitive ratio which considers sequences for which the costs of $\mathbb{A}$ and OPT are arbitrarily large. For this reason, throughout this paper we refer to the asymptotic competitive ratio as simply the competitive ratio.

The bin packing problem has provided some of the first-known explicit online algorithms. NEXTFIT is a simple algorithm that maintains at each step a single open bin. If an incoming item fits in the bin, it is placed there; otherwise, that bin is closed and a new bin is opened to accommodate the item. FIRSTFIT orders bins by their opening time and places an incoming item into the first bin which has enough space (opening a new bin if required). BESTFIT works similarly, except that it places the item into the bin with minimum available capacity which still has enough space for the item. It is known that Next Fit is 2-competitive, whereas FIRSTFIT and BESTFIT are both 1.7-competitive [17]. The best known online algorithm is HARMONIC++ which has a competitive ratio of 1.588 [21]. No online algorithm can have a competitive ratio better than 1.54037 [3], a result that holds for both deterministic and randomized algorithms.

Competitive analysis, due to its inherent comparison to the offline optimum, often leads to a more pessimistic performance evaluation of online algorithms than what observed in practice [8]. Different models have been proposed in order to address this issue, and one such approach is by allowing the online algorithm certain additional power. For example, the algorithm may be allowed to repack some items [14,15]. Alternatively, it may have access to lookahead [16], and, finally, may know the length of the input sequence [2] or the value of OPT [13]. The advice model is a generalization of the latter in which, *any* information can be passed to the algorithm in the form of *advice*. In this sense, we can think of the advice as generated by a benevolent offline oracle with access to the entire input; the online algorithm can exploit the advice so as to produce a better solution. In principle, there is a certain correlation between the number of advice bits and the quality of the resulting solution. For many problems, including bin packing, a large number of advice bits is required in order to achieve optimal solutions; however, this does not imply that one may not achieve efficient (albeit non-optimal) solutions with significantly smaller number of bits. In this paper, we study the impact of small-size advice (typically constant size) in improving the competitive ratio of bin packing algorithms. While our interest in studying the advice complexity stems from theoretical considerations, we emphasize that the advice setting may in fact have tangible applications. For instance, the advice model captures, among others, any relevant statistical information about the input that may be available through either preprocessing or historical data. We define the bin packing problem under the advice setting as follows:

Definition 1. *In the online bin packing problem with advice, the input is a sequence of items* $\sigma = \langle x_1, \ldots, x_n \rangle$, *with* $0 < x_i \leq 1$. *At time step* t, *an online algorithm must pack item* x_t *into a bin, and this decision is a function* *of $\Phi, x_1, \ldots, x_{t-1}$, where Φ is the content of the advice tape. An algorithm has advice complexity $s(n)$ if it accesses at most $s(n)$ bits of an advice tape Φ for any input of length n.*

Throughout the paper, for a given algorithm $\mathbb{A}$, we denote by $\mathbb{A}(\sigma)$ the number of bins used by $\mathbb{A}$ on sequence σ. Due to space limitations, we omit or sketch certain proofs (complete proofs can be found in the long version of the paper).

1.1   Previous Work and Our Contribution

The online advice model was first introduced by Böckenhauer *et al.* [6,7] and by Emek *et al.* [12]. Both papers were inspired by the work of Dobrev *et al.* [11]. In the model of Emek *et al.* an online algorithm receives a fixed number of bits of advice with each input item. Note that this model does not allow advice of sublinear size. In the model of Böckenhauer *et al.*, the advice is written on a read-only tape prior to the algorithm's execution, and the algorithm can read advice bits from that tape at will. The advice complexity has established itself as a prolific sub-field of online computation, and many online problems have been studied under the setting of online computation with advice (e.g., metrical task systems [12], job shop scheduling [7,18], the k-server problem [6,12,19], knapsack [5], buffer reordering management [1], and list update [9]).

In this paper, we study online bin packing under the advice-on-tape model. In this setting, Boyar *et al.* [10] proved tight bounds on the size of advice required to be optimal and showed that advice of super-linear size is necessary in order to attain optimality. They also proved that with advice of linear size, i.e., $\Theta(n)$ bits for a sequence of length n, one can achieve a competitive ratio of $4/3 + \epsilon$. This result was improved by Renault *et al.* [20] who showed that a competitive ratio arbitrary close to 1 can be achieved with $\Theta(n)$ bits. A related question is how many bits of advice are sufficient in order to outperform all online algorithms. Boyar *et al.* showed that advice of size $\Theta(\log n)$ is sufficient to achieve an algorithm with competitive ratio of 1.5, which is strictly better than the lower bound 1.54037 for online algorithms. They also proved that no algorithm is better than 9/8-competitive with advice of sub-linear size. A related problem, namely the minimum makespan problem on identical machines was studied in [20].

In our work, we address the power of small-sized advice in online bin packing. This is motivated by settings in which one may have some very limited information about the input, e.g., whether or not the input has many items of size beyond a certain threshold or some related statistical information. On the positive side, we prove that $O(1)$ advice suffices to outperform all online algorithms. More precisely, we first show that with only 16 bits of advice, we can achieve a competitive ratio of 1.530 (Section 2). Following a more complex approach, we show that constant-size advice suffices to go beyond the barrier of 1.5-competitiveness; more precisely, we achieve a competitive ratio arbitrarily close to 1.47012 (Section 3). This is, to date, the best upper bound for advice of sublinear size and demonstrates the significant impact of small-size advice on algorithmic performance. We should mention that the simple algorithm of

Section 2 reaches $1.5 + \epsilon$ with fewer advice bits than the complicated algorithm of Section 3. Last, we give a lower-bound construction that builds on ideas of [10] and which shows that advice of size $\Omega(n)$ is required to achieve a competitive ratio better than $7/6$, thus improving the previous lower bound of $9/8$ (Section 4).

In terms of techniques, for the upper bound of Section 2, we use information indirectly related to the ratio of "big" to "small" items; we show that this limited amount of information suffices to bring us arbitrarily close to the performance of algorithms that use logarithmic number of bits. For the more complicated upper bound of Section 3, we introduce two algorithms that, when combined, result in the desired upper bound. One of these algorithms uses a rounding technique to create close-to-optimal packings when there is an empty space of size ϵ or more in all bins of an optimal solution (ϵ is an arbitrary small positive value). The other algorithm achieves a competitive ratio of 1.3904 when all items are larger than $1/3$. Both algorithms use advice of constant size, i.e., independent of the length of sequence. Last, concerning the lower bound (Section 4), we base our construction on that of [10], using a better amortization scheme that leads to an improvement of the bound.

2   Constant-Size Advice Outperforms All Online Algorithms

In this section, we present an algorithm that achieves a competitive ratio of $1.5 + \epsilon$ and uses a constant number of bits of advice. Throughout the section, we distinguish items based on their sizes. An item is *huge* if it is larger than $2/3$, *critical* if it is in the range $(1/2, 2/3]$, *small* if it is in the range $(1/3, 1/2]$, and *tiny* if it is in the range $(0, 1/3]$.

Consider the algorithm RESERVECRITICAL [10] that works as follows. The algorithm treats huge items separately and places each of them in a single bin. Similarly, it places two small items in the same bin with no other items in said bin. The algorithm knows the number of critical items and reserves space of size $2/3$ for each of them (i.e., it opens a bin for each item and assumes the filled space of the bin is $2/3$). Critical items are placed in the reserved spaces. For tiny items, the algorithm uses FIRSTFIT to place them in critical bins with respect to their reserved spaces (and opens new bins as needed). To encode the number of critical bins in binary, $\Theta(\log n)$ advice bits are needed. As shown in [10], RESERVECRITICAL has a competitive ratio of 1.5. (Since our approach is related, in the long version of the paper, we provide a simpler proof of the result in [10].)

In what follows, we analyze another algorithm, called the REDBLUE algorithm, that receives an integer i, $0 \leq i < 2^k$ encoded in binary with k advice bits, where k is a constant independent of the length of the sequence. The value of i is determined by the packing of the RESERVECRITICAL algorithm. Let X and Y denote the number of bins in the packing of RESERVECRITICAL that include a critical item, and the number of bins opened for the tiny items, respectively.

The advice encodes an approximate value of $\frac{X}{X+Y}$, using k bits, by encoding the value of i such that

$$\beta = \frac{i}{2^k} \le \frac{X}{X+Y} < \frac{i+1}{2^k} = \beta + \frac{1}{2^k} . \tag{1}$$

Regardless of the value of β, REDBLUE always places each huge item in a single bin, and places small items in dedicated bins, with two such items per dedicated bin. In the following, we consider three (exhaustive) cases for β: $\beta > 1 - 1/2^{k/2}$, $\beta < 1/2^{k/2}$, and $1 - 1/2^{k/2} \le \beta \le 1/2^{k/2}$. For each case, we complete the definition of REDBLUE by describing how the critical and tiny items are packed.

Consider the first case: $\beta > 1 - 1/2^{k/2}$. For placing critical and tiny items, REDBLUE maintains a set of *blue* bins such that each bin has a reserved space of $2/3$ for critical items. To pack a critical item, REDBLUE packs it using FIRSTFIT among the set of blue bins, considering only the reserved space. To pack a tiny item, REDBLUE packs it using FIRSTFIT among the set of blue bins, considering, however, only the non-reserved space (of size $1/3$) of such bins. Any bin that FIRSTFIT opens for critical and tiny items will be blue, i.e., it has a reserved space of $2/3$ for critical and a space of $1/3$ for tiny items.

Lemma 1. *When $\beta > 1 - 1/2^{k/2}$, the competitive ratio of the REDBLUE algorithm is at most $1.5 + \frac{7.5}{2^{k/2}}$.*

Proof. From (1) and the statement of the lemma, we have:

$$\frac{Y}{X+Y} \le 1 - \beta < \frac{1}{2^{k/2}} \Rightarrow Y < \frac{1}{2^{k/2}} \cdot (X+Y). \tag{2}$$

Let B denote the set of blue bins. The first X bins in B are precisely the first X bins in the packing of RESERVECRITICAL, i.e., they include X critical items plus the same tiny items. Let Y' denote the number $|B| - X$. Then Y' bins in B only include tiny items (i.e., the reserved space is not occupied by a critical item); the level of all these bins, except possibly one, is at least $1/6$ (otherwise, FIRSTFIT could combine two in the same bin). Since the tiny items placed in these bins are the same as those placed in the last Y bins of the RESERVECRITICAL algorithm, we have $Y' \le 6Y + 1$; this is because the level of bins in the REDBLUE packing is at least $1/6$. Let H and S denote the number of huge and small items. From (2) and the fact that RESERVECRITICAL$(\sigma) = H + \lceil S/2 \rceil + X + Y \le 1.5 \text{OPT}(\sigma)$, we obtain REDBLUE$(\sigma) \le H + \lceil S/2 \rceil + X + 6Y + 1 < (1 + 5/2^{k/2}) \cdot 1.5 \text{OPT}(\sigma) + 1$.   □

Next, we consider the second case: $\beta < 1/2^{k/2}$. In this case, REDBLUE maintains a set of blue bins for critical items and a set of *red* bins for tiny items. The algorithm applies FIRSTFIT to pack critical items into the set of blue bins and tiny items into the set of red items. In this case, all the bins except the blue bins have a level of at least $2/3$; moreover, there are only a few blue bins. We can show that, on average, the level of all the bins (except 1 bin) is very close to $2/3$.

Lemma 2. *When $\beta < \frac{1}{2^{k/2}}$, the competitive ratio of* REDBLUE *is at most $3/2 + \frac{3}{2^k - 2}$.*

Next, we focus on the case $1 - 1/2^{k/2} \le \beta \le 1/2^{k/2}$. In this case, the algorithm maintains a set of blue bins such that each bin has a reserved space of $2/3$ for critical items. The remaining unreserved space of $1/3$ will be used for packing tiny items. The algorithm also maintains a set of red bins for packing tiny items.

We now explain precisely how REDBLUE packs critical and tiny items. For a critical item x, REDBLUE uses FIRSTFIT among the blue bins, and places x in the reserved space of such a bin. If x opens a new bin, the bin is declared blue. For a tiny item y, the algorithm applies FIRSTFIT to place y in either the unreserved space of a blue bin, or in a red bin. If the algorithm cannot place y in one of the existing bins, it opens a new bin for y. It declares the new bin as either a red or a blue bin as follows. Let B and R denote the number of blue and red bins, immediately before y is packed, respectively. The algorithm will then declare the new bin as a blue bin if $\frac{B+1}{B+R+1} \le \beta$; otherwise, it will declare the new bin as red. Note that, in this way, REDBLUE guarantees that $\frac{B_n}{B_n + R_n} \le \beta$, where B_n and R_n denote the number of blue and red bins after processing the entire sequence, respectively. It follows that the number of blue bins in the final packing of REDBLUE is equal to X, i.e, the number of critical items in the sequence. In other words, since β is a lower bound for the ratio $\frac{X}{X+Y}$, this strategy ensures that all bins declared as blue eventually receive a critical item.

Lemma 3. *When $1/2^{k/2} \le \beta \le 1 - 1/2^{k/2}$, the competitive ratio of* REDBLUE *is less than $1.5 + \frac{3}{2^{k/2} - 2}$.*

Proof. From (1) and the statement of the lemma, we have $\frac{X}{X+Y} < \beta + 1/2^k \Rightarrow X < \frac{\beta + 1/2^k}{1 - \beta - 1/2^k} Y$. In the given range for β, we have $\beta(1-\beta)2^k - \beta > 2^{k/2-1} - 1$. Hence,

$$\frac{1-\beta}{\beta} X < \left(1 + \frac{1}{\beta(1-\beta)2^k - \beta}\right) Y < \left(1 + \frac{1}{2^{k/2-1} - 1}\right) Y \qquad (3)$$

Let y_i be a tiny item for which REDBLUE opens the very last red bin in its packing. Let R_i and B_i denote the number of red and blue bins *after* placing y_i, respectively. From the statement of the algorithm, we have $\frac{B_i+1}{B_i+R_i} > \beta$, which implies that $R_i < \frac{(1-\beta)B_i+1}{\beta}$. Let R_n and B_n be the number of red and blue bins in the final packing of the algorithm. We have $R_n = R_i$ and $B_i \le B_n = X$. For the given range of β, in the final packing, all blue bins receive a critical item. Hence, $R_n \le \frac{1-\beta}{\beta} X + 1/\beta$. From the above inequality, we obtain $B_n + R_n \le X + Y + \left(\frac{2}{2^{k/2} - 2}\right) Y + 2^{k/2}$, and the cost of the algorithm can then be bounded as follows: REDBLUE$(\sigma) = H + \lceil S/2 \rceil + B_n + R_n \le H + \lceil S/2 \rceil + X + Y + 2Y/(2^{k/2}) - 2) + 2^{k/2} \le 1.5\,\text{OPT}(\sigma) + \frac{3}{2^{k/2} - 2}\,\text{OPT}(\sigma) + 2^{k/2}$, where we used (3) and the fact that RESERVECRITICAL$(\sigma) = H + \lceil S/2 \rceil + X + Y \le 1.5\,\text{OPT}(\sigma)$. Note also that $2^{k/2}$ is a constant independent of n.   □

Theorem 1. *For any $k \geq 4$, there is an online algorithm for bin packing with k bits of advice that has competitive ratio $1.5 + \frac{15}{2^{k/2+1}}$.*

Proof. From Lemmas 1, 2,3, the competitive ratio of the algorithm is no more than $1.5 + \max\left\{\frac{15}{2^{k/2+1}}, \frac{3}{2^k-2}, \frac{3}{2^{k/2}-2}\right\}$ which is $1.5 + \frac{15}{2^{k/2+1}}$ when $k \geq 4$.   □

In particular, for $k = 16$ bits of advice, we achieve a competitive ratio smaller than 1.530, which is strictly better than any online algorithm.

3   Beyond 1.5-competitiveness with $O(1)$ Advice Bits

We will present and analyze an online algorithm with constant number of advice bits that has a competitive ratio that is arbitrarily close to 1.47012. To this end, we will first introduce an algorithm for sequences in which all items are relatively large, namely larger than $1/3$. We will then use this algorithm as a subroutine in the final algorithm that handles arbitrary sequences.

3.1   Sequences with Items Larger than 1/3

Assume all items are larger than $1/3$. We will show that with only 1 bit of advice, we can achieve solutions which are 1.3904-competitive. For the remainder of this subsection, an item is said to be *small* if it is no larger than $1/2$, *large* if it has size larger than $1/2$ and is placed with a small item in the optimal packing, and *huge* if it is larger than $1/2$ and is alone in its bin in the optimal packing. We use S, L and H to denote the number of small, large and huge items, respectively. We can assume that the size of any huge item is no smaller than large items (otherwise they can be switched and thus obtain another optimal packing). The cost of OPT for the input sequence σ is then $\text{OPT}(\sigma) = H + L/2 + S/2$. We use $\text{OPT}_2(\sigma)$ to denote the number of bins in the optimal packing that include two items, i.e., $\text{OPT}_2(\sigma) = S/2 + L/2$.

The following is the main theorem of this subsection, and will also be used later in the proof of Lemma 10, in the context of general sequences.

Theorem 2. *For a sequence σ in which all items are strictly larger than $1/3$, there is an online algorithm with 1 bit of advice that opens at most $H + 1.3904 \cdot \text{OPT}_2(\sigma)$ bins.*

The following result is direct from Theorem 2, observing that $\text{OPT}(\sigma) = H + \text{OPT}_2(\sigma)$.

Corollary 1. *There is an algorithm for online bin packing with items larger than $1/3$ that uses 1 bit of advice and that has competitive ratio 1.3904.*

The single advice bit serves the purpose of determining the best algorithm among two purely online algorithms: ALMOSTBESTFIT (ABF) and CROSSBEST-FIT (CBF). ABF is similar to BESTFIT except that it opens a new bin for each item larger than $1/2$. CBF also applies BESTFIT, but it opens a new bin for each item smaller than or equal to $1/2$.

In order to prove Theorem 2, we consider three different cases and show that in each case, at least one of ABF and CBF is better than 1.3904-competitive. To define these cases, we consider two parameters α and β such that $0 \le \alpha \le 1$ and $1 \le \beta < 2$. We will determine the values of these parameters later in the proof. Note that in an optimal packing, L large items are matched with small items. We call such two items *partners*. Thus, the partner of a large item (respectively a small item) x is a small (respectively large) item which is placed in the same bin as x in the optimal packing. Let $X \le L$ denote the number of large items which have their partners among the forthcoming items (at the time they are placed). We consider the following three (exhaustive) cases: I) $L \le \frac{\beta-1}{2-\beta}S$, II) $L > \frac{\beta-1}{2-\beta}S$ and $X \ge \alpha L$, and III) $L > \frac{\beta-1}{2-\beta}S$ and $X < \alpha L$.

In the final packing of ABF, all small items (except potentially one of them) are placed with another item. With this observation, we can prove the following for Case I:

Lemma 4. *If $L \le \frac{\beta-1}{2-\beta}S$, ABF opens at most $H + \beta \cdot \text{OPT}_2(\sigma)$ bins.*

Next, we consider Case II.

Lemma 5. *If $L > \frac{\beta-1}{2-\beta}S$ and $X \ge \alpha L$ then ABF opens at most $H + (3/2 - \alpha/2) \cdot \text{OPT}_2(\sigma)$ bins.*

Proof. We claim that in the packing of ABF at least X small items are packed with large items. If this is true, then the number of bins opened by ABF is at most $H + L + (S - X)/2 \le H + L + (S - \alpha L)/2 = H + (2-\alpha)L/2 + S/2$. Note that the ratio $\frac{(2-\alpha)L/2 + S/2}{L/2 + S/2}$ is maximized when L as large as possible, namely when $L = S$. It follows that $\frac{(2-\alpha)L/2 + S/2}{L/2 + S/2} \le \frac{3-\alpha}{2}$, from which we obtain that $(2-\alpha)L/2 + S/2 \le \frac{3-\alpha}{2}\text{OPT}_2(\sigma)$. We thus conclude that $\text{ABF}(\sigma) \le H + (3/2 - \alpha/2)\text{OPT}_2(\sigma)$.

It remains to prove the claim. We maintain a mapping of size X as follows formed by X pairs of items. The mapping is initially formed by the X large items and their partners which appear later. We use $m(y)$ to denote the mapped item of an item y. The mapping is said to be *valid* if it has the following properties: i) for any pair $(x, m(x))$ in the mapping, x is larger than $1/2$ and $x + m(x) \le 1$; and ii) for any pair $(x, m(x))$ in the mapping, x appears earlier than $m(x)$ in the sequence. Note that the initial mapping is valid. We will show how to maintain a valid mapping of size X, upon the arrival and packing of each item, in such a way that all pairs of mapped items are placed in the same bin by the ABF algorithm.

Suppose that a new item y arrives. If y is larger than $1/2$, a new bin is opened for it and the mapping does not change. Next, suppose that y is small; moreover, suppose that the pair (z, y) is in the current mapping, for some item z. If y is placed with z in the same bin, then the mapping does not change. Assume y is placed with another item z' which is larger than z (by BESTFIT it cannot be placed with a smaller item). If z' is in the mapping, we replace

(z, y) with (z', y) (with a slight abuse of notation, we will say that an element r is in the mapping if there is an element q such that the pair (r, q) is in the mapping). Otherwise, since $z \le z'$ we have $z + m(z') \le 1$. In this case, we replace (z, y) and $(z', m(z'))$ with (z', y) and $(z, m(z'))$, respectively. The result is still a valid mapping. Finally, suppose that y is smaller than $1/2$ and it is not in the mapping. The mapping is not changed after packing y unless y is packed with an item z which is in the mapping. Note that z cannot be small; otherwise, it would have been placed with the large item that it is mapped to upon its arrival. Hence, z is a large item. In this case, we replace the pair $(z, m(z))$ with (z, y); this maintains a valid mapping. □

Finally, it remains to consider Case III. The proof of the following lemma uses techniques similar to the proof of Lemma 5.

Lemma 6. *Suppose $L > \frac{\beta - 1}{2 - \beta} S$ and $X < \alpha L$ then the number of bins opened by* CBF *is at most $H + (4 - 2(\alpha + \beta) + 2\alpha\beta) \cdot \mathrm{OPT}_2(\sigma)$.*

Proof of Theorem 2. From Lemmas 4, 5, 6, the competitive ratio of the best algorithm among ABF and CBF is at most $\max\{\beta, 3/2 - \alpha/2, 4 - 2(\alpha + \beta) + 2\alpha\beta\}$, where $0 \le \alpha \le 1$ and $1 \le \beta < 2$. The optimal choice is $\beta = (7 + \sqrt{17})/8$ and $\alpha = (5 - \sqrt{17})/4$ which gives a competitive ratio at most $\beta < 1.3904$. □

3.2 Arbitrary Sequences

We use the result of the previous section to show that advice of constant size suffices to achieve a competitive ratio of $1.47012 + \epsilon$ for any sequence and any arbitrarily small constant ϵ, $0 < \epsilon < 1/12$. To this end, we first define ϵ-*desirable* solutions.

Definition 2. *An ϵ-desirable packing of a sequence σ is a packing formed by a set of ϵ-desirable bins. A bin is ϵ-desirable and belongs to class 0 if there is an empty space of size at least ϵ in the bin. A bin is ϵ-desirable and belongs to class i ($i \in \{1, 2, 3\}$) if its empty space is less than ϵ and if it includes i items in the range $(1/i - \epsilon, 1/i]$.*

We begin with an outline of our approach. First, we will show that, for any packing that consists of X ϵ-desirable bins, there is an online algorithm DESIRABLEROUDING (DR) which opens $(1 + \epsilon)X$ bins and requires advice of size $f(\epsilon)$, where f is a function of ϵ (Lemma 7). Given an ϵ and an optimal offline packing of a sequence σ, we will define two new packings P_1 and P_2 in such a way that at least one of them provides a good approximation of the optimal packing, and the packings can be approximated in an online manner with constant advice. More precisely, P_1 is an ϵ-desirable packing of σ. The packing P_2 is comprised of two packings, P_{2a} and P_{2b}, of a partitioning of the items of σ. P_{2a} is a packing of the items with size at least $1/3$, and P_{2b} is an ϵ-desirable packing of the items with size no more than $1/3$. To approximate P_1, we use the DR algorithm. To approximate P_2, we use the algorithm from Section

3.1 so as to approximate P_{2a} and DR so as to approximate P_{2b}. One additional bit of advice can determine the best among the two online approximations of P_1 and P_2.

We now proceed with the technical details of the algorithm.

Lemma 7. *Consider an ϵ-desirable packing* OFF *of a sequence σ. There is an online algorithm* DR *with advice of size $O(2^{3.7/\epsilon} \cdot \log(1/\epsilon))$ that outputs a packing with at most $(1 + \epsilon)$ OFF(σ) bins, where* OFF(σ) *is the number of bins in the desirable packing.*

Proof Sketch. We give an outline of the proof. The full details are in the long version of the paper. Given an ϵ-desirable packing, the item sizes are rounded up so that there are m different item sizes or *item types*, where m is inversely proportional to ϵ^2. By applying this rounding scheme, there will be a constant (inversely proportional to ϵ^2) number of possible *bin types*, where the type of a bin is based on the number and types of the rounded items packed within. The advice indicates the approximate value for the fraction of bins from each bin type in the desirable packing. Each of these values are encoded in k bits, where k is function of ϵ. Provided with this advice, DR maintains similar ratios for the bins of each type that it opens. To accomplish this, instead of opening single bins, it opens a family of bins in which the bin types are pre-determined so as to maintain the same fraction of bin types as indicated by the advice. Each arriving item is packed into the appropriate reserved space based on the bin and item types. □

Lemma 7 suggests that we need ϵ-desirable packings that are good approximations of OPT(σ). Towards this direction, we need to distinguish between ϵ-*hard* and ϵ-*easy* bins as follows.

Definition 3. *We call a bin ϵ-hard if it contains two items of size larger than $1/3$ such that the total size of these two items is more than $1 - \epsilon$. Otherwise, we call the bin ϵ-easy.*

The following lemma implies that the set of ϵ-easy bins can be changed into a set of ϵ-desirable bins without much overhead. This is accomplished by removing items so as to make such bins desirable. New bins are opened for these removed items, in such a way that 3 bins account for one extra bin.

Lemma 8. *Given a set of items packed in m ϵ-easy bins, it is possible to obtain an ϵ-desirable packing of these items using at most $4/3 \cdot m + 2$ bins.*

We now define the packing P_1 and the online algorithm that approximates it. Let H and E denote the number of ϵ-hard and ϵ-easy bins in OPT(σ), respectively. Let also γ denote the ratio H/E.

To obtain P_1, we apply the procedure of Lemma 8 to transform the E ϵ-easy bins in OPT(σ) into at most $4/3E$ ϵ-desirable bins. Moreover, by applying a procedure similar to Lemma 8, we can transform the H ϵ-hard bins in

$\text{OPT}(\sigma)$ into at most $1.5H$ ϵ-desirable bins. To summarize, P_1 has at most $1.5H + 4/3E = (1.5\gamma + 4/3)E$ bins (omitting additive constants). From Lemma 7, the DR algorithm outputs a packing with $(1.5\gamma + 4/3 + \epsilon')E$ bins. Comparing this to $\text{OPT}(\sigma) = H + E = (1 + \gamma)E$, we get the following result.

Lemma 9. *There is an online algorithm that receives advice of constant size (dependant on ϵ) and achieves a competitive ratio of $\frac{9\gamma+8}{6\gamma+6} + \epsilon$.*

Next, we outline the packing P_2 and the online algorithm that approximates it. In particular, we will define the packings P_{2a} and P_{2b} (as described at the beginning of this section). For our analysis, we partition the set of ϵ-easy bins in the optimal packing into four groups depending on the number of items larger than $1/2$ or $1/3$ in these bins. Let E_1, E_2, E_3 and E_4 indicate the number of bins from these groups ($E_1 + E_2 + E_3 + E_4 = E$). With a similar classifying technique as in Lemmas 8, 9, we obtain P_2 with the following number of bins.

$$\underbrace{H + 1/2 \cdot E_1 + E_2 + E_3}_{P_{2a}:=\text{bins with items}>1/3} + \underbrace{2\epsilon'H + E_1 + 2/3 \cdot E_2 + 4/9 \cdot E_3 + 4/3 \cdot E_4}_{P_{2b}:=\text{desirable bins with items}\leq 1/3}$$

To approximate P_{2a}, since it consists of items of size larger than $1/3$, we can use the online algorithm with 1-bit of advice of Section 3.1. To approximate P_{2b}, since all the bins are ϵ-desirable, we use the online algorithm DR. This defines an online algorithm with at most $((1.3904 + 3\epsilon')\gamma + 1.8349 + \epsilon') \cdot E$ bins. The formal details can be found in the proof of the following lemma.

Lemma 10. *There is an online algorithm that receives advice of constant size (dependant on ϵ) and achieves a competitive ratio of $\frac{1.3904\gamma+1.8349}{\gamma+1} + \epsilon$.*

Theorem 3. *There is an online algorithm with advice of constant size (dependant on ϵ) that achieves a competitive ratio of at most $1.47012 + \epsilon$.*

Proof. We consider two cases depending on the value of γ. Define $\gamma^* = 5015/1096 \approx 4.7633$. If $\gamma \leq \gamma^*$, then we apply the algorithm of Lemma 9; this gives a ratio of at most $\frac{9\cdot\gamma^*+8}{6\cdot\gamma^*+6} + \epsilon < 1.470112 + \epsilon$. If $\gamma > \gamma^*$, then we apply the algorithm of Lemma 10; the competitive ratio is at most $\frac{1.3904\cdot\gamma^*+1.8349}{\gamma+1} + \epsilon < 1.47012 + \epsilon$. □

4   A 7/6 Lower Bound for Sublinear-Sized Advice

In this section, we prove that any online algorithm with $o(n)$ bits of advice has a competitive ratio of at least $7/6$. Our construction is inspired by the one given in [10], which showed a lower bound of $9/8$. Both lower bounds use a reduction from a variant of the *binary string guessing problem* (2-SGKH) [4,12]. In 2-SGKH, the online algorithm must guess an n-length bitstring bit-by-bit. The value of each bit is revealed after the algorithm makes its guess and the algorithm incurs a cost of 1 for each incorrect guess. In particular, we use the *binary string guessing problem with promise* (2-SGKH$_\beta$) that is parameterized by β. This problem is the same as 2-SGKH except that the input string is guaranteed to have exactly a β fraction of 0s (i.e., βn in total). [1]

Lemma 11. *Any deterministic algorithm for 2-SGKH$_\beta$ that is guaranteed to guess correctly more than αn bits, for $\max\{\beta, 1 - \beta\} < \alpha < 1$, requires at least $b(n) = (1 + (1 - \alpha)\log(1 - \alpha) + \alpha\log\alpha)n - e(\gamma n) - 1$ bits of advice, where $\gamma = \min\{\beta, 1 - \beta\}$ and $e(\gamma n) = \lceil\log(\gamma n + 1)\rceil + 2\lceil\log(\lceil\log(\gamma n + 1)\rceil + 1)\rceil + 1$.*

Given an instance $\mathcal{B}$ of the 2-SGKH$_{1/2}$ problem with a bitstring of length n, we construct a request sequence σ for the online bin packing problem with length $2n$ following [10]. (This is described fully in the long version of the paper.) The sequence consists of a *prefix* of $n/2$ items of size $1/2 + \epsilon$, a *central part* of n items of size less than $1/2$ and a *suffix* of $n/2$ items that are the exact complement of the $n/2$ smallest items in the central part. Among these n central items, we refer to the smallest $n/2$ items as *small* items and to the remaining items as *large* items. We observe that $\text{OPT}(\sigma) = n$. The $n/2$ small items are packed with their complements in the suffix; moreover, the remaining $n/2$ large items are packed each with an item of the prefix.

Let $\mathbb{B}$ denote an algorithm for the bin packing problem; we will show how to obtain an online algorithm $\mathbb{A}$ for 2-SGKH$_{1/2}$ that constructs σ and uses $\mathbb{B}$. $\mathbb{B}$ must open a bin for the $n/2$ items of the prefix of σ. The manner in which $\mathbb{B}$ packs each of n central items will determine the n guesses of $\mathbb{A}$. Let b_i be the i-th such item. Algorithm $\mathbb{B}$ has 3 options for packing b_i: (1) to open a new bin for b_i; (2) to pack b_i in a bin with an item from the prefix; or (3) to pack b_i in a bin with some item b_j, $j < i$. If $\mathbb{B}$ chooses option (1), the item is labeled as small and $\mathbb{A}$ guesses 0. If $\mathbb{B}$ chooses either option (2) or (3), the item is labeled as large and $\mathbb{A}$ guesses 1.

The following lemma relates the number of incorrect guesses (or number of mislabeled items) to the number of extra bins opened (in comparison to OPT). We will use the same accounting technique as in [10], but a new mapping of incorrect guesses to mislabellings, which leads to an improved bound. More precisely, we show that each extra bin corresponds to 3 mislabellings (as opposed to [10], in which the corresponding number equals 4). Let f_n denote the family of all request sequences σ constructed as described above, for all possible bitstrings of length n with exactly $n/2$ 0s.

Lemma 12. *Suppose that there is an algorithm $\mathbb{B}$ that uses $b(n)$ bits of advice and opens at most $\text{OPT}(\sigma) + c$ bins for all $\sigma \in f_n$. Then, there exists an algorithm for the 2-SGKH$_{1/2}$ problem that uses $b(n)$ bits of advice and makes at most $3c$ errors.*

We can now show that $\Omega(n)$ advice bits are necessary to obtain a competitive ratio better than $7/6$.

---

[1] Technically, the statement of Lemma 11 is very similar to Lemma 9 in [10]. We note, however, that the latter is correct only when the number of 0s is $n/2$. To avoid any ambiguity, the statement of Lemma 11 is parameterized by β, as opposed to Lemma 9 in [10].

52   S. Angelopoulos et al.

Theorem 4. *Any deterministic online algorithm with advice for the bin packing problem requires at least* $(1 + (1 - \alpha) \log(1 - \alpha) + \alpha \log \alpha)n - e(n/2) - 1$ *bits of advice to be* ρ*-competitive,* $1 < \rho < 7/6$, *where* $\alpha = 4 - 3\rho$ *and* $e(x) = \lceil \log(x + 1) \rceil + 2 \lceil \log(\lceil \log(x + 1) \rceil + 1) \rceil + 1$.

References

1. Adamaszek, A., Renault, M.P., Rosén, A., van Stee, R.: Reordering buffer management with advice. In: Kaklamanis, C., Pruhs, K. (eds.) WAOA 2013. LNCS, vol. 8447, pp. 132–143. Springer, Heidelberg (2014)
2. Ásgeirsson, E.I., Ayesta, U., Coffman, E.G., Etra, J., Momcilovic, P., Phillips, D.J., Vokhshoori, V., Wang, Z., Wolfe, J.: Closed on-line bin packing. Acta Cybernetica **15**(3), 361–367 (2002)
3. Balogh, J., Békési, J., Galambos, G.: New lower bounds for certain classes of bin packing algorithms. Theoretical Computer Science **440–441**, 1–13 (2012)
4. Böckenhauer, H., Hromkovic, J., Komm, D., Krug, S., Smula, J., Sprock, A.: The string guessing problem as a method to prove lower bounds on the advice complexity. Theoretical Computer Science **554**, 95–108 (2014)
5. Böckenhauer, H., Komm, D., Královic, R., Rossmanith, P.: The online knapsack problem: Advice and randomization. Theoretical Computer Science **527**, 61–72 (2014)
6. Böckenhauer, H.J., Komm, D., Královič, R., Královič, R.: On the advice complexity of the k-server problem. In: Aceto, L., Henzinger, M., Sgall, J. (eds.) ICALP 2011, Part I. LNCS, vol. 6755, pp. 207–218. Springer, Heidelberg (2011)
7. Böckenhauer, H.-J., Komm, D., Královič, R., Královič, R., Mömke, T.: On the advice complexity of online problems. In: Dong, Y., Du, D.-Z., Ibarra, O. (eds.) ISAAC 2009. LNCS, vol. 5878, pp. 331–340. Springer, Heidelberg (2009)
8. Borodin, A., El-Yaniv, R.: Online Computation and Competitive Analysis. Cambridge University Press (1998)
9. Boyar, J., Kamali, S., Larsen, K.S., López-Ortiz, A.: On the list update problem with advice. In: Dediu, A.-H., Martín-Vide, C., Sierra-Rodríguez, J.-L., Truthe, B. (eds.) LATA 2014. LNCS, vol. 8370, pp. 210–221. Springer, Heidelberg (2014)
10. Boyar, J., Kamali, S., Larsen, K.S., López-Ortiz, A.: Online bin packing with advice. In: Proc. 31st Symp. on Theoretical Aspects of Computer Science (STACS), pp. 174–186 (2014)
11. Dobrev, S., Královič, R., Pardubská, D.: Measuring the problem-relevant information in input. RAIRO - Theoretical Informatics and Applications **43**(3), 585–613 (2009)
12. Emek, Y., Fraigniaud, P., Korman, A., Rosén, A.: Online computation with advice. Theoretical Computer Science **412**(24), 2642–2656 (2011)
13. Epstein, L., Levin, A.: On bin packing with conflicts. SIAM J. Optimization **19**(3), 1270–1298 (2008)
14. Galambos, G., Woeginger, G.J.: Repacking helps in bounded space online bin packing. Computing **49**, 329–338 (1993)
15. Gambosi, G., Postiglione, A., Talamo, M.: Algorithms for the relaxed online bin-packing model. SIAM J. Computing **30**(5), 1532–1551 (2000)
16. Grove, E.F.: Online bin packing with lookahead. In: Proc. 6th Symp. on Discrete Algorithms (SODA), pp. 430–436 (1995)

17. Johnson, D.S., Demers, A.J., Ullman, J.D., Garey, M.R., Graham, R.L.: Worst-case performance bounds for simple one-dimensional packing algorithms. SIAM J. Computing **3**, 256–278 (1974)
18. Komm, D., Královič, R.: Advice complexity and barely random algorithms. RAIRO - Theoretical Informatics and Applications **45**(2), 249–267 (2011)
19. Renault, M.P., Rosén, A.: On online algorithms with advice for the k-server problem. Theory of Computing Systems **56**(1), 3–21 (2015)
20. Renault, M.P., Rosén, A., van Stee, R.: Online algorithms with advice for bin packing and scheduling problems. CoRR abs/1311.7589 (2013)
21. Seiden, S.S.: On the online bin packing problem. Journal of the ACM **49**, 640–671 (2002)

On the Approximability of Orthogonal Order Preserving Layout Adjustment

Sayan Bandyapadhyay, Santanu Bhowmick, and Kasturi Varadarajan[(⊠)]

Department of Computer Science, University of Iowa, Iowa City, USA
{sayan-bandyapadhyay,santanu-bhowmick,kasturi-varadarajan}@uiowa.edu

Abstract. Given an initial placement of a set of rectangles in the plane, we consider the problem of finding a disjoint placement of the rectangles that minimizes the area of the bounding box and preserves the orthogonal order i.e. maintains the sorted ordering of the rectangle centers along both x-axis and y-axis with respect to the initial placement. This problem is known as *Layout Adjustment for Disjoint Rectangles* (LADR). It was known that LADR is NP-hard, but only heuristics were known for it. We show that a certain decision version of LADR is APX-hard, and give a constant factor approximation for LADR.

1 Introduction

Graphs are often used to visualize relationships between entities in diverse fields such as software engineering (e.g. UML diagrams), VLSI (circuit schematics) and biology (e.g. biochemical pathways) [13]. For many such applications, treating graph nodes as points is insufficient, since each node may have a corresponding label explaining its significance. The presence of labels may lead to node overlapping. For the typical user, an uncluttered layout is more important than the amount of information presented [21]. For complex graphs, it is tedious to create meaningful layouts by hand, which has led to algorithms for layout generation.

Layout generation algorithms typically take a combinatorial description of a graph, and return a corresponding layout. Nodes are usually represented by boxes, and edges by lines connecting the boxes. For simplicity, the edges of the graph are ignored while creating the modified layout. In some interactive systems, modifications to the graph may happen in multiple stages. The layout must be adjusted after each alteration (if new nodes added overlap existing nodes), such that the display area is minimized. If we use layout creation algorithms after each iteration, we may get a layout that is completely different from the previous layout, which may destroy the 'mental map' of the user who is interacting with the system. Thus, we need an additional constraint in the form of maintaining some property of the layout, which would be equivalent to preserving the mental map. Eades et al. [6] defined *orthogonal ordering* as one of the key properties that should be maintained in an adjusted layout to preserve the user's mental

This material is based upon work supported by the National Science Foundation under Grant CCF-1318996.

F. Dehne et al. (Eds.): WADS 2015, LNCS 9214, pp. 54–65, 2015.
DOI: 10.1007/978-3-319-21840-3_5

map. Two layouts of a graph have the same orthogonal ordering if the horizontal and vertical ordering of the nodes are identical in both layouts.

We now state the problem studied in this paper, which involves laying out rectangles that represent the nodes in the graph being adjusted. We are given a set of rectangles R (each $r_i \in R$ is defined by an ordered pair, $r_i = (w_i, h_i)$, denoting its width and height respectively) and an initial layout λ^{in}. A layout consists of an assignment $\lambda : R \to \mathbb{R}^2$ of coordinates to the centers of rectangles in R. The goal is to find a layout in which no two rectangles intersect and orthogonal ordering of the rectangle centers w.r.t λ^{in} is maintained, while minimizing the area of the bounding box of the layout. We refer to this problem as *Layout Adjustment for Disjoint Rectangles* (LADR). Note that R is really a set of rectangle dimensions, and not a set of rectangles. Nevertheless, we will refer to R as a set of rectangles. See Section 2 for a more leisurely problem statement.

1.1   Previous Work

The concept of a mental map was introduced in [6], along with three quantitative models representing it - orthogonal ordering, proximity relations and topology. A framework for analyzing the various models of a mental map was presented in [4], which determined that orthogonal ordering constraint was the best metric for comparing different drawings of the same graph. A user study designed to evaluate human perceptions of similarity amongst two sets of drawings was given in [5], in which orthogonal ordering constraints received the highest rankings.

There has been a lot of work done using the concept of preserving mental maps. LADR was first introduced in [18], in which the authors described the *Force-Scan (FS)* algorithm. FS scans for overlapping nodes in both horizontal and vertical directions, and separates two intersecting nodes by "forcing" them apart along the line connecting the centers of the two nodes, while ensuring that the nodes being forced apart do not intersect any additional nodes in the layout. In [12], a modification of FS was presented (FS′), which resulted in a more compact layout than FS. Another version of FS algorithm, called the *Force-Transfer (FT)* algorithm, was given in [14]. For any two overlapping nodes, denote the vertical distance to be moved to remove the overlap as d_v, and let the horizontal distance for removing overlap be d_h. FT moves the overlapping node horizontally if $d_h < d_v$, else vertically, and experimentally, it has been shown that FT gives a layout of smaller area than FS and FS′.

FS, FS′ and FT belong to the family of force based layout algorithms. Spring based algorithms treat edges as springs obeying Hooke's Law, and the nodes are pushed apart or pulled in iteratively to balance the forces till an equilibrium is reached. A spring based algorithm *ODNLS*, which adjusts the attractive/repulsive force between two nodes dynamically, is proposed in [16], which preserves the orthogonal ordering of the input layout and typically returns a smaller overlap-free layout than the force-based family of algorithms. It is worth noting that none of the algorithms mentioned above give a provable worst-case guarantee on the quality of the output.

The hardness of preserving orthogonal constraints w.r.t various optimization metrics has also been well-studied. Brandes and Pampel [3] showed that it is NP-hard to determine if there exists an orthogonal-order preserving rectilinear drawing of a simple path, and extend the result for determination of uniform edge-length drawings of simple paths with same constraints. LADR was shown to be NP-hard by Hayashi et al. [12], using a reduction from $3SAT$.

1.2 Related Work

Algorithms for label placement and packing that do not account for orthogonal ordering have been extensively studied. The placement of labels corresponding to points on a map is a natural problem that arises in geographic information systems (GIS). In particular, placing labels on maps such that the label boundary coincides with the point feature has been a well-studied problem. A common objective in such label-placement problems is to maximize the number of features labelled, such that the labels are pairwise disjoint. We refer to [2, 15] as examples of this line of work.

Packing rectangles without orthogonality constraints has also been well-studied. One such problem is the strip packing problem, in which we want to pack a set of rectangles into a strip of given width while minimizing the height of the packing. It is known that the strip-packing problem is strongly NP-hard [17]. It can be easily seen that if the constraint for orthogonal order preservation is removed, then LADR can be reduced to multiple instances of strip packing problem. There has been extensive work done on strip packing [10, 19, 20], with the current best algorithm being a $5/3 + \varepsilon$-approximation by Harren et al. [9].

Another related packing problem is the two-dimensional geometric knapsack problem, defined as follows. The input consists of a set of weighted rectangles and a rectangular knapsack, and the goal is to find a subset of rectangles of maximum weight that can be placed in the knapsack such that no two rectangles have an overlap. The 2D-knapsack problem is known to be strongly NP-hard even when the input consists of a set of unweighted squares [17]. Recently, Adamaszek and Wiese [1] gave a quasi-polynomial time $(1 + \varepsilon)$ approximation scheme for this problem, with the assumption that the input consists of quasi-polynomially bounded integers.

1.3 Our Results

We point out an intimate connection between LADR and the problem of hitting segments using a minimum number of horizontal and vertical lines. In particular, the segments to be hit are the ones connecting each pair of rectangle centers in the input layout. The connection to the hitting set is described in Section 3. To our knowledge, this connection to hitting sets has not been observed in the literature. We exploit the connection to hitting set to prove hardness results for LADR in Section 4 that complement the NP-completeness result in [12]. We show that it is APX-hard to find a layout that minimizes the perimeter of the bounding box. We also show that if there is an approximate decision procedure that determines whether there is a layout that fits within a bounding box of specified dimensions, then $\mathbb{P} = \mathbb{NP}$. These hardness results hold even when the input rectangles are unit squares. The results for LADR follow from a hardness of approximation result that we show for a hitting set problem. The starting point of the latter is the result of Hassin and Megiddo [11] who show that it is NP-hard to determine if there is a set of k axis-parallel lines that hit a set of horizontal segments of unit length. The added difficulty that we need to overcome is that in our case, the set of segments that need to be hit cannot be arbitrarily constructed. Rather, the set consists of all segments induced by a set of arbitrarily constructed points. Due to space constraints, we defer most of the proofs to the full version of our paper.

It is possible to exploit this connection to hitting sets and use known algorithms for hitting sets (e.g. [8]) to devise an $O(1)$ approximation algorithm for LADR. Instead, we describe (in Section 5) a direct polynomial time algorithm for LADR that achieves a $4(1+o(1))$ approximation. This is the first polynomial time algorithm for LADR with a provable approximation guarantee. The algorithm involves solving a linear-programming relaxation of LADR followed by a simple rounding.

2   Preliminaries

We define a layout λ of a set of rectangles R as an assignment of coordinates to the center of each rectangle $r \in R$ i.e. $\lambda : R \to \mathbb{R}^2$. Our input for LADR consists of a set of rectangles R, and an initial layout λ^{in}. We will assume that λ^{in} is injective, i.e. no two rectangle centers coincide in the input layout. A rectangle r is defined by its horizontal width w_r and vertical height h_r, both of which are assumed to be integral. It is given that all rectangles are axis-parallel in λ^{in}, and rotation of rectangles is not allowed in any adjusted layout.

The coordinates of center of r in layout λ is denoted by $\lambda(r) = (x_r, y_r)$. For brevity, we denote the x-coordinate of $\lambda(r)$ by $\lambda_x(r)$, and the corresponding y-coordinate by $\lambda_y(r)$. The set of points $\{\lambda(r) : r \in R\}$ is denoted by $\lambda(R)$.

A pair of rectangles $r, r' \in R$ is said to intersect in a layout λ if and only if

$$|\lambda_x(r) - \lambda_x(r')| < \frac{w_r + w_{r'}}{2} \quad \text{and} \quad |\lambda_y(r) - \lambda_y(r')| < \frac{h_r + h_{r'}}{2}. \tag{1}$$

A layout λ is termed as a *disjoint* layout if no two rectangles in R intersect with each other. Let $W_l(\lambda)$ and $W_r(\lambda)$ denote the x-coordinates of the left and right sides of the smallest axis-parallel rectangle bounding the rectangles of R placed by λ, respectively. We then define the width of the layout, $W(\lambda) = W_r(\lambda) - W_l(\lambda)$. Similarly, let $H_t(\lambda)$ and $H_b(\lambda)$ define the y-coordinates of the top and bottom of the bounding rectangle, and the height of the layout is defined as $H(\lambda) = H_t(\lambda) - H_b(\lambda)$. The area of λ is thus defined as $A(\lambda) = H(\lambda) \times W(\lambda)$. The perimeter of λ is $2(H(\lambda) + W(\lambda))$.

Let λ and λ' be two layouts of R. Then, λ and λ' are defined to have the same *orthogonal ordering* if for any two rectangles $r, r' \in R$,

$$\lambda_x(r) < \lambda_x(r') \iff \lambda_x'(r) < \lambda_x'(r') \tag{2}$$

$$\lambda_y(r) < \lambda_y(r') \iff \lambda_y'(r) < \lambda_y'(r') \tag{3}$$

$$\lambda_x(r) = \lambda_x(r') \iff \lambda_x'(r) = \lambda_x'(r') \tag{4}$$

$$\lambda_y(r) = \lambda_y(r') \iff \lambda_y'(r) = \lambda_y'(r') \tag{5}$$

For any R and corresponding λ^{in}, the minimal area of a layout is defined as: $A^{\mathrm{min}} = \inf\{A(\lambda) : \lambda$ is a disjoint layout, λ has same orthogonal ordering as $\lambda^{\mathrm{in}}\}$ It should be noted that it may not be possible to attain a disjoint orthogonality preserving layout whose area is the same as A^{min} - we can only aim to get a layout whose area is arbitrarily close to A^{min}.

Let $\phi(p, p')$ be the segment whose endpoints are points p, p'. Then the set of segments induced by a set of points P is defined as $\Phi(P) = \{\phi(p, p') : p, p' \in P, p \neq p'\}$, denoted by Φ when P is clear from the context.

We also consider a simpler version of LADR where the set of rectangles R consists of unit squares. We call this version as the *Layout Adjustment for Disjoint Squares* problem, and refer to it as LADS for brevity.

3 Reduction of LADS to Hitting Set

We formally define a unit grid as follows. Let $f : \mathbb{R}^2 \to \mathbb{Z}^2$ be the function $f(x, y) = (\lfloor x \rfloor, \lfloor y \rfloor)$. The function f induces a partition of $\mathbb{R}^2$ into grid cells - grid cell (i, j) is the set $\{p \in \mathbb{R}^2 \mid f(p) = (i, j)\}$. We call this partition a unit grid on $\mathbb{R}^2$. The 'grid lines' are the vertical lines $x = \alpha$ and $y = \alpha$ for integer α.

Let S be the set of unit squares provided as input to LADS, having initial layout λ^{in}. Consider a disjoint, orthogonal order preserving layout λ for S. Let L be the subset consisting of those grid lines that intersect the minimum bounding box of $\lambda(S)$. Let ϕ be the line segment connecting the points $\lambda(s)$ and $\lambda(s')$, for some $s, s' \in S$. Since the layout λ is disjoint, $\lambda(s)$ and $\lambda(s')$ lie in different grid cells. Thus, there exists at least one line $\tau \in L$ that intersects ϕ. Motivated by this, we define a hitting set problem as follows.

We say a line τ hits a line segment ϕ if τ intersects the relative interior of ϕ but not either end point of ϕ. Thus, if ϕ is a horizontal line segment, then ϕ cannot be hit by a horizontal line $\tau \in L$. We thus define the Uniform Hitting Set (UHS) problem as follows:

Definition 1 (Uniform Hitting Set - Decision Problem). Given a set of segments Φ induced by a point set P and a non-negative integer k, is there a set of axis-parallel lines L that hit all segments in Φ, such that $|L| \leq k$?

Since the area of the minimum bounding box for $\lambda(S)$ is roughly the product of the number of horizontal grid lines intersecting it and the number of vertical grid lines intersecting it, we also need the following variant.

Definition 2 (Constrained Uniform Hitting Set - Decision Problem).
Given a set of line segments, Φ, induced by a set of points P, and non negative
integers r, c, is it possible to hit all segments in Φ with a set of lines L containing
at most r horizontal lines and c vertical lines ?

The term 'uniform' in the problem name refers to the fact that each segment
in Φ needs to be hit only once by a horizontal or vertical line. We denote the
problem thus defined as CUHS, and proceed to show its equivalence with a
constrained version of the layout adjustment problem.

Definition 3 (Constrained LADS - Decision Problem). Given n unit
squares S, initial layout λ^{in}, positive integers w, h and a constant $0 < \varepsilon < 1$,
is there a layout λ' having height $H(\lambda') \leq h + \varepsilon$ and width $W(\lambda') \leq w + \varepsilon$,
satisfying the following conditions?

1. λ' is a disjoint layout.
2. λ^{in} and λ' have the same orthogonal order.

We term the constrained version of layout adjustment problem as CLADS.
We now show how to transform a given instance of CLADS into an instance of
CUHS. We define Φ as the set of all line segments induced by points in $\lambda^{\text{in}}(S)$.

Lemma 1. *If there is a set of lines L containing at most r horizontal lines and
at most c vertical lines that hit all segments in Φ, then there is a disjoint layout
λ' that has the same orthogonality as λ^{in} and whose height and width is bounded
by $h + \varepsilon$ and $w + \varepsilon$, for any $\varepsilon > 0$. Here $h = r + 1$, $w = c + 1$.*

To solve LADS by multiple iterations of a procedure for solving CUHS, it
would be useful to guess the width of a disjoint layout with near-optimal area.
The following observation allows us to restrict our attention to layouts with
near integral width. That makes it possible to discretize LADS, by solving a
constrained version of LADS for all values of widths in $\{1, 2, \ldots, |S|\}$.

Lemma 2. *Any disjoint layout λ can be modified into a disjoint layout λ' having
the same height and orthogonal ordering as λ, such that $W(\lambda')(\leq W(\lambda))$ lies in
the interval $[w, w + \varepsilon]$, where $w \in \{1, 2, \ldots, |S|\}$ and $\varepsilon > 0$ is an arbitrarily small
constant.*

We can similarly modify a disjoint layout λ into an orthogonal order preserving disjoint layout λ' which has the same width, and whose height lies in the
interval $[h, h + \varepsilon]$ for some integer $h > 0$. Thus, combining the two methods, we
obtain the following corollary:

Corollary 3. *Any disjoint layout λ can be modified into an orthogonal order
preserving disjoint layout λ', such that $W(\lambda')(\leq W(\lambda))$ lies in the interval
$[w, w + \varepsilon]$ and $H(\lambda')(\leq H(\lambda))$ lies in the interval $[h, h + \varepsilon]$, where $w, h \in
\{1, 2, \ldots, |S|\}$ and $\varepsilon > 0$ is an arbitrarily small constant.*

Lemma 4. *For any $\varepsilon < 1/2$, if there is a disjoint layout λ' that has the same orthogonality as λ^{in} and whose height and width is bounded by $h + \varepsilon$ and $w + \varepsilon$ respectively, where h, w are positive integers, then there is a set of lines L containing at most c vertical lines and r horizontal lines that hit all segments in Φ. Here $r = h - 1$, $c = w - 1$.*

Lemmas 1 and 3 and Corollary 4 show the close connection between CLADS and CUHS. In subsequent sections, we exploit this connection to derive hardness results for CLADS.

4   Inapproximability of Layout Adjustment Problems

In this section, we prove APX-hardness of various layout adjustment problems. We consider a variant of LADS where instead of minimizing the area, we would like to minimize the perimeter of the output layout. We prove an inapproximability result for this problem which readily follows from APX-hardness of the Uniform Hitting Set problem. We also show that the decision problem Constrained LADS (CLADS) is NP-hard. Recall that in this problem, given an initial layout of n unit squares, positive integers w, h, and a constant $\varepsilon > 0$, the goal is to determine if there is an orthogonal order preserving layout having height and width at most $h + \varepsilon$ and $w + \varepsilon$ respectively. To be precise, we show a more general inapproximability result for this problem. We prove that there exists $0 < \xi < 1$ such that, given an instance of CLADS, it is NP-hard to determine whether there is an output layout of height and width at most $h + \varepsilon$ and $w + \varepsilon$ respectively, or there is no output layout of respective height and width at most $(1 + \xi)(h + \varepsilon)$ and $(1 + \xi)(w + \varepsilon)$. This result follows from the connection of CLADS with Constrained Uniform Hitting Set (CUHS) described in Section 3 and APX-hardness of CUHS. The APX-hardness of CUHS follows from the APX-hardness of UHS, to which we turn to next.

APX-Hardness of Hitting Set Problem. We consider the optimization version of UHS, in which given a set of points P, the goal is to find minimum number of vertical and horizontal lines that hit all segments in $\Phi(P)$. In this section, we prove that there exists some $0 < \xi < 1$ such that there is no polynomial time $(1 + \xi)$-factor approximation algorithm for UHS, unless $\mathbb{P} = \mathbb{NP}$. Note that the UHS problem we consider here is a special case of the hitting set problem where, given any set of segments S, the goal is to find a hitting set for S. This problem is known to be NP-hard. But, in case of UHS, given a set of points, we need to hit all the segments induced by the points. Thus the nontriviality in our result is to show that even this special case of hitting set is not only NP-hard, but also hard to approximate. To prove the result we reduce a version of maximum satisfiability problem (5-OCC-MAX-3SAT) to UHS. 5-OCC-MAX-3SAT is defined as follows. Given a set X of n boolean variables and a conjunction ϕ of m clauses such that each clause contains precisely three distinct literals and each variable is contained in exactly five clauses ($m = \frac{5n}{3}$), the goal is to find a binary assignment of the variables in X so that the maximum number of clauses of ϕ are satisfied. The following theorem follows from the work of Feige [7].

Theorem 5. *For some $\gamma > 0$, it is NP-hard to distinguish between an instance of 5-OCC-MAX-3SAT consisting of all satisfiable clauses, and one in which less than $(1 - \gamma)$-fraction of the clauses can be satisfied.*

The crux of the hardness result is to show the existence of a reduction from 5-OCC-MAX-3SAT to UHS having the following properties:

1. Any instance of 5-OCC-MAX-3SAT in which all the clauses can be satisfied, is reduced to an instance of UHS in which the line segments in $\Phi(P)$ can be hit using at most k lines, where k is a function of m and n.
2. Any instance of 5-OCC-MAX-3SAT in which less than $1 - \delta$ (for $0 < \delta \leq 1$) fraction of the clauses can be satisfied, is reduced to an instance of UHS in which more than $(1 + \frac{1}{55}\delta)k$ lines are needed to hit the segments in $\Phi(P)$.

The complete reduction appears in the full paper. The next theorem follows from the existence of such a reduction and from Theorem 5.

Theorem 6. *There is no polynomial time $(1 + \xi)$-factor approximation algorithm for UHS with $\xi \leq \frac{1}{55}\gamma$, unless $\mathbb{P} = \mathbb{NP}$, γ being the constant in Theorem 5.*

Now we consider the variant of LADS where we would like to minimize the perimeter $2(w + v)$ of the output layout, where w and v are the width and height of the layout respectively. We refer to this problem as Layout Adjustment for Disjoint Squares - Minimum Perimeter (LADS-MP). We note that in UHS we minimize the sum of the number of horizontal and vertical lines ($k = r + c$). Thus by Lemma 1 and Lemma 4 it follows that a solution for UHS gives a solution for LADS-MP (within an additive constant) and vice versa. Hence the following theorem easily follows from Theorem 6.

Theorem 7. *No polynomial time $(1 + \xi')$-factor approximation algorithm exists for LADS-MP with $\xi' = \frac{\xi}{4}$, unless $\mathbb{P} = \mathbb{NP}$, ξ being the constant in Theorem 6.*

Inapproximability of CUHS. We show that if there is a polynomial time approximate decision algorithm for Constrained Uniform Hitting Set - Decision Problem (CUHS), then $\mathbb{P} = \mathbb{NP}$. We use the inapproximability result of UHS for this purpose. See Definition 2 for the definition of CUHS. Now we have the following theorem whose proof follows from Theorem 6.

Theorem 8. *Suppose there is a polynomial time algorithm that, given $\Phi(P)$ and non-negative integers r, c as input to CUHS,*

(1) outputs "yes", if there is a set with at most c vertical and r horizontal lines that hits the segments in $\Phi(P)$; and
(2) outputs "no", if there is no hitting set for $\Phi(P)$ using at most $(1 + \xi)c$ vertical and $(1 + \xi)r$ horizontal lines, where ξ is the constant in Theorem 6.

Then $\mathbb{P} = \mathbb{NP}$.

Inapproximability of CLADS. We show that the existence of a polynomial time approximate decision algorithm for CLADS implies $\mathbb{P} = \mathbb{NP}$. See Definition 3 for the definition of CLADS. Now we have the following theorem whose proof follows from Theorem 8.

Theorem 9. *Suppose there is a polynomial time algorithm that, given S, λ^{in}, w, h, and ε as input to CLADS,*

(1) outputs "yes", if there is an output layout λ' with $H(\lambda') \le h + \varepsilon$ and $W(\lambda') \le w + \varepsilon$; and

(2) outputs "no", if there is no output layout λ' with $H(\lambda') \le (1 + \xi')(h + \varepsilon)$ and $W(\lambda') \le (1 + \xi')(w + \varepsilon)$, where $\xi' = \frac{\xi}{4}$ and ξ is the constant in Theorem 6.

Then $\mathbb{P} = \mathbb{NP}$.

5 Approximation Algorithm

In this section, we describe an approximation algorithm for LADR i.e. for a set R of axis-parallel rectangles having initial layout λ^{in}, we need to find a disjoint layout of minimum area that preserves the orthogonal ordering of λ^{in}. Let $W_{\max} = \max\{w_r \mid r \in R\}$ and $H_{\max} = \max\{h_r \mid r \in R\}$ be the maximum width and maximum height, respectively, amongst all rectangles in R. Lemma 2 showed that if the input consists of a set of squares S, any disjoint layout of S can be modified into a disjoint layout having same orthogonality such that its width is arbitrarily close to an integer from the set $\{1, \ldots, |S|\}$. It can be seen that Lemma 2 can be extended in a straightforward manner for a set of axis-parallel rectangles R i.e. any disjoint layout of R can be modified into a disjoint orthogonal-order preserving layout having a width that is arbitrarily close to an integer from the set $\{W_{\max}, W_{\max} + 1, \ldots, W_R\}$, where $W_R = \sum_{r \in R} w_r$. We henceforth state Corollary 3 in the context of LADR as follows.

Corollary 10. *Let $W_R = \sum_{r \in R} w_r$ and $H_R = \sum_{r \in R} h_r$ be the sum of the widths and sum of the heights of all the rectangles in R, respectively. Then, any disjoint layout λ of R can be modified into an orthogonal order preserving layout λ' of R, such that $W(\lambda')(\le W(\lambda))$ lies in the interval $[w, w + \varepsilon]$ and $H(\lambda')(\le H(\lambda))$ lies in the interval $[h, h + \varepsilon]$, where $w \in \{W_{\max}, W_{\max} + 1, \ldots, W_R\}$, $h \in \{H_{\max}, H_{\max} + 1, \ldots, H_R\}$ and $\varepsilon > 0$ is an arbitrarily small constant.*

Using Corollary 10, we know that for any disjoint layout λ of R, there is a corresponding disjoint layout λ' having the same orthogonal order as λ, whose height and width are arbitrarily close to an integer from a known set of integers. Hence, we look at all disjoint orthogonality preserving layouts in that range, and choose the one with the minimum area as our solution.

Given positive integers $w \in \{W_{\max}, W_{\max} + 1, \ldots, W_R\}$ and $h \in \{H_{\max}, H_{\max} + 1, \ldots, H_R\}$, we formulate as a LP the problem of whether there is an orthogonal order preserving layout λ with $W(\lambda) \leq w + \varepsilon$, $H(\lambda) \leq h + \varepsilon$. Here we fix some $0 < \varepsilon < 1$. Recall that a layout λ assigns a location $\lambda(r) = (x_r, y_r)$ for the center of each rectangle $r \in R$. The variables of our linear program are $\cup_{r \in R} \{x_r, y_r\}$. For any two rectangles $r, r' \in R$, $\lambda_x^{\mathrm{in}}(r) < \lambda_x^{\mathrm{in}}(r')$ implies that $x_r < x_{r'}'$. We add such a constraint for each pair of rectangles in R, both for x-coordinate and y-coordinate of the layout. Similarly, we add the constraint $x_r = x_r'$ for all pair of rectangles $r, r' \in R$ for which $\lambda_x^{\mathrm{in}}(r) = \lambda_x^{\mathrm{in}}(r')$. These constraints ensure orthogonality is preserved in the output layout.

We now look at constraints that ensure disjointness of the output layout. Let r and r' be two rectangles in the initial layout λ^{in}, having dimensions (w_r, h_r) and $(w_{r'}, h_{r'})$ respectively. We define $w(r, r') = \frac{w_r + w_{r'}}{2}$ and $h(r, r') = \frac{h_r + h_{r'}}{2}$. Let

$$x_{\mathtt{diff}}(r, r') = \begin{cases} x_r - x_{r'}, & \text{if } \lambda_x^{\mathrm{in}}(r') \leq \lambda_x^{\mathrm{in}}(r) \\ x_{r'} - x_r, & \text{otherwise} \end{cases}. \text{ We define } y_{\mathtt{diff}}(r, r') \text{ analogously.}$$

If r, r' are disjoint in some layout, then either their x-projections or their y-projections are disjoint in that layout. Equivalently, either the difference in x-coordinates of the centers of rectangles r, r' is at least $w(r, r')$, or the difference in y-coordinates of the centers is at least $h(r, r')$. We thus get the following LP:

$$x_r < x_{r'} \qquad \forall r, r' \in R : \lambda_x^{\mathrm{in}}(r) < \lambda_x^{\mathrm{in}}(r') \qquad (6)$$

$$x_r = x_{r'} \qquad \forall r, r' \in R : \lambda_x^{\mathrm{in}}(r) = \lambda_x^{\mathrm{in}}(r') \qquad (7)$$

$$y_r < y_{r'} \qquad \forall r, r' \in R : \lambda_y^{\mathrm{in}}(r) < \lambda_y^{\mathrm{in}}(r') \qquad (8)$$

$$y_r = y_{r'} \qquad \forall r, r' \in R : \lambda_y^{\mathrm{in}}(r) = \lambda_y^{\mathrm{in}}(r') \qquad (9)$$

$$\left(x_{r'} + \frac{w_{r'}}{2} \right) - \left(x_r - \frac{w_r}{2} \right) \leq w + \varepsilon \qquad \forall r, r' \in R : \lambda_x^{\mathrm{in}}(r) < \lambda_x^{\mathrm{in}}(r') \qquad (10)$$

$$\left(y_{r'} + \frac{h_{r'}}{2} \right) - \left(y_r - \frac{h_r}{2} \right) \leq h + \varepsilon \qquad \forall r, r' \in R : \lambda_y^{\mathrm{in}}(r) < \lambda_y^{\mathrm{in}}(r') \qquad (11)$$

$$\frac{x_{\mathtt{diff}}(r, r')}{w(r, r')} + \frac{y_{\mathtt{diff}}(r, r')}{h(r, r')} \geq 1 \qquad \forall r, r' \in R \qquad (12)$$

Inequalities (6) to (9) model the orthogonal ordering requirement for a layout, while Inequalities (10) to (11) restrict the width and height of the layout respectively. Since any two rectangles r, r' in a disjoint layout are separated by at least half the sum of their widths in the x-direction ($w(r, r')$) or at least half the sum of their heights in the y-direction ($h(r, r')$), Inequality (12) ensures that every such layout is a valid solution for the linear program. We incorporate the linear program into Algorithm 1 for solving LADR.

Lemma 11. *ApproxLADR(R, λ^{in}) returns a $4 + O(\varepsilon)$-approximation for LADR.*

Proof. Let $\lambda_{w,h}$ be any feasible layout returned by the LP in Line 4, for some value of w, h. Let r, r' be two rectangles in R, and assume that $\lambda_x^{\mathrm{in}}(r) > \lambda_x^{\mathrm{in}}(r')$, $\lambda_y^{\mathrm{in}}(r) > \lambda_y^{\mathrm{in}}(r')$. (The other cases are symmetric). By Inequality (12), either $\frac{x_{\mathtt{diff}}(r,r')}{w(r,r')} \geq \frac{1}{2}$ or $\frac{y_{\mathtt{diff}}(r,r')}{h(r,r')} \geq \frac{1}{2}$. Without loss of generality, assume its the former. Consider the layout $\lambda = 2\lambda_{w,h}$, as in Line 7. Hence, our assumption that

---

Algorithm 1. *ApproxLADR*(R, λ^{in})

Input: A set of rectangles R, and initial layout λ^{in}.
Output: A disjoint layout that has the same orthogonal order as λ^{in}.
1: **for** $w = W_{max}$ to W_R **do**
2:    **for** $h = H_{max}$ to H_R **do**
3:       **if** LP stated in Inequalities (6) to (12) is feasible **then**
4:          $\lambda_{w,h} \leftarrow$ Layout returned by solution of LP.
5:          **if** λ^{min} is undefined **or** $A(\lambda_{w,h}) < A(\lambda^{min})$ **then**
6:             $\lambda^{min} \leftarrow \lambda_{w,h}$
7: Define $\lambda(R) = 2 \cdot \lambda_{min}(R)$ i.e. $\lambda(r) = \left(2 * \lambda_x^{min}(r), 2 * \lambda_y^{min}(r)\right), \forall r \in R$
8: **return** The layout λ.

$\frac{x_{diff}(r,r')}{w(r,r')} \geq \frac{1}{2}$ implies that $\lambda_x(r) - \lambda_x(r') = 2x_r - 2x_{r'} \geq w(r,r')$, which satisfies the criteria for disjointness in Inequality (1). Since the final layout λ returned by the algorithm equals $2 \cdot \lambda_{w',h'}$ for some feasible layout $\lambda_{w',h'}$, λ is a disjoint layout that also satisfies the constraints for orthogonal ordering.

Let λ^* be any disjoint layout preserving the orthogonal ordering of λ^{in}. We may assume, by Corollary 10, that its width is in the interval $[w', w' + \varepsilon]$ and its height is in the interval $[h', h' + \varepsilon]$, for some integers $w' \in \{W_{max}, W_{max} + 1, \ldots, W_R\}$, $h' \in \{H_{max}, H_{max} + 1, \ldots, H_R\}$ and ε as fixed in the LP. Consider the iteration of the inner for loop in Algorithm 1 with $w = w'$ and $h = h'$. Since λ^* is a valid solution for the LP, the layout $\lambda_{w,h}$ computed in Line 4 (and hence λ^{min}) has an area that is at most $(w' + \varepsilon)(h' + \varepsilon)$. Algorithm 1 returns a layout $\lambda(R)$ obtained by multiplying each of the coordinates in λ^{min} by a factor of 2. Hence, the layout $\lambda(R)$ has width at most $2(w' + \varepsilon)$ and height at most $2(h' + \varepsilon)$, ensuring that $A(\lambda) \leq 4 * (w' + \varepsilon)(h' + \varepsilon)$.

We note that since W_R, H_R are not polynomial in the input size, the resultant algorithm is a pseudo-polynomial time algorithm. But by searching across exponentially increasing value of widths, and thereby losing a small approximation factor, we can obtain a $4(1 + o(1))$ polynomial time approximation for LADR. We also note that our approach can be used to get a $2(1 + o(1))$ approximation for the problem of finding a layout of rectangles that minimizes the perimeter. We conclude by summarizing our result as follows:

Theorem 12. *There is a polynomial time algorithm that returns a $4(1 + o(1))$-approximation for LADR i.e. given a set of rectangles R and an initial layout λ^{in}, it returns an orthogonal order preserving disjoint layout whose area is at most $4(1 + o(1))$ times the area attainable by any such layout.*

References

1. Adamaszek, A., Wiese, A.: A quasi-ptas for the two-dimensional geometric knapsack problem. In: SODA, pp. 1491–1505 (2015) 56
2. Agarwal, P.K., van Kreveld, M.J., Suri, S.: Label placement by maximum independent set in rectangles. Comput. Geom. **11**(3–4), 209–218 (1998) 56

3. Brandes, U., Pampel, B.: Orthogonal-ordering constraints are tough. J. Graph Algorithms Appl. **17**(1), 1–10 (2013) 56

4. Bridgeman, S.S., Tamassia, R.: Difference metrics for interactive orthogonal graph drawing algorithms. In: Whitesides, S.H. (ed.) GD 1998. LNCS, vol. 1547, pp. 57–71. Springer, Heidelberg (1999) 55

5. Bridgeman, S., Tamassia, R.: A user study in similarity measures for graph drawing. J. Graph Algorithms Appl. **6**(3), 225–254 (2002) 55

6. Eades, P., Lai, W., Misue, K., Sugiyama, K.: Preserving the mental map of a diagram. International Institute for Advanced Study of Social Information Science, Fujitsu Limited (1991) 54, 55

7. Feige, U.: A threshold of ln n for approximating set cover. J. ACM **45**(4), 634–652 (1998) 61

8. Gaur, D.R., Ibaraki, T., Krishnamurti, R.: Constant ratio approximation algorithms for the rectangle stabbing problem and the rectilinear partitioning problem. J. Algorithms **43**(1), 138–152 (2002) 57

9. Harren, R., Jansen, K., Prädel, L., van Stee, R.: A $(5/3 + \varepsilon)$-approximation for strip packing. Comput. Geom. **47**(2), 248–267 (2014) 56

10. Harren, R., van Stee, R.: Improved absolute approximation ratios for two-dimensional packing problems. In: Dinur, I., Jansen, K., Naor, J., Rolim, J. (eds.) Approximation, Randomization, and Combinatorial Optimization. LNCS, vol. 5687, pp. 177–189. Springer, Heidelberg (2009) 56

11. Hassin, R., Megiddo, N.: Approximation algorithms for hitting objects with straight lines. Discrete Appl. Math. **30**(1), 29–42 (1991) 57

12. Hayashi, K., Inoue, M., Masuzawa, T., Fujiwara, H.: A layout adjustment problem for disjoint rectangles preserving orthogonal order. In: Whitesides, S.H. (ed.) GD 1998. LNCS, vol. 1547, pp. 183–197. Springer, Heidelberg (1999) 55, 56

13. Herman, I., Melançon, G., Marshall, M.S.: Graph visualization and navigation in information visualization: A survey. IEEE Transactions on Visualization and Computer Graphics **6**(1), 24–43 (2000) 54

14. Huang, X., Lai, W., Sajeev, A., Gao, J.: A new algorithm for removing node overlapping in graph visualization. Information Sciences **177**(14), 2821–2844 (2007) 55

15. van Kreveld, M.J., Strijk, T., Wolff, A.: Point labeling with sliding labels. Comput. Geom. **13**(1), 21–47 (1999) 56

16. Li, W., Eades, P., Nikolov, N.S.: Using spring algorithms to remove node overlapping. APVIS. CRPIT **45**, 131–140 (2005) 55

17. Lodi, A., Martello, S., Monaci, M.: Two-dimensional packing problems: A survey. European Journal of Operational Research **141**(2), 241–252 (2002) 56

18. Misue, K., Eades, P., Lai, W., Sugiyama, K.: Layout adjustment and the mental map. Journal of visual languages and computing **6**(2), 183–210 (1995) 55

19. Schiermeyer, I.: Reverse-fit: A 2-optimal algorithm for packing rectangles. In: van Leeuwen, J. (ed.) ESA 1994. LNCS, vol. 855, pp. 290–299. Springer, Heidelberg (1994) 56

20. Steinberg, A.: A strip-packing algorithm with absolute performance bound 2. SIAM J. Comput. **26**(2), 401–409 (1997) 56

21. Storey, M.A.D., Müller, H.A.: Graph layout adjustment strategies. In: Brandenburg, F.J. (ed.) GD 1995. LNCS, vol. 1027, pp. 487–499. Springer, Heidelberg (1996) 54

An Optimal Algorithm for Plane Matchings in Multipartite Geometric Graphs

Ahmad Biniaz[1](✉), Anil Maheshwari[1], Subhas C. Nandy[2], and Michiel Smid[1]

[1] Carleton University, Ottawa, Canada
ahmad.biniaz@gmail.com
[2] Indian Statistical Institute, Kolkata, India

Abstract. Let P be a set of n points in general position in the plane which is partitioned into *color* classes. P is said to be *color-balanced* if the number of points of each color is at most $\lfloor n/2 \rfloor$. Given a color-balanced point set P, a *balanced cut* is a line which partitions P into two color-balanced point sets, each of size at most $2n/3+1$. A *colored matching* of P is a perfect matching in which every edge connects two points of distinct colors by a straight line segment. A *plane colored matching* is a colored matching which is non-crossing. In this paper, we present an algorithm which computes a balanced cut for P in linear time. Consequently, we present an algorithm which computes a plane colored matching of P optimally in $\Theta(n \log n)$ time.

1 Introduction

Let P be a set of n points in general position (no three points on a line) in the plane. Assume P is partitioned into *color* classes, i.e., each point in P is colored by one of the given colors. P is said to be *color-balanced* if the number of points of each color is at most $\lfloor n/2 \rfloor$. In other words, P is color-balanced if no color is in strict majority. For a color-balanced point set P, we define a *feasible cut* as a line ℓ which partitions P into two point sets Q_1 and Q_2 such that both Q_1 and Q_2 are color-balanced. In addition, if the number of points in each of Q_1 and Q_2 is at most $2n/3 + 1$, then ℓ is said to be a *balanced cut*. The well-known ham-sandwich cut (see [10]) is a balanced cut: given a set of $2m$ red points and $2m$ blue points in general position in the plane, a ham-sandwich cut is a line ℓ which partitions the point set into two sets, each of them having m red points and m blue points. Feasible cuts and balanced cuts are useful for convex partitioning of the plane and for computing plane structures, e.g., plane matchings and plane spanning trees.

Let n be an even number. Let $\{R, B\}$ be a partition of P such that $|R| = |B| = n/2$. Let $K_n(R, B)$ be the complete bipartite geometric graph on P which connects every point in R to every point in B by a straight-line edge. An *RB-matching* in P is a perfect matching in $K_n(R, B)$. Assume the points in R are colored red and the points in B are colored blue. An *RB-matching* in

Research supported by NSERC.

F. Dehne et al. (Eds.): WADS 2015, LNCS 9214, pp. 66–78, 2015.
DOI: 10.1007/978-3-319-21840-3_6

P is also referred to as a *red-blue matching* or a *bichromatic matching*. A *plane RB-matching* is an RB-matching in which no two edges cross. Let $\{P_1, \ldots, P_k\}$, where $k \geq 2$, be a partition of P. Let $K_n(P_1, \ldots, P_k)$ be the complete multipartite geometric graph on P which connects every point in P_i to every point in P_j by a straight-line edge, for all $1 \leq i < j \leq k$. Imagine the points in P to be colored, such that all the points in P_i have the same color, and for $i \neq j$, the points in P_i have a different color from the points in P_j. We say that P is a *k-colored* point set. A *colored matching* of P is a perfect matching in $K_n(P_1, \ldots, P_k)$. A *plane colored matching* of P is a perfect matching in $K_n(P_1, \ldots, P_k)$ in which no two edges cross. See Figure 1(a) (see the online version for colored figures).

In this paper we consider the problem of computing a balanced cut for a given color-balanced point set in general position in the plane. We show how to use balanced cuts to compute plane matchings in multipartite geometric graphs.

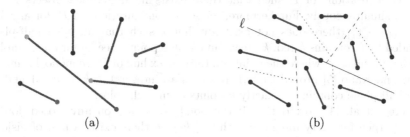

(a)   (b)

Fig. 1. (a) A plane colored matching. (b) Recursive ham sandwich cuts.

1.1 Previous Work

1.1.1 2-Colored Point Sets

Let P be a set of $n = 2m$ points in general position in the plane. Let $\{R, B\}$ be a partition of P such that $|R| = |B| = m$. Assume the points in R are colored red and the points in B are colored blue. It is well-known that $K_n(R, B)$ has a plane RB-matching [1]. In fact, a minimum weight RB-matching, i.e., a perfect matching that minimizes the total Euclidean length of the edges, is plane. A minimum weight RB-matching in $K_n(R, B)$ can be computed in $O(n^{2.5} \log n)$ time [13], or even in $O(n^{2+\epsilon})$ time [2]. Consequently, a plane RB-matching can be computed in $O(n^{2+\epsilon})$ time. As a plane RB-matching is not necessarily a minimum weight RB-matching, one may compute a plane RB-matching faster than computing a minimum weight RB-matching. Hershberger and Suri [8] presented an $O(n \log n)$ time algorithm for computing a plane RB-matching. They also proved a lower bound of $\Omega(n \log n)$ time for computing a plane RB-matching, by providing a reduction from sorting.

Alternatively, one can compute a plane RB-matching by recursively applying the ham sandwich theorem; see Figure 1(b). We say that a line ℓ *bisects* a point set R if both sides of ℓ have the same number of points of R; if $|R|$ is odd, then ℓ contains one point of R.

Theorem 1 (Ham Sandwich Theorem). *For a point set P in general position in the plane which is partitioned into sets R and B, there exists a line that simultaneously bisects R and B.*

A line ℓ that simultaneously bisects R and B can be computed in $O(|R| + |B|)$ time, assuming $R \cup B$ is in general position in the plane [10]. By recursively applying Theorem 1, we can compute a plane RB-matching in $\Theta(n \log n)$ time.

1.1.2 3-Colored Point Sets

Let P be a set of $n = 3m$ points in general position in the plane. Let $\{R, G, B\}$ be a partition of P such that $|R| = |G| = |B| = m$. Assume the points in R are colored red, the points in G are colored green, and the points in B are colored blue. A lot of research has been done to generalize the ham sandwich theorem to 3-colored point sets, see e.g. [4,5,9]. It is easy to see that there exist configurations of P such that there exists no line which bisects R, G, and B, simultaneously. Furthermore, for some configurations of P, for any $k \in \{1, \ldots, m - 1\}$, there does not exist any line ℓ such that an open half-plane bounded by ℓ contains k red, k green, and k blue points (see [5] for an example). For the special case, where the points on the convex hull of P are monochromatic, Bereg and Kano [5] proved that there exists an integer $1 \le k \le m - 1$ and an open half-plane containing exactly k points from each color.

Bereg et al. [4] proved that if the points of P are on any closed Jordan curve γ, then for every integer k with $0 \le k \le m$ there exists a pair of disjoint intervals on γ whose union contains exactly k points of each color. In addition, they showed that if m is even, then there exists a double wedge that contains exactly $m/2$ points of each color.

Now, let P be a 3-colored point set of size n in general position in the plane, with n even. Assume the points in P are colored red, green, and blue such that P is color-balanced. Let R, G, and B denote the set of red, green, and blue points, respectively. Note that $|R|$, $|G|$, and $|B|$ are at most $\lfloor n/2 \rfloor$, but, they are not necessarily equal. Kano et al. [9] proved the existence of a feasible cut, when the points on the convex hull of P are monochromatic.

Theorem 2 (Kano et al. [9]). *Let P be a 3-colored point set in general position in the plane, such that P is color-balanced and $|P|$ is even. If the points on the convex hull of P are monochromatic, then there exists a line ℓ which partitions P into Q_1 and Q_2 such that both Q_1 and Q_2 are color-balanced and have an even number of points and $2 \le |Q_i| \le |P| - 2$, for $i = 1, 2$.*

They also proved the existence of a plane perfect matching in $K_n(R, G, B)$ by recursively applying Theorem 2. Their proof is constructive. Although they did not analyze the running time, it can be shown that their algorithm runs in $O(n^2 \log n)$ time as follows. If the size of the largest color class is exactly $n/2$, then consider the points in the largest color class as R and the other points as B, then compute a plane RB-matching; and we are done. If there are two adjacent points of distinct colors on the convex hull, then match these two points and recurse on the remaining points. Otherwise, if the convex hull is monochromatic, pick a point $p \in P$ on the convex hull and sort the points in $P \setminus \{p\}$ around p. A line ℓ—partitioning the point set into two color-balanced point sets—is found by scanning the sorted list. Then recurse on each of the partitions. To find ℓ they spend $O(n \log n)$ time. The total running time of their algorithm is $O(n^2 \log n)$.

Based on the algorithm of Kano et al. [9], we can show that a plane perfect matching in $K_n(R, G, B)$ can be computed in $O(n \log^3 n)$ time. We can prove the existence of a feasible cut for P, even if the points on the convex hull of P are not monochromatic. To find feasible cuts recursively, we use the dynamic convex hull structure of Overmars and Leeuwen [11], which uses $O(\log^2 n)$ time for each insertion and deletion. Pick a point $p \in P$ on the convex hull of P and look for a point $q \in P \setminus \{p\}$, such that the line passing through p and q is a feasible cut. Search for q, alternatively, in clockwise and counterclockwise directions around p. To do this, we repeatedly check if the line passing through p and its (clockwise and counterclockwise in turn) neighbor on the convex hull, say r, is a feasible cut. If the line through p and r is not a feasible cut, then we delete r. At some point we find a feasible cut ℓ which divides P into Q_1 and Q_2. Add the two points on ℓ to either Q_1 or Q_2 such that they remain color-balanced. Let $|Q_1| = k$ and $|Q_2| \geq k$. In order to compute the data structure for Q_2, we use the current data structure and undo the deletions on the side of ℓ which contains Q_2. We rebuild the data structure for Q_1. Then, we recurse on Q_1 and Q_2. The running time can be expressed by $T(n) = T(n-k) + T(k) + O(k \log^2 n)$, where $k \leq n - k$. This recurrence solves to $O(n \log^3 n)$.

1.1.3 Multicolored Point Sets

Let $\{P_1, \ldots, P_k\}$, where $k \geq 2$, be a partition of P and $K_n(P_1, \ldots, P_k)$ be the complete multipartite geometric graph on P. A necessary and sufficient condition for the existence of a perfect matching in $K_n(P_1, \ldots, P_k)$ follows from the following result of Sitton [12].

Theorem 3 (Sitton [12]). *The size of a maximum matching in any complete multipartite graph $K_{n_1, \ldots, n_k}$, with $n = n_1 + \cdots + n_k$ vertices, where $n_1 \geq \cdots \geq n_k$, is*

$$|M_{max}| = \min \left\{ \sum_{i=2}^{k} n_i, \left\lfloor \frac{1}{2} \sum_{i=1}^{k} n_i \right\rfloor \right\}.$$

Theorem 3 implies that if n is even and $n_1 \leq \frac{n}{2}$, then $K_{n_1, \ldots, n_k}$ has a perfect matching. It is obvious that if $n_1 > \frac{n}{2}$, then $K_{n_1, \ldots, n_k}$ does not have any perfect matching. Therefore,

Corollary 1. *Let $k \geq 2$ and consider a partition $\{P_1, \ldots, P_k\}$ of a point set P, where $|P|$ is even. Then, $K_n(P_1, \ldots, P_k)$ has a colored matching if and only if P is color-balanced.*

Aichholzer et al. [3], and Kano et al. [9] show that the same condition as in Corollary 1 is necessary and sufficient for the existence of a plane colored matching in $K_n(P_1, \ldots, P_k)$:

Theorem 4 (Aichholzer et al. [3], and Kano et al. [9]). *Let $k \geq 2$ and consider a partition $\{P_1, \ldots, P_k\}$ of a point set P, where $|P|$ is even. Then, $K_n(P_1, \ldots, P_k)$ has a plane colored matching if and only if P is color-balanced.*

In fact, they show something stronger. Aichholzer et al. [3] show that a minimum weight colored matching in $K_n(P_1, \ldots, P_k)$, which minimizes the total Euclidean length of the edges, is plane. Gabow [7] gave an implementation of Edmonds' algorithm which computes a minimum weight matching in general graphs in $O(n(m + n \log n))$ time, where m is the number of edges in G. Since P is color-balanced, $K_n(P_1, \ldots, P_k)$ has $\Theta(n^2)$ edges. Thus, a minimum weight colored matching in $K_n(P_1, \ldots, P_k)$, and hence a plane colored matching in $K_n(P_1, \ldots, P_k)$, can be computed in $O(n^3)$ time. Kano et al. [9] extended their $O(n^2 \log n)$-time algorithm for the 3-colored point sets to the multicolored case.

Since the problem of computing a plane RB-matching in $K_n(R, B)$ is a special case of the problem of computing a plane colored matching in $K_n(P_1, \ldots, P_k)$, the $\Omega(n \log n)$ time lower bound for computing a plane RB-matching holds for computing a plane colored matching.

1.2   Our Contribution

Our main contribution, which is presented in Section 2, is the following: given any color-balanced point set P in general position in the plane, there exists a balanced cut for P. Further, we show that if n is even, then there exists a balanced cut which partitions P into two point sets each of even size, and such a balanced cut can be computed in linear time. In Section 3, we present a divide-and-conquer algorithm which computes a plane colored matching in $K_n(P_1, \ldots, P_k)$ in $\Theta(n \log n)$ time, by recursively finding balanced cuts in color-balanced subsets of P. In case P is not color-balanced, then $K_n(P_1, \ldots, P_k)$ does not admit a perfect matching; we describe how to find a plane colored matching with the maximum number of edges in Section 3.1. In addition, we show how to compute a maximum matching in any complete multipartite graph in linear time.

2   Balanced Cut Theorem

Given a color-balanced point set P with $n \geq 4$ points in general position in the plane, a *balanced cut* is a line which partitions P into two point sets Q_1 and Q_2, such that both Q_1 and Q_2 are color-balanced and $\max\{|Q_1|, |Q_2|\} \leq \frac{2n}{3} + 1$. Let $\{P_1, \ldots, P_k\}$ be a partition of P, where the points in P_i are colored C_i. In this section we prove the existence of a balanced cut for P. Moreover, we show how to find such a balance cut in $O(n)$ time.

If $k = 2$, the existence of a balanced cut follows from the ham sandwich cut theorem. If $k \geq 4$, we reduce the k-colored point set P to a three colored point set. Afterwards, we prove the statement for $k = 3$.

Lemma 1. *Let P be a color-balanced point set of size n in the plane with $k \geq 4$ colors. In $O(n)$ time P can be reduced to a color-balanced point set P' with 3 colors such that any balanced cut for P' is also a balanced cut for P.*

Proof. We repeatedly merge the color families in P until we get a color-balanced point set P' with three colors. Afterwards, we show that any balanced cut for P' is also a balanced cut for P.

Without loss of generality assume that $C_1, \ldots, C_k$ is a non-increasing order of the color classes according to the number of points in each color class. That is, $\lfloor |P|/2 \rfloor \geq |P_1| \geq \cdots \geq |P_k| \geq 1$ (note that P is color-balanced). In order to reduce the k-colored problem to a 3-colored problem, we repeatedly merge the two color families of the smallest cardinality. In each iteration we merge the stwo smallest color families, C_{k-1} and C_k, to get a new color class, C'_{k-1}, where $P'_{k-1} = P_{k-1} \cup P_k$. In order to prove that $P' = P_1 \cup \cdots \cup P_{k-2} \cup P'_{k-1}$ is color-balanced with respect to the coloring $C_1, \ldots, C_{k-2}, C'_{k-1}$ we have to show that $|P'_{k-1}| \leq \lfloor |P'|/2 \rfloor$. Note that before the merge we have $|P| = |P_1| + \cdots + |P_{k-2}| + |P_{k-1}| + |P_k|$, while after the merge we have $|P'| = |P_1| + \cdots + |P_{k-2}| + |P'_{k-1}|$, where $|P'_{k-1}| = |P_{k-1}| + |P_k|$. Since P_{k-1} and P_k are the two smallest and $k \geq 4$, $|P'_{k-1}| \leq |P_1| + \cdots + |P_{k-2}|$. This implies that after the merge we have $|P'_{k-1}| \leq \lfloor |P'|/2 \rfloor$. Thus P' is color-balanced. By repeatedly merging the points of the two smallest color families, at some point we get a 3-colored point set P' which is color-balanced. Without loss of generality assume that P' is colored by R, G, and B. Consider any balanced cut ℓ for P'; ℓ partitions P' into two sets Q_1 and Q_2, each of size at most $\frac{2}{3}n + 1$, such that the number points of each color in Q_i is at most $\lfloor |Q_i|/2 \rfloor$, where $i = 1, 2$. Note that the set of points in P colored C_j, for $1 \leq j \leq k$, is a subset of points in P' colored either R, G, or B. Thus, the number of points colored C_j in Q_i is at most $\lfloor |Q_i|/2 \rfloor$, where $j = 1, \ldots, k$ and $i = 1, 2$. Therefore, ℓ is a balanced cut for P.

In order to merge the color families, a monotone priority queue (see [6]) can be used, where the priority of each color C_j is the number of points colored C_j. The monotone priority queue offers **insert** and **extract-min** operations where the priority of an inserted element is greater than the priority of the last element extracted from the queue. We store the color families in a monotone priority queue of size $\frac{n}{2}$ (because all elements are in the range of 1 up to $\frac{n}{2}$). Afterwards, we perform a sequence of $O(k)$ **extract-min** and **insert** operations. Since $k \leq n$, the total time to merge k color families is $O(n)$. □

According to Lemma 1, from now on we assume that P is a color-balanced point set consisting of n points colored by three colors.

Lemma 2. *Let P be a color-balanced point set of $n \geq 4$ points in general position in the plane with three colors. In $O(n)$ time we can compute a line ℓ such that*

1. *ℓ does not contain any point of P.*
2. *ℓ partitions P into two point sets Q_1 and Q_2, where*
   (a) *both Q_1 and Q_2 are color-balanced,*
   (b) *both Q_1 and Q_2 contains at most $\frac{2}{3}n + 1$ points.*

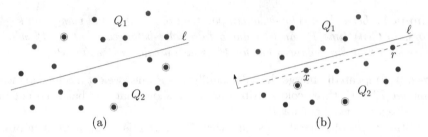

Fig. 2. Illustrating the balanced cut theorem. The blue points in X are surrounded by circles. The line ℓ is a balanced cut where: (a) $|R|$ is even, and (b) $|R|$ is odd.

Proof. Assume that the points in P are colored red, green, and blue. Let R, G, and B denote the set of red, green, and blue points, respectively. Without loss of generality assume that $1 \leq |B| \leq |G| \leq |R|$. Since P is color-balanced, $|R| \leq \lfloor \frac{n}{2} \rfloor$. Let X be an arbitrary subset of B such that $|X| = |R| - |G|$; note that $X = \emptyset$ when $|R| = |G|$, and $X = B$ when $|R| = \frac{n}{2}$ (where n is even). Let $Y = B - X$. Let ℓ be a ham sandwich cut for R and $G \cup X$ (pretending that the points in $G \cup X$ have the same color). Let Q_1 and Q_2 denote the set of points on each side of ℓ; see Figure 2(a). If $|R|$ is odd, then ℓ contains a point $r \in R$ and a point $x \in G \cup X$; see Figure 2(b). In this case without loss of generality assume that the number of blue points in Q_2 is at least the number of blue points in Q_1; slide ℓ slightly such that r and x lie in the same side as Q_2, i.e. Q_2 is changed to $Q_2 \cup \{r, x\}$. We prove that ℓ satisfies the statement of the theorem. The line ℓ does not contain any point of P and by the ham sandwich cut theorem it can be computed in $O(n)$ time.

Now we prove that both Q_1 and Q_2 are color-balanced. Let R_1, G_1, and B_1 be the set of red, green, and blue points in Q_1, where $X_1 = X \cap Q_1$, $Y_1 = Y \cap Q_1$, and $B_1 = X_1 \cup Y_1$. Similarly, define R_2, G_2, B_2, X_2, and Y_2 as subsets of Q_2. Since $|R| = |G \cup X|$ and ℓ bisects both R and $G \cup X$, we have $|R_1| = \lfloor |R|/2 \rfloor$ and $|G_1| + |X_1| = |R_1|$. In the case that $|R|$ is odd, we add the points on ℓ to Q_2 (assuming that $|B_2| \geq |B_1|$). Thus, in either case ($|R|$ is even or odd) we have $|R_2| = \lceil |R|/2 \rceil$ and $|G_2| + |X_2| = |R_2|$. Therefore,

$$|Q_1| \geq |R_1| + |G_1| + |X_1| = 2\lfloor |R|/2 \rfloor,$$
$$|Q_2| \geq |R_2| + |G_2| + |X_2| = 2\lceil |R|/2 \rceil. \tag{1}$$

Let t_1 and t_2 be the total number of red and green points in Q_1 and Q_2, respectively. Then, we have the following inequalities:

$$
\begin{aligned}
t_1 &= |R_1| + |G_1| \\
&= 2|R_1| - |X_1| \\
&\geq 2|R_1| - |X| \\
&= 2\lfloor |R|/2 \rfloor - (|R| - |G|) \\
&= \begin{cases} |G| & \text{if } R \text{ is even} \\ |G| - 1 & \text{if } R \text{ is odd,} \end{cases}
\end{aligned}
\qquad
\begin{aligned}
t_2 &= |R_2| + |G_2| \\
&= 2|R_2| - |X_2| \\
&\geq 2|R_2| - |X| \\
&= 2\lceil |R|/2 \rceil - (|R| - |G|) \\
&= \begin{cases} |G| & \text{if } R \text{ is even} \\ |G| + 1 & \text{if } R \text{ is odd.} \end{cases}
\end{aligned}
\tag{2}
$$

In addition, we have the following equations:

$$|Q_1| = t_1 + |B_1| \quad \text{and} \quad |Q_2| = t_2 + |B_2|. \tag{3}$$

Note that $|R_1| = \lfloor |R|/2 \rfloor$ and $|G_1| \leq |R_1|$, thus, by Inequality (1) we have $|R_1| \leq \lfloor |Q_1|/2 \rfloor$ and $|G_1| \leq \lfloor |Q_1|/2 \rfloor$. Similarly, $|R_2| \leq \lfloor |Q_2|/2 \rfloor$ and $|G_2| \leq \lfloor |Q_2|/2 \rfloor$. Therefore, in order to argue that Q_1 and Q_2 are color-balanced, it only remains to show that $|B_1| \leq \lfloor |Q_1|/2 \rfloor$ and $|B_2| \leq \lfloor |Q_2|/2 \rfloor$. Note that $|B_1|, |B_2| \leq |B|$ and by initial assumption $|B| \leq |G|$. We differentiate between two cases where $|R|$ is even and $|R|$ is odd. If $|R|$ is even, by Inequalities (2) we have $t_1, t_2 \geq |G|$. Therefore, by Equation (3), $|B_1| \leq \lfloor |Q_1|/2 \rfloor$ and $|B_2| \leq \lfloor |Q_2|/2 \rfloor$. If $|R|$ is odd, we slide ℓ towards Q_1; assuming that $|B_2| \geq |B_1|$. In addition, since $|B_1| + |B_2| = |B|$ and $|B| \geq 1$, $|B_2| \geq 1$. Thus, $|B_1| \leq |B| - 1 \leq |G| - 1$, while by Inequality (2), $t_1 \geq |G| - 1$. Therefore, Equality (3) implies that $|B_1| \leq \lfloor |Q_1|/2 \rfloor$. Similarly, by Inequality (2) we have $t_2 \geq |G| + 1$ while $|B_2| \leq |G|$. Thus, Equality (3) implies that $|B_2| \leq \lfloor |Q_2|/2 \rfloor$. Therefore, both Q_1 and Q_2 are color-balanced.

We complete the proof for $k = 3$ by providing the following upper bound on the size of Q_1 and Q_2. Since we assume that R is the largest color class, $|R| \geq \lceil \frac{n}{3} \rceil$. By Inequality (1), $\min\{|Q_1|, |Q_2|\} \geq 2\lfloor |R|/2 \rfloor$, which implies that

$$\max\{|Q_1|, |Q_2|\} \leq n - 2\lfloor \frac{|R|}{2} \rfloor \leq n - 2(\frac{|R|-1}{2}) \leq n - \frac{n}{3} + 1 = \frac{2n}{3} + 1.$$

$\square$

Therefore, by Lemma 1 and Lemma 2, we have proved the following theorem:

Theorem 5 (Balanced Cut Theorem). *Let P be a color-balanced point set of $n \geq 4$ points in general position in the plane. In $O(n)$ time we can compute a line ℓ such that*

1. *ℓ does not contain any point of P.*
2. *ℓ partitions P into two point sets Q_1 and Q_2, where*
   (a) *both Q_1 and Q_2 are color-balanced,*
   (b) *both Q_1 and Q_2 contains at most $\frac{2}{3}n + 1$ points.*

By Theorem 4, if P has even number of points and no color is in strict majority, then P admits a plane perfect matching. By Theorem 5, we partition P into two sets Q_1 and Q_2 such that in each of them no point is in strict majority. But, in order to apply the balanced cut theorem, recursively, to obtain a perfect matching on each side of the cut, we need both Q_1 and Q_2 to have an even number of points. Thus, we extend the result of Theorem 5 to a restricted version of the problem where $|P|$ is even and we are looking for a balanced cut which partitions P into Q_1 and Q_2 such that both $|Q_1|$ and $|Q_2|$ are even. The following theorem describes how to find such a balanced cut.

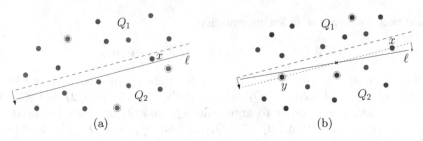

Fig. 3. Updating ℓ to make $|Q_1|$ and $|Q_2|$ even numbers, where: (a) ℓ passes over one point, and (b) ℓ passes over two points

Theorem 6. *Let P be a color-balanced point set of $n \geq 4$ points in general position in the plane with n even and three colors. In $O(n)$ time we can compute a line ℓ such that*

1. *ℓ does not contain any point of P.*
2. *ℓ partitions P into two point sets Q_1 and Q_2, where*
   (a) both Q_1 and Q_2 are color-balanced,
   (b) both Q_1 and Q_2 have even number of points,
   (c) both Q_1 and Q_2 contains at most $\frac{2}{3}n + 1$ points.

Proof. Let ℓ be the balanced cut obtained in the proof of Lemma 2, which divides P into Q_1 and Q_2. Note that ℓ does not contain any point of P. If $|Q_1|$ is even, subsequently $|Q_2|$ is even, thus ℓ satisfies the statement of the theorem and we are done. Assume that $|Q_1|$ and $|Q_2|$ are odd. Note that $|Q_1| = |R_1|+|G_1|+|X_1|+|Y_1|$ and $|Q_2| = |R_2| + |G_2| + |X_2| + |Y_2|$. Recall that $|R_1| = |G_1| + |X_1| = \lfloor |R/2| \rfloor$ and $|R_2| = |G_2|+|X_2| = \lceil |R/2| \rceil$. Thus, $|R_1|+|G_1|+|X_1|$ and $|R_2|+|G_2|+|X_2|$ are even. In order to make $|Q_1|$ and $|Q_2|$ to be odd numbers, both $|Y_1|$ and $|Y_2|$ have to be odd numbers. Thus, $|Y_1| \geq 1$ and $|Y_2| \geq 1$, which implies that

$$|Q_1| = |R_1| + |G_1| + |X_1| + |Y_1| \geq 2\lfloor |R/2| \rfloor + 1,$$
$$|Q_2| = |R_2| + |G_2| + |X_2| + |Y_2| \geq 2\lceil |R/2| \rceil + 1. \tag{4}$$

In addition,

$$|B_1| = |B| - (|X_2| + |Y_2|) \leq |B| - 1,$$
$$|B_2| = |B| - (|X_1| + |Y_1|) \leq |B| - 1. \tag{5}$$

Note that Q_1 is color-balanced. That is, $|R_1|, |G_1|, |B_1| \leq \lfloor |Q_1|/2 \rfloor$, where $|Q_1|$ is odd. Thus, by addition of one point (of any color) to Q_1, it still remain color-balanced. Therefore, we slide ℓ slightly towards Q_2 and stop as soon as it passes over a point $x \in Q_2$; see Figure 3(a). If ℓ passes over two points x and y, rotate ℓ slightly, such that x lies on the same side as Q_1 and y remains on the other side; see Figure 3(b). We prove that ℓ satisfies the statement of the theorem. It is obvious that updating the position of ℓ takes $O(n)$ time. Let $Q_1' = Q_1 \cup \{x\}$ and $Q_2' = Q_2 - \{x\}$. By the previous argument Q_1' is color-balanced. Now we show that Q_2' is color-balanced as well. Note that $|Q_2'| = |Q_2| - 1$, thus, by Inequality (4) we have

$$|Q_2'| \geq 2\lceil |R|/2 \rceil.$$

Let R_2', G_2', and B_2' be the set of red, green, and blue points in Q_2', and let t_2' be the total number of red and green points in Q_2'. Then,

$$|Q_2'| = t_2' + |B_2'|. \tag{6}$$

To prove that Q_2' is color-balanced we differentiate between three cases, where $x \in R_2$, $x \in G_2$, or $x \in B_2$:

- $x \in R_2$. In this case: (i) $|R_2'| = |R_2| - 1 = \lceil |R|/2 \rceil - 1 \leq \lfloor |Q_2'|/2 \rfloor$. (ii) $|G_2'| = |G_2| \leq |R_2| = \lceil |R|/2 \rceil \leq \lfloor |Q_2'|/2 \rfloor$. (iii) $t_2' = t_2 - 1 \geq |G| - 1$, while $|B_2'| = |B_2| \leq |B| - 1 \leq |G| - 1$; Inequality (6) implies that $|B_2'| \leq \lfloor |Q_2'|/2 \rfloor$.
- $x \in G_2$. In this case: (i) $|R_2'| = |R_2| = \lceil |R|/2 \rceil \leq \lfloor |Q_2'|/2 \rfloor$. (ii) $|G_2'| = |G_2| - 1 \leq |R_2| - 1 = \lceil |R|/2 \rceil - 1 \leq \lfloor |Q_2'|/2 \rfloor$. (iii) $t_2' = t_2 - 1 \geq |G| - 1$, while $|B_2'| = |B_2| \leq |B| - 1 \leq |G| - 1$; Inequality (6) implies that $|B_2'| \leq \lfloor |Q_2'|/2 \rfloor$.
- $x \in B_2$. In this case: (i) $|R_2'| = |R_2| = \lceil |R|/2 \rceil \leq \lfloor |Q_2'|/2 \rfloor$. (ii) $|G_2'| = |G_2| \leq |R_2| = \lceil |R|/2 \rceil \leq \lfloor |Q_2'|/2 \rfloor$. (iii) $t_2' = t_2 \geq |G|$, while $|B_2'| = |B_2| - 1 \leq |B| - 2 \leq |G| - 2$; Inequality (6) implies that $|B_2'| \leq \lfloor |Q_2'|/2 \rfloor$.

In all cases $|R_2'|, |G_2'|, |B_2'| \leq \lfloor |Q_2'|/2 \rfloor$, which imply that Q_2' is color-balanced.

As for the size condition,

$$\min\{|Q_1'|, |Q_2'|\} = \min\{|Q_1| + 1, |Q_2| - 1\} \geq 2\lfloor |R|/2 \rfloor,$$

where the last inequality resulted from Inequality (4). This implies that $\max\{|Q_1'|, |Q_2'|\} \leq \frac{2n}{3} + 1$. Thus, ℓ satisfies the statement of the theorem, where $Q_1 = Q_1'$ and $Q_2 = Q_2'$. $\qquad \square$

Note that both Theorem 6 and Theorem 2 prove the existence of a line ℓ which partitions a color-balanced point set P into two color-balanced point sets Q_1 and Q_2. But, there are two main differences: (i) Theorem 6 can be applied on any color-balanced point set P in general position. Theorem 2 is only applicable on color-balanced point sets in general position, where the points on the convex hull are monochromatic. (ii) Theorem 6 proves the existence of a balanced cut such that $\frac{n}{3} - 1 \leq |Q_i| \leq \frac{2n}{3} + 1$, while the cut computed by Theorem 2 is not necessarily balanced, as $2 \leq |Q_i| \leq n - 2$, where $i = 1, 2$. In addition, the balanced cut in Theorem 6 can be computed in $O(n)$ time, while the cut in Theorem 2 is computed in $O(n \log n)$ time.

3   Plane Colored Matching Algorithm

Let P be a color-balanced point set of n points in general position in the plane with respect to a partition $\{P_1, \ldots, P_k\}$, where n is even and $k \geq 2$. In this section we present an algorithm which computes a plane colored matching in $K_n(P_1, \ldots, P_k)$ in $\Theta(n \log n)$ time.

Let $\{C_1, \ldots, C_k\}$ be a set of k colors. Imagine all the points in P_i are colored C_i for all $1 \leq i \leq k$. Without loss of generality, assume that $|P_1| \geq |P_2| \geq \cdots \geq |P_k|$. If $k = 2$, then we can compute an RB-matching in $O(n \log n)$ time by recursively applying the ham sandwich theorem. If $k \geq 4$, as in Lemma 1, in $O(n)$ time, we compute a color-balanced point set P with three colors. Any plane colored matching for P with respect to the three colors, say (R, G, B), is also a plane colored matching with respect to the coloring $C_1, \ldots, C_k$. Hereafter, assume that P is a color-balanced point set which is colored by three colors.

By Theorem 6, in linear time we can find a line ℓ that partitions P into two sets Q_1 and Q_2, where both Q_1 and Q_2 are color-balanced with an even number of points, such that $\max\{|Q_1|, |Q_2|\} \leq \frac{2n}{3} + 1$. Since Q_1 and Q_2 are color-balanced, by Corollary 1, both Q_1 and Q_2 admit plane colored matchings. Let $M(Q_1)$ and $M(Q_2)$ be plane colored matchings in Q_1 and Q_2, respectively. Since Q_1 and Q_2 are separated by ℓ, $M(Q_1) \cup M(Q_2)$ is a plane colored matching for P. Thus, in order to compute a plane colored matching in P, one can compute plane colored matchings in Q_1 and Q_2 recursively, as described in Algorithm 1. The RGB-$matching$ function receives a colored point set P of n points, where n is even and the points of P are colored by three colors, and computes a plane colored matching in P. The $BalancedCut$ function partitions P into Q_1 and Q_2 where both are color-balanced and have even number of points.

Algorithm 1.. RGB-$matching(P)$

Input: a color-balanced point set P with respect to (R, G, B), where $|P|$ is even.
Output: a plane colored matching in P.

1: **if** P is 2-colored **then**
2:     **return** RB-$matching(P)$
3: **else**
4:     $\ell \leftarrow BalancedCut(P)$
5:     $Q_1 \leftarrow$ points of P to the left of ℓ
6:     $Q_2 \leftarrow$ points of P to the right of ℓ
7:     **return** RGB-$matching(Q_1) \cup RGB$-$matching(Q_2)$

Now we analyze the running time of the algorithm. If $k = 2$, then in $O(n \log n)$ time we can find a plane RB-matching for P. If $k \geq 4$, then by Lemma 1, in $O(n)$ time we reduce the k-colored problem to a 3-colored problem. Then, the function RGB-$matching$ computes a plane colored matching in P. Let $T(n)$ denote the running time of RGB-$matching$ on the 3-colored point set P, where $|P| = n$. As described in Theorem 5 and Theorem 6, in linear time we can find a balanced cut ℓ in line 4 in Algorithm 1. The recursive calls to RGB-$matching$ function in line 7 takes $T(|Q_1|)$ and $T(|Q_2|)$ time. Thus, the running time of RGB-$matching$ can be expressed by the following recurrence:

$$T(n) = T(|Q_1|) + T(|Q_2|) + O(n).$$

Since $|Q_1|, |Q_2| \leq \frac{2n}{3} + 1$ and $|Q_1| + |Q_2| = n$, this recurrence solves to $T(n) = O(n \log n)$.

Theorem 7. *Given a color-balanced point set P of size n in general position in the plane with n even, a plane colored matching in P can be computed in $\Theta(n \log n)$ time.*

3.1 Maximum Matching

If P is not color-balanced, then $K_n(P_1, \ldots, P_k)$ does not admit a perfect matching. In this case we compute a maximum matching.

Theorem 8. *Given a colored point set P of size n in general position in the plane, a maximum plane colored matching M in P can be computed optimally in $\Theta(n + |M| \log |M|)$ time.*

Theorem 9. *Given any complete multipartite graph $K_n(V_1, \ldots, V_k)$ on n vertices and $k \geq 2$, a maximum matching in $K_n(V_1, \ldots, V_k)$ can be computed optimally in $\Theta(n)$ time.*

Acknowledgments. We would like to thank Prosenjit Bose for pointing out the dynamic convex hull structure. We also would like to thank the members of the Computational Geometry Lab at Carleton University for several discussions.

References

1. The 1979 Putnam exam. In: Alexanderson, G.L., Klosinski, L.F., Larson, L.C. (eds.) The William Lowell Putnam Mathematical Competition Problems and Solutions: 1965–1984. Mathematical Association of America, USA (1985)
2. Agarwal, P.K., Efrat, A., Sharir, M.: Vertical decomposition of shallow levels in 3-dimensional arrangements and its applications. SIAM J. Comput. **29**(3), 912–953 (1999)
3. Aichholzer, O., Cabello, S., Monroy, R.F., Flores-Peñaloza, D., Hackl, T., Huemer, C., Hurtado, F., Wood, D.R.: Edge-removal and non-crossing configurations in geometric graphs. Disc. Math. & Theo. Comp. Sci. **12**(1), 75–86 (2010)
4. Bereg, S., Hurtado, F., Kano, M., Korman, M., Lara, D., Seara, C., Silveira, R.I., Urrutia, J., Verbeek, K.: Balanced partitions of 3-colored geometric sets in the plane. Discrete Applied Mathematics **181**, 21–32 (2015)
5. Bereg, S., Kano, M.: Balanced line for a 3-colored point set in the plane. Electr. J. Comb. **19**(1), P33 (2012)
6. Cherkassky, B.V., Goldberg, A.V., Silverstein, C.: Buckets, heaps, lists, and monotone priority queues. SIAM J. Comput. **28**(4), 1326–1346 (1999)
7. Gabow, H.N.: Data structures for weighted matching and nearest common ancestors with linking. In: Proceedings of the First Annual ACM-SIAM Symposium on Discrete Algorithms, pp. 434–443 (1990)
8. Hershberger, J., Suri, S.: Applications of a semi-dynamic convex hull algorithm. BIT **32**(2), 249–267 (1992)
9. Kano, M., Suzuki, K., Uno, M.: Properly colored geometric matchings and 3-trees without crossings on multicolored points in the plane. In: Akiyama, J., Ito, H., Sakai, T. (eds.) JCDCGG 2013. LNCS, vol. 8845, pp. 96–111. Springer, Heidelberg (2014)

10. Lo, C., Matousek, J., Steiger, W.L.: Algorithms for ham-sandwich cuts. Discrete & Computational Geometry **11**, 433–452 (1994)
11. Overmars, M.H., van Leeuwen, J.: Maintenance of configurations in the plane. J. Comput. Syst. Sci. **23**(2), 166–204 (1981)
12. Sitton, D.: Maximum matchings in complete multipartite graphs. Furman University Electronic Journal of Undergraduate Mathematics **2**, 6–16 (1996)
13. Vaidya, P.M.: Geometry helps in matching. SIAM J. Comput. **18**(6) (1989)

Generation of Colourings and Distinguishing Colourings of Graphs

William Bird[✉] and Wendy Myrvold

Department of Computer Science, University of Victoria, Victoria, BC, Canada
bbird@uvic.ca, wendym@uvic.ca

Abstract. A *colouring* of a graph X is an assignment of colours to the vertices of X. A *distinguishing colouring* of X is a colouring such that no non-trivial automorphism of X preserves all colours. The *distinguishing number* of X is the minimum number of colours in a distinguishing colouring. This research presents a new algorithm for the generation of all colourings and all distinguishing colourings of a graph X up to isomorphism, and presents computational data on the distinguishing numbers of vertex transitive graphs.

Keywords: Graph distinguishability · Combinatorial generation · Graph theory

1 Colourings of Graphs

A *k-colouring* of a graph X on n vertices is a sequence $C = c_1 c_2 \ldots c_n$ such that $c_i \in \{0, 1, \ldots, k-1\}$ for $1 \le i \le n$. In this article, the term 'colouring' does not carry any inherent restrictions on which vertices can receive which colours. The classical graph colouring problem in which adjacent vertices must receive different colours will be called 'proper colouring' for clarity.

A *k*-colouring C of a graph X is *distinguishing* if no automorphism of X besides the identity fixes the colour of every vertex of X. The minimum k such that there exists a k-distinguishing colouring of X is denoted by $D(X)$ and called the *distinguishing number* of X. In Figure 1, two colourings of C_5 (which has distinguishing number 3) are shown, one of which is 3-distinguishing and the other of which is not distinguishing.

The distinguishing number was introduced in an article by Albertson and Collins [1], which also proved several fundamental theorems about distinguishability. Subsequent research on distinguishability has led to results classifying the distinguishing number for certain families of graphs [3,6,7,17,22] and algorithms to evaluate the distinguishing number [2,22]. There is a paucity of data on the distinguishing numbers of graphs which are not among the families for which $D(X)$ has been classified analytically.

This article presents an algorithm to generate all k-colourings of a graph X up to isomorphism. The algorithm can easily be adapted to generate colourings with

W. Bird and W. Myrvold—Supported by NSERC.

F. Dehne et al. (Eds.): WADS 2015, LNCS 9214, pp. 79–90, 2015.
DOI: 10.1007/978-3-319-21840-3_7

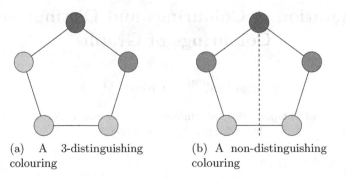

(a)   A   3-distinguishing   (b) A non-distinguishing
colouring   colouring

Fig. 1. Two colourings of the 5-cycle C_5. The left colouring is 3-distinguishing. The colouring on the right is fixed by a horizontal reflection, so it is not distinguishing.

restrictions, such as distinguishing colourings or proper colourings. Using the algorithm, along with various bounds on $D(X)$, we computed the distinguishing numbers of all vertex transitive graphs on up to 20 vertices. The computational results are summarized in Section 3.

2   Generating All k-colourings up to Isomorphism

This section presents a backtracking algorithm to generate k-colourings of a graph X up to isomorphism. The action of the automorphism group $\text{Aut}(X)$ partitions the set of all k-colourings into equivalence classes. Our algorithm generates the lexicographically minimum colouring in each equivalence class. Figure 2 shows the lexicographically minimum 2- and 3-colourings of the 4-cycle C_4. The algorithm is capable of handling extremely large group sizes (for the distinguishing number data summarized later in this article, the algorithm was run on groups as large as $20! \approx 2.4 \cdot 10^{18}$). Section 2.1 discusses one previous generation algorithm, which is limited to groups that can be stored explicitly in main memory. Section 2.2 introduces the Sims Table data structure, which is used by the algorithm to store the group, and Section 2.3 describes the algorithm itself. Finally, Section 2.4 describes the modification necessary to make the algorithm only generate distinguishing colourings.

The *Symmetric Group* on n symbols, denoted S_n, is the set of all permutations of $\{1, 2, \ldots, n\}$. The *automorphism group* of a graph X on the vertex set $V = \{v_1, v_2, \ldots, v_n\}$, denoted $\text{Aut}(X)$, is a subgroup of S_n which acts on V. In this article, permutations will be treated as bijective functions, not as a special class of object. Therefore, the group operation of permutation groups will be function composition, which is evaluated in right-to-left order.

2.1   The Permutation List Algorithm

Myrvold and Fowler, in [16], gave a backtracking algorithm to generate colourings of a graph up to isomorphism. In the original article, the main application of

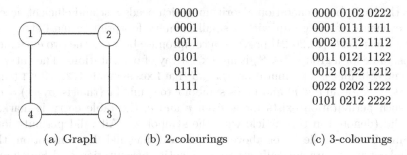

| (a) Graph | (b) 2-colourings | (c) 3-colourings |

Fig. 2. The 4-cycle C_4 with its lexicographically minimum 2- and 3-colourings

the algorithm is the generation of independent sets up to isomorphism, but the algorithm can also be used to generate k-colourings, k-distinguishing colourings, proper colourings and perfect matchings.

The algorithm in [16] takes a graph X and a listing of the permutations in the automorphism group $Aut(X)$ of X as input, and produces a listing of the lexicographically minimum k-colourings under the action of G. Since it operates on a list of the permutations in the group, this algorithm will be called the Permutation List (PL) algorithm in this article. The PL algorithm is relatively simple and has very high performance on graphs with a relatively small automorphism group for their size. As documented in [16], the PL algorithm outperforms traditional parent-child schemes for generation, which is why the PL algorithm was chosen as the baseline for comparison with our new algorithm.

The PL algorithm requires that all permutations in the automorphism group be stored explicitly in main memory. As a result, graphs whose automorphism group is too large to fit into memory cannot be processed. This limitation does not reflect the true complexity of storing the automorphism group, which can be represented much more compactly by a generating set. For example, Jerrum [11] gave a polynomial-time algorithm to produce a generating set containing at most $n - 1$ permutations for any permutation group $G \leq S_n$. To process groups too large for the PL algorithm, our new algorithm was developed using a *Sims Table* data structure, which requires $O(n^3)$ words of memory to represent a group $G \leq S_n$.

2.2   Computational Group Representation

Fundamentally, representations of a permutation group $G \leq S_n$ in a computer lie between two extremes. One extreme is storing a complete list of all group elements. This is relatively simple and allows every group element to be accessed easily, but requires $\Theta(n|G|)$ words of memory. The other extreme is storing a minimal generating set for the group, which requires at most $\Theta(n^2)$ words of memory [11] but does not give easy access to every element of the group (which must be formed by a product of generators).

As the basis for a generation algorithm, which needs fast and efficient access to all elements of the group with a small memory footprint, a data structure called a *Sims Table* [13,20,21] provides a compromise between the two extremes. A Sims Table for a group $G \leq S_n$ is an $n \times n$ array of permutations. The entry in row r and column c is a permutation $\sigma_{rc} \in G$ that fixes symbols $1, 2, \ldots, r\text{-}1$ (that is, $\sigma_{rc}(k) = k$ for all $k < r$) and maps symbol r to symbol c (that is, $\sigma_{rc}(r) = c$). If no such permutation exists for a given r and c, the table entry is marked as invalid (denoted in this article with the symbol $\times$). All valid permutations in a Sims Table will lie on or above the forward diagonal. By convention, the diagonal entries σ_{ii} are normally set to the identity permutation e, although any permutation meeting the definition is permissible. Figure 3 gives a Sims Table for an example graph.

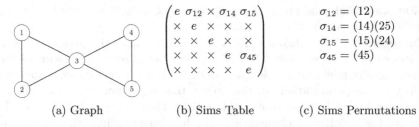

(a) Graph   (b) Sims Table   (c) Sims Permutations

Fig. 3. Sims Table and permutations for a sample graph

Given a permutation group $G \leq S_n$, the notation G_i will be used to denote the set

$$G_i = \{\pi \in G : \pi(j) = j \text{ for all } j < i\}$$

which is the pointwise stabilizer of all values less than i. It can be easily verified that each G_i is a group and

$$G = G_1 \geq G_2 \geq \ldots \geq G_n = \{e\}.$$

For each i, the group G_i is generated by the permutations in rows i through n of a Sims Table for G.

Sims' original publications [20,21] described a polynomial-time algorithm to create a Sims Table from a polynomial number of generators for G. The algorithm is usually called the 'Schreier-Sims algorithm' since it relies on a lemma of Schreier [10]. The term 'Sims Table' seems to originate with Knuth [12,13]. Several distinct formulations of the Schreier-Sims algorithm exist, each relying on the same mechanics as Sims' original results but differing in the specific structure of the algorithm [4]. To produce Sims Tables for the graphs studied in this article, pseudocode published by Kocay [14], based on a variant of the Schreier-Sims algorithm originated by Furst, Hopcroft and Luks [8], was used. One major advantage of a Sims Table compared to other generating sets is the unique factorization described by the following Lemma.

Lemma 1 (Sims [20]). *Given a Sims Table for $G \leq S_n$, every element $\pi \in G$ can be written uniquely in the form*

$$\pi = \sigma_{1j_1} \sigma_{2j_2} \cdots \sigma_{nj_n}.$$

Lemma 1 implies a simple algorithm to test whether a permutation $\pi \in S_n$ is a member of the group $G \leq S_n$. Specifically, if a decomposition $\pi = \beta\alpha$ is known such that $\beta = \sigma_{1j_1} \sigma_{2j_2} \cdots \sigma_{ij_i}$ for some i, and $\alpha(k) = k$ for all $k \leq i$, then $\pi \in G$ if and only if $\alpha \in G_{i+1}$. When $i = n$, α must be the identity, so the permutation β will give the factorization of π using the Sims table representation. Otherwise, let $k = \pi_{i+1}(i+1)$. If there is a permutation $\sigma_{i+1,k}$, the decomposition can be continued recursively by taking $\beta' = \beta\sigma_{i+1,k}$ and $\alpha' = \sigma_{i+1,k}^{-1}\alpha$. If no such permutation $\sigma_{i+1,k}$ exists in the Sims Table, then no factorization of π exists, and therefore $\pi \notin G$. Algorithm 1 gives pseudocode for this test. The arguments to the recursive TESTMEMBERSHIP function are a Sims Table T for G, the permutation π to be tested, and the current row i. For the initial call, i will be set to 1. At each recursive step, the algorithm determines which permutation σ_{ij_i} comprises the next element of the decomposition of π. If $\sigma_{ij_i} = \times$, the decomposition cannot exist and therefore π cannot be an element of G.

Algorithm 1. Test membership in G with a Sims Table

1: **procedure** TESTMEMBERSHIP(T, π, i)
2:     **if** $i = n$ **then return** true
3:     **if** $T[i][\pi(i)] = \times$ **then return** false
4:     $\sigma \leftarrow T[i][\pi(i)]$
5:     **return** TESTMEMBERSHIP$(T, \sigma^{-1}\pi, i+1)$
6: **end procedure**

2.3   Sims Table Generation Algorithm

A new colouring algorithm, which requires only $O(n^3)$ words of memory to generate all colourings of a group $G \leq S_n$, was developed by using a Sims Table representation of the group G. Besides providing a compact representation, the generating set contained in the Sims Table for G is very convenient for generating lexicographically minimum colourings. Our algorithm recursively generates partially-formed permutations in G and prefixes of colourings of G, backtracking when partial colourings are found to be infeasible.

Since all permutations below row i of a Sims Table for G must fix elements 1 through i, it is possible to avoid testing all permutations in G when verifying a potential minimum colouring C by only testing certain 'prefix' permutations $\pi = \sigma_{1j_1} \sigma_{2j_2} \cdots \sigma_{ij_i}$. If π maps C to a lexicographically greater colouring, all permutations prefixed by π will do so as well. Lemma 2 formalizes this condition.

Lemma 2 (Permutation Prefix Condition). *Let $G \leq S_n$, let T be a Sims Table for G and let $C = c_1 \ldots c_n$ be a colouring of G. Let $\pi = \sigma_{1i_1}\sigma_{2i_2}\ldots\sigma_{qi_q}$ be a permutation such that each σ_{ri_r} is a valid entry of T. If $c_{\pi(j)} = c_j$ for all $j < q$, and $c_{\pi(q)} > c_q$, then $C_{\pi\alpha} > C$ for all $\alpha \in G_{q+1}$.*

Algorithm 2. Test whether a colouring is lexicographically minimum

---

```
 1: procedure TESTMINIMUM(n, T, C, π, i)
 2:     if i = n + 1 then return true
 3:     for j ← i, i + 1, ..., n do
 4:         if T[i][j] ≠ × then
 5:             {Get the image of i under π ∘ T[i][j]}
 6:             q ← π(j)
 7:             if C[q] < C[i] then return false
 8:             else if C[q] > C[i] then return true
 9:             else
10:                 α ← π ∘ T[i][j]
11:                 if TESTMINIMUM(n, T, C, α, i + 1) = false then return false
12:             end if
13:         end if
14:     end for
15:     return true
16: end procedure
```

Lemma 2 implies a method to test whether a colouring C is minimum without necessarily testing C against all permutations in G. To check a given colouring $C = c_1 \ldots c_n$, the elements of G are generated by recursively building prefixes $\pi = \sigma_{1i_1}\sigma_{2i_2}\ldots\sigma_{qi_q}$ from the entries of the Sims Table. At each level q, if $c_{\pi(q)} < c_q$, the test returns **false** (since π maps C to a lexicographically smaller colouring). If $c_{\pi(q)} > c_q$, no permutation prefixed by π will map C to a smaller colouring by Lemma 2, so recursion backtracks. Otherwise, if $c_{\pi(q)} = c_q$, all permutations $\pi' = \pi\sigma_{q+1,j_{q+1}}$ are tested recursively. Algorithm 2 gives pseudocode for the test. The arguments to the TESTMINIMUM function in algorithm 2 are the number of elements n, a Sims Table for G in the form of a 2-dimensional array of permutations T, the colouring C, a prefix π and the first row i to test. The initial call to the recursive TESTMINIMUM function sets π to the identity permutation e and i to 1.

A naive method to generate all lexicographically minimum colourings simply generates all colourings and outputs those which Algorithm 2 reports as being minimum. A prefix condition for colourings, in a similar vein to Lemma 2, can be used to reduce the number of colourings considered. Lemma 3 gives the colouring prefix condition.

Lemma 3 (Colouring Prefix Condition). *Let $C_r = c_1, \ldots, c_r$ be a partial assignment of colours to elements of a group $G \leq S_n$. There is a lexicographically*

minimum colouring prefixed by C_i *only if there does not exist a permutation* $\pi \in G$ *such that* $\pi(r) < r$, $c_{\pi(r)} > c_r$ *and for all* $j < \pi(r)$,

$$\pi(j) < r, \ and$$

$$c_{\pi(j)} = c_j.$$

Our algorithm was developed as an outgrowth of Algorithm 2, with extra logic added to enforce Lemma 3. The algorithm uses the following data during recursion:

- The 2-dimensional array T stores a Sims Table for the group G.
- The current row i is tracked, and a permutation **perm** is maintained which is the product of entries from the first $i - 1$ rows of the Sims Table. At the beginning of recursion, i is set to 1 and **perm** is set to the identity permutation e.
- A vector **path** is used to store indices such that

$$\textbf{perm} = T[1][\textbf{path}[1]] \circ T[2][\textbf{path}[2]] \circ \ldots \circ T[i-1][\textbf{path}[i-1]].$$

- The colouring C is stored as a vector of length n, initially set to all zeroes.
- Since elements of C are not necessarily assigned values in left-to-right order, a boolean vector **fixed** is used to track the coloured elements. If $C[i]$ has been assigned a colour, **fixed**$[i]$ is set to **true**. Initially, every element of **fixed** is set to **false**.

Before recursion begins, $C[1]$ is set to each possible colour value (and the recursive process is run after each assignment). To start the recursive process, the variables described above are each assigned their initial values, and **fixed**$[1]$ is set to **true**. At each recursive step, it is assumed that the colour of each element before $C[i]$ is fixed, and that the partial colouring $C[1], C[2], \ldots, C[i]$ does not violate Lemma 3. The main operation at each recursive step is as follows:

1. The smallest index $j \geq \textbf{path}[i]$ such that $T[i][j] \neq \times$ is found. If no such value of j exists (that is, when $T[i][j] = \times$ for all $j \geq \textbf{path}[i]$), row i of the Sims table has been exhausted, so recursion returns to the previous level of the table. To do this, **perm** is first multiplied on the right by the inverse of $T[i-1][\textbf{path}[i-1]]$ to return it to its state before descending to level i, then **path**$[i]$ is set to 0, the value of **path**$[i-1]$ is incremented by 1, and finally, i is decremented. In the event that i is equal to -1 after decrementation, all permutations have been checked against C and no violations of either prefix condition have been found, so the colouring C is output as a minimum colouring.
2. The image q of i under $\textbf{perm} \circ T[i][j]$, which is equivalent to $\textbf{perm}[j]$, is found.
3. If $C[q] < C[i]$, Lemma 3 is violated for $c_1, \ldots, c_q$, so element q must be fixed to a colour greater than or equal to $C[i]$. Note that the invariant above implies that $q > i$. If position q is already fixed, it is impossible to change

the colour of $C[q]$, so the branch of recursion terminates. Otherwise, fixed[q] is set to true and new branches of recursion are started for each possible colour assignment to $C[q]$. After all branches terminate, the current branch is terminated.

4. If $C[q] > C[i]$, then by Lemma 2, no permutation prefixed by perm $\circ$ $T[i][j]$ will map C to any lexicographically smaller permutation, so $T[i][j]$ does not need to considered further and path[i] is set to $j + 1$ and recursion continues.

5. If $C[q] = C[i]$, then new branches of recursion are created for all assignments to $C[i]$ and $C[q]$ such that $C[q] \geq C[i]$. In branches where $C[q]$ continues to be equal to $C[i]$, i is incremented and perm is multiplied on the right by $T[i][j]$, since the permutation perm $\circ$ $T[i][j]$ has not yet been ruled out by Lemma 2. Branches where $C[q] > C[i]$ are handled according to step 4 above.

2.4   Generating k-distinguishing Colourings

A simple modification to the algorithm given in the previous section restricts the generated colourings to k-distinguishing colourings. When the current row i reaches n, the permutation perm will contain a permutation in G which fixes the current colouring C. As a result, C is not distinguishing and the active branch of recursion terminates. Since the recursive process reaches row n of the table if and only if every element of the current colouring C is fixed by the current permutation perm, this modification is sufficient to enforce the condition that C is distinguishing.

3   Finding the Distinguishing Number of Transitive Graphs

A graph X is *vertex transitive* if, for every pair of vertices $u, v \in V(X)$, there exists some $\pi \in \text{Aut}(X)$ such that $\pi(v) = u$. The automorphism group of a vertex transitive graph has a single orbit containing all vertices. Using a combination of bounds on the distinguishing number and the recursive colouring generation algorithm, the distinguishing number of every vertex transitive graph on up to 20 vertices was found. The distribution of distinguishing numbers by vertex count is summarized in Table 1.

The input graphs were taken from the database of all vertex transitive graphs published by Royle [18]. The nauty program [15] was used to find a generating set for the automorphism group of each graph, and an implementation of the Schreier-Sims algorithm, using the pseudocode given by Kocay in [14], was used to find a Sims Table for each group.

To find the distinguishing number of each graph X, upper and lower bounds on $D(X)$ were first evaluated. If the bounds differed, then the recursive colouring algorithm was used to search for a distinguishing colouring using progressively more colours until the upper bound was reached.

Four previously published upper bounds were evaluated [1,5,19,22], as well as a new upper bound (Theorem 1), based on the number of equivalence classes of all k-colourings. A new lower bound (Theorem 2) was also proven, and seems to be the first general lower bound on $D(X)$.

Theorem 1. *Let X be a graph on n vertices and $k \geq 1$. If q is the number of equivalence classes of the k^n k-colourings of X under the action of $Aut(X)$, and*

$$q < \frac{2k^n}{|Aut(X)|}$$

then

$$D(X) \leq k.$$

Proof. Consider a k-colouring C of X. Let S contain all permutations in $Aut(X)$ which fix C. Note that S is a subgroup of $Aut(X)$ since it is closed under permutation composition and for every $\alpha \in S$, the inverse permutation α^{-1} is also a member of S. Therefore, by Lagrange's Theorem, $|S|$ divides $|Aut(X)|$. Since each $\alpha \in S$ has the property $C_\alpha = C$, for any $\pi \in Aut(X)$, $C_{\pi\alpha} = C_\pi$, so the equivalence class of C under the action of $Aut(X)$ contains one colouring for each coset of S in $Aut(X)$.

If C is distinguishing, then $S = \{e\}$. Otherwise, $|S| \geq 2$ and the equivalence class of C has size

$$\frac{|Aut(X)|}{|S|} \leq \frac{|Aut(X)|}{2}.$$

The union of all equivalence classes must equal the complete set of k-colourings. If there are q equivalence classes of k-colourings with sizes $s_1, \ldots, s_q$, then

$$\sum_{i=1}^{q} s_i = k^n.$$

When no distinguishing colourings exist, every s_i must comply with the bound above, so

$$\sum_{i=1}^{q} s_i = k^n \leq q\frac{|Aut(X)|}{2}.$$

Therefore, if

$$q\frac{|Aut(X)|}{2} < k^n,$$

there must exist at least one k-distinguishing colouring of X.   □

Theorem 2. *Let X be a graph on n vertices. Let C be a k-colouring of the vertices of X with l_i equal to the number of vertices receiving colour i. If C is a distinguishing colouring of X, then*

$$|Aut(X)| \leq \binom{n}{l_1, l_2, \ldots, l_k}.$$

Proof. Suppose C is a distinguishing colouring. Then each permutation in Aut(X) carries C to a distinct colouring with the same distribution of colours. By the pigeonhole principle, the size of Aut(X) can be no greater than the number of such colourings, which is

$$\binom{n}{l_1, \ldots, l_k}.$$

□

Graphs composed of disjoint copies of a complete graph were difficult for our algorithm to process. By using our new Theorem 3, which analytically classifies the distinguishing number of such graphs, it was not necessary to perform a computational search on such graphs.

Theorem 3. *If X is a graph comprising q disjoint copies of K_n (for some $n \geq 1$), then $D(X)$ is equal to the least integer $k \geq n$ such that*

$$q \leq \binom{k}{n}.$$

Proof. It was proven in [1] that $D(K_n) = n$. Within each copy of K_n, for every pair v_1, v_2 of vertices there exists a permutation in Aut(X) which exchange v_1 and v_2 while fixing the rest of X. Therefore, any distinguishing colouring of X must assign distinct colours to each vertex within a copy of K_n. To prevent an automorphism from exchanging two copies of K_n, it is necessary for each copy of K_n to use a different subset of n-colours. A distinguishing colouring must therefore use at least k colours, with k defined as above, to allow each copy of K_n to receive a distinct set of colours.

A k-distinguishing colouring can be constructed by choosing q distinct n-subsets of $\{1, \ldots, k\}$ and assigning the colours in each subset to a copy of K_n.

□

The distinguishing number data for the set of all vertex transitive graphs on up to 20 vertices is summarized in Table 1. Several diagonal patterns can be observed in the table, corresponding to complete graphs and graphs consisting of copies of complete graphs. All of the graphs with distinguishing number six or more are graphs of this type (or their complement).

4   Future Research

Our generation algorithm can be adapted to produce all distinguishing colourings with k colours, all proper colourings with k colours and all independent sets up to isomorphism with minor modifications. A question for future work is whether it can be used as the basis for an algorithm to generate structures with more global restrictions, such as perfect matchings and vertex coverings. Additionally, although the two prefix conditions given in Section 2.3 are sufficient to make the

Table 1. Distinguishing numbers of all vertex transitive graphs on $1 - 20$ vertices

Number of Vertices \ Distinguishing Number	1	2	3	4	5	6	7	8	9	10	11	12	13	14	15	16	17	18	19	20
1	1																			
2		2																		
3			2																	
4			2	2																
5			1		2															
6		2	2	2		2														
7		2					2													
8		4	4	2	2			2												
9		4	1	2					2											
10		8	8	2		2				2										
11		6									2									
12		48	12	8	2		2					2								
13		12											2							
14		44	6		2			2						2						
15		38	2	2	2	2									2					
16		250	22	4	6				2							2				
17		34															2			
18		342	16	12	4		2			2								2		
19		58																	2	
20		1150	44	6	6	4					2									2
Total	1	2004	122	42	26	10	6	4	4	4	4	2	2	2	2	2	2	2	2	2

algorithm practical, there may be ways to refine both conditions further, or add extra conditions to further prune the search space.

The computational problem of generating all k-colourings is at least hard as counting all k-colourings, which is #P-Hard [9]. The complexity of generating distinguishing colourings is not yet known, although previous results have determined the complexity of several problems related to distinguishability [19]. The new bounds on $D(X)$ given in Section 3 rely on the structure of the equivalence classes of the k-colourings of X under the action of $\mathrm{Aut}(X)$. This provides an interesting link between distinguishing colourings and unrestricted colourings. Using similar techniques, it may be possible to derive much tighter bounds for specific families of groups (or families of graphs).

References

1. Albertson, M.O., Collins, K.L.: Symmetry breaking in graphs. Electronic Journal of Combinatorics **3**(R18), 1–17 (1996)
2. Arvind, V., Cheng, C.T., Devanur, N.R.: On computing the distinguishing numbers of planar graphs and beyond: A counting approach. SIAM Journal on Discrete Mathematics **22**(4), 1297–1324 (2008)
3. Bogstad, B., Cowen, L.: The distinguishing number of the hypercube. Discrete Mathematics **283**(1), 29–35 (2004)

4. Butler, G. (ed.): Fundamental Algorithms for Permutation Groups. LNCS, vol. 559. Springer, Heidelberg (1991)
5. Chan, M.: The maximum distinguishing number of a group. Electronic Journal of Combinatorics 13(R70), 1–8 (2006)
6. Chan, M.: The distinguishing number of the augmented cube and hypercube powers. Discrete Mathematics 308(11), 2330–2336 (2008)
7. Cheng, C.T.: On computing the distinguishing numbers of trees and forests. Electronic Journal of Combinatorics 13(R11), 1–12 (2006)
8. Furst, M., Hopcroft, J., Luks, E.: Polynomial-time algorithms for permutation groups. In: 21st Annual Symposium on Foundations of Computer Science, pp. 36–41, October 1980
9. Goldberg, L.: Automating Pólya theory: The computational complexity of the cycle index polynomial. Information and Computation 105(2), 268–288 (1993)
10. Hall, M.: The Theory of Groups. Macmillan, New York (1959)
11. Jerrum, M.: A compact representation for permutation groups. Journal of Algorithms 7(1), 60–78 (1986)
12. Knuth, D.E.: Efficient representation of perm groups. Combinatorica 11(1), 33–43 (1991)
13. Knuth, D.E.: The Art of Computer Programming, vol. 4A. Addison-Wesley, Reading (2011)
14. Kocay, W.: On writing isomorphism programs. Computational and Constructive Design Theory 368, 135–175 (1996)
15. McKay, B.: Practical graph isomorphism. Congressus Numerantium 30, 45–87 (1981)
16. Myrvold, W., Fowler, P.: Fast enumeration of all independent sets of a graph up to isomorphism. Journal of Combinatorial Mathematics and Combinatorial Computing 85, 173–194 (2013)
17. Potanka, K.S.: Groups, Graphs, and Symmetry-Breaking. Master's thesis, Virginia Polytechnic Institute and State University (1998)
18. Royle, G.F., Praeger, C.E.: Constructing the vertex-transitive graphs of order 24. Journal of Symbolic Computation 8(4), 309–326 (1989)
19. Russell, A., Sundaram, R.: A note on the asymptotics and computational complexity of graph distinguishability. Electronic Journal of Combinatorics 5(R23), 1–7 (1998)
20. Sims, C.C.: Computation with permutation groups. In: Proceedings of the Second ACM Symposium on Symbolic and Algebraic Manipulation, pp. 23–28. ACM (1971)
21. Sims, C.C.: Computational methods in the study of permutation groups. In: Leech, J. (ed.) Computational Problems in Abstract Algebra, pp. 169–183. Pergamon Press (1970)
22. Tymoczko, J.: Distinguishing numbers for graphs and groups. Electronic Journal of Combinatorics 11(R63), 1–13 (2004)

Strictly Implicit Priority Queues: On the Number of Moves and Worst-Case Time

Gerth Stølting Brodal, Jesper Sindahl Nielsen[(✉)], and Jakob Truelsen

MADALGO, Department of Computer Science, Aarhus University, Aarhus, Denmark
{gerth,jasn,jakobt}@cs.au.dk

Abstract. The binary heap of Williams (1964) is a simple priority queue characterized by only storing an array containing the elements and the number of elements n – here denoted a *strictly implicit* priority queue. We introduce two new strictly implicit priority queues. The first structure supports amortized $O(1)$ time INSERT and $O(\log n)$ time EXTRACT-MIN operations, where both operations require amortized $O(1)$ element moves. No previous implicit heap with $O(1)$ time INSERT supports both operations with $O(1)$ moves. The second structure supports worst-case $O(1)$ time INSERT and $O(\log n)$ time (and moves) EXTRACTMIN operations. Previous results were either amortized or needed $O(\log n)$ bits of additional state information between operations.

1 Introduction

In 1964 Williams presented "Algorithm 232" [12], commonly known as the binary heap. The binary heap is a priority queue data structure storing a dynamic set of n elements from a totally ordered universe, supporting the insertion of an element (INSERT) and the deletion of the minimum element (EXTRACTMIN) in worst-case $O(\log n)$ time. The binary heap structure is an *implicit* data structure, i.e., it consists of an array of length n storing the elements, and no information is stored between operations except for the array and the value n. Sometimes data structures storing $O(1)$ additional words are also called implicit. In this paper we restrict our attention to *strictly implicit* priority queues, i.e., data structures that do not store any additional information than the array of elements and the value n between operations.

Due to the $\Omega(n \log n)$ lower bound on comparison based sorting, either INSERT or EXTRACTMIN must take $\Omega(\log n)$ time, but not necessarily both. Carlson *et al.* [5] presented an implicit priority queue with worst-case $O(1)$ and $O(\log n)$ time INSERT and EXTRACTMIN operations, respectively. However, the structure is not strictly implicit since it needs to store $O(1)$ additional words. Harvey and Zatloukal [11] presented a strictly implicit priority structure achieving the same bounds, but amortized. No previous strictly implicit priority queue with matching worst-case time bounds is known.

Work supported in part by the Danish National Research Foundation grant DNRF84 through the Center for Massive Data Algorithmics.

© Springer International Publishing Switzerland 2015
F. Dehne et al. (Eds.): WADS 2015, LNCS 9214, pp. 91–102, 2015.
DOI: 10.1007/978-3-319-21840-3_8

Table 1. Selected previous and new results for implicit priority queues. The bounds are asymptotic, and $\star$ are amortized bounds.

	Insert	Extract-Min	Moves	Strict	Identical elements
Williams [12]	$\log n$	$\log n$	$\log n$	yes	yes
Carlsson et al. [5]	1	$\log n$	$\log n$	no	yes
Edelkamp et al. [7]	1	$\log n$	$\log n$	no	yes
Harvey and Zatloukal [11]	$\star\,1$	$\star\log n$	$\star\log n$	yes	yes
Franceschini and Munro [9]	$\star\log n$	$\star\log n$	$\star\,1$	yes	no
Section 2	$\star\,1$	$\star\log n$	$\star\,1$	yes	yes
Section 3	1	$\log n$	$\log n$	yes	no

A measurement often studied in implicit data structures and in-place algorithms is the number of element moves performed during the execution of a procedure. Franceschini showed how to sort n elements implicitly using $O(n \log n)$ comparisons and $O(n)$ moves [8], and Franceschini and Munro [9] presented implicit dictionaries with amortized $O(\log n)$ time updates with amortized $O(1)$ moves per update. The latter immediately implies an implicit priority queue with amortized $O(\log n)$ time INSERT and EXTRACTMIN operations performing amortized $O(1)$ moves per operation. No previous implicit priority queue with $O(1)$ time INSERT operations achieving $O(1)$ moves per operation is known.

For a more thorough survey of previous priority queue results, see [1].

Our Contribution. We present two strictly implicit priority queues. The first structure (Section 2) limits the number of moves to $O(1)$ per operation with amortized $O(1)$ and $O(\log n)$ time INSERT and EXTRACTMIN operations, respectively. However the bounds are all amortized and it remains an open problem to achieve these bounds in the worst case for strictly implicit priority queues. We note that this structure implies a different way of sorting in-place with $O(n \log n)$ comparisons and $O(n)$ moves. The second structure (Section 3) improves over [5,11] by achieving INSERT and EXTRACTMIN operations with worst-case $O(1)$ and $O(\log n)$ time (and moves), respectively. The structure in Section 3 assumes all elements to be distinct where as the structure in Section 2 also can be extended to support identical elements (see [4]). See Figure 1 for a comparison of new and previous results.

Preliminaries. We assume the *strictly implicit model* as defined in [3] where we are only allowed to store the number of elements n and an array containing the n elements. Comparisons are the only allowed operations on the elements. The number n is stored in a memory cell with $\Theta(\log n)$ bits (word size) and any operation usually found in a RAM is allowed for computations on n and intermediate values. The number of moves is the number of writes to the array storing the elements. That is, swapping two elements costs two moves.

A fundamental technique in the implicit model is to encode a 0/1-bit with a pair of distinct elements (x, y), where the pair encodes 1 if $x < y$ and 0 otherwise.

A binary heap is a complete binary tree structure where each node stores an element and the tree satisfies *heap order*, i.e., the element at a non-root node is larger than or equal to the element at the parent node. Binary heaps can be generalized to d-ary heaps [10], where the degree of each node is d rather than two. This implies $O(\log_d n)$ and $O(d \log_d n)$ time for INSERT and EXTRACTMIN, respectively, using $O(\log_d n)$ moves for both operations.

2   Amortized $O(1)$ Moves

In this section we describe a strictly implicit priority queue supporting amortized $O(1)$ time INSERT and amortized $O(\log n)$ time EXTRACTMIN. Both operations perform amortized $O(1)$ moves. In Sections 2.1-2.3 we assume elements are distinct. In [4] we describe how to handle identical elements.

Overview. The basic idea of our priority queue is the following (the details are presented in Section 2.1). The structure consists of four components: an insertion buffer B of size $O(\log^3 n)$; m insertion heaps $I_1, I_2, \ldots, I_m$ each of size $\Theta(\log^3 n)$, where $m = O(n/\log^3 n)$; a singles structure T, of size $O(n)$; and a binary heap Q, storing $\{1, 2, \ldots, m\}$ (integers encoded by pairs of elements) with the ordering $i \le j$ if and only if $\min I_i \le \min I_j$. Each I_i and B is a $\log n$-ary heap of size $O(\log^3 n)$. The table below summarizes the performance of each component:

	Insert		ExtractMin	
Structure	Time	Moves	Time	Moves
B, I_i	1	1	$\log n$	1
Q	$\log^2 n$	$\log^2 n$	$\log^2 n$	$\log^2 n$
T	$\log n$	1	$\log n$	1

It should be noted that the implicit dictionary of Franceschini and Munro [9] could be used for T, but we will give a more direct solution since we only need the restricted EXTRACTMIN operation for deletions.

The INSERT operation inserts new elements into B. If the size of B becomes $\Theta(\log^3 n)$, then m is incremented by one, B becomes I_m, m is inserted into Q, and B becomes a new empty $\log n$-ary heap. An EXTRACTMIN operation first identifies the minimum element in B, Q and T. If the overall minimum element e is in B or T, e is removed from B or T. If the minimum element e resided in I_i, where i is stored at the root of Q, then e and $\log^2 n$ further smallest elements are extracted from I_i (if I_i is not empty) and all except e inserted into T (T has cheap operations whereas Q does not, thus the expensive operation on Q is amortized over inexpensive ones in T), and i is deleted from and reinserted into Q with respect to the new minimum element in I_i. Finally e is returned.

For the analysis we see that INSERT takes $O(1)$ time and moves, except when converting B to a new I_m and inserting m into Q. The $O(\log^2 n)$ time and moves for this conversion is amortized over the insertions into B, which becomes amortized $O(1)$, since $|B| = \Omega(\log^2 n)$. For EXTRACTMIN we observe that an expensive deletion from Q only happens once for every $\log^2 n$-th element from I_i (the remaining ones from I_i are moved to T and deleted from T), and finally if there have been d EXTRACTMIN operations, then at most $d + m \log^2 n$ elements have been inserted into T, with a total cost of $O((d + m \log^2 n) \log n) = O(n + d \log n)$, since $m = O(n / \log^3 n)$.

2.1   The Implicit Structure

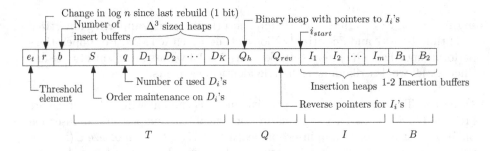

Fig. 1. The different structures ansd their layout in memory

We now give the details of our representation (see Figure 1). We select one element e_t as our *threshold element*, and denote elements greater than e_t as *dummy elements*. The current number of elements in the priority queue is denoted n. We fix an integer N that is an approximation of n, where $N \leq n < 4N$ and $N = 2^j$ for some j. Instead of storing N, we store a bit $r = \lfloor \log n \rfloor - \log N$, encoded by two dummy elements. We can then compute N as $N = 2^{\lfloor \log n \rfloor - r}$, where $\lfloor \log n \rfloor$ is the position of the most significant bit in the binary representation of n (which we assume is computable in constant time). The value r is easily maintained: When $\lfloor \log n \rfloor$ changes, r changes accordingly. We let $\Delta = \log(4N) = \lfloor \log n \rfloor + 2 - r$, i.e., Δ bits is sufficient to store an integer in the range $0..n$. We let $M = \lceil 4N/\Delta^3 \rceil$.

We maintain the invariant that the size of the insertion buffer B satisfies $1 \leq |B| \leq 2\Delta^3$, and that B is split into two parts B_1 and B_2, each being Δ-ary heaps (B_2 possibly empty), where $|B_1| = \min\{|B|, \Delta^3\}$ and $|B_2| = |B| - |B_1|$. We use two buffers to prevent expensive operation sequences that alternate inserting and deleting the same element. We store a bit b indicating if B_2 is nonempty, i.e., $b = 1$ if and only if $|B_2| \neq 0$. The bit b is encoded using two dummy elements. The structures $I_1, I_2, \ldots, I_m$ are Δ-ary heaps storing Δ^3 elements. The binary heap Q is stored using two arrays Q_h and Q_{rev} each of a fixed size $M \geq m$ and storing integers in the range $1..m$. Each value in both arrays is encoded using 2Δ dummy elements, i.e., Q is stored using $4M\Delta$ dummy elements. The first m entries of Q_h store the binary heap, whereas Q_{rev} acts as reverse pointers, i.e., if $Q_h[j] = i$ then $Q_{rev}[i] = j$. All operations on a regular binary heap take $O(\log n)$ time, but since each "read"/"write" from/to Q needs to decode/encode an integer the time increases by a factor 2Δ. It follows that Q supports INSERT and EXTRACTMIN in $O(\log^2 n)$ time, and FINDMIN in $O(\log n)$ time.

We now describe T and we need the following density maintenance result.

Lemma 1 ([2]). *There is a dynamic data structure storing n comparable elements in an array of length $(1 + \varepsilon)n$, supporting INSERT and EXTRACTMIN in amortized $O(\log^2 n)$ time and FINDPREDECESSOR in worst case $O(\log n)$ time. FINDPREDECESSOR does not modify the array.*

Corollary 1. *There is an implicit data structure storing n (key, index) pairs, while supporting INSERT and EXTRACTMIN in amortized $O(\log^3 n)$ time and moves, and FINDPREDECESSOR in $O(\log n)$ time in an array of length $\Delta(2+\varepsilon)n$.*

Proof. We use the structure from Lemma 1 to store pairs of a key and an index, where the index is encoded using 2Δ dummy elements. All additional space is filled with dummy elements. However comparisons are only made on keys and not indexes, which means we retain $O(\log n)$ time for FINDMIN. Since the stored elements are now an $O(\Delta) = \Theta(\log n)$ factor larger, the time for update operations becomes an $O(\log n)$ factor slower giving amortized $O(\log^3 n)$ time for INSERT and EXTRACTMIN. □

The singles structure T intuitively consists of a sorted list of the elements stored in T partitioned into buckets $D_1, \ldots, D_q$ of size at most Δ^3, where the minimum element e from bucket D_i is stored in a structure S from Corollary 1 as the pair (e, i). Each D_i is stored as a Δ-ary heap of size Δ^3, where empty slots are filled with dummy elements. Recall implicit heaps are complete trees, which means all dummy elements in D_i are stored consecutively after the last non-dummy element. In S we consider pairs (e, i) where $e > e_t$ to be empty spaces.

More specifically, the structure T consists of: q, S, $D_1, D_2, \ldots, D_K$, where $K = \lceil \frac{N}{16\Delta^3} \rceil \geq q$ is the number of D_i's available. The structure S uses $\lceil \frac{N}{4\Delta^2} \rceil$ elements and q uses 2Δ elements to encode a pointer. Each D_i uses Δ^3 elements.

The D_i's and S relate as follows. The number of D_i's is at most the maximum number of items that can be stored in S. Let $(e, i) \in S$, then $\forall x \in D_i : e < x$, and furthermore for any $(e', i') \in S$ with $e < e'$ we have $\forall x \in D_i : x < e'$. These invariants do not apply to dummy elements. Since D_i is a Δ-ary heap with Δ^3 elements we get $O(\log_\Delta \Delta^3) = O(1)$ time for INSERT and $O(\Delta \log_\Delta \Delta^3) = O(\Delta)$ for EXTRACTMIN on a D_i.

2.2 Operations

For both INSERT and EXTRACTMIN we need to know N, Δ, and whether there are one or two insert buffers as well as their sizes. First r is decoded and we compute $\Delta = 2 + \mathrm{msb}(n) - r$, where $\mathrm{msb}(n)$ is the position of the most significant bit in the binary representation of n (indexed from zero). From this we compute $N = 2^{\Delta-2}$, $K = \lceil N/(16\Delta^3) \rceil$, and $M = \lceil 4N/\Delta^3 \rceil$. By decoding b we get the number of insert buffers. To find the sizes of B_1 and B_2 we compute the value i_{start} which is the index of the first element in I_1. The size of B_1 is computed as follows. If $(n - i_{start}) \bmod \Delta^3 = 0$ then $|B_1| = \Delta^3$. If B_2 exists then B_1

starts at $n - 2\Delta^3$ and otherwise B_1 starts at $n - \Delta^3$. If B_2 exists and $(n - i_{start})$ mod $\Delta^3 = 0$ then $|B_2| = \Delta^3$, otherwise $|B_2| = (n - i_{start})$ mod Δ^3. Once all of this information is computed the actual operation can start. If $n = N + 1$ and an EXTRACTMIN operation is called, then the EXTRACTMIN procedure is executed and afterwards the structure is rebuilt as described in the paragraph below. Similarly if $n = 4N - 1$ before an INSERT operation the new element is appended and the data structure is rebuilt.

INSERT. If $|B_1| < \Delta^3$ the new element is inserted in B_1 by the standard insertion algorithm for Δ-ary heaps. If $|B_1| = \Delta^3$ and $|B_2| = 0$ and a new element is inserted the two elements in b are swapped to indicate that B_2 now exists. When $|B_1| = |B_2| = \Delta^3$ and a new element is inserted, B_1 becomes I_{m+1}, B_2 becomes B_1, $m + 1$ is inserted in Q (possibly requiring $O(\log n)$ values in Q_h and Q_{rev} to be updated in $O(\log^2 n)$ time). Finally the new element becomes B_2.

EXTRACTMIN. Searches for the minimum element e are performed in B_1, B_2, S, and Q. If e is in B_1 or B_2 it is deleted, the last element in the array is swapped with the now empty slot and the usual bubbling for heaps is performed. If B_2 disappears as a result, the bit b is updated accordingly. If B_1 disappears as a result, I_m becomes B_1, and m is removed from Q.

If e is in I_i then i is deleted from Q, e is extracted from I_i, and the last element in the array is inserted in I_i. The Δ^2 smallest elements in I_i are extracted and inserted into the *singles structure*: for each element a search in S is performed to find the range it belongs to, i.e. D_j, the structure it is to be inserted in. Then it is inserted in D_j (replacing a dummy element that is put in I_i, found by binary search). If $|D_j| = \Delta^3$ and $q = K$ the priority queue is rebuilt. Otherwise if $|D_j| = \Delta^3$, D_j is split in two by finding the median y of D_j using a linear time selection algorithm [6]. Elements $\geq y$ in D_j are swapped with the first $\Delta^3/2$ elements in D_q then D_j and D_q are made into Δ-ary heaps by repeated insertion. Then y is extracted from D_q and (y, q) is inserted in S. The dummy element pushed out of S by y is inserted in D_q. Finally q is incremented and we reinsert i into Q. Note that it does not matter if any of the elements in I_i are dummy elements, the invariants are still maintained.

If $(e, i) \in S$, the last element of the array is inserted into the singles structure, which pushes out a dummy element z. The minimum element y of D_i is extracted and z inserted instead. We replace e by y in S. If y is a dummy element, we update S as if (y, i) was removed. Finally e is returned. Note this might make B_1 or B_2 disappear as a result and the steps above are executed if needed.

Rebuilding. We let the new $N = n'/2$, where n' is n rounded to the nearest power of two. Using a linear time selection algorithm [6], find the element with rank $n - i_{start}$, this element is the new threshold element e_t, and it is put in the first position of the array. Following e_t are all the elements greater than e_t and they are followed by all the elements comparing less than e_t. We make sure to have at least $\Delta^3/2$ elements in B_1 and at most $\Delta^3/2$ elements in B_2 which dictates whether b encodes 0 or 1. The value q is initialized to 1. All the D_i structures are considered empty since they only contain dummy elements. The pointers in

Q_h and Q_{rev} are all reset to the value 0. All the I_i structures as well as B_1 (and possibly B_2) are made into Δ-ary heaps with the usual heap construction algorithm. For each I_j structure the Δ^2 smallest elements are inserted in the singles structure as described in the EXTRACTMIN procedure, and j is inserted into Q. The structure now satisfies all the invariants.

2.3 Analysis

In this subsection we give the analysis that leads to the following theorem.

Theorem 1. *There is a strictly implicit priority queue supporting* INSERT *in amortized* $O(1)$ *time,* EXTRACTMIN *in amortized* $O(\log n)$ *time. Both operations perform amortized* $O(1)$ *moves.*

INSERT. While $|B| < 2\Delta^3$, each insertion takes $O(1)$ time. When an insertion happens and $|B| = 2\Delta^3$, the insertion into Q requires $O(\log^2 n)$ time and moves. During a sequence of s insertions, this can at most happen $\lceil s/\Delta^3 \rceil$ times, since $|B|$ can only increase for values above Δ^3 by insertions, and each insertion at most causes $|B|$ to increase by one. The total cost for s insertions is $O(s + s/\Delta^3 \cdot \log^2 n) = O(s)$, i.e., amortized constant per insertion.

EXTRACTMIN. We first analyze the cost of updating the singles structure. Each operation on a D_i takes time $O(\Delta)$ and performs $O(1)$ moves. Locating an appropriate bucket using S takes $O(\log n)$ time and no moves. At least $\Omega(\Delta^3)$ operations must be performed on a bucket to trigger an expensive bucket split or bucket elimination in S. Since updating S takes $O(\log^3 n)$ time, the amortized cost for updating S is $O(1)$ moves per insertion and extraction from the singles structure. In total the operations on the singles structure require amortized $O(\log n)$ times and amortized $O(1)$ moves. For EXTRACTMIN the searches performed all take $O(\log n)$ comparisons and no moves. If B_1 disappears as a result of an extraction we know at least $\Omega(\Delta^3)$ extractions have occurred because a rebuild ensures $|B_1| \geq \Delta^3/2$. These extractions pay for extracting I_m from Q_h which takes $O(\log^2 n)$ time and moves, amortized this gives $O(1/\log n)$ additional time and moves. If the extracted element was in I_i for some i, then Δ^2 insertions occur in the singles structure each taking $O(\log n)$ time and $O(1)$ moves amortized. If that happens either $\Omega(\Delta^3)$ insertions or Δ^2 extractions have occurred: Suppose no elements from I_i have been inserted in the singles structure, then the reason there is a pointer to I_i in Q_h is due to $\Omega(\Delta^3)$ insertions. When inserting elements in the singles structure from I_i the number of elements inserted is Δ^2 and these must first be deleted. From this discussion it is evident that we have saved up $\Omega(\Delta^2)$ moves and $\Omega(\Delta^3)$ time, which pay for the expensive extraction. Finally if the minimum element was in S, then an extraction on a Δ-ary heap is performed which takes $O(\Delta)$ time and $O(1)$ moves, since its height is $O(1)$.

Rebuilding. The cost of rebuilding is $O(n)$, due to a selection and building heaps with $O(1)$ height. There are three reasons a rebuild might occur: (i) n became

$4N$, (ii) n became $N-1$, or (iii) An insertion into T would cause $q > K$. By the choice of N during a rebuild it is guaranteed that in the first and second case at least $\Omega(N)$ insertions or extractions occurred since the last rebuild, and we have thus saved up at least $\Omega(N)$ time and moves. For the last case we know that each extraction incur $O(1)$ insertions in the singles structure in an amortized sense. Since the singles structure accommodates $\Omega(N)$ elements and a rebuild ensures the singles structure has $o(n)$ non dummy elements (Lemma 2), at least $\Omega(N)$ extractions have occurred which pay for the rebuild.

Lemma 2. *Immediately after a rebuild $o(n)$ elements in the singles structure are non-dummy elements*

Proof. There are at most n/Δ^3 of the I_i structures and Δ^2 elements are inserted in the singles structure from each I_i, thus at most $n/\Delta = o(n)$ non-dummy elements reside in the singles structure after a rebuild.   □

The paragraphs above establish Theorem 1.

3   Worst Case Solution

In this section we present a strictly implicit priority queue supporting INSERT in worst-case $O(1)$ time and EXTRACTMIN in worst-case $O(\log n)$ time (and moves). The data structure requires all elements to be distinct. The main concept used is a variation on binomial trees. The priority queue is a forest of $O(\log n)$ such trees. We start with a discussion of the variant we call *relaxed binomial trees*, then we describe how to maintain a forest of these trees in an amortized sense, and finally we give the deamortization.

3.1   Relaxed Binomial Tree

Binomial trees are defined inductively: A single node is a binomial tree of size one and the node is also the root. A binomial tree of size 2^{i+1} is made by *linking* two binomial trees T_1 and T_2 both of size 2^i, such that one root becomes the rightmost child of the other root. We lay out in memory a binomial tree of size 2^i by a preorder traversal of the tree where children are visited in order of increasing size, i.e. $c_0, c_1, \ldots, c_{i-1}$. This layout is also described in [5]. See Figure 2 for an illustration of the layout. In a *relaxed binomial tree* (RBT) each nodes stores an element, satisfying the following order: Let p be a node with i children, and let c_j be a child of p. Let T_{c_j} denote the set of elements in the subtree rooted at c_j. We have the invariant that the element c_ℓ is less than either all elements in T_{c_ℓ} or less than all elements in $\bigcup_{j<\ell} T_{c_j}$ (see Figure 2). In particular we have the requirement that the root must store the smallest element in the tree. In each node we store a flag indicating in which direction the ordering is satisfied. Note that linking two adjacent RBTs of equal size can be done in $O(1)$ time: compare the keys of the two roots, if the lesser is to the right, swap the two nodes and finally update the flags to reflect the changes as just described.

For an unrelated technical purpose we also need to store whether a node is the root of a RBT. This information is encoded using three elements per node

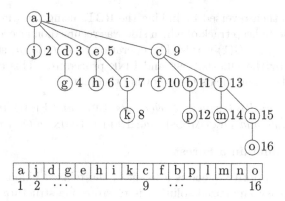

Fig. 2. An example of an RBT on 16 elements (a,b,...,o). The layout in memory of an RBT and a regular binomial tree is the same. Note here that node 9 has element c and is not the minimum of its subtree because node 11 has element b, but c is the minimum among the subtrees rooted at nodes 2, 3, and 5 (c_0, c_1, and c_2). Note also that node 5 is the minimum of its subtree but not the minimum among the trees rooted at nodes 2 and 3, which means only one state is valid. Finally node 3 is the minimum of both its own subtree and the subtree rooted at node 2, which means both states are valid for that node.

(allowing $3! = 6$ permutations, and we only need to differentiate between three states per node: "root", "minimum of its own subtree", or "minimum among strictly smaller subtrees").

To extract the minimum element of an RBT it is replaced by another element. The reason for replacing is that the forest of RBTs is implicitly maintained in an array and elements are removed from the right end, meaning only an element from the last RBT is removed. If the last RBT is of size 1, it is trivial to remove the element. If it is larger, then we *decompose* it. We first describe how to perform a DECOMPOSE operation which changes an RBT of size 2^i into i structures $T_{i-1}, \ldots, T_1, T_0$, where $|T_j| = 2^j$. Then we describe how to perform REPLACEMIN which takes one argument, a new element, and extracts the minimum element from an RBT and inserts the argument in the same structure.

A DECOMPOSE procedure is essentially reversing insertions. We describe a tail recursive procedure taking as argument a node r. If the structure is of size one, we are done. If the structure is of size 2^i the $(i-1)$th child, c_{i-1}, of r is inspected, if it is not the minimum of its own subtree, the element of c_{i-1} and r are swapped. The $(i-1)$th child should now encode "root", that way we have two trees of size 2^{i-1} and we recurse on the subtree to the right in the memory layout. This procedure terminates in $O(i)$ steps and gives $i+1$ structures of sizes $2^{i-1}, 2^{i-2}, \ldots, 2, 1,$ and 1 laid out in decreasing order of size (note there are two structures of size 1). This enables easy removal of a single element.

The REPLACEMIN operation works similarly to the DECOMPOSE, where instead of always recursing on the right, we recurse where the minimum element is the root. When the recursion ends, the minimum element is now in a structure of size 1, which is deleted and replaced by the new element. The decomposition is then reversed by linking the RBTs using the LINK procedure. Note it is possible to keep track of which side was recursed on at every level with $O(\log n)$ extra bits, i.e. $O(1)$ words. The operation takes $O(\log n)$ steps and correctness follows by the DECOMPOSE and LINK procedures. This concludes the description of RBTs and yields the following theorem.

Theorem 2. *On an RBT with $3 \cdot 2^i$ elements, LINK and FINDMIN can be supported in $O(1)$ time and DECOMPOSE and REPLACEMIN in $O(i)$ time.*

3.2   How to Maintain a Forest

As mentioned our priority queue is a forest of the relaxed binomial trees from Theorem 2. An easy amortized solution is to store one structure of size $3 \cdot 2^j$ for every set bit j in the binary representation of $\lfloor n/3 \rfloor$. During an insertion this could cause $O(\log n)$ LINK operations, but by a similar argument to that of binary counting, this yields $O(1)$ amortized insertion time. We are aiming for a worst case constant time solution so we maintain the invariant that there are at most 5 structures of size 2^i for $i = 0, 1, \ldots, \lfloor \log n \rfloor$. This enables us to postpone some of the LINK operations to appropriate times. We are storing $O(\log n)$ RBTs, but we do not store which sizes we have, this information must be decodable in constant time since we do not allow storing additional words. Recall that we need 3 elements per node in an RBT, thus in the following we let n be the number of elements and $N = \lfloor n/3 \rfloor$ be the number of nodes. We say a node is in node position k if the three elements in it are in positions $3k - 2$, $3k - 1$, and $3k$. This means there is a buffer of $0, 1$, or 2 elements at the end of the array. When a third element is inserted, the elements in the buffer become an RBT with a single node and the buffer is now empty. If an INSERT operation does not create a new node, the new element is simply appended to the buffer. We are not storing the structure of the forest (i.e. how many RBTs of size 2^j exists for each j), since that would require additional space. To be able to navigate the forest we need the following two lemmas.

Lemma 3. *There is a structure of size 2^i at node positions $k, k+1, \ldots, k+2^i-1$ if and only if the node at position k encodes "root", the node at position $k + 2^i$ encodes "root" and the node at position $k + 2^{i-1}$ encodes "not root".*

Proof. It is trivially true that the mentioned nodes encode "root", "root" and "not root" if an RBT with 2^i nodes is present in those locations.

We first observe there cannot be a structure of size 2^{i-1} starting at position k, since that would force the node at position $k + 2^{i-1}$ to encode "root". Also all structures between k and N must have less than 2^i elements, since both nodes at positions k and $k + 2^i$ encode "root". We now break the analysis in a few cases and the lemma follows from a proof by contradiction. Suppose there is a structure of size 2^{i-2} starting at k, then for the same reason as before there cannot be another one of size 2^{i-2}. Similarly, there can at most be one structure of size 2^{i-3} following that structure. Now we can bound the total number of nodes from position k onwards in the structure as: $2^{i-2} + 2^{i-3} + 5 \sum_{j=0}^{i-4} 2^j = 2^i - 5 < 2^i$, which is a contradiction. So there cannot be a structure of size 2^{i-2}

starting at position k. Note there can at most be three structures of size 2^{i-3} starting at position k, and we can again bound the total number of nodes as: $3 \cdot 2^{i-3} + 5 \sum_{j=0}^{i-4} 2^j = 2^i - 5 < 2^i$, again a contradiction. □

Lemma 4. *If there is an RBT with 2^i nodes the root is in position $N - 2^i k - x + 1$ for $k = 1, 2, 3, 4$ or 5 and $x = N$ mod 2^i.*

Proof. There are at most $5 \cdot 2^i - 5$ nodes in structures of size $\leq 2^{i-1}$. All structures of size $\geq 2^i$ contribute 0 to x, thus the number of nodes in structures with $\leq 2^{i-1}$ nodes must be x counting modulo 2^i. This gives exactly the five possibilities for where the first tree of size 2^i can be. □

We now describe how to perform an EXTRACTMIN. First, if there is no buffer (n mod $3 = 0$) then DECOMPOSE is executed on the smallest structure. We apply Lemma 4 iteratively for $i = 0$ to $\lfloor \log N \rfloor$ and use Lemma 3 to find structures of size 2^i. If there is a structure we call the FINDMIN procedure (i.e. inspect the element of the root node) and remember which structure the minimum element resides in. If the minimum element is in the buffer, it is deleted and the rightmost element is put in the empty position. If there is no buffer, we are guaranteed due to the first step that there is a structure with 1 node, which is now the buffer. On the structure with the minimum element REPLACEMIN is called with the rightmost element of the array. The running time is $O(\log n)$ for finding all the structures, $O(\log n)$ for decomposing the smallest structure and $O(\log n)$ for the REPLACEMIN procedure, in total we get $O(\log n)$ for EXTRACTMIN.

The INSERT procedure is simpler but the correctness proof is somewhat involved. A new element is inserted in the buffer, if the buffer becomes a node, then the *least significant bit i* of N is computed. If at least two structures of size 2^i exist (found using the two lemmas above), then they are linked and become one structure of size 2^{i+1}.

Lemma 5. *The INSERT and EXTRACTMIN procedures maintain that at most five structures of size 2^i exist for all $i \leq \lfloor \log n \rfloor$.*

Proof. Let $N_{\leq i}$ be the total number of nodes in structures of size $\leq 2^i$. Then the following is an invariant for $i = 0, 1, \ldots, \lfloor \log N \rfloor$.

$$N_{\leq i} + (2^{i+1} - ((N + 2^i) \bmod 2^{i+1})) \leq 6 \cdot 2^i - 1$$

The invariant states that $N_{\leq i}$ plus the number of inserts until we try to link two trees of size 2^i is at most $6 \cdot 2^i - 1$. Suppose that a new node is inserted and i is not the least significant bit of N then $N_{\leq i}$ increases by one and so does $(N + 2^i) \bmod 2^{i+1}$, which means the invariant is maintained. Suppose that i is the least significant bit in N (i.e. we try to link structures of size 2^i) and there are at least two structures of size 2^i, then the insertion makes $N_{\leq i}$ decrease by $2 \cdot 2^i - 1 = 2^{i+1} - 1$ and $2^{i+1} - (N + 2^i \bmod 2^{i+1})$ increases by $2^{i+1} - 1$, since $(N + 2^i) \bmod 2^{i+1}$ becomes zero, which means the invariant is maintained. Now suppose there is at most one structure of size 2^i and i is the least significant bit of N. We know by the invariant that $N_{\leq i-1} + (2^i - (N + 2^{i-1} \bmod 2^i)) \leq 6 \cdot 2^{i-1} - 1$

which implies $N_{\leq i-1} \leq 6 \cdot 2^{i-1} - 1 - 2^i + 2^{i-1} = 5 \cdot 2^{i-1} - 1$. Since we assumed there is at most one structure of size 2^i we get that $N_{\leq i} \leq 2^i + N_{\leq i-1} \leq 2^i + 5 \cdot 2^{i-1} - 1 = 3.5 \cdot 2^i - 1$. Since N mod $2^{i+1} = 2^i$ (i is the least significant bit of N) we have $N_{\leq i} + (2^{i+1} - (N + 2^i \text{ mod } 2^{i+1})) \leq 3.5 \cdot 2^i - 1 + 2^{i+1} = 5.5 \cdot 2^i - 1 < 6 \cdot 2^i - 1$.

The invariant is also maintained when deleting: for each i where $N_i > 0$ before the EXTRACTMIN, N_i decreases by one. For all i the second term increases by at most one, and possibly decreases by $2^{i+1} - 1$. Thus the invariant is maintained for all i where $N_i > 0$ before the procedure. If $N_i = 0$ before an EXTRACTMIN, we get $N_j = 2^{j+1} - 1$ for $j \leq i$. Since the second term can at most contribute 2^{j+1}, we get $N_j + (2^{j+1} - ((N + 2^j) \text{ mod } 2^{j+1})) \leq 2^{j+1} - 1 + 2^{j+1} \leq 6 \cdot 2^j - 1$, thus the invariant is maintained. $\square$

Correctness and running times of the procedures have now been established.

References

1. Brodal, G.S.: A survey on priority queues. In: Brodnik, A., López-Ortiz, A., Raman, V., Viola, A. (eds.) Ianfest-66. LNCS, vol. 8066, pp. 150–163. Springer, Heidelberg (2013)
2. Brodal, G.S., Fagerberg, R., Jacob, R.: Cache oblivious search trees via binary trees of small height. In: Proceedings of the Thirteenth Annual ACM-SIAM Symposium on Discrete Algorithms, pp. 39–48 (2002)
3. Brodal, G.S., Nielsen, J.S., Truelsen, J.: Finger search in the implicit model. In: Chao, K.-M., Hsu, T.-S., Lee, D.-T. (eds.) ISAAC 2012. LNCS, vol. 7676, pp. 527–536. Springer, Heidelberg (2012)
4. Brodal, G.S., Nielsen, J.S., Truelsen, J.: Strictly implicit priority queues: On the number of moves and worst-case time (2015). CoRR, abs/1505.00147
5. Carlsson, S., Munro, J.I., Poblete, P.V.: An implicit binomial queue with constant insertion time. In: Karlsson, R., Lingas, A. (eds.) SWAT 88. LNCS, vol. 318, pp. 1–13. Springer, Heidelberg (1988)
6. Carlsson, S., Sundström, M.: Linear-time in-place selection in less than 3n. In: Staples, J., Katoh, N., Eades, P., Moffat, A. (eds.) ISAAC 1995. LNCS, vol. 1004, pp. 244–253. Springer, Heidelberg (1995)
7. Edelkamp, S., Elmasry, A., Katajainen, J.: Ultimate binary heaps, Manuscript (2013)
8. Franceschini, G.: Sorting stably, in place, with $O(n \log n)$ comparisons and $O(n)$ moves. Theory of Computing Systems 40(4), 327–353 (2007)
9. Franceschini, G., Munro, J.I.: Implicit dictionaries with $O(1)$ modifications per update and fast search. In: Proceedings of the Seventeenth Annual ACM-SIAM Symposium on Discrete Algorithms, pp. 404–413 (2006)
10. Johnson, D.B.: Efficient algorithms for shortest paths in sparse networks. Journal of the ACM 24(1), 1–13 (1977)
11. Harvey, N.J.A., Zatloukal, K.C.: The post-order heap. In: 3rd International Conference on Fun with Algorithms (2004)
12. Williams, J.W.J.: Algorithm 232: Heapsort. Communications of the ACM 7(6), 347–348 (1964)

On Conflict-Free Multi-coloring

Andreas Bärtschi[1]($\boxtimes$) and Fabrizio Grandoni[2]

[1] Department of Computer Science, ETH Zürich, 8092 Zürich, Switzerland
baertschi@inf.ethz.ch
[2] IDSIA, University of Lugano, 6928 Manno, Switzerland
fabrizio@idsia.ch

Abstract. A *conflict-free coloring* of a hypergraph $H = (V, \mathcal{E})$ with $n = |V|$ vertices and $m = |\mathcal{E}|$ hyperedges (where $\mathcal{E} \subseteq 2^V$), is a coloring of the vertices V such that every hyperedge $E \in \mathcal{E}$ contains a vertex of "unique" color. Our goal is to minimize the total number of distinct colors. In its full generality, this problem is known as the conflict-free (hypergraph) coloring problem. It is known that $\Theta(\sqrt{m})$ colors might be needed in general.

In this paper we study the relaxation of the problem where one is allowed to assign multiple colors to the same node. The goal here is to substantially reduce the total number of colors, while keeping the number of colors per node as small as possible. By a simple adaptation of a result by Pach and Tardos [2009] on the single-color version of the problem, one obtains that only $O(\log^2 m)$ colors in total are sufficient (on every instance) if each node is allowed to use up to $O(\log m)$ colors.

By improving on the result of Pach and Tardos (under the assumption $n \ll m$), we show that the same result can be achieved with $O(\log m \cdot \log n)$ colors in total, and either $O(\log m)$ or $O(\log n \cdot \log \log m) \subseteq O(\log^2 n)$ colors per node. The latter coloring can be computed by a polynomial-time Las Vegas algorithm.

1 Introduction

Consider the following scenario motivated by wireless applications. We are given a collection of n transmitters, where each transmitter can transmit at a chosen frequency. Furthermore, we are given a collection of m receivers, where each receiver receives the signal of some subset of the transmitters. Each receiver can tune to a proper frequency, and it receives any message transmitted at that frequency if precisely one transmitter in its range is transmitting at that frequency (if two or more such transmitters do this, then interferences destroy the message). We have to choose frequencies such that each receiver can receive messages, and our goal is to minimize the total number of frequencies used altogether.

In its full generality, this is known as the conflict-free (hypergraph) coloring problem. We are given a hypergraph $H = (V, \mathcal{E})$, $\mathcal{E} \subseteq 2^V$, with n nodes and m

---

The second author is partially supported by the ERC Starting Grant NEWNET 279352.

F. Dehne et al. (Eds.): WADS 2015, LNCS 9214, pp. 103–114, 2015.
DOI: 10.1007/978-3-319-21840-3_9

hyperedges. A coloring of H with k colors is an assignment $c : V \rightarrow \{1, \ldots, k\}$ of an integer value (*color*) to each node. A coloring is *conflict-free* if for each hyperedge E there exists at least one node $v \in E$ such that $c(v) \neq c(u)$ for any other node $u \neq v$ with $u \in E$. Our goal is to find a conflict-free coloring with the minimum number of colors. The latter quantity $\chi_{\mathrm{cf}}(H)$ is the *conflict-free chromatic number* of H. Obviously, in the above scenario, nodes, hyperedges, and colors model transmitters, receivers, and frequencies, respectively.

Trivially, $\min\{n, m + 1\}$ frequencies are sufficient to achieve (in general) a conflict-free coloring. This result can be improved to $\Theta(\sqrt{m})$ [17][1]. The latter result is already tight: simply consider a complete graph on n nodes; a conflict-free coloring requires $n = \Theta(\sqrt{m})$ colors.

Our Results and Techniques. Motivated by the large (polynomial) number of colors needed to solve conflict-free coloring in general, in this paper we study a relaxation of the problem where we are allowed to use multiple colors at each node. This models a situation in which transmitters can transmit on multiple frequencies.

More formally, we study the following *conflict-free (hypergraph) multi-coloring* problem. Given a hypergraph $H = (V, \mathcal{E})$, $\mathcal{E} \subseteq 2^V$, with n nodes and m hyperedges, a multi-coloring of H with k colors is an assignment $C : V \rightarrow 2^{\{1, \ldots, k\}}$ of a subset of integer values (*colors*) to each node. A hyperedge $E \in \mathcal{E}$ is *conflict-free* if there exists at least one node $v \in E$ and one color $c(v) \in C(v)$ such that, for any other node $v \neq u \in E$ and any $c(u) \in C(u)$, one has $c(v) \neq c(u)$ (intuitively, some color appears exactly once in E). A multi-coloring is *conflict-free* if all hyperedges are conflict-free. Our goal is now two-fold: on one side, as before, we wish to minimize the total number of colors. At the same time, we would like to minimize the maximum number of colors assigned to each node. At high-level, we address the following main question: Is a small number of colors per node sufficient to drastically reduce the total number of colors?

We answer affirmatively to the above question. Indeed, one simple way to achieve a result of the above kind is via an adaptation of a result by Pach and Tardos [17] on the standard (single-color) version of the problem. Suppose that all hyperedges have size at least $2t - 1$ (for any integer $t \geq 1$). They show how to compute a conflict-free coloring with $O(t m^{1/t} \log m)$ colors in total using a simple (expected) polynomial-time Las-Vegas algorithm. The idea is to make $\Theta(\log m)$ copies of each node, and then apply the algorithm in [17]. The set of colors assigned to a given node is simply the union of the colors assigned to its copies. This way each node is assigned at most $O(\log m)$ colors, and the total number of colors is $O(\log^2 m)$. Pach and Tardos improve their result when the *dependencies* among hyperedges are limited, by means of a constructive version [16] of Lovász's Local Lemma (LLL) [10,19,22]. More formally, let $\Gamma \leq m - 1$

---

[1] Note that this is an absolute upper bound on the conflict-free chromatic number, while of course some hypergraphs might need fewer colors. All upper bounds in this paper are of this type.

denote the maximum number of different hyperedges that any hyperedge E intersects (the maximum *hyperedge degree* of H). In this case the result in [17] is refined to $O(t\Gamma^{1/t} \log \Gamma)$, and consequently one can obtain a conflict-free multi-coloring with $O(\log^2 \Gamma)$ colors in total, and $O(\log \Gamma)$ colors per node.

Our main result (which might be of independent interest) is an improvement on the $O(t\Gamma^{1/t} \log \Gamma)$ upper bound, under each of the following assumptions: (i) n is sufficiently smaller than Γ (see Section 2), (ii) hyperedges have size at most $O(\log \Gamma)$ (see Section 3):

Theorem 1. *There exists a polynomial-time Las Vegas algorithm for conflict-free coloring using $O(t\Gamma^{1/t} \log n) \subseteq O(tm^{1/t} \log n)$ colors, where $2t-1$ is a lower bound on the size of any hyperedge and Γ is the maximum hyperedge degree. If the maximum hyperedge size is $O(\log \Gamma)$, the number of colors can be reduced to $O(t\Gamma^{1/t})$.*

We remark that there are ranges of values of Γ, m and n such that we reduce the upper-bound on the conflict-free chromatic number by a factor $\Omega(\sqrt{n})$. For example, consider a hypergraph on n nodes for which we choose uniformly at random $m = n^{\sqrt{n}/\ln n}$ hyperedges of size $\sqrt{n}$. Then the probability that two hyperedges E, E' intersect is at least $1/\sqrt{n}$ (the probability that a fixed node is contained in E'). Hence in expectation a given hyperedge intersects at least $(m-1)/\sqrt{n}$ other hyperedges. Therefore we can assume $\Gamma \in \Omega(m/\sqrt{n})$. Since also $\Gamma \leq m-1$, we have $\sqrt{n} \in \Theta(\log m) = \Theta(\log \Gamma)$. In this case the result of Pach-Tardos gives the (up to constant factors) trivial bound of $O(\log^2 \Gamma) = O(n)$ colors, while our construction uses only $O(\log \Gamma) = O(\sqrt{n})$ colors.

Furthermore we can improve on Theorem 1 by a refined analysis in case of hypergraphs with hyperedge sizes bounded from below by $2t-1$ and from above by $O(t)$. For such almost-uniform hypergraphs we achieve a conflict-free coloring with $O(tm^{1/(t+1)})$ colors. This generalizes a result on uniform hypergraphs [14].

We next discuss the main ideas behind Theorem 1. The conflict-free coloring algorithm in [17] works as follows. There is a sequence of rounds. At round i we use a new color i and color each still uncolored node independently at random with some (fairly small) probability p. Observe that the color assigned to each node follows a geometric distribution.

Our main idea is to replace colors in the above approach with disjoint *color classes*, each one containing h colors. Then each node is independently assigned a color chosen uniformly at random in its color class. For our goal it is convenient to use a constant probability p and a large enough value of h. The rough idea is that, with large-enough probability, for each hyperdge E there is some round i where for the yet unassigned nodes $E' \subseteq E$ we have that $(h/|E'|)^{|E'|}$ is lower bounded by a polynomial in the maximum hyperedge degree Γ. Therefore with sufficiently large probability (with respect to $1/\Gamma$) some color in the i-th color class will appear only once in E' and hence in E, since color classes are disjoint.

By using node duplication as discussed before, one immediately obtains the following corollary for conflict-free multi-coloring.

Corollary 1. *There exists a polynomial-time Las Vegas algorithm for conflict-free multi-coloring using $O(\log \Gamma \cdot \log n)$ colors in total, and $O(\log \Gamma)$ colors per node, where Γ is the maximum hyperedge degree.*

Note that m (hence Γ) can be exponential in n. Therefore the upper bound $O(\log \Gamma)$ on the number of colors per node can be linear in n: this might be too much due to technological constraints. We were able to reduce the mentioned upper bound (for large enough Γ) by means of a more sophisticated algorithm, without increasing the total number of colors (see Section 3).

Theorem 2. *There exists a polynomial-time Las Vegas algorithm for conflict-free multi-coloring using $O(\log \Gamma \cdot \log n)$ colors in total, and $O(\log n \cdot \log \log \Gamma) \subseteq O(\log^2 n)$ colors per node, where Γ is the maximum hyperedge degree.*

The above refinement is obtained as follows: Observe that, using our result from Theorem 1, hyperedges of size $\Omega(\log \Gamma)$ can be conflict-free colored with a single color per node and $O(\log \Gamma \cdot \log n)$ colors in total. We partition the remaining hyperedges in $O(\log \log \Gamma)$ buckets of approximately uniform size. Hyperedges in each bucket are colored independently, using a novel set of colors each time. In each bucket we perform a node duplication which is strictly sufficient to achieve hyperedges of size $\Theta(\log \Gamma)$, and then apply a modified conflict-free coloring algorithm. As mentioned, due to the (approximate) uniformity of the hyperedge sizes, $O(t\Gamma^{1/t}) = O(\log \Gamma)$ colors are sufficient in each bucket (adding overall $O(\log \Gamma \cdot \log \log \Gamma) \subseteq O(\log \Gamma \cdot \log n)$ many colors to the total). For increasing value of the bucket size, on one hand the (potential) number of hyperedges increases, while on the other hand the number of node duplicates needed to reach the size $\Omega(\log \Gamma)$ decreases. The two phenomena compensate well. In particular, it is always sufficient to create $O(\log n)$ copies of each node (hence the total number of colors per node is $O(\log n \log \log \Gamma) \subseteq O(\log^2 n)$).

Our work also implies improved bounds for conflict hypergraphs induced by certain shapes in the plane. In particular, we can easily extend some known results for axis-parallel rectangles and disks to any shape with constant description complexity. Details are omitted from this extended abstract.

The following lower bounds show that our results are not very far from best possible, at least in some relevant cases.

Theorem 3. *Consider a complete r-uniform hypergraph on n nodes, with $r < n/2$. Then any conflict-free multi-coloring needs to use $\Omega(\log n)$ colors in total. Furthermore, any such coloring using $polylog(n)$ colors has to use $\Omega(\frac{\log n}{\log \log n})$ colors on some node.*

For intuition we give a proof for $r = 2$; the complete proof will be given in the full version of the paper. We can represent the multi-coloring of each node as a 0-1 vector, where the 1's indicate the colors assigned to that node. If two nodes u and v are labelled with the same vector, then the edge uv is not conflict-free. Suppose we use h_{tot} colors in total, and at most h_{max} colors per node. Then the number of 0-1 vectors is $O(\min\{h_{tot}^{h_{max}}, 2^{h_{tot}}\})$. As a consequence, we need $h_{tot} = \Omega(\log n)$ to have n distinct vectors. Similarly, if $h_{tot} = polylog(n)$, we need $h_{max} = \Omega(\log n/ \log \log n)$ to have n distinct vectors.

Table 1. Bounds on the conflict-free chromatic number of hypergraphs on n vertices, m edges and maximum hyperedge degree Γ

Constraint	Previous Results	Our Results
$\forall E \in \mathcal{E} : 2t - 1 \leq \lvert E \rvert$	$O(t\Gamma^{1/t} \log \Gamma)$ [17]	$O(t\Gamma^{1/t} \log n)$
$\forall E \in \mathcal{E} : 2t - 1 \leq \lvert E \rvert \in O(\log \Gamma)$		$O(t\Gamma^{1/t})$
$\forall E \in \mathcal{E} : \lvert E \rvert = r$	$\Omega(\frac{rm^{2/(r+2)}}{\log m}), O(rm^{\frac{2}{r+2}})$ [14]	
$\forall E \in \mathcal{E} : 2t - 1 \leq \lvert E \rvert \in O(t)$		$O(tm^{1/(t+1)})$

For comparison, consider a hypergraph on $m = \binom{n}{r}$ hyperedges of uniform size $r \leq n/e$. Observe that $\left(\frac{n}{r}\right)^r \leq m \leq \left(\frac{n \cdot e}{r}\right)^r$, hence $r \leq r(\ln n - \ln r) \leq \ln m$. The algorithm from Theorem 2 uses only one bucket and by this refined analysis assigns $O(\log m) = O(r \log n)$ colors in total and $O(\log n)$ colors per node. Hence for small r our algorithm is not far from best possible.

Related Work. An anonymous reviewer pointed us to the independent work of Bar-Yehuda, Goldreich and Itai [5] on the radio broadcast problem, which considers assigning time-slots to transmitters (rather than frequencies) in a periodic schedule. One can reinterpret time slots of their framework as colors in a multi-coloring, hence obtaining for our setting a randomized multi-coloring algorithm using $O(\log m \cdot \log \Delta)$ colors in total, where Δ is the maximum hyperedge size. Using Lovász's Local Lemma one can infer that this can be improved to $O(\log \Gamma \cdot \log \Delta)$ colors in total and $O(\log \Gamma)$ colors per node. Considering Theorem 2, the two results differ (i) for the total number of colors if $\log \Gamma \ll \Delta \ll n$ and (ii) for the number of colors per transmitter if $\Gamma \gg n$.

Other multi-coloring models for frequency assignment problems have been considered for standard graphs (for a survey, see e.g. [1]). We already mentioned a few results about the (single-color) conflict-free hypergraph coloring problem. Pach and Tardos [17] raised the question whether it is possible to get a coloring with $\tilde{O}(tm^{1/t})$ colors even when hyperedges have size at least t (rather than $2t - 1$). Kostochka et al. [14] have answered this in the negative, proving that there exists a r-uniform hypergraph H with m hyperedges (and *even $r \leq \ln m$*) such that $\chi_{cf}(H) \in \Omega(rm^{2/(r+2)}/\log m)$. They have also shown that for all r-uniform hypergraphs H, $\chi_{cf}(H) \in O(rm^{2/(r+2)})$. The known bounds on the conflict-free chromatic number are summarized in Table 1.

For obvious reasons related to the mentioned applications, it makes sense to consider the conflict-free coloring problem under geometric restrictions on the structure of the hypergraph. In particular, one can consider transmitters and receivers as points in an Euclidean space. Here each transmitter v reaches all the receivers E in a given geometric region around v (e.g., a circle or sphere centered at v). Indeed, the problem was first defined having such a geometric model in mind by Even et al. [11], and has further been studied by Smorodinsky [2,12,20], Pach [18] and Cheilaris [4,8] for various geometric hypergraphs, such as those induced by disks, rectangles or intervals. The problem has been studied in terms of approximation [13] and online algorithms [3,9]. For a comprehensive survey on this problem, see also [21].

Another recently studied conflict-free coloring problem is a chromatic variant of the art gallery problem, in which the hypergraph is induced by visibility regions of transmitters in a given polygon [6,7]. In this problem, the structure of the hypergraph depends on the placement of the transmitters, which is not prescribed, but rather can be chosen together with the coloring.

2   An Improved Conflict-Free Coloring Algorithm

In this section we describe the conflict free-coloring algorithm from Theorem 1. Recall that n denotes the number of nodes, m the number of hyperedges, and $\Gamma \leq m - 1$ the maximum number of hyperedges that intersect any given hyperedge E. Furthermore, the minimum hyperedge size is $2t - 1$.

Our proof proceeds as follows. We start by describing a simple randomized algorithm that assigns colors independently to each node. We remark that our algorithm framework contains the algorithm by Pach and Tardos [17] as a special case, however our choice of parameters is substantially different. Let B_E denote the (bad) event that a given hyperedge E is *not* conflict-free. We will show that $\Pr[B_E] \leq \frac{1}{e\Gamma}$. Since the event B_E depends on at most Γ other bad events B_F (namely those corresponding to hyperedges F intersecting E), we conclude from Lovasz Local Lemma (LLL) that our algorithm succeeds with positive probability[2]. We can therefore use the polynomial-time Las Vegas algorithm MT by Moser and Tardos [16] to construct the desired conflict-free coloring in expected polynomial time.

A Geometric Color Classes Algorithm. Consider the following Geometric Color Classes algorithm GCC. GCC has two parameters, a probability p and a positive integer h (to be fixed later). Let $C_1, C_2, \ldots, C_{\lceil \ln n \rceil}$ be pairwise disjoint subsets of h colors each (*color classes*). Our algorithm works in two steps:

Step 1 We independently assign a color class C_i to each node as follows. At each round $i = 1, \ldots, \lceil \ln n \rceil - 1$ we consider every node v that has not been assigned any color class yet, and we independently assign color class C_i to v with probability p. At the end of the process we assign the final color class $C_{\lceil \ln n \rceil}$ to the remaining unassigned nodes.

Step 2 For each node v we choose independently and uniformly at random one of the h colors from its assigned color class.

We next set the parameters p and h, and discuss some consequences of our choices that will turn out to be useful in the analysis of the algorithm. We choose $p = 1 - \frac{1}{e}$ and $h = 48t(2e\Gamma)^{1/t}$.

Remark 1. The assignment of nodes to color classes follows a *truncated* geometric distribution. In more detail, the probability that a node v is assigned to the

---

[2] Here we consider the refined version of LLL given by Shearer [19], however this is not crucial for us modulo updating a few constants.

color class C_i is $p(1 - p)^{i-1}$ for $i < \ln n$, and $(1 - p)^{\lceil \ln n \rceil - 1}$ for $i = \lceil \ln n \rceil$. In particular, the number X of nodes assigned to $C_{\lceil \ln n \rceil}$ in expectation is $\mathbb{E}[X] = n \cdot (1/e)^{\lceil \ln n \rceil - 1} \leq e$.

Remark 2. Consider h as a function $h(t)$ of t. Note that we can restrict the domain of $h(t)$ to $t \leq \ln \Gamma$: Since $t\Gamma^{1/t}$ achieves its minimum for $t = \ln \Gamma$, and since any hyperedge with more than $2\ln \Gamma - 1$ nodes has size at least $2\ln \Gamma - 1$, it is enough to show the claimed bound of $O(t\Gamma^{1/t} \log n)$ colors for $t \leq \ln \Gamma$. Over this domain $h(t)$ is monotonically decreasing because for $t \leq \ln \Gamma$ we have $h'(t) = 48(2e\Gamma)^{1/t} \cdot (t - \ln \Gamma - \ln 2 - 1)/t < 0$. We will make use of the fact that $t \leq t' \leq \ln \Gamma$ implies $h(t) \geq h(t')$.

Existence of a Good Coloring. In this section we will show that GCC, using $O(t\Gamma^{1/t} \log n)$ colors, finds a conflict-free coloring with positive probability. To this end, we prove that LLL is applicable to the randomized coloring given by GCC. Recall that in LLL one considers a set of (bad) events, each one happening with probability at most p, where each event is independent of all the others except for at most d of them. Then, if $epd \leq 1$, there is a nonzero probability that none of the events occur [19]. In our case the bad events are $\{B_E\}_{E \in \mathcal{E}}$, where B_E denotes the event that the hyperedge E is not conflict-free. Since E intersects at most Γ other hyperedges and colors are assigned independently, B_E is independent from all but Γ other events B_F (i.e., $d = \Gamma$). By LLL it is sufficient to show that $\Pr[B_E] \leq \frac{1}{e\Gamma}$.

Consider any given hyperedge E of size s. Recall that by assumption $s \geq 2t - 1$. We distinguish between the case that E is *small*, i.e., $s \leq 24\ln \Gamma$, and the case that E is *large*, i.e., $s > 24\ln \Gamma$.

Case of small hyperedges. Let E be a (small) hyperedge with $s = |E| \leq 24\ln \Gamma$ nodes. We can upper bound $\Pr[B_E]$ by means of the following coupling argument. Suppose that a node v is assigned the j-th color of the i-th color class C_i. Then we reassign to v the j-th color of C_1. Clearly this reassignment can only decrease the probability that each given hyperedge E is conflict-free. Therefore it is sufficient to upper bound $\Pr[B_E]$ under the assumption that all nodes in E are assigned to the same color class. Lemma 1 follows Kostochka et al. [14]:

Lemma 1. *Let E be a hyperedge of size s and let its nodes be colored uniformly at random with h colors. Then the probability that there is no unique color in E is $Pr[B_E] \leq \left(\frac{2s}{h}\right)^{\lceil s/2 \rceil}$.*

Lemma 2. *For any small hyperedge E, $Pr[B_E] \leq \frac{1}{2e\Gamma}$.*

Proof. By definition one has $s = |E|$ with $2t - 1 \leq s \leq 24\ln \Gamma$. First note that by Lemma 1 (and the mentioned coupling argument) we have

$$\Pr[B_E] \leq \left(\tfrac{2s}{h}\right)^{\lceil s/2 \rceil} = \left(\frac{2s}{48t(2e\Gamma)^{1/t}}\right)^{\lceil s/2 \rceil}. \tag{1}$$

Now we distinguish between two cases, depending on whether the size s of E is relatively close to t or not, i.e. whether $s \leq 24t$ or $t < \frac{s}{24}$.

Case 1: $s \leq 24t$. One has $2s \leq 48t$ and $\lceil s/2 \rceil \geq t$. Hence the right-hand side of (1) is bounded by

$$\left(\frac{2s}{48t(2e\Gamma)^{1/t}} \right)^{\lceil s/2 \rceil} \leq \left(\frac{1}{(2e\Gamma)^{1/t}} \right)^{\lceil s/2 \rceil} \leq \left(\frac{1}{(2e\Gamma)^{1/t}} \right)^{t} = \frac{1}{2e\Gamma}.$$

Case 2: $t < \frac{s}{24}$. Recall that by assumption we have $s \leq 24 \ln \Gamma \Leftrightarrow s = 24d \ln \Gamma$ for some $d \leq 1$. Thus we can write $t < d \ln \Gamma \leq \ln \Gamma$. By Remark 2, $h(t)$ is monotonically decreasing for $t \leq \ln \Gamma$ and thus $h = h(t) > h(d \ln \Gamma)$. Furthermore, $\lceil s/2 \rceil \geq 12d \ln \Gamma$. Putting everything together, the right-hand side of (1) is bounded by

$$\left(\frac{2s}{48t(2e\Gamma)^{1/t}} \right)^{\lceil s/2 \rceil} \leq \left(\frac{48d \ln \Gamma}{48d \ln \Gamma \cdot (2e\Gamma)^{1/(d \ln \Gamma)}} \right)^{12d \ln \Gamma} = \frac{1}{(2e\Gamma)^{12}} \leq \frac{1}{2e\Gamma}. \quad \square$$

Case of Large Hyperedges. Let E be a (large) hyperedge with $s = |E| > 24 \ln \Gamma$ nodes. To upper bound $\Pr[B_E]$, we show that with large enough probability E contains a subset of nodes $E' \subsetneq E$ of size $2t - 1 \leq |E'| \leq 24 \ln \Gamma$, whose nodes are assigned colors not appearing in $E \setminus E'$. This allows us to reuse the analysis for the case of small hyperedges (a coloring that is conflict-free on E' will also be conflict-free on E).

In more detail, consider the color classes assigned to the nodes of E by GCC, and denote them by $C'_1, C'_2, \ldots, C'_k$ (in the order given by the algorithm.) Recall that these color classes are pairwise disjoint. Denote by E_j the subset of nodes with color class C'_j. We show that there is either a *single* subset E_j of small size or that there is a *union* $E_{>k-l} := \bigcup_{j=0}^{l-1} E_{k-j}$ of the last l subsets, for some l, that has a small size. Depending on which case applies, we will use either $E' = E_j$ or $E' = E_{>k-l}$. Let us formally define these two events, for which we use mnemonic identifiers S (single color class) and U (union of color classes):

- $S =$ "There is an index j, $1 \leq j \leq l$, such that $2t - 1 \leq |E_j| \leq 24 \ln \Gamma$."
- $U =$ "There is an index l, $0 \leq l < k$, such that $2t - 1 \leq |E_{>k-l}| \leq 24 \ln \Gamma$."

Lemma 3. *For the events S and U as defined, we have $\Pr[\neg S \wedge \neg U] \leq \frac{1}{2e} \frac{1}{\Gamma}$.*

Proof. Assume neither S nor U occurs. Recall that $E_{>k-l} = \bigcup_{j=0}^{l-1} E_{k-j}$. Since by assumption U does not occur, there exists a unique l with $0 \leq l < k$ such that $E' := E_{>k-l}$ has a comparatively very small size $|E'| < 2t - 1$, while $E'' = E_{>k-l-1} = E' \cup E_{k-l}$ already has a large size $|E''| := a \ln \Gamma$ for some $a > 24$.

Since S does not occur, we must have $|E_{k-l}| > 24 \ln \Gamma$. Recall that by Remark 2 we can restrict ourselves to the case $t \leq \ln \Gamma$ and hence $|E'| < 2t - 1 < 2 \ln \Gamma$. Thus $|E_{k-l}| = |E'' \setminus E'| > (a - 2) \ln \Gamma$. We can conclude that

$$\Pr[\neg S \wedge \neg U] \leq \Pr[|E_{k-l}| \geq \ln \Gamma \cdot \max\{24, a - 2\}].$$

We distinguish two cases, depending on whether C'_{k-l} is the last color class $C_{\lceil \ln n \rceil}$ ever assigned by the algorithm or not.

First let us implicitly condition on the event $C'_{k-l} = C_{\lceil \ln n \rceil}$. By Remark 1, the number of all nodes X with assigned color class $C_{\lceil \ln n \rceil}$ is a sum of independent Bernoulli random variables with expectation $\mathbb{E}[X] \leq e$. Thus we can apply Chernoff bounds (see, e.g., [15]) to get

$$\Pr\left[|E_{k-l}| \geq 24 \ln \Gamma\right] \leq \Pr\left[X \geq 24 \ln \Gamma\right] \leq 2^{-24 \ln \Gamma} \leq \tfrac{1}{2e\Gamma}.$$

In the above inequalities we used the fact that $X \geq |E_{k-l}|$ and that Γ is sufficiently large.

Next we implicitly condition on the event $C'_{k-l} \neq C_{\lceil \ln n \rceil}$. Each of the nodes in E'' is chosen into E_{k-l} independently with probability $p = 1 - \tfrac{1}{e}$. Thus $|E_{k-l}|$ is a sum of independent Bernoulli random variables with expectation $\mathbb{E}\left[|E_{k-l}|\right] = p \cdot |E''| = \tfrac{(e-1)a}{e} \ln \Gamma$. Hence we can apply Chernoff bounds to get

$$\Pr\left[|E_{k-l}| \geq (a-2) \ln \Gamma\right] = \Pr\left[|E_{k-l}| \geq \left(1 + \tfrac{a-2e}{(e-1)a}\right) \tfrac{(e-1)a}{e} \ln \Gamma\right]$$

$$\leq e^{-\tfrac{(e-1)a}{e} \ln \Gamma \cdot \left(\tfrac{a-2e}{(e-1)a}\right)^2 \cdot \tfrac{1}{3}} = e^{-\ln \Gamma \tfrac{(a-2e)^2}{3e(e-1)a}} \leq e^{-\ln \Gamma \tfrac{(24-2e)^2}{3e(e-1)\cdot 24}} \leq \tfrac{1}{2e\Gamma}.$$

In the above inequalities we used the fact that $\tfrac{(a-2e)^2}{3e(e-1)a}$ is monotonically increasing in a for $a \geq 24$ and that Γ is sufficiently large. $\square$

Lemma 4. *For any large hyperedge E, $\Pr[B_E] \leq \tfrac{1}{e\Gamma}$.*

Proof. Using previous notation we get $\Pr[B_E] \leq \Pr\left[\neg S \wedge \neg U\right] + \Pr[B_E \mid S \vee U]$. By Lemma 3, $\Pr\left[\neg S \wedge \neg U\right] \leq \tfrac{1}{2e\Gamma}$. Given the event $S \vee U$, let E' be a corresponding subset of nodes of E. We recall that by definition $2t - 1 \leq |E'| \leq 24 \ln \Gamma$, and no color used for nodes in E' is also used for nodes in $E \setminus E'$. By the same analysis as in Lemma 2, the event $B_{E'}$ that there is no unique color among nodes E' has probability at most $\Pr[B_{E'}] \leq \tfrac{1}{2e\Gamma}$. Furthermore, when there is a unique color in E', then there is a unique color also in E, hence $\Pr[\neg B_{E'}] \leq \Pr[\neg B_E \mid S \vee U]$. Consequently, $\Pr[B_E \mid S \vee U] \leq \Pr[B_{E'}] \leq \tfrac{1}{2e\Gamma}$ and $\Pr[B_E] \leq \tfrac{2}{2e\Gamma} \leq \tfrac{1}{e\Gamma}$. $\square$

By Lemmas 2 and 4, and applying LLL, we obtain the following result.

Lemma 5. *Algorithm GCC computes a coloring using at most $O(t\Gamma^{1/t} \log n)$ colors. This coloring is conflict-free with positive probability.*

Computing a Conflict-Free Coloring. The probability that GCC computes a conflict-free coloring might be very small. For this reason, we rather use the Las Vegas algorithm ML in [16], adapted to our setting. In more detail, we start by coloring nodes according to GCC. Then, while there is some hyperedge E that is not conflict-free, we recolor the nodes in E using GCC (by resampling from the same product probability space as before, restricted to E). By the analysis in [16], this new algorithm GCC$^+$ computes a conflict-free coloring in expected time polynomial in n and m, provided that there exists a conflict-free coloring among the ones that can be returned by GCC. The latter condition holds by Lemma 5. The main part of Theorem 1 immediately follows.

3   A Refined Multi-coloring Algorithm

In this section we prove Theorem 2. This is achieved in two steps. First we describe and analyze a refined conflict-free coloring algorithm for the case that hyperedges have sizes in a small range. Then we present a non-trivial multi-coloring algorithm that exploits the new (and also the previous) coloring algorithm as a subroutine.

Hyperedges with Upper Bounded Size. Suppose that every hyperedge has size at most $k \cdot \ln \Gamma$, with $k \in o(\log n)$. Then we can modify the parameters in GCC to achieve an improved upper bound of $O(t\Gamma^{1/t}k)$ colors. In particular, for constant k the number of colors needed in the single-color case is $O(t\Gamma^{1/t})$ only.

In more detail, we set $p = 1$ and $h = 2kt(3e\Gamma)^{1/t}$ (i.e., we use only one color class of a size depending linearly on k). We denote this algorithm by $1C$. In order to prove that we have a conflict-free coloring with sufficiently large probability, it is sufficient to slightly adapt the proof of Lemma 2. In particular, in the case distinction we distinguish between hypedges of size $s \leq kt$ and hyperedges of size $kt < s \leq k \ln \Gamma$. The rest of the analysis is the same. We can also similarly modify the algorithm to make it run in expected polynomial time using the approach in [16]: let $1C^+$ denote this variant. Hence we obtain the following lemma, which shows the second part of Theorem 1.

Lemma 6. *There is a polynomial-time Las Vegas algorithm for conflict-free coloring using $O(t\Gamma^{1/t}k)$ colors, assuming that hyperedges have size at least $2t - 1$ and at most $k \ln \Gamma$.*

For $\Gamma \in \Theta(m)$, Lemma 6 yields a conflict-free coloring using $O(tm^{1/t}k)$ colors. Suppose that, additionally, all the hyperedges have size at most $k \cdot t$, with k a constant. Then we can improve the upper bound to $O(tm^{1/(t+1)})$ by using a deterministic preprocessing of the hypergraph similarly to [14]. Though this is not needed for our multi-coloring algorithm, we briefly present this result since it might be of some interest. Indeed, this generalizes the bound in [14] from uniform hypergraphs to hypergraphs with a constant factor gap between the minimum and maximum hyperedge size. The proof is omitted from this extended abstract.

Lemma 7. *There is a polynomial-time Las Vegas algorithm for conflict-free coloring using $O(tm^{1/(t+1)})$ colors, assuming that hyperedges have size at least $2t-1$ and at most $O(t)$.*

A Bucketing Multi-coloring Algorithm. We consider the following refined conflict-free multi-coloring algorithm. Let $q = \lceil \log_2(\ln \Gamma) \rceil$. Note that we have $q \in O(\log \log \Gamma)$. We partition the hyperedges into subsets $\mathcal{E}_0, \ldots, \mathcal{E}_q$, where for $i < q$ the subset $\mathcal{E}_i$ contains all hyperedges of size in $[2^i, 2^{i+1})$, while the last subset $\mathcal{E}_q$ contains all the remaining hyperedges (which have size $\geq 2^q \geq \ln \Gamma$). Then there is a sequence of rounds $i = 0, \ldots, q$. In round i the algorithm considers the sub-hypergraph induced by $\mathcal{E}_i$ (containing only the nodes V_i spanned by $\mathcal{E}_i$). If $i = q$, the algorithm colors nodes in V_i using algorithm GCC^+ from Theorem 1.

Otherwise, the algorithm splits each node in V_i into $\lceil \ln \Gamma_i/2^i \rceil$ copies, and colors such copies using algorithm $1C^+$ from Lemma 6. Here $\Gamma_i \leq |\mathcal{E}_i| - 1$ denotes the value of Γ in the considered sub-hypergraph. The algorithm uses a novel set of colors in each round. The final assignment of colors to a node v is simply the union of the colors assigned to any copy of v in any round. We next analyze this refined algorithm, proving Theorem 2.

Proof. (of Theorem 2) Consider the above Las Vegas algorithm. Its expected running is trivially polynomial. In each round i the algorithm obtains a conflict-free multi-coloring of hyperedges $\mathcal{E}_i$. Since each round uses different colors, the overall multi-coloring is conflict-free, too.

It remains to bound the total number of colors and the maximum number of colors per node. By Theorem 1, in round $i = q$ the algorithm uses one color per node and $O(\log \Gamma \log n)$ colors in total. In round $i < q$, the algorithm considers an instance with $m_i = |\mathcal{E}_i|$ hyperedges of size $\Theta(\log \Gamma_i)$ each (after node duplication). Applying Lemma 6 with $t = \Theta(\log \Gamma_i)$ and $k = O(1)$, the total number of colors used is $O(t\Gamma_i^{1/t}k) = O(\log \Gamma_i) \subseteq O(\log \Gamma)$. Furthermore, the number of extra colors used for each node is at most $O(\log(\Gamma_i)/2^i) = O(\log(m_i)/2^i) = O(\log(n^{2^{i+1}})/2^i) = O(\log n)$. Here we used the fact that hyperedges in $\mathcal{E}_i$ have size at most 2^{i+1}, hence there can be at most $O(n^{2^{i+1}})$ such hyperedges. Altogether, in rounds $i = 0, \ldots, q - 1$ the algorithm uses $O(\log \Gamma \log \log \Gamma) \subseteq O(\log \Gamma \log n)$ colors in total and $O(\log n \log \log \Gamma) \subseteq O(\log^2 n)$ colors per node. The claim follows.   □

Remark 3. In case hyperedges have size at most $O(\log \Gamma)$, the above algorithm (with a slight adaptation of q) uses only $O(\log \Gamma \cdot \log \log \Gamma)$ colors in total.

Acknowledgments. We would like to thank Parinya Chalermsook, Matus Mihalak, Thomas Tschager, and Peter Widmayer for helpful discussions.

References

1. Aardal, K.I., van Hoesel, S.P.M., Koster, A.M.C.A., Mannino, C., Sassano, A.: Models and solution techniques for frequency assignment problems. Annals of Operations Research **153**(1), 79–129 (2007)
2. Alon, N., Smorodinsky, S.: Conflict-free colorings of shallow discs. In: Proceedings of the 22nd Annual Symposium on Computational Geometry, SoCG, pp. 41–43 (2006)
3. Bar-Noy, A., Cheilaris, P., Olonetsky, S., Smorodinsky, S.: Online conflict-free colouring for hypergraphs. Combinatorics, Probability and Computing **19**, 493–516 (2010)
4. Bar-Noy, A., Cheilaris, P., Smorodinsky, S.: Deterministic conflict-free coloring for intervals: From offline to online. ACM Transactions on Algorithms **4**(4), 40:1–40:18 (2008)
5. Bar-Yehuda, R., Goldreich, O., Itai, A.: On the time-complexity of broadcast in multi-hop radio networks: An exponential gap between determinism and randomization. Journal of Computer and System Sciences **45**(1), 104–126 (1992)

6. Bärtschi, A., Ghosh, S.K., Mihalák, M., Tschager, T., Widmayer, P.: Improved bounds for the conflict-free chromatic art gallery problem. In: Proceedings of the 30th Annual Symposium on Computational Geometry, SoCG, pp. 144–153 (2014)

7. Bärtschi, A., Suri, S.: Conflict-free chromatic art gallery coverage. Algorithmica 68(1), 265–283 (2014)

8. Cheilaris, P., Gargano, L., Rescigno, A.A., Smorodinsky, S.: Strong conflict-free coloring for intervals. In: Chao, K.-M., Hsu, T., Lee, D.-T. (eds.) ISAAC 2012. LNCS, vol. 7676, pp. 4–13. Springer, Heidelberg (2012)

9. Chen, K., Fiat, A., Kaplan, H., Levy, M., Matoušek, J., Mossel, E., Pach, J., Sharir, M., Smorodinsky, S., Wagner, U., Welzl, E.: Online conflict-free coloring for intervals. SIAM Journal on Computing 36(5), 1342–1359 (2006)

10. Erdős, P., Lovász, L.: Problems and results on 3-chromatic hypergraphs and some related questions. In: Infinite and Finite Sets. Colloquia Mathematica Societatis János Bolyai, vol. 10, pp. 609–627 (1973)

11. Even, G., Lotker, Z., Ron, D., Smorodinsky, S.: Conflict-free colorings of simple geometric regions with applications to frequency assignment in cellular networks. SIAM Journal on Computing 33(1), 94–136 (2003)

12. Har-Peled, S., Smorodinsky, S.: Conflict-free coloring of points and simple regions in the plane. Discrete & Computational Geometry 34(1), 47–70 (2005)

13. Katz, M.J., Lev-Tov, N., Morgenstern, G.: Conflict-free coloring of points on a line with respect to a set of intervals. Computational Geometry 45(9), 508–514 (2012). CCCG 2007

14. Kostochka, A.V., Kumbhat, M., Łuczak, T.: Conflict-free colourings of uniform hypergraphs with few edges. Combinatorics, Probability and Computing 21(4), 611–622 (2012)

15. Mitzenmacher, M., Upfal, E.: Probability and Computing: Randomized Algorithms and Probabilistic Analysis, chap. 4.2: Deriving and Applying Chernoff Bounds. Cambridge University Press (2005)

16. Moser, R.A., Tardos, G.: A constructive proof of the general Lovász local lemma. Journal of the ACM 57(2), 11:1–11:15 (2010)

17. Pach, J., Tardos, G.: Conflict-free colourings of graphs and hypergraphs. Combinatorics, Probability and Computing 18(05), 819–834 (2009)

18. Pach, J., Tardos, G.: Coloring axis-parallel rectangles. Journal of Combinatorial Theory, Series A 117(6), 776–782 (2010)

19. Shearer, J.B.: On a problem of Spencer. Combinatorica 5(3), 241–245 (1985)

20. Smorodinsky, S.: On the chromatic number of geometric hypergraphs. SIAM Journal on Discrete Mathematics 21(3), 676–687 (2007)

21. Smorodinsky, S.: Conflict-Free Coloring and its Applications. Geometry-Intuitive, Discrete, and Convex, Bolyai Society Mathematical Studies 24, 331–389 (2014)

22. Spencer, J.: Asymptotic lower bounds for Ramsey functions. Discrete Mathematics 20, 69–76 (1977)

Semi-dynamic Connectivity in the Plane

Sergio Cabello[1]([⊠]) and Michael Kerber[2]

[1] Department of Mathematics, IMFM, and Department of Mathematics,
FMF, University of Ljubljana, Ljubljana, Slovenia
`sergio.cabello@fmf.uni-lj.si`
[2] Max-Planck-Institut für Informatik, Saarbrücken, Germany
`mkerber@mpi-inf.mpg.de`

Abstract. Motivated by a path planning problem we consider the following procedure. Assume that we have two points s and t in the plane and take $\mathcal{K} = \emptyset$. At each step we add to $\mathcal{K}$ a compact convex set that is disjoint from s and t. We must recognize when the union of the sets in $\mathcal{K}$ separates s and t, at which point the procedure terminates. We show how to add one set to $\mathcal{K}$ in $O(1 + k\alpha(n))$ amortized time plus the time needed to find all sets of $\mathcal{K}$ intersecting the newly added set, where n is the cardinality of $\mathcal{K}$, k is the number of sets in $\mathcal{K}$ intersecting the newly added set, and $\alpha(\cdot)$ is the inverse of the Ackermann function.

1 Introduction

Consider the *path planning problem* from robotics, also known as the *piano mover's problem* [9] [3, Ch.13]: Given an initial and a target configuration of a robot, the task is to decide whether the robot can move from the initial to the target configuration without colliding with itself or a surrounding object (and to find such a transformation if it exists). The problem is typically tackled by setting up a ***configuration space*** $\mathbb{X}$ where every robot position is encoded as a single point. Then $\mathbb{X}$ is partitioned into a ***free space*** $\mathbb{F} \subseteq \mathbb{X}$ of allowed configurations and its complement $\bar{\mathbb{F}} = \mathbb{X} \setminus \mathbb{F}$ denoting configurations that collide with obstacles. The initial and final state are denoted by two points s and t in $\mathbb{F}$, and the task is to decide whether s and t are in the same path-connected component of $\mathbb{F}$.

The following approach to solve the path planning problem is discussed by Wang, Chiang and Yap [13]. Assume for simplicity that the configuration space $\mathbb{X}$ is a unit cube in $\mathbb{R}^d$. For any given subcube, which we call ***box*** from now, we can decide whether the box is entirely contained in $\mathbb{F}$, entirely contained in $\bar{\mathbb{F}}$, or both contains points of $\mathbb{F}$ and $\bar{\mathbb{F}}$. We color a box green, red, or yellow, respectively, depending on the predicates outcome. Now, starting with the entire $\mathbb{X}$, we build a quadtree structure and keep subdividing yellow boxes into 2^d boxes of equal size until one of the following events occur:

(1) Points s and t lie in green boxes and are connected by a path that lies entirely in green boxes. Such a path is a solution to the path planning problem.
(2) Each path from s to t intersects some red box. In this case, no collision-free path from s to t can exist, and we say that the red boxes separate s and t.

© Springer International Publishing Switzerland 2015
F. Dehne et al. (Eds.): WADS 2015, LNCS 9214, pp. 115–126, 2015.
DOI: 10.1007/978-3-319-21840-3_10

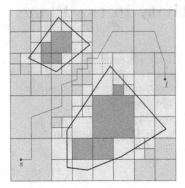 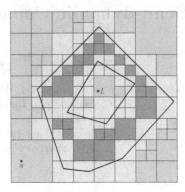

Fig. 1. Left: Configuration space with two (convex) holes. Right: Configuration space with an annulus-shaped obstacle.

Examples showing the events (1) and (2) are in Figure 1, left and right, respectively, where the next step of the subdivision is dashed. The described subdivision strategy is also used for the task of segmentation of digital images; see [1] and references therein. In that situation, the approach decides whether two pixels s and t belong to the same component of the image.

How quickly can we decide whether one of the two conditions is satisfied? Condition (1) can be easily checked by union-find [12]: just create a new element for each new green box and make unions to keep together adjacent green boxes, always checking whether the boxes containing s and t fall into the same set. That means that the amortized complexity of checking condition (1) is almost linear in the number of green boxes produced. For condition (2), the case seems less clear – an alternative way of phrasing the condition is to check whether the union of green and yellow boxes contains s and t in the same component. The union-find approach cannot directly be applied because yellow regions might turn into red and, therefore, the area covered by the boxes may shrink. In this paper, we discuss how to test the second condition in the planar case ($d = 2$).

We consider the following generalization of the problem. We have two points s and t in the plane. We get a set $\mathcal{K}$ of compact, convex sets in the plane iteratively, adding the sets one by one. Each of the sets added to $\mathcal{K}$ is disjoint from s and t. In the motivating problem, the red boxes would be the elements of $\mathcal{K}$. At the end of the insertion of a new compact convex set into $\mathcal{K}$, we want to know whether $\mathcal{K}$ separates s and t. Thus, we want a semi-dynamic data structure to store $\mathcal{K}$ that allows the insertion of new elements to $\mathcal{K}$ and decides whether $\mathcal{K}$ separates s and t.

We show that we can maintain $\mathcal{K}$ under insertions using a slightly more sophisticated union-find approach. The time to insert a new set K_u into $\mathcal{K}$ is the time we need to find all the k elements of $\mathcal{K}$ intersecting K_u, plus $O(k)$ union-find operations. In most cases, finding the elements intersecting K_u dominates the time complexity. However, in some applications with additional structure, as in our motivating scenario, the time needed for the union-find operations dominates the total running time. The idea is based on a classical parity argument saying that s and t are separated if and only if we can find a closed curve contained in the union of the elements of $\mathcal{K}$ that is crossed an odd number of times by the line segment ℓ from s to t. We maintain a union-find data structure for the sets of $\mathcal{K}$ and augment it by storing additional information about the parity of crossings with the line segment ℓ. Using this additional knowledge, we can quickly decide whether adding a new set to $\mathcal{K}$ forms a cycle that separates s and t, and the information can be maintained under union-find operations.

If in the motivating subdivision procedure we always subdivide a largest yellow box, we obtain $O(1)$ time per yellow box and $O(\alpha(n))$ amortized time per red box, where n is the number of red boxes and $\alpha(\cdot)$ is the inverse of the Ackermann function. The smooth quadtree discussed by Bennett and Yap [2] permits to subdivide boxes in an arbitrary order. Thus, we obtain the same asymptotic behavior for testing conditions (1) and (2).

Roadmap. In Section 2 we discuss a criterion to decide when $\mathcal{K}$ separates s and t in the static case. In Section 3 we extend this to the semi-dynamic case. In Section 4 we discuss the application to the motivating subdivision procedure.

Our aim is to provide a self-contained exposition. Some of the arguments are an adaptation of Cabello and Giannopoulos [4] to this simpler setting, others can be shorten substantially using machinery from Algebraic Topology.

2   Static Connectivity

Let $\mathcal{K}$ denote a finite family of compact convex sets in the plane, and let $\mathbb{K}$ denote their union. We use the notation $\bar{\mathbb{K}} = \mathbb{R}^2 \setminus \mathbb{K}$. Let s and t be points in $\bar{\mathbb{K}}$.

The set $\mathbb{K}$ **separates** s and t if they are in different path-connected components of $\bar{\mathbb{K}}$. Equivalently, $\mathbb{K}$ separates s and t if each path in the plane from s to t intersects $\mathbb{K}$. We also say that $\mathcal{K}$ separates s and t.

In the next subsection we discuss a criterion to decide when $\mathbb{K}$ separates s and t. The criterion is based on considering all polygonal paths contained in $\mathbb{K}$, and thus is computationally unfeasible. In Subsection 2.2 we discuss how this criterion can be checked in the intersection graph of $\mathcal{K}$, and thus obtain a discrete version suitable for computations.

We will consistently use Greek letters $\pi, \gamma, \tau, \ldots$ only for (polygonal) curves.

2.1   Topological Criterion for Separation

A polygonal curve π is **generic** (with respect to s and t) if π does not contain s nor t and the line segment from s to t does neither contain an endpoint of π nor a self-intersection of π. We will assume in our discussion that all the polygonal curves are generic. We can enforce this assumption making a rotation, so that ℓ is horizontal, and replacing the point s by $s' = s + (0, \varepsilon)$, for an infinitesimal $\varepsilon > 0$. We always use the same perturbed point s'. Since $\mathcal{K}$ is finite, separation of s and t with $\mathbb{K}$ is equivalent to separation of s' and t with $\mathbb{K}$. The computations can then be made using simulation of simplicity [7].

We fix ℓ as the line segment joining s' and t. The **crossing number** of ℓ with a polygonal curve π is the number of intersections of ℓ and π. We denote by $\mathrm{cr}_2(\ell, \pi)$ the modulo 2 value of the crossing number of ℓ and π. Thus, $\mathrm{cr}_2(\ell, \pi) = 1$ if and only if the crossing number is odd. For the whole paper, *any arithmetic involving* $\mathrm{cr}_2(\cdot, \cdot)$ *is done modulo 2*.

A polygonal curve π is **closed** if its endpoints coincide. It is **simple** if it does not have any self-intersections, except for the common endpoint in the case of closed polygonal paths.

Lemma 1. *The set* $\mathbb{K}$ *separates* s *and* t *if and only if there exists a closed polygonal curve* π *contained in* $\mathbb{K}$ *such that* $\mathrm{cr}_2(\ell, \pi) = 1$.

Proof. We use the following classical argument (e.g. [10, Sec. 2.1]): A simple closed polygonal curve π separates s' and t if and only if ℓ and π have an odd crossing number.

Assume that $\mathbb{K}$ contains a closed polygonal curve π such that ℓ and π have an odd crossing number. If π is not simple, we can split it at self-intersections to obtain simple, closed polygonal curves, and at least one of them has an odd crossing number with ℓ. This proves that $\mathbb{K}$ separates s and t.

Assume that $\mathbb{K}$ separates s and t, that is, s and t lie in different connected components of $\bar{\mathbb{K}}$. Since $\mathbb{K}$ is bounded, at least one of s or t lies in a bounded component A of $\bar{\mathbb{K}}$. The boundary curve of A is a simple closed curve in $\mathbb{K}$ that separates s and t. Using the convexity of elements in $\mathcal{K}$ and the compactness of $\mathbb{K}$, we can transform this separating curve into a simple polygonal curve in $\mathbb{K}$ that still separates s and t, and therefore has odd crossing number with ℓ. $\quad\square$

Corollary 1. *Let* K_u *and* K_v *be two compact convex sets of* $\mathcal{K}$. *For any two generic polygonal curves* π *and* π' *contained in* $K_u \cup K_v$ *with the same endpoints, we have* $\mathrm{cr}_2(\ell, \pi) = \mathrm{cr}_2(\ell, \pi')$.

Proof. A simple argument shows that a union of two compact convex sets $K_u \cup K_v$ cannot separate s and t. Therefore, Lemma 1 implies that any closed path γ contained in $K_u \cup K_v$ has $\mathrm{cr}_2(\ell, \gamma) = 0$. The concatenation of π and the reverse of π' is a closed path in $K_u \cup K_v$, so $\mathrm{cr}_2(\ell, \pi) + \mathrm{cr}_2(\ell, \pi') = 0$. $\quad\square$

2.2   Criterion on the Intersection Graph

Consider the **intersection graph** of $\mathcal{K}$ and denote it by G. Each element $K_v \in \mathcal{K}$ is a node of G; we will denote the node by v to match standard graph theory notation. There is an edge uv in G if and only if K_u and K_v intersect. The graph G is an abstract graph. Next we provide a geometric representation.

For each node v of G choose a point p_v in K_v. For each edge uv of G, let $\gamma(uv)$ be a polygonal path from p_u to p_v contained in the union $K_u \cup K_v$. Since K_u and K_v are convex and intersect, we can choose $\gamma(uv)$ with at most 2 segments. The collection of p_v's and $\gamma(uv)$'s is a drawing of G. (It is not necessarily a planar embedding because drawings of edges may cross.) For each walk $W = e_1 \ldots e_k$ in G, let $\gamma(W)$ be the polygonal path obtained by concatenating $\gamma(e_1), \ldots, \gamma(e_k)$. If W is a closed walk, then $\gamma(W)$ is a closed polygonal curve.

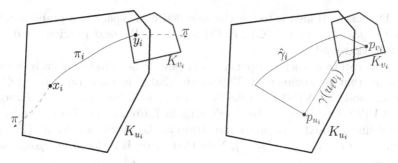

Fig. 2. Notation in the proof of Lemma 2

Lemma 2. *The set $\mathbb{K}$ separates s and t if and only if there exists a closed walk W in G such that $\mathrm{cr}_2(\ell, \gamma(W)) = 1$.*

Proof. Assume that $\mathbb{K}$ separates s and t. Because of Lemma 1, there is some polygonal curve π contained in $\mathbb{K}$ such that $\mathrm{cr}_2(\ell, \pi) = 1$. We break the path π into pieces such that each piece is contained in one set from $\mathcal{K}$. Let $\pi_1, \ldots, \pi_k$ be the resulting pieces, each of them a polygonal curve. For each piece π_i, let x_i and y_i be the endpoints of π_i, and let K_{u_i} be the element of $\mathcal{K}$ that contains π_i. Note that $y_i \in K_{u_{i+1}}$ and thus $u_i u_{i+1}$ is an edge of G. To avoid arithmetic in the subsubindex, define $v_i = u_{i+1}$ for all i. Note that $v_k = u_1$. Let W be the closed walk with edges $u_1 v_1, \ldots, u_k v_k$.

We claim that $\mathrm{cr}_2(\ell, \gamma(W)) = \mathrm{cr}_2(\ell, \pi) = 1$. To see this, consider for each piece π_i the polygonal curve $\hat{\gamma}_i$ from p_{u_i} to p_{v_i} obtained by concatenating the line segment from p_{u_i} to x_i, followed by π_i, and followed by the line segment from y_i to p_{v_i}. See Figure 2 for an example. For each piece π_i, the polygonal curves $\hat{\gamma}_i$ and $\gamma(u_i v_i)$ have the same endpoints and are contained in the union $K_{u_i} \cup K_{v_i}$. Because of Lemma 1, we have $\mathrm{cr}_2(\ell, \hat{\gamma}_i) = \mathrm{cr}_2(\ell, \gamma(u_i v_i))$. It follows that, if we define $\hat{\gamma}$ as the concatenation of $\hat{\gamma}_1, \ldots, \hat{\gamma}_k$, we have $\mathrm{cr}_2(\ell, \gamma(W)) = \mathrm{cr}_2(\ell, \hat{\gamma})$. Moreover, $\mathrm{cr}_2(\ell, \hat{\gamma}) = \mathrm{cr}_2(\ell, \pi)$ because $\hat{\gamma}$ is essentially π with spokes connecting x_i to p_{u_i}, where the number of crossings evens out. We conclude that $\mathrm{cr}_2(\ell, \gamma(W)) = \mathrm{cr}_2(\ell, \hat{\gamma}) = \mathrm{cr}_2(\ell, \pi) = 1$, finishing one direction of the proof.

For the other direction, assume that G has a closed walk W such that the crossing number of ℓ and $\gamma(W)$ is odd. Since the closed polygonal path $\gamma(W)$ is contained in $\mathbb{K}$ by construction, Lemma 1 implies that $\mathbb{K}$ separates s and t. $\square$

We extend Lemma 2 to a necessary and sufficient condition for s and t being disconnected that involves only a few cycles of G. Let T be any maximal spanning forest of G, that is, T contains a spanning tree of each connected component of G. For each edge e of $G - E(T)$, let $cycle(T, e)$ be the unique cycle in $T + e$, and let $\tau(T, e)$ be the curve $\gamma(cycle(T, e))$. That is, $\tau(T, e)$ is the polygonal curve describing $cycle(T, e)$ in the drawing.

Lemma 3. *Let T be a maximal spanning forest of G. The set $\mathbb{K}$ separates s and t if and only if there exists some edge $e \in E(G) \backslash E(T)$ such that $\mathrm{cr}_2(\ell, \tau(T, e)) = 1$.*

Proof. The essential idea is to use the so-called cycle space of a graph and the fact that $\{cycle(T, e) \mid e \in E(G) \setminus E(T)\}$ is a basis. We next provide the details using no background.

Since we can treat each component of G (and thus $\mathbb{K}$) independently, we will just assume that G is connected. This means that T is a spanning tree of G.

Fix any node $r \in V(G)$ and take the point $p_r \in K_r$ as a basepoint. For each node $v \in V(T)$, let $T[r, v]$ be the simple walk in T from r to v. For each edge uv of G we define a closed polygonal curve $\lambda(uv)$ as the concatenation of $\gamma(T[r, u])$, $\gamma(uv)$, and the reverse of $\gamma(T[r, v])$. Note that $\lambda(uv)$ is a closed polygonal path through p_r.

When $uv \notin E(T)$, $\lambda(uv)$ is $\tau(T, uv)$ concatenated with $\gamma(T[r, w])$ and its reverse, where w is the last common node of $T[r, u]$ and $T[r, v]$. This implies that

$$\forall uv \in E(G) \setminus E(T): \quad \mathrm{cr}_2(\ell, \tau(T, uv)) = \mathrm{cr}_2(\ell, \lambda(uv)). \tag{1}$$

When $uv \in E(T)$, $\lambda(uv)$ is a polygonal curve concatenated with its reverse, and therefore

$$\forall uv \in E(T): \quad \mathrm{cr}_2(\ell, \lambda(uv)) = 0. \tag{2}$$

Assume that the points s and t lie in different path-components of $\bar{\mathbb{K}}$. Because of Lemma 2, there exists some closed walk W in G with $\mathrm{cr}_2(\ell, \gamma(W)) = 1$. Let $u_1 v_1, \ldots, u_k v_k$ be the sequence of edges in W, where $u_1 = v_k$. Using arithmetic modulo 2, a simple calculation shows that

$$1 = \mathrm{cr}_2(\ell, \gamma(W)) = \sum_{i=1}^{k} \mathrm{cr}_2(\ell, \gamma(u_i v_i)) = \sum_{i=1}^{k} \mathrm{cr}_2(\ell, \lambda(u_i v_i)),$$

using that all segments on the right hand side involving r appear an even number of times and therefore cancel out. This means that, for some edge $u_i v_i$ of W, we have $\mathrm{cr}_2(\ell, \lambda(u_i v_i)) = 1$. This edge $u_i v_i$ cannot be in T because of (2). Therefore we have some edge $u_i v_i$ in $E(W)$, where $u_i v_i \notin E(T)$, with $\mathrm{cr}_2(\ell, \lambda(u_i v_i)) = 1$. Because of (1) we have $\mathrm{cr}_2(\ell, \tau(T, u_i v_i)) = \mathrm{cr}_2(\ell, \lambda(u_i v_i)) = 1$. This finishes the proof of one direction of the statement.

For the other direction, assume that there exists some edge $e \in E(G) \setminus E(T)$ such that $\mathrm{cr}_2(\ell, \tau(T, e)) = 1$. Taking $W = cycle(T, e)$ and using that $\tau(T, e) = \gamma(W)$ by definition, this means that W is a closed walk in G with $\mathrm{cr}_2(\ell, \gamma(W)) = 1$. It follows from Lemma 2 that $\mathbb{K}$ separates s and t.   □

3   Semi-dynamic Connectivity

In this section we discuss the separation of s and t under the addition of new sets to $\mathcal{K}$. We first describe a standard union-find data structure because we will build on it. Then we describe the setting and the notation we will use. It follows a description of the extension of the union-find data structure for our setting. Finally, we describe the data structure, its maintenance, and its correctness.

Algorithm FIND(u)
1.   **if** $u \neq parent(u)$ **then**
2. $parent(u) \leftarrow$ FIND($parent(u)$)
3.   **return** $parent(u)$

Algorithm UNION(u, v)
1. $\bar{u} \leftarrow$ FIND(u)
2. $\bar{v} \leftarrow$ FIND(v)
3.   **if** $rank(\bar{u}) > rank(\bar{v})$ **then**
4. $parent(\bar{v}) \leftarrow \bar{u}$
5.   **else** (* $rank(\bar{u}) \leq rank(\bar{v})$ *)
6. $parent(\bar{u}) \leftarrow \bar{v}$
7.   **if** $rank(\bar{u}) = rank(\bar{v})$ **then**
8. $rank(\bar{v}) \leftarrow rank(\bar{v}) + 1$

Fig. 3. The main two operations in the union-find data structure. u and v are nodes of the tree.

3.1   Preliminaries: Union-find

Here we review a standard union-find data structure and some of its properties. See [5, Chapter21], [6, Chapter5] or [8] for a comprehensive exposition.

A **union-find data structure** represents a disjoint set system supporting the operations MAKESET (create a new disjoint set with a single element), UNION (merge two sets), and FIND (return a representative of a given set). We can test whether two elements belong to the same set by testing whether the output of FIND for those two elements is the same. A common realization is to represent each disjoint set by a rooted tree in which each node holds one element of the set. The root of the tree holds the representative of the set. Each node has a pointer to its parent, while the root points to itself. Then FIND simply follows the parent pointer until it finds the root of the tree. The union operation merges two trees by making the root of one subtree a child of the root of the other. Thus, given two elements, we first locate the roots of their corresponding trees calling FIND, and then we proceed with the union.

Two optimizations are commonly used to obtain an efficient realization. *Union-by-rank* determines which root gets merged in a union operation: each root has a rank associated to it, in an union we simply make the root of lower rank a child of the root with larger rank, and we increase the rank of the root if both roots had the same rank. *Path compression* makes all nodes found on a search path from a node to its root direct children of the root. For later reference and modification, we include pseudocode in Figure 3. Combining these two optimizations, each operation has an amortized time complexity of $\alpha(n)$, where n is the number of elements in the set system and $\alpha(\cdot)$ is the extremely slow growing inverse Ackermann function. See references [5, Chapter21], [8] or [11] for an analysis of the time complexity.

3.2   Setting

Let s and t be two points in the plane. We have a finite family of convex sets $\mathcal{K}$, all of them disjoint from s and t. Following the previous notation, we denote by $\mathbb{K}$ the union of the sets in $\mathcal{K}$, and by G the intersection graph of $\mathcal{K}$.

Consider the addition of a new compact convex set K_u to $\mathcal{K}$. We use $\mathcal{K}_{\text{new}}$ for the resulting set, $\mathbb{K}_{\text{new}}$ for the union of its sets, and G_{new} for the intersection graph of $\mathcal{K}_{\text{new}}$.

The analysis of our data structure is based on a maximal spanning forest of the intersection graph of the convex sets. The definition of such spanning forest is iterative: Let $uv_1, \ldots, uv_k$ be an enumeration of the edges incident to u in G_{new}. That is, $K_{v_1}, \ldots, K_{v_k}$ are the sets of $\mathcal{K}$ intersecting the new set K_u. We consider adding the edges $uv_1, \ldots, uv_k$ to G one by one. We thus define G_0 as the union of G and a new vertex u for K_u. For each index $1 \leq j \leq k$, we define the graph $G_j = G_{j-1} + uv_j$. Note that $G_{\text{new}} = G_k$. The intermediate graphs $G_1, \ldots, G_{k-1}$ are not intersection graphs of $\mathcal{K}$ or $\mathcal{K}_{\text{new}}$, but something in between.

If at the time of adding uv_j the vertices u and v_j are already connected in the graph G_{j-1}, then we call uv_j a **cycle edge**. Otherwise, uv_j merges two components of G_{j-1} and we call it a **merge edge**. Whether an edge is a cycle edge or a merge edge depends on the order used in the addition of edges.

Let T be the maximal spanning forest of G. We define T_0 as the union of T and a new vertex u for K_u. For each index $1 \leq j \leq k$ we define

$$
T_j = \begin{cases} T_{j-1} & \text{if } uv_j \text{ is a cycle edge,} \\ T_{j-1} + uv_j & \text{if } uv_j \text{ is a merge edge.} \end{cases}
$$

It is easy to see by induction that, for each index $1 \leq j \leq k$, T_j is a maximal spanning forest of G_j. We define T_{new} as T_k. Thus T_{new} is a maximal spanning forest of $G_{\text{new}} = G_k$.

As it was done in Section 2.2, for each K_u we choose a point p_u in K_u and for each edge uv we choose a polygonal curve $\gamma(uv)$. These choices are made in the first appearance of the node or edge, and remain invariant from there onwards.

3.3 Augmented Union-find

We maintain a union-find data structure for the connected components of the graphs G_j. Recall that T_j is a maximal spanning forest of G_j. For each node v of G_j, we store a **parity bit**, denoted as $parity(v)$, with the following property:

- If v is the root of a union-find tree, then $parity(v) = 0$.
- If v has parent w in a union-find tree, then $parity(v) = \text{cr}_2(\ell, T_j[w, v])$. That is, we look at the parity of the crossing number of ℓ with the polygonal curve from p_v to p_w defined by the drawing of T_j.

For the rest of the paper, *any arithmetic involving parity bits is done modulo 2.*

We next argue that the correct parity bits can be maintained in the same complexity as the union-find operations, assuming that only certain unions are made. That is clear for MakeSet by giving the new node parity 0.

Consider the Find operation, which changes parent pointers due to path compression. Note that the graphs G_j and T_j do not change, but the union-find data structure does. Let u, v, w be nodes such that, in the union-find data

Algorithm FINDEXT(u)
1.   **if** $u \neq parent(u)$ **then**
2. $w \leftarrow parent(u)$
3. $r \leftarrow$ FIND(w)
4. $p(u) \leftarrow p(u) + p(w)$
5. $parent(u) \leftarrow r$
6.   **return** $parent(u)$

Algorithm UNIONEXT(u, v)
1. $\bar{u} \leftarrow$ FINDEXT(u)
2. $\bar{v} \leftarrow$ FINDEXT(v)
3. $b \leftarrow p(u) + p(v) + \mathbf{cr_2}(\ell, \gamma(uv))$
4.   **if** $rank(\bar{u}) > rank(\bar{v})$ **then**
5. $parent(\bar{v}) \leftarrow \bar{u}$
6. $p(\bar{v}) \leftarrow b$
7.   **else** (* $rank(\bar{u}) \leq rank(\bar{v})$ *)
8. $parent(\bar{u}) \leftarrow \bar{v}$
9. $p(\bar{u}) \leftarrow b$
10.   **if** $rank(\bar{u}) = rank(\bar{v})$ **then**
11. $rank(\bar{v}) = rank(\bar{v}) + 1$

Fig. 4. Extended find and union operations for nodes u and v. We write $p(\cdot)$ instead of $parity(\cdot)$ for a more compact notation.

structure, w is parent of v and v is parent of u. Note that

$$\mathbf{cr_2}(\ell, \gamma(T_j[u, w])) = \mathbf{cr_2}(\ell, \gamma(T_j[u, v])) + \mathbf{cr_2}(\ell, \gamma(T_j[v, w]))$$
$$= parity(u) + parity(v).$$

Therefore, when we update $parent(u) \leftarrow w$, we just have to set $parity(u) \leftarrow parity(u) + parity(v)$ to restore $parity(u)$ to its correct value.

We can now easily realize the augmented path compression. We define an extended function FINDEXT(u) that, for all nodes v from u to the root r of the tree containing u, sets $parent(u) = r$ and updates the value $parity(v)$ accordingly. Pseudocode is given in Figure 4 (left). It easily follows by induction that FINDEXT correctly maintains the parity bit of all elements.

Finally, we discuss the extension UNION to UNIONEXT. Its arguments are two nodes u and v_j such that uv_j is a merge edge and the union-find data structure stores the connectivity of G_{j-1}. Since uv_j is a merge edge, we have $T_j = T_{j-1} + uv_j$. This means that the sets K_u and K_{v_j} intersect but u and v_j were in different connected components of G_{j-1}. Like before, we first find the roots $\bar{u}$ and $\bar{v}$ of their trees using FINDEXT($\cdot$). After this it holds that $parity(u) = \mathbf{cr_2}(\ell, \gamma(T_{j-1}[\bar{u}, u]))$, and similarly $parity(v) = \mathbf{cr_2}(\ell, \gamma(T_{j-1}[\bar{v}, v]))$.

The walk $T_j[\bar{u}, \bar{v}]$ can be split into $T_{j-1}[\bar{u}, u]$, uv, and $T_{j-1}[v, \bar{v}]$. Thus,

$$\mathbf{cr_2}(\ell, \gamma(T_j[\bar{u}, \bar{v}])) = \mathbf{cr_2}(\ell, \gamma(T_{j-1}[u, \bar{u}])) + \mathbf{cr_2}(\ell, \gamma(uv)) + \mathbf{cr_2}(\ell, \gamma(T_{j-1}[v, \bar{v}]))$$
$$= parity(u) + \mathbf{cr_2}(\ell, \gamma(uv)) + parity(v).$$

The last values are either available through $parity(\cdot)$ or computable in constant time. If, for example, $\bar{u}$ gets $\bar{v}$ as its parent, then we have $parity(\bar{u}) = \mathbf{cr_2}(\ell, \gamma(T_j[u, v]))$. The other case is similar. We provide the resulting pseudocode in Figure 4 (right).

The properties of union-find imply that each of the extended operations, UNIONEXT and FINDEXT, has an amortized complexity of $\alpha(n)$, where n is the cardinality of $\mathcal{K}$.

3.4 Semi-dynamic Data Structure

We now describe the data structure to maintain $\mathcal{K}$. The data structure supports one operation: add a new compact convex set K_u to $\mathcal{K}$ and then report whether $\mathcal{K} \cup \{K_u\}$ separates s and t. We use the notation from Sections 3.2 and 3.3.

The data structure has the following elements:

- an augmented union-find data structure as described in Section 3.3;
- for each element K_v of $\mathcal{K}$, we store the point p_v;
- a semi-dynamic data structure $DS(\mathcal{K})$ that can find, for the new K_u, all the objects of $\mathcal{K}$ that intersect K_u.

The intersection graph G and the maximal spanning forest T are not kept. They are used only for the analysis.

We next describe how to insert K_u. We use the data structure $DS(\mathcal{K})$ to find the sets $K_{v_1}, \ldots, K_{v_k}$ of $\mathcal{K}$ that intersect K_u. We then insert K_u in the data structure $DS(\mathcal{K})$ to obtain $DS(\mathcal{K}_{\mathrm{new}})$. We choose a point p_u in K_u and create a new node u in the extended union-find data structure.

We then iterate over the edges $uv_1, \ldots, uv_k$. We first decide whether the considered edge uv_j is a merge edge or a cycle edge by checking whether $\mathrm{FINDEXT}(u)$ and $\mathrm{FINDEXT}(v_j)$ return the same representative. If uv_j is a merge edge, we just call $\mathrm{UNIONEXT}(u, v_j)$ and continue with the next step of the filtration.

Otherwise, uv_j is a cycle edge, and we proceed as follows. We want to check whether $\mathrm{cr}_2(\ell, \tau(T_j, uv_j)) = \mathrm{cr}_2(\ell, \tau(T_{j-1}, uv_j))$ is 1 or 0. For this, we use that u and v have already the same parent because of the calls $\mathrm{FINDEXT}(u)$ and $\mathrm{FINDEXT}(v)$. If we denote such a common parent by r, then

$$
\begin{aligned}
\mathrm{cr}_2(\ell, \tau(T_j, uv_j)) &= \mathrm{cr}_2(\ell, \gamma(T_{j-1}[u, v_j])) + \mathrm{cr}_2(\ell, \gamma(uv_j)) \\
&= \mathrm{cr}_2(\ell, \gamma(T_{j-1}[u, r])) + \mathrm{cr}_2(\ell, \gamma(T_{j-1}[v_j, r])) + \mathrm{cr}_2(\ell, \gamma(uv_j)) \\
&= parity(u) + parity(v_j) + \mathrm{cr}_2(\ell, \gamma(uv_j)).
\end{aligned}
$$

If $\mathrm{cr}_2(\ell, \tau(T_j, uv_j)) = 1$, then we conclude that $\mathbb{K}_{\mathrm{new}}$ separates s and t and we finish the algorithm. If $\mathrm{cr}_2(\ell, \tau(T_j, uv_j)) = 0$, we proceed to the next edge uv_{j+1}. This finishes the description of the algorithm. (Pseudocode for the insertion of K_u is given in the **appendix**.)

It follows from the invariants of the extended union-find discussed in Section 3.3, that we are correctly computing the value $\mathrm{cr}_2(\ell, \tau(T_j, uv_j))$. When $\mathrm{cr}_2(\ell, \tau(T_j, uv_j)) = 1$, then Lemma 3 implies that $\mathbb{K}_{\mathrm{new}}$ separates s and t. From that point on, we only need to remember that s and t are separated.

If $\mathrm{cr}_2(\ell, \tau(T_j, uv_j)) = 0$, then $\mathrm{cr}_2(\ell, \tau(T, uv_j))$ will remain 0 for all future maximal spanning forests T. This is so because the maximal spanning forest we maintain is monotone increasing: we only add vertices and edges, but never remove anything. Thus, we never need to check $\mathrm{cr}_2(\ell, \tau(T, uv_j))$ again later. In particular, if $\mathcal{K}$ did not separate s and t and we have $\mathrm{cr}_2(\ell, \tau(T_j, uv_j)) = 0$ for all j, then

$$
\forall vv' \in E(G_{\mathrm{new}}) \setminus E(T_{\mathrm{new}}): \quad \mathrm{cr}_2(\ell, \tau(T_{\mathrm{new}}, vv')) = 0.
$$

Since T_{new} is a maximal spanning forest of G_{new}, Lemma 3 implies that $\mathbb{K}_{\text{new}}$ does not separate s and t.

For each edge uv_j we are making 2 calls to FINDEXT, at most one call to UNIONEXT, and additional $O(1)$ work. This means that for each edge we spend $O(\alpha(n))$ amortized time, where n is the cardinality of $\mathcal{K}$. We also need the time needed to find the elements of $\mathcal{K}$ intersecting the new element K_u. We conclude.

Theorem 1. *Let s and t be two points in the plane. There is a semi-dynamic data structure to maintain a family $\mathcal{K}$ of n compact convex sets in the plane under insertions to decide whether $\mathcal{K}$ separates s can t. The insertion of a new set K_u in $\mathcal{K}$ that intersects k sets of $\mathcal{K}$ takes $O(1 + k\alpha(n))$ amortized time, plus the time needed to find the k elements of $\mathcal{K}$ intersecting K_u.*

Once s and t are separated by $\mathcal{K}$, the insertion of each new set can be carried out in constant time, since we only need to remember that $\mathcal{K}$ separates s and t.

4 Application to Dynamic Connectivity under Subdivision

We consider now the application discussed in the Introduction for $d = 2$.

We have two points s and t inside the unit square $\mathbb{X}$. Initially, the box $\mathbb{X}$ is colored yellow. In each iteration, we take a *largest* yellow box, subdivide it into 4 subboxes, and color each of them as red, yellow, or green depending on the outcome of some oracle. The boxes containing s or t are always colored yellow or green. We want to know at which point the red boxes separate s and t, meaning that each path from s to t contained in the unit square intersects some red box.

Boxes are assumed to contain their boundary, so that any two boxes intersect if their boundaries intersect, possibly only at a common vertex. For our arguments it is convenient to surround $\mathbb{X}$ with 8 red boxes of the same size as $\mathbb{X}$. This reduces the problem to finding certain curves within the red region. Without those additional squares, we should also consider boundary-to-boundary curves.

We maintain through the algorithm the intersection graph H of the yellow and red boxes. This intersection graph H has one node for each box that is yellow or red, and an edge between two nodes whenever the corresponding boxes intersect. The graph H is stored using an adjacency list representation [5, Chapter22]. The adjacency list of each vertex is stored as a doubly linked list. Moreover, for the appearance of a node v in the adjacency list of u, we keep a pointer to the appearance of u in the adjacency list of v. With this, we can perform the deletion of a node v in time proportional to its degree.

When we want to subdivide a yellow box K_u represented by a node u, we can locate its set of neighbors $N = N_H(u)$ in the graph H, delete u from the graph, subdivide K_u into four boxes, create the at most four new nodes representing the yellow and red boxes arising from the subdivision of K_u, check for intersection each of them against each of the nodes in N, and update the graph H accordingly. All this takes time $O(1 + |N|)$ time.

If we always subdivide a largest yellow box, there are at most 12 other boxes intersecting it. The same property is achieved is we maintain a smooth quadtree, as discussed by Bennett and Yap [2], and in such case we can subdivide boxes in an arbitrary order. This means that we can update the intersection graph H of yellow and red boxes in $O(1)$ time. Thus, we spend $O(1)$ time per subdivided yellow box and, for each red box, we get its neighboring red boxes in $O(1)$ time. Using Theorem 1 for the red boxes, and a normal union-find for the green boxes, as discussed in the Introduction, we obtain the following result.

Theorem 2. *Consider the subdivision procedure described in the Introduction where we always subdivide a largest yellow box. We can perform the subdivision until condition (1) or (2) occurs in $O(n\alpha(n))$ time, where n is the number of subdivisions performed.*

Acknowledgments. We thank Chee Yap for posing to us the problem about connectivity under subdivisions and reference [2]. The first author is supported by the Slovenian Research Agency, program P1-0297. The second author is supported by the Max Planck Center for Visual Computing and Communication.

References

1. Aizawa, K., Tanaka, S., Motomura, K., Kadowaki, R.: Algorithms for connected component labeling based on quadtrees. International Journal of Imaging Systems and Technology **19**(2), 158–166 (2009)
2. Bennett, H., Yap, C.: Amortized analysis of smooth quadtrees in all dimensions. In: Ravi, R., Gørtz, I.L. (eds.) SWAT 2014. LNCS, vol. 8503, pp. 38–49. Springer, Heidelberg (2014). doi: 10.1007/978-3-319-08404-6_4
3. de Berg, M., van Kreveld, M., Overmars, M., Schwarzkopf, O.: Computational Geometry: Algorithms and Applications, 2nd edn. Springer (2000)
4. Cabello, S., Giannopoulos, P.: The complexity of separating points in the plane. Algorithmica (to appear). doi: 10.1007/s00453-014-9965-6
5. Cormen, T.H., Leiverson, C.E., Rivest, R.L., Stein, C.: Introduction to Algorithms, 3rd edn. MIT Press (2009)
6. Dasgupta, S., Papadimitriou, C.H., Vazirani, U.V.: Algorithms. McGraw-Hill (2008)
7. Edelsbrunner, H., Mücke, E.P.: Simulation of simplicity: a technique to cope with degenerate cases in geometric algorithms. ACM Transactions on Graphics **9**(1), 66–104 (1990). doi: 10.1145/77635.77639
8. Erickson, J.: Algorithms notes: Maintaining disjoint sets ("union-find") (2015). http://web.engr.illinois.edu/jeffe/teaching/algorithms/
9. Lavalle, S.M.: Planning Algorithms. Cambridge University Press (2006)
10. Mohar, B., Thomassen, C.: Graphs on Surfaces. Johns Hopkins University Press (2001)
11. Seidel, R., Sharir, M.: Top-down analysis of path compression. SIAM Journal of Computing **34**(3), 515–525 (2005). doi: 10.1137/S0097539703439088
12. Tarjan, R.E.: Efficiency of a good but not linear set union algorithm. Journal of the ACM **22**(2), 215–225 (1975). doi: 10.1145/321879.321884
13. Wang, C., Chiang, Y.-J., Yap, C.: On soft predicates in subdivision motion planning. In: Proceedings of the Twenty-ninth Annual Symposium on Computational Geometry, SoCG 2013, pp. 349–358. ACM (2013). doi: 10.1145/2462356.2462386

Interval Selection in the Streaming Model

Sergio Cabello[1] and Pablo Pérez-Lantero[2]([✉])

[1] Department of Mathematics, FMF, University of Ljubljana, Ljubljana, Slovenia
`sergio.cabello@fmf.uni-lj.si`
[2] Escuela de Ingeniería Civil en Informática, Universidad de Valparaíso,
Valparaíso, Chile
`pablo.perez@uv.cl`

Abstract. A set of intervals is independent when the intervals are pairwise disjoint. In the interval selection problem, we are given a set $\mathbb{I}$ of intervals and we want to find an independent subset of intervals of largest cardinality, denoted $\alpha(\mathbb{I})$. We discuss the estimation of $\alpha(\mathbb{I})$ in the streaming model, where we only have one-time, sequential access to $\mathbb{I}$, the endpoints of the intervals lie in $\{1, \ldots, n\}$, and the amount of the memory is constrained.

For intervals of different sizes, we provide an algorithm that computes an estimate $\hat{\alpha}$ of $\alpha(\mathbb{I})$ that, with probability at least $2/3$, satisfies $\frac{1}{2}(1 - \varepsilon)\alpha(\mathbb{I}) \leq \hat{\alpha} \leq \alpha(\mathbb{I})$. For same-length intervals, we provide another algorithm that computes an estimate $\hat{\alpha}$ of $\alpha(\mathbb{I})$ that, with probability at least $2/3$, satisfies $\frac{2}{3}(1 - \varepsilon)\alpha(\mathbb{I}) \leq \hat{\alpha} \leq \alpha(\mathbb{I})$. The space used by our algorithms is bounded by a polynomial in ε^{-1} and $\log n$. We also show that no better estimations can be achieved using $o(n)$ bits.

1 Introduction

Several fundamental problems have been explored in the data streaming model [3, 14], where we have bounds on the amount of available memory, the data arrives sequentially, and we cannot afford to look at input data of the past, unless it was stored in our limited memory. This is effectively equivalent to assuming that we can only make one pass over the input data.

We consider the interval selection problem. A set of intervals is *independent* when all the intervals are pairwise disjoint. In the *interval selection problem*, the input is a set $\mathbb{I}$ of intervals and we want to find an independent subset of largest cardinality. Let $\alpha(\mathbb{I})$ denote this largest cardinality. There are actually two different problems: one problem is finding (or approximating) a largest independent subset, while the other problem is estimating $\alpha(\mathbb{I})$. In this paper, we focus on the estimation of $\alpha(\mathbb{I})$.

The **full version is online** at the arXiv repository [2].

S. Cabello—Supported by the Slovenian Research Agency, program P1-0297, projects J1-4106 and L7-5459; by the ESF EuroGIGA project (project GReGAS) of the European Science Foundation. Part of the work was done while visiting Universidad de Valparaíso.

P. Pérez-Lantero—Supported by project Millennium Nucleus Information and Coordination in Networks ICM/FIC RC130003 (Chile).

© Springer International Publishing Switzerland 2015
F. Dehne et al. (Eds.): WADS 2015, LNCS 9214, pp. 127–139, 2015.
DOI: 10.1007/978-3-319-21840-3_11

There are many natural reasons to consider the interval selection problem in the data streaming model. Firstly, it appears in many different contexts and several extensions have been studied [12]. Secondly, it is a natural generalization of the distinct elements problem: given a data stream of numbers, identify how many distinct numbers appeared in the stream. The distinct elements problem has a long tradition in data streams; see Kane, Nelson and Woodruff [11] for the latest result. Thirdly, there has been interest in understanding graph problems in the data stream model. However, several problems cannot be solved within the usual memory constraints, and other models have been proposed [6,7]. Finally, geometrically-defined graphs provide a rich family of graphs where certain graph problems may be solved within the traditional model. We advocate that graph problems should be considered for geometrically-defined graphs in the data stream model. The interval selection problem is one such case, since it is exactly finding a largest independent set in the intersection graph of the input intervals.

Previous Works. Emek, Halldórsson and Rosén [5] consider the interval selection problem with $O(\alpha(\mathbb{I}))$ space. They provide a 2-approximation algorithm for the case of arbitrary intervals and a (3/2)-approximation for the case of proper intervals, that is, when no interval contains another interval. Most importantly, they show that no better approximation factor can be achieved with sublinear space. Since any $O(1)$-approximation obviously requires $\Omega(\alpha(\mathbb{I}))$ space, their algorithms are optimal. They do not consider the problem of estimating $\alpha(\mathbb{I})$. Halldórsson et al. [7] consider maximum independent set in a different streaming model related to preemptive online algorithms.

Results. We consider the estimation of $\alpha(\mathbb{I})$ in the data streaming model for intervals with endpoints in $[n] = \{1, \ldots, n\}$. In this model it is common to assume that the input data, in our case the endpoints of the intervals, is from $[n]$.

(a) We provide an algorithm to obtain a value $\hat{\alpha}(\mathbb{I})$ such that $(1/2 - \varepsilon)\alpha(\mathbb{I}) \leq \hat{\alpha}(\mathbb{I}) \leq \alpha(\mathbb{I})$ with probability at least 2/3. The algorithm uses $O(\varepsilon^{-5} \log^6 n)$ space. This is explained in Section 3.
(b) For *same-length* intervals, we show how to find in $O(\varepsilon^{-2} \log(1/\varepsilon) + \log n)$ space an estimate $\hat{\alpha}(\mathbb{I})$ such that $(2/3 - \varepsilon)\alpha(\mathbb{I}) \leq \hat{\alpha}(\mathbb{I}) \leq \alpha(\mathbb{I})$ with probability at least 2/3. This is explained in Section 4.
(c) We provide lower bounds showing that the approximation ratios in (a) and (b) are essentially optimal, if we use $o(n)$ bits of space. Note that the lower bounds of Emek, Halldórsson and Rosén [5] hold for the interval selection problem but not for the estimation of $\alpha(\mathbb{I})$. This is explained in Section 5.

For the results in (a) and (b) we assume that a unit of memory can store values from $[n]$. As usual, the probability of error can be reduced by parallel repetition of the algorithm and taking the median of the results. The lower bounds of (c) are stated at bit level. Omitted details can be found in the *full version* [2]. In the full version we also develop new, approximate solutions to the interval selection problem, where we want to report a feasible solution using $O(\alpha(\mathbb{I}))$ space. Our algorithms for the interval selection problem match the optimal results by Emek et al. [5], but are much simpler.

2   Preliminaries

We assume that the input intervals are closed. Our algorithms can be easily adapted to handle other type of intervals. We will use the term 'interval' only for the input intervals. We will use the term 'window' for intervals constructed through the algorithm and 'segment' for intervals associated with the nodes of a segment tree. (This segment tree is explained later on.) The windows we consider may be of any type regarding the inclusion of endpoints. We assume that $0 < \varepsilon < 1/2$. Sometimes we use the notation $a = b \pm c$ for $b - c \le a \le b + c$.

A family of permutations $\mathcal{H} = \{h : [n] \to [n]\}$ is ε-**min-wise independent** if it satisfies

$$\forall X \subseteq [n], y \in X : \quad (1 - \varepsilon)/|X| \ \le \ \Pr_{h \in \mathcal{H}}[h(y) = \min h(X)] \ \le \ (1 + \varepsilon)/|X|.$$

Here, $h \in \mathcal{H}$ is chosen uniformly at random. The family of *all* permutations is 0-min-wise independent. However, there is no compact way to specify an arbitrary permutation. As discussed by Broder, Charikar and Mitzenmacher [1], the results of Indyk [9] can be used to construct a compact, computable family of permutations that is ε-min-wise independent. We prove this in the next lemma.

Lemma 1. *For every $\varepsilon \in (0, 1/2)$ and $n > 0$ there exists a family of permutations $\mathcal{H}(n, \varepsilon) = \{h : [n] \to [n]\}$ such that: (i) $\mathcal{H}(n, \varepsilon)$ has $n^{O(\log(1/\varepsilon))}$ permutations; (ii) $\mathcal{H}(n, \varepsilon)$ is ε-min-wise independent; (iii) an element of $\mathcal{H}(n, \varepsilon)$ can be chosen uniformly at random in $O(\log(1/\varepsilon))$ time; and (iv) for $h \in \mathcal{H}(n, \varepsilon)$ and $x, y \in [n]$, we can decide with $O(\log(1/\varepsilon))$ arithmetic operations whether $h(x) < h(y)$.*

Let us explain now how to use Lemma 1 to make a (nearly-uniform) random sample. We learned this idea from Datar and Muthukrishnan [4]. Consider any fixed subset $X \subseteq [n]$ and let $\mathcal{H} = \mathcal{H}(n, \varepsilon)$ be the family of permutations given in Lemma 1. An $\mathcal{H}$-**random** element s of X is obtained by choosing a permutation function $h \in \mathcal{H}$ uniformly at random, and setting $s = \arg\min\{h(x) \mid x \in X\}$. Although s is not chosen uniformly at random from X, from the definition of ε-min-wise independence we have

$$\forall Y \subseteq X : \quad (1 - \varepsilon)|Y|/|X| \ \le \ \Pr[s \in Y] \ \le \ (1 + \varepsilon)|Y|/|X|.$$

This means that, for a fixed Y, we can estimate the ratio $|Y|/|X|$ using $\mathcal{H}$-random samples from X repeatedly, and counting how many belong to Y.

Usually we use $\mathcal{H}$-random samples for the portion of the stream seen so far. We will also use $\mathcal{H}$ to make conditional sampling: we select $\mathcal{H}$-random samples until we get one satisfying a certain property. To analyze such a technique, the following result will be useful.

Lemma 2. *Let $Y \subseteq X \subseteq [n]$, $\varepsilon \in (0, 1/2)$, $\mathcal{H} = \mathcal{H}(n, \varepsilon)$ be the family of permutations of Lemma 1, and s a $\mathcal{H}$-random sample from X. Then*

$$\forall y \in Y : \quad (1 - 4\varepsilon)/|Y| \ \le \ \Pr[s = y \mid s \in Y] \ \le \ (1 + 4\varepsilon)/|Y|.$$

3   Size of Largest Independent Set of Intervals

Our idea is to carefully split the window $[1, n]$ into segments, and compute for each segment a 2-approximation using the algorithm of Emek et al. [5] or our new, simpler algorithm [2]. If each segment contains enough disjoint intervals from the input, then we do not do much error combining the results of the segments. We then have to estimate the number of segments in the partition of $[1, n]$ and the number of independent intervals in each segment. For the second estimation, it is useful that each segment does not contain too many independent intervals.

First, we describe the ingredients, independent of the streaming model, and discuss their properties. Then, we discuss how they can be computed in the streaming model.

3.1   Segments and Their Associated Information

Let T be a balanced segment tree on the n segments $[i, i+1)$, $i \in [n]$. Each leaf of T corresponds to a segment $[i, i+1)$ and the order of the leaves in T agrees with the order of their corresponding intervals along the real line. Each node v of T has an associated segment, denoted $S(v)$, that is the union of all segments stored at its descendants. It is easy to see that, for any internal node v with children v_ℓ and v_r, the segment $S(v)$ is the disjoint union of $S(v_\ell)$ and $S(v_r)$. We denote the root of T by r and have $S(r) = [1, n+1)$.

Let $\mathbb{S}$ be the set of $2n - 1$ segments associated with all nodes of T. Each $S \in \mathbb{S}$ contains the left endpoint and does not contain the right endpoint. For any segment $S \in \mathbb{S}$, where $S \neq S(r)$, let $\pi(S)$ be the "parent" segment of S: this is the segment stored at the parent of v, where $S(v) = S$.

For any $S \in \mathbb{S}$, let $\beta(S) = \alpha(\{I \in \mathbb{I} \mid I \subset S\})$. That is, we consider the restriction of the problem to intervals of $\mathbb{I}$ contained in S. Similarly, let $\hat{\beta}(S)$ be the size of a feasible solution computed for $\{I \in \mathbb{I} \mid I \subset S\}$ by using any 2-approximation algorithm [2,5]. Thus, $\beta(S) \geq \hat{\beta}(S) \geq \beta(S)/2$ for all $S \in \mathbb{S}$.

Lemma 3. *Let* $\mathbb{S}' \subset \mathbb{S}$ *be such that: (i)* $S(r)$ *is the disjoint union of the segments in* $\mathbb{S}'$, *and (ii) for each* $S \in \mathbb{S}'$, *we have* $\beta(\pi(S)) \geq 2\varepsilon^{-1}\lceil \log n \rceil$. *Then,*

$$\alpha(\mathbb{I}) \ \geq \ \sum_{S \in \mathbb{S}'} \hat{\beta}(S) \ \geq \ (1/2 - \varepsilon)\, \alpha(\mathbb{I}).$$

Proof. Since the segments in $\mathbb{S}'$ are disjoint because of hypothesis (i), we can merge the solutions giving $\beta(S)$ independent intervals, for all $S \in \mathbb{S}'$, to obtain a global feasible solution. We conclude that $\alpha(\mathbb{I}) \geq \sum_{S \in \mathbb{S}'} \beta(S) \geq \sum_{S \in \mathbb{S}'} \hat{\beta}(S)$.

Let $\tilde{\mathbb{S}}$ be the minimum subset of $\{\pi(S) \mid S \in \mathbb{S}'\}$ with the following property: for each $S \in \mathbb{S}'$, some segment $\tilde{S} \in \tilde{\mathbb{S}}$ is contained in the segment $\pi(S)$. Thus, each $\tilde{S} \in \tilde{\mathbb{S}}$ has some child in $\mathbb{S}'$ and no proper descendant in $\tilde{\mathbb{S}}$. For each $\tilde{S} \in \tilde{\mathbb{S}}$, let $\Pi_T(\tilde{S})$ be the path in T from the root to $\tilde{S}$. By construction, for each $S \in \mathbb{S}'$ there exists some $\tilde{S} \in \tilde{\mathbb{S}}$ such that the parent of S is on $\Pi_T(\tilde{S})$. By assumption (ii), for each $\tilde{S} \in \tilde{\mathbb{S}}$, we have $\beta(\tilde{S}) \geq 2\varepsilon^{-1}\lceil \log n \rceil$. Each $\tilde{S} \in \tilde{\mathbb{S}}$ is going to "pay" for the error we make in the sum at the segments whose parents belong to $\Pi_T(\tilde{S})$.

Let $\mathbb{J}^* \subseteq \mathbb{I}$ be an optimal solution to the interval selection problem. For each segment $S \in \mathbb{S}$, $\mathbb{J}^*$ has at most 2 intervals that intersect S but are not contained in S. Therefore, for all $S \in \mathbb{S}$ we have that

$$|\{J \in \mathbb{J}^* \mid J \cap S \neq \emptyset\}| \leq |\{J \in \mathbb{J}^* \mid J \subset S\}| + 2 \leq \beta(S) + 2. \qquad (1)$$

The segments in $\tilde{\mathbb{S}}$ are pairwise disjoint because in T none is a descendant of the other. This means that we can join solutions obtained inside the segments of $\tilde{\mathbb{S}}$ into a feasible solution. Combining this with hypothesis (ii) we get

$$|\mathbb{J}^*| \geq \sum_{\tilde{S} \in \tilde{\mathbb{S}}} \beta(\tilde{S}) \geq |\tilde{\mathbb{S}}| \cdot 2\varepsilon^{-1}\lceil \log n \rceil. \qquad (2)$$

For each $\tilde{S} \in \tilde{\mathbb{S}}$, the path $\Pi_T(\tilde{S})$ has at most $\lceil \log n \rceil$ vertices. Since each $S \in \mathbb{S}'$ has a parent in $\Pi_T(\tilde{S})$, for some $\tilde{S} \in \tilde{\mathbb{S}}$, we obtain from equation (2) that

$$|\mathbb{S}'| \leq 2\lceil \log n \rceil \cdot |\tilde{\mathbb{S}}| \leq 2\lceil \log n \rceil \cdot \frac{|\mathbb{J}^*|}{2\varepsilon^{-1}\lceil \log n \rceil} = \varepsilon \cdot |\mathbb{J}^*|. \qquad (3)$$

Using that $S(r) = \bigcup_{S' \in \mathbb{S}'} S'$, equations (1) and (3), and the fact that $\hat{\beta}(\cdot)$ is a 2-approximation of $\beta(\cdot)$, we obtain

$$|\mathbb{J}^*| \leq \sum_{S \in \mathbb{S}'} |\{J \in \mathbb{J}^* \mid J \cap S \neq \emptyset\}| \leq \sum_{S \in \mathbb{S}'} (\beta(S) + 2) = 2 \cdot |\mathbb{S}'| + \sum_{S \in \mathbb{S}'} \beta(S)$$

$$\leq 2\varepsilon \cdot |\mathbb{J}^*| + \sum_{S \in \mathbb{S}'} \beta(S) \leq 2\varepsilon \cdot |\mathbb{J}^*| + \sum_{S \in \mathbb{S}'} 2 \cdot \hat{\beta}(S).$$

The second inequality of the Theorem is obtained because $|\mathbb{J}^*| = \alpha(\mathbb{I})$. $\qquad \square$

We would like to find a set $\mathbb{S}'$ satisfying the hypothesis of Lemma 3. However, the definition should be local. The estimator $\hat{\beta}(S)$ is not suitable because for some segment $S \in \mathbb{S} \setminus \{S(r)\}$ it may happen that $\hat{\beta}(\pi(S)) < \hat{\beta}(S)$. We introduce another estimate that is an $O(\log n)$-approximation but is monotone non-decreasing along paths to the root. For each segment $S \in \mathbb{S}$, we define $\gamma(S)$ as the number of segments of $\mathbb{S}$ that are contained in S and contain some input interval.

Lemma 4. *For all $S \in \mathbb{S}$, we have the following properties:*

(i) $\gamma(S) \leq \gamma(\pi(S))$, if $S \neq S(r)$,
(ii) $\gamma(S) \leq \beta(S) \cdot \lceil \log n \rceil$,
(iii) $\gamma(S) \geq \beta(S)$, and
(iv) $\gamma(S)$ can be computed in $O(\gamma(S))$ space using the portion of the stream after the first interval contained in S.

Proof. Property (i) is obvious from the definition because any S' contained in S is also contained in the parent $\pi(S)$. For the rest of the proof, fix some $S \in \mathbb{S}$ and define $\mathbb{S}'$ as the segments of $\mathbb{S}$ that are contained in S and contain some input interval. Note that $\gamma(S) = |\mathbb{S}'|$. Let T_S be the subtree of T rooted at S.

For property (ii), note that T_S has at most $\lceil \log n \rceil$ levels. By the pigeonhole principle, there is some level L of T_S that contains at least $\gamma(S)/\lceil \log n \rceil$ different intervals of $\mathbb{S}'$. The segments of $\mathbb{S}'$ contained in level L are disjoint, and each of them contains some intervals of $\mathbb{I}$. Picking an interval from each $S' \in L$, we get a subset of intervals from $\mathbb{I}$ that are pairwise disjoint, and thus $\beta(S) \geq \gamma(S)/\lceil \log n \rceil$.

For property (iii), consider an optimal solution $\mathbb{J}^*$ for the interval selection problem in $\{I \in \mathbb{I} \mid I \subset S\}$. Thus $|\mathbb{J}^*| = \beta(S)$. For each interval $J \in \mathbb{J}^*$, let $S(J)$ be the smallest $S \in \mathbb{S}$ that contains J. Then $S(J) \in \mathbb{S}'$. Note that J contains the middle point of $S(J)$, as otherwise there would be a smaller segment in $\mathbb{S}$ containing J. This implies that the segments $S(J)$, $J \in \mathbb{J}^*$, are all distinct. (However, they are not necessarily disjoint.) We then have $\gamma(S) = |\mathbb{S}'| \geq |\{S(J) \mid J \in \mathbb{J}^*\}| = |\mathbb{J}^*| = \beta(S)$.

For property (iv), we store the elements of $\mathbb{S}'$ in a binary search tree. Whenever we obtain an interval I, we check whether the segments contained in S and containing I are already in the search tree and, if needed, update the structure. The space needed in a binary search tree is proportional to the number of elements stored and thus we need $O(\gamma(S))$ space.   □

A segment S of $\mathbb{S}$, $S \neq S(r)$, is **relevant** if the next two conditions are satisfied: $\gamma(\pi(S)) \geq 2\varepsilon^{-1}\lceil \log n \rceil^2$ and $1 \leq \gamma(S) < 2\varepsilon^{-1}\lceil \log n \rceil^2$. Let $\mathbb{S}_{rel} \subseteq \mathbb{S}$ be the set of relevant segments. If $\mathbb{S}_{rel}$ is empty, then we take $\mathbb{S}_{rel} = \{S(r)\}$.

Because of Lemma 4(i), $\gamma(\cdot)$ is non-decreasing along a leaf-to-root path in T. Combining lemmas 3 and 4, we obtain the following:

Lemma 5. *We have*

$$\alpha(\mathbb{I}) \geq \sum_{S \in \mathbb{S}_{rel}} \hat{\beta}(S) \geq (1/2 - \varepsilon)\,\alpha(\mathbb{I}).$$

Proof. (*Sketch*) We define

$$\mathbb{S}_0 = \{S \in \mathbb{S} \setminus \{S(r)\} \mid \gamma(S) = 0 \text{ and } \gamma(\pi(S)) \geq 2\varepsilon^{-1}\lceil \log n \rceil^2\},$$

and show that $\mathbb{S}' = \mathbb{S}_{rel} \cup \mathbb{S}_0$ satisfies the conditions of Lemma 3. We then use that $\gamma(S) = \hat{\beta}(S) = 0$ for all $S \in \mathbb{S}_0$.   □

Let N_{rel} be the number of relevant segments. A segment $S \in \mathbb{S}$ is **active** if $S = S(r)$ or its *parent* $\pi(S)$ contains some input interval. Let N_{act} be the number of active segments in $\mathbb{S}$. We are going to estimate N_{act}, the ratio N_{rel}/N_{act}, and the average value of $\hat{\beta}(S)$ over the relevant segments $S \in \mathbb{S}_{rel}$. With this, we will be able to estimate the sum considered in Lemma 5. The next section describes how the estimations are obtained in the data streaming model.

3.2 Algorithms in the Streaming Model

For each interval I, we use $\sigma_{\mathbb{S}}(I)$ for the sequence of segments from $\mathbb{S}$ that are active because of interval I, ordered *non-increasingly* by size. Thus, $\sigma_{\mathbb{S}}(I)$ contains $S(r)$ followed by the segments whose parents contain I. The selected ordering implies that a parent $\pi(S)$ appears before S, for all S in the sequence $\sigma_{\mathbb{S}}(I)$. Note that $\sigma_{\mathbb{S}}(I)$ has at most $2\lceil \log n \rceil$ elements because T is balanced.

Lemma 6. *There is an algorithm in the data stream model that in $O(\varepsilon^{-2}+\log n)$ space computes a value $\hat{N}_{act}$ such that*

$$\Pr\big[|N_{act} - \hat{N}_{act}| \le \varepsilon \cdot N_{act}\big] \ge 11/12.$$

Proof. The stream $\mathbb{I} = I_1, I_2, \ldots$ defines the stream $\sigma = \sigma_{\mathbb{S}}(I_1), \sigma_{\mathbb{S}}(I_2), \ldots$ of segments, that is $O(\log n)$ times longer. The segments appearing in σ are precisely the active segments. We have reduced the problem to the question of how many distinct elements appear in a stream of segments from $\mathbb{S}$. The result of Kane, Nelson and Woodruff [11] for distinct elements uses $O(\varepsilon^{-2}+\log|\mathbb{S}|) = O(\varepsilon^{-2}+\log n)$ space and computes a value $\hat{N}_{act}$ satisfying the claim. $\square$

Lemma 7. *There is an algorithm in the data stream model that uses $O(\varepsilon^{-4}\log^4 n)$ space and computes a value $\hat{N}_{rel}$ such that*

$$\Pr\big[|N_{rel} - \hat{N}_{rel}| \le \varepsilon \cdot N_{rel}\big] \ge 10/12.$$

Proof. (Sketch) The idea is the following. We estimate N_{act} by $\hat{N}_{act}$ using Lemma 6. We take a sample of active segments, and count how many of them are relevant. To get a representative sample, it is important to use a lower bound on N_{rel}/N_{act}. With this we can estimate $N_{rel} = (N_{rel}/N_{act}) \cdot N_{act}$ accurately.

In T, each relevant segment $S' \in \mathbb{S}_{rel}$ has $2\gamma(S') < 4\varepsilon^{-1}\lceil \log n\rceil^2$ active segments below it and at most $2\lceil\log n\rceil$ active segments whose parent is an ancestor of S'. This means that, for each relevant segment, there are at most $4\varepsilon^{-1}\lceil\log n\rceil^2 + 2\lceil\log n\rceil = O(\varepsilon^{-1}\log^2 n)$ active segments. Therefore $N_{rel}/N_{act} = \Omega(\varepsilon/\log^2 n)$.

We fix any injective mapping b between $\mathbb{S}$ and $[n^2]$ that can be easily computed, and consider a family $\mathcal{H} = \mathcal{H}(n^2, \varepsilon)$ of permutations $[n^2] \to [n^2]$ guaranteed by Lemma 1. For each $h \in \mathcal{H}$, the function $h \circ b$ gives an order among the elements of $\mathbb{S}$, and we use them to compute $\mathcal{H}$-random samples among the active segments.

We set an appropriate $k = \Theta(\varepsilon^{-3}\log^2 n)$, and choose permutations $h_1, \ldots, h_k \in \mathcal{H}$ uniformly and independently at random. For each h_j, where $j \in [k]$, let S_j be the *active* segment of $\mathbb{S}$ that minimizes $(h_j \circ b)(\cdot)$. Thus, $S_j = \arg\min\{h_j(b(S)) \mid S \in \mathbb{S} \text{ is active}\}$. The idea is that S_j is nearly a random active segment of $\mathbb{S}$. Therefore, if we define the random variable

$$X = \big|\{j \in \{1, \ldots, k\} \mid S_j \text{ is relevant}\}\big|$$

then $N_{rel}/N_{act} \approx X/k$. Below we discuss the computation of X.

To analyze the variable X, we define $p = \Pr_{h_j \in \mathcal{H}}[S_j$ is relevant]. Since S_j is selected among the active segments, the discussion after Lemma 1 implies $p = (1 \pm \varepsilon)N_{rel}/N_{act}$. Using the lower bound on N_{rel}/N_{act} and Chebyshev's inequality, we can prove that $\Pr[|X/k - p| \geq \varepsilon p] \leq 1/12$.

To finalize, we define the estimator $\hat{N}_{rel} = \hat{N}_{act}(X/k)$ of N_{rel}. The events $[|N_{act} - \hat{N}_{act}| \leq \varepsilon N_{act}]$ and $[|X/k - p| \leq \varepsilon p]$ occur simultaneously with probability at least $10/12$, and in such case it follows that $\hat{N}_{rel} = (1 \pm 7\varepsilon)N_{rel}$. Replacing ε by $\varepsilon/7$, the bound follows.

It remains to discuss how X can be computed. For each $j \in [k]$, we keep a variable that stores the current segment S_j for all the segments that are active so far, keep information about the choice of h_j, and keep information about $\gamma(S_j)$ and $\gamma(\pi(S_j))$, so that we can decide whether S_j is relevant.

Let $I_1, I_2, \ldots$ be the data stream of input intervals. We consider the stream of segments $\sigma = \sigma_{\mathbb{S}}(I_1), \sigma_{\mathbb{S}}(I_2), \ldots$. When handling a segment S of the stream σ, we have to update S_j when $h_j(b(S)) < h_j(b(S_j))$. Note that we can indeed maintain $\gamma(\pi(S_j))$ because S_j becomes active the first time that its parent contains some input interval. This is also the first time when $\gamma(\pi(S_j))$ becomes nonzero, and thus the forthcoming part of the stream has enough information to compute $\gamma(S_j)$ and $\gamma(\pi(S_j))$. (Here it is convenient that $\sigma_{\mathbb{S}}(I)$ gives segments in decreasing size.) To maintain $\gamma(S_j)$ and $\gamma(\pi(S_j))$, we use Lemma 4(iv). Since we are interested in knowing only whether $\gamma(S_j)$ and $\gamma(\pi(S_j))$ are smaller than $2\varepsilon^{-1}\lceil \log n \rceil^2$, we never need to store more than $O(\varepsilon^{-1} \log^2 n)$ segments. Therefore, we need in total $O(k\varepsilon^{-1} \log^2 n) = O(\varepsilon^{-4} \log^4 n)$ space.   □

Let $\rho = \left(\sum_{S \in \mathbb{S}_{rel}} \hat{\beta}(S) \right) / |\mathbb{S}_{rel}|$. The next result shows how to estimate ρ.

Lemma 8. *There is an algorithm in the data stream model that uses $O(\varepsilon^{-5} \log^6 n)$ space and computes a value $\hat{\rho}$ such that*

$$\Pr[|\rho - \hat{\rho}| \leq \varepsilon\rho] \geq 10/12.$$

Proof. (*Sketch*) Fix any injective mapping b between $\mathbb{S}$ and $[n^2]$, and consider a family $\mathcal{H} = \mathcal{H}(n^2, \varepsilon)$ of permutations $[n^2] \to [n^2]$ guaranteed by Lemma 1. Let $\mathbb{S}_{act}$ be the set of active segments, and consider a random variable Y_1 defined as follows. We repeatedly sample $h \in \mathcal{H}$ uniformly at random, until we get that $S_1 = \arg\min_{S \in \mathbb{S}_{act}} h(b(S))$ is a relevant segment, and set $Y_1 = \hat{\beta}(S_1)$. Because of Lemma 2, where $X = \mathbb{S}_{act}$ and $Y = \mathbb{S}_{rel}$, we have

$$\forall S \in \mathbb{S}_{rel}: \quad (1 - 4\varepsilon)/|\mathbb{S}_{rel}| \leq \Pr[S_1 = S] \leq (1 + 4\varepsilon)/|\mathbb{S}_{rel}|.$$

This can be used to show that $\mathbb{E}[Y_1] = (1 \pm 4\varepsilon) \cdot \rho$ and $\mathrm{Var}[Y_1] \leq O(\rho \cdot \varepsilon^{-1} \log^2 n)$. Note also that $\gamma(S) \geq 1$ implies $\hat{\beta}(S) \geq 1$ and therefore $\rho \geq 1$.

Consider an integer k to be chosen later. Let $Y_2, \ldots, Y_k$ be independent random variables with the same distribution that Y_1, and define the estimate $\hat{\rho} = (\sum_{i=1}^{k} Y_i)/k$. We can use Chebyshev's inequality and linearity to see that $\Pr[|\hat{\rho} - \mathbb{E}[Y_1]| \geq \varepsilon\rho] \leq O(k^{-1}\varepsilon^{-3} \log^2 n)$. Setting an appropriate $k = \Theta(\varepsilon^{-3} \log^2 n)$, we obtain $\Pr[|\hat{\rho} - \mathbb{E}[Y_1]| \geq \varepsilon\rho] \leq 1/12$.

We then proceed similar to the proof of Lemma 7. We choose an appropriate $k_0 = \Theta(k\varepsilon^{-1}\log^2 n) = \Theta(\varepsilon^{-4}\log^4 n)$. For each $j \in [k_0]$, take a function $h_j \in \mathcal{H}$ uniformly at random and select $S_j = \arg\min\{h(b(S)) \mid S \text{ is active}\}$. Let X be the number of relevant segments in $S_1, \ldots, S_{k_0}$, and let $p = \Pr[S_1 \in \mathbb{S}_{rel}]$. It can be proven that $\Pr[|X - k_0p| \geq k_0p/2] \leq 1/12$, which means that, with probability at least $11/12$, the sample $S_1, \ldots, S_{k_0}$ contains at least $(1/2)k_0p \geq k$ relevant segments. We can then use the first k of those relevant segments to compute the estimate $\hat{\rho}$, satisfying $\Pr[|\rho - \hat{\rho}| \leq 5\varepsilon\rho] \geq 10/12$.

It remains to show that we can compute $\hat{\rho}$ in the data stream model. Like before, for each $j \in [k_0]$, we have to maintain the segment S_j, information about the choice of h_j, information about $\gamma(S_j)$ and $\gamma(\pi(S_j))$, and the value $\hat{\beta}(S_j)$. Since $\hat{\beta}(S_j) \leq \beta(S_j) \leq \gamma(S_j)$ because of Lemma 4(iii), we need $O(\varepsilon^{-1}\log^2 n)$ space per index j. In total we need $O(k_0\varepsilon^{-1}\log^2 n) = O(\varepsilon^{-5}\log^6 n)$ space.   □

Theorem 1. *Let $\varepsilon \in (0, 1/2)$ and $\mathbb{I}$ be a set of intervals with endpoints in $[n]$ that arrive in a data stream. There is an algorithm that uses $O(\varepsilon^{-5}\log^6 n)$ space and computes a value $\hat{\alpha}$ such that*

$$\Pr\big[(1/2 - \varepsilon) \cdot \alpha(\mathbb{I}) \leq \hat{\alpha} \leq \alpha(\mathbb{I})\big] \geq 2/3.$$

Proof. (*Sketch*) We compute the estimates $\hat{N}_{rel}$ of Lemma 7 and $\hat{\rho}$ of Lemma 8. Define the estimate $\hat{\alpha} = \hat{N}_{rel} \cdot \hat{\rho}$. With probability at least $1 - 2/12 - 2/12 = 2/3$ the events $\big[|N_{rel} - \hat{N}_{rel}| \leq \varepsilon \cdot N_{rel}\big]$ and $\big[|\rho - \hat{\rho}| \leq \varepsilon\rho\big]$ simultaneously hold. When both events occur, we can use the definitions of N_{rel} and ρ together with Lemma 5, to prove that $(1 - \varepsilon)^2(1/2 - \varepsilon)\alpha(\mathbb{I}) \leq \hat{\alpha} \leq (1 + \varepsilon)^2\alpha(\mathbb{I})$. Rescaling ε and $\hat{\alpha}$ to avoid overestimation, the claimed approximation is obtained. The space bounds are those from Lemmas 7 and 8.   □

4   Size of Largest Independent Set for Same-Size Intervals

We consider the case when all the input intervals have the same length λ. The idea is based on the shifting technique of Hochbaum and Mass [8] with a grid of length 3λ and shifts of length λ. We observe that we can maintain an optimal solution restricted to a window of length 3λ because at most two disjoint intervals of length λ can fit in.

For $\ell \in \mathbb{R}$, let W_ℓ denote the window $[\ell, \ell + 3\lambda)$. For each $a \in \{0, 1, 2\}$ we define the partition of the real line $\mathbb{W}_a = \{W_{(a+3j)\lambda} \mid j \in \mathbb{Z}\}$ and let $\mathbb{I}_a$ be the set of input intervals contained in some window of $\mathbb{W}_a$. Since each interval of length λ is contained in exactly two windows of $\bigcup_a \mathbb{W}_a$, it follows that

$$\max\{\alpha(\mathbb{I}_0), \alpha(\mathbb{I}_1), \alpha(\mathbb{I}_2)\} \geq (2/3)\,\alpha(\mathbb{I}).$$

For $a = 0, 1, 2$, we will compute a value $\hat{\alpha}_a$ that $(1 + \varepsilon)$-approximates $\alpha(\mathbb{I}_a)$ with reasonable probability. We then return $\hat{\alpha} = \max\{\hat{\alpha}_0, \hat{\alpha}_1, \hat{\alpha}_2\}$, which catches a fraction at least $2(1 - \varepsilon)/3$ of $\alpha(\mathbb{I})$.

Lemma 9. *Let $a \in \{0, 1, 2\}$ and $\varepsilon \in (0, 1)$. There is an algorithm in the data stream model that in $O(\varepsilon^{-2} \log(1/\varepsilon) + \log n)$ space computes a value $\hat{\alpha}_a$ such that*

$$\Pr\left[|\alpha(\mathbb{I}_a) - \hat{\alpha}_a| \leq \varepsilon \cdot \alpha(\mathbb{I}_a)\right] \geq 8/9.$$

Proof. (*Sketch*) Let us fix some $a \in \{0, 1, 2\}$. We say that a window W of $\mathbb{W}_a$ is of type i if W contains *at least* i disjoint input intervals. Since the windows of $\mathbb{W}_a$ have length 3λ, they can be of type 0, 1 or 2. For $i = 0, 1, 2$, let γ_i be the number of windows of type i in $\mathbb{W}_a$. Then $\alpha(\mathbb{I}_a) = \gamma_1 + \gamma_2$.

We compute an estimate $\hat{\gamma}_1$ to γ_1 as follows. The stream of intervals $\mathbb{I} = I_1, I_2, \ldots$ defines the sequence of windows $W(\mathbb{I}) = W(I_1), W(I_2), \ldots$, where $W(I_i)$ denotes the window of $\mathbb{W}_a$ that contains I_i; if I_i is not contained in any window of $\mathbb{W}_a$, we then skip I_i. Then, γ_1 is the number of distinct elements in the sequence $W(\mathbb{I})$. Because of [11], we can compute using $O(\varepsilon^{-2} + \log n)$ space a value $\hat{\gamma}_1$ such that $\Pr[(1 - \varepsilon)\gamma_1 \leq \hat{\gamma}_1 \leq (1 + \varepsilon)\gamma_1] \geq 17/18$.

We next explain how to estimate the ratio $\gamma_2/\gamma_1 \leq 1$. Consider a family $\mathcal{H} = \mathcal{H}(n, \varepsilon)$ of permutations $[n] \to [n]$ guaranteed by Lemma 1, set an appropriate $k = \Theta(\varepsilon^{-2})$, and choose permutations $h_1, \ldots, h_k \in \mathcal{H}$ uniformly and independently at random. For each permutation h_j, where $j \in [k]$, let W_j be the window $[\ell, \ell + 3\lambda)$ of $\mathbb{W}_a$ that contains some input interval and minimizes $h_j(\ell)$. Thus

$$W_j = \arg\min\left\{h_j(\ell) \mid [\ell, \ell + 3\lambda) \in \mathbb{W}_a, \text{ some } I \in \mathbb{I} \text{ is contained in } [\ell, \ell + 3\lambda)\right\}.$$

The idea is that W_j is a nearly-uniform random window of $\mathbb{W}_a$, among those that contain some input interval. Therefore, if we define the random variable

$$M = \left|\{j \in \{1, \ldots, k\} \mid W_j \text{ is of type 2}\}\right|$$

then $M\gamma_1/k$ is roughly γ_2. We use Chebyshev's inequality on M and the choice of k to show that $M\gamma_1/k = \gamma_2 \pm \varepsilon\gamma_1$ with probability at least $17/18$. Note that we cannot guarantee an error smaller than $\varepsilon\gamma_2$ because γ_2 may be negligible in comparison to γ_1. However, since we want to estimate $\gamma_1 + \gamma_2$, making an error of $\varepsilon\gamma_1$ in the estimation of γ_2 is good enough. We then return $\hat{\gamma}_1\left(1 + \frac{M}{k}\right)$.

The computation of M can be done in $O(k \log(1/\varepsilon)) = O(\varepsilon^{-2} \log(1/\varepsilon))$ space as follows. For each $j \in [k]$, we keep information about the choice of h_j, keep a variable that stores the current window W_j for all the intervals that have been seen so far, and store the leftmost and rightmost intervals contained in W_j. Those two intervals tell us whether W_j is of type 1 or 2. $\qquad \square$

Theorem 2. *Let $\varepsilon \in (0, 1/2)$ and $\mathbb{I}$ be a set of intervals of length λ with endpoints in $[n]$ that arrive in a data stream. There is an algorithm that uses $O(\varepsilon^{-2} \log(1/\varepsilon) + \log n)$ space and computes a value $\hat{\alpha}$ such that*

$$\Pr\left[(2/3 - \varepsilon) \cdot \alpha(\mathbb{I}) \leq \hat{\alpha} \leq \alpha(\mathbb{I})\right] \geq 2/3.$$

Proof. For each $a = 0, 1, 2$ we compute the estimate $\hat{\alpha}_a$ to $\alpha(\mathbb{I}_a)$ with the algorithm described in Lemma 9. We then have that the three events $[|\alpha(\mathbb{I}_a) - \hat{\alpha}_a| \leq \varepsilon \cdot \alpha(\mathbb{I}_a)]$, $a = 0, 1, 2$, simultaneously occur with probability at least $2/3$. When the three events occur, it follows that $\frac{2}{3}(1 - \varepsilon) \cdot \alpha(\mathbb{I}) \leq \max\{\hat{\alpha}_0, \hat{\alpha}_1, \hat{\alpha}_2\} \leq (1 + \varepsilon)\alpha(\mathbb{I})$. Rescaling $\hat{\alpha}$ by $1/(1 + \varepsilon)$ and ε by $1/2$, the result is achieved. $\qquad \square$

5   Lower Bounds

Emek, Halldórsson and Rosén [5] showed that any streaming algorithm for the interval selection problem, where we have to report a feasible solution, cannot achieve an approximation ratio of r, for any constant $r < 2$, unless it uses $\Theta(n)$ bits. They also show that, for same-size intervals, one cannot obtain an approximation ratio below $3/2$. We are going to show that similar inapproximability results hold for estimating $\alpha(\mathbb{I})$.

Consider the problem INDEX: The input is a pair $(S, i) \in \{0, 1\}^n \times [n]$ and the output, denoted INDEX(S, i), is the i-th bit of S. One can think of S as a subset of $[n]$, and then INDEX(S, i) is asking whether element i is in the subset or not.

The one-way communication complexity of INDEX is well understood. In this scenario, Alice has S and Bob has i. Alice sends a message to Bob and then Bob has to compute INDEX(S, i). The key question is how long should be the message in the worst case so that Bob can compute INDEX(S, i) correctly with probability greater than, say, $2/3$. (Attaining probability $1/2$ is of course trivial.) To achieve this, the message of Alice must have $\Omega(n)$ bits in the worst case [10,13].

Theorem 3. *Let $c > 0$ be an arbitrary constant. Consider the problem of estimating $\alpha(\mathbb{I})$ for sets of same-length intervals $\mathbb{I}$ with endpoints in $[n]$. In the data streaming model, there is no algorithm that uses $o(n)$ bits of memory and computes an estimate $\hat{\alpha}$ such that*

$$\Pr\left[(2/3 + c)\,\alpha(\mathbb{I}) \le \hat{\alpha} \le \alpha(\mathbb{I})\right] \;\ge\; 2/3.$$

Proof. (*Sketch*) For simplicity, we use intervals with endpoints in $[3n]$ and mix closed and open intervals in the proof. Given an input (S, i) for INDEX, consider the following stream of intervals. Set $L = n + 2$. Let $\sigma_1(S)$ be a stream that, for each $j \in S$, contains the closed interval $[L + j, 2L + j]$. Let $\sigma_2(i)$ be the length-two stream with open intervals $(i, L + i)$ and $(2L + i, 3L + i)$. Finally, let $\sigma(S, i)$ be the concatenation of $\sigma_1(S)$ and $\sigma_2(i)$. See Figure 1 for an example. Let $\mathbb{I}$ be the intervals in $\sigma(S, i)$. It is straightforward to see that $\alpha(\mathbb{I})$ is 2 or 3. Moreover, $\alpha(\mathbb{I}) = 3$ if and only if INDEX$(S, i) = 1$.

Assume, for the sake of contradiction, that we have an algorithm in the data streaming model that uses $o(n)$ bits of space and computes a value $\hat{\alpha}$ that satisfies $\Pr\left[(2/3 + c)\,\alpha(\mathbb{I}) \le \hat{\alpha} \le \alpha(\mathbb{I})\right] \ge 2/3$. Then, Alice and Bob can solve INDEX(S, i) using $o(n)$ bits, as follows. Alice simulates the data stream algorithm on $\sigma_1(S)$ and sends to Bob a message encoding the state of the memory at the end of processing $\sigma_1(S)$. The message of Alice has $o(n)$ bits. Then, Bob continues the simulation on the last two items of $\sigma(S, i)$, that is, $\sigma_2(i)$. Bob has correctly computed the output of the algorithm on $\sigma(S, i)$, and therefore obtains $\hat{\alpha}$ so that $\Pr\left[(2/3 + c)\,\alpha(\mathbb{I}) \le \hat{\alpha} \le \alpha(\mathbb{I})\right] \ge 2/3$. If $\hat{\alpha} > 2$, then Bob returns the bit $\hat{\beta} = 1$. If $\hat{\alpha} \le 2$, then Bob returns $\hat{\beta} = 0$. This finishes the description of the protocol.

Analyzing separately the cases where INDEX$(S, i) = 1$ and INDEX$(S, i) = 0$, we obtain that $\Pr\left[\hat{\beta} = \text{INDEX}(S, i)\right] \ge 2/3$. Since Bob computes $\hat{\beta}$ after a message from Alice with $o(n)$ bits, this contradicts the lower bound of INDEX.   □

Fig. 1. Example showing $\sigma(S, i)$ for $n = 7$, $S = \{1, 3, 4, 6\}$, $L = 9$, and $i = 2$ in the proof of Theorem 3. The intervals are sorted from bottom to top in the order they appear in the data stream. The empty dots represent endpoints that are not included in the interval, while the full dots represent endpoints included in the interval.

For intervals of different sizes, we can use an alternative construction with the property that $\alpha(\mathbb{I})$ is either $k + 1$ or $2k + 1$. This means that we cannot get an approximation ratio arbitrarily close to 2.

Theorem 4. *Let $c > 0$ be an arbitrary constant. Consider the problem of estimating $\alpha(\mathbb{I})$ for sets of intervals $\mathbb{I}$ with endpoints in $[n]$. In the data streaming model, there is no algorithm that uses $o(n)$ bits of memory and computes an estimate $\hat{\alpha}$ such that*

$$\Pr\big[(1/2 + c)\,\alpha(\mathbb{I}) \le \hat{\alpha} \le \alpha(\mathbb{I})\big] \ge 2/3.$$

References

1. Broder, A.Z., Charikar, M., Mitzenmacher, M.: A derandomization using min-wise independent permutations. J. Discrete Algorithms **1**(1), 11–20 (2003)
2. Cabello, S., Pérez-Lantero, P.: Interval selection in the streaming model. arXiv preprint: 1501.02285 (2015). http://arxiv.org/abs/1501.02285
3. Chakrabarti, A., et al: CS49: Data stream algorithms lecture notes, fall (2011). http://www.cs.dartmouth.edu/ac/Teach/CS49-Fall11/
4. Datar, M., Muthukrishnan, S.M.: Estimating rarity and similarity over data stream windows. In: Möhring, R.H., Raman, R. (eds.) ESA 2002. LNCS, vol. 2461, pp. 323–334. Springer, Heidelberg (2002)
5. Emek, Y., Halldórsson, M.M., Rosén, A.: Space-constrained interval selection. In: Czumaj, A., Mehlhorn, K., Pitts, A., Wattenhofer, R. (eds.) ICALP 2012, Part I. LNCS, vol. 7391, pp. 302–313. Springer, Heidelberg (2012)
6. Feigenbaum, J., Kannan, S., McGregor, A., Suri, S., Zhang, J.: On graph problems in a semi-streaming model. Theor. Comput. Sci. **348**(2–3), 207–216 (2005)
7. Halldórsson, B.V., Halldórsson, M.M., Losievskaja, E., Szegedy, M.: Streaming algorithms for independent sets. In: Abramsky, S., Gavoille, C., Kirchner, C., Meyer auf der Heide, F., Spirakis, P.G. (eds.) ICALP 2010. LNCS, vol. 6198, pp. 641–652. Springer, Heidelberg (2010)
8. Hochbaum, D.S., Maass, W.: Approximation schemes for covering and packing problems in image processing and vlsi. J. ACM **32**(1), 130–136 (1985)
9. Indyk, P.: A small approximately min-wise independent family of hash functions. J. Algorithms **38**(1), 84–90 (2001)
10. Jayram, T.S., Kumar, R., Sivakumar, D.: The one-way communication complexity of hamming distance. Theory of Computing **4**(6), 129–135 (2008)

11. Kane, D.M., Nelson, J., Woodruff, D.P.: An optimal algorithm for the distinct elements problem. PODS 2010, pp. 41–52 (2010)
12. Kolen, A.W., Lenstra, J.K., Papadimitriou, C.H., Spieksma, F.C.: Interval scheduling: A survey. Naval Research Logistics (NRL) **54**(5), 530–543 (2007)
13. Kushilevitz, E., Nisan, N.: Communication Complexity. Cambridge University Press, New York (1997)
14. Muthukrishnan, S.: Data Streams: Algorithms and Applications. Foundations and trends in theoretical computer science. Now Publishers (2005)

On the Bounded-Hop Range
Assignment Problem

Paz Carmi$^{(\boxtimes)}$, Lilach Chaitman-Yerushalmi, and Ohad Trabelsi

Department of Computer Science,
Ben-Gurion University of the Negev, Beersheba, Israel
carmip@gmail.com

Abstract. We study the problem of assigning transmission ranges to
radio stations in the plane such that any pair of stations can communicate within a bounded number of hops h and the cost of the network is
minimized. The cost of transmitting in a range r is proportional to r^α,
where $\alpha \geq 1$.

We consider two settings of this problem: collinear station locations
and arbitrary locations. For the case of collinear stations, we introduce
the pioneer polynomial-time exact algorithm for any $\alpha \geq 1$ and constant
h, and thus conclude that the 1D version of the problem, where h is a
constant, is in P. For an arbitrary h, not necessarily a constant, and
$\alpha = 1$, we propose a 1.5-approximation algorithm. This improves the
previously best known approximation ratio of 2.

For the case of stations placed arbitrarily in the plane, we present a
$(6 + \varepsilon)$-approximation algorithm, for any $\varepsilon > 0$. This improves the previously best known approximation ratio of $4(9^{h-2})/(\sqrt[h]{2} - 1)$. Moreover,
we show a $(1.5 + \varepsilon)$-approximation algorithm for a case where deviation
of one hop ($h + 1$ hops in total) is acceptable.

1 Introduction

A *wireless ad-hoc network* prevails in scenarios where a fixed wired infrastructure
is not available, either because it is physically impossible or not economically practical. A wireless ad-hoc network is a decentralized network that consists of independent radio stations communicating over radio channels without relying on any
existing infrastructure. Each station is able to transmit a signal over a fixed range,
and any other station within this transmission range receives the message. Communication with stations outside the transmission range is achieved by *multi-hop*
transmission. The twenty-first century witnesses widespread deployment of wireless networks for both professional and private applications. This field continuously experiences technological progress and market growth. For a comprehensive
survey of this field see [15].

Let S be a set of points in the Euclidean plane representing radio stations.
A *range assignment* for S is a function $\rho : S \rightarrow \mathbb{R}^+$ that assigns each point

The research is partially supported by the Lynn and William Frankel Center for
Computer Science and by grant 680/11 from the Israel Science Foundation (ISF).

© Springer International Publishing Switzerland 2015
F. Dehne et al. (Eds.): WADS 2015, LNCS 9214, pp. 140–151, 2015.
DOI: 10.1007/978-3-319-21840-3_12

a transmission range (radius). The cost of a range assignment is defined as $cost(\rho) = \sum_{v \in S}(\rho(v))^{\alpha}$, for some real constant $\alpha \geq 1$. In the case of $\alpha \in (1, 6]$, the *cost* represents the power consumption of the network, where α varies depending on different environmental factors [15]. The linear setting of the problem ($\alpha = 1$) corresponds to minimizing the sum of ranges (radii).

A range assignment ρ induces a *directed communication graph* $G_\rho = (S, E_\rho)$, where $E_\rho = \{(u, v) : \rho(u) \geq |uv|\}$ and $|uv|$ denotes the Euclidean distance between u and v. There is a variety of *minimum cost range assignment* problems that aim to find a minimum cost range assignment ρ, provided that the induced communication graph admits a specified constraint, mainly regarding its connectivity. This class of problems has been considered extensively in many settings, for different values of α and with respect to various required constraints. We refer to a range assignment as *feasible* if its induced graph satisfies the constraint of the addressed problem.

In this paper we consider the Bounded-Hop Minimum Cost (Power) Range Assignment (hMINPOWER) problem, whose objective is to compute a minimum cost range assignment under the constraint that the induced graph contains a directed path between any two nodes with at most h edges for a given $1 \leq h \leq n - 1$, where $\alpha \geq 1$. Some of our work concentrate on the specific linear case where $\alpha = 1$. We refer to the hMINPOWER problem under this linear model as the hMINRANGE problem. Minimization of the radii sum ($\alpha = 1$) has been considered also in the context of the unbounded-hop version [2,4] and other range assignment problems, such as a set of circles connectivity [5] and circle coverage [1,13,14]. This linear model may be appropriate also for power consumption in future systems, as predicted in [14], where the transmitting stations do not transmit in all directions simultaneously, but rather focus the transmission energy in a narrow angle beam whose direction changes according to the needs of the network.

In [11], Kirousis et al. considered the 1D hMINPOWER problem for the case with $h = n - 1$, where n is the number of input points, i.e., the unbounded-hop version, and showed an $O(n^4)$-time exact algorithm. Later, Das et al. [9] improved the running time to $O(n^3)$. Finally, Carmi and Chaitman-Yerushalmi [4] proposed an $O(n^2)$-time exact algorithm. Unlike the unbounded case, the complexity of the 1D hMINPOWER problem for $h < n - 1$ has not been known. The best known approximation algorithm is due to Clementi et al. [7] that ensures a 2-approximation ratio, for any arbitrary $1 \leq h < n$.

In Section 2, we present the first polynomial-time exact algorithm for the hMINPOWER problem, for any constant $1 \leq h < n$. For any arbitrary $1 \leq h < n$, not necessarily a constant, we show an algorithm that outputs a feasible range assignment for the hMINRANGE problem of cost 1.5 times the optimal cost. This improves the previously best known approximation ratio for the problem, which is 2.

While the 1D version of the hMINPOWER problem, where h is a constant, can be solved optimally, no exact algorithm is known for the hMINPOWER problem in higher dimensions. The unbounded case, i.e., $h = n - 1$, has been proven to be NP-hard [6,11] for any $d \geq 2$ and $\alpha \geq 1$, however, a 2-approximation algorithm exists due to Kirousis et al. [11]. For $\alpha = 1$, Ambühl et al. [2] gave a 1.5-approximation, and lately Carmi et al. [4] gave an algorithm with improved approximation ratio of $1.5 - c$, for a suitable constant $c > 0$. In the case where the hop bound is a constant $1 \leq h < n$, Calinescu et al. [3] provided an $(O((\log n)/h), O((\log n)^\alpha))$ bi-criteria approximation algorithm for any $\alpha \geq 1$. Namely, their algorithm outputs an assignment whose cost is bounded by $O(\log n)$ times the optimal cost and the induced network has a hop-diameter bounded by $O(\log n)$ times h. Kantor and Peleg introduced in [10] the first constant-approximation (though exponential in h) for the problem (for general metrics) with approximation ratio of $(1/\sqrt[h]{2} - 1)^\alpha (1 + 3^\alpha)(3^{\alpha+1})^{h-2}$. For the Euclidean hMINRANGE problem, this ratio equals $4(9^{h-2})/(\sqrt[h]{2} - 1)$.

In Section 3, we present two approximation algorithms for the hMINRANGE problem in the plane. The first algorithm admits a $(3/2 + \varepsilon, 1 + 1/h)$ bi-criteria approximation. Namely, it outputs an assignment whose cost is bounded by $(3/2 + \varepsilon)$ times the optimal cost and the induced network has a hop-diameter bounded by $h + 1$, rather than h. The second algorithm has an approximation ratio of $(6+\varepsilon)$, while it guarantees a hop-diameter of exactly h. Both results hold for any small enough positive constant ε and use the PTAS for the bounded-hop MST problem by Laue et al. [12].

2   The hMinPower Problem in 1D

Consider a set $S = \{s_1, \ldots s_n\}$ of points on a line, representing stations. For simplicity, we assume that the line is horizontal and for every $i < j$, s_i lies to the left of s_j. Given a range assignment $\rho : S \to \mathbb{R}^+$, we say that a station s_i reaches s_j, and denote it by $i \to_\rho j$, if $\rho(s_i) \geq |s_i s_j|$. In addition, we say that a station s_i reaches s_j in at most h hops, and denote it by $i \xrightarrow{h}_\rho j$, if there is a path in G_ρ from s_i to s_j with at most h edges.

We present a polynomial-time exact algorithm for the hMINPOWER problem for every $\alpha \geq 1$ and a constant h. In addition, we show a 1.5-approximation algorithm for the hMINRANGE problem for any arbitrary h, not necessarily a constant. The later improves the previously best known 2-approximation algorithm [7] for the linear model of the hMINRANGE problem.

2.1   Exact Algorithm for the hMinPower Problem

Our algorithm uses a *dynamic programming* approach inspired by the polynomial-time approximation scheme in [12]. We begin with a simple observation.

Observation 1. *A range assignment ρ for S is feasible if and only if every $s_k \in S$ satisfies $k \xrightarrow{h}_\rho 1$ and $k \xrightarrow{h}_\rho n$.*

Our algorithm finds a minimum cost range assignment under the constraint that every $s_k \in S$ reaches both s_1 and s_n in at most h hops. By Observation 1, this assignment admits an optimal solution for the hMINPOWER problem.

Consider the communication graph induced by an optimal range assignment. Each s_i, for $i \in \{1, n\}$, implies a *shortest-paths tree* rooted at s_i, which we refer as T_i. Each tree T_i associates each station with a value between 0 to h that corresponds to its level in the tree T_i. Namely, a station of level j in T_i, referred as a j-level$_i$ station, is a station whose hop-distance to s_i in T_i is j. Clearly, s_i is a single station of level 0 in T_i (0-level$_i$), see Figure 1. Therefore, each station is associated with two levels with respect to the two trees T_1 and T_n. Our algorithm uses *dynamic programming* approach in order to determine for every station its level, j, and its distance from the closest $(j-1)$-level station, in both T_1 and T_n. This implies the range of the station, which is the maximal among the two distances.

Next, we describe a hierarchical decomposition of the input set that is later used to define our subproblems. Consider the input S and a bounding interval I_S containing all points of S in its interior (and no point on its boundary). We divide I_S into two intervals sharing a common boundary point located between $s_{\lceil n/2 \rceil}$ and $s_{\lceil n/2 \rceil + 1}$. Thus, each interval contains at most $\lceil n/2 \rceil$ points. We continue with dividing the two intervals recursively by the same method until the intervals contain a single point. We view the obtained intervals as nodes of a binary *split tree* with I_S as a root. The children of an interval node correspond to the two intervals obtained by the above partition. Note that this tree has $O(n)$ nodes.

We define a set of *subproblems* on each interval I in the split-tree. Each subproblem on I represents a different sequence of "guessed" distances from I's boundaries to the closest j-level station in T_i for each $i \in \{1, n\}$ and $0 \leq j \leq h-1$. Formally, each subproblem is specified by the following parameters (see Figure 1):

- an interval in the split-tree;
- for every $i \in \{1, n\}$ and $0 \leq j \leq h-1$, a function $in_{(i,j)}$ that assigns each boundary point of the interval the distance to the closest j-level$_i$ station inside the interval; and
- for every $i \in \{1, n\}$ and $0 \leq j \leq h-1$, a function $out_{(i,j)}$ that assigns the left (resp. right) boundary point of the interval the distance to the closest j-level$_i$ station, outside, on the left (resp. right) of the interval;

For an interval I in the split-tree, a pair $(i, j) \in \{1, n\} \times \{0, ..., h-1\}$, and a boundary point p of I, there are $O(n)$ possible values for each $in_{(i,j)}(p)$ and $out_{(i,j)}(p)$ that correspond to distances between I's boundaries and stations. Note that the distance may be ∞. Thus, the number of subproblems on a given interval is $O(n^{4h})$.

The *cost* of a subproblem on an interval I corresponds to $\sum_{s \in I} (\rho(s))^\alpha$, where ρ is an optimal assignment with respect to the "guessed" distances. The algorithm performs a bottom-up traversal of the split tree, and at each step it solves the set of subproblems on the current interval and stores their costs in a table.

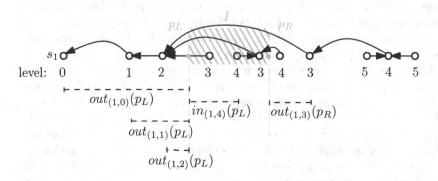

Fig. 1. The tree T_1 is depicted in black with the associated levels of all stations. The gray interval I has a left boundary point p_L and a right boundary point p_R. The values of the $in_{(i,j)}$ and $out_{i,j}$ functions, depicted as dashed segments, correspond to a subproblem on I of optimal cost (since it complies with the correct levels in T_1).

Finally, it returns the minimum value among all table entries corresponding to subproblems on I_S.

Computing the Subproblems.

Base Subproblems. The base subproblems correspond to leaves of the split-tree, i.e., intervals containing a single station s. Their cost is determined only by the range assigned to s. For every $i \in \{1, n\}$, s should be associated with exactly one level $0 \le j_i \le h$. Thus, legal subproblems must satisfy the following property:

Property 1. For every $i \in \{1, n\}$, there exists at most one $0 \le j_i \le h$ that satisfies: for each boundary point p of the interval,

1. if $j_i < h$, then $in_{(i,j_i)}(p) = |p\,s|$; and
2. for every $k \ne j_i$, $in_{(i,k)}(p) = \infty$.

Moreover, if s is s_1 (resp. s_n), then j_1 (resp. j_n) must be 0.

If Property 1 is satisfied, then for every $i \in \{1, n\}$, we consider s as a j_i-level$_i$ station in T_i and thus it must reach the closest $(j_i - 1)$-level$_i$ station. Therefore, the cost of the subproblem stored in the table is

$$\max_{i \in \{1,n\}} \{d_i\}, \text{ where } d_i = \left\{ \begin{array}{ll} 0, & s = s_i \\ \left(\min_{boundary\ p} \{|sp| + out_{(i,j_i-1)}(p)\} \right)^{\alpha}, & otherwise \end{array} \right\}.$$

Otherwise (Property 1 is not satisfied), the cost is ∞.

General Subproblems. The cost of a subproblem on an interval I with two children I_1 and I_2 is computed as follows. Assume w.l.o.g. that I_1 is to the left of I_2. Let p_L (resp. p_R) denote the left (resp. right) boundary point of I and let p_M be the division point, i.e., the common boundary point of I_1 and I_2. The algorithm considers all pairs of subproblems on I_1 and I_2 with associated functions $out^1_{(i,j)}, in^1_{(i,j)}$ and $out^2_{(i,j)}, in^2_{(i,j)}$, respectively, that satisfy the following property:

Property 2. For every $i \in \{1, n\}$ and $0 \leq j \leq h - 1$,

1. the subproblem on I complies with those on I_1 and I_2:

$$out_{(i,j)}(p_L) = out^1_{(i,j)}(p_L),$$
$$out_{(i,j)}(p_R) = out^2_{(i,j)}(p_R),$$
$$in_{(i,j)}(p_L) = \min\{in^1_{(i,j)}(p_L), |p_L p_M| + in^2_{(i,j)}(p_M)\},$$
$$in_{(i,j)}(p_R) = \min\{in^2_{(i,j)}(p_R), |p_R p_M| + in^1_{(i,j)}(p_M)\}; \text{ and}$$

2. the subproblems on I_1 and I_2 comply with each other:

$$out^2_{(i,j)}(p_M) = \min\{in^1_{(i,j)}(p_M), |p_M p_L| + out^1_{(i,j)}(p_L)\},$$
$$out^1_{(i,j)}(p_M) = \min\{in^2_{(i,j)}(p_M), |p_M p_R| + out^2_{(i,j)}(p_R)\};$$

Then, the algorithm observes the table entries of all such pairs, chooses the pair that sums to minimal cost and sets this sum to be the cost of the current subproblem. Note that every pair of subproblems on I_1 and I_2 complies with exactly one subproblem on I. Therefore, the time required for computing all subproblems for I is $O((n^{4h})^2)$. Since there are $O(n)$ intervals, the total running time is $O(n^{8h+1})$ and Theorem 2 follows.

Theorem 2. *Given a set S of points on a line, the hMINPOWER problem for every $\alpha \geq 1$ can be solved in $O(n^{8h+1})$ time.*

2.2   1.5-Approximation Algorithm for the hMinRange Problem for an Arbitrary h

In this section we present a 1.5-approximation algorithm for the hMINRANGE problem in 1D for an arbitrary h. This result improves the 2-approximation algorithm suggested in [8] for the linear model of the hMINRANGE problem. For consistency, we follow the definitions from [8] with only slight changes.

Definition 1. *For any $1 \leq i < j \leq n$,*

$$\overrightarrow{\text{ALL}}_h(i,j) = \min_{\rho}\{cost(\rho) \mid \forall k \in [i,j], \ k \xrightarrow{h}_\rho j\};$$

$$\overleftarrow{\text{ALL}}_h(i,j) = \min_{\rho}\{cost(\rho) \mid \forall k \in [i,j], \ k \xrightarrow{h}_\rho i\}.$$

Definition 2. *We say that a station s_i is a* Base *if $i \rightarrow_\rho 1$ and $i \rightarrow_\rho n$. A range assignment ρ is of type* B *if there is at least one* Base *and for every $1 \leq i \leq n$, there exists a* Base *s_b such that $i \xrightarrow{h-1}_\rho b$. The cost of a minimum assignment for S of type* B *is denoted by $\mathsf{BASES}_h(S)$. The cost of a minimum assignment for S of type* B *subject to the constraint that s_i is a* Base *is denoted by $\mathsf{BASES}_h(S, i)$. We also define $\mathsf{BASES}_h^r(S, i)$ (resp. $\mathsf{BASES}_h^l(S, i)$) as the cost of a minimum assignment for S of type* B *subject to the constraint that s_i is the rightmost (resp. leftmost)* Base.

According to [8], the values $\overrightarrow{\mathsf{ALL}}_h(i, j)$, $\overleftarrow{\mathsf{ALL}}_h(i, j)$, $\mathsf{BASES}_h(S)$, $\mathsf{BASES}_h^r(S, i)$ and $\mathsf{BASES}_h^l(S, i)$ over all $1 \leq i < j \leq n$ can be computed in $O(hn^3)$ time.

Corollary 1. *The values $\mathsf{BASES}_h(S, i)$ over all $1 \leq i \leq n$ can be computed in $O(hn^3)$ time.*

Proof. Consider the minimum range assignment of type B for S under the constraint that s_i is a Base. Note that for every station s_j with $1 \leq j < i$ there exists a Base s_b with $1 \leq b \leq i$, such that s_j reaches s_b in at most $h - 1$ hops. Symmetrically, for every station s_j with $i < j \leq n$ there exists a Base s_b with $i \leq b \leq n$, such that s_j reaches s_b in at most $h - 1$ hops. Let ρ be the assignment associated with the cost $\mathsf{BASES}_h^r(S, i)$ (resp. $\mathsf{BASES}_h^l(S, i)$), then for every $i < j \leq n$ (resp. $1 \leq j < i$), it holds that $j \xrightarrow{h-1}_\rho i$. Therefore,

$$\mathsf{BASES}_h(S, i) = \mathsf{BASES}_h^r(S, i) - \overleftarrow{\mathsf{ALL}}_{h-1}(i, n) +$$
$$\mathsf{BASES}_h^l(S, i) - \overrightarrow{\mathsf{ALL}}_{h-1}(1, i) - \max\{|s_1 s_i|, |s_i s_n|\},$$

where the subtraction of $\max\{|s_1 s_i|, |s_i s_n|\}$ is due to the double addition of s_i's range to the total cost. □

Let s_m be the closest station to the midpoint of the segment $\overline{s_1 s_n}$. Lemma 1 states an invariant that holds for any feasible assignment for S.

Lemma 1. *For every $1 \leq i \leq n$, s_i reaches s_m or a* Base *in at most $h - 1$ hops.*

Proof. Consider a station s_i and assume w.l.o.g. that s_i is to the right of s_m, i.e., $i > m$. If s_i reaches s_m in exactly h hops, then it also reaches s_1 in exactly h hops. This means that the last edge in the corresponding path from s_i to s_1 is necessarily (s_j, s_1) for $j > m$. Thus, s_j reaches s_1 in one hop which implies that s_j also reaches s_n in one hop, namely, s_j is a Base. Since s_i reaches s_j in $h - 1$ hops, the lemma follows. □

Consider an optimal assignment to the $h\mathrm{MINRANGE}$ problem, ρ^*, and its cost OPT. By Lemma 1, applying one modification on ρ^* of setting s_m to be a Base yields an assignment of type B having s_m as a Base. Therefore,

$$\mathsf{BASES}_h(S, m) \leq OPT + \max\{|s_1 s_m|, |s_m s_n|\}.$$

Assume w.l.o.g. that s_m is to the left of the midpoint, then

$$\max\{|s_1 s_m|, |s_m s_n|\} \leq (|s_1 s_n| + |s_m s_{m+1}|)/2.$$

By Lemma 3 (in Section 3), for every $e \in MST(S)$ it holds that $OPT \geq wt(MST(S)) + wt(e)$; especially, $OPT \geq |s_1 s_n| + |s_m s_{m+1}|$. Therefore, $\mathsf{BASES}_h(S, m) \leq 1.5 \cdot OPT$ and Theorem 3 follows.

Theorem 3. *Given a set S of points on a line, a 1.5-approximation for the hMINRANGE problem and an arbitrary h can be computed in $O(hn^3)$ time.*

3   The hMinRange Problem in the Plane

In this section we introduce two approximation algorithms for the hMINRANGE problem in the plane. First we present a $(3/2 + \varepsilon, 1 + \frac{1}{h})$ cost-hop bi-criteria approximation, i.e., the algorithm outputs an assignment whose cost is bounded by $(3/2 + \varepsilon)$ times the optimal cost and the induced graph has a hop-diameter bounded by $h + 1$. Later, we show an algorithm that does not exceed the h hop bound, at the cost of increasing the approximation ratio to $(6 + \varepsilon)$. Both algorithms use the PTAS by Laue et al., denoted by HMST, for the bounded-hop MST (hHOPMST) problem [12]. This problem receives as an input a set S, a root node $r \in S$ and a hop bound h and outputs a minimum weight tree rooted at r, in which every node is connected to r with a path containing at most h edges. The output of the HMST algorithm for an input (S, r, h) is denoted by HMST(S,r,h).

Throughout this section, let S denote the input set. We denote by ρ^* the optimal assignment for the hMINRANGE problem and by OPT_ρ its cost. In addition, we denote by T_r^* the optimal assignment for the hHOPMST problem with a root $r \in S$ and by OPT_{T_r} its weight. We begin with bounding the weight of any feasible solution for the hMINRANGE problem with respect to an optimal solution of the hHOPMST problem, rooted at $r \in S$.

The following lemmas introduce some lower bounds on OPT_ρ, used later to analyze the approximation ratio of our algorithms.

Lemma 2. *For any $r \in S$, $OPT_\rho > OPT_{T_r}$.*

Proof. Let T_r^ρ be the shortest paths tree (with respect to number of edges) to r, induced by ρ^*, having r as its root. That is, for each $v \in V$, we compute the directed shortest-hop path from v to r in G_{ρ^*}. This construction yields that (i) $wt(T_r^\rho) < cost(\rho^*)$; (ii) for all $v \in V$ there is a path from v to r in T_r^ρ consisting of at most h edges. Since T_r^* is the tree of minimum weight among all trees rooted at r with bounded hop h, i.e., $wt(T_r^*) \leq wt(T_r^\rho)$, the lemma follows.   □

Lemma 3. *Let e be an edge in $MST(S)$, then*

$$OPT_\rho \geq wt(MST(S)) + wt(e).$$

Proof. Consider the two connected components of $MST(S) \backslash e$. To connected them there must be a station $s \in S$ with a range of at least $wt(e)$. Moreover, since all points in $S \backslash \{s\}$ must reach s, by the same arguments as in the proof of Lemma 2, their ranges must sum up to at least $wt(MST(S))$.   □

3.1 Bi-criteria Approximation Algorithm for the hMinRange Problem

In this section we show a $(3/2 + \varepsilon, 1 + \frac{1}{h})$ cost-hop bi-criteria approximation algorithm for the hMINRANGE problem. We start by describing our algorithm, then we prove the feasibility of the received range assignment and bound its cost.

Algorithm Description. Given a set S of points in $\mathbb{R}^2$ and a constant $h > 0$, the algorithm finds a point $r \in S$ that minimizes the enclosing circle centered at r (containing all points in S). Let d_r be the radius of this minimum enclosing circle centered at r, and let T be HMST(S,r,h). Moreover, let ρ_T be the range assignment induced by T by directing all edges toward r. The algorithm sets ρ to be ρ_T with one modification that is $\rho(r) = d_r$ and returns the assignment ρ.

Algorithm Analysis. The weight of T is at most $(1 + \varepsilon)OPT_{Tr}$, which by Lemma 2 is smaller than $(1 + \varepsilon)OPT_\rho$. Note that $cost(\rho_T) = wt(T)$ and thus, $cost(\rho_T) < (1 + \varepsilon)OPT_\rho$. Next we bound the additional range d_r which the algorithm sets to r.

Consider the longest path in $MST(S)$ and its midpoint m. Obviously, the minimum enclosing circle centered at m has a radius of at most $wt(MST(S)/2$, however, m may lie on an edge of $MST(S)$. Let $c \in S$ be the closest station to m, then the radius d_c of the minimum enclosing circle centered at c is at most $(wt(MST(S)) + wt(e_m))/2$, where e_m is the largest edge in $MST(S)$. By the choice of r, $d_r \leq d_c$ and by Lemma 3, $OPT_\rho \geq wt(MST(S)) + wt(e_m)$; thus, $d_r \leq OPT_\rho/2$.

Altogether we receive

$$cost(\rho) \leq (1 + \varepsilon)OPT_\rho + \frac{OPT_\rho}{2} = (\frac{3}{2} + \varepsilon)OPT_\rho.$$

As for the feasibility, any point has a path with at most h edges to r in G_ρ, and r can reach any point in one hop. Thus, we have the following theorem.

Theorem 4. *Given a set S of points in the plane and an integer h, a polynomial $(3/2 + \varepsilon, 1 + 1/h)$ cost-hop approximation for the hMINRANGE can be computed in polynomial time, for any small $\varepsilon > 0$.*

3.2 $(6 + \varepsilon)$-Approximation for the hMinRange Problem

In this section we present an approximation algorithm that outputs an assignment ρ of cost at most $(6 + \varepsilon)OPT_\rho$, for any small $\varepsilon > 0$. We start by giving the algorithm description, illustrated in Figure 2.

Let l and r be the two farthest points in S, w.l.o.g. assume that both are on the x-axis and l is to the left of r. Let b be the perpendicular bisector to the segment $\overline{lr}$ and let H_l and H_r be the half planes to the left and to the right of b, respectively.

Algorithm 1.

Input: A set of points S in $\mathbb{R}^2$ and an integer $h > 0$
Output: A range assignment ρ
 1: let $T_l := HMST(S, l, h)$, $T_r := HMST(S, r, h)$;
 2: let ρ_l (resp. ρ_r) be the range assignment induced by T_l (resp. T_r), by directing
      the edges of T_l (resp. T_r) towards l (resp. r).
 3: **for** every $(q, l) \in T_l$ **do**
 4:     **if** q has a descendant $q_d \in H_r$ in T_l **then**
 5:         set $\rho_l(q) := |lr|$;
 6: **for** every $(q, r) \in T_r$ **do**
 7:     **if** q has a descendant $q_d \in H_l$ **then**
 8:         set $\rho_r(q) := |lr|$;
 9: **for** every $v \in S$ **do**
10:     $\rho(v) := max\{\rho_l(v), \rho_r(v)\}$;
11: **return** ρ;

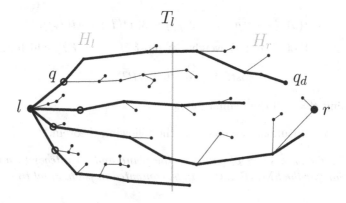

Fig. 2. An illustration of the tree T_l from Algorithm 1. The points q to which a range $|lr|$ is assigned are depicted as circles. The paths of T_l that are charged for those ranges are depicted in bold.

The correctness of the algorithm is proved in the following two lemmas. In Lemma 4 we show that the algorithm returns a feasible range assignment ρ and Lemma 5 proves that ρ admits the required approximation ratio.

Lemma 4. *The induced graph G_ρ has a path with at most h edges between any two points in S.*

Proof. Let $u, v \in S$ and assume w.l.o.g. that $u \in H_r$. Since the assignment ρ_l is induced by the bounded-hop tree T_l, there exists a path, $P_{u,l}$, from u to l in G_{ρ_l} with at most h hops. Let $(q, l) \in T_l$ be the (last) edge in $P_{u,l}$. Then, since $u \in H_r$ is a descendant of q in T_l, the algorithm set $\rho_l(q)$ to be $|lr|$. Thus, $\rho(q) = |lr|$, i.e., the maximum distance between any two points in S. Since the prefix of $P_{u,l}$ from u to q contains at most $h - 1$ edges, and q can reach any point in S and in particular v, we have that G_ρ has a path with at most h edges from u to v, and the lemma follows.

$\square$

Lemma 5. *For any choice of* $\varepsilon > 0$, $cost(\rho) \leq (6 + \varepsilon)OPT_\rho$.

Proof. At the end of step 2 of the algorithm, we have

$$cost(\rho_l) + cost(\rho_r) = wt(T_l) + wt(T_r).$$

In the rest of the algorithm, additional $|lr|$ ranges are assigned to points q adjacent to l or r that have a descendant q_d beyond the bisector. Consider such a point q adjacent to l (the arguments for points adjacent to r are symmetric). The path from q_d to l in T_l weights at least $|lr|/2$. We can charge this path twice to achieve the required range assigned to q. Since any two neighbors of l have no common descendants and the paths from their descendants to l in T_l are vertex disjoint, we do not charge any path more than once (see Figure 2). Therefore, we have

$$cost(\rho) \leq cost(\rho_l) + cost(\rho_r) \leq 3(wt(T_l) + wt(T_r)).$$

By Lemma 2, each of T_l and T_r weights at most $(1 + \varepsilon')OPT_\rho$ which implies

$$cost(\rho) \leq (6 + 6\varepsilon')OPT_\rho,$$

and by picking $\varepsilon' = \varepsilon/6$ the lemma follows.

$\square$

By the above 2 lemmas, we conclude the following theorem.

Theorem 5. *Given a set* S *of points in the plane and an integer* h, *a* $(6 + \varepsilon)$-*approximation for the* hMINRANGE *can be computed in polynomial time, for any* $\varepsilon > 0$.

References

1. Alt, H., Arkin, E.M., Brönnimann, H., Erickson, J., Fekete, S.P., Knauer, C., Lenchner, J., Mitchell, J.S.B., Whittlesey, K.: Minimum-cost coverage of point sets by disks. In: Proceedings of the Twenty-second Annual Symposium on Computational Geometry, SCG 2006, pp. 449–458 (2006)
2. Ambühl, C., Clementi, A.E.F., Penna, P., Rossi, G., Silvestri, R.: On the approximability of the range assignment problem on radio networks in presence of selfish agents. Theor. Comput. Sci. **343**(1–2), 27–41 (2005)
3. Călinescu, G., Kapoor, S., Sarwat, M.: Bounded-hops power assignment in ad hoc wireless networks. Discrete Applied Mathematics **154**(9), 1358–1371 (2006)
4. Carmi, P., Chaitman-Yerushalmi, L.: On the minimum cost range assignment problem. CoRR, abs/1502.04533 (2015)
5. Chambers, E.W., Fekete, S.P., Hoffmann, H.-F., Marinakis, D., Mitchell, J.S.B., Srinivasan, V., Stege, U., Whitesides, S.: Connecting a set of circles with minimum sum of radii. In: Dehne, F., Iacono, J., Sack, J.-R. (eds.) WADS 2011. LNCS, vol. 6844, pp. 183–194. Springer, Heidelberg (2011)

6. Clementi, A.E.F., Penna, P., Silvestri, R.: On the power assignment problem in radio networks. Mob. Netw. Appl. **9**(2), 125–140 (2004)
7. Clementi, A.E.F., Ferreira, A., Penna, P., Pérénnes, S., Silvestri, R.: The minimum range assignment problem on linear radio networks. In: Paterson, M. (ed.) ESA 2000. LNCS, vol. 1879, pp. 143–154. Springer, Heidelberg (2000)
8. Clementi, A.E.F., Di Ianni, M., Silvestri, R.: The minimum broadcast range assignment problem on linear multi-hop wireless networks. Theoretical Computer Science **299**(13), 751–761 (2003)
9. Das, G.K., Ghosh, S.C., Nandy, S.C.: Improved algorithm for minimum cost range assignment problem for linear radio networks. Int. J. Found. Comput. Sci. **18**(3), 619–635 (2007)
10. Kantor, E., Peleg, D.: Approximate hierarchical facility location and applications to the bounded depth steiner tree and range assignment problems. J. Discrete Algorithms **7**(3), 341–362 (2009)
11. Kirousis, L., Kranakis, E., Krizanc, D., Pelc, A.: Power consumption in packet radio networks. Theoretical Computer Science **243**(1–2), 289–305 (2000)
12. Laue, S., Matijevic, D.: Approximating k-hop minimum spanning trees in euclidean metrics. Inf. Process. Lett. **107**(3–4), 96–101 (2008)
13. Lev-Tov, N., Peleg, D.: Exact algorithms and approximation schemes for base station placement problems. In: Penttonen, M., Schmidt, E.M. (eds.) SWAT 2002. LNCS, vol. 2368, pp. 90–99. Springer, Heidelberg (2002)
14. Lev-Tov, N., Peleg, D.: Polynomial time approximation schemes for base station coverage with minimum total radii. Computer Networks **47**(4), 489–501 (2005)
15. Pahlavan, K.: Wireless information networks. John Wiley, Hoboken (2005)

Greedy Is an Almost Optimal Deque

Parinya Chalermsook[1], Mayank Goswami[1], László Kozma[2($\boxtimes$)],
Kurt Mehlhorn[1], and Thatchaphol Saranurak[3]

[1] Max-Planck Institute for Informatics, 66123 Saarbrücken, Germany
[2] Department of Computer Science,
Saarland University, 66123 Saarbrücken, Germany
kozma@cs.uni-saarland.de
[3] KTH Royal Institute of Technology, 11428 Stockholm, Sweden

Abstract. In this paper we extend the geometric binary search tree (BST) model of Demaine, Harmon, Iacono, Kane, and Pătraşcu (DHIKP) to accommodate for insertions and deletions. Within this extended model, we study the online GREEDY BST algorithm introduced by DHIKP. GREEDY BST is known to be equivalent to a maximally greedy (but inherently offline) algorithm introduced independently by Lucas in 1988 and Munro in 2000, conjectured to be dynamically optimal.

With the application of forbidden-submatrix theory, we prove a quasilinear upper bound on the performance of GREEDY BST on deque sequences. It has been conjectured (Tarjan, 1985) that splay trees (Sleator and Tarjan, 1983) can serve such sequences in linear time. Currently neither splay trees, nor other general-purpose BST algorithms are known to fulfill this requirement. As a special case, we show that GREEDY BST can serve output-restricted deque sequences in linear time. A similar result is known for splay trees (Tarjan, 1985; Elmasry, 2004).

As a further application of the insert-delete model, we give a simple proof that, given a set U of permutations of $[n]$, the access cost of any BST algorithm is $\Omega(\log |U| + n)$ on "most" of the permutations from U. In particular, this implies that the access cost for a random permutation of $[n]$ is $\Omega(n \log n)$ with high probability.

Besides the splay tree noted before, GREEDY BST has recently emerged as a plausible candidate for dynamic optimality. Compared to splay trees, much less effort has gone into analyzing GREEDY BST. Our work is intended as a step towards a full understanding of GREEDY BST, and we remark that forbidden-submatrix arguments seem particularly well suited for carrying out this program.

1 Introduction

Binary search trees (BST) are among the most popular and most thoroughly studied data structures for the dictionary problem. There remain however, several outstanding open questions related to the BST model. In particular, what

---

T. Saranurak—Work mostly done while at Saarland University.

F. Dehne et al. (Eds.): WADS 2015, LNCS 9214, pp. 152–165, 2015.
DOI: 10.1007/978-3-319-21840-3_13

is the best way to adapt a BST in an online fashion, in reaction to a sequence of operations (e.g. access, insert, and delete), and what are the theoretical limits of such an adaptation? Does there exist a "one-size-fits-all" BST algorithm, asymptotically as efficient as any other dynamic BST algorithm, regardless of the input sequence?

Splay trees have been proposed by Sleator and Tarjan [13] as an efficient BST algorithm, and were shown to be competitive with any *static* BST (besides a number of other attractive properties, such as the *balance, working set*, and *static finger* properties). Furthermore, Sleator and Tarjan conjectured splay trees to be competitive with any *dynamic* BST algorithm; this is the famous *dynamic optimality conjecture* [13]. An easier, but similarly unresolved, question asks whether such a dynamically optimal algorithm exists at all. We refer to [7] for a survey of work related to the conjecture.

A different BST algorithm (later called GREEDYFUTURE) has been proposed independently by Lucas [8] and by Munro [9]. GREEDYFUTURE is an offline algorithm: it anticipates future accesses, preparing for them according to a greedy strategy. In a breakthrough result, Demaine, Harmon, Iacono, Kane, and Pătraşcu (DHIKP) transformed GREEDYFUTURE into an online algorithm (called here GREEDY BST), and presented a geometric view of BST that facilitates the analysis of access costs (while abstracting away many details of the BST model).

At present, our understanding of both splay trees and GREEDY BST is incomplete. For splay trees, besides the above-mentioned four properties (essentially subsumed[1] by a single statement called the access lemma), a few other corollaries of dynamic optimality have been shown, including the *sequential access* [15] and the *dynamic finger* [1,2] theorems. The only known proof of the latter result uses very sophisticated arguments, which makes one pessimistic about the possibility of proving even stronger statements.

A further property conjectured for splay trees is a linear cost on deque sequences (stated as the "deque conjecture" by Tarjan [15] in 1985). Informally, a deque sequence consists of insert and delete operations at *minimum* or *maximum* elements of the current dictionary. Upper bounds for the cost of splay on a sequence of n deque operations are $O(n\alpha(n))$ by Sundar [14] and $O(n\alpha^*(n))$ by Pettie [10]. Here α is the *extremely* slowly growing inverse Ackermann function, and α^* is its iterated version. A linear bound for splay trees on *output-restricted* deque sequences (i.e. where deletes occur only at minima) has been shown by Tarjan [15], and later improved by Elmasry [5].

In general, our understanding of GREEDY BST is even more limited. Fox [6] has shown that GREEDY BST satisfies the access lemma and the sequential access theorem, but no other nontrivial bounds appear to be known. One might optimistically ascribe this to a (relative) lack of trying, rather than to insurmountable technical obstacles. This motivates our attempt at the deque conjecture for GREEDY BST.

---

[1] Apart from a technicality for working set, that poses no problem in the case of splay trees and GREEDY BST.

As mentioned earlier, a deque sequence consists of insert and delete operations. In the tree-view, e.g. for splay trees, such operations have a straightforward implementation. Unfortunately, the geometric view in which GREEDY BST can be most naturally expressed only concerns with accesses. Thus, prior to our work there was no way to formulate the deque conjecture in a managable way for GREEDY BST.

Our Contributions. We augment the geometric model of DHIKP to allow insert and delete operations (exemplified by the extension of the GREEDY BST algorithm), and we show the offline and online equivalence of a sequence of operations in geometric view with the corresponding sequence in tree-view. This extended model allows us to formulate the deque conjecture for GREEDY BST. We transcribe the geometric view of GREEDY BST in matrix form, and we apply the forbidden-submatrix technique to derive the quasilinear bound $O(m2^{\alpha(m,m+n)} + n)$ on the cost of GREEDY BST, while serving a deque sequence of length m on keys from $[n]$.

We also prove an $O(m+n)$ upper bound for the special case of output-restricted deque sequences. We find this proof considerably simpler than the corresponding proofs for splay trees, and we observe that a slight modification of the argument gives a new (and perhaps simpler) proof of the sequential access theorem for GREEDY BST.

As a further application of the insert-delete model we show through a reduction to sorting that for *any* BST algorithm, most representatives from a set U of permutations on $[n]$ have an access cost of $\Omega(\log |U|+n)$. In particular, this implies that a random permutation of $[n]$ has access cost $\Omega(n \log n)$ with high probability. A similar result has been shown by Wilber [16] for random access sequences (that might not be permutations). Our proof is self-contained, not relying on Wilber's BST lower bound. Permutation access sequences are important, since it is known that the existence of a BST algorithm that is constant-competitive on permutations implies the existence of a dynamically optimal algorithm (on arbitrary access sequences).

Related Work. A linear cost for deque sequences is achieved by the multi-splay algorithm [4, Thm 3] in the special case when the initial tree is empty; by contrast, the results in this paper make no assumption on the initial tree.

Most relevant to our work is the deque bound of Pettie for splay trees [10]. That result relies on bounds for Davenport-Schinzel sequences, which can be reformulated in the forbidden-submatrix framework. Indeed, the use of forbidden-submatrix theory for proving data structure bounds was pioneered by Pettie, who reproved the sequential access theorem for splay trees [11] (among other data structure results). Our application of forbidden-submatrix theory is somewhat simpler and perhaps more intuitive: the geometric view of GREEDY BST seems particularly suitable for these types of arguments, as the structure of BST accesses is readily available in a matrix form, without the need for an extra "transcribing" step.

2   Geometric Formulation of BST with Insertion/Deletion

In this section we extend the model of DHIKP [3] to allow for insertions and deletions. After defining our geometric model, we prove the equivalence of the arboreal (i.e. tree-view) and the geometric views of BSTs.

2.1   Rotations and Updates

Definition 1 (Valid Reconfiguration). *Given a BST* T_1, *a (connected) subtree* τ *of* T_1 *containing the root, and a tree* τ' *on the same nodes as* τ, *except that one node may be missing or newly added, we say that* T_1 *can be reconfigured by an operation* $\tau \to \tau'$ *to another BST* T_2 *if* T_2 *is identical to* T_1 *except for* τ *being replaced by* τ', *meaning that the child pointers of elements not in* τ *do not change. The cost of the reconfiguration is* $\max\{|\tau|, |\tau'|\}$.

This definition differs from [3, Def.2] in that τ' need not be defined on the same nodes as τ. Note that, according to the definition, if an operation $\tau \to \tau'$ changes a child pointer of an element x, then $x \in \tau$. See Figure 1 for examples.

Definition 2 (Execution of Update Sequence). *Given an update sequence*

$$S = \langle (s_1, \mathsf{op}_1), (s_2, \mathsf{op}_2), \ldots, (s_m, \mathsf{op}_m) \rangle, \text{ where } \mathsf{op}_i \in \{\mathsf{access}, \mathsf{insert}, \mathsf{delete}\},$$

we say that a BST algorithm executes S *by an execution* $E = \langle T_0, \tau_1 \to \tau'_1, \ldots, \tau_m \to \tau'_m \rangle$ *if all reconfigurations* $\tau_t \to \tau'_t$ *transforming* T_{t-1} *to* T_t *are valid, and for all* t

- *if* $\mathsf{op}_t = \mathsf{access}$, *then* $s_t \in \tau_t$ *and* $\tau'_t = \tau_t$ *as a set,*
- *if* $\mathsf{op}_t = \mathsf{insert}$, *then* $\tau'_t = \{s_t\} \dot\cup \tau_t$ *as a set,*
- *if* $\mathsf{op}_t = \mathsf{delete}$, *then* $\tau_t = \{s_t\} \dot\cup \tau'_t$ *as a set.*

We also say that E *executes* S. *The* cost of execution of E *is the sum over all reconfiguration costs. If an element* $x \in \tau_t \cup \tau'_t$, *we say that* x *is* touched *at time* t.

We assume that we work over the set $[n]$. Each element can be inserted or deleted many times, but insertions and deletions on the same element must be alternating. We also assume that every element is accessed or updated at least once.

2.2   Valid Sets

Definition 3 (Geometric View of Update Sequence). *The geometric view of an update sequence* S *is a point set* $P(S) = A(S) \dot\cup I(S) \dot\cup D(S)$ *in the integer grid* $[n] \times [m]$ *consisting of access points* $A(S) = \{(s_t, t) \mid \mathsf{op}_t = \mathsf{access}\}$, *insertion points* $I(S) = \{(s_t, t) \mid \mathsf{op}_t = \mathsf{insert}\}$, *and deletion points* $D(S) = \{(s_t, t) \mid \mathsf{op}_t = \mathsf{delete}\}$. *Update points are* $U(S) = I(S) \dot\cup D(S)$.

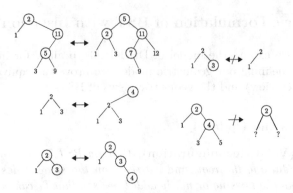

Fig. 1. (left) Examples of valid insert/delete operations. Circled elements indicate τ and τ'; (right) Examples of invalid operations: τ does not contain root (above) and τ' cannot link all pendant trees (below)

We usually omit the parameter S and simply write A, I, D, U when the choice of S is clear from context. We denote the x-coordinate and t-coordinate of a point p by (p_x, p_t). By element x, we mean the column x. By time t, we mean the row t.

Definition 4 (Valid Point). *Given a point set $P(S)$ in the integer grid $[n]\times[m]$, let p be a point (p may not be in $P(S)$), and let $p', p'' \in U(S)$ denote the update points nearest to p, below (resp. above) p, i.e. $p'_x = p''_x = p_x$, and $p'_t < p_t < p''_t$. One or both of p' and p'' might not exist. We say that p is* valid *in $P(S)$, iff:*

- *$p \notin U(S)$, $p' \in I(S)$ (or does not exist), and $p'' \in D(S)$ (or does not exist), or*
- *$p \in I(S)$, $p' \in D(S)$ (or does not exist), and $p'' \in D(S)$ (or does not exist), or*
- *$p \in D(S)$, $p' \in I(S)$ (or does not exist), and $p'' \in I(S)$ (or does not exist).*

Let T_t denote the resulting tree at time t during an execution of the BST algorithm E on the update sequence S. Observe that Definition 4 allows elements to be accessed or deleted without having been inserted before. Such elements are (implicitly) in the initial tree T_0.

Fact 5. *A point x can be touched at time t iff (x, t) is valid.*

Suppose that (x, t) is valid. If (x, t) is a deletion point, then x is in T_{t-1} but not T_t, and it is touched. If (x, t) is an insertion point, then x is in T_t but not T_{t-1}, and it is touched. If (x, t) is not an update point, then x is in both trees, and might or might not be touched. See Figure 2 for an illustration.

Definition 6 (Predecessor/Successor of a Point). *Given $P(S)$, the predecessor $pred(p)$ of a point p is the largest element x' smaller than p_x such that (x', p_t) is valid. We also write $pred(p) = (x', p_t)$ as a point. The successor $succ(p)$ of p is symmetrically defined.*

Definition 7 (Valid Set). *A point set $P \supseteq P(S)$ is valid iff every point $p \in P$ is valid.*

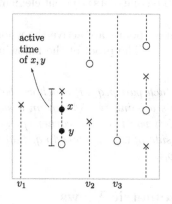

active time of x, y

v_1 v_2 v_3

Fig. 2. A point set with insert ($\bigcirc$) and delete ($\times$) points. Dashed lines indicate valid points. Observe that $succ(x) = v_3$, $succ(y) = v_2$, and $pred(x) = pred(y) = v_1$.

For any node x in a tree T, let $pred_T(x)$ and $succ_T(x)$ denote the predecessor, respectively successor of x in T. The following lemma shows that points in a valid set, and their predecessor and successor, are associated with nodes in the tree at the corresponding time.

Lemma 8. *Let $P \supseteq P(S)$ be a valid point set, and E executes S. For any $p \in U(S)$, we have $pred(p) = pred_{T_{p_t}}(p_x)$ and $succ(p) = succ_{T_{p_t}}(p_x)$.*

Proof. Let $x' = pred(p)$ and hence (x', p_t) is valid by definition. By Fact 5, x' can be touched at time p_t. Since x' is not an updated element, we have $x' \in T_{p_t}$. Moreover, x' is the closest element on the left of p_x at this time. So $x' = pred_{T_{p_t}}(p_x)$. The proof for successor is symmetric. $\square$

Definition 9 (Active Time of Points). *Let p be a point in a valid point set $P \supseteq P(S)$. The active time $act(p)$ of p is the maximal consecutive interval of time $[t_{ins}(p), t_{del}(p)]$ containing p_t such that, for all $t \in act(p)$, (p_x, t) is valid. We call $t_{ins}(p)$ insertion time of p, and $t_{del}(p)$ deletion time of p.*

2.3   Arboreally Satisfied Set

Definition 10 (Geometric View of BST Execution). *The geometric view of a BST execution $E = \langle T_0, \tau_1 \to \tau'_1, \ldots, \tau_m \to \tau'_m \rangle$ of some update sequence S is the point set $P(E) = \{(x, t) \mid x \in \tau_t \cup \tau'_t\}$ in the integer grid, indicating which element is touched at which time. Note that $P(E) \supseteq P(S)$.*

Definition 11 (Arboreally Satisfied Set). *A valid point set $P \supseteq P(S)$ is* (arboreally) satisfied *iff the following holds:*

- *For each pair $p, q \in P$ that are both active from time p_t to q_t (called an active pair), either both p and q lie in the same vertical/horizontal line, or there is a point $r \in \square_{pq} \cap P \setminus \{p, q\}$. If r is on the bottommost row of $\square_{pq}$, then r cannot be a deletion point. If r is on the topmost row of $\square_{pq}$, then r cannot be an insertion point.*
- *For each update point $p \in U$, if both $pred(p)$ and $succ(p)$ exist, then either $pred(p)$ or $succ(p)$ is also in P.*

The first condition is almost the same as the one in [3, Def.2.3] but focused only on *active pairs* (they are active from p_t to q_t), and with additional technical condition due to update points. The second condition says that if the updated element is not the current minimum/maximum, then one of its adjacent elements must be touched.

Note that if there are no update points, then all points are active the whole time and our definition is equivalent to [3, Def.2.3]. The proof of the following fact is omitted in this version of the paper.

Fact 12. *Suppose that P is satisfied. Then, for each pair $p, q \in P$ which are both active from time p_t to q_t and $p_t < q_t$, there exists a point in $P \setminus \{p, q\}$ on a side of $\square_{pq}$ incident to p, that is either a non-deletion point, or the corner (p_x, q_t). Similarly, there exists a point in $P \setminus \{p, q\}$ on a side of $\square_{pq}$ incident to q, that is either a non-insertion point, or the corner (q_x, p_t).*

3   Equivalence of Arboreal and Geometric Views

In this section we prove the following theorem:

Theorem 13. *A point set P is satisfied iff $P = P(E)$ for some BST execution E.*

The first direction of the proof involves considering a BST algorithm and showing that it generates a satisfied point set (tree to geometry). The second direction is showing how to convert a satisfied point set to a BST algorithm (geometry to tree).

3.1   Tree to Geometry

Lemma 14. *Let x and z be elements with consecutive values in a BST T, with $x < z$. Then one of x and z is an ancestor of the other.*

Proof. Suppose not. Then the lowest common ancestor of x and z is another element y. We know $x < y < z$ which is a contradiction. □

Lemma 15. *Suppose that y is not the minimum or maximum element in a BST T. To insert or delete y in T, either $pred_T(y)$ or $succ_T(y)$ must be touched.*

Lemma 16. *For any execution E, a point set $P(E)$ is satisfied.*

Proof. There are two conditions that need to be checked.

For the first condition, let p, q be a pair of points in $P(E)$ active from time p_t to q_t. Suppose that p, q violate the condition. Hence, they are not vertically or horizontally aligned. We assume that $p_t < q_t$ and $p_x < q_x$. Since p_x and q_x are active at time p_t, by Fact 5 and the statement below the fact, they exist in the tree T_{p_t}. Hence, a lowest common ancestor a of p_x and q_x in T_{p_t} is well-defined. There are two cases.

If $a = p_x$, then p_x is an ancestor of q_x. Since $\square_{pq}$ is not satisfied, q_x is not touched from time p_t to $q_t - 1$ and p_x remains an ancestor of q_x right before time q_t. Thus, to touch q_x at time q_t, p_x must be touched, and so $(p_x, q_t) \in \square_{pq}$. Only insertion point can be in the topmost row of unsatisfied $\square_{pq}$. So (p_x, q_t)

an insertion point. But this implies that p and q are not active pair, which is a contradiction.

If $a \neq p_x$, then a must be touched at time p_t. As a has value between p_x and q_x, we have $(a, p_t) \in \square_{pq}$. Since $\square_{pq}$ is not satisfied, (a, p_t) is a deletion point and, moreover, p_x must be its predecessor. Hence p_x becomes an ancestor of q_x right after time p_t and we can use the previous argument again.

For the second condition, suppose that $p \in U$ is an update point. That is, we update p_x in the BST T_{p_t}. If both $pred(p)$ and $succ(p)$ exist, then p_x is not a minimum or maximum in T_{p_t}. By Lemma 15, either $pred_{T_{p_t}}(p_x)$ or $succ_{T_{p_t}}(p_x)$ is touched at time p_t. By Lemma 8, $pred_{T_{p_t}}(p_x) = pred(p)$ and $succ_{T_{p_t}}(p_x) = succ(p)$, and we are done. $\qquad\square$

3.2   Geometry to Tree

Now we show how to convert a valid point set to an offline algorithm first. We need the following lemma, which is essentially a converse of Lemma 15, saying that if we touch either $pred_T(y)$ or $succ_T(y)$, then we can insert or delete y. The proofs of the following two statements are deferred to the full version of the paper.

Lemma 17. *Suppose either $pred_T(y)$ or $succ_T(y)$ is in a subtree τ containing the root of T, or y is the minimum or maximum element in T. Then (i) any reconfiguration $\tau \to \tau'$, where $\tau' = \tau \dot\cup \{y\}$ as a set, is valid, and (ii) any reconfiguration $\tau \to \tau'$, where $\tau = \tau' \dot\cup \{y\}$ as a set, is valid.*

Lemma 18 (Offline Equivalence). *For any satisfied set X, there is a point set $P(E) = X$ for some execution E. We call E a tree view of X.*

By Lemma 16 and 18, this concludes the proof of Theorem 13.

Observe that if $X = P(E)$, the quantity $|X|$ is exactly the execution cost of E.

3.3   Geometry to Tree: Online

The discussion in § 3.2 assumes that a satisfied set X is available all at once, and we show that there exists an execution E (i.e. an offline BST algorithm) whose point set $P(E)$ is exactly X.

We call an *online geometric algorithm* an algorithm that, given a geometric update sequence $P(S) \subseteq [n] \times [m]$, outputs a satisfied superset $P \supseteq P(S)$, with the condition that both the input and output are revealed *row-by-row* (i.e. the decision on which points to touch can depend only on the current and preceding rows of the input). We remark that GREEDY BST (as extended in § 4) is such an algorithm.

Analogously, by an online BST algorithm we mean a procedure that, given an initial set $S_0 \subseteq [n]$, and an update sequence S, outputs an execution E, with the condition that both the input and output are revealed *item-by-item* (i.e. the decision on which reconfiguration to perform can depend only on the current and preceding update operations).

Theorem 19 (Online Equivalence). *For any online geometric algorithm $\mathcal{A}$, there exists an online BST algorithm $\mathcal{A}'$ such that, on any update sequence, the cost of $\mathcal{A}'$ is bounded by a constant times the cost of $\mathcal{A}$.*

The proof of Theorem 19 is an adaptation of the proof of Lemma 2.3 in [3] to the new geometric setting, and is analogous to the proof of Lemma 18. We omit the proof in this extended abstract.

4   Defining GREEDY BST with Insertion/Deletion

GREEDY BST is an online algorithm for constructing a satisfied set given an update sequence S. At each time t, GREEDY BST minimally satisfies the point set up to time t. Having defined satisfied sets when there are update points, we naturally obtain the extension of GREEDY BST that can handle insertions and deletions.

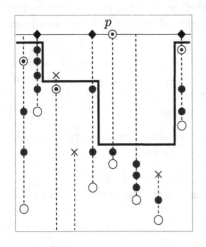

We develop some notation for describing the algorithm. A rectangle $\square_{pq}$ is *unsatisfied* if there is no other point in the proper (closed) rectangle formed by points p and q. We say that p and q are an *active pair* if they are active from time p_t to q_t. The *stair* of point p is denoted by $stair(p) = \{p\} \cup \{q \mid \square_{pq}$ is unsatisfied rectangle formed by an active pair p and q where q is below $p\}$. The *stair* of element x at time t is the stair of the point (x, t). *Satisfying/touching $stair(x, t)$* means visiting/touching, at time t, the elements of points in the stair: $\{(q_x, t) \mid q \in stair(x, t)\}$. These elements are then added to the row at time t.

Fig. 3. A GREEDY BST execution with insert ($\bigcirc$), delete ($\times$), access ($\odot$), touched ($\bullet$) points, and touched points at time p_t ($\blacklozenge$). Thick line shows stair of p. Observe that a non-(min/max) insert or delete must access a neighbor as well.

Fact 20. *Touching the stair $stair(p)$ is to minimally satisfy the point p.*

Therefore, when GREEDY BST gets an access point p, it touches only $stair(p)$. For an update point p, if p is not the minimum or maximum, then GREEDY BST chooses the smaller set between $stair(p) \cup stair(pred(p))$ and $stair(p) \cup stair(succ(p))$. This is because of the second condition of satisfied set. If p is the minimum or maximum, then GREEDY BST just touches $stair(p)$. The execution of GREEDY BST is illustrated in Figure 3.

The following observation is useful for deque sequences. For insertion point p, observe that $stair(p) = \{p\}$ because the active time of p begins at time p_t itself (for any point q below p, p and q are not an active pair by definition).

Fact 21. *To insert p such that p is the minimum or maximum,* GREEDY BST *touches only p.*

5   Performance of GREEDY BST on Deque Sequences

Definition 22 (Deque Sequence). *An update sequence is a* deque sequence *if it has only insertions and deletions at the current minimum or maximum element, and no access operations.*

Definition 23 (Output-restricted). *A deque sequence is* output-restricted *if it has deletions only at minimum elements.*

Theorem 24. *The cost of executing a deque sequence on* $[n]$ *of length m by* GREEDY BST *is at most* $O(m2^{\alpha(m,n+m)} + n)$, *where* α *is the inverse Ackermann function.*

Theorem 25. *The cost of executing an output-restricted deque sequence on* $[n]$ *of length m by* GREEDY BST *is at most* $24m + 12n$.

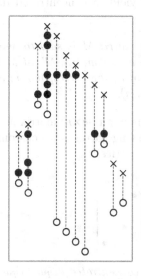

Fig. 4. Sample execution of GREEDY BST on a concentrated deque sequence with insert (○), delete (×), and touched (●) points. Dashed lines show the active times of elements.

Remark. The bound in Theorem 25 refers to the cost of the online *geometric* GREEDY BST. In the online tree-view equivalent the constants can be larger, hinging on the details of Theorem 19, but the bound remains of the form $O(m + n)$.

The rest of this section is devoted to the proofs of Theorems 24 and 25.

5.1   Concentrated Deque Sequences

We first reduce the analysis of GREEDY BST on any deque sequence to that on a special type of deque sequence that we call a concentrated deque sequence. Recall that in a deque sequence we can delete only the current minimum or maximum. We define two sets of elements as follows: let L_t be the set of elements which are deleted (from the left) before time t when they were the minimum at their deletion time, and R_t be the set of elements which are deleted (from the right) before time t when they were *not the minimum* at their deletion time. Observe that $L_t \cap R_t = \emptyset$.

Definition 26 (Concentrated Deque Sequence). *A deque sequence is concentrated if, for any time t, if the inserted element x is the minimum, then $y < x$ for all $y \in L_t$, and if x is the maximum, then $x < y$ for all $y \in R_t$.*

Note that the definition implies that each element in a concentrated deque sequence can be inserted and deleted at most once. The proof of the following lemma is deferred to the full paper.

Lemma 27. *For any deque sequence S, there is a concentrated deque sequence S' such that the execution of any BST algorithm on S' and S have the same cost.*

5.2   GREEDY BST on a Concentrated Deque Sequence

Now we analyze the performance of GREEDY BST on concentrated deque sequences (see Figure 4 for an example). Because of Lemma 27, we can view the points touched by GREEDY BST as an $(m \times (n + m))$ binary matrix (i.e. with entries 0 and 1), with all touched points represented as ones, and all other grid elements as zeroes. Notice that the number of columns is $n + m$ instead of n because of the reduction in Lemma 27 which allows each element to be inserted and deleted at most once. We further observe that if a deque sequence is output-restricted, then the transformation of Lemma 27 yields a concentrated deque sequences that is similarly output-restricted.

Definition 28 (Forbidden Pattern). *A binary matrix M is said to avoid a binary matrix P (called a pattern) if there exists no submatrix M' of M with same dimensions as P, such that for all 1-entries of P, the corresponding entry in M' is 1 (the 0-entries of P are "don't care" values).*

We denote by $\mathrm{Ex}(P, m, n)$ the largest number of 1s in an $(m \times n)$ matrix M that avoids pattern P. In this work, we refer to the following patterns (as customary, we write dots for 1-entries and empty spaces for 0-entries).

$$P_5 = \begin{pmatrix} \bullet & & \bullet & & \bullet \\ & \bullet & & \bullet & \end{pmatrix} \quad and \quad P_4 = \begin{pmatrix} \bullet & & \\ & & \bullet \\ \bullet & & \\ & & \bullet \end{pmatrix}$$

Lemma 29. *The execution of GREEDY BST on concentrated deque sequences avoids the pattern P_5.*

Proof. Suppose that P_5 appears in the GREEDY BST execution, and name the touched points matched to the 1-entries in P_5 from left to right as a, b, c, d, and e.

Let $t > b_t$ be smallest such that (c_x, t) is touched. Then $t \le c_t$ and either b or d must have been deleted within the time interval $[b_t, t]$. Otherwise, any update point in the interval $[b_t, t]$ is outside the interval $[b_x, d_x]$ and c_x is "hidden" by b and d (it cannot be on the stair of any update point).

Assume w.l.o.g. that b is deleted. If b is deleted by a minimum-delete, then a cannot be touched. If b is deleted by a maximum-delete, then e cannot be touched. This is because the sequence is concentrated. $\square$

Lemma 30. *The execution of* GREEDY BST *on concentrated output-restricted deque sequences avoids the pattern* P_4.

Proof: Suppose that P_4 appears in the GREEDY BST execution, and name the touched points matched to the 1-entries in P_4 from left to right as a, b, c, and d. We claim that in order to touch c, there has to be a deletion point in the interval $[b_x, d_x]$ in the time interval $[d_t, c_t]$. Otherwise, any deletion point in the time interval $[d_t, c_t]$ is left of b_x (as deletes happen only at the minimum). Furthermore, all insertion points in the time interval $[b_t, c_t]$ must be outside of $[b_x, d_x]$ (since both b and d are active at time b_t). We remind that insertion touches nothing else besides the insertion point itself. This means that c cannot be touched: it is "hidden" to deletion points on the left of b_x by b.

Denote the deletion point in the rectangle $[b_x, d_x] \times [d_t, c_t]$ as d'. Observe that a is to the left of and above d', and since we only delete minima, a is not active at time d'_t. In order to be touched, a must become active after d'_t via an insertion, contradicting that the sequence is concentrated. $\square$

Fact 31. ([**12, Thm3.4**]). $\mathrm{Ex}(P_5, u, v) = O(u2^{\alpha(u,v)} + v)$.

Fact 32. ([**12, Thm1.5(5)**]). $\mathrm{Ex}(P_4, u, v) < 12(u + v)$.

Proof of Theorem 24: By Lemma 27, it is enough to analyze the cost of GREEDY BST on concentrated deque sequences. This cost is bounded by $O(m2^{\alpha(m,m+n)} + n)$ using Lemma 29 and Fact 31.

Proof of Theorem 25: By Lemma 27, it is enough to analyze the cost of GREEDY BST on concentrated deque sequences. This cost is bounded by $24m + 12n$ using Lemma 30 and Fact 32.

Remark. The proof of Theorem 25 can be minimally adjusted to prove the sequential access theorem for GREEDY BST. Sequential access can be simulated as a sequence of minimum-deletions. In this way we undercount the cost by exactly one touched point above each access, which adds a linear term to the bound.

6   A Lower Bound on Accessing a Set of Permutations

Let U be a set of permutations on $[n]$. In this section we prove the following theorem:

Theorem 33. *Fix a BST algorithm $\mathcal{A}$ and a constant $\epsilon < 1$. There exists $U' \subseteq U$ of size $|U'| \geq (1 - \frac{1}{|U|^\epsilon})|U|$ such that $\mathcal{A}$ requires $\Omega(\log|U| + n)$ access cost on any permutation in U'.*

Proof. The proof utilizes the geometric view of insertions, and uses two reductions. We first claim that there exists an algorithm $\mathcal{B}$ that is capable of insertions such that the cost of $\mathcal{A}$ to access a permutation π is no less than the cost of $\mathcal{B}$ to insert π. Note that since $\mathcal{A}$ is accessing π, all the points are active by definition. We will describe $\mathcal{B}$ in the geometric view simply by requiring that upon inserting $\pi(t)$ at time t, $\mathcal{B}$ touches all the points that $\mathcal{A}$ touches while accessing $\pi(t)$ at time t. Note that $\mathcal{A}$ touches at least all the points in $stair(\pi(t), t)$, and $\mathcal{B}$ is required only to touch either $pred(\pi(t))$ and its stair, or $succ(\pi(t))$ and its stair (Definition 11). Since $pred(\pi(t))$ belongs to $stair(\pi(t), t)$, one easily sees that $stair(pred(p)) \subset stair(\pi(t), t)$, and this defines a valid insertion algorithm.

We now reduce $\mathcal{B}$ to an algorithm for sorting π. Just by a traversal of the tree maintained by $\mathcal{B}$ at time n, we can produce the sorted order of π after incurring a cost of $O(n)$. However, we know that to sort a set U of permutations, any (comparison-based) sorting algorithm must require $\Omega(\log|U| + n)$ comparisons on at least a $1 - \frac{1}{|U|^\epsilon}$ fraction of the permutations in U. To see this, note that the decision tree of any sorting algorithm must have at least $|U|$ leaves (note that here we are assuming the weaker hypothesis that $\mathcal{A}$ and hence the sorting algorithm, are only designed to work on U; they may fail outside U). The number of leaves at height at most $(1 - \epsilon) \log|U|$ is at most $|U|^{1-\epsilon}$, and hence at least a $1 - \frac{1}{|U|^\epsilon}$ fraction require at least $(1 - \epsilon) \log|U| = \Omega(\log|U|)$ comparisons. Adding the trivial bound of $\Omega(n)$ to scan the input permutation gives us the desired bound.

Remark. Upper bounds proved for our model do not directly translate into bounds for algorithms. For example, when a new maximum is inserted, this can be done at a cost of one by making the element the root of the tree, respectively, only touching the element inserted. Note that this requires the promise that the element inserted is actually a new maximum. A slight extension makes the model algorithmic. This is best described in tree-view. We put all nodes of the tree in in-order into a doubly-linked list. Then, in the case of an insertion one can actually stop the search once the predecessor or the successor of the new element has been reached in the search because by also comparing the new element with the neighboring list element, one can verify that a node contains the predecessor or successor. Thus at the cost of a constant factor, bounds proved for our model are algorithmic. □

Acknowledgments. We thank an anonymous reviewer for valuable comments.

References

1. Cole, R.: On the dynamic finger conjecture for splay trees. part ii: The proof. SIAM Journal on Computing **30**(1), 44–85 (2000)

2. Cole, R., Mishra, B., Schmidt, J., Siegel, A.: On the dynamic finger conjecture for splay trees. part i: Splay sorting log n-block sequences. SIAM J. Comput. **30**(1), 1–43 (2000)
3. Demaine, E.D., Harmon, D., Iacono, J., Kane, D.M., Patrascu, M.: The geometry of binary search trees. In: SODA 2009, pp. 496–505 (2009)
4. Derryberry, J., Sleator, D., Wang, C.C.: Properties of multi-splay trees (2009)
5. Elmasry, A.: On the sequential access theorem and deque conjecture for splay trees. Theor. Comput. Sci. **314**(3), 459–466 (2004)
6. Fox, K.: Upper bounds for maximally greedy binary search trees. In: Dehne, F., Iacono, J., Sack, J.-R. (eds.) WADS 2011. LNCS, vol. 6844, pp. 411–422. Springer, Heidelberg (2011)
7. Iacono, J.: In pursuit of the dynamic optimality conjecture. In: Brodnik, A., López-Ortiz, A., Raman, V., Viola, A. (eds.) Ianfest-66. LNCS, vol. 8066, pp. 236–250. Springer, Heidelberg (2013)
8. Lucas, J.M.: Canonical forms for competitive binary search tree algorithms. Tech. Rep. DCS-TR-250, Rutgers University (1988)
9. Munro, J.I.: On the competitiveness of linear search. In: Paterson, M. (ed.) ESA 2000. LNCS, vol. 1879, pp. 338–345. Springer, Heidelberg (2000)
10. Pettie, S.: Splay trees, davenport-schinzel sequences, and the deque conjecture. In: SODA 2008, pp. 1115–1124 (2008)
11. Pettie, S.: Applications of forbidden 0–1 matrices to search tree and path compression-based data structures. In: SODA 2010, pp. 1457–1467 (2010)
12. Pettie, S.: Generalized davenport-schinzel sequences and their 0–1 matrix counterparts. Journal of Combinatorial Theory, Series A **118**(6), 1863–1895 (2011)
13. Sleator, D.D., Tarjan, R.E.: Self-adjusting binary search trees. J. ACM **32**(3), 652–686 (1985)
14. Sundar, R.: On the deque conjecture for the splay algorithm. Combinatorica **12**(1), 95–124 (1992)
15. Tarjan, R.E.: Sequential access in splay trees takes linear time. Combinatorica **5**(4), 367–378 (1985)
16. Wilber, R.: Lower bounds for accessing binary search trees with rotations. SIAM Journal on Computing **18**(1), 56–67 (1989)

A New Approach for Contact Graph Representations and Its Applications

Yi-Jun Chang and Hsu-Chun Yen[✉]

Department of Electrical Engineering, National Taiwan University,
Taipei, 106 Taiwan, Republic of China
yen@cc.ee.ntu.edu.tw

Abstract. A contact graph representation is a classical graph drawing style in which vertices are represented by geometric objects such that edges correspond to contacts between objects. Based on a characterization of stretchable systems of pseudo segments, we present a new approach for constructing a wide range of contact graph representations. Using Courcelle's theorem, some useful fixed-parameter tractability results are derived. Our approach can also be applied to giving quick proofs for some existing results of contact graph representations. We feel that the technique developed in the paper gives new insight to the study of contact representations of plane graphs.

1 Introduction

A *contact graph representation* is a classical graph drawing style in which vertices are represented by interior-disjoint geometric objects such that edges correspond to contacts between those objects. Following the well-known Koebe's circle packing theorem [12] that every planar graph can be drawn as touching circles, a variety of contact representations have been proposed and studied over the years. Parameters that differentiate one contact representation from another include the object shape (circle, triangle, . . . , etc) and the contact style (point vs. side contact, for instance).

Several quality measures arise naturally in designing contact graph representations. It is intuitive that one should avoid the presence of holes and minimize the size of the unused areas if at all possible. From the aesthetic and cognitive viewpoints, convex polygons are more pleasing to the eye than non-convex ones. To simplify the complexity of the drawing, it is also desirable that the number of sides of polygons (i.e., the polygonal complexity) be minimized.

Motivated by applications in floor-planning, cartographic design, and data visualization, *rectangular duals*, where all vertices are represented by axis-aligned rectangles such that the drawing forms a tiling of a rectangle (as a result, the drawing contains no hole), have received extensive investigation in both VLSI

H.-C. Yen—Research supported in part by Ministry of Science and Technology of Taiwan under Grant MOST-103-2221-E-002-154-MY3.

F. Dehne et al. (Eds.): WADS 2015, LNCS 9214, pp. 166–177, 2015.
DOI: 10.1007/978-3-319-21840-3_14

design and graph drawing communities. *Rectilinear duals*, which generalize rectangular duals, allow vertices to be drawn as rectilinear polygons. See, e.g., [2].

In reality, it is not uncommon to encounter objects displayed as non-
rectilinear polygons, see, e.g., [8], which deals with table cartograms with segments drawn not in an axis-aligned fashion. As it is mathematically more difficult
to deal with a non-rectilinear situation, only a scarcity of results were available in
this setting. The most notable in non-rectilinear setting is the so-called *triangle
contact representations*. A contact representation that forms a triangular tilling
of a triangle is called a *proper touching triangle graph* (proper-TTG) representation. The investigation of proper-TTG representations has been reported in
two recent articles, i.e., [11] and [9]. In [11], a fixed-parameter tractable decision
algorithm was proposed for triconnected planar graphs, and an inductive construction approach was used to show triconnected cubic planar graphs to admit
proper-TTG representations. In [9], strongly-connected outerplanar graphs, a
subclass of biconnected outerplanar graphs, were shown to have proper-TTG
representations iff the graph has at most 2 internal faces. Touching triangle representations without boundary constraints have been studied in [10]. See Fig. 1
for a showcase of some triangle contact representations.

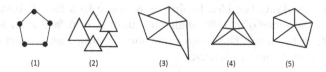

Fig. 1. A variety of triangle contact representations: (1) the 5-cycle, (2) the point-
side triangle contact representation, (3) the TTG representation without any boundary
constraint, (4) the proper-TTG representation, (5) the 3-sided convex polygonal dual

Our goal in this paper is to present a new technique for constructing a wide
range of contact representations, in particular tackling the non-rectilinear situation. In particular, the contributions of our work include the following:

- A very general drawing style called the *convex polygonal dual*, which sub-
  sumes well-studied drawing styles like the proper-TTG representation and
  the rectangular dual, is proposed.
- We characterize graphs admitting *straight-line convex t-gon representations*
  and *straight-line t-gon representations*, which can be regarded as a primal
  version of convex polygonal duals. This extends the main result of [1].
- Based on the above result, a characterization for a plane graph to admit a
  t-sided convex polygonal dual is presented.
- Using Courcelle's theorem, we derive some useful fixed-parameter tractability results for convex polygonal duals.
- To show that our approach is useful, we give quick alternative proofs for the
  following existing results:
  - Maximal plane graphs admit 6-sided convex polygonal duals [7].
  - Triconnected cubic plane graphs admit proper-TTG representations [11].

2   Preliminary

A graph is *planar* iff it can be drawn in the Euclidean plane without crossings. A *plane graph* is a planar graph with a fixed combinatorial embedding and a designated outer face. We write $f_O(G)$ (or f_O if the underlying graph G is understood) to denote the outer face of a plane graph $G = (V, E)$. Given a face f, $V(f)$ (resp., $E(f)$) denotes the set of nodes (resp., edges) along the boundary of f. We call a vertex (resp., an edge) in $V(f_O)$ (resp., $E(f_O)$) a *boundary* vertex (resp., edge).

Some definitions presented below can be seen as an extension or generalization of similar ones in [1].

Definition 1. *Given a biconnected plane graph $G = (V, E)$ such that all degree 2 vertices are in $V(f_O(G))$, a t-flat angle assignment (t-FAA, for short) is a mapping from a subset of $V \setminus \{v | v \in V(f_O(G))\}$ to inner faces of G such that:*

1. *Each vertex is assigned at most once;*
2. *Each inner face F is assigned at least $|V(F)| - t$ times;*
3. *for each mapping associating a vertex v to a face F, we have $v \in V(F)$.*

Intuitively speaking, the idea behind assigning v to a face F in a t-FAA is to capture the presence of a 180° angle surrounding v in face F in a drawing. Condition (2) is to ensure that each inner face is drawn as a convex polygon which has at most t convex corners.

Definition 2. *A straight line t-gon representation (t-SLR, for short) is a planar drawing such that:*

1. *each inner face is a polygon of at most t sides, and*
2. *the outer face is a convex polygon.*

A straight line convex t-gon representation (t-convex-SLR, for short) is a t-SLR with an additional constraint that each inner face is convex.

FAAs are also closely related to the so-called *contact systems of pseudo-segments* [5], each of which is a set of non-crossing Jordan arcs where any two of them intersect in at most one point, and each intersecting point is internal to at most one arc. A contact system is *stretchable* if there exists a homeomorphism transforming the contact system into a drawing where each arc is a straight line. Stretchable contact systems of pseudo-segments were characterized in [5] based on the notion of *extremal points*.

Definition 3. *A point p is an extremal point of a contact system S of pseudo-segments if the following three conditions are satisfied:*

1. *p is an endpoint of a pseudo-segment in S.*
2. *p is not interior to any pseudo-segment in S.*
3. *p is incident to the unbounded region of S.*

Theorem 1 ([5]). *A contact system S of pseudo-segments is stretchable iff each of its subsystems (i.e., subsets of pseudo-segments) S' of cardinality greater than 1 has at least 3 extremal points.*

It is not difficult to see that a t-FAA of a plane graph naturally defines a contact system of pseudo-segments in which each pseudo-segment is associated with a path $e_1, ..., e_{k-1}$ ($e_i = (v_i, v_{i+1})$, $1 \le i \le k - 1$) between two vertices v_1 and v_k such that $k \ge 2$ and $\forall 1 < j < k$, (1) v_j is assigned to a face containing e_{j-1} and e_j, and (2) v_1 and v_k are unassigned or assigned to a face not containing e_1 and e_{k-1}, respectively. Such a pseudo-segment is said to be *induced* by edge e_j, where $1 \le j \le k - 1$. Note that an edge induces exactly one pseudo-segment.

Given a plane graph G, the *inner* (also known as *internal* or *interior*) region of a cycle C is the region enclosed by C, and the *outer region* of C is the region outside of C. The inner and outer regions of C are written as $in(C)$ and $out(C)$, respectively. The edges and vertices located along C are neither in the inner region nor in the outer region of C. For ease of explanation, we write S_C to denote the set of pseudo-segments induced by C w.r.t. a given t-FAA.

It is clear that a graph admits a t-FAA corresponding to a stretchable contact system of pseudo-segments iff it admits a t-convex-SLR. With respect to a t-FAA, we call a corner of an inner face a *combinatorial convex corner* if it is not assigned to the face. For a more detailed exposition, the reader is referred to [1].

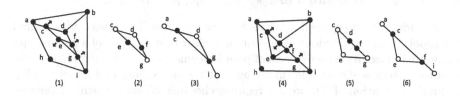

Fig. 2. Illustrations of concepts introduced in Section 3

3   Characterizing t-sided Convex Polygonal Duals

A *t-sided convex polygonal dual* is a side-contact representation of a plane graph in which all vertices are represented by convex polygons of at most t sides such that the drawing forms a tiling of a convex polygon. The goal in this section is to give a combinatorial characterization for plane graphs admitting such drawings.

In what follows we first derive a characterization for a graph admitting a t-convex-SLR based on the notion of t-FAAs.

Definition 4. *Let C be a cycle in a biconnected plane graph G whose degree 2 vertices are all in $V(f_O)$, and let v be a vertex in C. Given a t-FAA, we call v free in C if one of the following conditions is satisfied:*

1 v *is unassigned, or*
2 v *is assigned to a face F in $out(C)$, and F is not the only face to which v is incident in $out(C)$.*

Moreover, v is strongly-free *if Condition 1 above is replaced by*

1' v *is unassigned, and v is either in the outer face or incident to more than one face in $out(C)$*

Intuitively speaking, a free vertex (strongly-free vertex) of a cycle C indicates a corner (convex corner) in $in(C)$. Fig. 2(2) is a cycle C in Fig. 2(1), which is drawn in 5-convex-SLR. The vertices c, d, and g are strongly-free vertices of that cycle. Fig. 2(3) shows the set of pseudo-segments S_C for cycle C. The vertices a, d, and i are the extremal points in S_C. Fig. 2(4) is a 6-SLR. As we shall prove in the following theorem, the FAA described in Fig. 2(4) cannot be a convex-SLR since the cycle (c, d, f, g, e) only has 2 strongly-free vertices c and g. Note that the vertex e is free but not strongly-free. In any drawing realizing that FAA, e must be a concave corner in the face interior to the cycle (a, c, e, g, i, h).

The following key theorem, one of the main contributions of this paper, characterizes graphs admitting t-SLR and t-convex-SLR in terms of FAAs.

Theorem 2. *Let G be a biconnected plane graph whose degree 2 vertices are all in $V(f_O)$. G admits a t-convex-SLR (resp., t-SLR) iff there exists a t-FAA such that each cycle has at least 3 strongly-free (resp., free) vertices.*

Proof. (Idea) Due to space limitation, here we only give the intuitive idea behind the proof. From our previous discussion, it is clear that a t-FAA of a plane graph naturally induces a contact system of pseudo-segments. For deciding whether the contact system is stretchable (implying that the plane graph admits a t-SLR), a direct application of Theorem 1 requires checking *all* sub-systems of pseudo-segments for the availability of 3 extremal points. The current theorem shows a simpler characterization, i.e., examining only subsets of pseudo-segments of the form S_C for some cycle C is sufficient. Furthermore, we are able to relate the availability of 3 extremal points of pseudo-segments of S_C to the presence of at least 3 free vertices along cycle C. See Fig. 2(2, 3, 5, 6) for instance.

If each face is further required to be a convex polygon, we need to prevent a vertex from causing a face to be a concave polygon, like the vertex e in the cycle depicted in Fig. 2(5). It turns out that adding the constraint forcing each free vertex to be incident to more than one face in $out(C)$ (see Condition (1') in Definition 4) leads to a necessary and sufficient characterization.   □

It is easy to extend Theorem 2 to all biconnected plane graphs by modifying the definition of FAAs to handle degree 2 inner vertices. However, as the situation would not be encountered throughout the paper, we omit it in order to reduce complication.

To give a characterization of convex polygonal duals, in what follows we establish a link between t-sided convex polygonal duals and its primal counterpart, t-convex-SLRs.

Given a plane graph G, one may hope to find some sort of a "dual" graph G^* such that any t-convex-SLR of G^* is also a t-sided convex polygonal dual of G. Unfortunately, this kind of a reduction strategy turns out to be more complex than it appears on the surface, as the polygon associated with a vertex $v \in f_O(G)$ may touch the boundary of the t-sided convex polygonal dual of G on $0, 1, \ldots,$ $t - 1$ sides (see Fig. 3(4)). As an attempt to resolve such a difficulty, we define the G^* associated with a graph G as follows:

Definition 5. *Given a plane graph G and an integer t, the graph G^* is defined to be the result of the following construction steps:*

1. *Add a new vertex s in the unbounded face of G, and add an edge between s and each vertex in the boundary face.*
2. *Take the dual, and the new outer face is designated to the one corresponding to s.*
3. *Subdivide each edge into a path of $t - 1$ edges in the boundary face.*

See Fig. 3(1)-(2) for a graph G and the corresponding G^* (for $t = 3$), respectively.

The following result is then straightforward.

Theorem 3. *A plane graph G admits a t-sided convex polygonal dual iff there is a graph G', resulting from contracting some edges along the boundary of G^*, that admits a t-convex-SLR.*

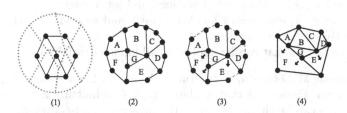

(1)   (2)   (3)   (4)

Fig. 3. (1) A graph G, (2) its associated G^*, (3) applying edge contractions to the dashed edges along the boundary of G^*, (4) a 3-convex-SLR of G^* which is also a 3-sided convex polygonal dual of G. In (3) and (4), flat angle assignments are annotated by arrows.

Theorems 2 and 3 relate the problem of finding a convex polygonal dual to finding a set of edges to be contracted and a corner labeling satisfying some constraints. In comparison with previous techniques designed for contact graph representations, the greatest advantage of Theorem 3 is that it turns a geometry problem to a purely graph-theoretic one. This, in conjunction with Theorem 2, allows us to get rid of any tedious and laborious geometric construction process when designing algorithms for contact graph representations.

By offering the possibility of contracting boundary edges of G^*, polygons associated with vertices in $f_O(G)$ can touch the boundary of the convex polygonal dual of G on $0, 1, \ldots, t-1$ sides. See Fig. 3(3)-(4). In Fig. 3(4), for instance, faces B, C and F touch the boundary on 0, 1, or 2 sides, respectively. Note that an edge contraction has the same effect of a corner assignment in $f_O(G^*)$. Therefore, we can assume that no assignment occurs in $f_O(G^*)$ since edge contraction already handles it.

(Remark) Theorem 2 is of independent interest as it improves the main result in [1] (i.e., Thm 2.10 in [1]) in the following way: (i) We check only simple cycles instead of all outline cycles; (ii) the result holds for all t-FAAs instead of 3-FAAs only; and (iii) we are able to deal with both polygons and convex polygons.

4   Fixed-Parameter Tractability Results

Monadic second-order logic (MSO), a fragment of second-order logic, only allows quantification over unary relations (i.e., sets). Among numerous applications of MSO, the study of graph structures has benefited in recent years from the advance of the theory of MSO. A powerful algorithmic meta-theorem for investigating graph structures in the logical framework is *Courcelle's theorem*, which says that any graph property expressible in MSO_2 is linear time solvable for graphs of bounded treewidth [3,4], where MSO_2 on graphs includes the following ingredients:

- Variables: vertices, edges, set of vertices, and set of edges.
- Relations: $\in$, $=$, edge-vertex incidence (INC), and adjacency (ADJ).
- Connectives: $\vee, \wedge, \neg, \rightarrow$.
- Quantifiers: $\forall, \exists$ that can be applied to all kinds of variables.

For more about MSO on graph structures, the reader is referred to, e.g., [4,6].

Recall from Theorem 3 that a plane graph G admitting a t-sided convex polygonal dual can be characterized by the presence of a t-convex-SLR of G' (a graph resulting from applying some edge contraction in f_O of G^*), and the latter can be further captured by t-FAAs (Theorem 2). Like many graph structures expressible in MSO_2, it turns out that such a characterization can be formulated in the framework of MSO_2. More precisely, we have:

Theorem 4. *Given a plane graph G, one can construct a graph $\tilde{G}$ along with a designated set of vertices F_{IN}, a designated vertex F_O, and a formula φ in MSO_2 such that G has a t-sided convex polygonal dual iff $(\tilde{G}, F_{IN}, F_O) \models \varphi$.*

Proof. (Sketch) First note that the parameter t in a t-sided convex polygonal of a graph is considered a fixed constant.

The $\tilde{G}$ is constructed from G^* using the following procedure: (1) add a new vertex for each face in G^*; (2) for each newly added vertex v and its associated face F, for all $u \in V(F)$, add edge $\{u, v\}$. In setting up φ, we allow some designated vertices, edges, set of vertices, and set of edges to be associated with free variables. We use F_{IN} to denote the designated set of vertices in $V(\tilde{G})$ corresponding to inner faces in G^*, and F_O to denote the designated vertex in $V(\tilde{G})$ corresponding to the outer face in G^*.

We define the formula $\text{CORNER}(e) \equiv (\exists u, v)[\text{INC}(e, u) \wedge \text{INC}(e, v) \wedge (u \in F_{IN}) \wedge (v \notin F_{IN})]$ which is true iff e is an edge incident to a vertex in $V(G^*)$ and a vertex in F_{IN}.

We use a subset U of $\{e \in E(\tilde{G}) | \text{CORNER}(e)\}$ to encode a t-FAA and a subset R of edges in the outer face of G^* to encode edge contraction. Since each $e \in E(\tilde{G})$ such that $\text{CORNER}(e)$ corresponds to a corner in G^*, when e is a corner of an inner face not located along the boundary of the drawing, it represents a flat angle assignment.

Our goal is to define φ as $(\exists U, R)t\text{-VALIDFAA}(U, R)$, where $t\text{-VALIDFAA}(U, R)$ is true iff U, together with R, represents a t-FAA such that each cycle has at least 3 strongly-free vertices.

$$t\text{-VALIDFAA}(U, R) \equiv t\text{-FAA}(U, R) \wedge (\forall C)\{\text{CYCLE}(C, R) \rightarrow \bigcup_{k=0,\ldots,3}$$

$$[(3 - k)\text{-BOUNDARYCORNERS}(C, R) \wedge (\exists v_1, \ldots, v_k) \bigwedge_{i=1,\ldots,k} \text{SFREE}(v_i, C, U)]\}$$

$t\text{-FAA}(U, R)$ is used to capture Definition 1. $\text{CYCLE}(C, R)$ is to ensure that C is a cycle after applying edge contraction R. $i\text{-BOUNDARYCORNERS}(C, R)$ is true iff the number of vertices in $V(C) \cap V(f_O)$ remains at least i after applying the edge contraction R (these vertices are strongly-free boundary vertices in C). $\text{SFREE}(v, C, U)$ is true iff v is strongly-free and non-boundary in C under the FAA U. Note that $\bigwedge_{i=1,\ldots,k} \text{SFREE}(v_i, C, U)$ is vacuously true if $k = 0$.   □

We are in a position to give our main result in this section.

Theorem 5. *For any t, it can be decided in polynomial time whether a plane graph G admits a t-sided convex polygonal dual if there is a constant k such that:*

1. *tree-width of $G \leq k$, or*
2. *For all $v \in V(G)$ such that $\deg(v) > 3$, there is a path linking v to the outer face of length $\leq k$.*

Proof. (Idea) In view of Theorem 4 and Courcelle's theorem, it suffices to show that each of the two conditions implies that the graph $\tilde{G}$ constructed in the proof of Theorem 4 is of bounded tree-width.

For the first condition, the tree-width of $\tilde{G}$ can be shown to be bounded by $O(k^2)$. The proof for the second condition is a little bit tricky. The underlying idea is that a triangle in G^* whose nearby faces are all triangles is unimportant and is irrelevant to the decision of whether the graph G admits a t-sided convex polygonal dual. In this spirit, we devise a modification to G^* which results in a bounded tree-width $\tilde{G}$ by eliminating all unimportant portions carefully such that the property of having t-convex-SLR is preserved.   □

Theorem 5 implies polynomial time algorithms for many important graph classes appearing frequently in the literature [9,11]. We have:

Corollary 1. *Deciding whether a plane graph admits a t-sided convex polygonal dual is solvable in polynomial time for graphs of max degree 3, partial 3-trees, and k-outerplane graphs.*

5   Further Applications of our Technique

In addition to the fixed-parameter tractability results in the previous section, in this section we give short proofs for some interesting existing results using the technique we have developed. First, we give a simple proof for a result of [7]:

Theorem 6 ([7]). *Each maximal plane graph admits a 6-sided convex polygonal dual.*

Proof. The first step in our alternative proof relies on a result of [2], showing that maximal plane graphs admit rectilinear duals using only upside-down T-modules and their degenerated modules. Fig. 4(1.1) lists the set of allowed modules, while modules listed in Fig. 4(1.2) are not allowed.

Given a maximal plane graph G (see Fig. 4(2.1)), an FAA of G^* is constructed naturally according to a rectilinear dual as shown in Fig. 4(2.2). To be precise, we make an assignment at each $180°$ corner not in the boundary of the drawing. Note that concave corners in any module does not correspond to a vertex in G^*. As such an FAA may not lead to a stretchable drawing, we do some adjustments by unassigning some vertices according to the rules specified in Fig. 4(3).

It is easy to see that the resulting FAA is a 6-FAA. Prior to the adjustment, it is a 6-FAA since only a convex corner of a module can be a combinatorial convex corner, and since each module has at most 6 convex corners. Though each adjustment increases the number of combinatorial convex corner of a face by one, we can apply it only when we have a nearby convex corner that is not a vertex in G^*.

What is left to be done is to show that each cycle has at least 3 strongly-free vertices. Let C be any cycle in G^*. Consider the sub-drawing, which is a rectilinear polygon, of C in the rectilinear dual. Let $\overline{ab}$ and $\overline{cd}$ be its highest and lowest horizontal segments, respectively, as shown in Fig. 4(4). It is immediate that a and b are strongly-free vertices of C. Suppose that there is no strongly-free vertex on $\overline{cd}$. Then, c and d must be bends in the drawing (i.e. not a vertex in G^*), and no adjustment is applied on $\overline{cd}$. This implies that there is no line segment touching $\overline{cd}$ from $out(C)$, meaning that there is a non-convex polygon F in $out(C)$ incident to $\overline{cd}$, which is a contradiction to the allowed set of modules (i.e., upside-down T-modules and their degeneracies). Therefore, we conclude that there is a strongly-free vertex in $\overline{cd}$, and hence C has at least 3 strongly-free vertices.

See Fig. 4(2.3) for an example of an FAA after adjustment, and see Fig. 4(2.4) for the resulting convex polygonal dual.   □

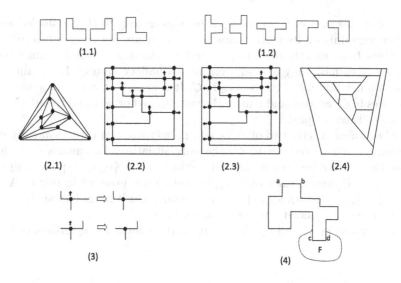

Fig. 4. Illustrations for the proof of Theorem 6

As each Hamiltonian maximal plane graph admits a rectilinear dual using only L-shape and rectangular modules, following a similar approach, our technique can be utilized to give a simple proof for the following:

Theorem 7 ([13]). *Each Hamiltonian maximal plane graph admits a 5-sided convex polygonal dual.*

Next, we showcase a quick proof for the main result of [11]:

Theorem 8 ([11]). *Each triconnected cubic plane graph admits a proper-TTG representation.*

Proof. A proper-TTG representation is just a 3-sided convex polygonal dual whose boundary is a triangle.

Let G be a triconnected cubic plane graph, and we construct its associated G^* as described in Section 3. We let $f_O(G) = (v_1, v_2, \ldots, v_s)$ be the outer face of G. Note that we must have $s \geq 3$ since G is simple. It is easy to see that the subgraph H of G^* induced by the faces corresponding to vertices in $V(G) \setminus V(f_O(G))$ (the shaded area in Fig. 5) is biconnected, since otherwise G is not triconnected.

We contract most of the boundary edges, only leaving a boundary edge for each of F_1, F_2, and F_3, where F_i is the face in G^* corresponding to v_i. We let the 3-FAA contain only $u_i \rightarrow F_i, i \in \{1, 2, 3\}$, where $u_i \in V(G^*)$ is the shared non-boundary vertex of F_i and F_{i+1}. See Fig. 5 for an illustration. We claim that our edge contraction and FAA work. It is immediate that the assignment is a 3-FAA such that the boundary in the resulting drawing is a triangle whose three corners are c_1, c_2 and c_3 in Fig. 5. What remains to be done is to verify that each cycle C has 3 free vertices:

- If C contains none of c_1, c_2 and c_3, it belongs entirely to H (the shaded area). Then, certainly all its vertices are free, as they are not assigned to $in(C)$.
- If C contains exactly one of c_1, c_2 and c_3, the one it contains must be c_3 (since c_1, c_2 have only one adjacent non-boundary vertex). Let x and y be the two neighboring vertices of c_3 in C. It is clear that x, c_3 and y are 3 free vertices in C, Since x and y are either unassigned or assigned to $out(C)$, and since c_3 is unassigned.
- If C contains exactly two of c_1, c_2 and c_3, as these two corners already contribute two free vertices to C, the only situation that makes C to have less than 3 free vertices is that all vertices in $C \setminus \{c_1, c_2, c_3\}$ are assigning to $in(C)$. However, since only u_1, u_2 and u_3 are involved in our FAA (i.e. $V(C) \subseteq \{c_1, c_2, c_3, u_1, u_2, u_3\}$), we can assure that it never happen by examining a small bounded amount of possibilities.
- If C contains c_1, c_2 and c_3, these three corners form 3 free vertices of C. $\quad\square$

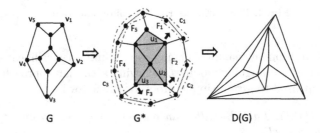

G   G*   D(G)

Fig. 5. Illustrations for the proof of Theorem 8

Adapting our approach, the laborious process of explicitly assigning positions for each point to construct a drawing, which inevitably appears in many works on contact graph representations in non-rectilinear situation, can be prevented.

6   Conclusion and Future Work

We have proposed a new approach for tackling a wide range of problems of contact graph representations. In addition to the facilitation of Courcelle's Theorem in the framework of MSO_2 to yield some fixed-parameter tractability results, the usefulness of this new technique is further amplified through several short proofs of some interesting existing results. Some intriguing problems still remain:

- Is there a general approach to deal with the case when holes are allowed? Also, how about other types of contact styles?
- As the huge constant involved in Courcelle's Theorem makes the FPT algorithm practically unusable, it would be helpful to have a practically usable solution.

- The problem of deciding whether a graph admits a t-sided polygonal dual is still not known to be NP-complete in general.
- The work of [9] showed that a special subclass of outerplanar graphs enjoys proper-TTG representations. Is it possible to extend the result to a broader graph class such as the entire class of outerplanar graphs?
- Though we already have a characterization for graphs admitting a t-sided convex polygonal dual in general, to extend Theorem 8, it will be nice to have a simpler characterization and a simpler polynomial time recognition algorithm for the biconnected cubic plane graphs admitting a 3-sided convex polygonal dual (or proper-TTG representation).
- In view of Theorems 6 and 7, it will be interesting to see more results linking rectilinear contact representations to non-rectilinear ones.

References

1. Aerts, N., Felsner, S.: Straight line triangle representations. In: Wismath, S., Wolff, A. (eds.) GD 2013. LNCS, vol. 8242, pp. 119–130. Springer, Heidelberg (2013)
2. Alam, M.J., Biedl, T., Felsner, S., Kaufmann, M., Kobourov, S.G., Ueckert, T.: Computing Cartograms with Optimal Complexity. Discrete & Computational Geometry **50**(3), 784–810 (2013)
3. Courcelle, B.: The monadic second-order logic of graphs. I. Recognizable sets of finite graphs. Information and Computation **85**(1), 12–75 (1990)
4. Courcelle, B., Engelfriet, J.: Graph Structure and Monadic Second-Order Logic: A Language-Theoretic Approach. Cambridge University Press (2012)
5. De Fraysseix, H., de Mendez, P.O.: Barycentric systems and stretchability. Discrete Applied Mathematics **155**(9), 1079–1095 (2007)
6. Downey, R., Fellows, M.: Fundamentals of Parameterized Complexity. Springer (2013)
7. Duncan, C., Gansner, E., Hu, Y., Kaufmann, M., Kobourov, S.: Optimal Polygonal Representation of Planar Graphs. Algorithmica **63**(3), 672–691 (2012)
8. Evans, W., Felsner, S., Kaufmann, M., Kobourov, S.G., Mondal, D., Nishat, R.I., Verbeek, K.: Table cartograms. In: Bodlaender, H.L., Italiano, G.F. (eds.) ESA 2013. LNCS, vol. 8125, pp. 421–432. Springer, Heidelberg (2013)
9. Fowler, J.J.: Strongly-connected outerplanar graphs with proper touching triangle representations. In: Wismath, S., Wolff, A. (eds.) GD 2013. LNCS, vol. 8242, pp. 155–160. Springer, Heidelberg (2013)
10. Gansner, E.R., Hu, Y., Kobourov, S.G.: On touching triangle graphs. In: Brandes, U., Cornelsen, S. (eds.) GD 2010. LNCS, vol. 6502, pp. 250–261. Springer, Heidelberg (2011)
11. Kobourov, S.G., Mondal, D., Nishat, R.I.: Touching triangle representations for 3-connected planar graphs. In: Didimo, W., Patrignani, M. (eds.) GD 2012. LNCS, vol. 7704, pp. 199–210. Springer, Heidelberg (2013)
12. Koebe, P.: Kontaktprobleme der konformen Abbil-dung. Ber. Verh. Sachs. Akademie der Wissenschaften Leipzig, Math.-Phys. Klasse **88**, 141–164 (1936)
13. Ueckerdt, T.: Geometric Representations of Graphs with low Polygonal Complexity. PhD thesis, Technische Universität Berlin (2011)

Dealing with 4-Variables by Resolution: An Improved MaxSAT Algorithm

Jianer Chen[1,2], Chao Xu[1], and Jianxin Wang[1]([⊠])

[1] School of Information Science and Engineering, Central South University,
Changsha, People's Republic of China
jxwang@mail.csu.edu.cn
[2] Department of Computer Science and Engineering, Texas A&M University,
College Station, USA

Abstract. We study techniques for solving the MAXIMUM SATISFIABIL-
ITY problem (MAXSAT). Our focus is on variables of degree 4. We identify cases for degree-4 variables and show how the resolution principle
and the kernelization techniques can be nicely integrated to achieve more
efficient algorithms for MAXSAT. As a result, we present a parameterized algorithm of time $O^*(1.3248^k)$ for MAXSAT, improving the previous
best upper bound $O^*(1.358^k)$ by Bliznets and Golovnev.

1 Introduction

The SATISFIABILITY problem (SAT) and its optimization version, the MAXI-
MUM SATISFIABILITY problem (MAXSAT) are of fundamental importance in
computer science [4], in particular in the study of approximation algorithms [12].
Since the problems are NP-hard [10], different algorithmic approaches, including
heuristic algorithms ([11,17]), approximation algorithms ([2,19]), and exact and
parameterized algorithms ([5,6,16]), have been extensively studied.

The main result of the current paper is an improved parameterized algorithm
for the MAXSAT problem, which is formally defined as follows.

> MAXSAT: Given a CNF formula F and an integer k (the *parameter*), is
> there an assignment to the variables in F that satisfies at least k clauses
> in F?

It is known that the MAXSAT problem is fixed-parameter tractable, i.e.,
it is solvable in time $O^*(f(k))$.[1] The research on parameterized algorithms for
MAXSAT has an impressive list, as shown in Figure 1.

Most algorithms for SAT and MAXSAT are based on the branch-and-bound
process [11]. The *Strong Exponential Time Hypothesis* [8,13] indicates, to some
extent, a popular opinion that branch-and-bound is perhaps unavoidable in solving the SAT problem and its variations.

Supported by the National Natural Science Foundation of China under Grants
(61173051, 61232001, 61472449, 61420106009, 71221061).

[1] Following the current convention in exact and parameterized algorithms, we use the
notation $O^*(f(k))$ to denote the bound $f(k)n^{O(1)}$, where n is the instance size.

F. Dehne et al. (Eds.): WADS 2015, LNCS 9214, pp. 178–188, 2015.
DOI: 10.1007/978-3-319-21840-3_15

Bound	Reference	Year
$O^*(1.618^k)$	Mahajan, Raman [14]	1999
$O^*(1.400^k)$	Niedermeier, Rossmanith [15]	1999
$O^*(1.381^k)$	Bansal, Raman [3]	1999
$O^*(1.370^k)$	Chen, Kanj [6]	2002
$O^*(1.358^k)$	Bliznets, Golovnev [5]	2012
$O^*(1.325^k)$	this paper	2015

Fig. 1. Progress in MaxSAT algorithms

Therefore, how to branch more effectively in algorithms solving SAT and MaxSAT has become critical. For MaxSAT, it is well-known that branching on variables of high degrees in a formula will be sufficiently effective. On the other hand, variables of degree bounded by 2 can be handled efficiently based on the resolution principle [9]. Recently, Bliznets and Golovnev [5] proposed new strategies for branching on variables of degree 3 effectively and improved Chen-Kanj's algorithm [6], which had stood as the best MaxSAT algorithm for 10 years.

The next bottleneck is on degree-4 variables (case 3.10 in [6], Theorem 5, step 10 in [5]). Degree-4 variables seem neither to have a large enough degree to support direct branchings efficiently, nor to have structures simple enough to yield efficient case-by-case manipulations.

A contribution of this paper is to show how the resolution principle [9] can be used in handling degree-4 variables in solving the MaxSAT problem. The resolution principle is a quite powerful tool in solving the SAT problem [9], because it preserves the satisfiability of the formula. Unfortunately, resolutions cannot be used directly in solving MaxSAT in general because the underlying formula is not assumed to be satisfiable.

The current paper identifies cases for degree-4 variables and shows how the resolution principle can be applied efficiently on these cases (R-Rules 6-7). This technique eliminates the structures that slow down the branching process. Observing that resolutions on high degree variables may significantly increase the size of a formula, we integrate resolutions nicely with the technique of polynomial-time kernelization in parameterized algorithms [6]. Therefore, the resolution principle can be used whenever it is applicable – once the formula size gets too large, we simply use the kernelization algorithm to reduce the formula size. In fact, one of our reduction rules (R-Rule 7) decreases the number of variables while keeping the parameter value unchanged. This rule is valid since it can be applied at most polynomial many times while the kernelization of MaxSAT keeps the formula size from going too large.

A nice approach suggested by Bliznets and Golovnev [5] is to transform solving a class of special instances of MaxSAT into solving the SET-COVER problem. However, the method proposed in [5] is not efficient enough to achieve our bound. For this, we introduce a new branching rule that is sufficiently efficient and further reduces the instances to an even more restricted form. In particular, we eliminate all clauses of size 2 and 3. This restricted instance allows us to apply more powerful techniques in randomized algorithms and in derandomization [19] to derive tighter lower bounds on the instances of MaxSAT, so that we can use more effectively the existing algorithm for Set-Cover [18].

We start with some preliminary concepts and definitions.

A (Boolean) *variable* x can be assigned value either 1 (TRUE) or 0 (FALSE). A variable x has two corresponding literals: the *positive literal* x and the *negative literal* $\bar{x}$, called the *literals* of x. The variable x is called the *variable for the literals* x *and* $\bar{x}$. A *clause* C is a disjunction of a set of literals. Let $C_1 = zC_2$ indicate that the clause C_1 consists of the literal z plus all literals in the clause C_2, and use C_1C_2 to denote the clause consisting of all literals that are in either C_1 or C_2, or both. Assume that a literal can appear in a clause at most once. A clause C is *satisfied* by an assignment if at least one literal in C gets a value 1. A *CNF formula* F is a conjunction of clauses, which is *satisfied* by an assignment if all clauses in F are satisfied by the assignment. We will always denote by n (resp. m) the number of variables (resp. clauses) in a given formula.

A literal z is an (i,j)-*literal* if z appears i times and $\bar{z}$ appears j times in F. A variable x is an (i,j)-*variable* if the literal x is an (i,j)-literal. A variable x has degree h, called an h-*variable*, if x is an (i,j)-variable and $h = i + j$. A variable is an h^+-*variable* if its degree is at least h.

The *size* of a clause C is the number of literals in C. A clause is an h-*clause* if its size is h, and an h^+-*clause* if its size is at least h. A clause is *unit* if its size is 1 and is *non-unit* if its size is larger than 1. The *size* of a CNF formula F is equal to the sum of the sizes of the clauses in F.

A *resolvent* on a variable x in a formula F is a clause of the form CD such that xC and $\bar{x}D$ are clauses in F. The *resolution* on the variable x in F is the conjunction of all resolvents on x.

2 Reduction Rules

An instance (F, k) of the MaxSAT problem asks whether there is an assignment to the variables in a given CNF formula F that satisfies at least k clauses in F. A *reduction rule* transforms, in polynomial time, an instance (F, k) of MaxSAT into another instance (F', k') with $k \geq k'$ such that (F, k) is a Yes-instance if and only if (F', k') is a Yes-instance.

We present a set of reduction rules, R-Rules 1-7. An R-Rule j is applied only when none of R-Rules i with $i < j$ is applicable. The first three reduction rules are from [6]. Let $F_{z=1}$ (resp. $F_{z=0}$) be the formula obtained from F with the literal assignment $z = 1$ (resp. $z = 0$).

R-Rule 1 ([6]). $(F \wedge (x\bar{x}C), k) \rightarrow (F, k-1)$, $(F \wedge (x) \wedge (\bar{x}), k) \rightarrow (F, k-1)$.

R-Rule 2 ([6]). For an (i,j)-literal z such that there are at least j unit clauses (z) in F, $(F, k) \rightarrow (F_{z=1}, k-i)$.

Assume that R-Rule 2 is not applicable to F, then F has no *pure literals*, i.e., literals whose negation does not appear in F. Thus, all variables are 2^+-variables.

Under this condition, we can process 2-variables based on the resolution principle [9], whose correctness can be easily verified.

R-Rule 3 ([6]). For a 2-variable x, $(F \wedge (xC_1) \wedge (\bar{x}C_2), k) \rightarrow (F \wedge (C_1 C_2), k - 1)$.

In case none of R-Rules 1-3 is applicable, every variable is a 3^+-variable. Moreover, for each $(i, 1)$-literal z, there is no unit clause (z), and for each $(i, 2)$-literal z, there is at most one unit clause (z). Now we show two reduction rules from [5] (Simplification Rule 5, Corollary 1).

R-Rule 4 ([5]). For a $(2, 1)$-literal z and an arbitrary literal y, $(F \wedge (zy) \wedge (zC_2) \wedge (\bar{z}C_3), k) \rightarrow (F \wedge (yC_3) \wedge (\bar{y}C_2 C_3), k - 1)$.

R-Rule 5 ([5]). For a formula $F_0 = F \wedge (zC_1) \wedge (zC_2) \wedge (\bar{z}C_3)$, where z is a $(2, 1)$-literal in F_0 and $C_1 \cup C_2 \cup C_3$ contains both y and $\bar{y}$ for some variable y, $(F \wedge (zC_1) \wedge (zC_2) \wedge (\bar{z}C_3), k) \rightarrow (F \wedge (C_1 C_3) \wedge (C_2 C_3), k - 1)$.

Therefore, in case none of R-Rules 1-5 is applicable, for each $(2, 1)$-literal z, the two clauses containing z are 3^+-clauses. Now, we introduce two new reduction rules that are based on the resolution principle.

R-Rule 6. If there exist an $(i, 1)$-literal z and a $(j, 1)$-literal y in F_1 such that $F_1 = F \wedge (zC_1) \wedge \cdots \wedge (zC_i) \wedge (\bar{z}yC)$, then $(F_1 = F \wedge (zC_1) \wedge \cdots \wedge (zC_i) \wedge (\bar{z}yC), k) \rightarrow (F_2 = F \wedge (yCC_1) \wedge \cdots \wedge (yCC_i), k - 1)$.

Lemma 1. *R-Rule 6 transforms instance* (F_1, k) *into* $(F_2, k-1)$ *such that* (F_1, k) *is a Yes-instance if and only if* $(F_2, k - 1)$ *is a Yes-instance.*

Proof. (sketch) It can be shown that there is an optimal assignment to F_1 that satisfies all $i + 1$ clauses $(zC_1) \wedge \cdots \wedge (zC_i) \wedge (\bar{z}yC)$. Similarly, there is an optimal assignment to F_2 that satisfies all i clauses $(yCC_1) \wedge \cdots \wedge (yCC_i)$. These plus the resolution principle give immediately that (F_1, k) is a Yes-instance if and only if $(F_2, k - 1)$ is a Yes-instance. □

Based on the resolution principle, our last reduction rule deals with a $(2, 2)$-variable, which does not decrease the parameter value k, but reduces the number of variables by eliminating the $(2, 2)$-variable.

R-Rule 7. Let z be a $(2, 2)$-literal in a formula $F_1 = F \wedge (zy_1 C_1) \wedge (zy_2 C_2) \wedge (\bar{z}y_3 C_3) \wedge (\bar{z}y_4 C_4)$, such that each y_h is an $(i_h, 1)$-literal for some i_h. Then, $(F_1 = F \wedge (zy_1 C_1) \wedge (zy_2 C_2) \wedge (\bar{z}y_3 C_3) \wedge (\bar{z}y_4 C_4), k) \rightarrow (F_2 = F \wedge (y_1 y_3 C_1 C_3) \wedge (y_2 y_3 C_2 C_3) \wedge (y_1 y_4 C_1 C_4) \wedge (y_2 y_4 C_2 C_4), k)$.

Lemma 2. *R-Rule 7 transforms the instance* (F_1, k) *into* (F_2, k) *such that* (F_1, k) *is a Yes-instance if and only if* (F_2, k) *is a Yes-instance.*

Proof. (sketch) The proof is similar to that for Lemma 1. It can be proved that there is an optimal assignment to F_1 that satisfies all 4 clauses $(zy_1 C_1)$, $(zy_2 C_2)$, $(\bar{z}y_3 C_3)$, $(\bar{z}y_4 C_4)$, and there is an optimal assignment to F_2 that satisfies all 4 clauses $(y_1 y_3 C_1 C_3)$, $(y_2 y_3 C_2 C_3)$, $(y_1 y_4 C_1 C_4)$, $(y_2 y_4 C_2 C_4)$. These two facts plus the resolution principle give that (F_1, k) is a Yes-instance if and only if (F_2, k) is a Yes-instance. □

We remark that instead of decreasing the parameter value k, R-Rule 7 decreases the number of variables. It may increase the size of the formula. However, whenever the size of the formula F in an instance (F, k) gets too large, we can simply apply the polynomial-time kernelization algorithm in [6] that will reduce the formula size and bound the size by $2k^2$.

3   Branching Rules

In a typical branch-and-bound algorithm, a branching step on an instance (F, k) produces, in polynomial time, a collection $\{(F_1, k - d_1), \ldots, (F_r, k - d_r)\}$ of instances of MAXSAT, such that (F, k) is a Yes-instance if and only if at least one of $(F_1, k - d_1)$, $\ldots$, $(F_r, k - d_r)$ is a Yes-instance. Such a branching step is called a $(d_1, \ldots, d_r)$-branching, the vector $t = (d_1, \ldots, d_r)$ is called the branching vector for the branching, and each instance $(F_i, k - d_i)$, $1 \leq i \leq r$, is called a branch of the branching. It can be shown ([7]) that the polynomial $p_t(x) = x^k - x^{k-d_1} - \cdots - x^{k-d_r}$, has a unique positive root, denoted as $\rho(t)$, and $\rho(t) \geq 1$. We say that the t_1-branching is inferior to the t_2-branching if $\rho(t_1) > \rho(t_2)$. It is well-known that for a parameterized algorithm based on the branch-and-bound process, if the root of every branching step in the algorithm is bounded by a constant $c \geq 1$, then the algorithm runs in time $O^*(c^k)$.

If any of R-Rules 1-7 is applicable on a formula F, we apply the rule, which either decreases the parameter value k (R-Rules 1-6) or reduces the number of variables without increasing the parameter value (R-Rule 7). A formula F is reduced if none of R-Rules 1-7 is applicable on F. It is obvious that each of R-Rules 1-7 takes polynomial time, and these rules can be applied at most polynomial many times (this holds true for R-Rule 7 because MAXSAT problem has a kernel of size $2k^2$ [6]). Thus, with a polynomial-time preprocessing, we can always reduce a given instance into a reduced instance. Therefore, we can assume that the formula F before the branch-and-bound process is always reduced.

Now, we present a series of branching rules (B-Rules). Again a B-Rule j is applied only when none of B-Rules i with $i < j$ is applicable.

For an instance (F, k), and an (i, j)-literal z in F, by "branching on z", we mean to construct two instances $(F_{z=1}, k - i)$ and $(F_{z=0}, k - j)$.

As well known, branching on a high degree variable is efficient enough.

Lemma 3. (B-Rule 1) If a reduced formula F contains a 6^+-variable x or a $(3, 2)$-literal x, then branch on x. The branching is not inferior to the $(3, 2)$-branching.

We also note a result for branching on 3-variables (Theorem 2 in [5]):

Lemma 4. (B-Rule 2) ([5]) If a reduced formula F contains a 3-variable, then we can make a branching that is not inferior to the $(6, 1)$-branching, and thus it is not inferior to the $(3, 2)$-branching.

Note that if B-Rules 1-2 are not applicable on a reduced formula F, then F contains only $(4, 1)$-, $(3, 1)$-, and $(2, 2)$-literals and their negations. An $(i, 1)$-literal z in a formula F is an $(i, 1)$-*singleton* (resp. $(i, 1)$-nonsingleton) if the clause containing $\bar{z}$ is a unit (resp. non-unit) clause.

Lemma 5. (B-Rule 3) *Given a reduced formula F, if a literal z is an $(i, 1)$-nonsingleton such that $\bar{z}$ is contained in a non-unit clause $(\bar{z}y_1 \cdots y_h)$, then branch with (B1) $z = 1$; (B2) $z = y_1 = \cdots = y_h = 0$. The branching is not inferior to the $(3, 2)$-branching.*

After Lemma 5, we can assume that all $(i, 1)$-literals are $(i, 1)$-singletons. A literal is a *singleton* if it is an $(i, 1)$-singleton for some i.

Lemma 6. (B-Rule 4) *Given a reduced formula F, if an $(i, 1)$-literal z is contained in a 2-clause (zy), then branch with: (B1) $z = 1$; and (B2) $z = 0$ and $y = 1$. The branching is not inferior to the $(3, 2)$-branching.*

By Lemma 6 and R-Rule 2, every $(i, 1)$-literal is in a 3^+-clause.

The next nine branching rules are dealing with $(2, 2)$-literals, which present the most difficult cases for our algorithm. The first three rules are easy to prove. Note that each variable is either a 4-variable or a $(4, 1)$-variable. For two clauses that both contain a $(2, 2)$-literal z and literals of another variable, B-Rules 5-6 solve the cases.

Lemma 7. (B-Rule 5) *Given a reduced formula F, if a $(2, 2)$-literal z is contained in two clauses (zy_1C_1) and (zy_2C_2), where y_1 and y_2 are literals of the same 4-variable y, then branch with: (B1) $z = 0$; and (B2) $z = 1$ followed by an application of R-Rule 2 or 3. The branching is not inferior to the $(3, 2)$-branching.*

Since B-Rule 5 is not applicable, if any $(2, 2)$-literal z is contained in two clauses (zy_1C_1) and (zy_2C_2), where y_1 and y_2 are literals of the same variable, then y_1 and y_2 must be $(4, 1)$-singletons, so $y_1 = y_2$.

Lemma 8. (B-Rule 6) *Given a reduced formula F, if two clauses both contain literals z and y, where z is a $(2, 2)$-literal, then branch with: (B1) $y = 0$; and (B2) $y = 1$ followed by an application of R-Rule 2. The branching is not inferior to the $(3, 2)$-branching.*

Next, B-Rule 7 deals with unit clauses containing $(2, 2)$-literals.

Lemma 9. (B-Rule 7) *Given a reduced formula F, if there is a $(2, 2)$-literal z with two clauses (zC_1) and (zC_2) such that $(\bar{z})$ is a unit clause, then branch with: (B1) $z = 1$, $C_1C_2 = 0$; and (B2) $z = 0$. The branching is not inferior to the $(3, 2)$-branching.*

If B-Rule 7 is not applicable, then $(2, 2)$-literals are only in 2^+-clauses. Now, we show an important branching on $(2, 2)$-variables.

Lemma 10. (B-Rule 8) *Given a reduced formula F, if one clause contains an $(i, 1)$-literal y_1 and a $(2, 2)$-literal z, and another contains z and a $(2, 2)$-literal y_2, then branch with:* (B1) $y_2 = 1$, *then apply R-Rule 6; and* (B2) $y_2 = 0$. *The branching is not inferior to the $(3, 2)$-branching.*

Denote by (zC_1), (zC_2), $(\bar{z}D_1)$ and $(\bar{z}D_2)$ the four clauses containing the literals of a $(2, 2)$-variable z. Since R-Rule 7 and B-Rules 7-8 are not applicable, $C_1C_2D_1D_2$ must satisfy one of the two cases: (1) all literals in $C_1C_2D_1D_2$ are $(2, 2)$-literals; and (2) one of C_1C_2 and D_1D_2 contains only singletons and the other contains only $(2, 2)$-literals. Based on this observation, we introduce two new terminologies for $(2, 2)$-literals.

Definition 1. A $(2, 2)$-literal z is *skewed* if for z_1, which is either z or $\bar{z}$, all other literals in the clauses containing z_1 are singletons and all literals in the clauses containing $\bar{z}_1$ are $(2, 2)$-literals. A $(2, 2)$-literal z is *evened* if the four clauses containing either z or $\bar{z}$ contain only $(2, 2)$-literals.

If B-Rules 1-8 are not applicable, then formula F contains only $(3, 1)$-singletons, $(4, 1)$-singletons, skewed $(2, 2)$-literals, and evened $(2, 2)$-literals.

Lemma 11. (B-Rule 9) *If an evened $(2, 2)$-literal z is in a 2-clause in a reduced formula F, then pick any literal $y \neq \bar{z}$ in a clause containing $\bar{z}$, and branch with:* (B1) $y = 1$, *then apply R-Rule 2 or 4; and* (B2) $y = 0$. *The branching is not inferior to the $(3, 2)$-branching.*

If B-Rule 9 is not applicable, then every $(2, 2)$-literal in a 2-clause is skewed. Combined with the fact that B-Rule 4 is not applicable, this guarantees that every literal in a 2-clause is a skewed $(2, 2)$-literal. The next branching rule is to deal with literals in 2-clauses.

Lemma 12. (B-Rule 10) *For a given 2-clause (zy), let the two clauses containing $\bar{z}$ be $(\bar{z}C_1)$ and $(\bar{z}C_2)$. Branch with:* (B1) $y = 1$; (B2) $y = 0$, $z = 1$; *and* (B3) $y = z = C_1 = C_2 = 0$. *The branching is not inferior to the $(8, 4, 2)$-branching, which is not inferior to the $(3, 2)$-branching.*

If B-Rule 10 is not applicable, then all 2^+-clauses are 3^+-clauses. The next branching rule solves all skewed $(2,2)$-literals.

Lemma 13. (B-Rule 11) *Given a reduced formula F, if a clause (zyC_1) contains two $(2, 2)$-literals z and y, where y is a skewed $(2, 2)$-literal and the other clause containing z is (zC_2), then branch with:* (B1) $z = 0$; (B2) $z = 1$, $yC_1 = 0$; *and* (B3) $z = 1$, $C_2 = 0$. *The branching is not inferior to the $(6, 5, 2)$-branching, which is not inferior to the $(3, 2)$-branching.*

Let y be a skewed $(2, 2)$-literal. By Lemma 13, if B-Rule 11 is not applicable, a clause C containing y cannot contain other $(2, 2)$-literals. Therefore, all other literals in the clause C are singletons. Note that $\bar{y}$ is also a skewed $(2, 2)$-literal, so all other literals in a clause containing $\bar{y}$ are also singletons. However, in this case, R-Rule 7 would have become applicable. Therefore, if B-Rule 11 is not applicable, then a reduced formula F contains no skewed $(2, 2)$-literals. Thus, the formula F contains only $(4, 1)$-singletons, $(3, 1)$-singletons, and evened $(2, 2)$-literals.

Lemma 14. (B-Rule 12) *Given a reduced formula F, if the clauses containing an evened $(2, 2)$-literal z are (zy_1C_1) and (zy_2C_2), and there is a third clause $(y_1\bar{y}_2C_3)$, then branch on z and in the branch $z = 1$ also apply R-Rule 6. The branching is not inferior to the $(3, 2)$-branching.*

With Lemma 14, we are ready to eliminate all $(2, 2)$-literals.

Lemma 15. (B-Rule 13) *For clauses (zy_1C_1), $(\bar{z}y_2C_2)$, (y_1D_1), (y_2D_2) in a reduced formula, where (y_1D_1) could be the same as (y_2D_2), and z is an evened $(2, 2)$-literal, branch with: (B1) $z = 1$, $y_1 = 0$, then apply B-Rule 2; (B2) $z = y_1 = 1$, $D_1 = 0$; (B3) $z = 0$, $y_2 = 0$, then apply B-Rule 2; and (B4) $z = 0$, $y_2 = 1$, $D_2 = 0$. The branching is not inferior to the $(10, 10, 6, 6, 5, 5)$-branching, which is not inferior to the $(3, 2)$-branching.*

Remark. Note that the clauses (zy_1C_1), $(\bar{z}y_2C_2)$, (y_1D_1), and (y_2D_2) contain only $(2, 2)$-literals. Thus, when branch (B1) (resp. (B3)) assigns values to z and y_1 (resp. y_2), new 3-variables are created so B-Rule 2 becomes applicable. This verifies the validity of these branches in B-Rule 13.

If the branching rule B-Rule 13 is not applicable, then all literals in a reduced formula F are either $(3, 1)$-singletons or $(4, 1)$-singletons, or their negations. Moreover, all non-unit clauses are 3^+-clauses. The following branching rule will further eliminate all 3-clauses.

Lemma 16. (B-Rule 14) *If a reduced formula F contains a 3-clause $(z_1z_2z_3)$, then branch with: (B1) $z_1 = 1$; (B2) $z_1 = 0$, $z_2 = 1$; and (B3) $z_1 = z_2 = 0$, $z_3 = 1$. The branching is not inferior to the $(3, 4, 5)$-branching, which is not inferior to the $(3, 2)$-branching.*

Summarizing all Lemmas 3-16, we conclude that if none of the reduction rules R-Rules 1-7 and the branching rules B-Rules 1-14 is applicable, then all literals are $(i, 1)$-singletons or their negations, where i is either 3 or 4, and all non-unit clauses are 4^+-clauses.

4   An $O^*(1.3248^k)$-Time Algorithm for MAXSAT

An instance (F, k) is a *simplified instance* if every variable in F is either a 3-singleton or a 4-singleton, and each non-unit clause in F is a 4^+-clause. By Lemmas 3-16, for any instance (F, k) of the MAXSAT problem, we apply the branching rules B-Rules 1-14, which are all not inferior to the $(3, 2)$-branching, until the formula F becomes a simplified instance.

The MAXSAT problem on simplified instances can be solved by reducing the problem to the MIN SET-COVER problem [5]. We first refine this method to get an algorithm of time $O^*(1.3226^k)$ (compared to that of time $O^*(1.3574^k)$ in [5]), based on an observation derived from a classical result of Yannakakis [19].

Lemma 17. *If* $m + n/2 \geq 1.829k$, *then for a simplified instance* (F, k), *there is an assignment that satisfies at least* k *clauses in* F, *and the assignment can be constructed in polynomial time.*

Proof. Since every variable x_i in F is a singleton, there are exactly n unit clauses $(\bar{x}_i)$, $1 \leq i \leq n$, and $m - n$ non-unit clauses. Set $p = 0.1795$, and assign each variable x_i with value 1 with a probability p. Therefore, each unit clause $(\bar{x}_i)$ is satisfied with a probability $1 - p$. Since each non-unit clause contains at least 4 positive literals, the assignment satisfies a non-unit clause with a probability at least $1 - (1-p)^4$. Therefore, the expected number of satisfied clauses under this random assignment is at least

$$n(1-p) + (m-n)(1 - (1-p)^4) = n(1-p) + (m + \frac{n}{2})(1 - (1-p)^4)$$
$$-\frac{3n}{2}(1 - (1-p)^4) \geq (m + \frac{n}{2})(1 - (1-p)^4) \geq 1.829k(1 - (1-p)^4) \geq k.$$

Now a polynomial-time deranandomization process (see [19]) can construct an assignment satisfying at least k clauses in the formula F. □

Therefore, we only need to consider simplified instances (F, k) satisfying $m + n/2 < 1.829k$. We follow the approach proposed in [5] and reduce the simplified instance (F, k) of MaxSAT to an instance C_F of the MIN SET-COVER problem. Each non-unit clause C_h in F corresponds to an element a_{C_h} in the universal set U_F, and each variable x_i in F corresponds to a set S_{x_i} in C_F such that the set S_{x_i} contains the element a_{C_h} if and only if the literal x_i is in the clause C_h. Thus, the collection C_F consists of n sets S_{x_i}, $1 \leq i \leq n$, the universal set U_F has $m - n$ elements (note that there are exactly n unit clauses $(\bar{x}_i)$), and we are looking for the minimum number of sets in C_F that cover all elements in the universal set U_F. As observed in [5], for a given instance (F, k) of MaxSAT, at least one optimal assignment satisfies all non-unit clauses.

Lemma 18. *From any minimum set cover* C' *for the collection* C_F, *an optimal assignment to the formula* F *in the simplified instance* (F, k) *of* MaxSAT *can be constructed in polynomial time.*

Then, we solve such a simplified instance using a result in [18].

Theorem 1. *The* MaxSAT *problem on simplified instances can be solved in time* $O^*(1.3226^k)$.

Proof. MIN SET-COVER on C_F is solvable in time $O^*(1.29^{0.6|U_F|+0.9|S_F|})$ [18], where $|U_F| = m - n$, $|S_F| = n$, and by Lemma 17, $m + n/2 \leq 1.829k$. Thus, $O^*(1.29^{0.6(m-n)+0.9n}) \leq O^*(1.29^{0.6 \times 1.829k}) = O^*(1.3226^k)$. □

In summary, we present our algorithm in Figure 2.

Theorem 2. *The algorithm* **MaxSAT-Solver** *solves the* MaxSAT *problem in time* $O^*(1.3248^k)$.

Algorithm MaxSAT-Solver(F, k)

INPUT: an instance (F, k) of MaxSAT, where F is a CNF formula

OUTPUT: an assignment that satisfies at least k clauses, or report non-exist

1. apply R-Rules 1-7, in order, repeatedly until (F, k) is reduced;
2. **if** $k \leq 1$ **then** directly solve the problem and return;
3. **if** (F, k) is a simplified instance
   then solve the problem in time $O^*(1.3226^k)$; return;
4. apply the first B-Rule that is applicable to (F, k);
   recursively solve the instance in each of the branches.

Fig. 2. The main algorithm for MaxSAT

Proof. Each leaf of the search tree $\mathcal{T}$ for an execution of the algorithm corresponds to an execution of step 2 or 3, which, by Theorem 1, takes time $O^*(1.3226^k)$ on an instance (F, k). Each internal node of $\mathcal{T}$ corresponds to applying one of B-Rules 1-14 in step 4, which, by Lemmas 3-16, is not inferior to the $(3, 2)$-branching that has its root ≤ 1.3248. As a result, the algorithm **Max-SAT-Solver** solves MaxSAT in time $O^*(1.3248^k)$.   □

5   Conclusion

We presented an $O^*(1.3248^k)$-time algorithm for MaxSAT, improving the best previous bound $O^*(1.358^k)$ [5]. We showed how the resolution principle is used effectively to eliminate instance structures that cause inefficient branchings. We presented techniques to show how MaxSAT on simplified instances is more effectively reduced to SET-COVER, leading to a more efficient algorithm for simplified MaxSAT instances.

The *Exponential Time Hypothesis* [8,13] conjectures that there is a fixed constant $c_0 > 1$ such that the MaxSAT problem cannot be solved in time $O^*(c_0^k)$. Therefore, there is a limit for improving the constant c for upper bound $O^*(c^k)$ for MaxSAT. Naturally, it will become increasingly harder to further reduce the value of c, which perhaps requires more careful and tedious analysis on more complicated instance structures. On the other hand, our algorithm does not require much more detailed structure analysis but reaches the most significant improvement, improving the base c by 0.033 over the previous best result [5], compared with the two recent improvements [5,6] that improve c by no more than 0.012.

Further improvement over our algorithm seems to require new techniques and new ideas. Our bound $O^*(1.3248^k)$ is "tight" in the sense that all our branching rules, except B-Rules 2 and 13, have their roots equal to 1.3248. Besides handling degree-4 variables more efficiently, we will need to deal with $(5, 1)$- and $(3, 2)$-literals, introducing more complicated instance structures that have not been considered in the literature, yet.

Finally, our results imply improvements on two variations of MaxSAT: (1) the upper bound for the MaxSAT ABOVE GUARANTEED VALUE problem is improved from $O^*(6.9158^{k'})$ [1] to $O^*(1.3248^{6k'}) = O^*(5.41^{k'})$, using methods in [14], where we find an assignment satisfying at least $\lceil m/2 \rceil + k'$ clauses; and (2) our upper bound $O^*(1.325^k)$ also improves the previous best exact algorithm of time $O^*(1.325^m)$ [6], because $k \leq m$.

References

1. Alber, J., Gramm, J., Niedermeier, R.: Faster exact algorithms for hard problems: a parameterized point of view. Discrete Mathematics **229**, 3–27 (2001)
2. Asano, T., Williamson, D.: Improved approximation algorithms for MAX-SAT. Journal of Algorithms **42**, 173–202 (2002)
3. Bansal, N., Raman, V.: Upper bounds for MaxSat: further improved. In: Aggarwal, A.K., Pandu Rangan, C. (eds.) ISAAC 1999. LNCS, vol. 1741, pp. 247–258. Springer, Heidelberg (1999)
4. Biere, A., Heule, M., van Maaren, H., Walsh, T.: Handbook of Satisfiability. Frontiers in Artificial Intelligence and Applications, vol. 185. IOS Press (2009)
5. Bliznets, I., Golovnev, A.: A new algorithm for parameterized MAX-SAT. In: Thilikos, D.M., Woeginger, G.J. (eds.) IPEC 2012. LNCS, vol. 7535, pp. 37–48. Springer, Heidelberg (2012)
6. Chen, J., Kanj, I.: Improved exact algorithms for Max-SAT. Discrete Applied Mathematics **142**, 17–27 (2004)
7. Chen, J., Kanj, I., Jia, W.: Vertex cover: further observations and further improvements. Journal of Algorithms **41**, 280–301 (2001)
8. Cai, L., Juedes, D.: On the existence of subexponential parameterized algorithms. Journal of Computer and System Sciences **67**, 789–807 (2003)
9. Davis, M., Putnam, H.: A computing procedure for quantification theory. Journal of the ACM **7**, 201–215 (1960)
10. Garey, M., Johnson, D.: Computers and Intractability: A Guide to the Theory of NP-Completeness. W.H.Freeman and Company, New York (1979)
11. Gu, J., Purdom, P., Wah, W.: Algorithms for the satisfiability (SAT) problem: a survey. In: Satisfiability Problem: Theory and Applications. DIMACS Series in Discrete Mathematics and Theoretical Computer Science, AMS, pp. 19–152 (1997)
12. Hochbaum, D. (ed.): Approximation Algorithms for NP-Hard problems. PWS Publishing Company, Boston (1997)
13. Impagliazzo, R., Paturi, R.: On the complexity of k-SAT. Journal of Computer and System Sciences **62**, 367–375 (2001)
14. Mahajan, M., Raman, V.: Parameterizing above guaranteed values: MaxSat and MaxCut. J. Algorithms **31**(2), 335–354 (1999)
15. Niedermeier, R., Rossmanith, P.: New upper bounds for maximum satisfiability. J. Algorithms **36**(1), 63–88 (2000)
16. Schöning, U.: Algorithmics in exponential time. In: Diekert, V., Durand, B. (eds.) STACS 2005. LNCS, vol. 3404, pp. 36–43. Springer, Heidelberg (2005)
17. The Ninth Evaluation of Max-SAT Solvers, Vienna, Austria, July 14–17, 2014
18. van Rooij, J., Bodlaender, H.: Exact algorithms for dominating set. Discrete Applied Mathematics **159**(17), 2147–2164 (2011)
19. Yannakakis, M.: On the approximation of maximum satisfiability. Journal of Algorithms **17**, 475–502 (1994)

Select with Groups of 3 or 4

Ke Chen[✉] and Adrian Dumitrescu

Department of Computer Science, University of Wisconsin-Milwaukee,
Milwaukee 53201-0784, USA
{kechen,dumitres}@uwm.edu

Abstract. We revisit the selection problem, namely that of computing the ith order statistic of n given elements, in particular the classical deterministic algorithm by grouping and partition due to Blum, Floyd, Pratt, Rivest, and Tarjan (1973). While the original algorithm uses groups of odd size at least 5 and runs in linear time, it has been perpetuated in the literature that using groups of 3 or 4 will force the worst-case running time to become superlinear, namely $\Omega(n \log n)$. We first point out that the arguments existent in the literature justifying the superlinear worst-case running time fall short of proving this claim. We further prove that it is possible to use group size 3 or 4 while maintaining the worst case linear running time. To this end we introduce two simple variants of the classical algorithm, the repeated step algorithm and the shifting target algorithm, both running in linear time.

Keywords: Median selection · ith order statistic · Comparison algorithm

1 Introduction

Together with sorting, selection is one of the most widely used procedure in computer algorithms. Indeed, it is easy to find hundreds if not thousands of algorithms (documented in at least as many research articles) that use selection as a subroutine. A classical example is [24].

Given a sequence A of n numbers (usually stored in an array), and an integer (target) parameter $1 \le i \le n$, the selection problem asks to find the ith smallest element in A. Trivially sorting solves the selection problem, but if one aims at a linear time algorithm, a higher level of sophistication is needed. A now classical approach for selection [6,14,18,27,29] from the 1970s is to use an element in A as a pivot to partition A into two smaller subsequences and recurse on one of them with a (possibly different) selection parameter i.

The time complexity of this kind of algorithms is sensitive to the pivots used. For example, if a good pivot is used, many elements in A can be discarded; while if a bad pivot is used, in the worst case, the size of the problem may be only reduced by a constant, leading to a quadratic worst-case running time. But choosing a good pivot can be time consuming.

Randomly choosing the pivots yields a well-known randomized algorithm with expected linear running time (see e.g., [7, Ch. 9.2], [22, Ch. 13.5], or [25, Ch. 3.4]), however its worst case running time is quadratic in n.

© Springer International Publishing Switzerland 2015
F. Dehne et al. (Eds.): WADS 2015, LNCS 9214, pp. 189–199, 2015.
DOI: 10.1007/978-3-319-21840-3_16

The first deterministic linear time selection algorithm SELECT (called PICK by the authors), in fact a theoretical breakthrough at the time, was introduced by Blum et al. [6]. By using the median of medians of small (constant size) disjoint groups of A, good pivots that guarantee reducing the size of the problem by a constant fraction can be chosen with low costs. The authors [6, page 451, proof of Theorem 1] required the group size to be at least 5 for the SELECT algorithm to run in linear time. It has been perpetuated in the literature the idea that SELECT with groups of 3 or 4 does not run in linear time: an exercise of the book by Cormen et al. [7, page 223, exercise 9.3-1] asks the readers to argue that "SELECT does not run in linear time if groups of 3 are used".

We first point out that the argument for the $\Omega(n \log n)$ lower bound in the solution to this exercise [8, page 23] is incomplete by failing to provide an input sequence with one third of the elements being discarded in each recursive call in both the current sequence and its sequence of medians; the difficulty in completing the argument lies in the fact that these two sequences are not disjoint thus cannot be constructed or controlled independently. The question whether the original SELECT algorithm runs in linear time with groups of 3 remains open at the time of this writing.

Further, we show that this restriction on the group size is unnecessary, namely that group sizes 3 or 4 can be used to obtain a deterministic linear time algorithm for the selection problem. Since selecting the median in smaller groups is easier to implement and requires fewer comparisons (e.g., 3 comparisons for group size 3 versus 6 comparisons for group size 5), it is attractive to have linear time selection algorithms that use smaller groups. Our main result concerning selection with small group size is summarized in the following theorem.

Theorem 1. *There exist suitable variants of* SELECT *with groups of* 3 *and* 4 *running in* $O(n)$ *time.*

Historical background. The interest in selection algorithms has remained high over the years with many exciting developments (e.g., lower bounds, parallel algorithms, etc) taking place; we only cite a few here [2,5,9,11–17,19,20,26, 28,29]. We also refer the reader to the dedicated book chapters on selection in [1,3,7,10,22,23] and the recent article [21].

Outline. In Section 2, the classical SELECT algorithm is introduced (rephrased) under standard simplifying assumptions. In Section 3, we introduce a variant of SELECT, the *repeated step* algorithm, which runs in linear time with both group size 3 and 4. With groups of 3, the algorithm executes a certain step, "group by 3 and find the medians of the groups", twice in a row. In Section 4, we introduce another variant of SELECT, the *shifting target* algorithm, a linear time selection algorithm with group size 4. In each iteration, upper or lower medians are used based on the current rank of the target, and the shift in the target parameter i is controlled over three consecutive iterations. In Section 5, we briefly introduce three other variants of SELECT with group size 4, including one due to Zwick [30], all running in linear time. We also put forward a conjecture on the running time of the original SELECT algorithm from [6] with groups of 3 and 4. In Section 6, we present experimental results comparing the running times of our algorithms (with group size 3 and 4) to the running time of the original SELECT algorithm (with group size 5).

2   Preliminaries

Without affecting the results, the following two standard simplifying assumptions are convenient: (i) the input sequence A contains n distinct numbers; and (ii) the floor and ceiling functions are omitted in the descriptions of the algorithms and their analyses. We also assume that all the grouping steps are carried out using the "natural" order, i.e., given a sequence $A = \{a_1, a_2, \ldots, a_n\}$, "arrange A into groups of size m" means that group 1 contains $a_1, a_2, \ldots, a_m$, group 2 contains $a_{m+1}, a_{m+2}, \ldots, a_{2m}$ and so on. Under these assumptions, SELECT with groups of 5 (from [6]) can be described as follows (using this group size has become increasingly popular, see e.g., [7, Ch. 9.2]):

1. If $n \leq 5$, sort A and return the ith smallest number.
2. Arrange A into groups of size 5. Let M be the sequence of medians of these $n/5$ groups. Select the median of M recursively, let it be m.
3. Partition A into two subsequences $A_1 = \{x | x < m\}$ and $A_2 = \{x | x > m\}$ (the order of elements is preserved). If $i = |A_1| + 1$, return m. If $i < |A_1| + 1$, go to step 1 with $A \leftarrow A_1$ and $n \leftarrow |A_1|$. If $i > |A_1| + 1$, go to step 1 with $A \leftarrow A_2$, $n \leftarrow |A_2|$ and $i \leftarrow i - |A_1| - 1$.

Denote the worst case running time of the recursive selection algorithm on an n-element input by $T(n)$. As shown in Figure 1, at least $(n/5)/2 * 3 = 3n/10$ elements are discarded at each iteration, which yields the recurrence

$$T(n) \leq T(n/5) + T(7n/10) + O(n).$$

Since the coefficients sum to $1/5 + 7/10 = 9/10 < 1$, the recursion solves to $T(n) = \Theta(n)$ (as it is well-known).

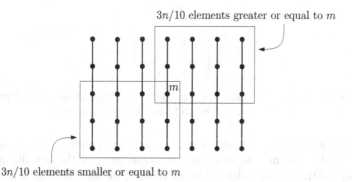

$3n/10$ elements greater or equal to m

$3n/10$ elements smaller or equal to m

Fig. 1. One iteration of the SELECT algorithm with group size 5. At least $3n/10$ elements can be discarded.

3   The Repeated Step Algorithm

Using group size 3 directly in the SELECT algorithm in [6] yields

$$T(n) \leq T(n/3) + T(2n/3) + O(n), \tag{1}$$

which solves to $T(n) = O(n \log n)$. Here a large portion (at least one third) of A is discarded in each iteration but the cost of finding such a good pivot is too high, namely $T(n/3)$. The idea of our *repeated step* algorithm, inspired by the algorithm in [4], is to find a weaker pivot in a faster manner by performing the operation "group by 3 and find the medians" twice in a row (as illustrated in Figure 2).

Algorithm

1. If $n \leq 3$, sort A and return the it smallest number.
2. Arrange A into groups of size 3. Let M be the sequence of medians of these $n/3$ groups.
3. Arrange M into groups of size 3. Let M' be the sequence of medians of these $n/9$ groups.
4. Select the median of M' recursively, let it be m.
5. Partition A into two subsequences $A_1 = \{x | x < m\}$ and $A_2 = \{x | x > m\}$. If $i = |A_1|+1$, return m. If $i < |A_1|+1$, go to step 1 with $A \leftarrow A_1$ and $n \leftarrow |A_1|$. If $i > |A_1| + 1$, go to step 1 with $A \leftarrow A_2$, $n \leftarrow |A_2|$ and $i \leftarrow i - |A_1| - 1$.

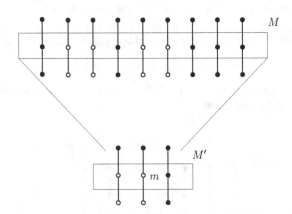

Fig. 2. One iteration of the *repeated step* algorithm with groups of 3. Empty disks represent elements that are guaranteed to be smaller or equal to m.

Analysis. Since elements are discarded if and only if they are too large or too small to be the ith smallest element, the correctness of the algorithm follows. Regarding the time complexity of this algorithm, we have the following lemma:

Lemma 1. *The repeated step algorithm with groups of 3 runs in $\Theta(n)$ time on an n-element input.*

Proof. By finding the median of medians of medians instead of the median of medians, the cost of selecting the pivot m reduces from $T(n/3) + O(n)$ to $T(n/9) + O(n)$. We need to determine how well m partitions A in the worst case. In step 4, m is guaranteed to be greater or equal to $(n/9)/2 * 2 = n/9$ elements in M. Each element in M is a median of a group of size 3 in A, so it is greater or equal to 2 elements in its group. All the groups of A are disjoint, thus m is at least greater or equal to $2n/9$ elements in A. Similarly, m is at least smaller or equal to $2n/9$ elements in A. Thus, in the last step, at least $2n/9$ elements can be discarded. The recursive call in step 4 takes $T(n/9)$ time. So the resulting recurrence is

$$T(n) \leq T(n/9) + T(7n/9) + O(n),$$

and since the coefficients on the right side sum to $8/9 < 1$, we have $T(n) = \Theta(n)$, as required.

Note that grouping by 3 twice and finding the median of medians of medians is different from grouping by 9 and finding the median of medians. The number of comparisons required for grouping by 3 twice is $3n/3 + 3n/9 = 12n/9$ while for grouping by 9 the number is $14n/9$ (14 comparisons for selecting 5th out of 9). The number of elements guaranteed to be discard is also different. For grouping by 3 twice, at least $2n/9$ elements can be discarded. For grouping by 9, this number is $5n/18$.

4   The Shifting Target Algorithm

In the SELECT algorithm introduced in [6], the group size is restricted to odd numbers in order to avoid the calculation of the average of the upper and lower median. For group size of 4, depending on the choice of upper, lower or average median, there are three possible partial orders to be considered (see Figure 3).

Fig. 3. Three partial orders of 4 elements based on the upper (left), lower (middle) and average (right) medians. The empty square represents the average of the upper and lower median which is not necessarily part of the 4-element sequence.

If the upper (or lower) median is always used, only $(n/4)/2*2 = n/4$ elements are guaranteed to be discarded in each iteration (see Figure 4) which gives the recurrence

$$T(n) \leq T(n/4) + T(3n/4) + O(n). \tag{2}$$

The term $T(n/4)$ is for the recursive call to find the median of all $n/4$ medians. This recursion solves to $T(n) = O(n \log n)$. Even if we use the average of the two medians, the recursion remains the same since only 2 elements from each of the $(n/4)/2 = n/8$ groups are guaranteed to be discarded.

Observe that if the target parameter satisfies $i \leq n/2$ (resp., $i \geq n/2$), using the lower (resp., upper) median gives a better chance to discard more elements and thus obtain a better recurrence; detailed calculations are given in the proof of Lemma 2. Inspired by this idea, we propose the *shifting target* algorithm as follows:

Algorithm

1. If $n \leq 4$, sort A and return the ith smallest number.
2. Arrange A into groups of size 4. Let M be the sequence of medians of these $n/4$ groups. If $i \leq n/2$, the lower medians are used; otherwise the upper medians are used. Select the median of M recursively, let it be m.
3. Partition A into two subsequences $A_1 = \{x | x < m\}$ and $A_2 = \{x | x > m\}$. If $i = |A_1| + 1$, return m. If $i < |A_1| + 1$, go to step 1 with $A \leftarrow A_1$ and $n \leftarrow |A_1|$. If $i > |A_1| + 1$, go to step 1 with $A \leftarrow A_2$, $n \leftarrow |A_2|$ and $i \leftarrow i - |A_1| - 1$.

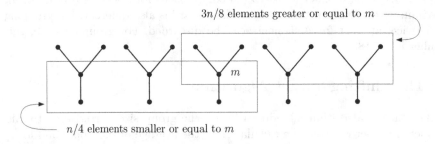

$3n/8$ elements greater or equal to m

m

$n/4$ elements smaller or equal to m

Fig. 4. Group size 4 with lower medians used

Analysis. Regarding the time complexity, we have the following lemma:

Lemma 2. *The shifting target algorithm with group size 4 runs in $\Theta(n)$ time on an n-element input.*

Proof. Assume first that $i \leq n/4$ in some iteration so the lower medians are used. Recall that m is guaranteed to be greater or equal to $(n/4)/2 * 2 = n/4$ numbers in A. So either m is the ith smallest element in A or at least $(n/4)/2 * 3 = 3n/8$ largest numbers are discarded (see Figure 4), hence the worst-case running time recurrence is

$$T(n) \leq T(n/4) + T(5n/8) + O(n). \tag{3}$$

Observe that in this case the coefficients on the right side sum to $7/8 < 1$, yielding a linear solution, as required.

Now consider the case $n/4 < i \leq n/2$, so the lower medians are used. If $|A_1| \geq i$, i.e., the rank of m is higher than i, again at least $(n/4)/2 * 3 = 3n/8$ largest numbers are discarded and (3) applies. Otherwise, suppose that only

$t = |A_1| \geq (n/4)/2 * 2 = n/4$ smallest numbers are discarded. Then in the next iteration, $i' = i - t$, $n' = n - t$.

If $i' \leq n'/4$, at least $3n'/8$ numbers are discarded. The first iteration satisfies recurrence (2) and we can use recurrence (3) to bound the term $T(3n/4)$ from above. We deduce that in two iterations the worst case running time satisfies the recurrence:

$$\begin{aligned} T(n) &\leq T(n/4) + T(3n/4) + O(n) \\ &\leq T(n/4) + T((3n/4)/4) + T((3n/4) * 5/8) + O(n) \\ &= T(n/4) + T(3n/16) + T(15n/32) + O(n). \end{aligned} \qquad (4)$$

Observe that the coefficients on the right side sum to $29/32 < 1$, yielding a linear solution, as required. Subsequently, we can therefore assume that $i' \geq n'/4$. We have

$$\begin{aligned} i'/n' &= (i - t)/(n - t) \\ &\leq (i - n/4)/(n - n/4) \\ &\leq (n/2 - n/4)/(n - n/4) \\ &= 1/3. \end{aligned}$$

Since $1/4 < i'/n' \leq 1/3 \leq 1/2$, the lower medians will be used. As described above, if at least $3n'/8$ largest numbers are discarded, in two iterations, the worst case running time satisfies the same recurrence (4).

So suppose that only $t' \geq (n'/4)/2 * 2 = n'/4$ smallest numbers are discarded. Let $i'' = i' - t'$, $n'' = n' - t'$. We have

$$\begin{aligned} i''/n'' &= (i' - t')/(n' - t') \\ &\leq (i' - n'/4)/(n' - n'/4) \\ &\leq (n'/3 - n'/4)/(n' - n'/4) \\ &= 1/9. \end{aligned}$$

Since $i''/n'' < 1/4$, in the next iteration, at least $3n''/8$ numbers will be discarded. The first two iterations satisfy recurrence (2) and we can use recurrence (3) to bound the term $T(9n/16)$ from above. We deduce that in three iterations the worst case running time satisfies the recurrence:

$$\begin{aligned} T(n) &\leq T(n/4) + T(3n/4) + O(n) \\ &\leq T(n/4) + T((3n/4)/4) + T((3n/4) * 3/4) + O(n) \\ &= T(n/4) + T(3n/16) + T(9n/16) + O(n) \\ &\leq T(n/4) + T(3n/16) + T((9n/16)/4) + T((9n/16) * 5/8) + O(n) \\ &= T(n/4) + T(3n/16) + T(9n/64) + T(45n/128) + O(n). \end{aligned}$$

The sum of the coefficients on the right side is $119/128 < 1$, so again the solution is $T(n) = \Theta(n)$.

By symmetry, the analysis also holds for the case $i \geq n/2$, and the proof of Lemma 2 is complete.

5   Other Variants

A similar idea of repeating the group step (from Section 3) also applies to the case of groups of 4 and yields

$$T(n) \le T(n/16) + T(7n/8) + O(n),$$

and thereby another linear time selection algorithm with group size 4.

Yet another variant of SELECT with group size 4 (we refer to it as the hybrid algorithm), can be obtained by using the ideas of both algorithms together, i.e., repeat the grouping by 4 step twice in a row while M contains the lower medians and M' contains the upper medians (or vice versa). Recursively selecting the median m of M' takes time $T(n/16)$. Notice that m is greater or equal to at least $(n/16)/2 * 3 = 3n/32$ elements in M of which each is greater or equal to 2 elements in its group in A. So m is greater or equal to at least $3n/16$ elements of A. Also, m is smaller or equal to at least $(n/16)/2 * 2 = n/16$ elements in M of which each is smaller or equal to 3 elements in its group of A. So m is smaller or equal to at least $3n/16$ elements of A, thus the resulting recurrence is

$$T(n) \le T(n/16) + T(13n/16) + O(n),$$

again with a linear solution, as desired.

Zwick's variant. The fact that the SELECT algorithm can be modified so that it works with groups of 4 in linear time was observed prior to this writing. The following variant, from 2010, is due to Zwick [30]. Split the elements in A into quartets. Find the 2nd smallest element of each quartet (i.e., the lower median), and let M be this subset of $n/4$ elements. Recursively find the $(3/5)(n/4)$th smallest element m of M. Now $(3/5)(n/4)$ groups of A have 2 elements smaller or equal to m, so m is greater or equal to at least $2(3/5)(n/4) = 3n/10$ elements in A. Similarly, $(2/5)(n/4)$ groups of A have 3 elements greater or equal to m, so m is smaller or equal to at least $3(2/5)(n/4) = 3n/10$ elements in A. Thus, the remaining recursive call involves at most $7n/10$ elements, and the resulting recurrence is

$$T(n) \le T(n/4) + T(7n/10) + O(n).$$

Since $1/4 + 7/10 < 1$, the solution is linear.

Comment. The question whether the original selection algorithm introduced in [6] (outlined in Section 2) runs in linear time with group size 3 and 4 remains open. Although the recurrences

$$T(n) \le T(n/3) + T(2n/3) + O(n), \text{ and}$$
$$T(n) \le T(n/4) + T(3n/4) + O(n)$$

(see (1) and (2)) for its worst-case running time with these group sizes both solve to $T(n) = O(n \log n)$, we believe that they only give non-tight upper bounds on the worst case scenarios. In any case, and against popular belief we think that $\Theta(n \log n)$ is *not* the answer:

Conjecture 1. The SELECT algorithm introduced by Blum et al. [6] runs in $o(n \log n)$ time with groups of 3 or 4.

6 Experimental Results

To compare our algorithms with the original SELECT algorithm, we first derive upper bounds on the exact numbers of comparisons for each variant in the same manner as in Section 2 of [6]. Sharper upper bounds are possible by taking extra care in avoiding comparisons with known outcomes against the pivot; however, for simplicity of implementation we opted to forego this saving. In order to avoid the overhead of repeated array copying, all the five algorithms were implemented in-place, in the sense that, with the exception of the recursion, only $O(1)$ extra space is used in addition to the input array. This requires minor modifications of the algorithms; however, their running time analyses remain unchanged.

Let now $T(n)$ denote the total number of comparisons performed. For the original SELECT algorithm with group size 5, we have

$$T(n) \leq T(n/5) + T(7n/10) + 6n/5 + n,$$

in which $6n/5$ is for computing the $n/5$ medians (recall that each takes at most 6 comparisons) and n is for partitioning the sequence using the selected pivot. Solving the recurrence yields $T(n) \leq 22n$. Similarly, for the repeated step algorithm, we have

$$T(n) \leq T(n/9) + T(7n/9) + 3n/3 + 3n/9 + n,$$

and consequently, $T(n) \leq 21n$. For the hybrid algorithm, we have

$$T(n) \leq T(n/16) + T(13n/16) + 4n/4 + 4n/16 + n,$$

and consequently, $T(n) \leq 18n$. For Zwick's algorithm, we have

$$T(n) \leq T(n/4) + T(7n/10) + 4n/4 + n,$$

and consequently, $T(n) \leq 40n$. For the shifting target algorithm, the analysis is more involved; it yields $T(n) \leq 66n$.

We carried out 1000 experiments[1] on selecting medians in arrays of 10 million randomly permuted distinct integers. The results are summarized in the following table:

Algorithm	Number of Comparisons	Average Running Time
Hybrid algorithm	$\leq 18n$	434.4ms
Repeated step algorithm	$\leq 21n$	442.8ms
Original algorithm	$\leq 22n$	523.7ms
Zwick's algorithm	$\leq 40n$	620.7ms
Shifting target algorithm	$\leq 66n$	619.5ms

---

[1] The experiments were performed on a laptop with 64bits operating system, 4GB memory and Intel® Core™ i5-2410M 2.3GHz processor.

The C code used can be downloaded at https://pantherfile.uwm.edu/kechen/ linear_selection_small_group/small_group_experiment/src/.

We observed that the experimental results agree with the worst-case estimates in the number of comparisons, i.e., showing roughly the same speed ranking. Note also that the optimizations introduced in Section 3 of [6] are applicable to reduce the constant factors computed here. However, as the authors of [6] stated, "The optimized algorithm is full of red tape, and could not in practice be implemented efficiently,...".

References

1. Aho, A.V., Hopcroft, J.E., Ullman, J.D.: Data Structures and Algorithms. Addison-Wesley, Reading (1983)
2. Ajtai, M., Komlós, J., Steiger, W.L., Szemerédi, E.: Optimal parallel selection has complexity $O(\log \log n)$. Journal of Computer and System Sciences **38**(1), 125–133 (1989)
3. Baase, S.: Computer Algorithms: Introduction to Design and Analysis, 2nd edn. Addison-Wesley, Reading (1988)
4. Battiato, S., Cantone, D., Catalano, D., Cincotti, G., Hofri, M.: An efficient algorithm for the approximate median selection problem. In: Bongiovanni, G., Petreschi, R., Gambosi, G. (eds.) CIAC 2000. LNCS, vol. 1767, p. 226. Springer, Heidelberg (2000)
5. Bent, S.W., John, J.W.: Finding the median requires $2n$ comparisons. In: Proceedings of the 17th Annual ACM Symposium on Theory of Computing (STOC 1985), pp. 213–216. ACM (1985)
6. Blum, M., Floyd, R.W., Pratt, V., Rivest, R.L., Tarjan, R.E.: Time bounds for selection. Journal of Computer and System Sciences **7**(4), 448–461 (1973)
7. Cormen, T.H., Leiserson, C.E., Rivest, R.L., Stein, C.: Introduction to Algorithms, 3rd edn. MIT Press, Cambridge (2009)
8. Cormen, T.H., Lee, C., Lin, E.: Instructor's Manual, to accompany Introduction to Algorithms, 3rd edn. MIT Press, Cambridge (2009)
9. Cunto, W., Munro, J.I.: Average case selection. Journal of ACM **36**(2), 270–279 (1989)
10. Dasgupta, S., Papadimitriou, C., Vazirani, U.: Algorithms. Mc Graw Hill, New York (2008)
11. Dor, D., Håstad, J., Ulfberg, S., Zwick, U.: On lower bounds for selecting the median. SIAM Journal on Discrete Mathematics **14**(3), 299–311 (2001)
12. Dor, D., Zwick, U.: Finding the αnth largest element. Combinatorica **16**(1), 41–58 (1996)
13. Dor, D., Zwick, U.: Selecting the median. SIAM Journal on Computing **28**(5), 1722–1758 (1999)
14. Floyd, R.W., Rivest, R.L.: Expected time bounds for selection. Communications of ACM **18**(3), 165–172 (1975)
15. Fussenegger, F., Gabow, H.N.: A counting approach to lower bounds for selection problems. Journal of ACM **26**(2), 227–238 (1979)
16. Hadian, A., Sobel, M.: Selecting the t-th largest using binary errorless comparisons. Combinatorial Theory and Its Applications **4**, 585–599 (1969)
17. Hoare, C.A.R.: Algorithm 63 (PARTITION) and algorithm 65 (FIND). Communications of the ACM **4**(7), 321–322 (1961)

18. Hyafil, L.: Bounds for selection. SIAM Journal on Computing 5(1), 109–114 (1976)
19. John, J.W.: A new lower bound for the set-partitioning problem. SIAM Journal on Computing 17(4), 640–647 (1988)
20. Kirkpatrick, D.G.: A unified lower bound for selection and set partitioning problems. Journal of ACM 28(1), 150–165 (1981)
21. Kirkpatrick, D.: Closing a long-standing complexity gap for selection: $V_3(42) = 50$. In: Brodnik, A., López-Ortiz, A., Raman, V., Viola, A. (eds.) Ianfest-66. LNCS, vol. 8066, pp. 61–76. Springer, Heidelberg (2013)
22. Kleinberg, J., Tardos, É.: Algorithm Design. Pearson & Addison-Wesley, Boston (2006)
23. Knuth, D.E.: The Art of Computer Programming. Sorting and Searching, vol. 3, 2nd edn. Addison-Wesley, Reading (1998)
24. Megiddo, N.: Partitioning with two lines in the plane. Journal of Algorithms 6(3), 430–433 (1985)
25. Mitzenmacher, M., Upfal, E.: Probability and Computing: Randomized Algorithms and Probabilistic Analysis. Cambridge University Press (2005)
26. Paterson, M.: Progress in selection. In: Karlsson, R., Lingas, A. (eds.) SWAT 1996. LNCS, vol. 1097, pp. 368–379. Springer, Heidelberg (1996)
27. Schönhage, A., Paterson, M., Pippenger, N.: Finding the median. Journal of Computer and System Sciences 13(2), 184–199 (1976)
28. Yao, A., Yao, F.: On the average-case complexity of selecting the kth best. SIAM Journal on Computing 11(3), 428–447 (1982)
29. Yap, C.K.: New upper bounds for selection. Communications of the ACM 19(9), 501–508 (1976)
30. Zwick, U.: Personal communication, September 2014

Approximating Nearest Neighbor Distances

Michael B. Cohen[1], Brittany Terese Fasy[2], Gary L. Miller[3], Amir Nayyeri[4], Donald R. Sheehy[5][✉], and Ameya Velingker[3]

[1] Massachusetts Institute of Technology, Cambridge, USA
[2] Montana State University, Bozeman, USA
[3] Carnegie Mellon University, Pittsburgh, USA
[4] Oregon State University, Corvallis, USA
[5] University of Connecticut, Mansfield, USA
don.r.sheehy@gmail.com

Abstract. Several researchers proposed using non-Euclidean metrics on point sets in Euclidean space for clustering noisy data. Almost always, a distance function is desired that recognizes the closeness of the points in the same cluster, even if the Euclidean cluster diameter is large. Therefore, it is preferred to assign smaller costs to the paths that stay close to the input points.

In this paper, we consider a natural metric with this property, which we call the nearest neighbor metric. Given a point set P and a path γ, this metric is the integral of the distance to P along γ. We describe a $(3+\varepsilon)$-approximation algorithm and a more intricate $(1 + \varepsilon)$-approximation algorithm to compute the nearest neighbor metric. Both approximation algorithms work in near-linear time. The former uses shortest paths on a sparse graph defined over the input points. The latter uses a sparse sample of the ambient space, to find good approximate geodesic paths.

1 Introduction

Many problems lie at the interface of computational geometry, machine learning, and data analysis, including, but not limited to: clustering, manifold learning, geometric inference, and nonlinear dimensionality reduction. Although the input to these problems is often a Euclidean point cloud, a different distance measure may be more *intrinsic* to the data, other than the metric inherited from the Euclidean space. In particular, we are interested in a distance that recognizes the closeness of two points in the same cluster, even if their Euclidean distance is large, and, conversely, recognizes a large distance between points in different clusters, even if the Euclidean distance is small. For example, in Figure 1, the distance between a and b must be larger than the distance between b and c.

There are at least two seemingly different approaches to define a non-Euclidean metric on a finite set of points in $\mathbb{R}^d$. The first approach is to form a graph metric on the point set. An example of a potential graph is the kth nearest neighbor graph, where an edge between two points exists if and only if they are

---

Partially supported by the NSF grant CCF-1065106.

F. Dehne et al. (Eds.): WADS 2015, LNCS 9214, pp. 200–211, 2015.
DOI: 10.1007/978-3-319-21840-3_17

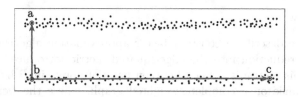

Fig. 1. The intrinsic density-based distance should recognize two points within the same cluster as cloesr than two points in different clusters, regardless of the actual Euclidean distance

both in the k nearest neighbor set of the other. In this graph, the edge weights may be a constant or the Euclidean distances. In this paper, we consider the complete graph, where the edge lengths are a power of their Euclidean lengths. We are particularly interested in the squared length, which we refer to as the edge-squared metric.

The second approach is to endow all of $\mathbb{R}^d$ with a new metric. We start with a cost function $c : \mathbb{R}^d \to \mathbb{R}$ that takes the point cloud into account. Then, the length of a path $\gamma : [0, 1] \to \mathbb{R}^d$ is the integral of the cost function along the path.

$$\ell_c(\gamma) = \int_\gamma c(s)ds = \int_0^1 c(\gamma(t)) \left| \frac{d\gamma}{dt}(t) \right| dt. \tag{1}$$

The distance between two points $x, y \in \mathbb{R}^d$ is then the length of the shortest path between them:

$$\mathbf{d}_c(x, y) = \inf_\gamma \ell_c(\gamma), \tag{2}$$

where the infimum is over paths that start at x and end at y. Note that the constant function, $c(x) = 1$ for all $x \in \mathbb{R}^d$, gives the Euclidean metric; whereas, other functions allow space to be stretched in various ways.

In order to reinforce paths within clusters, one would like to assign smaller lengths to paths that stay close to the point cloud. Therefore, the simplest natural cost function on $\mathbb{R}^d$ is the distance to the point cloud. More precisely, given a finite point set P the cost $c(x)$ for $x \in \mathbb{R}^d$ is chosen to be $\mathrm{N}(x)$, the Euclidean distance from x to $\mathrm{NN}(x)$, where $\mathrm{NN}(x)$ denotes the nearest point to x in P. The *nearest neighbor length* (N-length) $\ell_N(\gamma)$ of a curve is given by (1), where we set $c(x) = \mathrm{N}(x)$ for all points $x \in C$. We refer to the corresponding metric given by (2) as the *nearest neighbor metric* or simply the N-distance.

In this paper, we investigate approximation algorithms for N-distance computation. We describe a $(3 + \varepsilon)$-approximation algorithm and a $(1 + \varepsilon)$-approximation algorithm. The former comes from comparing the nearest neighbor metric with the edge-squared metric. The latter is a tighter approximation that samples the ambient space to find good approximate geodesics.

1.1 Overview

In Section 3, we describe a constant factor approximation algorithm obtained via an elegant reduction into the edge-squared metric introduced by [BRS11] and [VB03]. This metric is defined between pairs of points in P by considering the graph distance on a complete weighted graph, where the weight of each edge is the square of its Euclidean length. We show that the N-distance and edge-squared metric are equivalent up to a factor of three (after a scaling by a factor of four). As a result, because spanners for the edge-squared metric can be computed in nearly linear time [LSV06], we obtain a $(3 + \varepsilon)$-approximation algorithm for computing N-distance.

Theorem 1. *Let P be a set of points in $\mathbb{R}^d$, and let $x, y \in P$. The nearest neighbor distance between x and y can be approximated within a $(3 + \varepsilon)$ factor in $O(n \log n + n \varepsilon^{-d})$ time, for any $0 < \varepsilon \leq 1$.*

In Section 4, we describe a $(1 + \varepsilon)$-approximation algorithm for the N-distance that works in time $\varepsilon^{-O(d)} n \log n$. Our algorithm computes a discretization of the space for points that are sufficiently far from P. Nevertheless, the sub-paths that are close to P are computed exactly. We can adapt our algorithm to work for any Lipschitz cost function that is bounded away from zero; thus, the algorithm can be applied to many different scenarios.

Theorem 2. *For any finite set of points $P \subset \mathbb{R}^d$ and any fixed number $0 < \varepsilon < 1$, the shortest N-distance between any pair of points of the space can be $(1 + \varepsilon)$-approximated in time $O(\varepsilon^{-O(d)} n \log n)$.*

1.2 Related Work

Computing the distance between a pair of points with respect to a cost function encompasses several significant problems that have been considered by different research communities for at least a few centuries. As early as 1696, Johann Bernoulli introduced the *brachistochrone* curve, the shortest path in the presence of gravity, as "an honest, challenging problem, whose possible solution will bestow fame and remain as a lasting monument" [Ber96]. With six solutions to his problem published just one year after it was posed, this event marked the birth of the field of *calculus of variations*.

Rowe and Ross [RR90] as well as Kime and Hespanha [KH03] consider the problem of computing anisotropic shortest paths on a terrain. An anisotropic path cost takes into account the (possibly weighted) length of the path and the direction of travel. Note that this problem can be translated into the problem of computing a shortest path between two compact subspaces of $\mathbb{R}^6$ under a certain cost function

When c is a piecewise constant function, the problem is known as the weighted region problem [MP91]. Mitchell and Papadimitriou [MP91] gave a linear-time approximation scheme in the plane and list the problem for more general cost functions as an open problem (See Section 10, problem number

(3)). Further work on this problem has led to fast approximations for shortest paths on terrains [AMS05].

Similar distances have been used in semi-supervised machine learning under the name density-based distance (DBDs) [SO05]. The goal here is to place points that can be connected through dense regions in the same cluster. Several approaches [VB03, BCH04] have been suggested that first estimate the density and then discretize space in a similar manner to that of Tsitsiklis [Tsi95], however, they do not provide any analysis on the complexity of the discretized space. Another approach is to search for shortest paths among a sample [BRS11] and this approach was shown to give good approximations to sufficiently long paths [HDI14]. The nearest neighbor metric can be viewed as a special case of density-based distance when the underlying density is the nearest neighbor density estimator.

2 Preliminaries

2.1 Metrics

In this paper, we consider three metrics. Each metric is defined by a length function on a set of paths between two points of the space. The distance between two points is the length of the shortest path between them.

Euclidean metric. This is the most natural metric defined by the Euclidean length. We use $\ell(\gamma)$ to denote the Euclidean length of a curve γ; $\ell(\gamma)$ can also be defined by setting $c(x) = 1$ for all $x \in \mathbb{R}^d$ in (1). We use $\mathbf{d}(x, y)$ to denote the distance between two points $x, y \in \mathbb{R}^d$ based on the Euclidean metric.

Nearest neighbor metric. As mentioned above, the nearest neighbor length of a curve with respect to a set of points P, is defined by setting $c(\cdot)$ to be $\mathrm{N}(\cdot)$ in (1). The nearest neighbor length of a curve γ is denoted by $\ell_\mathrm{N}(\gamma)$, and the distance between two points $x, y \in \mathbb{R}^d$ with respect to the nearest neighbor metric is denoted by $\mathbf{d}_\mathrm{N}(x, y)$.

Edge-squared metric. Finally, the edge-squared metric is defined as the shortest path metric on a complete graph on a point set P, where the length of each edge is its Euclidean length squared. The length of a path γ in this graph is naturally the total length of its edges and it is denoted by $\ell_\mathrm{sq}(\gamma)$. The edge-squared distance between two points $x, y \in P$ is the length of the shortest path and is denoted by $\mathbf{d}_\mathrm{sq}(x, y)$.

2.2 Voronoi Diagrams and Delaunay Triangulations

Let P be a finite set of points, called *sites*, in $\mathbb{R}^d$, for some $d \geq 1$. The Delaunay triangulation $\mathrm{Del}(P)$ is a decomposition of the convex closure of P into simplices such that for each simplex $\sigma \in \mathrm{Del}(P)$, the Delaunay empty circle property is satisfied; that is, there exists a sphere C such that the vertices of σ are on the boundary of C and $\text{int}(C) \cap P$ is empty. The Voronoi diagram, denoted $\text{Vor}(P)$, is the dual to $\text{Del}(P)$. We define the in-ball of a Voronoi cell with site p to be the maximal ball centered at p that is contained in the cell. The inradius of a Voronoi cell is the radius of its in-ball. We refer the reader to [DBVKOS00] for more details.

3 N-Distance Versus Edge-Squared Distance

In this section, we show that the nearest neighbor distance of two points $x, y \in P$ can be approximated within a factor of three by looking at their edge-squared distance. More precisely, $\mathbf{d}_{\text{sq}}(x, y)/4 \geq \mathbf{d}_{\text{N}}(x, y) \geq \mathbf{d}_{\text{sq}}(x, y)/12$ (see Lemma 1 and Lemma 3).

As a consequence, a constant factor approximation of the N-distance can be obtained via computing shortest paths on a weighted graph, in nearly-quadratic time. This approximation algorithm becomes more efficient, if the shortest paths are computed on a Euclidean spanner of the points, which is computable in nearly linear time [Hp11]. A result of Lukovszki et al. (Theorem 16(ii) of [LSV06]) confirms that a $(1 + \varepsilon)$-Euclidean spanner is a $(1 + \varepsilon)^2$-spanner for the edge squared metric. Therefore, we obtain Theorem 1.

3.1   The Upper Bound

We show that the edge-squared distance between any pair of points $x, y \in P$ (with respect to the point set P) is always larger than four times the N-distance between x and y (with respect to P). To this end, we consider any shortest path with respect to the edge-squared measure and observe that its N-length is an upper bound on the N-distance between its endpoints.

Lemma 1. *Let $P = \{p_1, p_2, \ldots, p_n\}$ be a set of points in $\mathbb{R}^d$, and let $\mathbf{d}_{\text{N}}$ and $\mathbf{d}_{\text{sq}}$ be the associated nearest neighbor and edge-squared distances, respectively. Then, for any distinct points $x, y \in P$, we have that $\mathbf{d}_{\text{N}}(x, y) \leq \frac{1}{4}\mathbf{d}_{\text{sq}}(x, y)$.*

3.2   The Lower Bound

Next, we show that the edge-squared distance between any pair of points from P cannot be larger than twelve times their N-distance. To this end, we break a shortest path of the N-distance into segments in a certain manner, and shadow the endpoints of each segment into their closest point of P to obtain a short edge-squared path. The following definition formalizes our method of discretizing paths.

Definition 1. *Let $P = \{p_1, p_2, \cdots, p_n\}$ be a set of points in $\mathbb{R}^d$, and let $x, y \in P$. Let $\gamma : [0, 1] \to \mathbb{R}^d$ be an (x, y)-path that is internally disjoint from P. A sequence $0 < t_0 \leq t_1 \leq \cdots \leq t_k < 1$ is a proper breaking sequence of γ if it has the following properties:*

1. *The nearest neighbors of* $\gamma(t_0)$ *and* $\gamma(t_k)$ *in* P *are* x *and* y, *respectively.*
2. *For all* $1 \leq i \leq k$, *we have* $\ell(\gamma[t_{i-1}, t_i]) = \frac{1}{2}(N(\gamma(t_{i-1})) + N(\gamma(t_i)))$

The following lemma guarantees the existence of breaking sequences.

Lemma 2. *Let* $P = \{p_1, p_2, \cdots, p_n\}$ *be a set of points in* $\mathbb{R}^d$, *and let* $x, y \in P$. *Let* γ *be a path from* x *to* y *that is internally disjoint from* P. *There exists a proper breaking sequence of* γ.

Given a path γ that realizes the nearest neighbor distance between two points x and y, in the proof of the following lemma we show how to obtain another (x, y)-path with bounded edge-squared length. The proof heavily relies on the idea of breaking sequences.

Lemma 3. *Let* $P = \{p_1, p_2, \cdots, p_n\}$ *be a set of points in* $\mathbb{R}^d$, *and let* $\mathbf{d}_N$ *and* $\mathbf{d}_{sq}$ *be the associated nearest neighbor and edge-squared distances, respectively. Then, for any distinct points* $x, y \in P$, $\mathbf{d}_N(x, y) \geq \frac{1}{12}\mathbf{d}_{sq}(x, y)$.

4   A $(1 + \varepsilon)$-Approximation of the N-Metric

In this section, we describe a polynomial time approximation scheme to compute the N-distance between a pair of points from a finite set $P \subset \mathbb{R}^d$. The running time of our algorithm is $\varepsilon^{-O(d)} n \log n$ for n points in d-dimensional space. We start with Section 4.1, which describes an exact algorithm for the simple case in which P consists of just one site. Section 4.2 describes how to obtain a piecewise linear path using infinitely many Steiner points, the technical details of which may be found in the full version [CFM+15]. Section 4.3 combines ideas from 4.2 and 4.1 to cut down the required Steiner points to a finite number. Finally, Section 4.4 describes how to generate the necessary Steiner points.

4.1   Nearest Neighbor Distance with One Site

We describe a method for computing $\mathbf{d}_N$ for the special case that P is a single point using complex analysis. This case will be important since distances will go to zero at an input point and thus we must be more careful at input points. Far from input points, we use a piecewise constant approximation for the nearest neighbor function, and near input points, we use exact distances. More than likely this case has been solved by others since the solution is so elegant. We refer the interested reader to [Str] for more general methods to solve similar problems in the field of calculus of variations.

Suppose we want to compute $\mathbf{d}_N(x, y)$ where $P = \{(0, 0)\}$. Writing $(x, y) \in \mathbb{C}$ in polar coordinates as $z = re^{i\theta}$, we define the *quadratic transformation* $f \colon \mathbb{C} \to \mathcal{R}$ by

$$f(z) = z^2/2 = (r^2/2)e^{i2\theta},$$

where $\mathcal{R}$ is the two-fold Riemann surface; see Figure 2. The important point here is that the image is a double covering of $\mathbb{C}$. For example, the points 1 and -1

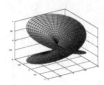

Fig. 2. To make the complex function one-to-one, one needs to extend the complex plane to the two-fold cover called the two-fold Riemann Surface

are mapped to different copies of 1/2. Therefore, on the Riemann surface, the distance between 1 and −1 is one and the shortest path goes through the origin. More generally, given any two nonzero points p and q on the surface, the minimum angle between them (measured with respect to the origin) will be between 0 and 2π. Moreover, if this angle is $\geq \pi$, then the shortest path between them will consist of the two straight lines $[p, 0]$ and $[0, q]$. Otherwise, the line $[p, q]$ will be a line on the surface and, thus, the geodesic from p to q.

Let $\mathbf{d}_{\mathcal{R}}$ denote the distance on the Riemann surface. We next show that for a single point, the nearest neighbor geodesic is identical to the geodesic on the Riemann surface.

Lemma 4. *Let $\gamma\colon [0, 1] \to \mathbb{C}$ be a curve. Then, the image of γ under f, denoted by $f \circ \gamma$ satisfies the following property:*

$$\mathbf{d}_{\mathcal{R}}(f \circ \gamma) = \ell_{\mathrm{N}}(\gamma).$$

Proof. Suppose $\gamma\colon [0, 1] \to \mathbb{C}$ is any piecewise differentiable curve, and let $\alpha := f \circ \gamma$. The N-length $\ell_{\mathrm{N}}(\gamma)$ of γ is the finite sum of the N-length of all differentiable pieces of γ. If the path γ goes through the origin, we further break the path at the origin so that α is also differentiable. Thus, it suffices to consider $(a, b) \subset [0, 1]$ so that $\gamma[a, b]$ is a differentiable piece of γ. Then, we have

$$\ell_{\mathrm{N}}(\gamma[a, b]) = \int_a^b |\gamma(t)||\gamma'(t)|\, dt \qquad |\cdot| \text{ is modulus.}$$

$$= \int_a^b |\gamma(t)\gamma'(t)|\, dt \qquad \text{Modulus commutes with product.}$$

$$= \int_a^b |\alpha'(t)|\, dt \qquad \text{Chain rule.}$$

$$= \ell_{\mathcal{R}}(\alpha[f(a), f(b)]).$$

Corollary 1 (Reduction to Euclidean Distances on a Riemann Surface). *Given three points x, y, and p in $\mathbb{R}^d$ such that $p = \mathrm{NN}(x) = \mathrm{NN}(y)$, the nearest neighbor geodesic G from x to y satisfies the following properties:*

1. *G is in the plane determined by x, y, p.*
2. *(a) If the angle formed by x, p, y is $\pi/2$ or more, then G consists of the two straight segments $\overline{xp}$ and $\overline{py}$.*

(b) *Otherwise, G is the preimage of the straight line from $f(x)$ to $f(y)$, where f is the quadratic map in the plane given by x, y, p to the Riemann Surface.*

4.2   Approximating with Steiner Points

Assume $P \subset \mathbb{R}^d$, $x, y \in P$, and let γ be an arbitrary (x, y)-path. We show how to approximate γ with a piecewise linear path through a collection of Steiner points in $\mathbb{R}^d$. To obtain an accurate estimation of γ, we require the Steiner points to be sufficiently dense. The following definition formalizes this density with a parameter δ.

Definition 2 (δ-sample). *Let $P = \{p_1, p_2, \cdots, p_n\}$ be a set of points in $\mathbb{R}^d$, and let $D \subseteq \mathbb{R}^d$. For a real number $0 < \delta < 1$, a δ-sample is a (possibly infinite) set of points $T \subseteq D$ such that if $z \in D \setminus P$, then $\mathbf{d}(z, T) \leq \delta \cdot \mathrm{N}(z)$.*

The following lemma guarantees that an accurate estimation of γ can be computed using a δ-sample. Its proof may be found in the full paper [CFM+15].

Lemma 5. *Let $P = \{p_1, p_2, \cdots, p_n\}$ be a set of points in $\mathbb{R}^d$, and let S be a δ-sample, and let $0 < \delta < 1/10$. Then, for any pair of points $x, y \in P$, there is a piecewise linear path $\eta = (x, s_1, \ldots, s_k, y)$, where $s_1, \ldots, s_k \in S$, such that:*

$$\ell_{\mathrm{N}}(\eta) \leq (1 + C_1 \delta^{2/3}) \mathbf{d}_{\mathrm{N}}(x, y),$$

and, for all $1 \leq i \leq k - 1$,

$$\ell_{\mathrm{N}}((s_i, s_{i+1})) \leq C_2 \cdot \delta^{2/3} \cdot \mathrm{N}(s_i).$$

C_1 *and* C_2 *are universal constants.*

4.3   The Approximation Graph

So far we have shown that any shortest path can be approximated using a δ-sample that is composed of infinitely many points. In addition, we know how to compute the exact N-distance between any pair of points if they reside in the same Voronoi cell of $\mathrm{Vor}(P)$. Here, we combine these two ideas to be able to approximate any shortest path using only a finite number of Steiner points. The high-level idea is to use the Steiner point approximation while γ passes through regions that are far from P and switch to the exact distance computation as soon as γ is sufficiently close to one of the points in P.

Let $P = \{p_1, p_2, \cdots, p_n\}$ be a set of points in $\mathbb{R}^d$, and let B be any convex body that contains P. Fix $\delta \in (0, 1)$, and for any $1 \leq i \leq n$, let $r_i = r_P(p_i)$ be the inradius of the Voronoi cell with site p_i. Also, let $u_i = (1 - \delta^{2/3}) r_i$. Finally, let S be a δ-sample on the domain $B \setminus \bigcup_{1 \leq i \leq n} B(p_i, u_i)$.

Definition 3 (Approximation Graph). *The approximation graph* $\mathcal{A} = \mathcal{A}(P, \{u_1, \ldots, u_n\}, S, \delta) = (V_{\mathcal{A}}, E_{\mathcal{A}})$ *is a weighted undirected graph, with weight function* $w : E_{\mathcal{A}} \rightarrow \mathbb{R}^+$. *The vertices in* $V_{\mathcal{A}}$ *are in one to one correspondence with the points in* $S \cup P$; *for simplicity we use the same notation to refer to corresponding elements in* $S \cup P$ *and* $V_{\mathcal{A}}$. *The set* $E_{\mathcal{A}}$ *is composed of three types of edges:*

1. *If* $s_1, s_2 \in S$ *and* $s_1, s_2 \in B(p_i, r_i)$ *for any* p_i, *then* $(s_1, s_2) \in E_{\mathcal{A}}$ *and* $w(s_1, s_2) = \mathbf{d}_N(s_1, s_2)$. *We compute this distance using Corollary 1.*
2. *Otherwise, if* $s_1, s_2 \in S$ *and* $\ell(s_1, s_2) \leq C_2 \delta^{2/3} \max(N(s_1), N(s_2))$, *where* C_2 *is the constant of Lemma 5, then* $(s_1, s_2) \in E_{\mathcal{A}}$ *and* $w(s_1, s_2) = \max(N(s_1), N(s_2)) \cdot \ell(s_1, s_2)$.
3. *If* $s_1 \in S$ *and* $s_1 \in B(p_i, r_i)$ *then* $(p_i, s_1) \in E_{\mathcal{A}}$ *and* $w(p_i, s_1) = \mathbf{d}_N(p_i, s_1) = (\mathbf{d}(p_i, s_1))^2/2$; *see Corollary 1.*

For $x, y \in V_{\mathcal{A}}$ *let* $\mathbf{d}_{\mathcal{A}}(x, y)$ *denote the length of the shortest path from* x *to* y *in the graph* $\mathcal{A}$.

The following lemma guarantees that the shortest paths in the approximation graph are sufficiently accurate estimations. Its proof my be found in the full paper [CFM+15].

Lemma 6. *Let* $\{u_1, \ldots, u_n\}$, S *and* δ *be defined as above. Let* $\mathcal{A}(P, \{u_1, \ldots, u_n\}, S, \delta)$ *be the approximation graph for* P. *For any pair of points* $x, y \in P$ *we have:*

$$(1 - C_2 \delta^{2/3}) \cdot \mathbf{d}_N(x, y) \leq \mathbf{d}_{\mathcal{A}}(x, y) \leq (1 + C_4 \delta^{2/3}) \cdot \mathbf{d}_N(x, y),$$

where C_2 *and* C_4 *are constants computable in* $O(1)$ *time.*

4.4   Construction of Steiner Points

The only remaining piece that we need to obtain an approximation scheme is an algorithm for computing a δ-sample. For this section, given a point set T and $x \in T$, let $r_T(x)$ denote the inradius of the Voronoi cell of $\text{Vor}(T)$ that contains x. Also, given a set T and an arbitrary point x (not necessarily in T), let $\mathbf{f}_T(x)$ denote the distance from x to its *second* nearest neighbor in T.

We can apply existing algorithms for generating meshes and well-spaced points to compute a δ-sample on $\mathcal{D} \setminus \bigcup_i B(p_i, u_i)$, where $\mathcal{D} \subseteq \mathbb{R}^d$ is a domain, and $u_i = (1 - \delta^{2/3}) r_P(p_i)$. The procedure consists of two steps:

1. Use the algorithm of [MSV13] to construct a well-spaced point set M (along with its associated approximate Delaunay graph) with aspect ratio τ in time $2^{O(d)}(n \log n + |M|)$.
2. Then over-refine M to S for the sizing function $g(x) = \frac{2\delta}{11\tau} \mathbf{f}_P(x)$ (while maintaining aspect ratio τ) in time $2^{O(d)}|S|$ by using the algorithm of Section 3.7 in [She11]. (see also [HOMS10] for an earlier use of this technique)

In the above algorithm, we will choose τ to be a fixed constant, say, $\tau = 6$. Both of the meshing algorithms listed above are chosen for their theoretical guarantees on running time. In practice, one could use any quality Delaunay meshing algorithm, popular choices include Triangle [She96] in $\mathbb{R}^2$ and Tetgen [Si11] or CGAL [ART+12] in $\mathbb{R}^3$.

From the guarantees in ([She11]), we know that

$$|S| = O\left(\int_{\mathcal{D}} \frac{dx}{g(x)^d}\right) = \delta^{-O(d)} n \log \Delta, \tag{3}$$

where Δ is the *spread* of P, i.e., the ratio of the largest distance between two points in P to the smallest distance between two points in P.

Now, it remains to show that the point set S is indeed a δ-sample on $\mathcal{D} \setminus \bigcup_i B(p_i, u_i)$. This is provided by the following lemma, whose proof may be found in the full paper [CFM+15].

Lemma 7. S is a δ-sample on $\mathcal{D} \setminus \bigcup_i B(p_i, u_i)$.

Now, we calculate the number of edges that will be present in the approximation graph defined in the previous section. For this, we require a few lemmas.

Lemma 8. Let $A = B(p_i, r_P(p_i)) \setminus B(p_i, u_i)$ be an annulus around p_i. Then, $|A \cap S| = \delta^{-O(d)}$.

Proof. By the meshing guarantees of [She11], we know that for any point $s \in A \cap S$, $B(s, t)$ does not contain a point from $S \setminus \{s\}$ for $t = \Omega(r_S(s)) = \Omega(\delta \cdot r_P(p))$. Thus, the desired result follows using a simple sphere packing argument.

Lemma 9. If $s \in S$, then $|B(s, C_2 \delta^{2/3} N(s)) \cap S| = \delta^{-O(d)}$, where C_2 is the constant in Lemma 5.

Proof. As in the previous lemma, meshing guarantees tell us that for any $s' \in B(s, C_2 \delta^{2/3} N(s))$, we have that $B(s', t)$ does not contain a point from $S \setminus \{s'\}$ for $t = \Omega(\delta \cdot N(s')) = \Omega(\delta \cdot N(s))$. Thus, we again obtain the desired result from a sphere packing argument.

From the above lemmas, we see that $\mathcal{A}$ is composed of $|S| = \delta^{-O(d)} n \log \Delta$ vertices and $n \delta^{-O(d)} + |S| \cdot \delta^{-O(d)} = |S| \cdot \delta^{-O(d)}$ edges.

Remark. Note that the right hand side of (3) is in terms of the spread, a non-combinatorial quantity. Indeed, one can construct examples of P for which the integral in (3) is not bounded from above by any function of n. However, for many classes of inputs, one can obtain a tighter analysis. In particular, if P satisfies a property known as *well-paced*, one can show that the resulting set S will satisfy $|S| = 2^{O(d)} n$ (see [MPS08, She12]).

In a more general setting (without requiring that P is well-paced), one can modify the algorithms to produce output in the form of a *hierarchical mesh* [MPS11]. This then produces an output of size $2^{O(d)}n$, and $(1 + \varepsilon)$-approximation algorithm for the nearest neighbor metric can be suitably modified so that the underlying approximation graph uses a hierarchical set of points instead of a full δ-sample. However, we ignore the details here for the sake of simplicity of exposition.

The above remark, along with the edge count of $\mathcal{A}$ and the running time guarantees from [MSV13], yields Theorem 2, the main theorem of this section.

5   Discussion

Motivated by estimating geodesic distances within subsets of $\mathbb{R}^n$, we consider two distance metrics in this paper: the N-distance and the edge-squared distance. The main focus of this paper is to find an approximation of the N-distance. One possible drawback of our $(1 + \varepsilon)$-approximation algorithm is its exponential dependency on d. To alleviate this dependency a natural approach is using a Johnson-Lindenstrauss type projection. Thereby, we would like to ask which properties are preserved under random projections such as those in Johnson-Lindenstrauss transforms.

We are currently working on implementing the approximation algorithm presented in Section 3. We hope to show that this approximation is fast in practice as well as in theory.

Acknowledgement. The authors would like to thank Larry Wasserman for helpful discussions.

References

[AMS05] Aleksandrov, L., Maheshwari, A., Sack, J.-R.: Determining approximate shortest paths on weighted polyhedral surfaces. J. ACM **52**(1), 25–53 (2005)

[ART+12] Alliez, P., Rineau, L., Tayeb, S., Tournois, J., Yvinec, M.: 3D mesh generation. In: CGAL User and Reference Manual. CGAL Editorial Board, 4.1 edn. (2012)

[BCH04] Bousquet, O., Chapelle, O., Hein, M.: Measure based regularization. In: 16th NIPS (2004)

[Ber96] Bernoulli, J.: Branchistochrone problem. Acta Eruditorum, June 1696

[BRS11] Bijral, A.S., Ratliff, N.D., Srebro, N.: Semi-supervised learning with density based distances. In: Cozman, F.G., Pfeffer, A. (eds.) UAI, pp. 43–50. AUAI Press (2011)

[CFM+15] Cohen, M.B., Fasy, B.T., Miller, G.L., Nayyeri, A., Sheehy, D., Velingker, A.: Approximating nearest neighbor distances, 2015. CoRR, abs/1502.08048

[DBVKOS00] De Berg, M., Van Kreveld, M., Overmars, M., Schwarzkopf, O.C.: Computational Geometry. Springer (2000)

[HDI14] Hwang, S.J., Damelin, S.B., Hero, A.O., III: Shortest path through random points (2014). arXiv/1202.0045v3

[HOMS10]   Hudson, B., Oudot, S.Y., Miller, G.L., Sheehy, D.R.: Topological inference via meshing. In: SOCG: Proceedings of the 26th ACM Symposium on Computational Geometry (2010)

[Hp11]   Har-peled, S.: Geometric Approximation Algorithms. American Mathematical Society, Boston (2011)

[KH03]   Kim, J., Hespanha, J.P.: Discrete approximations to continuous shortest-path: Application to minimum-risk path planning for groups of uavs. In: 42nd IEEE ICDC, January 2003

[LSV06]   Lukovszki, Tamás, Schindelhauer, Christian, Volbert, Klaus: Resource efficient maintenance of wireless network topologies. Journal of Universal Computer Science 12(9), 1292–1311 (2006)

[MP91]   Joseph, S.B.: Mitchell and Christos H. Papadimitriou. The weighted region problem: finding shortest paths through a weighted planar subdivision. J. ACM 38(1), 18–73 (1991)

[MPS08]   Miller, G.L., Phillips, T., Sheehy, D.R.: Linear-size meshes. In: CCCG: Canadian Conference in Computational Geometry (2008)

[MPS11]   Miller, G.L., Phillips, T., Sheehy, D.R.: Beating the spread: Time-optimal point meshing. In: SOCG: Proceedings of the 27th ACM Symposium on Computational Geometry (2011)

[MSV13]   Miller, G.L., Sheehy, D.R., Velingker, A.: A fast algorithm for well-spaced points and approximate delaunay graphs. In: 29th SOCG. SoCG 2013, pp. 289–298. ACM, New York (2013)

[RR90]   Rowe, Neil, Ross, Ron: Optimal grid-free path planning across arbitrarily contoured terrain with anisotropic friction and gravity effects. IEEE Transactions on Robotics and Automation 6(5), 540–553 (1990)

[She96]   Shewchuk, J.R.: Triangle: Engineering a 2D quality mesh generator and Delaunay triangulator. In: Lin, M.C., Manocha, Dinesh (eds.) FCRC-WS 1996 and WACG 1996. LNCS, vol. 1148, pp. 203–222. Springer, Heidelberg (1996)

[She11]   Sheehy, D.: Mesh Generation and Geometric Persistent Homology. PhD thesis, Carnegie Mellon University, Pittsburgh, July 2011. CMU CS Tech Report CMU-CS-11-121

[She12]   Sheehy, Donald R.: New Bounds on the Size of Optimal Meshes. Computer Graphics Forum 31(5), 1627–1635 (2012)

[Si11]   Si, H.: TetGen: A quality tetrahedral mesh generator and a 3D Delaunay triangulator, January 2011. http://tetgen.org/

[SO05]   Sajama and Orlitsky, A.: Estimating and computing density based distance metrics. In: ICML 2005, pp. 760–767. ACM, New York (2005)

[Str]   Strain, J.: Calculus of variation. http://math.berkeley.edu/strain/170. S13/cov.pdf

[Tsi95]   Tsitsiklis, John N.: Efficient algorithms for globally optimal trajectories. IEEE Transactions on Automatic Control 40, 1528–1538 (1995)

[VB03]   Vincent, P., Bengio, Y.: Density sensitive metrics and kernels. In: Snowbird Workshop (2003)

Linearity Is Strictly More Powerful Than Contiguity for Encoding Graphs

Christophe Crespelle[1,2,6]([⊠]), Tien-Nam Le[3],
Kevin Perrot[4,5], and Thi Ha Duong Phan[6]

[1] Université Claude Bernard Lyon 1, Villeurbanne, France
[2] CNRS, DANTE/INRIA, LIP UMR CNRS 5668, ENS de Lyon, Université de Lyon,
Lyon, France
christophe.crespelle@inria.fr
[3] ENS de Lyon, Université de Lyon, Lyon, France
tien-nam.le@ens-lyon.fr
[4] Universidad de Chile, DIM, CMM UMR CNRS 2807, Santiago, Chile
[5] Aix-Marseille Université CNRS, LIF UMR 7279,, 13288 Marseille, France
kevin.perrot@lif.univ-mrs.fr
[6] Institute of Mathematics, Vietnam Academy of Science and Technology,
18 Hoang Quoc Viet, Hanoi, Vietnam
phanhaduong@math.ac.vn

Abstract. Linearity and contiguity are two parameters devoted to
graph encoding. Linearity is a generalisation of contiguity in the sense
that every encoding achieving contiguity k induces an encoding achieving linearity k, both encoding having size $\Theta(k.n)$, where n is the number
of vertices of G. In this paper, we prove that linearity is a strictly more
powerful encoding than contiguity, i.e. there exists some graph family
such that the linearity is asymptotically negligible in front of the contiguity. We prove this by answering an open question asking for the worst
case linearity of a cograph on n vertices: we provide an $O(\log n / \log \log n)$
upper bound which matches the previously known lower bound.

1 Introduction

One of the most widely used operation in graph algorithms is the *neighbourhood
query*: given a vertex x of a graph G, one wants to obtain the list of neighbours
of x in G. The classical data structure that allows to do so is the adjacency
lists. It stores a graph G in $O(n + m)$ space, where n is the number of vertices
of G and m its number of edges, and answers a neighbourhood query on any

This work was partially funded by a grant from Région Rhône-Alpes and by the
delegation program of CNRS.

This work was partially funded by the Vietnam Institute for Advanced Study in
Mathematics (VIASM) and by the Vietnam National Fondation for Science and
Technology Developement (NAFOSTED).

This work was partially funded by Fondecyt Postdoctoral grant 3140527 and Núcleo
Milenio Información y Coordinación en Redes (ACGO).

© Springer International Publishing Switzerland 2015
F. Dehne et al. (Eds.): WADS 2015, LNCS 9214, pp. 212–223, 2015.
DOI: 10.1007/978-3-319-21840-3_18

vertex x in $O(d)$ time, where d is the degree of vertex x. This time complexity is optimal, as long as one wants to produce the list of neighbours of x. On the other hand, in the last decades, huge amounts of data organized in the form of graphs or networks have appeared in many contexts such as genomic, biology, physics, linguistics, computer science, transportation and industry. In the same time, the need, for industrials and academics, to algorithmically treat this data in order to extract relevant information has grown in the same proportions. For these applications dealing with very large graphs, a space complexity of $O(n+m)$ is often very limiting. Therefore, as pointed out by [13], finding compact representations of a graph providing optimal time neighbourhood queries is a crucial issue in practice. Such representations allow to store the graph entirely in memory while preserving the complexity of algorithms using neighbourhood queries. The conjunction of these two advantages has great impact on the running time of algorithms managing large amount of data.

One possible way to store a graph G in a very compact way and preserve the complexity of neighbourhood queries is to find an order σ on the vertices of G such that the neighbourhood of each vertex x of G is an interval in σ. In this way, one can store the order σ on the vertices of G and assign two pointers to each vertex: one toward its first neighbour in σ and one toward its last neighbour in σ. Therefore, one can answer adjacency queries on vertex x simply by listing the vertices appearing in σ between its first and last pointer. It must be clear that such an order on the vertices of G does not exist for all graphs G. Nevertheless, this idea turns out to be quite efficient in practice and some compression techniques are precisely based on it [1–4,11]: they try to find orders of the vertices that group the neighbourhoods together, as much as possible.

Then, a natural way to relax the constraints of the problem so that it admits a solution for a larger class of graphs is to allow the neighbourhood of each vertex to be split in at most k intervals in order σ. The minimum value of k which makes possible to encode the graph G in this way is a parameter called *contiguity* [9] and denoted by $cont(G)$. Another natural way of generalization is to use at most k orders $\sigma_1, \ldots, \sigma_k$ on the vertices of G such that the neighbourhood of each vertex is the union of exactly one interval taken in each of the k orders. This defines a parameter called the *linearity* of G [6], denoted $lin(G)$. The additional flexibility offered by linearity (using k orders instead of just 1) results in a greater power of encoding, in the sense that if a graph G admits an encoding by contiguity k, using one linear order σ and at most k intervals for each vertex, it is straightforward to obtain an encoding of G by linearity k: take k copies of σ and assign to each vertex one of its k intervals in each of the k copies of σ.

As one can expect, this greater power of encoding requires an extra cost: the size of an encoding by linearity k, which uses k orders, is greater than the size of an encoding by contiguity k, which uses only 1 order. Nevertheless, very interestingly, the sizes of these two encodings are equivalent up to a multiplicative constant. Indeed, storing an encoding by contiguity k requires to store a linear ordering of the n vertices of G, i.e. a list of n integers, and the bounds of each of the k intervals for each vertex, i.e. $2kn$ integers, the total size of the encoding being $(2k + 1)n$ integers. On the other hand, the linearity encoding also requires to store $2kn$ integers for the bounds of the k intervals of each vertex, but it needs k linear orderings of the vertices instead of just one, that is kn integers. Thus, the total size of an encoding by linearity k is $3kn$ integers instead of $(2k + 1)n$ for contiguity k and therefore the two encodings have equivalent sizes.

Then the question naturally arises to know whether there are some graphs for which the linearity is significantly less than the contiguity. More formally, does there exist some graph family for which the linearity is asymptotically negligible in front of the contiguity? Or are these two parameters equivalent up to a multiplicative constant? This is the question we address here. Our results show that linearity is strictly more powerful than contiguity.

Related Work. Only little is known about contiguity and linearity of graphs. In the context of $0 - 1$ matrices, [9,14] studied closed contiguity and showed that deciding whether an arbitrary graph has closed contiguity at most k is NP-complete for any fixed $k \geq 2$. For arbitrary graphs again, [8] (Corollary 3.4) gave an upper bound on the value of closed contiguity which is $n/4 + O(\sqrt{n \log n})$. Regarding graphs with bounded contiguity or linearity, only the class of graphs having contiguity 1, or equivalently linearity 1, has been characterized, as being the class of proper (or unit) interval graphs [12]. For interval graphs and permutation graphs, [6] showed that both contiguity and linearity can be up to $\Omega(\log n / \log \log n)$. For cographs, a subclass of permutation graphs, [7] showed that the contiguity can even been up to $\Omega(\log n)$ and is always $O(\log n)$, implying that both bounds are tight. The $O(\log n)$ upper bound consequently applies for the linearity (of cographs) as well, but [7] only provides an $\Omega(\log n / \log \log n)$ lower bound.

Our Results. Our main result (Corollary 1) is to exhibit a family of graphs G_h, $h \geq 1$, such that the linearity of G_h is asymptotically negligible in front of the contiguity of G_h. In order to do so, we prove (Theorem 1) that the linearity of a cograph G on n vertices is always $O(\log n / \log \log n)$. It turns out that this bound is tight, as it matches the previously known lower bound on the worst-case linearity of a cograph [7].

2   Preliminaries

All graphs considered here are finite, undirected, simple and loopless. In the following, G is a graph, V (or $V(G)$) is its vertex set and E (or $E(G)$) is its edge set. We use the notation $G = (V, E)$ and n stands for the cardinality $|V|$ of $V(G)$. An edge between vertices x and y will be arbitrarily denoted by xy or yx. The (open) neighbourhood of x is denoted by $N(x)$ (or $N_G(x)$) and its closed neighbourhood by $N[x] = N(x) \cup \{x\}$. The subgraph of G induced by the set of vertices $X \subseteq V$ is denoted by $G[X] = (X, \{xy \in E \mid x, y \in X\})$.

For a rooted tree T and a node $u \in T$, the depth of u in T is the number of edges in the path from the root of T to u (the root has depth 0). The *height* of T, denoted by $h(T)$, is the greatest depth of its leaves. We employ the usual terminology for *children, father, ancestors* and *descendants* of a node u in T (the two later notions including u itself), and denote by $\mathcal{C}(u)$ the set of children of u. The subtree of T rooted at u, denoted by T_u, is the tree induced by node u and all its descendants in T. A *monotonic path* C of a rooted tree T is a path such that there exists some node $u \in C$ such that all nodes of C are ancestors of u. The unique node of C which has no parent in C is called the root of the monotonic path.

In the following, the notion of *minors* of rooted trees is central. This is a special case of minors of graphs (see e.g. [10]), for which we give a simplified definition in the context of rooted trees. The *contraction of edge uv* in a rooted tree T, where u is the parent of v, consists in removing v from T and assigning its children (if any) to node u.

Definition 1 (Minor). *A rooted tree T' is a minor of a rooted tree T if it can be obtained from T by a sequence of edge contractions.*

There are actually two notions of linearity depending on whether one uses the open neighbourhood $N(x)$ or closed neighbourhood $N[x]$.

Definition 2 (p-line-model). *A closed p-line-model (resp. open p-line-model) of a graph $G = (V, E)$ is a tuple $(\sigma_1, \ldots, \sigma_p)$ of linear orders on V such that $\forall v \in V, \exists (I_1, \ldots, I_p)$ such that $\forall i \in [\![1, p]\!]$, I_i is an interval of σ_i and $N[x] = \bigcup_{1 \le i \le p} I_i$ (resp. $N(x) = \bigcup_{1 \le i \le p} I_i$).*
The closed linearity (resp. open linearity) of G, denoted by $cl(G)$ (resp. $ol(G)$), is the minimum integer p such that there exists a closed p-line-model (resp. open p-line-model) of G.

Remark 1. *In the definition of a p-line-model, the set of vertices of the intervals I_i assigned to a vertex x are not necessarily disjoint. They are only required to cover the neighbourhood of x while being included in it.*

In all the paper, we abusively extend the notion of linearity to cotrees, referring to the linearity of their associated cograph. Moreover, we consider only closed linearity but, from the inequalities below, the bounds we obtain (which hold up to multiplicative constants) also hold for the open linearity. Then, for the sake of clarity, as we will not use the open notion, in the following, we denote $lin(G)$ instead of $cl(G)$.

Lemma 1. *For an arbitrary graph G, we have the following inequalities: $cl(G) - 1 \le ol(G) \le 2cl(G)$.*

There are several characterizations of the class of *cographs*. They are often defined as the graphs that do not admit the P_4 (path on 4 vertices) as induced subgraph. Equivalently, they are the graphs obtained from a single vertex under the closure of the parallel composition and the series composition. The parallel composition of two graphs $G_1 = (V_1, E_1)$ and $G_2 = (V_2, E_2)$ is the disjoint union of G_1 and G_2, i.e., the graph $G_{par} = (V_1 \cup V_2, E_1 \cup E_2)$. The series composition of two graphs G_1 and G_2 is the disjoint union of G_1 and G_2 plus all possible edges from a vertex of G_1 to one of G_2, i.e., the graph $G_{ser}(V_1 \cup V_2, E_1 \cup E_2 \cup \{xy \mid x \in V_1, y \in V_2\})$. These operations can naturally be extended to a finite number of graphs.

This gives a very nice representation of a cograph G by a tree whose leaves are the vertices of the graph and whose internal nodes (non-leaf nodes) are labelled P, for parallel, or S, for series, corresponding to the operations used in the construction of G. It is always possible to find such a labelled tree T representing G such that every internal node has at least two children, no two parallel nodes are adjacent in T and no two series nodes are adjacent. This tree T is unique [5] and is called the *cotree* of G. Note that the subtree T_u rooted at some node u of cotree T also defines a cograph, denoted G_u, and then $V(G_u)$ is the set of leaves of T_u. The adjacencies between vertices of a cograph can easily be read on its cotree, in the following way.

Remark 2. *Two vertices x and y of a cograph G having cotree T are adjacent iff the least common ancestor u of leaves x and y in T is a series node. Otherwise, if u is a parallel node, x and y are not adjacent.*

For a graph encoding scheme Enc and a graph G, we denote $|Enc(G)|$ the minimum size of an encoding of G based on Enc. We now give a formal definition for an encoding scheme to be strictly more powerful than another one.

Definition 3 (Strictly more powerful encoding). *Let Enc_1 and Enc_2 be two graph encoding schemes. We say that Enc_2 is at least as powerful as Enc_1 iff there exists $\alpha > 0$ such that for all graphs G, $|Enc_2(G)| \leq \alpha|Enc_1(G)|$. Moreover, we say that Enc_2 is strictly more powerful than Enc_1 iff Enc_2 is at least as powerful as Enc_1 and the converse is not true.*

Note that, Enc_1 is not at least as powerful as Enc_2 iff there exists a series of graphs G_h, $h \geq 1$, such that $|Enc_1(G_h)|/|Enc_2(G_h)|$ tends to infinity when h tends to infinity. In the introduction, we showed that the encoding schemes $LinEnc$ and $ContEnc$ based on linearity and contiguity respectively are such that, for any graph G on n vertices, we have $2\,n\,cont(G) \leq |ContEnc(G)| \leq 3\,n\,cont(G)$ and $|LinEnc(G)| = 3\,n\,lin(G)$. Since $lin(G) \leq cont(G)$, this gives $|LinEnc(G)| \leq \frac{3}{2}|ContEnc(G)|$. In addition, the previous inequalities also imply that $\frac{2}{3}cont(G)/lin(G) \leq |ContEnc(G)|/|LinEnc(G)| \leq cont(G)/lin(G)$. Altogether, we obtain the following remark.

Remark 3. *Linearity is an encoding at least as powerful as contiguity according to Definition 3. Moreover, it is strictly more powerful iff there exists a series of graphs G_h, $h \geq 1$, such that $|cont(G_h)|/|lin(G_h)|$ tends to infinity when h tends to infinity.*

3   Linearity of a Cograph and Factorial Rank of Its Cotree

In this section, we show that the linearity of a cograph is upper bounded by the size of some maximal structure contained in its cotree, more precisely by the height of a maximal double factorial tree (defined below), which we call the factorial rank of a cotree. This result is interesting by itself as it provides a structural explanation of the difficulty of encoding a cograph by linearity. For our concern, the interesting point is that the number of leaves of a double factorial tree of height h is $\Omega(h!)$. Combined with this fact, the result presented in this section (Lemma 2) will allow us to derive in next section the desired $O(\log n / \log \log n)$ upper bound on the linearity of cographs. We start by some necessary definitions.

Definition 4 (Double factorial tree). *The* double factorial tree F^h *of height h is defined inductively as the tree whose root has $2h + 1$ children u, whose subtrees F_u are precisely F^{h-1}, F^0 being the unique tree of height 0 (i.e., made of a single leaf node).*

Definition 5 (Factorial rank). *The* factorial rank *of a rooted tree T denoted $factrank(T)$, is the maximum height of a double factorial tree being a minor of T, that is:*
$$factrank(T) = \max\{h(T') \mid T' \text{ is a double factorial tree and a minor of } T\}.$$

We extend the notion of factorial rank to a node, referring to the factorial rank of its subtree. The case where the children of node u all have factorial rank strictly less than the one of u will play a key role.

Definition 6 (Minimally of factorial rank k). *Let u be a node of a tree T. If u has factorial rank k and if all the children of u have factorial rank at most $k - 1$, we say that u is* minimally of factorial rank k.

We are now ready to state the result of this section, which claims that the linearity of a cograph is linearly bounded by the factorial rank of its cotree.

Lemma 2. *Let T be a cotree and let $u \in T$ of factorial rank $k \geq 0$. Then, $lin(G_u) \leq 2k + 1$. Moreover, if $k \geq 1$ and u is minimally of factorial rank k, then $lin(G_u) \leq 2k$.*

Sketch of Proof. We prove the result by induction. We consider an integer $k \geq 1$ such that: all nodes of factorial rank $j \leq k - 1$ have linearity at most $2j + 1$; and all nodes which are minimally of factorial rank k (i.e., whose children have factorial rank at most $k - 1$) have linearity at most $2k$. Then, we show that any node u of factorial rank k (not necessarily minimally) can be encoded using one more order (i.e. $2k + 1$) and that adding again one more order (i.e. using $2k + 2$ orders), we can also encode any node v which is minimally of factorial rank $k + 1$.

Node u of Factorial Rank k. In order to describe a $2k + 1$-line-model of G_u we need to distinguish different parts of T_u. Let U_k be the subset of nodes of T_u having factorial rank k and consider the set $U_{kmin} = \{u_1, u_2, \ldots, u_l\} \subsetneq U_k$ of its minimal elements for the ancestor relationship (i.e. the lowest in the cotree). Note that $|U_{kmin}| = l \leq 2k$, as otherwise u would be of factorial rank $k + 1$

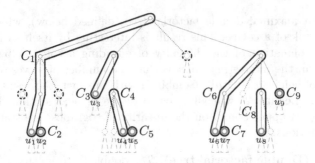

Fig. 1. Example of partition into monotonic paths in the case where u is of factorial rank k. The three dot circled nodes of $U_{\leq k-1}$ form the set $U^1_{\leq k-1}$.

(since it would have $2k + 1$ independent descendants of rank $2k$). By definition, all the children of the nodes of U_{kmin} have factorial rank at most $k-1$, and then the nodes of U_{kmin} are minimally of rank k. By induction hypothesis, it follows that for all $i \in [\![1, l]\!]$, u_i admits a $2k$-line-model for which we denote $\sigma_j(u_i)$, with $1 \leq j \leq 2k$, its $2k$ orders. We denote T'_u the subtree of T_u induced by the set of nodes U_k (by definition, $U_{kmin} \subseteq T'_u$). We also denote $U_{\leq k-1}$ the set of nodes of $T_u \setminus T'_u$ whose parent is in $T'_u \setminus U_{kmin}$. Nodes of $U_{\leq k-1}$ have, by definition, rank at most $k - 1$ and it follows from the induction hypothesis that they admit a $(2k - 1)$-line-model. Then, for a node $w \in U_{\leq k-1}$, we again denote $\sigma_j(w)$, with $1 \leq j \leq 2k-1$, the $2k-1$ orders of such a model. In addition , we use a partition $\mathcal{P}$ of the nodes of T'_u into l monotonic paths C_i such that for all $i \in [\![1, l]\!]$, $u_i \in C_i$ (see Figure 1). Partition $\mathcal{P}$ naturally induces a generalised partition (some parts may be empty) of $U_{\leq k-1}$ whose parts are the subset of nodes $U^i_{\leq k-1}$ of $U_{\leq k-1}$ whose parent belongs to $C_i \setminus \{u_i\}$.

We can now describe the $2k+1$ orders $(\sigma_j)_{1 \leq j \leq 2k+1}$ of the model we build for G_u. Importantly, note that $V(G_w)$, $w \in U_{kmin} \cup U_{\leq k-1}$, is a partition of $V(G_u)$. In our construction, $V(G_w)$ will always be an interval of σ_j for all $w \in U_{kmin} \cup U_{\leq k-1}$ and all $j \in [\![1, 2k+1]\!]$. Then, the description of σ_j is in two steps: we first give the order, denoted π_j, in which the intervals of nodes $w \in U_{kmin} \cup U_{\leq k-1}$ appear in σ_j and then, for each w, we give the order, denoted σ^w_j, in which the vertices of G_w appear in this interval. The description of orders π_j will be done by choosing a local order on the children of each node of $U_k \setminus U_{kmin}$. Then π_j is defined as the unique order on $U_{kmin} \cup U_{\leq k-1}$ respecting all the chosen local orders, i.e. such that for any $v, v' \in U_{kmin} \cup U_{\leq k-1}$, if v and v' has the same parent z and if v comes before v' in the order chosen on children of z, then all descendants of v comes before all descendants of v' in π_j.

To fully describe the $2k + 1$-line-model of u, we must also assign to each vertex x one interval of its neighbours in each of the orders of the model, in such a way that these intervals entirely cover the neighbourhood of x. In order to help our analysis, we distinguish between the *external neighbourhood* of node x, which is $N[x] \setminus V(G_w)$, where w is the unique node of $U_{kmin} \cup U_{\leq k-1}$ being an ancestor of leaf x in T_u, and its *internal neighbourhood* $N[x] \cap V(G_w)$.

Our construction mainly focusses on the $2k$ first orders of the model, which we use to encode the majority of adjacencies of G_u, order σ_{2k+1} being used to encode the remaining ones.

For $j \in [\![1, 2k]\!]$, the purpose of order σ_j is to satisfy the external neighbourhoods of vertices of G_w for $w \in \{u_j\} \cup U^j_{\leq k-1}$. It entirely succeeds to do so for u_j and encodes only half of the external neighbourhoods of $V(G_w)$ for nodes $w \in U^j_{\leq k-1}$, the other half being encoded in σ_{2k+1}. Then, for each $w \in \{u_j\} \cup U^j_{\leq k-1}$, the internal neighbourhoods of vertices of G_w are encoded in the remaining $2k - 1$ orders of $(\sigma_j)_{1 \leq j \leq 2k}$. It is enough for $w \in U^j_{\leq k-1}$, since they admit a $2k - 1$-line-model by recursion hypothesis, but one order is missing for u_j which is minimally of linearity k and is then only guaranteed to admit a $2k$-line model by recursion hypothesis. Again, the missing order will be found in σ_{2k+1}.

External Neighbourhoods and Choice of π_j's. Let us now show how to choose the order π_j used for defining σ_j such that, as claimed above, most of the external adjacencies of vertices of G_w, for $w \in \{u_j\} \cup U^j_{\leq k-1}$, will be satisfied in σ_j. We choose π_j the order induced by the following local orders on the children of nodes $u' \in U_k \setminus U_{kmin}$: if u' is a series node (resp. parallel node) and a strict ancestor of u_i, then the child of u' which is an ancestor of u_j is placed first (resp. last) in the order on the children of u' (the order on the other children of u' does not matter), in all other cases, the order on the children of u' does not matter. This way, the external neighbourhood of vertices of G_{u_j} is an interval at the end of σ_j (the interval following G_{u_j}) and this is the interval assigned to vertices of G_{u_j} in σ_j. For nodes $w \in U^j_{\leq k-1}$ whose parent (which is a strict ancestor of u_j by definition) is a parallel node, the situation is the same. But for nodes $w \in U^j_{\leq k-1}$ whose parent w' is series, their external neighbourhood is split into two intervals of σ_j: one following $V(G_w)$, which is the one we assign to vertices of G_w in σ_j, and one preceding $V(G_w)$, denoted $I_{<w}$, which is constituted by the leaves of T_u descending from the children of w' that precede w in the order chosen for π_j.

This is where we need order σ_{2k+1} and the partition of T'_u into paths C_i introduced earlier. To define order π_{2k+1}, for any node $u' \in U_k \setminus U_{kmin}$, we use the same order on the children of u' as the one used for π_i, with $i \in [\![1, l]\!]$ such that $u' \in C_i$. This ensures that for any node $w \in U_{\leq k-1}$ whose parent w' is a series node of C_i, the interval $I_{<w}$ of external neighbours which was not covered in order σ_i (note that since $w' \in C_i$ then $w \in U^i_{\leq k-1}$) will also be an interval of σ_{2k+1}. This is precisely the interval we assign to vertices of G_w in σ_{2k+1}, which is possible as their internal neighbourhood will be entirely satisfied in the $2k$ first orders, as described below.

Internal Neighbourhoods and Choice of σ_j^w's. The orders σ_j^w used for the vertices of G_w, with $w \in U_{kmin} \cup U_{\leq k-1}$, in order σ_j, with $j \in [\![1, 2k]\!]$ are chosen as follows. For a node $w \in U_{\leq k-1}$ whose parent belong to path C_i of the partition, if $j < i$ (resp. if $j > i$) then we use the order $\sigma_j(w)$ (resp. $\sigma_{j-1}(w)$), and the interval of σ_j associated to the vertices of G_w is the same as the one associated to them in $\sigma_j(w)$ (resp. $\sigma_{j-1}(w)$). Otherwise, if $j = i$ the order chosen for vertices of G_w does not matter as σ_j is used only for satisfying their external neighbourhood, see above. Proceeding this way, the internal neighbourhoods of vertices of G_w are entirely satisfied in orders $(\sigma_j)_{j \in [\![1,2k]\!]}$. For a node $u_i \in U_{kmin}$, if $j \neq i$, the order chosen on the vertices of G_{u_i} is $\sigma_j(u_i)$ and the interval associated to vertices of G_{u_i} in σ_j is the same as the one associated to them in $\sigma_j(u_i)$. Otherwise, if $j = i$, the order chosen for vertices of G_{u_i} does not matter again as σ_j is used only for satisfying their external neighbourhood. Then, only $2k - 1$ orders among the $2k$ first ones are used to encode the internal neighbourhoods of G_{u_i}, while the recursion hypothesis only guarantees that $lin(G_{u_i}) \leq 2k$. For this reason, we chose the order on the vertices of G_{u_i} in σ_{2k+1} as being $\sigma_i(u_i)$, the one which was not used until now, and the interval associated to vertices of G_{u_i} in σ_{2k+1} is the same as the one associated to them in $\sigma_i(u_i)$. This is possible as the external neighbourhood of vertices of G_{u_i} has already been entirely satisfied before, in order σ_i. Then, all adjacencies are satisfied and $lin(G_u) \leq 2k + 1$.

Node v Minimally of Factorial Rank $k+1$. The only interesting case is when v is a series node (the result is straightforward when v is parallel), then we denote $v_1, v_2, \ldots, v_l$, with $l \in \mathbb{N}$, the children of v, which have factorial rank at most k by definition. From what precedes, each of them v_i admit a $(2k + 1)$-line-model denoted $(\sigma_j(v_i))_{j \in [\![1,2k+1]\!]}$. A remarkable property of this $(2k + 1)$-line-model, which we have constructed above, is that for any vertex x, there exists an index j, later denoted $ind(x)$, such that the interval associated to x in $\sigma_j(v_i)$ contains the last vertex of $\sigma_j(v_i)$. Based on this, the model $(\sigma_j)_{1 \leq j \leq 2k+2}$ we build for G_v is as follows. For $j \in [\![1, 2k + 1]\!]$, order σ_j is the concatenation of orders $\sigma_j(v_i)$ in the order from $i = 1$ to $i = l$. For any vertex x of G_{v_i}, if $j \neq ind(x)$, the interval associated to x in σ_j is the same as the one associated to x in $\sigma_j(v_i)$; and if $j = ind(x)$, as the interval associated to x in $\sigma_{ind(x)}(v_i)$ contains the last vertex of $\sigma_{ind(x)}(v_i)$, in the order $\sigma_{ind(x)}$ of the model of G_v, we extend this interval on the right by including the vertices of $G_{v_{i'}}$ for all $i' > i$. As v is a series node, all these vertices are indeed adjacent to x, as well as all the vertices of $G_{v_{i'}}$ for all $i' < i$, which are the only adjacencies of x that are not covered in the orders $(\sigma_j)_{1 \leq j \leq 2k+1}$. We use order σ_{2k+2} to cover these adjacencies in the following way. For each node v_i, we choose an arbitrary order on the vertices of G_{v_i} and concatenate them in the order from $i = 1$ to $i = l$. Then, to any vertex x of G_{v_i}, we associate the interval made by all the vertices of $G_{v_{i'}}$ for all $i' < i$. This completes the $2k + 2$-model of v and the proof of the lemma. □

4   Main Results

The first result we derive from Lemma 2 is a tight upper bound on the worst-case linearity of cographs on n vertices. Until now, the best known upper bound [7] was $O(\log n)$, and [7] also exhibits some cograph families having a linearity up to $\Omega(\log n / \log \log n)$. Here, we show a new upper bound of $O(\log n / \log \log n)$ that matches the lower bound of [7]. This is a direct consequence of Lemma 2 and of the fact that a double factorial tree of height h has $\Omega(h!)$ vertices.

Theorem 1. *For any cograph G on n vertices, we have $lin(G) = O(\log n / \log \log n)$, and this upper bound is tight.*

Proof. Let T denote the cotree of G and $k = factrank(T)$. From Lemma 2, the linearity of G is in $O(k)$. Let us now show that $k = O(\log n / \log \log n)$, which will conclude this proof. According to the definition of factorial rank, G has at least as many vertices as the double factorial tree of height k, which has $\prod_{i=0}^{k}(2i+1)$ vertices. It follows from Stirling's approximation of factorial that

$$n \geq \prod_{i=0}^{k}(2i+1) = \frac{(2(k+1))!}{2^{k+1}(k+1)!} \geq \frac{2\sqrt{\pi}}{e}\left(\frac{2(k+1)}{e}\right)^{k+1}$$

and consequently

$$\log n \geq (k+1)\left(\log(k+1) + \log\left(\frac{2}{e}\right)\right) + \log\left(\frac{2\sqrt{\pi}}{e}\right) \geq (k+1)\left(\log(k+1) - 1\right).$$

As $x \geq y > 1$ implies $\frac{x}{\log x} \geq \frac{y}{\log y}$, we have

$$\frac{\log n}{\log \log n} \geq \frac{(k+1)\left(\log(k+1) - 1\right)}{\log(k+1) + \log\left(\log(k+1) - 1\right)}$$

and it follows that $k = O(\log n / \log \log n)$.

And finally, as [7] exhibits some cographs having linearity $\Omega(\log n / \log \log n)$, consequently, the upper bound provided by the lemma is tight. □

We now prove the main result aimed by this paper: linearity is a strictly more powerful encoding than contiguity, which means, from Remark 3, that there exists some graph families for which the linearity is asymptotically negligible in front of the contiguity (hereafter denoted $cont(G)$ for a graph G).

Corollary 1. *There exists a series of graphs G_h, $h \geq 1$, such that $cont(G_h)/lin(G_h)$ tends to infnity when h tends to infinity.*

Proof. For $h \geq 1$, let G_h be the connected cograph whose cotree is a complete binary tree of height h and let $n = 2^h$ denote the number of vertices of G_h. It is proven in [7] that $cont(G_h) = \Theta(\log n)$ and that $lin(G_h) = \Omega(\log n / \log \log n)$. Then, Theorem 1 above implies that $lin(G_h) = \Theta(\log n / \log \log n)$ and therefore $cont(G_h)/lin(G_h) = \Theta(\log \log n)$, which achieves the proof. □

5   Perspectives

In this paper, we showed that linearity provides a strictly more powerful encoding for graphs than contiguity does, meaning that the ratio between the contiguity and the linearity of a graph is not bounded by a constant. From a practical point of view, the meaning of our result is that using several orders, instead of just one, for grouping neighbourhoods of vertices can greatly enhance compression rates in some cases.

We obtained this result by exhibiting a graph family, namely a subfamily of cographs, for which the ratio between the contiguity and the linearity tends to infinity as fast as $\Omega(\log \log n)$, with n the number of vertices in the graph. As a by-product of our proof, but meaningful in itself, we also showed tight bounds for the worst-case linearity of cographs on n vertices; tight bounds were previously known for contiguity. Several questions naturally arises from these results and others.

Open Question 1. *What is the worst case contiguity and the worst-case linearity of arbitrary graphs?*

It is straightforward to see that both of these values are bounded by $n/2$. Conversely, since there are $2^{n(n-1)/2}$ graphs on n labelled vertices and since contiguity and linearity do not depend on the labels of the vertices, then both encodings must use at least n^2 bits for graphs on n vertices. Moreover, when the value of the parameter is k, the size of the corresponding encoding is $O(k\,n)$ integers, that is $O(k\,n \log n)$ bits. Consequently, both parameters must be at least $\Omega(n/\log n)$ in the worst case. For contiguity, [8] gave an upper bound asymptotically equivalent to $n/4$. Is $\Omega(n)$ indeed the worst-case contiguity of a graph? Is the worst-case for linearity the same as the one for contiguity? Another appealing question which is closely related is the following.

Open Question 2. *For arbitrary graphs, what is the maximum gap between contiguity and linearity?*

In other words, let $(G_n)_{n \geq 1}$ be a family of graphs on n vertices and let $f(n) = cont(G_n)/lin(G_n)$. Can $f(n)$ tends to infinity faster than $\Omega(\log \log n)$? What is the maximum asymptotic growth possible for $f(n)$? Answering those questions would be both theoretically and practically of key interest for the field of graph encoding.

References

1. Apostolico, A., Drovandi, G.: Graph compression by BFS. Algorithms **2**(3), 1031–1044 (2009)
2. Boldi, P., Vigna, S.: The webgraph framework I: compression techniques. In: WWW 2004, pp. 595–602. ACM (2004)
3. Boldi, P., Vigna, S.: Codes for the world wide web. Internet Mathematics **2**(4), 407–429 (2005)
4. Boldi, P., Santini, M., Vigna, S.: Permuting web and social graphs. Internet Mathematics **6**(3), 257–283 (2009)
5. Corneil, D., Lerchs, H., Burlingham, L.: Complement reducible graphs. Discrete Applied Mathematics **3**(3), 163–174 (1981)

6. Crespelle, C., Gambette, P.: Efficient neighborhood encoding for interval graphs and permutation graphs and $O(n)$ breadth-first search. In: Fiala, J., Kratochvíl, J., Miller, M. (eds.) IWOCA 2009. LNCS, vol. 5874, pp. 146–157. Springer, Heidelberg (2009)

7. Crespelle, C., Gambette, P.: (nearly-)tight bounds on the contiguity and linearity of cographs. Theoretical Computer Science **522**, 1–12 (2014)

8. Gavoille, C., Peleg, D.: The compactness of interval routing. SIAM Journal on Discrete Mathematics **12**(4), 459–473 (1999)

9. Goldberg, P., Golumbic, M., Kaplan, H., Shamir, R.: Four strikes against physical mapping of DNA. Journal of Computational Biology **2**(1), 139–152 (1995)

10. Lovász, L.: Graph minor theory. Bulletin of the American Mathematical Society **43**(1), 75–86 (2006)

11. Maserrat, H., Pei, J.: Neighbor query friendly compression of social networks. In: KDD 2010, pp. 533–542. ACM (2010)

12. Roberts, F.: Representations of Indifference Relations. Ph.D. thesis, Stanford University (1968)

13. Turan, G.: On the succinct representation of graphs. Discr. Appl. Math. **8**, 289–294 (1984)

14. Wang, R., Lau, F., Zhao, Y.: Hamiltonicity of regular graphs and blocks of consecutive ones in symmetric matrices. Discr. Appl. Math. **155**(17), 2312–2320 (2007)

On the Complexity of an Unregulated Traffic Crossing

Philip Dasler[✉] and David M. Mount

Department of Computer Science, University of Maryland,
College Park, MD 20742, USA
{daslerpc,mount}@cs.umd.edu

Abstract. One of the most challenging aspects of traffic coordination involves traffic intersections. In this paper we consider two formulations of a simple and fundamental geometric optimization problem involving coordinating the motion of vehicles through an intersection.

We are given a set of n vehicles in the plane, each modeled as a unit length line segment that moves monotonically, either horizontally or vertically, subject to a maximum speed limit. Each vehicle is described by a start and goal position and a start time and deadline. The question is whether, subject to the speed limit, there exists a collision-free motion plan so that each vehicle travels from its start position to its goal position prior to its deadline.

We present three results. We begin by showing that this problem is NP-complete with a reduction from 3-SAT. Second, we consider a constrained version in which cars traveling horizontally can alter their speeds while cars traveling vertically cannot. We present a simple algorithm that solves this problem in $O(n \log n)$ time. Finally, we provide a solution to the discrete version of the problem and prove its asymptotic optimality in terms of the maximum delay of a vehicle.

1 Introduction

As autonomous and semi-autonomous vehicles become more prevalent, there is an emerging interest in algorithms for controlling and coordinating their motions to improve traffic flow. The steady development of motor vehicle technology will enable cars of the near future to assume an ever increasing role in the decision making and control of the vehicle itself. In the foreseeable future, cars will have the ability to communicate with one another in order to better coordinate their motion. This motivates a number of interesting algorithmic problems. One of the most challenging aspects of traffic coordination involves traffic intersections. In this paper we consider two formulations of a simple and fundamental geometric optimization problem involving coordinating the motion of vehicles through an intersection.

Supported by the National Science Foundation under grant CCF-1117259 and the Office of Naval Research under grant N00014-08-1-1015.

© Springer International Publishing Switzerland 2015
F. Dehne et al. (Eds.): WADS 2015, LNCS 9214, pp. 224–235, 2015.
DOI: 10.1007/978-3-319-21840-3_19

Traffic congestion is a complex and pervasive problem with significant economic ramifications. Practical engineering solutions will require consideration of myriad issues, including the physical limitations of vehicle motion and road conditions, the complexities and dynamics of traffic and urban navigation, external issues such as accidents and break-downs, and human factors. We are motivated by the question of whether the field of algorithm design can contribute positively to such solutions. We aim to identify fundamental optimization problems that are simple enough to be analyzed formally, but realistic enough to contribute to the eventual design of actual traffic management systems. In this paper, we focus on a problem, the *traffic crossing problem*, that involves coordinating the motions of a set of vehicles moving through a system of intersections. In urban settings, road intersections are *regulated* by traffic lights or stop/yield signs. Much like an asynchronous semaphore, a traffic light locks the entire intersection preventing cross traffic from entering it, even when there is adequate space to do so. Some studies have proposed a less exclusive approach in which vehicles communicate either with one another or with a local controller that allows vehicles, possibly moving in different directions, to pass through the intersection simultaneously if it can be ascertained (perhaps with a small adjustment in velocities) that the motion is collision-free (see, e.g., [9]). Even though such systems may be beyond the present-day automotive technology, the approach can be applied to controlling the motion of parcels and vehicles in automated warehouses [17].

Prior work on autonomous vehicle control has generally taken a high-level view (e.g., traffic routing [5,6,15,18]) or a low-level view (e.g., control theory, kinematics, etc. [10,14]). We propose a mid-level view, focusing on the control of vehicles over the course of minutes rather than hours or microseconds, respectively. The work by Fiorini and Shiller on velocity obstacles [11] considers motion coordination in a decentralized context, in which a single agent is attempting to avoid other moving objects. Much closer to our approach is work on *autonomous intersection management* (AIM) [2,4,7-9,16]. This work, however, largely focuses on the application of multi-agent techniques and deals with many real-world issues. As a consequence, formal complexity bounds are not proved. Berger and Klein consider a dynamic motion-panning problem in a similar vein to ours, which is loosely based on the video game *Frogger* [3]. Their work is based, at least in part, on the work of Arkin, Mitchell, and Polishchuk [1] in which a group of circular agents must cross a field of polygonal obstacles. These obstacles are dynamic, but their motion is fixed and known *a priori*.

We consider a simple problem formulation of the traffic crossing problem, but one that we feel captures the essential computational challenges of coordinating crosswise motion through an intersection. Vehicles are modeled as line segments moving monotonically along axis-parallel lines (traffic lanes) in the plane. Vehicles can alter their speed, subject to a maximum speed limit, but they cannot reverse direction. The objective is to plan the collision-free motion of these segments as they move to their goal positions.

After a formal definition of our traffic crossing problem in Section 2, we present three results. First, we show in Section 3 that this problem is NP-complete.

(While this is a negative result, it shows that this problem is of a lower complexity class than similar PSPACE-complete motion-planning problems, like sliding-block problems [12].) Second, in Section 4 we consider a constrained version in which cars traveling vertically travel at a fixed speed. This variant is motivated by a scenario in which traffic moving in one direction (e.g., a major highway) has priority over crossing traffic (e.g., a small road). We present a simple algorithm that solves this problem in $O(n \log n)$ time.

Finally, we consider the problem in a discrete setting in Section 5, which simplifies the description of the algorithms while still capturing many of the interesting scheduling elements of the problem. As part of this consideration, we provide a solution to the problem that limits the maximum delay of any vehicle and prove that this solution is asymptotically optimal.

2   Problem Definition

The Traffic Crossing Problem is one in which several vehicles must cross an intersection simultaneously. For a successful crossing, all vehicles must reach the opposite side of the intersection without colliding, and they must do so in a reasonable amount of time. Formally, a traffic crossing is defined as a tuple $C = (V, \delta_{max})$. This tuple is comprised of a set of n vehicles V which exist in $\mathbb{R}^2$ and a global speed limit $\delta_{max} \in \mathbb{R}^+$, where $\mathbb{R}^+$ denotes the set of nonnegative reals. Each vehicle is modeled as a vertical or horizontal open line segment that moves parallel to its orientation. Like a car on a road, each vehicle moves monotonically, but its speed may vary between zero and the speed limit. A vehicle's position is specified by its leading point (relative to its direction).

Each vehicle $v_i \in V$ is defined as a set of properties, $v_i = \{l_i, p_i^\vdash, p_i^\dashv, t_i^\vdash, t_i^\dashv\}^1$, where l_i is the vehicle's length, $p_i^\vdash$ and $p_i^\dashv$ are its initial and goal positions, respectively, and $t_i^\vdash$ and $t_i^\dashv$ are its start time and deadline for reaching its goal position.

The set V and the global speed limit δ_{max} define the problem and remain invariant throughout. Our objective is to determine whether there exists a collision-free motion of the vehicles that respects the speed limit and satisfies the goal deadlines. Such a motion is described by a set of functions, called speed profiles, that define the instantaneous speed of the vehicles at time t.

This set of functions is defined as $D = \{\delta_i(t) \mid i \in [1, n], \forall t, 0 \le \delta_i(t) \le \delta_{max}\}$. A set D of speed profiles is *valid* if no vehicle (1) moves prior to its start time or after its deadline, (2) violates the speed limit or travels in reverse (3) collides with another vehicle or (4) fails to reach its goal prior to its deadline.

A traffic crossing C is solvable if there exists a valid set of speed profiles D.

3   Hardness of Traffic Crossing

Determining whether a given instance of the traffic crossing problem is solvable is NP-complete. We show its NP-hardness by proving the following theorem:

---

[1] The notational use of $\vdash$ and $\dashv$ set above a variable (e.g., $\alpha^\vdash$) represents the beginning and end of a closed interval, respectively (e.g., start and end times).

Theorem 1. *Given a Boolean formula F in 3-CNF, there exists a traffic crossing $C = (V, \delta)$, computable in polynomial time, such that F is satisfiable if and only if there exists a valid set of speed profiles D for C.*

The input to the reduction is a boolean formula F in 3-CNF (i.e., an instance of 3-SAT). Let $\{z_1, \ldots, z_n\}$ denote its variables and $\{c_1, \ldots, c_m\}$ denote its clauses. Each variable z_i in F is represented by a pair of vehicles whose motion is constrained to one of two possible states by intersecting their paths with a perpendicular pair of vehicles. This constraining mechanism (seen in Fig. 1) is the core concept around which all mechanisms in the reduction are built. It allows us to represent logical values, to transmit these values throughout the construction, and to check these values for clause satisfaction.

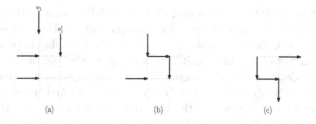

            (a)   (b)   (c)

Fig. 1. (a) An example of transferring values at $t_i^\vdash$. v_1 and v_1' are **true** and **false**, respectively. (b) At time $t_i^\vdash + 1$, the upper horizontal vehicle will take on the value of v_1' while the lower takes the value of v_1. (c) $= t_i^\vdash + 2$.

All vehicles in the reduction are of unit length and (barring a few special cases) their deadlines are set so that they can reach their goal position with at most one unit time delay. More formally, $t_i^\dashv - t_i^\vdash - 1 = \frac{(\|p_i^\dashv - p_i^\vdash\|)}{\delta_{max}}$. In general, the delay may take multiple forms (e.g., the vehicle could take a delay of 1 at any point during its travel or spread the delay out by traveling slower than δ_{max}), but the mechanism described above constrains the delay to only one of two types: a delay of exactly 0 or 1 taken immediately at the vehicle's start time.

For each clause $c_i \in F$, a mechanism is created that forces a collision if, and only if, all three literals are **false**. This mechanism checks the positive and negative literals separately, then combines the results in order to determine whether the clause is satisfied.

These mechanisms each require only a constant number of vehicles, resulting in a reduction complexity on the order of $O(n + m)$, where n and m are the number of variables and clauses, respectively[2].

3.1 Membership in NP

Lemma 1. *The Traffic Crossing Problem is in NP.*

---

[2] Detailed descriptions of these mechanisms have been omitted due to space constraints, but can be found in the arXiv version of this paper.

First, observe that for each pair of orthogonal vehicles, v_i, v_j, their paths cross at a single intersection. The certificate provides a priority for each such pair, specifying which vehicle crosses through the intersection first. Next, it can be shown that if there exists a valid set of speed profiles for an instance of the problem, then there exists another valid set where vehicles move at the maximum speed and are subject to the constraints in the certificate. Finally, when proving the validity of a solution provided by the certificate, only a number of events polynomial in n must be processed and the number of bits of precision required is polynomial in the number of bits in the input plus $\log n$. For the sake of space, the formal proof has been omitted.

4 A Solution to the One-Sided Problem

While the generalized Traffic Crossing Problem is NP-complete, it is possible to solve a constrained version of the problem more efficiently. The complexity of the generalized Traffic Crossing Problem arises from the interplay between horizontal and vertical vehicles, which results in a complex cascade of constraints. To break this interdependency, the vertically traveling vehicles are given priority, allowing them to continue through the intersection at a fixed speed. In this variant, called the *one-sided problem*, the horizontal vehicles can plan their motion with complete information and without fear of complex constraint chains.

First, we assume that the vertically traveling vehicles are invariant and are all traveling at the same speed, s_n. With vertical vehicle motion now fixed, there is no way for horizontal vehicles to affect each other and movement profiles for each can be found in isolation from the others. Finally, we assume that all vehicles are of length l and in general position.

For the purpose of illustration we begin with a simplified version of the problem and then, over the course of three cases, relax the restrictions until we are left with a solution to the original problem under the fixed, one-sided policy described above. These three cases are:

Intersection Between One-Way Highways
- Vertical vehicles approach from the North only.
- Horizontal vehicles approach from the West only.
- Each vehicle is in its own lane (i.e., no two vehicles are collinear).

Intersection Between a One-Way Street and a Two-Way Highway
- Vertical vehicles approach from the North and the South.
- Horizontal vehicles approach from the West only.
- There is a single horizontal lane (i.e., all horizontal vehicles are collinear) and one or more vertical lanes.

Intersection Between Two-Way Highways
- Vertical vehicles approach from the North and the South.
- Horizontal vehicles approach from the West and the East.
- There are k horizontal lanes, one or more vertical lanes, and vehicles may share lanes.

4.1   Intersection Between One-Way Highways

Formally, vehicles from the North are in the subset $N \subset V$ and their direction of travel is $d_n = (0, -1)$, where as vehicles from the West are in the subset $W \subset V$ with a direction of travel of $d_w = (1, 0)$. Again, our only task is to find valid speed profiles for vehicles coming from the West.

To begin, the problem space is transformed so that the vehicles in W are represented as points rather than line segments. This makes movement planning simpler while maintaining the geometric properties of the original space. Every vehicle in W is contracted from left to right, until it is reduced to its leading point. In response, the vehicles in N are expanded, transforming each into a square obstacle with sides of length l and with their left edges coincident with the original line segments.

Given the global speed limit δ_{max}, there are regions in front of each obstacle in which a collision is inevitable (this concept is similar to the obstacle avoidance work done in [13]). These triangular zones (referred to as *collision zones*) are based on the speed constraints of the vehicles and are formed by a downward extension of the leading edge of each obstacle. The leftmost point of this edge is extended vertically and the rightmost point is extended at a slope derived from the ratio between δ_{max} and the obstacle speed. As one last concession to clarity, we scale the axes of our problem space so that this ratio becomes 1. Formally, a collision zone Z_O for the obstacle O is the set of all points p, such that there is no path originating at p with a piecewise slope in the interval $[1, \infty]$ that does not intersect O.

Expanding the vehicles in N into rectangular obstacles may cause some to overlap, producing larger obstacles and, consequently, larger collision zones. This merger and generation of collision zones is done through a standard sweep line algorithm and occurs in $O(n \log n)$ steps, where n is the number of obstacles, as described below.

Merging Obstacles and Growing Collision Zones. This process is done using a horizontal sweep line moving from top to bottom. While the following is a relatively standard application of a sweep line algorithm, it is included for the sake of completeness. First, the event list is populated with the horizontal edges of every obstacle, in top-to-bottom order, requiring $O(n \log n)$ time for $O(n)$ obstacles. The sweep line status stores a set of intervals representing the interiors of disallowed regions (e.g., the inside of an obstacle or collision zone). Each interval holds three pieces of information: the location of its left edge, a sorted list of the right edges of any obstacles within the interval, and the slopes of these right edges. These slopes will be either infinite (i.e., the edges are vertical) or will have a slope of 1.

In addition to horizontal edge positions, the event list must keep track of three other events which deal with the termination of the sloped edges of the collision zones. These edges begin at the bottom right edge of an obstacle and terminate in one of three ways: against the top of another obstacle, against the right edge of another obstacle, or by reaching the left edge of an interval. The first case is already in the event list as the top edges were added at the start of this process. The remaining two cases are added as the sweep line progresses through the obstacles.

The initial population of the event list occurs in $O(n \log n)$. As the sweep line progresses through the obstacle space, it adds and removes the right edges of obstacles to the appropriate intervals. These lists of edges are built incrementally in sorted order, requiring only $O(\log n)$ time. Finally, as there is a constant number of possible events per obstacle (a single top edge, a single bottom edge, and a single termination of its sloped edge), there are at most $O(n)$ events to be processed. Thus, the sweep line processes the obstacle space in $O(n \log n)$ time.

Movement Planning. Currently, vehicles only move horizontally and obstacles only move vertically. Instead, we will treat the obstacles as static objects and add a corresponding vertical component to the vehicles' motion. To find a movement plan, a vehicle moves through the obstacle space at maximum speed (giving it a slope of 1 under our scaled axes) until either reaching its goal or encountering an obstacle. If the goal is reached, the plan is complete. If an obstacle is encountered, the vehicle travels vertically until it is no longer blocked (this vertical motion corresponds to stopping and waiting for the obstacle to pass). Once the path is clear, the vehicle continues at maximum speed.

The path created by the above behavior can be found with another line sweep. The sweep line in this case is perpendicular to the vehicles' trajectories (giving it a slope of -1), moves from the upper right to the lower left, and determines how obstacles occlude one another, as seen from the vehicles' perspective. These occlusions reveal which obstacles are encountered and how the vehicle must move in order to follow the strategy laid out above.

4.2   Intersection Between a One-Way Street and a Two-Way Highway

In this case, vertical vehicles approach from the North and the South while horizontal vehicles travel in a single lane.

To account for the bidirectional vertical vehicles we fold the space along the horizontal lane. This rotates the northbound traffic to an equivalent southbound set of vehicles (see Fig. 2). This only requires a $O(n)$ transformation. Using the plane sweep algorithm above yields a combined obstacle space.

Finally, we must prevent the vehicles from rear-ending each other. Once the lead vehicle has found a motion plan through the obstacles, it creates a new set of constraints for the vehicles behind it. The monotonic path of the lead vehicle is stored in a binary search tree, allowing for easy collision queries. As each vehicle finds its own path through the obstacles, this search tree is updated to appropriately constrain subsequent vehicles [3].

In the end, we can still account for shared lanes without a running time greater than $O(n \log n)$.

---

[3] Details of how this is done can be found in the arXiv version of this paper.

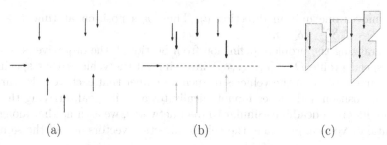

Fig. 2. (a) An example of bidirectional cross-traffic. (b) To account for how these vehicles interact when they reach a horizontal lane, we can fold the space along the lane, rotating one set of vehicles about it. (c) Then, we run the same space transformation and obstacle merger detailed above.

4.3 Intersection Between Two-Way Highways

Finally, this case combines the two above, allowing for bidirectional movement horizontally and vertically, with multiple lanes along each axis, and the possibility of collinear vehicles.

The vehicles approaching from the East are independent of those approaching from the West, presenting a symmetric problem that can be solved with the techniques discussed above. The addition of horizontal lanes, however, impacts the running time of the algorithm. Previously, the bidirectional vertical traffic was accounted for by folding the obstacle space along a single horizontal lane, but in this case, because the position of the vertical vehicles relative to each other is different at any given lane, the folding must occur individually for each lane. Thus, the algorithm runs in $O(kn \log n)$, for k horizontal lanes. In general, we assume that k is a relatively small constant.

5   Traffic Crossing in the Discrete Setting

In this section we consider the problem in a simple discrete setting, significantly simplifying the description of the algorithms and freeing us from a number of cumbersome continuous issues while still capturing the most salient elements of the original traffic-crossing problem. We assume that each vehicle occupies a point on the integer grid in the plane, $\mathbb{Z}^2$. Time advances discretely in unit increments, and at each time step a vehicle may either advance to the next grid point or remain where it is. A collision occurs if two vehicles occupy the same grid point.

The *discrete traffic crossing problem* is defined in much the same manner as in the continuous case. The problem is presented as a set V of n vehicles on the integer grid. Each vehicle v_i is represented by its initial and goal positions $p_i^\vdash$ and $p_i^\dashv$, respectively, both in $\mathbb{Z}^2$. Also given are a starting time $t_i^\vdash$ and deadline $t_i^\dashv$, both in $\mathbb{Z}^+$ (where $\mathbb{Z}^+$ denotes the set of nonnegative integers). A vehicle's direction d_i is a unit length vector directed from its initial position to its goal, which is either horizontal or vertical. Time proceeds in unit increments starting at zero. The motion of v_i is specified as a function of time, $\delta_i(t) \in \{0,1\}$. Setting $\delta_i(t) = 0$ means that at time t vehicle i remains stationary, and $\delta_i(t) = 1$ means that it moves one unit in direction d_i. Thus, v_i's position at time $t \geq 0$ is $p_i(t) = p_i^\vdash + d_i \sum_{x=0}^{t} \delta_i(x)$.

Generalizing the problem definition from Section 2, the objective is to compute a speed profile $D = \langle \delta_1, \ldots, \delta_n \rangle$ involving all the vehicles that specifies a collision-free motion of the vehicles in such a manner that each vehicle starts at its initial position and moves monotonically towards its goal, arriving there at or before its given deadline. Similar to road networks, we assume that along any horizontal or vertical grid line, the vehicle direction vectors are all the same.

5.1   Maximum Delay

Because we will be largely interested in establishing approximation bounds in this section, we will depart from the decision problem and consider a natural optimization problem instead, namely, minimizing the maximum delay experienced by any vehicle, defined formally as follows. For each vehicle we consider only its initial and goal positions, and let us assume that all vehicles share the same starting time at $t = 0$. A vehicle v_i experiences a *delay* at time t if it does not move at this time (that is, $\delta_i(t) = 0$). The *total delay* experienced by a vehicle is the total number of time instances where it experiences a delay until the end of the motion simulation. The *maximum delay* of the system is the maximum total delay experienced by any vehicle.

While we will omit a formal proof, it is not hard to demonstrate that the NP-hardness reduction of Section 3 can be transformed to one showing that it is NP-hard to minimize maximum delay in the discrete setting. (Intuitively, the reason is that the reduction involves purely discrete quantities: integer vehicle coordinates and starting times, vehicles of unit length, and unit speed limit. The system described in the reduction is feasible if and only if the maximum delay is at most five time units.) However, it is interesting to note that the question of whether there exists a solution involving at most single unit delay can be solved efficiently. This is stated in the following result.

Theorem 2. *There exists an $O(nm)$ time algorithm that, given an instance of the discrete traffic crossing problem with n vehicles where each vehicle encounters at most m intersections, determines whether there exists a solution with maximum delay of at most one time unit.*

Due to space limitations, we have omitted the proof, but the algorithm involves a straightforward reduction to 2-SAT. The key insight is that each vehicle can be in one of two states, *not-delayed* or *delayed*. Since all potential collisions involve pairs of vehicles, we can express the feasibility of a single unit delay solution through an instance of 2-SAT.

5.2   The Parity Heuristic

In the discrete setting it is possible to describe a simple common-sense heuristic. Intuitively, each intersection will alternate in allowing horizontal and vertical traffic to pass. Such a strategy might be far from optimal because each time a vehicle arrives at an intersection, it might suffer one more unit of delay. To address this, whenever a delay is imminent, we will choose which vehicle to delay in a manner that will avoid cross traffic at all future intersections. Define the *parity* of a grid point $p = (p_x, p_y)$ to be $(p_x + p_y)$ mod 2. Given a horizontally moving vehicle v_i and a time t, we say that v_i is *on-parity* at t if the parity of its position at time t equals t mod 2. Otherwise, it is *off-parity*. Vertically moving vehicles are just the opposite, being *on-parity* if the parity of their position is *not* equal to t mod 2. Observe that if two vehicles arrive at an intersection at the same time, one moving vertically and one horizontally, exactly one of them is on-parity. This vehicle is given the right of way, as summarized below.

Parity Heuristic: If two vehicles are about to arrive at the same intersection at the same time t, the vehicle that is on-parity proceeds, and the other vehicle waits one time unit (after which it will be on-parity, and will proceed).

The parity heuristic has a number of appealing properties. First, once all the vehicles in the system are on-parity, every vehicle may proceed at full speed without the possibility of further collisions. Second, the heuristic is not (locally) wasteful in the sense that it does not introduce a delay into the system unless a collision is imminent. Finally, the rule is scalable to large traffic systems, since a traffic controller at an intersection need only know the current time and the vehicles that are about to enter the intersection.

5.3   Steady-State Analysis of the Parity Heuristic

Delays may be much larger than a single time unit under the parity heuristic. (For example, a sequence of k consecutive vehicles traveling horizontally that encounters a similar sequence of k vertical vehicles will result in a cascade of delays, spreading each into an alternating sequence of length $2k$.) This is not surprising given the very simple nature of the heuristic. It is not difficult to construct counterexamples in which the maximum delay of the parity heuristic is arbitrarily large relative to an optimal solution. We will show, however, that the parity heuristic is asymptotically optimal in a uniform, steady-state scenario (to be made precise below).

Consider a traffic crossing pattern on the grid. Let m_x and m_y denote the numbers of vertical and horizontal lanes, respectively. Each lane is assigned a direction arbitrarily (up or down for vertical lanes and left or right for horizontal). Let R denote a $W \times W$ square region of the grid containing all the intersections (see Fig. 3(a)). In order to study the behavior of the system in steady-state, we will imagine that R is embedded on a torus, so that vehicles that leave R on one side reappear instantly in the same lane on the other side (see Fig. 3(b)). Equivalently, we can think of this as a system of infinite size by tiling the plane with identical copies (see Fig. 3(c)). We assume that W is even.

If the system is sufficiently dense, the maximum delay of the system will generally grow as a function of time. Given a scheduling algorithm and a discrete traffic crossing, define its *delay rate* to be the maximum delay after t time units divided by t. Define the *asymptotic delay rate* to be the limit supremum of the delay rate for $t \to \infty$. Our objective is to show that, given a suitably uniform traffic crossing instance on the torus, the asymptotic delay rate of the parity algorithm is optimal.

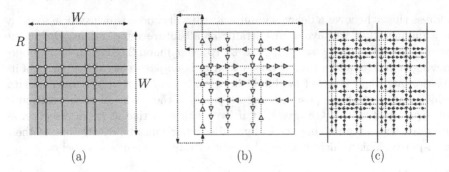

Fig. 3. Analysis of the Parity Heuristic

We say that a traffic crossing on the torus is *uniform* if every lane (within the square R) has an equal number of vehicles traveling on this lane. Letting n' denote this quantity, the total number of vehicles in the system is $n = n'(m_x + m_y)$. (The total number of positions possible is $W(m_x + m_y) - m_x m_y$, and so $n' \leq W - m_x m_y/(m_x + m_y)$.) The initial positions of the vehicles within each of the lanes is arbitrary. Let $p = n'/W$ denote the density of vehicles within each lane. Let $\rho_\infty^{\mathrm{par}} = \rho_\infty^{\mathrm{par}}(W, p, m_x, m_y)$ denote the worst-case asymptotic delay rate of the parity heuristic on any uniform discrete traffic crossing instance of the form described above, and let $\rho_\infty^{\mathrm{opt}} = \rho_\infty^{\mathrm{opt}}(W, p, m_x, m_y)$ denote the worst-case asymptotic delay for an optimum scheduler.

Our approach will be to relate the asymptotic performance of parity and the optimum to a parameter that describes the inherent denseness of the system. Define $\chi = \max(0, 2p - 1)$ to be the *congestion* of the system. Observe that $0 \leq \chi \leq 1$, where $\chi = 0$ means that the density is at most $1/2$ and $\chi = 1$ corresponds to placing vehicles at every available point on every lane (which is not really possible given that $n' < W$). To demonstrate that the parity heuristic is asymptotically optimal in this setting, it can be shown that $\rho_\infty^{\mathrm{par}} \leq \chi/(1+\chi) \leq \rho_\infty^{\mathrm{opt}}$. This is a consequence of the following two lemmas, whose proofs are omitted due to space constraints.

Lemma 2. *Given any uniform traffic crossing instance on the torus with congestion χ, $\rho_\infty^{\mathrm{par}} \leq \chi/(1 + \chi)$.*

Lemma 3. *Given any uniform traffic crossing instance on the torus with congestion χ, $\rho_\infty^{\mathrm{opt}} \geq \chi/(1 + \chi)$.*

While the proofs are somewhat technical, the intuition behind them is relatively straightforward. If $\chi = 0$, then while local delays may occur, there is sufficient capacity in the system for them to dissipate over time, and hence the asymptotic delay rate tends to zero as well. On the other hand, if $\chi > 0$, then due to uniformity and the cyclic nature of the system, delays will and must grow at a predictable rate. As an immediate consequence of the above lemmas, we have the following main result of this section.

Theorem 3. *Given a uniform traffic crossing instance on the torus, the asymptotic delay rate of the parity heuristic is optimal.*

References

1. Arkin, E.M., Mitchell, J.S.B., Polishchuk, V.: Maximum thick paths in static and dynamic environments. Comput. Geom. Theory Appl. **43**(3), 279–294 (2010)
2. Au, T.-C., Stone, P.: Motion planning algorithms for autonomous intersection management. In: Bridging the Gap Between Task and Motion Planning (2010)
3. Berger, F., Klein, R.: A traveller's problem. In: Proc. 26th Annu. Sympos. Comput. Geom., SoCG 2010, pp. 176–182. ACM, New York (2010)
4. Carlino, D., Boyles, S.D., Stone, P.: Auction-based autonomous intersection management. In: 2013 16th International IEEE Conference on Intelligent Transportation Systems-(ITSC), pp. 529–534. IEEE (2013)
5. Clarke, G., Wright, J.W.: Scheduling of vehicles from a central depot to a number of delivery points. Operations Res. **12**(4), 568–581 (1964)
6. Dantzig, G.B., Ramser, J.H.: The truck dispatching problem. Management Sci. **6**(1), 80–91 (1959)
7. Dresner, K., Stone, P.: Multiagent traffic management: a reservation-based intersection control mechanism. In: Proc. Third Internat. Joint Conf. on Auton. Agents and Multi. Agent Syst., pp. 530–537. IEEE Computer Society (2004)
8. Dresner, K., Stone, P.: Multiagent traffic management: an improved intersection control mechanism. In: Proc. Fourth Internat. Joint Conf. on Auton. Agents and Multi. Agent Syst., pp. 471–477. ACM (2005)
9. Dresner, K.M., Stone, P.: A multiagent approach to autonomous intersection management. J. Artif. Intell. Res. **31**, 591–656 (2008)
10. Fenton, R.E., Melocik, G.C., Olson, K.W.: On the steering of automated vehicles: Theory and experiment. IEEE Trans. Autom. Control **21**(3), 306–315 (1976)
11. Fiorini, P., Shiller, Z.: Motion planning in dynamic environments using velocity obstacles. Internat. J. Robot. Res. **17**(7), 760–772 (1998)
12. Hearn, R.A., Demaine, E.D.: Pspace-completeness of sliding-block puzzles and other problems through the nondeterministic constraint logic model of computation. Theo. Comp. Sci. **343**(1–2), 72–96 (2005)
13. Petti, S., Fraichard, T.: Safe motion planning in dynamic environments. In: 2005 IEEE/RSJ International Conference on Intelligent Robots and Systems, 2005. (IROS 2005), pp. 2210–2215, August 2005
14. Rajamani, R.: Vehicle Dynamics and Control. Springer Science & Business Media, December 2011
15. Solomon, M.M.: Algorithms for the vehicle routing and scheduling problems with time window constraints. Operations Res. **35**(2), 254–265 (1987)
16. Van Middlesworth, M., Dresner, K., Stone, P.: Replacing the stop sign: unmanaged intersection control for autonomous vehicles. In: Proc. Seventh Internat. Joint Conf. on Auton. Agents and Multi. Agent Syst., pp. 1413–1416. International Foundation for Autonomous Agents and Multiagent Systems (2008)
17. Wurman, P.R., D'Andrea, R., Mountz, M.: Coordinating hundreds of cooperative, autonomous vehicles in warehouses. The AI magazine **29**(1), 9–19 (2008)
18. Yu, J., LaValle, S.M.: Multi-agent path planning and network flow. In: Frazzoli, E., Lozano-Perez, T., Roy, N., Rus, D. (eds.) Algorithmic Foundations of Robotics X. Springer Tracts in Advanced Robotics, vol. 86, pp. 157–173. Springer, Heidelberg (2013). http://dx.doi.org/10.1007/978-3-642-36279-8_10

Finding Pairwise Intersections
Inside a Query Range

Mark de Berg[1], Joachim Gudmundsson[2], and Ali D. Mehrabi[1]([⊠])

[1] Department of Computer Science, TU Eindhoven, Eindhoven, The Netherlands
`amehrabi@win.tue.nl`
[2] School of IT, University of Sydney, Sydney, Australia

Abstract. We study the following problem: preprocess a set $\mathcal{O}$ of
objects into a data structure that allows us to efficiently report all pairs
of objects from $\mathcal{O}$ that intersect inside an axis-aligned query range Q.
We present data structures of size $O(n \operatorname{polylog} n)$ and with query time
$O((k + 1) \operatorname{polylog} n)$, where k is the number of reported pairs, for two
classes of objects in the plane: axis-aligned rectangles and objects with
small union complexity. For the 3-dimensional case where the objects and
the query range are axis-aligned boxes in $\mathbb{R}^3$, we present a data structure
of size $O(n\sqrt{n} \operatorname{polylog} n)$ and query time $O((\sqrt{n} + k) \operatorname{polylog} n)$. When
the objects and query are fat, we obtain $O((k + 1) \operatorname{polylog} n)$ query time
using $O(n \operatorname{polylog} n)$ storage.

1   Introduction

The study of geometric data structures is an important subarea within computational geometry, and range searching forms one of the most widely studied
topics within this area [1,11]. In a range-searching query, the goal is to report
or count all points from a given set $\mathcal{O}$ that lie inside a query range Q. The
more general version, where $\mathcal{O}$ contains other objects than just points and the
goal is to report all objects intersecting Q, is often called intersection searching
and it has been studied extensively as well. A common characteristic of almost
all range-searching and intersection-searching problems studied so far, is that
whether an object $o_i \in \mathcal{O}$ should be reported (or counted) depends only on o_i
and Q. In this paper we study a range-searching variant where we are interested
in reporting *pairs* of objects that satisfy a certain criterion. In particular, we
want to preprocess a set $\mathcal{O} = \{o_1, \ldots, o_n\}$ of n objects in the plane such that,
given a query range Q, we can efficiently report all pairs of objects o_i, o_j that
intersect inside Q.

Our motivation for studying these problems is the following. Suppose we are
given a collection of n discrete trajectories representing the movements of, say,

M. de Berg and A.D. Mehrabi were supported by the Netherlands Organization for
Scientific Research (NWO) under grants 024.002.003 and 612.001.118, respectively.
J. Gudmundsson was supported by the Australian Research Council (project num-
bers FT100100755 and DP150101134).

F. Dehne et al. (Eds.): WADS 2015, LNCS 9214, pp. 236–248, 2015.
DOI: 10.1007/978-3-319-21840-3_20

people. Each trajectory is a sequence of locations (points in the plane) with a corresponding time stamp; for discrete trajectories the movement in between consecutive locations is not considered. The query we are interested in is: which pairs of people met inside a given rectangular query region Q? A natural way to define that two people meet is to require that they are within a given distance D from each other. When we restrict our attention to a fixed time instance, we can place a disk of radius $D/2$ around the location of each person and the question becomes: which pairs of disks intersect within Q? When we consider the ℓ_∞ metric, we get the same problem but now for squares instead of disks. A more general version of the query also specifies a time interval I: which pairs of people met within a region Q' during time interval I? To deal with the fact that the time stamps may not be synchronized for the different trajectories, we assume that each location is valid for some interval of time. If we then model time as the third dimension and consider distances in the ℓ_∞ metric, we get the question: which pairs of boxes (which are the product of a square around a location and a time interval) intersect with the query box $Q := Q' \times I$?

An obvious approach to our problem is to precompute all intersections between the objects and store the intersections in a suitable intersection-searching data structure. This may give fast query times, but in the worst case any two objects intersect, so $\Omega(n^2)$ is a lower bound on the storage for this approach. The main question is thus: can we achieve fast query times with a data structure that uses subquadratic (and preferably near-linear) storage in the worst case?

Rahul *et al.* [13] answered this question affirmatively when Q is an axis-aligned rectangle in the plane and the objects are axis-aligned line segments. Their data structure uses $O(n \log n)$ storage and answers queries in time $O(\log n + k)$, where k is the number of answers. Our contribution is to obtain similar results for a broader class of objects than those of [13], namely axis-aligned rectangles and objects with small union complexity. For axis-aligned rectangles our data structure uses $O(n \log n)$ storage and has $O(\log n \log^* n + k \log n)$ query time,[1] where k is the number of reported pairs of objects. Our data structure for classes of objects with small union complexity—disks and other types of fat objects are examples—uses $O(U(n) \log n)$ storage, where $U(n)$ is maximum union complexity of n objects from the given class, and it has $O((k + 1) \log^2 n)$ query time. We also consider a 3-dimensional version of the problem, where the range Q and the objects in $\mathcal{O}$ are axis-aligned boxes. Here our data structure uses $O(n\sqrt{n} \log n)$ storage and $O((\sqrt{n} + k) \log^2 n)$ query time. When the query range and the objects are fat, we improve this to $O(n \log^2 n)$ storage and $O((k + 1) \log^2 n)$ query time.

2   Axis-Aligned Objects

In this section we study the case where the set $\mathcal{O}$ is a set of n axis-aligned rectangles in the plane or boxes in $\mathbb{R}^3$. Our approach for these cases is the same and uses the following two-step query process.

---

[1] Here $\log^* n$ denotes the iterated logarithm.

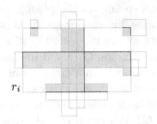

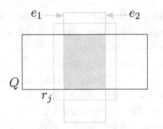

Fig. 1. Gray areas are intersections of other rectangles with r_i, black segments indicate witness segments

Fig. 2. Example of Case B-3-I

1. Compute a *seed set* $\mathcal{O}^*(Q) \subseteq \mathcal{O}$ of objects such that the following holds: for any two objects o_i, o_j in $\mathcal{O}$ such that o_i and o_j intersect inside Q, at least one of o_i, o_j is in $\mathcal{O}^*(Q)$.
2. For each seed object $o_i \in \mathcal{O}^*(Q)$, perform an intersection query with the range $o_i \cap Q$ in the set $\mathcal{O}$, to find all objects $o_j \neq o_i$ intersecting o_i inside Q.

For this approach to be efficient, $\mathcal{O}^*(Q)$ should not contain too many objects that do not give an answer in Step 2. For the planar case we will ensure $|\mathcal{O}^*(Q)| = O(1 + k)$, where k is the number of pairs of objects intersecting inside Q, while for the 3-dimensional case we will have $|\mathcal{O}^*(Q)| = O(\sqrt{n} + k)$.

2.1 The Planar Case

Let $\mathcal{O} = \{r_1, \ldots, r_n\}$ be a set of axis-aligned rectangles in the plane. The key to our approach is to be able to efficiently find the seed set $\mathcal{O}^*(Q)$. To this end, during the preprocessing we compute a set W of axis-aligned *witness segments*. For each rectangle $r_i \in \mathcal{O}$ we define at most ten witness segments, two for each edge of r_i and two in the interior of r_i, as follows—see also Fig. 1.

Let e be an edge of r_i, and consider the set $S(e) := e \cap (\cup_{j \neq i} r_j)$, that is, the part of e covered by the other rectangles. The set $S(e)$ consists of a number of sub-edges of e. If e is vertical then we add the topmost and bottommost sub-edge from $S(e)$ (if any) to W; if e is horizontal we add the leftmost and rightmost sub-edge to W. The two witness segments in the interior of r_i are defined as follows. Suppose there are vertical edges (belonging to other rectangles r_j) completely crossing r_i from top to bottom. Then we put $e' \cap r_i$ into W, where e' is the rightmost such crossing edge. Similarly, we put into W the topmost horizontal edge e'' completely crossing r_i from left to right. Our data structure to find the seed set $\mathcal{O}^*(Q)$ now consists of the following components.

- We store the witness set W in a data structure $\mathcal{D}_1$ that allows us to report the witness segments that intersect the query rectangle Q.
- We store the vertical edges of the rectangles in $\mathcal{O}$ in a data structure $\mathcal{D}_2$ that allows us to decide if the set $\mathsf{V}(Q)$ of edges that completely cross a query

rectangle Q from top to bottom, is non-empty. The data structure should also be able to report all (rectangles corresponding to) the edges in $V(Q)$.

- We store the horizontal edges of the rectangles in $\mathcal{O}$ in a data structure $\mathcal{D}_3$ that allows us to decide if the set $H(Q)$ of edges that completely cross a query rectangle Q from left to right, is non-empty.
- We store the set $\mathcal{O}$ in a data structure $\mathcal{D}_4$ that allows us to report the rectangles that contain a query point q.

Step 1 of the query procedure, where we compute $\mathcal{O}^*(Q)$, proceeds as follows.

1(i) Perform a query in $\mathcal{D}_1$ to find all witness segments intersecting Q. For each reported witness segment, insert the corresponding rectangle into $\mathcal{O}^*(Q)$.

1(ii) Perform queries in $\mathcal{D}_2$ and $\mathcal{D}_3$ to decide if the sets $V(Q)$ and $H(Q)$ are both non-empty. If so, report all rectangles corresponding to edges in $V(Q)$ and put them into $\mathcal{O}^*(Q)$.

1(iii) For each corner point q of Q, perform a query in $\mathcal{D}_4$ to report all rectangles in $\mathcal{O}$ that contain q, and put them into $\mathcal{O}^*(Q)$.

Lemma 1. *Let r_i, r_j be two rectangles in $\mathcal{O}$ such that $(r_i \cap r_j) \cap Q \neq \emptyset$. Then at least one of r_i, r_j is put into $\mathcal{O}^*(Q)$ by the above query procedure.*

Proof. Let $I := (r_i \cap r_j) \cap Q$. Each edge of I is either contributed by r_i or r_j, or by Q. Let $E(I)$ denote the (possibly empty) set of edges of r_i and r_j that contribute an edge to I. We distinguish two cases, with various subcases.

CASE A: At least one edge $e \in E(I)$ has an endpoint, v, inside Q. Now the witness sub-edge on e closest to v must intersect Q and, hence, the corresponding rectangle will be put into $\mathcal{O}^*(Q)$ in Step 1(i).

CASE B: All edges in $E(I)$ cross Q completely. We now have several subcases.

CASE B-1: $|E(I)| \leqslant 1$. Now Q contributes at least three edges to I, so at least one corner of I is a corner of Q. Hence, both r_i and r_j are put into $\mathcal{O}^*(Q)$ in Step 1(iii).

CASE B-2: $|E(I)| \geqslant 3$. Since each edge of $E(I)$ crosses Q completely and $|E(I)| \geqslant 3$, both $V(Q)$ and $H(Q)$ are non-empty. Thus at least one of r_i and r_j is put into $\mathcal{O}^*(Q)$ in Step 1(ii).

CASE B-3: $|E(I)| = 2$. Let e_1 and e_2 denote the segments in $E(I)$. If one of e_1, e_2 is vertical and the other is horizontal, we can use the argument from Case B-2. It remains to handle the case where e_1 and e_2 have the same orientation, say vertical.

CASE B-3-I: Edges e_1 and e_2 belong to the same rectangle, say r_i, as in Fig. 2. If e_1 has an endpoint, v, inside r_j, then e_1 has a witness sub-edge starting at v that intersects Q, so r_i is put into $\mathcal{O}^*(Q)$ in Step 1(i). If r_j contains a corner of Q then r_j will be put into $\mathcal{O}^*(Q)$ in Step 1(iii). In the remaining case the right edge of r_j crosses Q and there are vertical edges completely crossing r_j (namely e_1 and e_2). Hence, the rightmost edge completely crossing r_j, which is a witness for r_j, intersects Q. Thus r_j is put into $\mathcal{O}^*(Q)$ in Step 1(i).

CASE B-3-II: Edge e_1 is an edge of r_i and e_2 is an edge of r_j (or vice versa). Assume without loss of generality that the y-coordinate of the top endpoint of e_1 is less than or equal to the y-coordinate of the top endpoint of e_2. Then the top endpoint, v, of e_1 must lie in r_j, and so e_1 has a witness sub-edge starting at v that intersects Q. Hence, r_i is put into $\mathcal{O}^*(Q)$ in Step 1(i). □

In the second part of the query procedure we need to report, for each rectangle r_i in the seed set $\mathcal{O}^*(Q)$, the rectangles $r_j \in \mathcal{O}$ intersecting $r_i \cap Q$. Thus we store $\mathcal{O}$ in a data structure $\mathcal{D}_5$ that can report all rectangles intersecting a query rectangle. Putting everything together we obtain the following theorem.

Theorem 1. *Let $\mathcal{O}$ be a set of n axis-aligned rectangles in the plane. There is a data structure that uses $O(n \log n)$ storage and can report, for any axis-aligned query rectangle Q, all pairs of rectangles r_i, r_j in $\mathcal{O}$ such that r_i intersects r_j inside Q in $O(\log n \log^* n + k \log n)$ time, where k denotes the number of answers.*

Proof. For the data structure $\mathcal{D}_1$ on the set W we use the data structure developed by Edelsbrunner *et al.* [9], which uses $O(n \log n)$ preprocessing time and storage, and has $O(\log n + \#\text{answers})$ query time. For data structure $\mathcal{D}_2$ (and, similarly, $\mathcal{D}_3$) we note that a vertical segment $s_i := x_i \times [y_i, y_i']$ crosses $Q := [x_Q, x_Q'] \times [y_Q, y_Q']$ if and only if the point (x_i, y_i, y_i') lies in the range $[x_Q, x_Q'] \times [-\infty, y_Q] \times [y_Q', \infty]$. Hence, we can use the data structure of Subramanian and Ramaswamy [14], which uses $O(n \log n)$ storage and has $O(\log n \log^* n + \#\text{answers})$ query time. For data structure $\mathcal{D}_4$ we use the point-enclosure data structure developed by Chazelle [4], which uses $O(n)$ storage and can be used to report all rectangles in $\mathcal{O}$ containing a query point in $O(\log n + \#\text{answers})$ time.

Note that $|\mathcal{O}^*(Q)| \leqslant 2k + 4$ where k is the total number of reported pairs. Indeed, each rectangle in $\mathcal{O}^*(Q)$ intersects at least one other rectangle inside Q and for every reported pair we put at most two rectangles into the seed set; the extra term "+4" is because in Step 1(iii) we may report at most one rectangle per corner of Q that does not have an intersection inside Q. Hence, the time for Step 1 is $O(\log n \log^* n + |\mathcal{O}^*(Q)|) = O(\log n \log^* n + k)$.

It remains to analyze Step 2 of the query procedure, where we need to find for a given $r_i \in \mathcal{O}^*(Q)$ all $r_j \in \mathcal{O}$ such that $r_i \cap Q$ intersects r_j. First notice that a rectangle r_j intersects a rectangle $r_i' := r_i \cap Q$ if and only if (i) a corner of r_j is inside r_i', or (ii) a corner of r_i' is inside r_j, or (iii) an edge of r_j intersects an edge of r_i'. Thus $\mathcal{D}_5$ consists of three components: All r_j satisfying (i) can be found in $O(\log n + \#\text{answers})$ time using a range tree with fractional cascading [3], which uses $O(n \log n)$ storage. All r_j satisfying (ii) and (iii) can be found using, respectively, the data structure by Chazelle [4] and the one by Edelsbrunner *et al.* [9]. Thus the running time of Step 2 is $\sum_{r_i \in \mathcal{O}^*(Q)} O(\log n + k_i)$, where k_i denotes the number of rectangles in $\mathcal{O}$ that intersect r_i inside Q, and so the total time for Step 2 is $O((k + 1) \log n)$. □

2.2   The 3-Dimensional Case

We now study the case where the set $\mathcal{O}$ of objects and the query range Q are axis-aligned boxes in $\mathbb{R}^3$. We first present a solution for the general case, and then an improved solution for the special case where the input as well as the query are cubes. Both solutions use the same query strategy as above: we first find a seed set $\mathcal{O}^*(Q)$ that contains at least one object o_i from every pair that intersects inside Q and then we find all other objects intersecting o_i inside Q.

The General Case. Let $\mathcal{O} := \{b_1, \ldots, b_n\}$ be a set of axis-aligned boxes. The pairs of boxes b_i, b_j intersecting inside Q come in three types: (i) $b_i \cap b_j$ fully contains Q, (ii) $b_i \cap b_j$ lies completely inside Q, (iii) $b_i \cap b_j$ intersects a face of Q.

Type (i) is easy to handle without using seeds sets: we simply store $\mathcal{O}$ in a data structure for 3-dimensional point-enclosure queries [4], which allows us to report all boxes $b_i \in \mathcal{O}$ containing a query point in $O(\log^2 n + \#\text{answers})$ time. If we query this structure with a corner q of Q and report all pairs of boxes containing q then we have found all intersecting pairs of Type (i).

Lemma 2. *We can find all intersecting pairs of boxes of Type (i) in $O(\log^2 n + k)$ time, where k is the number of such pairs, with a structure of size $O(n \log n)$.*

For Type (ii) we proceed as follows. Note that a vertex of $b_i \cap b_j$ is either a vertex of b_i or b_j, or it is the intersection of an edge e of one of these two boxes and a face f of the other box. To handle the first case we create a set W of witness points, which contains for each box b_i all its vertices that are contained in at least one other box. We store W in a data structure for 3-dimensional orthogonal range reporting [14]. In the query phase we then query this data structure with Q, and put all boxes corresponding to the witness vertices inside Q into the seed set $\mathcal{O}^*(Q)$. For the second case we show next how to find the intersecting pairs e, f where e is a vertical edge (that is, parallel to the z-axis) and f is a horizontal face (that is, parallel to the xy-plane); the intersecting pairs with other orientations can be found in a similar way.

Let E be the set of vertical edges of the boxes in $\mathcal{O}$ and let F be the set of horizontal faces. We sort F by z-coordinate—we assume for simplicity that all z-coordinates of the faces are distinct—and partition F into $O(\sqrt{n})$ *clusters*: the cluster F_1 contains the first $\sqrt{n}$ faces in the sorted order, the second cluster F_2 contains the next $\sqrt{n}$ faces, and so on. We call the range between the minimum and maximum z-coordinate in a cluster its *z-range*. For each cluster F_i we store, besides its z-range and the set F_i itself, the following information. Let $E_i \subseteq E$ be the subset of edges that intersect at least one face in F_i, and let $\overline{E_i}$ denote the set of points obtained by projecting the edges in E_i onto the xy-plane. We store $\overline{E_i}$ in a data structure $\mathcal{D}(\overline{E_i})$ for 2-dimensional orthogonal range reporting. Note that for a query box Q whose z-range contains the z-range of F_i we have: an edge $e \in E$ intersects at least one face $f \in F_i$ inside Q if and only if $e \in E_i$ and $\overline{e}$ lies in $\overline{Q}$, the projection of Q onto the xy-plane.

A query with a box $Q = [x_1 : x_2] \times [y_1 : y_2] \times [z_1 : z_2]$ is now answered as follows. We first find the clusters F_i and F_j whose z-range contains z_1 and z_2, respectively, and we put (the boxes corresponding to) the faces in these clusters into the seed set $\mathcal{O}^*(Q)$. Next we perform, for each $i < t < j$, a query with the projected range $\overline{Q}$ in the data structure $\mathcal{D}(\overline{E_t})$. For each of the reported points $\overline{e}$ we put the box corresponding to the edge e into the seed set $\mathcal{O}^*(Q)$. Finally, we remove any duplicates from the seed set. This leads to the following lemma.

Lemma 3. *Using a data structure of size $O(n\sqrt{n}\log n)$ we can find in time $O(\sqrt{n}\log n + k)$ a seed set $\mathcal{O}^*(Q)$ of $O(\sqrt{n}+k)$ boxes containing at least one box from every intersecting pair of Type (ii), where k is the number of such pairs.*

It remains to handle the Type (iii) pairs, in which $b_i \cap b_j$ intersects a face of Q. We describe how to find the pairs such that $b_i \cap b_j$ intersects the bottom face of Q; the pairs intersecting the other faces can be found in a similar way.

We first sort the z-coordinates of the horizontal faces of the boxes in $\mathcal{O}$. For $1 \leqslant i \leqslant 2\sqrt{n}$, let h_i be a horizontal plane containing the $(i\sqrt{n})$-th horizontal face. These planes partition $\mathbb{R}^3$ into $O(\sqrt{n})$ horizontal slabs $\Sigma_0, \ldots, \Sigma_{2\sqrt{n}+1}$. We call a box $b \in \mathcal{O}$ *short* at Σ_i if it has a horizontal face inside Σ_i, and we call it *long* if it completely crosses Σ_i. For each Σ_i, we store the short boxes in a list. We store the projections of the long boxes onto the xy-plane in a data structure $\mathcal{D}(\Sigma_i)$ for the 2-dimensional version of the problem, namely the structure of Theorem 1.

A query with the bottom face of Q is now answered as follows. We first find the slab Σ_i containing the face. We put all short boxes of Σ_i into our seed set $\mathcal{O}^*(Q)$. We then perform a query with $\overline{Q}$, the projection of Q onto the xy-plane, in the data structure $\mathcal{D}(\Sigma_i)$. For each answer we get from this 2-dimensional query—that is, each pair of projections intersecting inside $\overline{Q}$—we directly report the corresponding pair of long boxes. (There is no need to go through the seed set for these pairs.) This leads to the following lemma for the Type (iii) pairs.

Lemma 4. *Using a data structure of size $O(n\sqrt{n}\log n)$ we can find in time $O(\sqrt{n} + k\log n\log^* n)$ a seed set $\mathcal{O}^*(Q)$ of $O(\sqrt{n})$ boxes plus a collection $B(Q)$ of pairs of boxes intersecting inside Q such that, for each pair of Type (iii) boxes, either at least one of these boxes is in $\mathcal{O}^*(Q)$ or b_i, b_j is a pair in $B(Q)$.*

In Step 2 of our query procedure we need to report all boxes $b_j \in \mathcal{O}$ intersecting a query box $B := Q \cap b_i$, where $b_i \in \mathcal{O}^*(Q)$. Note that B intersects b_j if (i) B contains a vertex of b_j, or (ii) a vertex of B is contained in b_j, or (iii) an edge e of B intersects a face of b_j, or (iv) a face f of B intersects an edge of b_j. All r_j satisfying (i) and (ii) can be found using a 3D range tree with fractional cascading [3] and the 3D point-enclosure data structure of [4], respectively. For (iii), assume e is parallel to the z-axis and consider the faces of b_j parallel to the xy-plane. Then we can use a 2-level structure whose first level is a tree on the z-coordinates of the faces, and whose second-level structures are 2D point-enclosure structures [4] on the projections onto the xy-plane. For (iv), assume f is parallel to the xy-plane and consider the edges of b_j parallel to the z-axis. Then we can use 2-level structure whose first level is a segment tree on the z-ranges of the edges, and whose second-level structures are 2D range trees (with fractional cascading). The components that we need for Step 2 together need $O(n\log^2 n)$ storage and querying takes $O(\log^2 n + \#\text{answers})$ time.

Putting everything together we obtain the following theorem.

Theorem 2. *Let $\mathcal{O}$ be a set of n axis-aligned boxes in $\mathbb{R}^3$. Then there is a data structure that uses $O(n\sqrt{n}\log n)$ storage and that allows us to report, for any axis-aligned query box Q, all pairs of boxes b_i, b_j in $\mathcal{O}$ such that b_i intersects b_j inside Q in $O((\sqrt{n}+k)\log^2 n)$ time, where k denotes the number of answers.*

Fat Boxes. Next we obtain better bounds when the boxes in $\mathcal{O}$ and the query box Q are fat, that is, when their *aspect ratio*—the ratio between the length of the longest edge and the length of the shortest edge—is bounded by a constant α. First we consider the case of cubes.

Let $\mathcal{O} := \{c_1, \cdots, c_n\}$ be a set of n cubes in $\mathbb{R}^3$ and let Q be the query cube. We compute a set W of witness points for each cube c_i, as follows. Let e be an edge of c_i, and consider the set $S(e) := e \cap (\cup_{j \neq i} c_j)$, that is, the part of e covered by the other cubes. We put the two extreme points from $S(e)$—in other words, the two points closest to the endpoints of e—into W. Similarly, we assign each face f of c_i at most four witness points, namely points from $S(f) := f \cap (\cup_{j \neq i} c_j)$ that are extreme in the axis-aligned directions parallel to f. For example, if f is parallel to the xy-plane, then we take points of maximum and minimum x-coordinate in $S(f)$ and points of maximum and minimum y-coordinate in $S(f)$ as witnesses. We store W in a data structure $\mathcal{D}_1$ for orthogonal range queries, and we store $\mathcal{O}$ in a data structure $\mathcal{D}_2$ for point-enclosure queries.

To compute $\mathcal{O}^*(Q)$ in the first phase of the query procedure, we query $\mathcal{D}_1$ to find all witness points inside Q and for each reported witness point, we insert the corresponding cube into $\mathcal{O}^*(Q)$. Furthermore, for each corner point q of Q, we query $\mathcal{D}_2$ to find the cubes in $\mathcal{O}$ that contain q, and we put them into $\mathcal{O}^*(Q)$.

Lemma 5. *Let c_i, c_j be two cubes in $\mathcal{O}$ such that $(c_i \cap c_j) \cap Q \neq \emptyset$. Then at least one of c_i, c_j is put into $\mathcal{O}^*(Q)$ by the above query procedure.*

Proof. Suppose $c_i \cap c_j$ intersects Q, and assume without loss of generality that c_i is not larger than c_j. If c_i or c_j contains a corner q of Q then the corresponding cube will be put into the seed set when we perform a point-enclosure query with q, so assume c_i and c_j do not contain a corner. We have two cases.

CASE A: c_i does not intersect any edge of Q. Because c_i and Q are cubes, this implies that c_i is contained in Q or c_i intersects exactly one face of Q. Assume that c_i intersects the bottom face of Q; the cases where c_i intersects another face and where c_i is contained in Q can be handled similarly. We claim that at least one of the vertical faces of c_i contributes a witness point inside Q. To see this, observe that c_j will intersect at least one vertical face, f, of c_i inside Q, since c_j intersects c_i inside Q and c_i is not larger than c_j. Hence, the witness point on f with maximum z-coordinate will be inside Q. Thus c_i will be put into $\mathcal{O}^*(Q)$.

CASE B: c_i intersects one edge of Q. (If c_i intersects more than one edge of Q then it would contain a corner of Q.) Assume without loss of generality that c_i intersects the bottom edge of the front face of Q; see Fig. 3. Observe that if c_j intersects the top face of c_i then the witness point of the face with minimum x-coordinate is inside Q. Similarly, if c_j intersects the back face of c_i (the face parallel to the yz-plane and with minimum x-coordinate) then the witness point

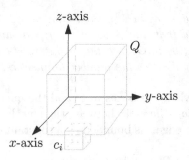

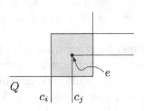

Fig. 3. Case B in the proof of Lemma 5; c_j is not shown

Fig. 4. Cross-section of Q, c_i, and c_j with a plane parallel to the xz-plane. The gray area indicates $Q \cap c_i$ in the cross-section.

of the face with maximum z-coordinate is inside Q. Otherwise, as illustrated in Fig 4, c_j must have an edge e parallel to the y-axis that intersects c_i inside Q, and one of the witness points on e will be inside Q—note that e lies fully inside Q because c_j does not contain a corner of Q.   □

To adapt the above solution to boxes of aspect ratio at most α, we cover each box $b_i \in \mathcal{O}$ by $O(\alpha^2)$ cubes, and preprocess the resulting collection $\widetilde{\mathcal{O}}$ of cubes as described above, making sure we do not introduce witness points for pairs of cubes used in the covering of the same box b_i. To perform a query, we cover Q by $O(\alpha^2)$ query cubes and compute a seed set for each query cube. We take the union of these seed sets, replace the cubes from $\widetilde{\mathcal{O}}$ in the seed set by the corresponding boxes in $\mathcal{O}$, and filter out duplicates. This gives us our seed set $\mathcal{O}^*(Q)$ for the second phase of the query procedure.

In the second phase we take each $b_i \in \mathcal{O}^*(Q)$ and report all $b_j \in \mathcal{O}$ intersecting $b_i \cap Q$, using the data structure $\mathcal{D}^*$ described just before Theorem 2. We obtain the following theorem.

Theorem 3. *Let $\mathcal{O}$ be a set of n axis-aligned boxes in $\mathbb{R}^3$ of aspect ratio at most α. Then there is a data structure that uses $O(\alpha^2 n \log^2 n)$ storage and that allows us to report, for any axis-aligned query box Q of aspect ratio at most α, all pairs of cubes c_i, c_j in $\mathcal{O}$ such that c_i intersects c_j inside Q in $O(\alpha^2(k+1)\log^2 n)$ time, where k denotes the number of answers.*

Proof. The data structures $\mathcal{D}_1$ and $\mathcal{D}_2$ can be implemented such that they use $O(n \log n)$ storage, and have $O(\log n \log^* n + \#\text{answers})$ and $O(\log^2 n + \#\text{answers})$ query time, respectively [4,14]. Since Step 2 of the query procedure is the same as the second step of query procedure of Subsection 2.2 we can use the data structures that we designed there, which need $O(n \log^2 n)$ storage and have $O(\log^2 n + \#\text{answers})$ query time. The conversion of boxes of aspect ratio α to cubes give an additional factor $O(\alpha^2)$.   □

3   Objects with Small Union Complexity in the Plane

In the previous section we presented efficient solutions for the case where $\mathcal{O}$ consists of axis-aligned rectangles. In this section we obtain results for classes of constant-complexity objects (which may have curved boundaries) with small union complexity. More precisely, we need that $U(n)$, the maximum union complexity of any set of n objects from the class, is small. This is for instance the case for disks (where $U(m) = O(m)$ [12]) and for locally fat objects (where $U(m) = m2^{O(\log^* m)}$ [2]).

In Step 2 of the query algorithm of the previous section, we performed a range query with $o_i \cap Q$ for each $o_i \in \mathcal{O}^*(Q)$. When we are dealing with arbitrary objects, this will be expensive, so we modify our query procedure.

1. Compute a seed set $\mathcal{O}^*(Q) \subseteq \mathcal{O}$ of objects such that, for any two objects o_i, o_j in $\mathcal{O}$ intersecting inside Q, both o_i and o_j are in $\mathcal{O}^*(Q)$.
2. Compute all intersecting pairs of objects in the set $\{o_i \cap Q : o_i \in \mathcal{O}^*(Q)\}$ by a plane-sweep algorithm.

Next we describe how to efficiently find $\mathcal{O}^*(Q)$, which should contain all objects intersecting at least one other object inside Q, when the union complexity $U(n)$ is small. For each object $o_i \in \mathcal{O}$ we define $o_i^* := \bigcup_{o_j \in \mathcal{O}, j \neq i}(o_i \cap o_j)$ as the union of all intersections between o_i and all other objects in $\mathcal{O}$. Let $|o_i^*|$ denote the complexity (that is, number of vertices and edges) of o_i^*.

Lemma 6. $\sum_{i=1}^{n} |o_i^*| = O(U(n))$.

Proof. Consider the arrangement induced by the objects in $\mathcal{O}$. We define the *level* of a vertex v in this arrangement as the number of objects from $\mathcal{O}$ that contain v in their interior. We claim that every vertex of any o_i^* is a level-0 or level-1 vertex. Indeed, a level-k vertex for $k > 1$ is in the interior of more than one object, which implies it cannot be a vertex of any o_i^*.

Since the level-0 vertices are exactly the vertices of the union of $\mathcal{O}$, the total number of level-0 vertices is $U(n)$. It follows from the Clarkson-Shor technique [7] that the number of level-1 vertices is $O(U(n))$ as well. The lemma now follows, because each level-0 or level-1 vertex contributes to at most two different o_i^*'s.   □

Our goal in Step 1 is to find all objects o_i such that o_i^* intersects Q. To this end consider the connected components of o_i^*. If o_i^* intersects Q then one of these components lies completely inside Q or an edge of Q intersects o_i^*.

Lemma 7. *We can find all o_i^* that have a component completely inside Q in $O(\log n + k)$ time, where k is the number of pairs of objects that intersect inside Q, with a data structure that uses $O(U(n) \log n)$ storage.*

Proof. For each o_i, take an arbitrary representative point inside each component of o_i^*, and store all the representative points in a structure for orthogonal range reporting. By Lemma 6 we store $O(U(n))$ points, and so the structure for orthogonal range reporting uses $O(U(n) \log n)$ storage.

The query time is $O(\log n + t)$, where t is the number of representative points inside Q. This implies the query time is $O(\log n + k)$, because if o_i^* has t_i representative points inside Q then o_i intersects $\Omega(t_i)$ other objects inside Q. This is true because the objects have constant complexity, so a single object o_j cannot generate more than a constant number of components of o_i^*.   □

Next we describe a data structure for reporting all o_i^* intersecting a vertical edge of Q; the horizontal edges of Q can be handled similarly. The data structure is a balanced binary tree $\mathcal{T}$, whose leaves are in one-to-one correspondence to the objects in $\mathcal{O}$. For an (internal or leaf) node ν in $\mathcal{T}$, let $\mathcal{T}(\nu)$ denote the subtree rooted at ν and let $\mathcal{O}(\nu)$ denote the set of objects corresponding to the leaves of $\mathcal{T}(\nu)$. Define $\mathcal{U}(\nu) := \cup_{o_i \in \mathcal{O}(\nu)} o_i^*$. At node ν, we store a point-location data structure [8] on the trapezoidal map of $\mathcal{U}(\nu)$. (If the objects are curved, then the "trapezoids" may have curved top and bottom edges.)

Lemma 8. *The tree $\mathcal{T}$ uses $O(U(n) \log n)$ storage and allows us to report all o_i^* intersecting a vertical edge s of Q in $O((t+1) \log^2 n)$ time, where t is the number of answers.*

Proof. To report all o_i^* intersecting s we walk down $\mathcal{T}$, only visiting the nodes ν such that s intersects $\mathcal{U}(\nu)$. This way we end up in the leaves corresponding to the o_i^* intersecting s. To decide if we have to visit a child ν of an already visited node, we do a point location with both endpoints of s in the trapezoidal map of $\mathcal{U}(\nu)$. Now s intersects $\mathcal{U}(\nu)$ if and only if one of these endpoints lies in a trapezoid inside $\mathcal{U}(\nu)$ and/or the two endpoints lie in different trapezoids. Thus we spend $O(\log n)$ time for the decision. Since we visit $O(t \log n)$ nodes, the total query time is as claimed.

To analyze the storage we claim that the sum of the complexities of $\mathcal{U}(\nu)$ over all nodes ν at any fixed height of $\mathcal{T}$ is $O(U(n))$. The bound on the storage then follows because the point-location data structures take linear space [8] and the height of $\mathcal{T}$ is $O(\log n)$. It remains to prove the claim. Consider a node ν at a given height h in $\mathcal{T}$. It can be argued that each vertex in $\mathcal{U}(\nu)$ is either a level-0 or level-1 vertex of the arrangement induced by the objects in $\mathcal{O}(\nu)$, or a vertex of o_i^*, for some o_i in $\mathcal{O}(\nu)$. The proof of the claim then follows from the following two facts. First, the number of vertices of the former type is $O(U(|\mathcal{O}(\nu)|))$, which sums to $O(U(n))$ over all nodes at height h. Second, by Lemma 6 the number of vertices of the latter type over all nodes at height h sums to $O(U(n))$.   □

Theorem 4. *Let $\mathcal{O}$ be a set of n constant-complexity objects in the plane from a class of objects such that the maximum union complexity of any m objects from the class is $U(m)$. Then there is a data structure that uses $O(U(n) \log n)$ storage and that allows us to report for any axis-aligned query rectangle Q, in $O((k+1) \log^2 n)$ time all pairs of objects o_i, o_j in $\mathcal{O}$ such that o_i intersects o_j inside Q, where k denotes the number of answers.*

4   Concluding Remarks

We presented data structures for finding intersecting pairs of objects inside a query rectangle. An obvious open problem is whether our bounds can be improved. In particular, one would hope that better solutions are possible for 3-dimensional boxes, where we obtained $O((k + \sqrt{n})\operatorname{polylog} n)$ query time with $O(n\sqrt{n}\log n)$ storage. (We can reduce the query time to $O((k + m)\operatorname{polylog} n)$, for any $1 \leqslant m \leqslant \sqrt{n}$, but at the cost of increasing the storage to $O((n^2/m)\operatorname{polylog} n)$.)

Two settings where we have not been able to obtain efficient solutions are when the objects are balls in $\mathbb{R}^3$, and when they are arbitrary segments in the plane. Especially the latter case is challenging. Indeed, suppose $\mathcal{O}$ consists of $n/2$ horizontal lines and $n/2$ lines of slope 1. Suppose furthermore that the query is a vertical line ℓ and that we only want to check if ℓ contains at least one intersection. A data structure for this can be used to solve the following 3SUM-hard problem: given three sets of parallel lines, decide if there is a triple intersection [10]. Thus it is unlikely that we can obtain a solution with sublinear query time and subquadratic preprocessing time. However, storage is not the same as preprocessing time. This raises the following question: is it possible to obtain sublinear query time with subquadratic storage?

References

1. Agarwal, P.K., Erickson, J.: Geometric range searching and its relatives. Contemporary Mathematics **223**, 1–56 (1999)
2. Aronov, B., de Berg, M., Ezra, E., Sharir, M.: Improved bounds for the union of locally fat objects in the plane. SIAM J. Comput. **43**(2), 543–572 (2014)
3. de Berg, M., Cheong, O., van Kreveld, M., Overmars, M.: Computational Geometry: Algorithms and Applications, 3rd edn. Springer (2008)
4. Chazelle, B.: Filtering search: A new approach to query-answering. SIAM J. Comput. **15**, 703–724 (1986)
5. Chazelle, B.: A functional approach to data structures and its use in multidimensional searching. SIAM J. Comput. **17**, 427–462 (1988)
6. Chazelle, B., Edelsbrunner, H., Guibas, L.J., Sharir, M.: Algorithms for bichromatic line-segment problems and polyhedral terrains. Algorithmica **11**, 116–132 (1994)
7. Clarkson, K.L., Shor, P.W.: Applications of random sampling in computational geometry. II. Discr. Comput. Geom. **4**, 387–421 (1989)
8. Edelsbrunner, H., Guibas, L.J., Stolfi, J.: Optimal point location in a monotone subdivision. SIAM J. Comput. **15**, 317–340 (1986)
9. Edelsbrunner, H., Overmars, M.H., Seidel, R.: Some methods of computational geometry applied to computer graphics. Comput. Vision, Graphics and Image Proc. **28**, 92–108 (1984)
10. Gajentaan, A., Overmars, M.H.: On a class of $O(n^2)$ problems in computational geometry. Comput. Geom. Theory Appl. **5**, 165–185 (1995)
11. Goodman, J.E., O'Rourke, J.: Range Searching. Handbook of Discrete and Computational Geometry, 2nd edn., Chapter 36 (2004)

12. Keden, K., Livne, R., Pach, J., Sharir, M.: On the union of Jordan regions and collision-free translational motion amidst polygonal obstacles. Discr. Comput. Geom. 1, 59–71 (1986)
13. Rahul, S., Das, A.S., Rajan, K.S., Srinatan, K.: Range-aggregate queries involving geometric aggregation operations. In: Proc. Workshop on Alg. Comp. vol. 1, pp. 122–133 (2011)
14. Subramanian, S., Ramaswamy, S.: The P-range tree: A new data structure for range searching in secondary memory. In: Proc. 6th ACM-SIAM Symp. Discr. Alg., pp. 378–387 (1995)

Cache-Oblivious Iterated Predecessor Queries via Range Coalescing

Erik D. Demaine, Vineet Gopal, and William Hasenplaugh$^{(\boxtimes)}$

Massachusetts Institute of Technology, Cambridge, MA 02139, USA
{edemaine,whasenpl}@mit.edu, vineetg@alum.mit.edu
http://toc.csail.mit.edu/

Abstract. In this paper we develop an optimal cache-oblivious data structure that solves the iterated predecessor problem. Given k static sorted lists $L_1, L_2, \ldots, L_k$ of average length n and a query value q, the *iterated predecessor* problem is to find the largest element in each list which is less than q. Our solution to this problem, called "range coalescing", requires $O(\log_{B+1} n + k/B)$ memory transfers for a query on a cache of block size B, which is information-theoretically optimal. The range-coalescing data structure consumes $O(kn)$ space, and preprocessing requires only $O(kn/B)$ memory transfers with high probability, given a tall cache of size $M = \Omega(B^2)$.

1 Introduction

The *predecessor* problem is to find the largest item in a given sorted list L that is less than a query value $q \in \mathbb{R}$. The *iterated predecessor* problem is to find the predecessor for a query q in each of a set of k static lists $L_1, L_2, \ldots, L_k$, each of average size n. A naive solution involves individual binary searches over all k lists, which would require $O(k \lg n)$ time in the worst-case. However, Chazelle and Guibas [6] showed that the lists can be preprocessed to support iterated predecessor queries in $O(\lg n + k)$ time, with linear preprocessing time and linear space via their technique *fractional cascading*.

In this project, we will demonstrate that the iterated predecessor problem can also be solved using a technique called *range coalescing* in $O(\lg n + k)$ time. Range coalescing is *cache-oblivious* [7], using only $O(\log_{B+1} n + k/B)$ memory transfers per query in the worst case.[1] Furthermore, range coalescing requires only linear space and the preprocessing requires $O(kn)$ time and $O(kn/B)$ memory transfers.

The essence of range coalescing, as the name suggests, is to coalesce ranges of the query space into n "bins", each of which could generate $O(k)$ different results depending on the specific value of q within that range. Figure 1 illustrates how the smallest element in each bin is the "splitter" for the bin, so named because they collectively "split" all of the elements into bins of contiguous value ranges.

---

[1] Throughout this paper we will use the notation $\lg n$ to mean $\log_2 n$ and $\ln n$ to mean the natural logarithm.

© Springer International Publishing Switzerland 2015
F. Dehne et al. (Eds.): WADS 2015, LNCS 9214, pp. 249–262, 2015.
DOI: 10.1007/978-3-319-21840-3_21

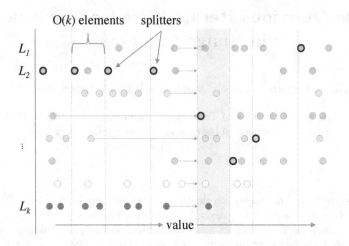

Fig. 1. Range coalescing data structure for the iterated predecessor problem with the value of the element on the x-axis and each row representing a list in $\{L_1, L_2, \ldots, L_k\}$. Each vertical line delineates the $O(k)$ elements in each bin. The elements with a heavy black border are the splitters for each bin — the smallest item in the bin is the splitter. An example bin is highlighted by the blue vertical bar. The elements to the left of the bar with rightward arrows are the predecessors of the splitter from the blue bin in each list.

In addition, the predecessor of the splitter from each list is included in order to service predecessor requests that are smaller in value than the smallest element in the bin from each list. Figure 2 gives an example of how the data in a bin is stored. The elements from each list are stored in sorted subsequences of varying lengths, but of total length $O(k)$. Each bin stores $O(k)$ different values and thus the total data structure is linear space.

A query q on a bin D walks through the $O(k)$ elements in D in a single pass. Within D are k subsequences of each list in sorted order as depicted in Figure 2, so we merely take the largest element from the ith subsequence which is less than q as our predecessor answer from the ith list. A more detailed description of the query process can be found in Section 4.

Throughout this paper we will let M be the size of the cache and B be the size of the cache blocks on a hypothetical external-memory machine. We will consider solutions to the iterated predecessor problem, in which we are given k n-length lists $L_1, L_2, \ldots, L_k$ and we preprocess them to improve the query time.

In Section 2 we discuss some simple known results from the study of cache-oblivious algorithms and data structures which are useful in subsequent analysis. Section 3 gives an overview of solutions to the iterated predecessor problem which are not cache-oblivious, but nonetheless serve as a reasonable baseline for our work. We present range coalescing in Section 4 and show that they answer queries cache-obliviously using $O(\log_{B+1} n + k/B)$ memory transfers. Section 5 demonstrates that the preprocessing for a range coalescing data structure requires

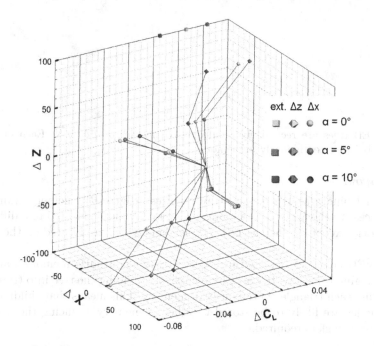

Fig. 2. An example of how a bin is constructed in a range coalescing data structure. The elements that fall in the range of the blue bar and the predecessor of the splitter (with heavy black border) from each list is represented in the bin. In particular, the subsequences of those elements from each list are stored together and concatenated so that the subsequence from L_1 comes first and so on until the subsequence from L_k.

only $O(kn/B)$ with high probability. Section 6 describes our implementations of each solution described herein and an experimental methodology for testing the performance of each. Finally, Section 7 describes some limitations of range coalescing, providing opportunities for future work.

2   Cache-Oblivious Tools

This section describes some cache-oblivious primitive operations that are known in the literature and useful to build up solutions to the iterated predecessor problem in this paper.

Array Scanning

Accessing a random element in an length n array requires $O(1)$ memory transfers. However, if we access the entire array in order, we can achieve $O(n/B)$ memory transfers, where B is the size of a block in cache. Each memory transfer brings B elements into cache, so we must make at most $O(n/B)$ transfers.

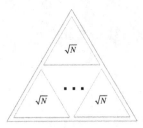

Fig. 3. vEB trees are recursively divided into triangles of size $\sqrt{N}$. Each of these triangles is stores contiguously in memory.

vEB Layout

Traditional binary search on an array requires $O(\lg(n/B))$ memory transfers. Every access to the array is a random access, and could be located in a different cache block, except for the last $O(\lg B)$ elements which are located on the same block.

The vEB layout as depicted in Figure 3 tries to optimize memory transfers by rearranging the array. It works by recursively dividing the tree into triangles, and storing each triangle contiguously in memory. This means that children and parent nodes are likely to be stored together in memory, reducing the number of memory transfers required.

Lemma 1. *A query on an binary tree in the vEB layout requires $O(\log_{B+1} n)$ memory transfers.*

Proof. Triangles of size S are recursively divided into smaller triangles of size $\sqrt{S}$. Lets examine the largest triangle that has at most B elements. This triangle must have at least $\sqrt{B+1}$ elements, so its height must be at least $\frac{1}{2}\lg(B+1)$. This entire triangle can be loaded into one cache block. There are at most $\lg n / \frac{1}{2}\lg(B+1) = 2\log_{B+1} n$ of these triangles along a root to leaf path, so we only need to make $O(\log_{B+1} n)$ memory transfers to find an element. □

The vEB layout has been the basis for several known cache-oblivious algorithms, including B-trees [2], funnel sort [7], and priority queues [1,5]. It is also used in our solution for range coalescing.

3   Known Solutions

We examine several simple solutions to the iterated predecessor problem. These solutions provide a good background for understanding range coalescing and serve as our implementation baselines in Section 6.

Sequential Binary Search

The simplest solution to the iterative predecessor problem is to do a binary search on each of the k unmodified lists $L_1, L_2, \ldots, L_k$, and write down the output from each. Since each binary search requires $O(\lg(n/B))$ memory transfers, this solution requires a total of $O(k \lg(n/B))$ memory transfers. The total space usage is optimal, $O(kn)$.

Sequential vEB Binary Search

Using the vEB layout described previously, we can do binary searches using only $O(\log_{B+1} n)$. If we use a vEB layout for each of the k lists, we can perform the searches in $O(k \log_{B+1} n)$. The total space usage is optimal, $O(kn)$.

Fractional Cascading

Fractional cascading [6] is the incumbent solution for the in-memory iterated predecessor problem. An external memory-oriented extension of fractional cascading described below achieves a runtime of $O(\log_{B+1} n + k)$.

  The main idea behind fractional cascading is to use the query result from each list to perform a search on the next list in constant time. One needs only to do a binary search on the first list to prime the pipeline. To do this, we store pointers in each list to its predecessor and successor in the next list. This gives us a constant-sized range to search through in the next list.

  If we did this naively, this would be of no benefit — the predecessor and successor of the ith list could span the entirety of the $i + 1$st list. However, by altering the lists slightly, we can achieve constant time per remaining search. Starting with the last list, we insert every other element into the previous list. We do this for each list. This ensures that the range between predecessor and successor is at most a constant value.

  We store the initial list in the vEB layout to minimize memory transfers for the initial binary searches. Using this method, we can perform a query using $O(\log_{B+1} n + k)$ memory transfers. The total space usage is optimal, $O(kn)$.

Quadratic Storage

A brute-force fast solution involves storing one sorted kn-length list of all elements from all lists using the vEB layout. Accompanying each element in the list is a k-length sublist with a copy of its k predecessors, one from each list in $\{L_1, L_2, \ldots, L_k\}$. We can iterate over this contiguous sublist using $O(k/B)$ memory transfers, so the total number of memory transfers is $O(\log_{B+1}(kn) + k/B)$. However, the total space usage is $O(nk^2)$, since we must store a k-length sublist for each element.

4   Range Coalescing

In this section we describe how an iterated predecessor query can be satisfied cache-obliviously using $O(\log_{B+1} n + k/B)$ memory transfers using a ***range coalescing*** data structure. We describe how a range coalescing data structure is built cache-obliviously from a set of sorted lists $L_1, L_2, \ldots, L_k$ each of size n using only $O(kn/B)$ memory transfers with high probability in Section 5.

  Let H be a range coalescing data structure built from a set of n-length sorted lists $L_1, L_2, \ldots, L_k$. H is composed of n bins, each of size $O(k)$, which partition the space of possible query values using a sorted list of n splitters S, as depicted

```
QUERY(H, q)
1   ⟨D, s⟩ = vEB(H.S, q)
2   j = 1
3   for i = 1 to D.size − 1
4       if D_i < q
5           Z_j = D_i
6       if D_{i+1} < s
7           j = j + 1
8   return Z
```

Fig. 4. Pseudocode of the QUERY method for a range coalescing data structure H. H contains a sorted array S of splitters organized using a van Emde Boas layout [3]. The function vEB returns a bin D, organized as an array, and a splitter s, which is the predecessor of q in S. The bin D is walked in a linear fashion, overwriting potential predecessors in the output array Z and incrementing the output position whenever the subsequence of the next list begins. The jth subsequence begins with the one and only element from L_j that is smaller than the splitter s and each bin is appended with $-\infty$ to handle the boundary condition when D_{i+1} is compared with the splitter s.

in Figure 1. Figure 2 illustrates how a bin concatenates k sequences of elements, each of which is a subsequence of each constituent list from $\{L_1, L_2, \ldots, L_k\}$. The first element of the ith subsequence in the jth bin is the predecessor of the splitter S_j in L_i and is strictly smaller than S_j by construction, a fact that we will exploit to implicitly denote the beginning of each subsequence.

Lemma 2. *A range coalescing data structure H built from a set of n-length sorted lists $L_1, L_2, \ldots, L_k$ consumes $O(kn)$ space.*

Proof. The elements from all n-length lists $L_1, L_2, \ldots, L_k$ are partitioned into n bins. In addition, each bin has exactly one element for each of the k lists which is smaller in value than the splitter for the bin. Thus, each of the n bins has $O(k)$ elements and the data structure has $O(kn)$ space.   □

Iterated predecessor queries

This section describes the process by which a range coalescing data structure answers iterated predecessor queries and demonstrates that the process incurs $O(\log_{B+1} n + k/B)$ memory transfers with high probability. Figure 4 gives pseudocode for the procedure QUERY, which takes a range coalescing data structure H and a query q and returns an ordered list of results which correspond to the predecessors of q for each constituent list in $\{L_1, L_2, \ldots, L_k\}$.

While it may be that the function QUERY is correct by inspection, we leave nothing to chance and prove it here.

Lemma 3. *Given a range coalescing data structure H and a query value q, the function QUERY(H,q) returns the correct answer.*

Proof. Consider the jth bin, with corresponding splitter S_j, which is used to satisfy all queries $q \in [S_j, S_{j+1})$. By construction, the jth bin contains all elements $\{l \in \cup_{i=1}^k L_i \text{ s.t. } l \in [S_j, S_{j+1})\}$ in addition to the predecessor of S_j from each list in $\{L_1, L_2, \ldots, L_k\}$. Thus, the jth bin contains the k correct answers — the predecessors of q in each constituent list in $\{L_1, L_2, \ldots, L_k\}$. We also see that each subsequence has exactly one element that is less than the splitter S_j, which allows us to know which subsequence we are in during the course of the scan — each element falling below the splitter denotes the beginning of a new subsequence. Furthermore, since the subsequences are stored in sorted order, we know that the predecessor result for a particular list L_i corresponds to the largest element less than q in L_i's subsequence. $\square$

Now we bound the number of memory transfers incurred by QUERY by walking through the pseudocode in Figure 4.

Theorem 1. *An iterated predecessor query* QUERY(H, q) *on a range coalescing data structure H built from a set of n-length sorted lists $L_1, L_2, \ldots, L_k$ incurs $O(\log_{B+1} n + k/B)$ memory transfers on a processor with cache blocks of size B.*

Proof. We use the cache-oblivious search tree structure described by Bender, Demaine and Farach-Colton [3] on line 1 of Figure 4 to find the bin corresponding to the predecessor in the sorted n-length splitter list S using $O(\log_{B+1} n)$ memory transfers. After we find the splitter and the corresponding bin D, we merely scan through D once and write out the answers in a continuous stream to the array *output*. Thus, we incur a read stream and a write stream, each of which is $O(k)$ elements and $O(k/B)$ memory transfers. $\square$

5   Preprocessing

This section describes how a range coalescing data structure is built cache-obliviously from a set of k n-length sorted lists $L_1, L_2, \ldots, L_k$ and bounds the number of memory transfers incurred by the process. We do this in four steps. First, we give a suboptimal deterministic strategy for finding the "splitters" — the values that partition the query space such that each partition has $O(k)$ elements from the constituent lists in $\{L_1, L_2, \ldots, L_k\}$. Second, we demonstrate that the elements from each list can be assembled in the bin corresponding to each splitter using $O(kn/B)$ memory transfers. Finally, we give two randomized algorithm for finding the splitters when $k < \ln^2 n$ and $k \geq \ln^2 n$, respectively, each of which incurs $O(kn/B)$ memory transfers.

Preprocessing suboptimally

In this section, we show how to find an n-length sorted splitter array S, such that $O(k)$ elements from $\mathcal{L} = \cup_{i=1}^k L_i$ fall in each range $[S_j, S_{j+1})$ $\forall$ $1 \leq j \leq n$. If we assume that all elements in $\mathcal{L}$ are unique, we can merely merge all the elements and take every kth element in the merged list as the splitters.[2] We can use a cache-oblivious k-merger [7] to merge the elements using $O((kn/B)\log_{M/B}(k/B) + k)$ memory transfers if $k \leq \sqrt[3]{n}$ and $O((kn/B)\log_{M/B}(kn/B))$ memory transfers otherwise.

Bin construction

This section demonstrates how we can build the $O(k)$-sized bin corresponding to each splitter in the array S using $O(kn/B)$ memory transfers in the worst case. If we were to merely build each of the n bins in sequence, each of which could incur as many as $2k$ memory transfers since k may be larger than M, we could incur as many as $O(kn)$ memory transfers overall. This is unacceptable. Instead, we will build the bins using a Z-order traversal [8] of the 2D space spanned by the cross-product of bin number and list number, notated as the bin number $\times$ list number iteration space.

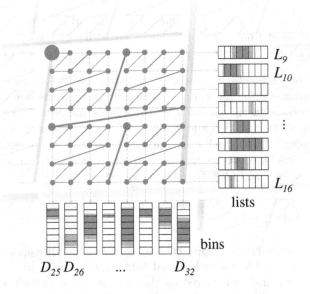

Fig. 5. Example 2^r by 2^r (for $r = 3$) region of a Z-order traversal of the bin number $\times$ list number iteration space. During the course of the execution of this example region there are 8 lists and 8 bins active. The blue regions represent cache blocks which are partially read (lists $L_9, L_{10}, \ldots, L_{16}$) and partially written (bins $D_{25}, D_{26}, \ldots, D_{32}$). The orange blocks are those which are fully read or written, respectively.

Theorem 2. *Given a sorted list of $O(n)$ splitters S and k sorted n-length lists $L_1, L_2, \ldots, L_k$, the bins for a range coalescing data structure can be constructed*

---

[2] We can extend the value of each element with the list number in order to make them unique, since the elements from any particular list L_i are unique. Note that if each list contained a value l and the value l from the L_i was chosen as the jth splitter S_j, we do not compromise the correctness of the query, since the next smaller value than S_j from each list in $\{L_1, L_2, \ldots, L_k\}$ is contained in the jth bin.

deterministically and cache-obliviously using $O(kn/B)$ memory transfers on a processor with a cache of size $M = \Omega(B^2)$ and cache blocks of size B.

Proof. Please see Appendix.   □

Finding splitters for small k

In this section, we show how to find the splitters which partition all of the elements into bins, each of which stores the answers to any query which falls in the range of values between the associated splitter and the splitter for the next bin of larger value. When k is less than $\ln^2 n$, we can randomly select elements to be splitters with probability $1/k$ using $O(kn/B)$ memory transfers by streaming the lists and writing out the samples to separate lists. Next, we subdivide bins that are too large, potentially generating extra splitters. We will show that this process generates $O(kn/B)$ memory transfers with high probability.[3] First, we establish two straightforward yet useful lemmas.

Lemma 4. *Consider a coin with heads probability $1/k$. In kn flips we will see between $n/2$ and $2n$ heads with probability at least $1 - (kn)^{-c}$ for some $c > 1$.*

Proof. The proof follows from an application of Hoeffding's inequality, for sufficiently large n and the assumption that $k < \ln^2 n$.   □

Lemma 5. *The largest number of elements with value between two splitters selected randomly with probability $1/k$ is $(1 + \varepsilon)k \ln(kn)$ with probability at least $1 - (kn)^{-\varepsilon}$ for any $\varepsilon > 0$.*

Proof. We can think of a bin as being created by successive coin flips with probability of heads equal to $1/k$: every time tails comes up, the bin grows by one. Thus, the probability that a bin is of a particular size R is at most $(1 - 1/k)^R/k$. Summing from $R = R'$ to ∞, we can bound the probability that a particular bin has size at least R', $\sum_{R=R'}^{\infty} (1 - 1/k)^R/k = (1 - 1/k)^{R'}$. Letting $R' = (1 + \varepsilon)k \ln(kn)$ and taking the union bound across at most kn different bins, the proof follows.   □

Consider the process of splitting a bin which exceeds $2(1 + \varepsilon)k$ elements. We use $\lfloor R/2(1 + \varepsilon)k \rfloor$ applications of a cache-oblivious selection algorithm [7] to subdivide large bins of size R into bins of size at most $2(1 + \varepsilon)k$ elements using $O(\lfloor R/2(1 + \varepsilon)k \rfloor R/B)$ memory transfers.[4]

Theorem 3. *The total number of memory transfers S required to subdivide all m bins to be less than $2(1 + \varepsilon)k$ elements each is $O(kn/B)$, assuming that the largest bin is at most $(1 + \varepsilon)k \ln(kn)$ elements and $n/2 \le m \le 2n$.*

---

[3] In this context, **with high probability** means with probability greater than $1 - N^{-c}$ where N is the total number of elements in the problem and $c > 1$ is some constant.

[4] For convenience, we assume $R > B$. There is no need to make the bins smaller than B elements, since processing a bin incurs at least one memory reference.

Proof. Please see Appendix.   □

We need to verify that the process of subdivision does not unduly increase the number of bins.

Lemma 6. *After subdivision, we will have $O(n)$ bins.*

Proof. Initially, there are $O(n)$ splitters with high probability by Lemma 4. The process of subdividing bins generates bins with size at least k, thus we can create at most n extra bins through subdivision.   □

Finding splitters for large k

When k is at least $\ln^2 n$, we use an oversampling technique as used in sample sort [4] to find a set of splitters and bound the size of all bins. In particular, we start by randomly sampling elements as candidate splitters with probability $1/\ln k$, which can be accomplished by streaming each list and writing out the samples to a separate candidate list with $O(kn/B)$ memory transfers. Then, we merge these candidates using and a cache-oblivious k-merger [7] using $O((kn/B \ln k) \log_{M/B}(k/B) + k) = O(kn/B)$ memory transfers, assuming $M = \Omega(B^2)$ and $n = \Omega(B)$. Finally, we take every n evenly space elements from this sorted list as our set of splitters.

Theorem 4. *Given an oversampling rate of $k/\ln k$, the largest resulting bin has at least $2(1 + \varepsilon)k$ elements with probability less than $(kn)^{-\varepsilon}$.*

Proof. Please see Appendix.   □

6   Implementation and Experimentation

We implemented each of the simple solutions described in Section 3, and compared their performance to range coalescing on a variety of data sets.

Each solution was implemented in C++, and compiled and tested on an Intel i7 processor with 3MB of L3 cache. We implemented the full merge range coalescing solution described in Section 5, instead of the randomized solution. Before each test, k lists each of length n were generated using a uniform distribution of integers from 0 to 1 million. These lists were passed in as input to each of the solutions. The initialization times and average query times of each solution were recorded.

Range coalescing performed significantly better in practice than other linear-storage solutions. It performed better on both small and large datasets. Average query times are shown in Figure 6 and Figure 7. As the datasets got larger, the effects of range coalescing became more evident. For $n = 50$ and $k = 1000$, range coalescing did 5 times better than a simple binary search, whereas for $n = 5000$ and $k = 1000$, range coalescing performed 18 times better.

However, range coalescing requires much more time for preprocessing. It takes about 20 times longer to initialize than the vEB search, and 3 times longer than fractional cascading. We believe these results could be improved upon - we did not implement the linear time randomized preprocessing method described in Section 5, leaving it instead to future work.

The average query time for the quadratic storage solution is better than all linear storage solutions. However, it requires $O(nk^2)$ time to initialize. For $k = 1000, n = 100$, this is 42 times longer than the preprocessing for the range coalescing solution.

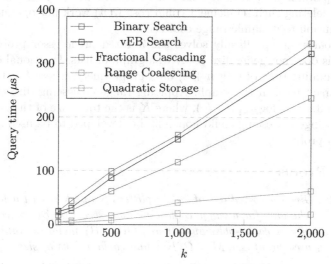

Fig. 6. Query times vs k (for fixed n=1000)

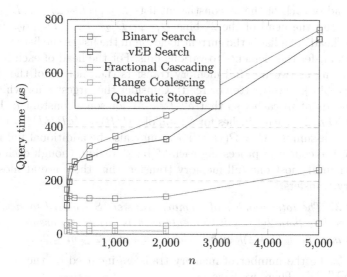

Fig. 7. Query times vs n (for fixed k=1000)

7   Future Work

We have presented an optimal cache oblivious solution for the static iterated predecessor query. There are several areas in which this can be extended. Range coalescing does not currently support dynamic operations like insert or delete. For instance, it is not obvious how one would avoid the adversarial behavior of repeatedly adding and deleting an element. An element can appear in many bins if the corresponding list does not have other elements in those bins. Repeatedly adding and deleting such an element could cost $\Omega(n)$ work per operation given a naive extension to dynamic range coalescing.

Range coalescing specifically solves the iterated predecessor problem on k lists, but this does not generalize easily to a graph of lists. Fractional cascading achieves a running time of $O(\lg n + k)$ on a graph query, where k is the length of the traversed path in the graph. Applying range coalescing directly to this problem results in $O(\log_{B+1} n + K)$, where K is the total size of the graph. The concepts of range coalescing could hopefully be developed to be used as a black box for such problems.

Omitted Proofs

Theorem 2. *Given a sorted list of $O(n)$ splitters S and k sorted n-length lists $L_1, L_2, \ldots, L_k$, the bins for a range coalescing data structure can be constructed deterministically and cache-obliviously using $O(kn/B)$ memory transfers on a processor with a cache of size $M = \Omega(B^2)$ and cache blocks of size B.*

Proof. Consider a 2^r by 2^r naturally aligned region in the bin number $\times$ list number iteration space, like the one depicted in Figure 5.[5] The cache need only keep the head of each of the 2^r constituent lists $L_{i2^r+1}, L_{i2^r+2}, \ldots, L_{(i+1)2^r}$ for some $i > 0$ and the head of the 2^r bins $D_{j2^r+1}, D_{j2^r+2}, \ldots, D_{(j+1)2^r}$ for some $j > 0$. The "head" of a list is the current location in the list as the list is streamed linearly to transfer the elements to various bins. For the head of each list, there can be as many as two non-full memory transfers (i.e., not all of the elements on the cache block were written or read). Consider the largest r for which these $2 \cdot 2^r$ list heads fit in cache, so that $a2^r B = M$ for some constant a. In total, there will be $kn/2^{2r}$ cache flushes for a total of $(M/B) \cdot kn(aB/M)^2 = O(kn/B)$ cache blocks, assuming $M = \Omega(B^2)$. There may also be additional full memory transfers in the course of processing each 2^r by 2^r region, though each element may appear in at most one full memory transfer, thus there are at most kn/B full memory transfers.   □

Theorem 3. *The total number of memory transfers S required to subdivide all m bins to be less than $2(1 + \varepsilon)k$ elements each is $O(kn/B)$, assuming that the largest bin is at most $(1 + \varepsilon)k \ln(kn)$ elements and $n/2 \le m \le 2n$.*

Proof. Let x_i be the number of memory transfers incurred by the ith bin and thus $S = \sum_{i=1}^{m} X_i$. Then, we have

---
[5] A *naturally aligned* region of size c by c is one which begins with some index $i \equiv 1$ (mod c) in one dimension and $j \equiv 1$ (mod c) in the other.

$$E[x_i] \leq a \sum_{R=0}^{\infty} \frac{1}{k}\left(1 - \frac{1}{k}\right)^R \left\lfloor \frac{R}{2(1+\varepsilon)k} \right\rfloor \frac{R}{B}$$

$$\leq a \sum_{R=0}^{\infty} \left(1 - \frac{1}{k}\right)^R \frac{R^2}{2(1+\varepsilon)k^2 B}$$

$$\leq a \frac{k}{B}$$

for some constant $a > 0$. Also, we see that

$$x_i \leq \left\lfloor \frac{(1+\varepsilon)k\ln(kn)}{2(1+\varepsilon)k} \right\rfloor \frac{(1+\varepsilon)k\ln(kn)}{B}$$

$$\leq \frac{k}{2B}(1+\varepsilon)\ln^2(kn)$$

for all i since the largest bin is assumed to have at most $(1+\varepsilon)k\ln(kn)$ elements. Let $t = (k/2B)(1+\varepsilon)\ln^2(kn)$ and note that each random variable in $\{x_1, x_2, \ldots, x_m\}$ has support in the range $[0, t]$. A Hoeffding bound on S gives us $\Pr\left\{S - E[S] \geq t\sqrt{\varepsilon m \ln(kn)}\right\}$

$$\leq \exp\left(-2\frac{\varepsilon m \ln(kn)t^2}{mt^2}\right)$$

$$\leq \exp(-2\varepsilon \ln(kn))$$

$$\leq (kn)^{-\varepsilon}.$$

Then, for sufficiently large kn, $S = O(kn/B)$ with high probability. □

Theorem 4. *Given an oversampling rate of $k/\ln k$, the largest resulting bin has at least $2(1+\varepsilon)k$ elements with probability less than $(kn)^{-\varepsilon}$.*

Proof. Let R be the size of the largest bin. By Theorem B.3 of [4], for sufficiently large kn, we have that

$$\Pr\{R > 2(1+\varepsilon)k\} \leq kn\exp\left(-(1+\varepsilon)\left(\frac{1+2\varepsilon}{2(1+\varepsilon)}\right)^2 \frac{k}{\ln k}\right)$$

$$\leq kn\exp\left(-(1+\varepsilon)\frac{k}{4\ln k}\right)$$

$$\leq kn\exp\left(-(1+\varepsilon)\frac{\ln^2\frac{kn}{\ln^2 n}}{4\ln\ln^2 n}\right)$$

$$\leq kn\exp\left(-(1+\varepsilon)\frac{\ln^2(kn) - o(\ln^2(kn))}{8\ln\ln n}\right)$$

$$\leq kn\exp(-(1+\varepsilon)\ln(kn))$$

$$\leq (kn)^{-\varepsilon}.$$

□

Acknowledgments. This work was begun during the open-problem sessions of the MIT class 6.851: Advanced Data Structures taught by E. Demaine in Spring 2014. We thank the other participants for making a creative and productive environment.

References

1. Arge, L., Bender, M.A., Demaine, E.D., Holland-Minkley, B., Munro, J.I.: Cache-oblivious priority queue and graph algorithm applications. In: STOC (2002)
2. Bayer, R., McCreight, E.M.: Organization and maintenance of large ordered indexes. Acta Informatica (1972)
3. Bender, M.A., Demaine, E., Farach-Colton, M.: Cache-oblivious B-trees. In: FOCS, pp. 399–409 (2000)
4. Blelloch, G.E., Leiserson, C.E., Maggs, B.M., Plaxton, C.G., Smith, S.J., Zagha, M.: A comparison of sorting algorithms for the connection machine CM-2. In: SPAA (1991)
5. Brodal, G.S., Fagerberg, R.: Funnel heap - a cache oblivious priority queue. In: Bose, P., Morin, P. (eds.) ISAAC 2002. LNCS, vol. 2518, pp. 219–228. Springer, Heidelberg (2002)
6. Chazelle, B., Guibas, L.: Fractional cascading: 1. a data structuring technique. Algorithmica (1986)
7. Frigo, M., Leiserson, C.E., Prokop, H., Ramachandran, S.: Cache-oblivious algorithms. In: FOCS (1999)
8. Morton, G.: A computer oriented geodetic data base; and a new technique in file sequencing. Tech. rep., IBM (1966)

Polylogarithmic Fully Retroactive Priority Queues via Hierarchical Checkpointing

Erik D. Demaine$^{(\boxtimes)}$, Tim Kaler, Quanquan Liu, Aaron Sidford, and Adam Yedidia

MIT CSAIL, Cambridge, MA, USA
{edemaine,tfk,quanquan}@mit.edu

Abstract. Since the introduction of retroactive data structures at SODA 2004 [1], a major open question has been the difference between partial retroactivity (where updates can be made in the past) and full retroactivity (where queries can also be made in the past). In particular, for priority queues, partial retroactivity is possible in $O(\log m)$ time per operation on a m-operation timeline, but the best previously known fully retroactive priority queue has cost $\Theta(\sqrt{m} \log m)$ time per operation.

We address this open problem by providing a general logarithmic-overhead transformation from partial to full retroactivity called "hierarchical checkpointing," provided that the given data structure is "time-fusible" (multiple structures with disjoint timespans can be fused into a timeline supporting queries of the present). As an application, we construct a fully retroactive priority queue which can insert an element, delete the minimum element, and find the minimum element, at any point in time, in $O(\log^2 m)$ amortized time per update and $O(\log^2 m \log \log m)$ time per query, using $O(m \log m)$ space. Our data structure also supports the operation of determining the time at which an element was deleted in $O(\log^2 m)$ time.

1 Introduction

Retroactivity. We can think of a data structure as being defined by a sequence of updates $u_1, u_2, \ldots, u_m$ applied to its initial (empty) state. Traditional data structures "live in the present" in the sense that the user can only append updates to this sequence, and ask queries about the final state of the data structure resulting from the entire update sequence. **Retroactive data structures**, introduced at SODA 2004 [1], allow for updates to be inserted or deleted in the middle of the sequence, instead of just the end. Effectively, this feature enables the user to travel back in time and make a retroactive change to the data structure (similar to the movie *Back to the Future*). Thus we refer to the mutable update sequence as the **timeline**.

We distinguish two forms of retroactivity. In **partial retroactivity**, queries can be made only of the final version resulting from all of the updates in the timeline; effectively, retroactive updates must be propagated all the way through the timeline in order to answer such queries correctly. In **full retroactivity**,

© Springer International Publishing Switzerland 2015
F. Dehne et al. (Eds.): WADS 2015, LNCS 9214, pp. 263–275, 2015.
DOI: 10.1007/978-3-319-21840-3_22

queries can be made about the data structure at any time, corresponding to the result from a prefix of the timeline. In short, both forms of retroactivity enable modifying the past, and full retroactivity enables querying the past.

Known Results. In some settings, retroactivity is easy to achieve. If updates commute with each other and have inverses, then retroactive updates can be moved to the end of the timeline, making partial (but not full) retroactivity easy. If updates are inserts and deletes, and the queries fall under Bentley and Saxe's decomposable search problems, then full retroactivity is possible with an $O(\log m)$ factor overhead [1].

Retroactivity becomes challenging when updates can have non-trivial interactions. Here one retroactive update can have a propagated effect on potentially all later updates. In the extreme, when the data structure is a general-purpose computer, a retroactive update can require an $\Omega(m)$ factor overhead [1].

The more interesting middle ground is when the updates have some but limited influence on each other—a common scenario in many classic data structures. For example, logarithmic fully retroactive stacks (with push/pop), queues (with enqueue/dequeue), deques (with all four), union-find, dictionaries, and predecessor/successor structures all have logarithmic fully retroactive data structures [1,2]. Of these results, predecessor/successor was the most challenging; the original paper [1] solved partial retroactivity in $O(\log m)$ but full retroactivity in $O(\log^2 m)$, which was later improved to $O(\log m)$ by Giora and Kaplan [2]. This problem is equivalent to dynamic rectilinear ray shooting, which was in fact the original motivation for defining retroactivity.

Challenges. A key open problem in retroactivity, posed at SODA 2004, is whether there is a difference in difficulty between obtaining partial versus full retroactivity. The only known upper bound on the separation is a conversion from partial to full retroactivity with $O(\sqrt{m})$ factor overhead [1]. Essentially, this conversion maintains $\Theta(\sqrt{m})$ checkpoints of the timeline using a partially retroactive data structure, and to query in between, replays the necessary $O(\sqrt{m})$ intervening updates. On the other hand, the only known data structural problem with a polynomial separation between the best partially retroactive and best fully retroactive data structures is priority queues (with insert and delete-min operations). The logarithmic partially retroactive priority queue [1] is one of the most sophisticated retroactive data structures, propagating potentially linear-length chain reactions in just logarithmic time. However, the existing approach appeared limited to partial retroactivity. Until now, the fastest known fully retroactive priority queue was the $O(\sqrt{m}\log m)$ bound that follows from the general conversion.

Our Results. In this paper, we solve this 11-year-old open problem by constructing the first polylogarithmic fully retroactive priority queue. Specifically, our data structure supports inserting an element, deleting the minimum element, and finding the minimum element, at any time in the timeline, in $O(\log^2 m)$ amortized time per update and $O(\log^2 m \log \log m)$ time per query, using $O(m \log m)$ space.

We also show how to support another natural query over the timeline: finding the time at which a given element gets deleted as the minimum (or finding that it remains in the structure in the present).

More importantly, we present a new general transformation from partial to full retroactivity with only a logarithmic factor overhead. This result shows a strong upper bound on the separation between partial versus full retroactivity, but it requires one additional assumption. Specifically, we call a (partially retroactive) data structure *time-fusible* if, given two such data structures representing two different timelines (contained in disjoint time intervals), it is possible to form a new (read-only) data structure representing the concatenation of those timelines. Roughly speaking, this assumption lets us apply the $O(\sqrt{m})$ checkpointing idea recursively in a binary tree structure built over the timeline, storing a partially retroactive data structure for the sub-timeline represented by each rooted subtree. Hence we call the transformation *hierarchical checkpointing*. A retroactive query can then be answered by fusing $O(\log m)$ structures and asking a query about the present.

Our fully retroactive priority queue data structure is an application of this general technique. With some modifications, we show how to fuse two of the logarithmic partially retroactive priority queues from [1] in polylogarithmic time. Applying the general technique gives us a polylogarithmic bound on fully retroactive priority queues, but with worse bounds than those stated above. By a more careful analysis tailored to priority queues, we show how to further tune the hierarchical checkpointing analysis to improve the running time by a logarithmic factor and get the claimed bounds of $\tilde{O}(\log^2 m)$.

Organization. We organize the sections of this paper as follows. Section 2 introduces our hierarchical checkpointing framework in greater detail. Section 3 describes time-fusible partially retroactive priority queues whose timelines may be fused together in polylogarthmic time. Section 4 applies the technique of hierarchical checkpointing to obtain a fully retroactive priority queue with polylogarthmic overheads.

2   Hierarchical Checkpointing

In this section, we present our hierarchical checkpointing technique for transforming a time-fusible partially retroactive data structure into one that is fully retroactive while incurring only polylogarithmic overheads. In later sections, these results will be employed to design a fully retroactive priority queue with polylogarithmic overheads.

We begin by defining in Section 2.1 the notion of time fusibility for retroactive data structures. Then in Section 2.2 we describe the hierarchical checkpoint procedure and prove its correctness.

2.1   Definitions

Here we discuss the properties of partially retroactive data structures and the conditions necessary to use hierarchical checkpointing to obtain full retroactivity.

We define a **retroactive update** operation to be the insertion or deletion of a data structure operation at a particular time. These operations are:

- INSERT-OP(o, t): insert a data structure update operation o into the retroactive structure's timeline at time t.
- DELETE-OP(o, t): delete a data structure update operation o from the retroactive structure's timeline at time t.

We define a **retroactive query** operation to be one that can determine some aspect of the state of the retroactive data structure at some point in time. We use GET-VIEW(t) as the canonical query procedure when we describe our transformation.

- GET-VIEW(t): returns some aspect of the state of the retroactive data structure at time t.

For partially retroactive structures, query operations can only be performed in the present (i.e. $t = \infty$). Fully retroactive data structures, however, may be queried at any time t. It turns out, that a collection of partially retroactive data structures can be used to support fully retroactive query operations when it is possible to "fuse" their timelines. Formally, we say a partially retroactive data structure is **time fusible** if it has the following properties:

1. It supports a function, FUSE(d_1, d_2), that fuses the timelines of two instances d_1 and d_2 of the partially retroactive data structure, producing a version of the data structure that allows read-only queries and reflects the updates in both d_1 and d_2. FUSE(d_1, d_2) need only support fusion between structures containing updates that span disjoint and adjacent intervals of the timeline.
2. Sequences of operations made on it exhibit substring closure; in other words, given a valid sequence of operations, any contiguous subsequence of operations on the structure is also valid.

2.2   The Data Structure

In this section we describe how to transform a time-fusible partially retroactive data structure into one that is fully retroactive using our hierarchical checkpointing framework. Specifically, we obtain a fully retroactive data structure with $O(T(m) \log m + Q(m, k))$ query time and $O(A(m) \log^2 m)$ amortized update time, where $T(m)$ and $A(m)$ represent the merge and update time, respectively, in the original partially retroactive data structure, and $Q(m, k)$ is the query time of a time-fused structure consisting of k fusions and containing m updates.

The first step of our transformation is to build a **checkpoint tree** — a balanced binary search tree in which each node of the tree contains a partially retroactive data structure consisting of all the updates in the subtree rooted at that node. Our checkpoint tree is similar to a segment tree [3] in that each partially retroactive data structure can be viewed as a segment with endpoints given by the first and last chronological update in the structure. The structures in the leaves of our checkpoint tree each contain only one update, and the leaves are sorted by the time of their one update. The update operations INSERT-OP(o, t) or DELETE-OP(o, t) can be performed on the fully retroactive structure by inserting into or deleting the update, o, from all of the partially retroactive structures in the search path. A query can be performed at time t by merging $O(\log n)$ disjoint partially retroactive structures obtained from the balanced binary tree such that the fused structure contains all updates in the time span $(-\infty, t]$.

Theorem 1. *Given a partially retroactive data structure that is time fusible, we may construct a fully retroactive version of the data structure using hierarchical checkpointing. This data structure will have an $O(A(m) \log^2 m)$ amortized update time and $O(T(m) \log m + Q(m, k))$ query time.*

We prove Theorem 1 in two parts below.

Lemma 1. *Our hierarchical checkpointing method produces a fully retroactive data structure with $O(A(m) \log^2 m)$ amortized update time.*

Proof. Let F be a fully retroactive data structure based on a time-fusible partially retroactive data structure P. Suppose that m updates have been inserted into F and that the update operation for P runs in $O(A(m))$ time.

We utilize a scapegoat tree [4] to represent the checkpoint tree for F. The checkpoint tree contains all updates to the fully retroactive structure at its leaves ordered by time. Each internal node, x, is associated with an instance of P that reflects the application of all updates in its subtree. To perform INSERT-OP(o, t) or DELETE-OP(o, t), we insert the update as a leaf in the checkpoint tree, and apply the update to the instances of P associated with nodes along the roof to leaf path in $O(A(m) \log m)$ time.

To rebalance the checkpoint tree, the tree rooted at the scapegoat node is rebuilt. We begin by obtaining a sorted list of the k updates ordered by time by performing an in-order walk of the subtree. We create a balanced binary tree with these k updates at the leaves, and initialize an empty instance of P for each internal node of the subtree. Then, we insert the update at each leaf into each of its $O(\log k)$ ancestors. Because applying an update to an instance of P takes $O(A(k))$ time, the total time required to rebuild a subtree containing k updates is $O(A(k) \log k)$. The total cost of an INSERT-OP or DELETE-OP operation for the fully retroactive structures is then the sum of the cost of an insertion or deletion and the amortized cost of rebuilding, $O(A(m) \log^2 m)$ amortized. ∎

Lemma 2. *Our hierarchical checkpointing method produces a fully retroactive data structure with $O(T(m) \log m + Q(m, k))$ query time.*

Proof. Suppose that $T(m)$ is the time it takes to fuse any two instances of P, and $Q(m, k)$ is the time it takes to query an instance of P, where m is the total number of updates in P, and k is the number of components that were used to create the fused structure.

To perform GET-VIEW(t), we first traverse the checkpoint tree to identify the $O(\log m)$ disjoint subtrees that represent the time interval $(-\infty, t]$. The time-fusible partially retroactive structures associated with these subtrees are then fused in-order, resulting in a single structure representing the interval $(-\infty, t]$. We can fuse $O(\log m)$ P structures in $O(T(m) \log m)$ time. Querying this structure then takes $O(Q(m, k))$ time. Therefore, the total runtime of GET-VIEW(t) is $O(T \log m + Q(m, k))$.

3   Time-Fusible Partially Retroactive Priority Queue

In this section we present a partially retroactive priority queue that supports a polylogarithmic fusion operation. Specifically, we describe an algorithm that fuses $k = O(\log m)$ partially retroactive priority queues containing m updates in $O(k \log k \log m)$ time. This time-fusible partially retroactive priority queue enables the use of hierarchical checkpointing to obtain a fully retroactive priority queue with polylogarithmic overheads.

3.1   Partially Retroactive Priority Queues

We begin with an informal review of a partially retroactive priority queue data structure. To simplify our exposition, we treat the partially retroactive priority queue from [1] as a black box and maintain 2 auxillary data structures: Q_{now} containing the set of all keys remaining in the priority queue at time $t = \infty$, and Q_{del} containing all keys that were removed from the priority queue at some point in the past.

We assume that the partially retroactive priority queue returns, following each retroactive update, the keys which should be inserted or deleted from Q_{now} and Q_{del}. If a priority queue is empty at time t, then a delete-min operation will, by convention, insert a placeholder key of infinite weight into Q_{del}. It is known that, following a retroactive update at time t, it is only necessary to insert or delete a single key into Q_{now} and Q_{del} [1]. We can, therefore, synchronize our auxillary data structures Q_{now} and Q_{del} with the partially retroactive priority queue in $O(\log m)$ time. A proof of this claim and an in-depth description of the partially retroactive priority queue data structure can be found in [1, 5.4].

The auxillary Q_{now} and Q_{del} structures are maintained using weight-balanced B-trees [5–7] which for a balance factor $d > 4$ have the following properties:

- Insertion and deletion operations on a B-tree containing m elements take $O(\log m)$ time.
- For all non-root nodes u at height h the weight $w(u)$ of the subtree rooted at u is bounded as follows: $d^h/2 \le w(u) \le 2d^h$.
- The root r of a height-h tree has bounded weight $w(r)$: $d^{h-1} \le w(r) \le 2d^h$.
- Tree-split and concatenate operations on a size-m tree take $O(\log m)$ time.

- A height-h' subtree T' of a height-h weight-balanced B-tree T can be deleted to form the weight-balanced B-tree $T - T'$ in $O(d(h - h'))$ time.

A weight-balanced B-tree data structure possessing these properties is described in [6,7]. Specifically, we apply the result of [6] with balance factor $d = 8$ to maintain Q_{now} and Q_{del}.

3.2   Fusion Algorithm

Before describing our algorithm for fusion, let us better understand the structure of the problem by proving a mathematical relationship between two partially retroactive priority queues that represent two fusible (i.e. disjoint and adjacent) intervals of time.

Lemma 3. *Consider two partially retroactive priority queues Q_1 and Q_2 whose update times lie in the intervals $[a, b)$ and $[b, c)$ respectively. Then, the partially retroactive priority queue Q_3 containing all updates in Q_1 and Q_2 in the interval $[a, c)$ has the property that*

$$Q_{3,now} = Q_{2,now} \cup max\text{-}A\{Q_{1,now} \cup Q_{2,del}\} \tag{1}$$

$$Q_{3,del} = Q_{1,del} \cup min\text{-}D\{Q_{1,now} \cup Q_{2,del}\} \tag{2}$$

where $A = |Q_{1,now}| - |Q_{2,del}|$, $D = |Q_{2,del}|$ and max-$C\{S\}$ denotes the C largest elements in the set S.

Using Lemma 3 we can construct a time-fused representation of Q_3 from Q_1 and Q_2 in polylogarithmic time. We will represent each of $Q_{3,now}$ and $Q_{3,del}$ as a list of trees obtained via tree-split operations consistent with the application of Equation 1 and Equation 2. We say that a time-fusible partially retroactive priority queue has order k, and use the superscript notation Q^k, if Q_{now}^k and Q_{del}^k are represented as lists of at most k trees.

In Figure 1 we provide the pseudocode for FUSE which fuses two partially retroactive priority queues Q_1^k, Q_2^k to obtain Q_3^{3k}. Step 1 computes the value of A from Lemma 3, and step 2 concatenates the list of trees representing $Q_{1,now}^k$ and and $Q_{2,del}^k$ to form a list L containing $2k$ trees. Step 3 computes a "split-key" x that is greater than A elements contained in trees of L. Next each tree in L is split in step 4 by performing a tree-split operation to divide each tree T_i into a tree $T_{i,<}$ containing all keys in T_i that are less than x and $T_{i,>}$ containing all keys in T_i that are greater than x. The trees $T_{i,>}$ for $i = 1, 2, \ldots, 2k$ combined with the trees in $Q_{2,now}$ contain the elements satisfying the relation of Equation (1) in Lemma 3, and similarly the trees in $Q_{1,del}$ and in $T_{i,<}$ for $i = 1, 2, \ldots, 2k$ contain the elements satisfying the relation of Equation (2).

The following theorem proves that FUSE fuses two partially retroactive priority queues of order k in $O(k \log m)$ time.

Theorem 2. *Consider two partially retroactive priority queues Q_1^k and Q_2^k with order k containing m operations. Then FUSE(Q_1^k, Q_2^k) runs in $O(k \log m)$ time.*

GETSPLITKEY($s, T_1, \ldots, T_k$)

1. If $N = \sum_i |T_i| < C$ (for constant C), sort $\bigcup_i T_i$ and return the sth element.
2. If $s < N/2$, set $s = N - s$ and "invert" the order of each T_i.
3. For each T_i, pick a leftmost subtree T_{m_i} containing keys in the range $(-\infty, m_i)$ where m_i has an order statistic in T_i contained in the range $(|T_i|/256, |T_i|/4)$.
4. Assign each m_i the weight $w_i = |T_i|$. Using weighted selection, select the $N/4$th element m_j among $m_1, m_2, \ldots, m_k$
5. For $m_i \leq m_j$, let $T_i' = T_i - T_{m_i}$. For $m_i > m_j$, let $T_i' = T_i$.
6. Set $s_{new} = s - \sum_i (|T_i| - |T_i'|)$ and return GETSPLITKEY($s_{new}, T_1', \ldots, T_k'$).

(a)

FUSE(Q_1^k, Q_2^k)

1. $A = |Q_{1,now}| - |Q_{2,del}|$
2. Form a list of $2k$ trees $L = T_1, \ldots, T_{2k}$ by concatenating the list of k trees representing $Q_{1,now}$ with the k trees representing $Q_{2,del}$.
3. $x = $ GETSPLITKEY($A, T_1, \ldots, T_{2k}$)
4. For $i = 1, 2, \ldots 2k$, split the tree T_i on the key x to obtain 2 trees $T_{i,>}$ and $T_{i,<}$.
5. $Q_{3,now} = Q_{2,now} + T_{1,>}, \ldots, T_{2k,>}$
6. $Q_{3,del} = Q_{1,del} + T_{1,<}, \ldots, T_{2k,<}$
7. Return Q_3

(b)

Fig. 1. Pseudocode for (a) the GETSPLITKEY operation; and (b) the FUSE operation. GETSPLITKEY takes a key s and a list of k binary trees, and returns a key x such that s keys in $T_1, T_2, \ldots, T_k$ are less than x

Proof. We first show that GETSPLITKEY runs in $O(k \log m)$ time. Our algorithm for finding the split key is an adaptation of the approach of Frederickson and Johnson to compute order statistics for sorted arrays [8].

Steps 1, 2, and 4 of GETSPLITKEY run in $O(k)$ time (step 4 uses linear-time weighted selection from [9]).

Step 3 finds a leftmost subtree T_{m_i} whose contents are contained in the range $(-\infty, m_i)$ and where the order statistic of m_i is in the range $(|T_i|/256, |T_i|/4)$. We show that step 3 runs in $O(k)$ time by showing that for each T_i such a subtree exists at a distance of at most 2 from the root. Consider a height-h weight-balanced B-tree with balance factor d, root node r, and an internal node u at height $h - 2$. The weight-balance criteria for B-trees provided in Section 3.1 implies that the ratio $w(u)/w(r)$ is bounded in the range $(1/256, 1/4)$. The key m_i can, therefore, be found in $O(1)$ time by selecting the maximum key from the leftmost height-$(h - 2)$ subtree of T_i.

Step 5 deletes the subtree T_{m_i} from T_i if $m_i \leq m_j$. The difference in the heights of T_{m_i} and T_i is at most 2, which allows $T - T_{m_i}$ to be obtained in $O(d)$ time while preserving weight-balance. For $d = 8$, this step runs in $O(k)$ time. Note that the subtrees deleted in this step contain elements whose order statistic is strictly less than $N/2$ and thus these subtrees can not contain the sth order statistic. To prove this we show that the order statistic of m_j, computed in step 4, is less than $N/2$. The key m_j is selected in step 4 such that $3N/4$ elements are contained in trees T_i for which $m_i > m_j$. For each such i, the key m_i is smaller than at least $3|T_i|/4$ of the elements in T_i. The key m_j is, therefore, smaller than at least $9N/16$ elements, and thus has an order statistic less than $N/2$.

Step 6 updates the value of s to reflect the reduced problem size, and recursively calls GETSPLITKEY. To bound the depth of the recursion, it is sufficient to show that step 5 eliminates a constant fraction of the elements. Since a total of $N/4$ elements are contained in trees T_i for which $m_i \leq m_j$, and at least $|T_i|/256$ elements in T_i are smaller than m_i, step 5 eliminates at least $N/1024$ elements. The recursion depth is, therefore, bounded by $O(\log N)$. Since $N = O(m)$, the total runtime of GETSPLITKEY is $O(k \log m)$.

Next let us analyze the FUSE operation. Steps 1-2 and 5-6 of FUSE can be performed in $O(k)$ time. Step 3 to compute the split key runs in $O(k \log m)$ time, and step 4 may be performed in $O(k \log m)$ time by performing an $O(\log m)$ time tree split operation on each of k trees. The runtime of FUSE is bounded by the time to compute the split key, and therefore is $O(k \log m)$.

The bound proved in Theorem 2 depends on the order k of the two time-fusible partially retroactive priority queues Q_1^k, Q_2^k being merged. It turns out, that the fusion of k partially retroactive priority queues can be constructed efficiently while being represented using only $O(k)$ trees by combining trees in Q_{now} and Q_{del} that originated from a split operation on a common tree. The ability to perform such a reduction relies on the following lemma.

Lemma 4. *Let $Q_1, \ldots, Q_k$ denote k partially retroactive priority queues each with disjoint time intervals that increase monotonically with k. Let Q_* be a priority queue containing the updates in $Q_1, \ldots, Q_k$ applied consecutively. Then $Q_{*,now}$ and $Q_{*,del}$ consist of contiguous intervals of $Q_{i,now}$ and $Q_{i,del}$, i.e.*

$$Q_{*,now} = \cup_{i \in S_{now}} Q_{i,now}[a_i, b_i] \cup_{i \in S_{del}} Q_{i,del}[a_i', b_i'] \tag{3}$$

$$Q_{*,del} = \cup_{i \in T_{now}} Q_{i,now}[c_i, d_i] \cup_{i \in T_{del}} Q_{i,del}[c_i', d_i'] \tag{4}$$

for some sets $S_{now}, S_{del}, T_{now}, T_{del} \subseteq \{1, \ldots, k\}$ and elements $a_i, a_i', b_i, b_i', c_i, c_i', d_i, d_i'$ where for a set S and $a, b \in S$ we let $S[a, b] = \{x \in S : a \leq x \leq b\}$.

The preceding lemma allows us to tweak the fusion algorithm to guarantee that the order of the fusion of k time-fusible partially retroactive priority queues is bounded by $2k$. This is accomplished by adding a post-processing step POST-FUSE immediately after the fusion procedure FUSE. After obtaining the fusion Q_3 of Q_1 and Q_2, the trees representing $Q_{3,now}$ are checked in POSTFUSE to identify pairs of split-trees that were obtained by splitting a common tree. By Lemma 4 the union of these intervals span disjoint intervals and these pairs of trees can, therefore, be concatenated in logarithmic time.

Lemma 5. *The fusion of k time-fusible partially retroactive priority queues has order bounded by $2k$ and runs in $O(k \log m)$ time when using the POSTFUSE procedure.*

To combine the results of this section, we prove the following theorem.

Theorem 3. *Consider* $k = O(\log m)$ *time-fusible partially retroactive priority queues. The time to fuse these* k *data structures is bounded by* $O(k \log k \log m)$, *and the time required to query this structure is* $O(\log^2 m)$.

Proof. We arrange the k time-fusible structures at the leaves of a balanced height-$\log k$ merge tree. By Lemma 5 the sum of the orders of time-fusible partially retroactive priority queues at level i in the merge tree is $O(k)$. The total work to perform fusions at level i is, therefore, $O(k \log m)$ Since there are $\log \log m$ levels in the merge tree the total time is $O(k \log m \log \log m)$. To query the fused structure we perform a query on each of the $O(\log m)$ trees representing Q_{now} which can be done in $O(\log^2 m)$ time.

4   Fully Retroactive Priority Queue

In this section we describe the design of a fully retroactive priority queue that uses hierarchical checkpointing. We begin in Section 4.1 by showing how to apply our technique of hierachical checkpointing using the time-fusible partially retroactive priority queue of Section 3. This yields a fully retroactive priority queue that supports retroactive updates in $O(\log^3 m)$ amortized time, retroactive queries in $O(\log^2 m \log \log m)$ time, and FIND-DELETION-TIME in $O(\log^3 m \log \log m)$ time. Next, in Section 4.2, we optimize our application of hierarchical checkpointing for priority queues to obtain $O(\log^2 m)$ amortized time updates, and $O(\log^2 m)$ time FIND-DELETION-TIME queries.

4.1   Obtaining Full Retroactivity Using Hierarchical Checkpointing

Here we analyze the fully retroactive priority queue obtained by a straightforward application of hierarchical checkpointing. The time-fusible partially retroactive priority queue described in Section 3 meets the prerequisites of Theorem 1 needed to perform the partial-to-full transformation. Consequently we can directly apply this theorem to obtain a fully retroactive priority queue which follows the structure laid out in Section 2. A checkpoint tree contains all retroactive updates ordered by time, and each internal node maintains a time-fusible partially retroactive priority queue that contains the updates within its subtree.

The checkpoint-tree data structure used in this fully retroactive priority queue is shown in Figure 2(a) after 16 retroactive operations have been performed. In this example, the checkpoint tree has 16 leaves each corresponding to a retroactive operation on the priority queue. The time-fusible partially retroactive priority queue data structure described in Section 3 is used to represent the partial checkpoints in a checkpoint tree. Each internal node, $Q_{[a,b)}$, maintains a time-fusible partially retroactive priority queue that contains all retroactive operations in its subtree (i.e. all operations occurring at times $t \in [a, b)$).

The GET-VIEW(t) operation is illustrated in Figure 2(b). A checkpoint representing the priority queue at time $t = 10$ is constructed by combining 3 partial checkpoints from the checkpoint tree. The time-fusible partially retroactive priority queues $Q_{[0,8)}$, $Q_{[8,10)}$, and $Q_{[10,11)}$ that are highlighted in Figure 2 are

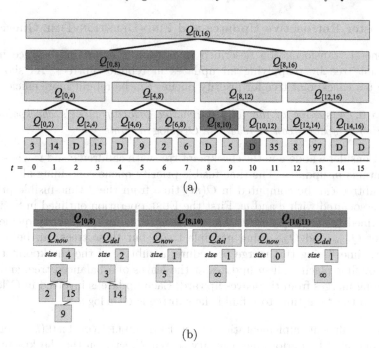

Fig. 2. Hierarchical checkpointing for fully retroactive priority queue. Illustration of the checkpoint tree for a fully retroactive priority queue with 16 operations.

collected and then merged to obtain obtain a partially retroactive priority queue containing all updates in in the interval $[-\infty, 10]$.

Theorem 4. *There exists a fully retroactive priority queue that supports retroactive updates in $O(\log^3 m)$ amortized time, queries in $O(\log^2 m \log \log m)$, and the operation,* FIND-DELETION-TIME, *in $O(\log^3 m \log \log m)$ time.*

Proof. The time-fusible partially retroactive priority queue described in Section 3 supports retroactive updates in $O(\log m)$ time. Applying Lemma 1 with $A(m) = \log m$ shows that retroactive updates run in $O(\log^3 m)$ amortized time. By Theorem 3, the time to merge $O(\log m)$ time-fusible partially retroactive priority queues is bounded by $O(\log^2 m \log \log m)$. Similarly, the time to query this merged structure is bounded by $O(\log^2 m)$ since the merged priority queue has order $O(\log m)$. Applying Lemma 2 with $T(m) = O(\log^2 m)$ and $Q(m) = O(\log^2 m \log \log m)$ shows that retroactive queries run in $O(\log^2 m \log \log m)$ time. Finally, the FIND-DELETION-TIME(x) operation can be performed via binary search to identify the first time t for which the key x is not in the queue. This involves $O(\log m)$ retroactive queries showing that FIND-DELETION-TIME runs in $O(\log^3 m \log \log m)$ time.

4.2   Faster Retroactive Updates and FIND-DELETION-TIME Queries

The general transformation described in Section 2 maintains balance in the checkpoint tree by reapplying all updates in rebuilt subtrees. As shown in Lemma 6 a checkpoint tree for priority queues can be rebuilt more efficiently.

Lemma 6. *A subtree of the fully retroactive priority queue's checkpoint tree containing m updates can be rebuilt in $O(m \log m)$ time.*

Proof. Consider a node u in the checkpoint tree with children v and w whose subtree contains m updates. The time-fusible priority queue containing all updates in u's subtree can be computed in $O(m)$ time from the 2 time-fusible priority queues associated with v and w. First the FUSE operation outlined in Section 2 is performed to merge v and w. The resulting time-fusible priority queue may represent Q_{now} and Q_{del} using multiple trees, but these trees can be merged in $O(m)$ time. Using this merge procedure, a subtree of the checkpoint tree is rebuilt by first placing all m updates at the leaves of a balanced tree, and then performing merges from the leaves upward. Each update is involved in $O(\log m)$ merges, so the total time to rebuild the subtree is $O(m \log m)$.

A more efficient implementation of the FIND-DELETION-TIME(k) operation can be obtained by performing a binary search directly on the checkpoint tree. The high-level idea is to perform a binary search for the time of deletion by keeping track of the current number of surviving keys that are less than or equal to k at any particular time. Due to space limitations, this result is stated in Lemma 7 without proof.

Lemma 7. *The FIND-DELETION-TIME operation which performs a binary search directly on the checkpoint tree data structure runs in $O(\log^2 m)$ time.*

Theorem 5. *The fully retroactive priority queue performs updates in $O(\log^2 m)$ amortized time when using a checkpoint tree with the memoized subtree rebuilding procedure, and performs FIND-DELETION-TIME operations in $O(\log^2 m)$ time.*

Acknowledgments. This research was initiated during the open-problem sessions of the MIT class 6.851: Advanced Data Structures taught by E. Demaine in Spring 2014. We thank Adam Hesterberg, Ofir Nachum, and other members of the class for helpful discussion regarding this problem.

References

1. Demaine, E.D., Iacono, J., Langerman, S.: Retroactive data structures. ACM Transactions on Algorithms **3**, 13 (2007). Originally in SODA 2003
2. Giora, Y., Kaplan, H.: Optimal dynamic vertical ray shooting in rectilinear planar subdivisions. ACM Transactions on Algorithms (TALG) **5**, 28 (2009)
3. Bentley, J.L.: Solutions to klees rectangle problems. Technical report, Carnegie Mellon University (1977)

4. Galperin, I., Rivest, R.L.: Scapegoat trees. In: Proceedings of the Fourth Annual ACM-SIAM Symposium on Discrete algorithms, pp. 165–174. Society for Industrial and Applied Mathematics (1993)
5. Sleator, D.D., Tarjan, R.E.: Self-adjusting binary search trees. Journal of the ACM (JACM) **32**, 652–686 (1985)
6. Bender, M.A., Demaine, E.D., Farach-Colton, M.: Cache-oblivious b-trees. In: Proceedings of the 41st Annual Symposium on Foundations of Computer Science, pp. 399–409. IEEE (2000)
7. Arge, L., Vitter, J.S.: Optimal dynamic interval management in external memory. In: Proceedings of the 37th Annual Symposium on Foundations of Computer Science, pp. 560–569. IEEE (1996)
8. Frederickson, G.N., Johnson, D.B.: The complexity of selection and ranking in $X+Y$ and matrices with sorted columns. Journal of Computer and System Sciences **24**, 197–208 (1982)
9. Rauh, A., Arce, G.: A fast weighted median algorithm based on quickselect. In: 2010 17th IEEE International Conference on Image Processing (ICIP), pp. 105–108 (2010)

On the Minimum Eccentricity
Shortest Path Problem

Feodor F. Dragan and Arne Leitert[✉]

Department of Computer Science, Kent State University, Kent, Ohio, USA
{dragan,aleitert}@cs.kent.edu

Abstract. In this paper, we introduce and investigate the *Minimum Eccentricity Shortest Path (MESP)* problem in unweighted graphs. It asks for a given graph to find a shortest path with minimum eccentricity. We demonstrate that:

- a minimum eccentricity shortest path plays a crucial role in obtaining the best to date approximation algorithm for a minimum distortion embedding of a graph into the line;
- the MESP-problem is NP-hard on general graphs;
- a 2-approximation, a 3-approximation, and an 8-approximation for the MESP-problem can be computed in $\mathcal{O}(n^3)$ time, in $\mathcal{O}(nm)$ time, and in linear time, respectively;
- a shortest path of minimum eccentricity k in general graphs can be computed in $\mathcal{O}(n^{2k+2}m)$ time;
- the MESP-problem can be solved in linear time for trees.

1 Introduction

All graphs occurring in this paper are connected, finite, unweighted, undirected, loopless and without multiple edges. For a graph $G = (V, E)$, we use $n = |V|$ and $m = |E|$ to denote the cardinality of the vertex set and the edge set of G. For a vertex v of G, $N_G(v) = \{u \in V \mid uv \in E\}$ is called the *open neighborhood*, and $N_G[v] = N_G(v) \cup \{v\}$ the *closed neighborhood* of v.

The *length* of a path from a vertex v to a vertex u is the number of edges in the path. The *distance* $d_G(u, v)$ of two vertices u and v is the length of a shortest path connecting u and v. The distance between a vertex v and a set $S \subseteq V$ is defined as $d_G(v, S) = \min_{u \in S} d_G(u, v)$. The *eccentricity* $\mathrm{ecc}_G(v)$ of a vertex v is $\max_{u \in V} d_G(u, v)$. For a set $S \subseteq V$, its eccentricity is $\mathrm{ecc}_G(S) = \max_{u \in V} d_G(u, S)$.

In this paper, we investigate the following problem.

Definition 1 (Minimum Eccentricity Shortest Path Problem). *For a given a graph G, find a shortest path P such that for each shortest path Q, $\mathrm{ecc}_G(P) \leq \mathrm{ecc}_G(Q)$.*

© Springer International Publishing Switzerland 2015
F. Dehne et al. (Eds.): WADS 2015, LNCS 9214, pp. 276–288, 2015.
DOI: 10.1007/978-3-319-21840-3_23

Although this problem might be of an independent interest (it may arise in determining a "most accessible" speedy linear route in a network and can find applications in communication networks, transportation planning, water resource management and fluid transportation), our interest in this problem stems from the role it plays in obtaining the best to date approximation algorithm for a minimum distortion embedding of a graph into the line. In Section 2, we demonstrate that every graph G with a shortest path of eccentricity k admits an embedding f of G into the line with distortion at most $(8k+2)\,\mathrm{ld}(G)$, where $\mathrm{ld}(G)$ is the minimum line-distortion of G. Furthermore, if a shortest path of G of eccentricity k is given in advance, then such an embedding f can be found in linear time.

This fact augments the importance of investigating the *Minimum Eccentricity Shortest Path* problem (MESP-problem) in graphs. Fast algorithms for it will imply fast approximation algorithms for the minimum line distortion problem. Existence of low eccentricity shortest paths in special graph classes will imply low approximation bounds for those classes. For example, all AT-free graphs (and hence all interval, permutation, cocomparability graphs) enjoy a shortest path of eccentricity at most 1 [3], all convex bipartite graphs enjoy a shortest path of eccentricity at most 2 [5].

We prove also that for every graph G with $\mathrm{ld}(G) = \lambda$, the minimum eccentricity of a shortest path of G is at most $\lfloor \frac{\lambda}{2} \rfloor$. Hence, one gets an efficient embedding of G into the line with distortion at most $\mathcal{O}(\lambda^2)$.

In Section 3, we show that the MESP-problem is NP-hard on general graphs and that a shortest path of minimum eccentricity k in general graphs, can be computed in $\mathcal{O}(n^{2k+2}m)$ time. In Section 4, we design for the MESP-problem on general graphs a 2-approximation algorithm that runs in $\mathcal{O}(n^3)$ time, a 3-approximation algorithm that runs in $\mathcal{O}(nm)$ time and an 8-approximation algorithm that runs in linear time. In Section 5, we demonstrate that the MESP-problem can be solved in linear time for trees and distance-hereditary graphs, and in polynomial time for chordal graphs and dually chordal graphs.

Note that our Minimum Eccentricity Shortest Path problem is close but different from the *Central Path* problem in graphs introduced in [16]. It asks for a given graph G to find a path P (not necessarily shortest) such that any other path of G has eccentricity at least $\mathrm{ecc}_G(P)$. The Central Path problem generalizes the Hamiltonian Path problem and therefore is NP-hard even for chordal graphs [15]. Our problem is polynomial time solvable for chordal graphs.

In what follows we will need the following additional notions and notations.

The *diameter* of a graph G is $\mathrm{diam}(G) = \max_{u,v \in V} d_G(u,v)$. The diameter $\mathrm{diam}_G(S)$ of a set $S \subseteq V$ is defined as $\max_{u,v \in S} d_G(u,v)$. A pair of vertices x, y of G is called a *diametral pair* if $d_G(u,v) = \mathrm{diam}(G)$. In this case, every shortest path connecting x and y is called a *diametral path*.

A path P of a graph G is called a *k-dominating path* of G if $\mathrm{ecc}_G(P) \leq k$. In this case, we say also that P *k-dominates* each vertex of G. A pair of vertices x, y of G is called a *k-dominating pair* if every path connecting x and y has eccentricity at most k.

For a vertex s, let $L_i^{(s)} = \{v \mid d_G(s,v) = i\}$ denote the vertices with distance i from s. We will also refer to $L_i^{(s)}$ as the i-th layer.

2   Motivation Through the Line-Distortion of a Graph

Computing a minimum distortion embedding of a given n-vertex graph G into the line ℓ was recently identified as a fundamental algorithmic problem with important applications in various areas of computer science, like computer vision [17], as well as in computational chemistry and biology (see [12,13]). The *minimum line distortion* problem asks, for a given graph $G = (V, E)$, to find a mapping f of vertices V of G into points of ℓ with minimum number λ such that $d_G(x,y) \leq |f(x) - f(y)| \leq \lambda d_G(x,y)$ for every $x, y \in V$. The parameter λ is called the *minimum line-distortion* of G and denoted by $\mathrm{ld}(G)$. The embedding f is called *non-contractive* since $d_G(x,y) \leq |f(x) - f(y)|$ for every $x, y \in V$.

In [2], Bădoiu et al. showed that this problem is hard to approximate within a constant factor. They gave an exponential-time exact algorithm and a polynomial-time $\mathcal{O}(n^{1/2})$-approximation algorithm for arbitrary unweighted input graphs, along with a polynomial-time $\mathcal{O}(n^{1/3})$-approximation algorithm for unweighted trees. In another paper [1], Bădoiu et al. showed that the problem is hard to approximate by a factor $\mathcal{O}(n^{1/12})$, even for weighted trees. They also gave a better polynomial-time approximation algorithm for general weighted graphs, along with a polynomial-time algorithm that approximates the minimum line-distortion λ embedding of a weighted tree by a factor that is polynomial in λ.

Fast exponential-time exact algorithms for computing the line-distortion of a graph were proposed in [7,8]. Fomin et al. [8] showed that a minimum distortion embedding of an unweighted graph into the line can be found in time $5^{n+o(n)}$. Fellows et al. [7] gave an $\mathcal{O}(n\lambda^4(2\lambda+1)^{2\lambda})$ time algorithm that for an unweighted graph G and integer λ either constructs an embedding of G into the line with distortion at most λ, or concludes that no such embedding exists. They extended their approach also to weighted graphs obtaining an $\mathcal{O}(n\lambda^{4W}(2\lambda+1)^{2\lambda W})$ time algorithm, where W is the largest edge weight. Thus, the problem of minimum distortion embedding of a given n-vertex graph G into the line ℓ is Fixed Parameter Tractable.

Heggernes et al. [10,11] initiated the study of minimum distortion embeddings into the line of specific graph classes. In particular, they gave polynomial-time algorithms for the problem on bipartite permutation graphs and on threshold graphs [11]. Furthermore, in [10], Heggernes et al. showed that the problem of computing a minimum distortion embedding of a given graph into the line remains NP-hard even when the input graph is restricted to a bipartite, cobipartite, or split graph, implying that it is NP-hard also on chordal, cocomparability, and AT-free graphs. They also gave polynomial-time constant-factor approximation algorithms for split and cocomparability graphs.

Recently, in [5], a more general result for unweighted graphs was proven: for every class of graphs with path-length bounded by a constant, there exists an efficient constant-factor approximation algorithm for the minimum line-distortion problem. As a byproduct, an efficient algorithm was obtained which for each unweighted graph G with $\mathrm{ld}(G) = \lambda$ constructs an embedding with distortion at most $\mathcal{O}(\lambda^2)$. Furthermore, for AT-free graphs, a linear time 8-approximation algorithm for the minimum line-distortion problem was obtained. Note that AT-free graphs contain all cocomparability graphs and hence all interval, permutation and trapezoid graphs.

In this section we simplify and improve the result of [5]. We show that a minimum eccentricity shortest path plays a crucial role in obtaining the best to date approximation algorithm for the minimum line-distortion problem.

We will need the following simple "local density" lemma.

Lemma 1. *For every vertex set $S \subseteq V$ of an arbitrary graph $G = (V, E)$,*

$$|S| - 1 \le \mathrm{diam}_G(S)\,\mathrm{ld}(G).$$

Proof. Consider an embedding f^* of G into the line ℓ with distortion $\mathrm{ld}(G)$. Let a and b be the leftmost and the rightmost, respectively, in ℓ vertices of S, i.e., $f^*(a) = \min\{f^*(v) \mid v \in S\}$ and $f^*(b) = \max\{f^*(v) \mid v \in S\}$. Consider a shortest path P in G between a and b. Since for each edge xy of G (and hence of P) $|f^*(x) - f^*(y)| \le \mathrm{ld}(G)$ holds, we get $f^*(b) - f^*(a) \le d_G(a, b)\,\mathrm{ld}(G) \le \mathrm{diam}_G(S)\,\mathrm{ld}(G)$. On the other hand, since all vertices of S are mapped to points of ℓ between $f^*(a)$ and $f^*(b)$, we have $f^*(b) - f^*(a) \ge |S| - 1$. $\square$

The main result of this section is the following.

Theorem 1. *Every graph G with a shortest path of eccentricity k admits an embedding f of G into the line with distortion at most $(8k+2)\,\mathrm{ld}(G)$. If a shortest path of G of eccentricity k is given in advance, then such an embedding f can be found in linear time.*

Proof. Let $P = (x_0, x_1, \ldots, x_i, \ldots, x_j, \ldots, x_q)$ be a shortest path of G of eccentricity k. Build a $BFS(P, G)$-tree T of G (i.e., a Breadth-First-Search tree of G started at path P). Denote by $\{X_0, X_1, \ldots, X_q\}$ the decomposition of the vertex set V of G obtained from T by removing the edges of P. That is, X_i is the vertex set of a subtree (branch) of T growing from vertex x_i of P. See Fig. 1(a) for an illustration. Since eccentricity of P is k, we have $d_G(v, x_i) \le k$ for every $i \in \{1, \ldots, q\}$ and every $v \in X_i$.

We define an embedding f of G into the line ℓ by performing a preorder traversal of the vertices of T starting at vertex x_0 and visiting first vertices of X_i and then vertices of X_{i+1}, $i = 0, \ldots, q-1$. We place vertices of G on the line in that order, and also, for each $i \in \{0, \ldots, q-1\}$, we leave a space of length $d_T(v_i, v_{i+1})$ between any two vertices v_i and v_{i+1} placed next to each other (this can be done during the preorder traversal). Alternatively, f can be defined by creating a twice around tour of the tree T, which visits vertices of X_i prior to vertices of X_{i+1}, $i = 0, \ldots, q-1$, and then returns to x_0 from x_q along edges of P. Following vertices of T from x_0 to x_q as shown in Fig. 1(b) (i.e., using upper part of the twice around tour), $f(v)$ can be defined as the first appearance of vertex v in that subtour (see Fig. 1(c)).

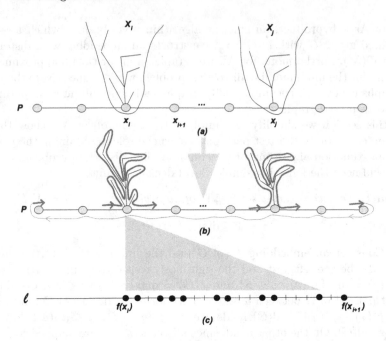

Fig. 1. Illustration to the proof of Theorem 1. (a) The decomposition $\{X_0, X_1, \ldots, X_q\}$ of the vertex set V of G. (b) The upper part of the twice around tour. (c) An embedding f obtained from following the upper part of the twice around tour.

We claim that f is a (non-contractive) embedding with distortion at most $(8k+2)\,\mathrm{ld}(G)$. It is sufficient to show that $d_G(x,y) \leq |f(x) - f(y)|$ for every two vertices of G that are placed by f next to each other in ℓ and that $|f(v) - f(u)| \leq (8k+2)\,\mathrm{ld}(G)$ for every edge uv of G (see, e.g., [2,11]).

Let x, y be arbitrary two vertices of G that are placed by f next to each other in ℓ. By construction, we know that $|f(x) - f(y)| = d_T(x,y)$. Since $d_G(x,y) \leq d_T(x,y)$, we get also $d_G(x,y) \leq |f(x) - f(y)|$, i.e., f is non-contractive.

Consider now an arbitrary edge uv of G and assume $u \in X_i$ and $v \in X_j$ $(i \leq j)$. Note that $d_P(x_i, x_j) = j - i \leq 2k+1$, since P is a shortest path of G and $d_P(x_i, x_j) = d_G(x_i, x_j) \leq d_G(x_i, u) + 1 + d_G(x_j, v) \leq 2k+1$. Set $S = \bigcup_{h=i}^{j} X_h$. For any two vertices $x, y \in S$, $d_G(x,y) \leq d_G(x, P) + 2k+1 + d_G(y, P) \leq k + 2k+1 + k = 4k+1$ holds. Hence, $\mathrm{diam}_G(S) \leq 4k+1$. Consider subtree T_S of T induced by S. Clearly, T_S is connected and has $|S| - 1$ edges. Therefore, $f(v) - f(u) \leq 2(|S| - 1)$ since each edge of T_S contributes to $f(v) - f(u)$ at most 2 units. Now, by Lemma 1, $f(v) - f(u) \leq 2(|S| - 1) \leq 2\,\mathrm{diam}_G(S)\,\mathrm{ld}(G) \leq (8k+2)\,\mathrm{ld}(G)$. □

Recall that a pair x, y of vertices of a graph G forms a k-dominating pair if every path connecting x and y in G has eccentricity at most k. It turns out that the following result is true.

Proposition 1. *If the minimum line-distortion of a graph G is λ, then G has a $\lfloor \frac{\lambda}{2} \rfloor$-dominating pair.*

Proof. Let f be an optimal line embedding for G. This embedding has a first vertex v_1 and a last vertex v_n. Let u be an arbitrary vertex and P an arbitrary path from v_1 to v_n. If u is not on this path, there is an edge $v_i v_j$ of P with $f(v_i) < f(u) < f(v_j)$. Without loss of generality, we can say that $f(u) - f(v_i) \leq \lfloor (f(v_j) - f(v_i))/2 \rfloor \leq \lfloor \frac{\lambda}{2} \rfloor$. Thus, each vertex is $\lfloor \frac{\lambda}{2} \rfloor$-dominated by each path from v_1 to v_n, i.e., v_1, v_n is a $\lfloor \frac{\lambda}{2} \rfloor$-dominating pair. $\square$

Corollary 1. *For every graph G with $\mathrm{ld}(G) = \lambda$, the minimum eccentricity of a shortest path of G is at most $\lfloor \frac{\lambda}{2} \rfloor$.*

Theorem 1 and Corollary 1 stress the importance of investigating the *Minimum Eccentricity Shortest Path* problem (MESP-problem) in graphs. As we will show later, although the MESP-problem is NP-hard on general graphs, there are much better (than for the minimum line distortion problem) approximation algorithms for it. We design for the MESP-problem on general graphs a 2-approximation algorithm that runs in $\mathcal{O}(n^3)$ time, a 3-approximation algorithm that runs in $\mathcal{O}(nm)$ time and an 8-approximation algorithm that runs in linear time.

Combining Theorem 1 and Corollary 1 with those approximation results, we reproduce a result of [2] and [5].

Corollary 2 ([2,5]). *For every graph G with $\mathrm{ld}(G) = \lambda$, an embedding into the line with distortion at most $\mathcal{O}(\lambda^2)$ can be found in polynomial time.*

It should be noted that, since the difference between the minimum eccentricity of a shortest path and the line-distortion of a graph can be very large (close to n), the result in Theorem 1 seems to be stronger. Furthermore, one version of our algorithm (that uses an 8-approximation algorithm for the MESP-problem) runs in total linear time.

3   NP-Completeness Result

In this section, we will show that in general it is NP-complete to find a minimum eccentricity shortest path. For this, we define the decision version of this problem (k-ESP) as follows: Given a graph G and an integer k, does G contain a shortest path P with eccentricity at most k?

Theorem 2. *The decision version of the minimum eccentricity shortest path problem is NP-complete.*

Proof. We will proof this by reducing SAT to k-ESP.

Let $\mathcal{I}$ be an instance of SAT with the variables $\mathcal{P} = \{p_1, \ldots, p_n\}$ and the clauses $\mathcal{C} = \{c_1, \ldots, c_m\}$. We assume $\mathcal{I}$ is a formula given in CNF. Also, let $k = \max\{n, m\}$. We create a graph G as shown in Figure 2. For each variable p_i

create two vertices, one representing p_i and one representing $\neg p_i$. Create one vertex c_i for every clause c_i. Additionally, create two vertices u_0, u_n and, for each i with $0 \leq i \leq n$, a vertex v_i.

Connect each variable vertex p_i and $\neg p_i$ with v_{i-1} and v_i directly with an edge. Connect each clause with the variables containing it with a path of length k. Also connect v_0 with u_0 and v_n with u_n with a path of length k.

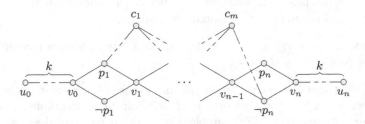

Fig. 2. Reduction from SAT to k-DSP. Illustration to the proof of Theorem 2.

Note that every shortest path in G not containing v_0 and v_n has an eccentricity larger than k. Also, a shortest path from v_0 to v_n has length $2n$ ($d_G(v_{i-1}, v_i) = 2$, passing p_i or $\neg p_i$). Since $k \geq n$, no shortest path from v_0 to v_n is passing a vertex c_i; in this case the minimal length would be $2k + 2$. Additionally, note that for all vertices in G except the vertices which represent clauses, the distance to a vertex v_i with $0 \leq i \leq n$ is at most k.

We will now show that $\mathcal{I}$ is satisfiable if and only if G has a shortest path with eccentricity k.

First assume $\mathcal{I}$ is satisfiable. Let $f \colon \mathcal{P} \to \{T, F\}$ be a satisfying assignment for the variables. As shortest path P we choose a shortest path from v_0 to v_n. Thus, we have to chose between p_i and $\neg p_i$. We will chose p_i if and only if $f(p_i) = T$. Because $\mathcal{I}$ is satisfiable, there is a p_i for each c_j such that either $f(p_i) = T$ and $d_G(c_j, p_i) = k$, or $f(p_i) = F$ and $d_G(c_j, \neg p_i) = k$. Thus, P has eccentricity k.

Next consider a shortest path P in G of eccentricity k. As mentioned above, P contains either p_i or $\neg p_i$. Now we define $f \colon \mathcal{P} \to \{T, F\}$ as follows:

$$f(p_i) = \begin{cases} T & \text{if } p_i \in P, \\ F & \text{else, i. e. } \neg p_i \in P. \end{cases}$$

Because P has eccentricity k and only vertices representing a variable in the clause c_j are at distance k to vertex c_j, f is a satisfying assignment for $\mathcal{I}$.   □

V. B. LE[1] pointed out that, by slightly modifying the created graph, it can be shown that the problem remains NP-complete even if the graph has a bounded vertex-degree of 3.

---

[1] University of Rostock, Germany.

Note that the factor k in this reduction depends on the input size. In [14] it was already mentioned that for $k = 1$ the problem can be solved in $\mathcal{O}(n^3m)$ time by modifying an algorithm given in [4]. There, the problem was called *Dominating Shortest Path* problem. In the full version of this paper, we show that the k-ESP problem can be solved in $\mathcal{O}(n^{2k+2}m)$ time for every fixed $k \geq 0$.

We can slightly modify the MESP problem such that a start vertex s and an end vertex t of the path are given. This is, for a given a graph G and two vertices s and t, find a shortest (s,t)-path P such that for each shortest (s,t)-path Q, $\mathrm{ecc}_G(P) \leq \mathrm{ecc}_G(Q)$. We call this the (s,t)-MESP problem. From the reduction above, it follows that the decision version of this problem is NP-complete, too.

Corollary 3. *The decision version of the (s,t)-MESP problem is NP-complete.*

4   Approximation Algorithms

In this section we will present different approximation algorithms. The algorithms differ in their approximation factor and runtime. Base for them are the following two lemmas.

Lemma 2. *In a graph G, let P be a shortest path from s to t of eccentricity at most k. For each layer $L_i^{(s)}$ there is a vertex $p_i \in P$ such that the distance from p_i to each vertex $v \in L_i^{(s)}$ is at most $2k$. Additionally, $p_i \in L_i^{(s)}$ if $i \leq d_G(s,t)$, and $p_i = t$ if $i \geq d_G(s,t)$.*

Proof. For each vertex v, let $p(v) \in P$ be a vertex with $d_G(p(v), v) \leq k$.

For each $i \leq d_G(s,t)$, let $p_i \in P \cap L_i^{(s)}$ be the vertex in P with distance i to s. For an arbitrary vertex $v \in L_i^{(s)}$, let $j = d_G(s, p(v))$. Because $\mathrm{ecc}_G(P) \leq k$ and P is a shortest path, $|i - j| \leq k$. Thus, $d_G(p_i, v) \leq d_G(p_i, p(v)) + d_G(p(v), v) \leq 2k$.

Let $L' = \{v \mid d_G(s,v) \geq d_G(s,t)\}$. Because P has eccentricity at most k, $d_G(p,t) \leq k$ for all $p \in \{p(v) \mid v \in L'\}$. Therefore, $d_G(t,v) \leq 2k$ for all $v \in L'$. $\square$

Lemma 3. *If G has a shortest path of eccentricity at most k from s to t, then every path Q with $s \in Q$ and $d_G(s,t) \leq \max_{v \in Q} d_G(s,v)$ has eccentricity at most $3k$.*

Proof. Let P be a shortest path from s to t with $\mathrm{ecc}_G(P) \leq k$ and Q an arbitrary path with $s \in Q$ and $d_G(s,t) \leq \max_{v \in Q} d_G(s,v)$. Without loss of generality, we can assume that Q starts at s. Also let u be an arbitrary vertex. Since $\mathrm{ecc}_G(P) \leq k$, there is a vertex $p \in P$ with $d_G(u,p) \leq k$. Because $d_G(s,t) \leq \max_{v \in Q} d_G(s,v)$, there is a vertex $q \in Q$ with $d_G(s,p) = d_G(s,q)$. By Lemma 2, the distance between p and q is at most $2k$. Thus, the distance from q to u is at most $3k$. $\square$

Corollary 4. *For a given graph G and two vertices s and t, each shortest (s,t)-path is a 3-approximation for the (s,t)-MESP problem.*

Theorem 3. *Algorithm 1 calculates a 3-approximation for the MESP problem in $\mathcal{O}(nm)$ time.*

Algorithm 1. A 3-approximation for the MESP problem.

Input: A graph $G = (V, E)$.

Output: A shortest path with eccentricity at most $3k$, where k is the minimum eccentricity of all paths in G.

1 **foreach** $s \in V$ **do**

2   Find a vertex v for which the distance to s is maximal. Also find a shortest path $P(s)$ from s to v.

3   Calculate $k(s) = \mathrm{ecc}_G(P(s))$.

4 Among all computed paths $P(s)$, select one for which $k(s)$ is minimal.

Proof. Assume a given graph G has a shortest path P from s to t with $\mathrm{ecc}_G(P) = k$ and s is the vertex selected by the loop in line 1. Let v be a vertex such that $d_G(s, v)$ is maximal (line 2). Because $d_G(s, v)$ is maximal, $d_G(s, t) \leq d_G(s, v)$. Thus, by Lemma 3, each path from s to v has eccentricity at most $3k$, i.e. $k(s) \leq 3k$ (line 3). Therefore, the eccentricity of the path selected in line 4 is also at most $3k$.

It is easy to see that line 2 and line 3 run in $\mathcal{O}(m)$ time for a given s. Therefore, the overall runtime for the algorithm is $\mathcal{O}(nm)$. □

Theorem 4. *Algorithm 2 calculates a 2-approximation for the MESP problem in $\mathcal{O}(n^3)$ time.*

Proof (Correctness). Assume a given graph G has a shortest path P from s to t with $\mathrm{ecc}_G(P) = k$ and s is the vertex selected by the loop in line 2. Let Q be a shortest path from s to v. We say the *layer-wise eccentricity* of Q is ϕ if for each layer $L_i^{(s)}$ ($i \leq d_G(s, v)$) there is a vertex $q_i \in Q \cap L_i^{(s)}$ with $\max\{d_G(q_i, u) \mid u \in L_i^{(s)}\} \leq \phi$.

We will now show that lines 4 to 8 calculate for each v the minimal $\phi(v)$ such that there is a shortest path Q from s to v with a layer-wise eccentricity $\phi(v)$.

By induction assume this is true for all vertices $u \in L_j^{(s)}$ with $j \leq i - 1$. Now let v be an arbitrary vertex in $L_i^{(s)}$. Line 6 calculates the maximal distance $\phi(v)$ from v to all other vertices in $L_i^{(s)}$. Since v is the only vertex in $Q \cap L_i^{(s)}$ for every shortest path Q from s to v, the layer-wise eccentricity of each Q is at least $\phi(v)$. Let u be a neighbour of v in the previous layer. By induction $\phi(u)$ is optimal. Therefore, $\phi(v) := \max\{\min_{u \in N_G^-[v]} \phi(u), \phi(v)\}$ (line 7) is optimal for v.

Since line 9 selects the vertex u with the smallest $\phi(u)$ as parent for v, each path Q from s to v in $T(s)$ has an optimal layer-wise eccentricity of $\phi(v)$. Line 8 calculates the maximal distance from v to all vertices in $\{u \mid d_G(s, u) \geq d_G(s, v)\}$. Thus, $\mathrm{ecc}_G(Q) \leq \phi'(v)$ and line 10 and 11 select a shortest path which has an eccentricity at most $\phi'(v)$.

By Lemma 2, we know that P has a layer-wise eccentricity of at most $2k$. Thus, the path Q from s to t in $T(s)$ has a layer-wise eccentricity of at most $2k$. Additionally, Lemma 2 says that t $2k$-dominates all vertices in $\{v \mid d_G(s, v) \geq$

$d_G(s,t)\}$. Therefore, $\mathrm{ecc}_G(Q) \leq 2k$. Thus, the path selected in line 11 is a shortest path with eccentricity at most $2k$. □

Proof (Complexity). Line 1 runs in $\mathcal{O}(nm)$ time. If the distances are stored in an array, they can be later accessed in constant time. Therefore, line 6 and line 8 run in $\mathcal{O}(n)$ time for a given s and v or in $\mathcal{O}(n^3)$ time overall. For a given s, line 7 runs in $\mathcal{O}(m)$ time and therefore has an overall runtime of $\mathcal{O}(nm)$. Line 9 has an overall runtime of $\mathcal{O}(nm)$, line 11 takes $\mathcal{O}(n^2)$ time, and line 10 runs in $\mathcal{O}(n)$ time. Adding all together, the total runtime is $\mathcal{O}(n^3)$. □

Algorithm 2. A 2-approximation for the MESP problem.

Input: A graph $G = (V, E)$.
Output: A shortest path with eccentricity at most $2k$, where k is the minimum eccentricity of all paths in G.

1 Calculate the distances $d_G(u, v)$ for all vertex pairs u and v, including
   $L_i^{(u)} = \{v \in V \mid d_G(u, v) = i\}$ with $0 \leq i \leq \mathrm{ecc}_G(u)$ for each u.
2 **foreach** $s \in V$ **do**
3 | Set $\phi(s) := 0$.
4 | **for** $i := 1$ **to** $\mathrm{ecc}_G(s)$ **do**
5 | | **foreach** $v \in L_i^{(s)}$ **do**
6 | | | Set $\phi(v) := \max_{u \in L_i^{(s)}} d_G(u, v)$.
7 | | | Let $N_G^-(v) = L_{i-1}^{(s)} \cap N_G(v)$ denote the neighbours of v in the previous layer. Set $\phi(v) := \max\{\min_{u \in N_G^-(v)} \phi(u), \phi(v)\}$.
8 | | | Set $\phi^+(v) := \max\{d_G(u, v) \mid d_G(s, u) \geq i\}$.
9 | Calculate a BFS-tree $T(s)$ starting from s. If multiple vertices u are possible as parent for a vertex v, select one with the smallest $\phi(u)$.
10 | Let t be the vertex for which $\phi'(t) := \max\{\phi(t), \phi^+(t)\}$ is minimal. Set $k(s) := \phi'(t)$.
11 Among all computed pairs s and t, select a pair (and corresponding path in $T(s)$) for which $k(s)$ is minimal.

Algorithm 1 and 2 both iterate over all vertices of the graph to find the best start vertex. Lemma 4 will show that a constant factor approximation can be found with a simple algorithm which starts at an arbitrary vertex. However, the approximation factor will be much higher.

Lemma 4. *Let G be a graph having a shortest path of eccentricity k. Let x be a vertex most distant from some arbitrary vertex, and y be a vertex most distant from x. Then, x, y is a $8k$-dominating pair of G.*

Proof. Let p be an end vertex of a shortest path of eccentricity k in a given graph G. By Lemma 2, the diameter in G of each layer $L_i^{(p)}$ is at most $4k$. Assume, x is most distant from an arbitrary vertex s.

If there is a layer containing both s and x, then $d_G(s,x) \leq 4k$. By the choice of x, each vertex of G is within distance at most $4k$ from s, hence, within distance at most $8k$ from x. Evidently, in this case, x, y is a $8k$-dominating pair of G.

Assume now, without loss of generality, that $x \in L_i^{(p)}$ and $s \in L_l^{(p)}$ with $i < l$. Consider an arbitrary vertex v of G which belongs to a layer with an index smaller than i. We show that $d_G(x,v) \leq 8k$. As $L_i^{(p)}$ separates v from s, a shortest path $P(s,v)$ of G between s and v must have a vertex u in $L_i^{(p)}$. We have $d_G(s,x) \geq d_G(s,v) = d_G(s,u) + d_G(u,v)$ and, by the triangle inequality, $d_G(s,x) \leq d_G(s,u) + d_G(u,x)$. Hence, $d_G(u,v) \leq d_G(u,x)$ and, since both u and x belong to same layer $L_i^{(p)}$, $d_G(u,x) \leq 4k$. That is, $d_G(x,v) \leq d_G(x,u) + d_G(u,v) \leq 2d_G(u,x) \leq 8k$.

If $d_G(x,y) \leq 8k$ then, by the choice of y, each vertex of G is within distance at most $8k$ from x. Hence, x, y is a $8k$-dominating pair of G. So, assume that $d_G(x,y) > 8k$, i.e., the layer $L_j^{(p)}$ with $i < j$ contains y. Repeating the arguments of the previous paragraph, we can show that $d_G(y,v) \leq 8k$ for every vertex v that belongs to a layer with an index greater than j.

Consider now an arbitrary path P of G connecting vertices x and y. P has a vertex in every layer $L_h^{(p)}$ with $i \leq h \leq j$. Hence, for each vertex v of G that belongs to layer $L_h^{(p)}$ ($i \leq h \leq j$), there is a vertex $u \in P \cap L_h^{(p)}$ such that $d_G(v,u) \leq 4k$. As $d_G(v,x) \leq 8k$ for each vertex v from $L_{i'}^{(p)}$ with $i' < i$ and $d_G(v,y) \leq 8k$ for each vertex v from $L_{j'}^{(p)}$ with $j' > j$, we conclude that $\operatorname{ecc}_G(P) \leq 8k$. $\square$

Corollary 5. *An 8-approximation for the MESP problem can be calculated in linear time.*

5   MESP for Certain Graph Classes

So far, we investigated the MESP problem in general graphs. Next, we will show that the problem is solvable in linear or polynomial time for certain graph classes.

Lemma 5. *If a tree has a shortest path of eccentricity k, then any diametral path has eccentricity at most k.*

Proof. In a tree T, let P be a shortest path from s to t with $\operatorname{ecc}_G(P) = k$ and D be a diametral path from x to y. Assume P and D do not intersect. Then there is a vertex $u \in P$ with minimal distance to D and a vertex $z \in D$ with minimal distance to P. Thus, the paths from u to x and from u to y contain z. Because $d_T(x,P) \leq k$, $d_T(y,P) \leq k$, and $d_T(u,z) > 0$, we have $d_T(z,x) < k$ and $d_T(z,y) < k$. Therefore, $d_T(x,y) < 2k$. Each diametral path of length l in a tree contains a vertex c with $\operatorname{ecc}_T(c) = \lceil l/2 \rceil$ [9]. Thus, $\operatorname{ecc}_G(D) \leq k$.

Next, assume P and D intersect. Then there is a vertex $x' \in P \cap D$ with $d_T(x,x') = d_T(x,P) \leq k$ and $y' \in P \cap D$ with $d_T(y,y') = d_T(y,P) \leq k$. Assume there is a vertex v with $d_T(v,D) > k$. Thus, there is a vertex $v' \in P \setminus D$ with $d_T(v,v') \leq k$ and, without loss of generality, $d_T(s,v') < d_T(s,x')$. Therefore,

x' is the vertex in D with minimal distance to v. It follows that $d_T(y,v) = d_T(y,x') + d_T(x',v) > d_T(y,x') + d_T(x',x) = d_T(y,x)$. This contradicts with D being a diametral path.   □

Recall that a diametral path in a tree can be found as follows: Select an arbitrary vertex v. Find a most distant vertex x from v and then a most distant vertex y from x. The path from x to y is a diametral path. Thus, it follows from Lemma 5:

Theorem 5. *The MESP problem can be solved for trees in linear time.*

In [6] we show that the MESP problem can be solved in linear time for distance-hereditary graphs and in polynomial time for chordal graphs and dually chordal graphs.

References

1. Bădoiu, M., Chuzhoy, J., Indyk, P., Sidiropoulos, A.: Low-distortion embeddings of general metrics into the line. In: Proceedings of the 37th Annual ACM Symposium on Theory of Computing (STOC 2005), pp. 225–233. ACM (2005), Baltimore
2. Bădoiu, M., Dhamdhere, K., Gupta, A., Rabinovich, Y., Raecke, H., Ravi, R., Sidiropoulos, A.: Approximation algorithms for low-distortion embeddings into low-dimensional spaces. In: Proceedings of the ACM/SIAM Symposium on Discrete Algorithms (2005)
3. Corneil, D.G., Olariu, S., Stewart, L.: Linear Time Algorithms for Dominating Pairs in Asteroidal Triple-free Graphs. SIAM J. Computing **28**, 292–302 (1997)
4. Deogun, J.S., Kratsch, D.: Diametral path graphs. In: Nagl, M. (ed.) WG 1995. LNCS, vol. 1017, pp. 344–357. Springer, Heidelberg (1995)
5. Leitert, A., Dragan, F.F., Köhler, E.: Line-distortion, bandwidth and path-length of a graph. In: Ravi, R., Gørtz, I.L. (eds.) SWAT 2014. LNCS, vol. 8503, pp. 158–169. Springer, Heidelberg (2014)
6. Dragan, F.F., Leitert, A.: Minimum eccentricity shortest paths in some structured graph classes. In: WG 2015: 41st International Workshop on Graph-Theoretic Concepts in Computer Science, June 17–19, 2015, Munich, Germany, Lecture Notes in Computer Science (2015) (to appear)
7. Fellows, M.R., Fomin, F.V., Lokshtanov, D., Losievskaja, E., Rosamond, F.A., Saurabh, S.: Distortion is fixed parameter tractable. In: Albers, S., Marchetti-Spaccamela, A., Matias, Y., Nikoletseas, S., Thomas, W. (eds.) ICALP 2009, Part I. LNCS, vol. 5555, pp. 463–474. Springer, Heidelberg (2009)
8. Fomin, F.V., Lokshtanov, D., Saurabh, S.: An exact algorithm for minimum distortion embedding. Theor. Comput. Sci. **412**, 3530–3536 (2011)
9. Handler, G.Y.: Minimax location of a facility in an undirected tree graph. Transportation Science **7**, 287–293 (1973)
10. Heggernes, P., Meister, D.: Hardness and approximation of minimum distortion embeddings. Information Processing Letters **110**, 312–316 (2010)
11. Heggernes, P., Meister, D., Proskurowski, A.: Computing minimum distortion embeddings into a path of bipartite permutation graphs and threshold graphs. Theoretical Computer Science **412**, 1275–1297 (2011)

12. Indyk, P.: Algorithmic applications of low-distortion geometric embeddings. In: Proceedings of FOCS 2001, pp. 10–35. IEEE (2005)
13. Indyk, P., Matousek, J.: Low-distortion embeddings of finite metric spaces, Handbook of Discrete and Computational Geometry, 2nd edn., pp. 177–196. CRC Press (2004)
14. Kratsch, D.: Domination and total domination on asteroidal triple-free graphs. Discrete Applied Mathematics **99**, 111–123 (2000)
15. Müller, H.: Hamiltonian circuits in chordal bipartite graphs. Discrete Mathematics **156**, 291–298 (1996)
16. Slater, P.J.: Locating central paths in a graph. Transportation Science **16**, 1–18 (1982)
17. Tenenbaum, J.B., de Silva, V., Langford, J.C.: A global geometric framework for nonlinear dimensionality reduction. Science **290**, 2319–2323 (2000)

Convex Polygons in Geometric Triangulations

Adrian Dumitrescu[1] and Csaba D. Tóth[2,3](✉)

[1] University of Wisconsin–Milwaukee, Milwaukee, USA
dumitres@uwm.edu
[2] California State University Northridge, Los Angeles, CA, USA
[3] Tufts University, Medford, MA, USA
csaba.toth@csun.edu

Abstract. We show that the maximum number of convex polygons in a triangulation of n points in the plane is $O(1.5029^n)$. This improves an earlier bound of $O(1.6181^n)$ established by van Kreveld, Löffler, and Pach (2012) and almost matches the current best lower bound of $\Omega(1.5028^n)$ due to the same authors. We show how to compute efficiently the number of convex polygons in a given a planar straight-line graph with n vertices.

1 Introduction

Convex polygons. According to the celebrated Erdős-Szekeres theorem [13], every set of n points in the plane, no three on a line, contains $\Omega(\log n)$ points in convex position, and, apart from the constant factor, this bound is the best possible. The minimum and maximum *number* of subsets in convex position contained in an n-element point set have also been investigated [17]. When the n points are in convex position, then trivially all the $2^n - 1$ nonempty subsets are also in convex position. Erdős [12] proved that the minimum number of subsets in convex position is $\exp(\Theta(\log^2 n))$.

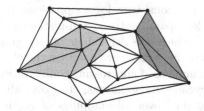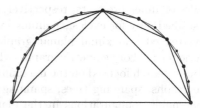

Fig. 1. Left: A (geometric) triangulation on 19 points; the two shaded convex polygons are subgraphs of the triangulation. Right: A triangulation on $2^4 + 1 = 17$ points in convex position, whose dual graph is a full binary tree with 8 leaves.

Recently, van Kreveld et al. [15] posed analogous problems concerning the number of convex polygons in a triangulation of n points in the plane; see Fig. 1 (left). They proved that the maximum number of convex polygons in a triangulation of n points, no three on a line, is between $\Omega(1.5028^n)$ and $O(1.6181^n)$.

© Springer International Publishing Switzerland 2015
F. Dehne et al. (Eds.): WADS 2015, LNCS 9214, pp. 289–300, 2015.
DOI: 10.1007/978-3-319-21840-3_24

Their lower bound comes from a balanced binary triangulations on $2^4 + 1 = 17$ points shown in Fig. 1 (right). At the other end of the spectrum, Löffler et al. [16] showed that the *minimum* number of convex polygons in an n-vertex triangulation is $\Theta(n)$. Here we study the maximum number of convex polygons contained in an n-vertex triangulation. This number is known [15] to be exponential in n, and our interest is in the base of the exponent: what is the infimum of $a > 0$ such that every n-vertex triangulation contains $O(a^n)$ convex polygons?

Throughout this paper we consider planar point sets $S \subset \mathbb{R}^2$ with no 3 points collinear. A *(geometric) triangulation* of a set $S \subset \mathbb{R}^2$ is a plane straight-line graph with vertex set S such that all bounded faces are triangles that jointly tile the convex hull of S.

Our results. We first prove that the maximum number of convex polygons in an n-vertex triangulation is attained, up to an $O(n)$-factor, for point sets in convex position. Consequently, determining the maximum becomes a purely combinatorial problem. We then show that the maximum number of convex polygons in a triangulation of n points in the plane is $O(1.5029^n)$. This improves an earlier bound of $O(1.6181^n)$ established by van Kreveld, Löffler, and Pach [15] and almost matches the current best lower bound of $\Omega(1.5028^n)$ due to the same authors (Theorem 3 and Corollary 1 in Subsection 2.4). In deriving the new upper bound, we start with a careful analysis of a balanced binary triangulation indicated in Fig. 1 (right), and then extend the analysis to *all* triangulations on n points in convex position. Given a planar straight-line graph G with n vertices, we show how to compute efficiently the number of convex polygons in G (Theorem 4 in Section 3). Most proofs are omitted from this extended abstract due to space limitations, and are available in the full paper [10].

Related work. We derive new upper and lower bounds on the maximum and minimum number of convex cycles in straight-line triangulations with n points in the plane. Both subgraphs we consider can be defined geometrically (in terms of angles or inner products, respectively). Previously, analogous problems have been studied only for cycles, spanning cycles, spanning trees, and matchings [7] in n-vertex edge-maximal planar graphs—which are defined in purely graph theoretic terms. For geometric graphs, where the vertices are points in the plane, previous research focused on the maximum number of noncrossing configurations (plane graphs, spanning trees, spanning cycles, triangulations, etc.) over all n-element point configurations in the plane (i.e., over all mappings of K_n into $\mathbb{R}^2$) [1,2,8,14,18,20–23]; see also [9,24]. Early upper bounds in this area were obtained by multiplying the maximum number of triangulations on n point in the plane with the maximum number of desired configurations in an n-vertex triangulation, since every planar straight-line graph can be augmented into a triangulation.

2   Convex Polygons

Section outline. We reduce the problem of determining the maximum number of convex polygons in an n-vertex triangulation (up to polynomial factors) to triangulations of n points in convex position (Theorem 1, Section 2.1). We further reduce the problem to counting convex *paths* between two adjacent vertices in a triangulation (Lemma 2, Subsection 2.2). We first analyze the number of convex paths in a balanced binary triangulation, which gives the current best lower bound [15] (Theorem 2, Subsection 2.3). The new insight gained from this analysis is then generalized to derive an upper bound for all n-vertex triangulations (Theorem 3 and Corollary 1, Subsection 2.4).

2.1   Reduction to Convex Position

For a triangulation T of n points in the plane, let $C(T)$ denote the number of convex polygons in T. For an integer $n \geq 3$, let $C(n)$ be the maximum of $C(T)$ over all triangulations T of n points in the plane; and let $C_x(n)$ be the maximum of $C(T)$ over all triangulations T of n points *in convex position*. It is clear that $C_x(n) \leq C(n)$ for every integer $n \geq 3$. The main result of this section is the following.

Theorem 1. *For every integer $n \geq 3$, we have $C(n) \leq (2n - 5)C_x(n)$.*

Theorem 1 is an immediate consequence of the following lemma.

Lemma 1. *Let T be a triangulation on a set S of n points in the plane, and let f be a bounded face of T. Then there exists a triangulation T' on a set S' of n points in convex position such that the number of convex polygons in T whose interior contains the face f is at most $C(T')$.*

Proof. We construct a point set S' in convex position, a triangulation T' on S', and then give an injective map from the set of convex polygons in T that contain f into the set of convex polygons of T'.

Let o be a point in the interior of the face f, and let O be a circle centered at o that contains all points in S. Refer to Fig. 2. For each point $p \in S$, let p' be the intersection point of the ray $\overrightarrow{op}$ with O. Let $S' = \{p' : p \in S\}$.

We now construct a plane graph T' on the point set S'. For two points $p', q' \in S'$, there is an edge $p'q'$ in T' iff there is an empty triangle $\triangle oab$ such that ab is contained in an edge of T, point p lies on segment oa, and q lies on ob. Note that no two edges in T' cross each other. Indeed, suppose to the contrary that edges $p_1'q_1'$ and $p_2'q_2'$ cross in T'. By construction, there are empty triangles $\triangle oa_1b_1$ and $\triangle oa_2b_2$ that induce $p_1'q_1'$ and $p_2'q_2'$, respectively. We may assume w.l.o.g. that both $\triangle oa_1b_1$ and $\triangle oa_2b_2$ are oriented counterclockwise. Since a_1b_1 and a_2b_2 do not cross (they may be collinear), either segment ob_2 lies in $\triangle oa_1b_1$ or segment oa_1 lies in $\triangle oa_2b_2$. That is, one of $\triangle oa_1b_1$ and $\triangle oa_2b_2$ contains a point from S, contradicting our assumption that both triangles are empty.

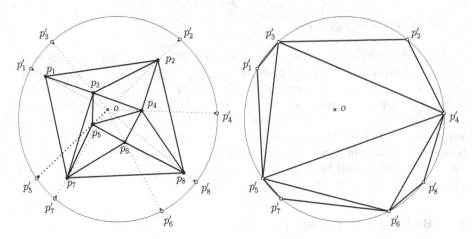

Fig. 2. A triangulation on the point set $\{p_1, \ldots, p_8\}$ (left) is mapped to a triangulation on the point set $\{p'_1, \ldots, p'_8\}$ in convex position (right)

Finally, we define an injective map from the convex polygons of T that contain o into the convex polygons of T'. To define this map, we first map every edge of T to a path in T'. Let pq be an edge in T induced by a triangle $\triangle opq$ oriented counterclockwise. We map the edge pq to the path $(p', r'_1, \ldots, r'_k, q')$, where $(r_1, \ldots, r_k)$ is the sequence of all points in S lying in the interior of $\triangle opq$ in counterclockwise order around o. A convex polygon $A = (p_1, \ldots, p_k)$ containing o in T is mapped to the convex polygon A' in T' obtained by concatenating the images of the edges $p_1 p_2, \ldots, p_{k-1} p_k$, and $p_k p_1$.

It remains to show that the above mapping is injective on the convex polygons of T that contain o. Consider a convex polygon $A' = (p'_1, \ldots, p'_k)$ in T' that is the image of some convex polygon in T. Then its preimage A must be a convex polygon in T that contains $\{p_1, \ldots, p_k\}$ on its boundary or in its interior. Hence A must be the boundary of the convex hull of $\{p_1, \ldots, p_k\}$, that is, A' has a unique preimage.   □

Proof of Theorem 1. Let T be a triangulation with n vertices. Every n-vertex triangulation has $2n - 4$ faces (including the outer face), and hence at most $2n - 5$ bounded faces. By Lemma 1, each bounded face f of T lies in the interior of at most $C_x(n)$ convex polygons contained in T. Summing over all bounded faces f, the number of convex polygons in T is bounded by $C(T) \leq (2n - 5)C_x(n)$, as required.   □

2.2   Reduction to Convex Paths

A *convex path* is a polygonal chain $(p_1, \ldots, p_m)$ that makes a right turn at each interior vertex $p_2, \ldots, p_{m-1}$. Let $P(n)$ denote the maximum number of convex paths between two adjacent vertices in a triangulation of n points in convex position. A convex path from a to b is either a direct path consisting of a single

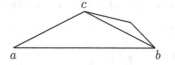 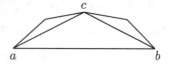

Fig. 3. Convex paths in a triangulation. Left: $P(4) = P(2) P(3) + 1 = 3$. Right: $P(5) = P(3) P(3) + 1 = 5$.

segment ab, or a path that can be decomposed into two convex subpaths sharing a common endpoint c, where $\triangle abc$ is a counterclockwise triangle incident to ab; see Fig. 3.

Thus $P(n)$ satisfies the following recurrence for $n \geq 3$, with initial values $P(2) = 1$ and $P(3) = 2$.

$$P(n) = \max_{\substack{n_1+n_2=n+1 \\ n_1,n_2 \geq 2}} \{P(n_1) P(n_2) + 1\} \tag{1}$$

Remark. The values of $P(n)$ for $2 \leq n \leq 18$ are shown in Table 1. It is worth noting that $P(n)$ need not be equal to $P(\lfloor \frac{n+1}{2} \rfloor) P(\lceil \frac{n+1}{2} \rceil) + 1$; for instance, $P(7) = P(3) P(5) + 1 > P(4) P(4) + 1$. That is, the balanced partition of a convex n-gon into two subpolygons does not always maximize $P(n)$. However, we have $P(n) = P(\frac{n+1}{2}) P(\frac{n+1}{2}) + 1$ for $n = 2^k + 1$ and $k = 1, 2, 3, 4$; these are the values relevant for the (perfectly) balanced binary triangulation discussed in Subsection 2.3.

Let ab be a hull edge of a triangulation T on n points in convex position. Suppose that ab is incident to counterclockwise triangle $\triangle abc$. The edges ac and bc decompose T into three triangulations T_1, $\triangle abc$ and T_2, of size n_1, 3 and n_2, where $n_1 + n_2 = n + 1$. A convex polygon in T is either (i) contained in T_1; or (ii) contained in T_2; or (iii) the union of ab and a convex path from a to b that passes through c; see Fig. 3. Consequently, $C_x(n)$, the maximum number of convex polygons contained in a triangulation of n points in convex position, satisfies the following recurrence:

$$C_x(n) = \max_{\substack{n_1+n_2=n+1 \\ n_1,n_2 \geq 2}} \{P(n_1) P(n_2) + C_x(n_1) + C_x(n_2)\} \tag{2}$$

for $n \geq 3$, with initial values $C_x(2) = 0$ and $C_x(3) = 1$. The values of $C_x(n)$ for $2 \leq n \leq 9$ are displayed in Table 1.

Table 1. $P(n)$ and $C_x(n)$ for small n

n	2	3	4	5	6	7	8	9	10	11	12	13	14	15	16	17	18
$P(n)$	1	2	3	5	7	11	16	26	36	56	81	131	183	287	417	677	937
$C_x(n)$	0	1	3	6	11	18	29	45									

Lemma 2. *We have* $C_x(n) \leq \sum_{k=2}^{n-1} P(k)$. *Consequently,* $C_x(n) \leq n\,P(n)$.

Proof. We first prove the inductive inequality:

$$C_x(n) \leq P(n-1) + C_x(n-1). \tag{3}$$

Let T be an arbitrary triangulation of a set S of n points in the plane. Consider the dual graph T^* of T, with a vertex for each triangle in T and an edge for every pair of triangles sharing an edge. It is well known that if the n points are in convex position, then T^* is a tree. Let $\varDelta abc$ be a triangle corresponding to a leaf in T^*, sharing a unique edge, say $e = ab$, with other triangles in T.

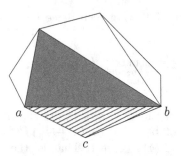

Fig. 4. Proof of Lemma 2

We distinguish two types of convex polygons contained in T: (i) those containing both edges ac and cb, and (ii) those containing neither ac nor cb. Observe that the number of convex polygons of type (i) is at most $P(n-1)$, since any such polygon can be decomposed into the path (b, c, a) and another path connecting a and b in the subgraph of T induced by $S \setminus \{c\}$. Similarly, the number of convex polygons of type (ii) is at most $C_x(n-1)$, since they are contained in the subgraph of T induced by $S \setminus \{c\}$. Altogether we have $C_x(n) \leq P(n-1) + C_x(n-1)$ and (3) is established.

Summing up inequality (3) for $n, n-1, \ldots, 3$ yields $C_x(n) \leq \sum_{k=2}^{n-1} P(k)$, as required. Since $P(k) \leq P(k+1)$, for every $k \geq 2$, it immediately follows that $C_x(n) \leq n\,P(n)$, for every $n \geq 3$, as desired.   □

2.3   Analysis of Balanced Binary Triangulations

We briefly review the lower bound construction of van Kreveld, Löffler and Pach [15, Sec. 3.1]. For a constant $k \in \mathbb{N}$, let T_k be the triangulation on $n = 2^k + 1$ points, say, on a circular arc, such that the dual graph T_k^* is a balanced binary tree; see Fig. 1 (right). The authors constructed a triangulation of $n = m2^k + 1$ points, for $m \in \mathbb{N}$, by concatenating m copies of T_k along a common circular arc, where consecutive copies share a vertex, and by triangulating the convex hull of the m chords arbitrarily to obtain a triangulation of the n points. They settled on $k = 4$.

Denote by λ_k the number of convex paths between the diametrical pair of vertices in T_k. As noted in [15], λ_k satisfies the following recurrence:

$$\lambda_{k+1} = \lambda_k^2 + 1, \text{ for } k \geq 0, \qquad \lambda_0 = 1. \tag{4}$$

The values of λ_k for $0 \leq k \leq 5$ are shown in Table 2. Note that $\lambda_k = P(2^k+1)$ for these values. Obviously (4) implies that the sequence $(\lambda_k)^{1/2^k}$ is strictly increasing. Van Kreveld et al. [15] proved that $\lambda_4 \geq 1.5028^{2^4}$. By the product rule, this gives $C(n) \geq C_x(n) \geq (\lambda_4)^{n/16} = \left(\lambda_4^{1/16}\right)^n \geq 1.5028^n$, for every $n = 2^4m + 1$: a lower bound construction is obtained by concatenating m triangulations, each with 17 vertices with $P(17)$ convex a-to-b paths, where consecutive copies share a vertex, and triangulating the convex hull of the m chords ab arbitrarily.

Table 2. The values of λ_k for small k

k	0	1	2	3	4	5
λ_k	1	2	5	26	677	458330

As noted above, $\lambda_k \geq 1.5028^{2^k}$ for every $k \geq 4$. In this section (Theorem 2), we establish an almost matching upper bound $\lambda_k \leq 1.50284^{2^k}$, or equivalently, $(\lambda_k)^{1/2^k} \leq 1.50284$ for every $k \geq 0$. We start by bounding λ_k from above by a product. To this end we frequently use the standard inequality $1+x \leq e^x$, where e is the base of the natural logarithm.

Lemma 3. *For $k \in \mathbb{N}$, we have*

$$\lambda_k \leq 2^{2^{k-1}} \prod_{i=1}^{k-1} \left(1 + \frac{1}{2^{2^i}}\right)^{2^{k-1-i}}. \tag{5}$$

The following sequence is instrumental for manipulating the exponents in (5). Let

$$\alpha_k = 2^k + k + 1 \quad \text{for } k \geq 1. \tag{6}$$

That is, $\alpha_1 = 4$, $\alpha_2 = 7$, $\alpha_3 = 12$, $\alpha_4 = 21$, $\alpha_5 = 38$, etc. The way this sequence appears will be evident in Lemma 4, and subsequently, in the proof of Theorem 3. The following lemma is proved by induction.

Lemma 4. *For $k \in \mathbb{N}$, we have*

$$\lambda_k \leq 2^{2^{k-1}} \exp\left(2^k \sum_{i=1}^{k-1} 2^{-\alpha_i}\right). \tag{7}$$

Taking roots (i.e., the $1/2^k$ root) in (7) yields a first rough approximation:

$$(\lambda_k)^{1/2^k} \leq 2^{2^{k-1}/2^k} \exp\left(2^k/2^k \sum_{i=1}^{k-1} 2^{-\alpha_i}\right) \leq 2^{1/2} \exp\left(\sum_{i=1}^{\infty} 2^{-\alpha_i}\right) \leq 1.5180,$$

To obtain a sharper estimate, we keep the first few terms in the sequence as they are, and only introduce approximations for latter terms.

Theorem 2. *For every $k \in \mathbb{N}$, we have $\lambda_k \leq 1.50284^{2^k}$. Consequently, for every $n = m2^k + 1$ points, the triangulation obtained by extending (via concatenation) the balanced triangulation on $2^k + 1$ points in convex position has at most $O(1.50284^n)$ convex polygons.*

2.4 Upper Bound for Triangulations of Convex Polygons

In this section we show that the maximum number of convex polygons present in a triangulation on n points in convex position, $C(n)$, is $O(1.50285^n)$. In the main step, a complex proof by induction yields the following.

Theorem 3. *Let $n \geq 2$ where $2^k + 1 \leq n \leq 2^{k+1}$. Then*

$$P(n)^{\frac{1}{n-1}} \leq (P(17))^{1/16} \exp\left(\sum_{i=4}^{k-1} 2^{-\alpha_i}\right) = 677^{1/16} \exp\left(\sum_{i=4}^{k-1} 2^{-\alpha_i}\right). \quad (8)$$

Proof. We prove the inequality by induction on n. The base cases $2 \leq n \leq 32$ are verified by direct calculation:

$$\max_{2 \leq n \leq 16} P(n)^{\frac{1}{n-1}} = P(9)^{1/8} = 26^{1/8} = 1.50269\ldots.$$

$$\max_{17 \leq n \leq 32} P(n)^{\frac{1}{n-1}} = P(17)^{1/16} = 677^{1/16} = 1.50283\ldots.$$

Assume now that $n \geq 33$, hence $k \geq 5$, and that the required inequality holds for all smaller n. We will show that for all pairs $n_1, n_2 \geq 2$ with $n_1 + n_2 = n+1$, the expression $P(n_1) P(n_2) + 1$ is bounded from above as required. Since $n_1 + n_2 = n+1$, we have $n_1, n_2 \leq n-1$, thus using the induction hypothesis for n_1 and n_2 is justified. It suffices to consider pairs with $n_1 \leq n_2$. We distinguish two cases:

Case 1: $2 \leq n_1 \leq 16$. Since $n \geq 33$, we have $18 \leq n_2 \leq n - 1$. By the induction hypothesis we have

$$P(n_2)^{1/(n_2-1)} \leq 677^{1/16} \exp\left(\sum_{i=4}^{k-1} 2^{-\alpha_i}\right).$$

Further,

$$P(n) \leq P(n_1) P(n_2) + 1$$

$$\leq P(n_1)\, 677^{\frac{n_2-1}{16}} \exp\left((n_2 - 1) \sum_{i=4}^{k-1} 2^{-\alpha_i}\right) + 1$$

$$\leq P(n_1)\, 677^{\frac{n_2-1}{16}} \exp\left((n_2 - 1) \sum_{i=4}^{k-1} 2^{-\alpha_i}\right) \left(1 + (P(n_1))^{-1}\, 677^{-\frac{n_2-1}{16}}\right)$$

$$\leq P(n_1)\, 677^{\frac{n_2-1}{16}} \exp\left((n_2 - 1) \sum_{i=4}^{k-1} 2^{-\alpha_i}\right) \exp\left((P(n_1))^{-1}\, 677^{-\frac{n_2-1}{16}}\right).$$

To settle Case 1, it suffices to show that

$$P(n_1)\, 677^{\frac{n_2-1}{16}} \exp\left((n_2 - 1) \sum_{i=4}^{k-1} 2^{-\alpha_i}\right) \exp\left((P(n_1))^{-1}\, 677^{-\frac{n_2-1}{16}}\right) \leq$$

$$\leq 677^{\frac{n-1}{16}} \exp\left((n-1)\sum_{i=4}^{k-1} 2^{-\alpha_i}\right),$$

or equivalently,

$$P(n_1)\exp\left((P(n_1))^{-1}\,677^{-\frac{n_2-1}{16}}\right) \leq 677^{\frac{n_1-1}{16}} \exp\left((n_1-1)\sum_{i=4}^{k-1} 2^{-\alpha_i}\right). \qquad (9)$$

We have $n_1 + n_2 = n + 1$, hence $n_2 - 1 = n - n_1 \geq 33 - n_1$. To verify (9) it suffices to verify that the following inequality holds for $2 \leq n_1 \leq 16$.

$$P(n_1)\exp\left((P(n_1))^{-1}\,677^{-\frac{33-n_1}{16}}\right) \leq 677^{\frac{n_1-1}{16}}. \qquad (10)$$

Indeed, (10) would imply

$$P(n_1)\exp\left((P(n_1))^{-1}\,677^{-\frac{n_2-1}{16}}\right) \leq P(n_1)\exp\left((P(n_1))^{-1}\,677^{-\frac{33-n_1}{16}}\right)$$

$$\leq 677^{\frac{n_1-1}{16}} \leq 677^{\frac{n_1-1}{16}}\exp\left((n_1-1)\sum_{i=4}^{k-1} 2^{-\alpha_i}\right),$$

as required by (9). Finally, (10) can be deduced via the following fact: For $2 \leq n \leq 16$, we have

$$P(n)\exp\left((P(n))^{-1}\,677^{-\frac{33-n}{16}}\right) \leq 677^{\frac{n-1}{16}}. \qquad (11)$$

Case 2: $n_1 \geq 17$. We distinguish two subcases, $n \leq 2^k + 2$ and $n \geq 2^k + 3$. Due to space constraints, the proofs of the two subcases are omitted (the reader is referred to [10]). $\qquad \square$

Corollary 1. $C(n) = O(1.50285^n)$.

Proof. Note that

$$677^{1/16}\exp\left(\sum_{i=4}^{\infty} 2^{-\alpha_i}\right) \leq 1.50284.$$

By Theorem 3 and the above inequality we obtain

$$P(n)^{\frac{1}{n}} \leq P(n)^{\frac{1}{n-1}} \leq 677^{\frac{1}{16}}\exp\left(\sum_{i=4}^{k-1} 2^{-\alpha_i}\right) \leq 677^{\frac{1}{16}}\exp\left(\sum_{i=4}^{\infty} 2^{-\alpha_i}\right) \leq 1.50284.$$

Further, by Lemma 2, we have $C_x(n) \leq n\,P(n)$. Consequently, Theorem 1 yields

$$C(n) \leq (2n-5)\,C_x(n) \leq 2n^2\,P(n) \leq 2n^2 \cdot 1.50284^n = O(1.50285^n),$$

as required. $\qquad \square$

3   Algorithmic Aspects

The number of crossing-free structures (matchings, spanning trees, spanning cycles, triangulations) on a set of n points in the plane is known to be exponential in n [8, 14, 18, 21–23]. It is a challenging problem to determine the number of configurations faster than listing all such configurations (i.e., count faster than enumerate). Exponential-time algorithms have been recently developed for triangulations [4], planar graphs [19], and matchings [25] that count these structures exponentially faster than the number of structures. It is worth pointing out that counting (exactly) matchings, spanning trees, spanning cycles, and triangulations, can be done in polynomial time in non-trivial cases by a result of Alvarez et al. [3].

Given a planar straight-line graph G with n vertices, we show how to compute in polynomial time the number of convex polygons in G. In particular, convex polygons can be counted in polynomial time in a given triangulation.

Theorem 4. *Given a planar straight-line graph G with n vertices, the number of convex polygons in G can be computed in $O(n^4)$ time. The convex polygons can be enumerated in an additional $O(1)$-time per edge.*

Computing the number of convex polygons in a given graph. Let $G = (V, E)$ be a planar straight line graph. For counting and enumerating convex cycles in G, we adapt a dynamic programming approach by Eppstein et al. [11], originally developed for finding the subsets of an n-element point set in the plane in convex position optimizing various parameters, e.g., the area or the perimeter of the convex hull. The dynamic program relies on the following two observations:

1. Introduce a canonical notation for the convex polygons in G. Assume, by rotating G if necessary, that no two vertices have the same x- or y-coordinates. Order the vertices of G by their x-coordinates. Now every convex polygon $\xi = (v_1, v_2, \ldots, v_t)$ can be labeled such that v_1 is the leftmost vertex, and the vertices are in counterclockwise order.

2. Consider the triangle (v_1, v_i, v_{i+1}), for $1 < i < t$, in the convex polygon $\xi = (v_1, v_2, \ldots, v_t)$. The triangle $\triangle v_1 v_i v_{i+1}$ decomposes ξ into two convex arcs[1] $(v_1, \ldots, v_i)$ and $(v_{i+1}, \ldots, v_t, v_1)$. The convex arc $(v_1, \ldots, v_i)$ lies in the closed region $R(v_1, v_i, v_{i+1})$ on the right of the vertical line through v_1, and right of both directed lines $\overrightarrow{v_1 v_i}$ and $\overrightarrow{v_{i+1} v_i}$ (Fig. 5). Importantly, the region $R(v_1, v_i, v_{i+1})$ is defined in terms of only three vertices, irrespective of any interior vertices of the arc $(v_1, \ldots, v_i)$.

For every ordered triple of vertices $(a, b, c) \in V^3$ and every integer $3 \le k \le n$, we compute the following function by dynamic programming. Let $f_k(a, b, c)$ denote the number of counterclockwise convex arcs $(v_1, \ldots, v_k)$ with k vertices such that $a = v_1$ is the leftmost vertex, $b = v_{k-1}$ and $c = v_k$.

---

[1] A convex arc is a polygonal arc that lies on the boundary of a convex polygon.

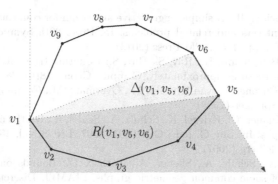

Fig. 5. A convex polygon $\xi = (v_1, \ldots, v_9)$ where v_1 is the leftmost vertex. Region $R(v_1, v_5, v_6)$ is shaded.

Observe that if $v_1 v_k$ is an edge of G, then this edge completes all $f_k(a, b, c)$ convex arcs into a convex polygon in G. The initial values $f_3(a, b, c)$ can be computed in $O(n^3)$ time by examining all triples $(a, b, c) \in V^3$. If (a, b, c) is a counterclockwise 2-edge path in G, where a is the leftmost vertex, then $f_3(a, b, c) = 1$, otherwise $f_3(a, b, c) = 0$. In the induction step, we compute $f_k(a, b, c)$ for all $(a, b, c) \in V^3$ based on the values $f_{k-1}(a, b, c)$. It is enough to consider counterclockwise triples (a, b, c), where a is the leftmost vertex and $bc \in E$. For such a triple (a, b, c) we have $f_k(a, b, c) = \sum_v f_{k-1}(a, v, b)$ where the sum is over all vertices $v \in V$ that lie in the region $R(a, b, c)$. For any other triple (a, b, c), we have $f_k(a, b, c) = 0$.

Note that for $k = 4, \ldots, n$, the value of $f_k(a, b, c)$ is the sum of at most $\deg(b) - 1$ terms. Consequently for every $k = 4, \ldots, n$, all nonzero values of $f_k(a, b, c)$ can be computed in

$$O\left(n \cdot \sum_{v \in V} \deg^2(v)\right) = O(n^3)$$

time. The total running time over all k is $O(n^4)$. Finally, the total number of convex polygons is obtained by summing all values $f_k(a, b, c)$ for which $ac \in E$, again in $O(n^4)$ time. Note that $f_k(a, b, c)$ counts the number of convex polygons in T with k vertices, leftmost vertex a, and containing a counterclockwise convex arc (b, c, a), hence each convex polygon is counted precisely once.

References

1. Aichholzer, O., Hackl, T., Vogtenhuber, B., Huemer, C., Hurtado, F., Krasser, H.: On the number of plane geometric graphs. Graphs Combin. **23**(1), 67–84 (2007)
2. Ajtai, M., Chvátal, V., Newborn, M., Szemerédi, E.: Crossing-free subgraphs. Ann. Discrete Math. **12**, 9–12 (1982)
3. Alvarez, V., Bringmann, K., Curticapean, R., Ray, S.: Counting crossing-free structures. In: Proc. 28th Sympos. on Comput. Geom. (SOCG), pp. 61–68. ACM Press (2012). arXiv:1312.4628

4. Alvarez, V., Seidel, R.: A simple aggregative algorithm for counting triangulations of planar point sets and related problems. In: Proc. 29th Sympos. on Comput. Geom. (SOCG), pp. 1–8. ACM Press (2013)

5. Alvarez, V., Bringmann, K., Ray, S., Ray, S.: Counting triangulations and other crossing-free structures approximately. Comput. Geom. **48**(5), 386–397 (2015)

6. de Berg, M., Cheong, O., van Kreveld, M., Overmars, M.: Computational Geometry, 3rd edn. Springer (2008)

7. Buchin, K., Knauer, C., Kriegel, K., Schulz, A., Seidel, R.: On the number of cycles in planar graphs. In: Lin, G. (ed.) COCOON 2007. LNCS, vol. 4598, pp. 97–107. Springer, Heidelberg (2007)

8. Dumitrescu, A., Schulz, A., Sheffer, A., Tóth, C.D.: Bounds on the maximum multiplicity of some common geometric graphs. SIAM J. Discrete Math. **27**(2), 802–826 (2013)

9. Dumitrescu, A., Tóth, C.D.: Computational Geometry Column 54. SIGACT News Bulletin **43**(4), 90–97 (2012)

10. Dumitrescu, A., Tóth, C.D.: Convex polygons in geometric triangulations, November 2014. arXiv:1411.1303

11. Eppstein, D., Overmars, M., Rote, G., Woeginger, G.: Finding minimum area k-gons. Discrete Comput. Geom. **7**(1), 45–58 (1992)

12. Erdős, P.: Some more problems on elementary geometry. Austral. Math. Soc. Gaz. **5**, 52–54 (1978)

13. Erdős, P., Szekeres, G.: A combinatorial problem in geometry. Compos. Math. **2**, 463–470 (1935)

14. García, A., Noy, M., Tejel, A.: Lower bounds on the number of crossing-free subgraphs of K_N. Comput. Geom. **16**(4), 211–221 (2000)

15. van Kreveld, M., Löffler, M., Pach, J.: How Many potatoes are in a mesh? In: Chao, K.-M., Hsu, T., Lee, D.-T. (eds.) ISAAC 2012. LNCS, vol. 7676, pp. 166–176. Springer, Heidelberg (2012)

16. Löffler, M., Schulz, A., Tóth, C.D.: Counting carambolas. In: Proc. 25th Canadian Conf. on Comput. Geom. (CCCG), Waterloo, ON, pp. 163–168 (2013)

17. Morris, W., Soltan, V.: The Erdős-Szekeres problem on points in convex position–a survey. Bull. AMS **37**, 437–458 (2000)

18. Razen, A., Snoeyink, J., Welzl, E.: Number of crossing-free geometric graphs vs. triangulations, Electron. Notes. Discrete Math. **31**, 195–200 (2008)

19. Razen, A., Welzl, E.: Counting plane graphs with exponential speed-up. In: Calude, C.S., Rozenberg, G., Salomaa, A. (eds.) Rainbow of Computer Science. LNCS, vol. 6570, pp. 36–46. Springer, Heidelberg (2011)

20. Sharir, M., Sheffer, A.: Counting triangulations of planar point sets. Electron. J. Combin. **18**, P70 (2011)

21. Sharir, M., Sheffer, A.: Counting plane graphs: cross-graph charging schemes. Combin., Probab. Comput. **22**(6), 935–954 (2013)

22. Sharir, M., Sheffer, A., Welzl, E.: Counting plane graphs: perfect matchings, spanning cycles, and Kasteleyn's technique. J. Combin. Theory, Ser. A **120**(4), 777–794 (2013)

23. Sharir, M., Welzl, E.: On the number of crossing-free matchings, cycles, and partitions. SIAM J. Comput. **36**(3), 695–720 (2006)

24. Sheffer, A.: Numbers of plane graphs (version of May, 2015). https://adamsheffer.wordpress.com/numbers-of-plane-graphs/

25. Wettstein, M.: Counting and enumerating crossing-free geometric graphs. In: Proc. 30th Sympos. on Comput. Geom. (SOCG), pp. 1–10. ACM Press (2014)

Straight-Line Drawability of a Planar Graph Plus an Edge

Peter Eades[1], Seok-Hee Hong[1](✉), Giuseppe Liotta[2],
Naoki Katoh[3], and Sheung-Hung Poon[4]

[1] University of Sydney, Sydney, Australia
{peter.eades,seokhee.hong}@sydney.edu.au
[2] University of Perugia, Perugia, Italy
liotta@diei.unipg.it
[3] Kwansei Gakuin University, Nishinomiya, Japan
naoki.katoh@kwansei.ac.jp
[4] Institut Teknologi Brunei, Gadong, Brunei
sheung.hung.poon@gmail.com

Abstract. We investigate straight-line drawings of topological graphs that consist of a planar graph plus one edge, also called almost-planar graphs. We present a characterization of such graphs that admit a straight-line drawing. The characterization enables a linear-time testing algorithm to determine whether an almost-planar graph admits a straight-line drawing, and a linear-time drawing algorithm that constructs such a drawing, if it exists. We also show that some almost-planar graphs require exponential area for a straight-line drawing.

1 Introduction

This paper investigates straight-line drawings of *almost-planar* graphs, that is, graphs that become planar after the deletion of just one edge. Our work is partly motivated by the classical *planarization* approach [1] to graph drawing. This method takes as input a graph G, deletes a small number of edges to give a planar subgraph G^-, and then constructs a planar topological embedding (i.e., a plane graph) of G^-. Then the deleted edges are re-inserted, one by one, to give a topological embedding of the original graph G. Finally, a drawing algorithm is applied to the topological embedding. A number of variations on this basic approach give a number of graph drawing algorithms (see, e.g., [1]). This paper is concerned with the final step of creating a drawing from the topological embedding.

This research began at the *Blue Mountains Workshop on Geometric Graph Theory*, August, 2010, in Australia, and supported by the University of Sydney IPDF funding and the ARC (Australian Research Council). Hong is supported by ARC Future Fellowship. Liotta is also supported by the Italian Ministry of Education, University, and Research (MIUR) under PRIN 2012C4E3KT AMANDA. For the full version of this paper with proofs, see [6].

© Springer International Publishing Switzerland 2015
F. Dehne et al. (Eds.): WADS 2015, LNCS 9214, pp. 301–313, 2015.
DOI: 10.1007/978-3-319-21840-3_25

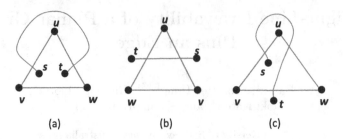

Fig. 1. (a) An almost-planar topological graph G; (b) a straight-line drawing of G that preserves its embedding on the sphere but not on the plane; (c) An almost-planar topological graph for which an embedding preserving straight-line drawing does not exist

Minimizing the number of edge crossings is an NP-hard problem even when the given graph is almost-planar [3]. However, Gutwenger *et al.* [8] present an elegant polynomial-time solution to the following simpler problem: Given a graph G and an edge e such that $G^- = G - e$ is planar, find a planar topological embedding of G^- that minimizes the number of edge crossings when re-inserting e in G.

While the output of the algorithm of Gutwenger *et al.* [8] has the minimum number of edge crossings, it may not give rise to a straight-line planar drawing. In this paper we study the following problem: Let G be a topological graph consisting of a planar graph plus an edge e. We want to test whether G admits a straight-line drawing that preserves the given embedding.

It is important to remark that by "preserving the embedding" we mean that the straight-line drawing must preserve the cyclic order of the edges around each vertex and around each crossing. In other words, we want to preserve a given embedding on the *sphere*. Note that the problem is different if, in addition to preserving the cyclic order of the edges around the vertices and the crossings, we also want the preservation of a given external boundary; in other words the problem is different if we want to maintain a given embedding on the *plane* instead of on the sphere. For example, consider the graph of Fig. 1(a). If we regard this as a topological graph on the sphere, then it has an embedding preserving straight-line drawing, as shown in Fig. 1(b). However, the drawing in Fig. 1(a) has a different external face to Fig. 1(b). It is easy to show that there is no straight-line drawing with the same external face as in Fig. 1(a). For a contrast, Fig. 1(c) shows a topological graph G that does not have a straight-line drawing that preserves the embedding on the sphere.

In this paper we mostly focus on spherical topologies but, as a byproduct, we obtain a result for topologies on the plane that may be of independent interest. Namely, the main results of this paper are as follows.

- We characterize those almost-planar topological graphs that admit a straight-line drawing that preserves a given embedding on the sphere. The characterization gives rise to a linear-time testing algorithm.

- We characterize those almost-planar topological graphs that admit a straight-line drawing that preserves a given embedding on the plane.
- We present a drawing algorithm that constructs straight-line drawings when such drawings exist. This drawing algorithm runs in linear time; however, the model of computation used is the real RAM, and the drawings that are produced have exponentially bad resolution. We show that, in the worst case, the exponentially bad resolution is inevitable.

Our results also contribute to the rapidly increasing literature about topological graphs that are "nearly" plane, in some sense. An interesting example is the class of *1-plane graphs*, that is, topological graphs with at most one crossing per edge. Thomassen [13] gives a "Fáry-type theorem" for 1-plane graphs, that is, a characterization of 1-plane topological graphs that admit a straight-line drawing. Hong et al. [9] present a linear-time algorithm that constructs a straight-line 1-planar drawing of 1-plane graph, if it exists. More generally, Nagamochi [12] investigates straight-line drawability of a wide class of topological non-planar topological graphs. He presents Fáry-type theorems as well as polynomial-time testing and drawing algorithms. This paper considers graphs that are "nearly plane" in the sense that deletion of a single edge yields a planar graph. Such graphs are variously called "1-skew graphs" or "almost-planar" graphs in the literature. Our characterization can be regarded as a Fáry-type theorem for almost-planar graphs.

Section 2 gives notation and terminology. The characterization of almost-planar topological graphs on the sphere that admit an embedding preserving straight-line drawing is given in Section 3. The extension of this characterization to topological graphs on the plane and the exponential area lower bound are described in Section 4. Open problems can be found in Section 5. Because of page limits, we omit many proofs; the omitted proofs can be found in [6].

2   Preliminaries

A *topological graph* $G = (V, E)$ is a representation of a simple graph on a given surface, where each vertex is represented by a point and each edge is represented by a simple Jordan arc between the points representing its endpoints. If the given surface is the sphere, then we say that G is an $\mathbb{S}^2$-*topological graph*; if the given surface is the plane, then we say that G is an $\mathbb{R}^2$-*topological graph*. Two edges of a topological graph *cross* if they have a point in common, other than their endpoints. The point in common is called a *crossing*. We assume that a topological graph satisfies the following non-degeneracy conditions: (i) an edge does not contain a vertex other than its endpoints; (ii) edges must not meet tangentially; (iii) no three edges share a crossing; and (iv) an edge does not cross an incident edge.

An $\mathbb{S}^2$-*embedding* of a graph is an equivalence class of $\mathbb{S}^2$-topological graphs under homeomorphisms of the sphere. An $\mathbb{S}^2$-topological graph has no unbounded face; in fact an $\mathbb{S}^2$-embedding is uniquely determined merely by the clockwise order of edges around each vertex and each edge crossing. An $\mathbb{R}^2$-*embedding* of a graph is an equivalence class of $\mathbb{R}^2$-topological graphs under homeomorphisms of the plane. Note that one face of an $\mathbb{R}^2$-topological graph in the plane is unbounded; this is the *external face*.

The concepts of $\mathbb{R}^2$-embedding and $\mathbb{S}^2$-embedding are very closely related. Each $\mathbb{S}^2$-topological graph gives rise to a representation of the same graph on the plane, by a stereographic projection about an interior point of a chosen face. This chosen face becomes the external face of the $\mathbb{R}^2$-topological graph. Thus we can regard an $\mathbb{R}^2$-embedding to be an $\mathbb{S}^2$-embedding in which one specific face is chosen to be the external face. Further, each $\mathbb{R}^2$-topological graph gives rise to a representation of the same graph on the sphere, by a simple projection.

A topological graph (either on the plane or on the sphere) is *planar* if no two edges cross. A topological graph is *almost-planar* if it has an edge (s,t) whose removal makes it planar. An *almost-planar* $\mathbb{R}^2$-*embedding* ($\mathbb{S}^2$-embedding) of a graph is an equivalence class of almost-planar $\mathbb{R}^2$-topological graphs ($\mathbb{S}^2$-topological graphs) under homeomorphisms of the plane (sphere).

Throughout this paper, $G = (V, E)$ denotes an almost-planar topological graph ($\mathbb{S}^2$ or $\mathbb{R}^2$) and (s,t) denotes an edge of G whose deletion makes G planar. The embedding obtained by deleting the edge (s,t) is denoted by $\hat{G}$. More generally, we use the convention that the notation $\hat{X}$ normally denotes X without the edge (s,t).

Let G be an $\mathbb{S}^2$-topological graph and let G' be an $\mathbb{R}^2$-topological graph with the same underlying simple graph. We say that G' *preserves the* $\mathbb{S}^2$-*embedding of* G if for each vertex and for each crossing they have the same cyclic order of incident edges. Further, let G be an $\mathbb{R}^2$-topological graph and let G' be an $\mathbb{R}^2$-topological graph with the same underlying simple graph. We say that G' *preserves the* $\mathbb{R}^2$-*embedding of* G if for each vertex and for each crossing they have the same cyclic order of incident edges and the same external face. A *straight-line drawing* of a graph is an $\mathbb{R}^2$-topological graph whose edges are represented by straight-line segments.

3   Straight-Line Drawability of an Almost-Planar $\mathbb{S}^2$-Embedding

In this section we state our main theorem. Let G be a topological graph with a given almost-planar $\mathbb{S}^2$-embedding. Suppose that α is a crossing between edges (s,t) and (u,v) in G. If the clockwise order of vertices around α is $\langle s, u, t, v \rangle$, then u is a *left* vertex and v is a *right* vertex (with respect to the ordered pair (s,t) and the crossing α). We say that a vertex of G is *inconsistent* if it is both left and right, and *consistent* otherwise. For example, vertex v in Fig. 1(c) is inconsistent: it is a left vertex with respect the first crossing along (s,t), and it is a right vertex with respect to the final crossing along (s,t).

Theorem 1. *An almost-planar* $\mathbb{S}^2$-*topological graph* G *with* n *vertices admits an* $\mathbb{S}^2$-*embedding preserving straight-line drawing if and only if every vertex of* G *is consistent. This condition can be tested in* $O(n)$ *time.*

The necessity of every vertex being consistent is straightforward. The proof of sufficiency involves many technicalities and it occupies most of the remainder of this paper. Namely, we prove the sufficiency of the condition in Theorem 1 by the following steps.

Augmentation: We show that we can add edges to an almost-planar $\mathbb{S}^2$-topological graph to form a maximal almost-planar graph, without changing the property that every vertex is consistent. Let G' be the augmented $\mathbb{S}^2$-topological graph (subsection 3.1).

Choice of an external face: We find a face f_o of G' such that if the $\mathbb{S}^2$-embedding of G' is projected on the plane with f_0 as the external face, G' satisfies an additional property that we call *face consistency* (subsection 3.2).

Split the augmented graph: After having projected G' on the plane with f_o as the external face, we split the $\mathbb{R}^2$-embedding of G' into the "inner graph" and the "outer graph". The inner graph and outer graph share a cycle called the "separating cycle" (subsection 3.3).

Straight-line drawing computation: We draw the outer graph leaving a convex shaped "hole" for the inner graph; the boundary of this hole is the separating cycle. Then we draw the "inner graph", whose external face is the separating cycle, such that it fits exactly into the convex shaped "hole" (subsections 3.4 and 3.5).

Before presenting more details of the proof of sufficiency, we observe that the condition stated in Theorem 1 can be tested in linear time. By regarding crossing points as dummy vertices, we can apply the usual data structures for plane graphs to almost-planar graphs (see [5], for example). A simple traversal of the crossing points along the edge (s, t) can be used to compute the left and the right vertices. Since the number of crossing points in an almost-planar graph is linear, these data structures can be applied without asymptotically increasing total time complexity.

3.1   Augmentation

Let G be an $\mathbb{S}^2$-topological graph. An $\mathbb{S}^2$-*embedding preserving augmentation* of G is an $\mathbb{S}^2$-topological graph G' obtained by adding edges (and no vertices) to G such that for each vertex (for each crossing) of G', the cyclic order of the edges of $G' \cap G$ around the vertex (around the crossing) is the same in G' and in G. An almost-planar topological graph is *maximal* if the addition of any edge would result in a topological graph that is not almost planar. The following lemma describes a technique to compute an $\mathbb{S}^2$-embedding preserving augmentation of an almost-planar $\mathbb{S}^2$-topological graph that gives rise to a maximal almost-planar graph. The proof is reported in [6].

Lemma 1. *Let G be an almost-planar $\mathbb{S}^2$-topological graph with n vertices. If G satisfies the vertex consistency condition, then there exists a maximal almost-planar $\mathbb{S}^2$-embedding preserving augmentation G' of G such that G' satisfies the vertex consistency condition. Also, such augmentation can be computed in $O(n)$ time.*

Some remarks about maximal almost-planar graphs are in [6].

Lemma 2. *If G is a maximal almost-planar topological graph, then either $\hat{G}$ is a maximal planar graph (that is, every face of $\hat{G}$ has size 3); or every face of $\hat{G}$ has size 3, except exactly one face f_4 which has the following properties: (i)f_4 has size 4; (ii)f_4 induces a clique in G; and (iii) both s and t are on f_4.*

3.2   Choice of an External Face

The augmentation step results in a maximal almost-planar $\mathbb{S}^2$-topological graph G' in which every vertex is consistent. Next, we want to identify a face f_o of G' such that if we choose f_0 to be the external face, then G' becomes an $\mathbb{R}^2$-topological graph that has an embedding preserving straight-line drawing in the plane. To identify such a face, we need some further terminology.

Let G be an almost-planar topological graph. Let $\hat{G}$ denote $G - (s, t)$. We denote the set of left (resp. right) vertices of G by V_L (resp. V_R). We denote the subgraph of $\hat{G}$ induced by $V_L \cup \{s, t\}$ (resp. $V_R \cup \{s, t\}$) by $\hat{G}_L$ (resp. $\hat{G}_R$). The union of $\hat{G}_L$ and $\hat{G}_R$ is $\hat{G}_{LR}$, and G_{LR} denotes the topological subgraph of G formed from $\hat{G}_{LR}$ by adding the edge (s, t). Note that G_{LR} and $\hat{G}_{LR}$ are *not* necessarily induced subgraphs of $\hat{G}$. A face of G_{LR} is *inconsistent* if it contains a left vertex and a right vertex, and *consistent* otherwise. In fact G_{LR} has exactly one inconsistent face, as stated in the next Lemma.

Lemma 3. *Let G be an $\mathbb{S}^2$-topological graph in which every vertex is consistent. Then G_{LR} has exactly one inconsistent face.*

We now proceed as follows. Let G be an $\mathbb{S}^2$-topological graph in which every vertex is consistent and let G' be a maximal almost-planar $\mathbb{S}^2$-embedding preserving augmentation of G constructed by using Lemma 1. We project G on the plane such that the only inconsistent face of G_{LR} is its external face. The following lemma is a consequence of the discussion above and of Lemma 3.

Lemma 4. *Let G be a maximal almost-planar $\mathbb{S}^2$-topological graph in which every vertex is consistent. There exists an $\mathbb{R}^2$-topological graph G' that preserves the $\mathbb{S}^2$-embedding of G and such that: (i) every internal face of $\hat{G}'$ consists of three vertices (i.e. it is a triangle); (ii) every internal face of G'_{LR} is consistent.*

Examples of an almost-planar $\mathbb{R}^2$-topological graph G and of its subgraphs $\hat{G}_L$, $\hat{G}_R$, and G_{LR} are given in [6].

3.3   Splitting the Augmented Graph

For the remainder of Section 3, we assume that G is a maximal almost-planar $\mathbb{R}^2$-topological graph; that is, that the augmentation and choice of an outer face have been done. Next we divide G into the "inner graph" and the "outer graph".

Denote the induced subgraph of $\hat{G}$ on the vertex set $V_L \cup V_R \cup \{s, t\}$ by $\hat{G}_{LR}^+$. Note that $\hat{G}_{LR}$ is a subgraph of $\hat{G}_{LR}^+$, but these graphs may not be the same; in particular, $\hat{G}^+_{LR}$ may have edges with a left endpoint and a right endpoint that do not cross (s, t); such an edge is called a *cap* edge. Although $\hat{G}$ is internally triangulated by Lemma 4, $\hat{G}^+_{LR}$ may have *non-triangular* inner faces. Nevertheless, the outside face of $\hat{G}^+_{LR}$ is a simple cycle. (See [6] for proof and an illustration.)

Lemma 5. *If G is a maximal almost-planar $\mathbb{R}^2$-topological graph such that every internal face of G_{LR} is consistent, then the external face of $\hat{G}^+_{LR}$ is a simple cycle.*

We call the external face of $\hat{G}^+_{LR}$ the *separating cycle* of the graph G. The topological subgraph consisting of the separating cycle as well as all vertices and edges that lie outside the separating cycle is the *outer graph* G_{out}. (An example of an outer graph is in [6].) The *inner graph* consists of $\hat{G}^+_{LR}$ with the addition of some dummy edges. Namely, for every face f of $\hat{G}^+_{LR}$ that is not a triangle, we perform a *fan triangulation*; that is, we choose a vertex u in f with degree 2 in f, and add dummy edges incident with u to triangulate f. The graph formed by fan triangulating every non-triangular internal face of $\hat{G}^+_{LR}$ is the *inner graph* $\hat{G}_{in}$.

Note that the vertices of G that are neither left vertices nor right vertices and that are inside the separating cycle belong to neither the inner nor the outer graph. At the end of next section we show how to reinsert these vertices and their incident edges into the drawing.

3.4   Drawing the Outer Graph

Since G is maximal almost-planar, by using Lemma 2 we can show that G_{out} is triconnected as long as the separating cycle has no chord. But since G_{in} contains the subgraph of $\hat{G}$ induced by the separating cycle, every chord on the separating cycle is in G_{in} and not in G_{out}. Thus G_{out} is triconnected. We use the linear-time convex drawing algorithm of Chiba et al. [4] to draw G_{out} such that every face in the drawing is a convex polygon. This drawing of the outer graph has a convex polygonal drawing of the separating cycle, which we shall call the *separating polygon*. In the next section we show how to draw the inner graph such that its outside face (i.e. the separating cycle) is the separating polygon.

3.5   Drawing the Inner Graph

The overall approach for drawing the inner graph is described as follows. For each edge e of the separating cycle, we define a "side graph" S_e; intuitively, S_e consists on vertices and edges that are "close" to e. There may be two special side graphs, that contain cap edges (that is, edges that join a left vertex and a right vertex but do not cross (s, t)); these side graphs are "cap graphs". Each side graph has a block-cutvertex tree T_e. We root T_e at the block (biconnected component) that contains the edge e. The algorithm first draws the root block for each side graph, then proceeds from the root to the leaves of these trees, drawing the blocks one by one. Cap graphs are drawn with a different algorithm from that used for other side graphs.

Each non-root block B with parent cutverex c in T_e is associated with circular arc $\gamma(B)$, and two regions, called a "safe wedge" $\omega(B)$ and a "safe moon" $\mu(c)$; these are defined precisely below. We draw all the vertices of B and its descendants in $\mu(c)$, with all vertices except the parent cutvertex lying on $\gamma(B)$ inside $\mu(c) \cap \omega(B)$. Every edge with exactly one endpoint in B and its descendants lies inside $\omega(B)$.

First the root blocks are drawn, and then the algorithm proceeds by repeating the following steps until every vertex of every side graph is drawn. (1) Choose a "safe block" B from the child blocks of drawn vertices; (2) Compute the "safe moon" $\mu(c)$, the "safe wedge" $\omega(B)$, and the circular arc $\gamma(B)$; (3) Draw each vertex of B except c on $\gamma(B)$.

Side Graphs and Cap Graphs. To define "side graphs" and "cap graphs", we need to first define a certain closed walk in the inner graph. Denote the edges that cross (s, t) by $e_0, e_1, \ldots, e_{p-1}$, ordered from s to t by their crossing points along (s, t). Suppose that $e_i = (\ell_i, r_i)$ for $0 \leq i \leq p - 1$, where ℓ_i is a left vertex and r_i is a right vertex. Note that cyclic list $(s, \ell_0, \ell_1, \ldots, \ell_{p-1}, t, r_{p-1}, \ldots, r_0)$ may contain repeated vertices.

Now let W be the sublist of $(s, \ell_0, \ell_1, \ldots, \ell_{p-1}, t, r_{p-1}, \ldots, r_0)$ obtained by replacing each contiguous subsequence of the same vertex by a single occurrence of that vertex. Note that W may contain repeated vertices, but these repeats are not contiguous. Namely, W is a closed spanning walk of $\hat{G}_{LR}$.

Now let $e = (u, v)$ be an edge of the separating cycle, with u before v in clockwise order around the separating cycle. Note that both u and v are elements of the closed walk W. Suppose that the clockwise sequence of vertices in W between u and v is $(u = u_1, u_2, \ldots, u_k = v)$. If u occurs more than once in W, then we choose u_1 to be the first occurrence of u in clockwise order after s; similarly choose u_k. The *side graph* S_e is the induced subgraph of G on $\{u_1, u_2, \ldots, u_k\}$.

If S_e contains both left and right vertices then it is a *cap graph*. Note that a cap graph contains either s or t; one can show that s and t are not in the same cap graph. An example of the closed walk W with side graphs and cap graphs in [6].

Drawing the Root Blocks of Side Graphs. Next we show how to draw the root block B_e^* of the side graph S_e. The edge e is drawn as a side λ_e of the separating polygon. We define a circular arc $\gamma(B_e^*)$ through the endpoints of λ_e, with radius chosen such that the maximum distance from λ_e to $\gamma(B_e^*)$ is ϵ_1. We will show how to choose ϵ_1 later; for the moment, we assume that ϵ_1 is very small in comparison to the length of the smallest edge of the separating polygon. The convex region bounded by λ_e and $\gamma(B_e^*)$ is called the *pillow* of e.

Suppose that B_e^* of S_e has $a + 1$ vertices, which occur in clockwise order on the closed walk W as $w_0, w_1, \ldots, w_a$. Since B_e^* is biconnected, this sequence is a Hamilton path of S_e. We compute $a + 1$ equally spaced points $\alpha(w_0), \alpha(w_1), \ldots, \alpha(w_a)$ on λ_e as in Fig. 2(a). Let $\zeta(w_i)$ denote the line through $\alpha(w_i)$ orthogonal to λ_e, as in Fig. 2(a).

If S_e is not a cap graph, then we simply place vertex w_i in B_e^* at the point $\beta(w_i)$ where $\zeta(w_i)$ intersects the circular arc $\gamma(B_e^*)$ ($0 \leq i \leq a$). Note that the

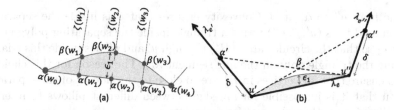

Fig. 2. (a) A pillow. (b) Defining ϵ_1

edges of S_e (which are chords on the Hamilton path $(w_0, w_1, \ldots, w_a)$) lie within the pillow of e.

If S_e is a cap graph, then we place vertex w_i on the line $\zeta(w_i)$, but not necessarily at $\beta(w_i)$. First we define an acyclic directed graph as follows. We direct edges along the Hamilton path $(w_0, w_1, \ldots, w_a)$ from w_0 to w_a, and direct other edges so that the result is a directed acyclic graph $\overrightarrow{B_e^*}$ with a source at w_0 and a sink at w_a. Note that $\overrightarrow{B_e^*}$ is a *leveled planar graph* with one vertex on each level [11]. One can use the algorithm in [7] to draw $\overrightarrow{B_e^*}$ so that there are no edge crossings, vertex w_i lies on the line $\zeta(w_i)$, and the external face is a given polygon. We choose the external face to be the convex hull of λ_e and the points $\beta(w_i)$, $0 \leq i \leq a$. Note that the vertices $w_0, w_1, \ldots, w_a$ are in monotonic order in the direction of the edge (w_0, w_k). The general picture after the drawing of the root blocks is illustrated in Fig. 3.

Fig. 3. The general picture with pillows

Next we show how to choose ϵ_1. Let δ denote d/n, where d is the minimum length of a side of the separating polygon, and n is the number of vertices in the graph. Suppose that $\lambda_{e'}$, λ_e, and $\lambda_{e''}$ are three consecutive sides of the separating polygon, as in Fig. 2(b); we show how to choose ϵ_1 for the edge e. Suppose that the endpoints of e are u' and u'', and α' and α'' are points on $\lambda_{e'}$ and $\lambda_{e''}$ distant δ from u' and u'' respectively. Suppose that the line from u' to α'' meets the line from u'' to α' at β. Convexity ensures that β is inside the separating polygon, and thus (u', β, u'') forms a triangle inside the separating polygon. We choose ϵ_1 so that the circular arc $\gamma(B_e^*)$ through u' and u'' lies inside this triangle (meeting the triangle only on the line segment λ_e). The reason for this choice of ϵ_1 is to ensure that all vertices in B_e^* are so close to the side λ_e of the separating polygon that it is impossible for an edge between different pillows to intersect with pillows other than those at its endpoints.

Safe Blocks. To describe the algorithm for drawing the non-root blocks, we need some terminology. Suppose that c is a cutvertex in the side graph S_e, and $B = (V_B, E_B)$ is a child block of c. Suppose that c is a left vertex. In the clockwise order of edges in G around c, there is an edge $e_1 \notin E_B$, followed by a number of edges in E_B, followed by an edge $e_2 \notin E_B$, as illustrated in Fig. 4(a). We say that e_1 and e_2 are the *bounding* edges of B. Note that a bounding edge either crosses (s, t), or has s or t as an endpoint.

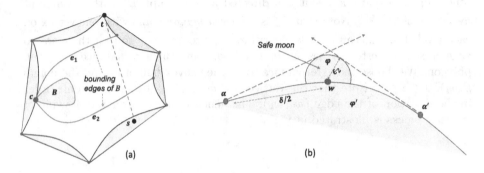

Fig. 4. (a) Bounding edges of a block. (b)The safe moon $\mu(u)$ at u

At any stage of the drawing algorithm, a block may be *safe* or *unsafe*. A block B is *safe* if the following properties hold: (i) The parent cutvertex c (that is, the parent of B in the block-cutvertex tree) has been drawn, and the other vertices in B are not drawn; (ii) Suppose that the boundary edges of B are $e_1 = (c, u_1)$ and $e_2 = (c, u_2)$; let u'_1 and u'_2 be the vertices which are the least already-drawn ancestors of u_1 and u_2 respectively in their respective block-cutvertex trees. Then we require that $u'_1 \neq u'_2$.

Lemma 6. *If there is an undrawn vertex, then there is a safe block.*

Safe Moon. Suppose that w is a parent cutvertex for a safe block B; for the moment we assume that w is not on the separating cycle. Suppose that the parent block of w is B'; then w has been drawn on the circular arc $\gamma(B')$; denote the circular disc defined by $\gamma(B')$ by ϕ'. Let ϕ be a circular disc of radius ϵ_2 with centre at w. We show how to choose ϵ_2 later; for the moment we can assume that ϵ_2 is very small in comparison to the radius of $\gamma(B')$. The *safe moon* $\mu(w)$ for w is the interior of $\phi - \phi'$; see Fig.4(b).

Now we show how to choose ϵ_2. Again let δ denote d/n, where d denotes the minimum length of a side of the separating polygon, and n is the number of vertices in the graph. Now consider two points α and α' at distance $\frac{\delta}{2}$ from u. We choose ϵ_2 small enough that: (i) $\mu(w)$ at u does not intersect the tangents to $\gamma(B')$ at α and α'; (ii) $\mu(w)$ does not intersect the line through s and t. Small adjustments to this choice of $\mu(w)$ are required for the cases where w is on the separating cycle, and where w is an endpoint of $\gamma(B')$.

A consequence of the definition of safe moon is the following: Let w_1 and w_2 be vertices on the circular arcs $\gamma(B_1)$ and $\gamma(B_2)$ for two blocks B_1 and B_2 that have been drawn. Let α_1 be a point in $\mu(w_1)$ and α_2 be a point in $\mu(w_2)$; the line segment between α_1 and α_2 does not intersect any safe moon other than $\mu(w_1)$ and $\mu(w_2)$.

Safe Wedges. Suppose that the boundary edges of a non-root block B are $e_1 = (c, u_1)$ and $e_2 = (c, u_2)$; let u_1' and u_2' be the vertices which are the least drawn ancestors of u_1 and u_2 respectively in their respective block-cutvertex trees. Since B is safe, $u_1' \neq u_2'$. For each point α_1 (resp. α_2) in $\mu(u_1')$ (resp. $\mu(u_2')$), consider the wedge $\omega(\alpha_1, \alpha_2)$ formed by the rays from c through α_1 and α_2. The *safe wedge* $\omega(B)$ of B is the intersection of all such wedges $\omega(\alpha_1, \alpha_2)$ with the safe moon of c. This is illustrated in Fig. 5(a).

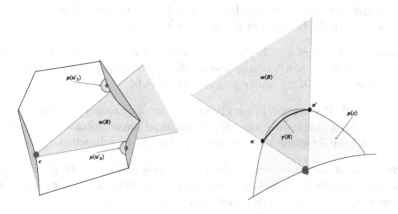

Fig. 5. (a) A safe wedge. (b) The circular arc $\gamma(B)$.

The Circular Arc $\gamma(B)$. Suppose that B is a non-root block. We give a location to each vertex in B except the parent cutvertex c (which is already drawn). These vertices are drawn on a circular arc $\gamma(B)$, defined as follows. Suppose that the boundaries of $\mu(c)$ and $\omega(B)$ intersect at points α and α' as shown in Fig 5(b). Then $\gamma(B)$ is a circular arc that passes through α and α'. The radius of $\gamma(B)$ is chosen so that it lies inside $\mu(c)$, and it is distant at most ϵ_1 from the straight line between α and α'. Here ϵ_1 is chosen in exactly the same way as for the root block.

Putting it All Together. In the construction of the inner graph in subsection 3.3, all vertices that are neither left nor right are removed, and the resulting non-triangular faces are fan-triangulated. These vertices can be drawn as follows. Each fan-triangulated face, after removal of the dummy edges, is *star-shaped*. The non-aligned vertices (neither left nor right) that came from this face form a triangulation inside the face. Thus we can use the linear-time algorithm of Hong and Nagamochi [10] to construct a straight-line drawing replacing the non-aligned vertices. This concludes the proof of sufficiency of Theorem 1.

4   Concluding Remarks

Assuming the real RAM model of computation, it can be proved that all algorithmic steps presented in the previous section can be executed in $O(n)$ time, where n is the number of vertices of G.

Theorem 2. *Let G be an almost-planar $\mathbb{S}^2$-topological graph with n vertices such that every vertex of G is consistent. There exists an $O(n)$ time algorithm that computes an $\mathbb{S}^2$-embedding preserving straight-line drawing of G.*

The real RAM model of computation allows for exponentially bad resolution in the drawing. The next theorem shows that such exponentially bad resolution is inevitable in the worst case. The construction of the family of almost-planar graphs for Theorem 3 is based on a family of upward planar digraphs first described by Di Battista et al. [2]. (See [6] for details.)

Theorem 3. *For each $k \geq 1$, there is an almost-planar $\mathbb{S}^2$-topological graph G_k with $2k + 1$ vertices, such that any $\mathbb{S}^2$-embedding preserving straight-line drawing of G_k requires area $\Omega(2^k)$ under any resolution rule.*

We conclude this section by observing that the arguments used to prove Theorem 1 lead to a characterization of the maximal almost-planar $\mathbb{R}^2$-topological graphs that have $\mathbb{R}^2$-embedding preserving straight-line drawings.

Theorem 4. *A maximal almost-planar $\mathbb{R}^2$-topological graph G admits an $\mathbb{R}^2$-embedding preserving straight-line drawing of G if and only if every vertex of G is consistent, and every internal face of G_{LR} is consistent.*

Namely, the sufficiency of Theorem 4 is proved in Sections 3.3, 3.4, and 3.5. The proof that the conditions of Theorem 4 are also necessary is in [6].

5   Open Problems

We mention two open problems that are naturally suggested by the research in this paper. The first open problem is about characterizing those almost-planar $\mathbb{R}^2$-topological graphs that admit an embedding preserving straight-line drawing. Theorem 4 provides such a characterization for the family of maximal almost-planar graphs.

The second open problem is about extending Theorem 1 to k-skew graphs with $k > 1$. A topological graph $G = (V, E)$ is k-skew if there is a set $E' \subset E$ of edges such that $G^- = (V, E - E')$ has no crossings where $|E'| \leq k$. Many graphs that arise in practice are k-skew for small values of k; this paper gives drawing algorithms for the case $k = 1$. For each edge $e \in E'$, one could define "left vertex relative to e" and "right vertex relative to e", extending the definitions of left and right in this paper. However, it is not difficult to find a topological 2-skew graph in which all vertices are consistent with respect to the 2 "crossing" edges, but do not admit a straight-line drawing. It would be interesting to characterize k-skew graphs that admit a straight-line drawing for $k > 1$.

References

1. Di Battista, G., Eades, P., Tamassia, R., Tollis, I.G.: Graph Drawing: Algorithms for the Visualization of Graphs. Prentice-Hall (1999)
2. Di Battista, G., Tamassia, R., Tollis, I.G.: Area requirement and symmetry display of planar upward drawings. Discrete & Computational Geometry **7**, 381–401 (1992)
3. Cabello, S., Mohar, B.: Adding one edge to planar graphs makes crossing number and 1-planarity hard. SIAM J. Comput. **42**(5), 1803–1829 (2013)
4. Chiba, N., Yamanouchi, T., Nishizeki, T.: Linear algorithms for convex drawings of planar graphs. In: Progress in Graph Theory, pp. 153–173. Academic Press, London (1984)
5. Chrobak, M., Eppstein, D.: Planar orientations with low out-degree and compaction of adjacency matrices. Theor. Comput. Sci. **86**(2), 243–266 (1991)
6. Eades, P., Hong, S.H., Liotta, G., Katoh, N., Poon, S.H.: Straight-line drawability of a planar graph plus an edge (2015). ArXiv, 1504.06540
7. Eades, P., Feng, Q.-W., Lin, X., Nagamochi, H.: Straight-line drawing algorithms for hierarchical graphs and clustered graphs. Algorithmica **44**(1), 1–32 (2006)
8. Gutwenger, C., Mutzel, P., Weiskircher, R.: Inserting an edge into a planar graph. Algorithmica **41**(4), 289–308 (2005)
9. Hong, S.H., Eades, P., Liotta, G., Poon, S.H.: Fáry's theorem for 1-planar graphs. In: Gudmundsson, J., Mestre, J., Viglas, T. (eds.) COCOON 2012. LNCS, vol. 7434, pp. 335–346. Springer, Heidelberg (2012)
10. Hong, S.H., Nagamochi, H.: Convex drawings of graphs with non-convex boundary constraints. Discrete Applied Mathematics **156**(12), 2368–2380 (2008)
11. Jünger, M., Leipert, S.: Level planar embedding in linear time. J. Graph Algorithms Appl. **6**(1), 67–113 (2002)
12. Nagamochi, H.: Straight-line drawability of embedded graphs. Technical Report 2013-005, Graduate School of Informatics, Kyoto University (2013)
13. Thomassen, C.: Rectilinear drawings of graphs. Journal of Graph Theory **12**(3), 335–341 (1988)

Solving Problems on Graphs of High Rank-Width

Eduard Eiben, Robert Ganian[(✉)], and Stefan Szeider

Algorithms and Complexity Group, Institute of Computer Graphics and Algorithms,
TU Wien, Vienna, Austria
{eduard.eiben,rganian}@gmail.com, stefan@szeider.net

Abstract. A modulator of a graph G to a specified graph class $\mathcal{H}$ is a set of vertices whose deletion puts G into $\mathcal{H}$. The cardinality of a modulator to various graph classes has long been used as a structural parameter which can be exploited to obtain FPT algorithms for a range of hard problems. Here we investigate what happens when a graph contains a modulator which is large but "well-structured" (in the sense of having bounded rank-width). Can such modulators still be exploited to obtain efficient algorithms? And is it even possible to find such modulators efficiently?

We first show that the parameters derived from such well-structured modulators are strictly more general than the cardinality of modulators and rank-width itself. Then, we develop an FPT algorithm for finding such well-structured modulators to any graph class which can be characterized by a finite set of forbidden induced subgraphs. We proceed by showing how well-structured modulators can be used to obtain efficient parameterized algorithms for MINIMUM VERTEX COVER and MAXIMUM CLIQUE. Finally, we use the concept of well-structured modulators to develop an algorithmic meta-theorem for efficiently deciding problems expressible in Monadic Second Order (MSO) logic, and prove that this result is tight in the sense that it cannot be generalized to LinEMSO problems.

1 Introduction

Many important graph problems are known to be NP-hard, and yet admit efficient solutions in practice due to the inherent structure of instances. The parameterized complexity paradigm [9,22] allows a more refined analysis of the complexity of various problems and hence enables the design of more efficient algorithms. In particular, given an instance of size n and a numerical parameter k which captures some property of the instance, one asks whether the instance can be solved in time $f(k) \cdot n^{\mathcal{O}(1)}$. Parameterized problems which admit such an algorithm are called *fixed parameter tractable* (FPT), and the algorithms themselves are often called FPT *algorithms*.

Given the above, it is natural to ask what kind of structure can be exploited to obtain FPT algorithms for a wide range of natural graph problems. There

---

Supported by the Austrian Science Fund (FWF), project P26696.

F. Dehne et al. (Eds.): WADS 2015, LNCS 9214, pp. 314–326, 2015.
DOI: 10.1007/978-3-319-21840-3_26

are two very successful, mutually incomparable approaches which tackle this question.

A. *Width measures.* Treewidth has become an extremely successful structural parameter with a wide range of applications in many fields of computer science. However, treewidth is not suitable for use in dense graphs. This led to the development of algorithms that use the parameter clique-width [6], which can be viewed as a relaxation of treewidth towards dense graphs. However, while there are efficient theoretical algorithms for computing tree-decompositions, this is not the case for decompositions for clique-width. This shortcoming has later been overcome by the notion of rank-width [23], which improves upon clique-width by allowing the efficient computation of rank-decompositions while retaining all of the positive algorithmic results previously obtained for clique-width.

B. *Modulators.* A modulator is a vertex set whose deletion places the considered graph into some specified graph class. A substantial amount of research has been placed into finding as well as exploiting small modulators to various graph classes [2,10]. Popular notions such as vertex cover and feedback vertex set are also special cases of modulators (to the classes of edgeless graphs and forests, respectively). One advantage of parameterizing by the size of modulators is that it allows us to build on the vast array of research of polynomial-time algorithms on specific graph classes (see, for instance, [5,21]). In other fields of computer science, modulators are often called *backdoors* and have been successfully used to obtain efficient algorithms for, e.g., Satisfiability and Constraint Satisfaction [12].

Our primary goal in this paper is to push the boundaries of tractability for a wide range of problems above the state of the art for both of these approaches. We summarize our contributions below.

1. We introduce a family of "hybrid" parameters that combine approaches A and B.

Given a graph G and a fixed graph class $\mathcal{H}$, the new parameters capture (roughly speaking) the minimum rank-width of any modulator of G into $\mathcal{H}$. We call this the *well-structure number* of G or $wsn^{\mathcal{H}}(G)$. The formal definition of the parameter also relies on the notion of *split decompositions* [7] and is provided in Section 3, where we also prove that for any graph class $\mathcal{H}$ of unbounded rank-width, $wsn^{\mathcal{H}}$ is not larger and in many cases much smaller than both rank-width and the size of a modulator to $\mathcal{H}$.

2. We develop an FPT algorithm for computing $wsn^{\mathcal{H}}$.

As with most structural parameters, virtually all algorithmic applications of the well-structure number rely on having access to an appropriate decomposition. In Section 4 we provide an FPT algorithm for computing $wsn^{\mathcal{H}}$ along with the corresponding decomposition for any graph class $\mathcal{H}$ which can be characterized by a finite set of forbidden induced subgraphs (*obstructions*). This is achieved by building on the polynomial algorithm for computing split-decompositions [16] in combination with the FPT algorithm for computing rank-width [18].

3. We design FPT algorithms for Minimum Vertex Cover (MINVC) and Maximum Clique (MAXCLQ) parameterized by $wsn^{\mathcal{H}}$.

Specifically, in Section 5 we show that for any graph class $\mathcal{H}$ (which can be characterized by a finite set of obstructions) such that the problem is polynomial-time tractable on $\mathcal{H}$, the problem becomes fixed parameter tractable when parameterized by $wsn^{\mathcal{H}}$. We also give an overview of possible choices of $\mathcal{H}$ for MINVC and MAXCLQ.

4. We develop a *meta-theorem* to obtain FPT algorithms for problems definable in Monadic Second Order (MSO) logic [6] parameterized by $wsn^{\mathcal{H}}$.

The meta-theorem requires that the problem is FPT when parameterized by the cardinality of a modulator to $\mathcal{H}$. We prove that this condition is not only sufficient but also necessary, in the sense that the weaker condition of polynomial-time tractability on $\mathcal{H}$ used for MINVC and MAXCLQ is not sufficient for FPT-time MSO model checking. Formal statements and proofs can be found in Section 6.

5. We show that, in general, solving LinEMSO problems [6,11] is not FPT when parameterized by $wsn^{\mathcal{H}}$.

In particular, in the concluding Section 7 we give a proof that these problems are in general paraNP-hard when parameterized by $wsn^{\mathcal{H}}$ under the same conditions as those used for MSO model checking.

2   Preliminaries

The set of natural numbers (that is, positive integers) will be denoted by $\mathbb{N}$. For $i \in \mathbb{N}$ we write $[i]$ to denote the set $\{1, \ldots, i\}$. If $\sim$ is an equivalence relation over a set A, then for $a \in A$ we use $[a]_{\sim}$ to denote the equivalence class containing a.

Graphs We will use standard graph theoretic terminology and notation (cf. [8]). All graphs considered in this document are simple and undirected.

Given a graph $G = (V(G), E(G))$ and $A \subseteq V(G)$, we denote by $N(A)$ the set of neighbors of A in $V(G) \setminus A$; if A contains a single vertex v, we use $N(v)$ instead of $N(\{v\})$. We use V and E as shorthand for $V(G)$ and $E(G)$, respectively, when the graph is clear from context. Two vertex sets A, B are *overlapping* if $A \cap B, A \setminus B, B \setminus A$ are all nonempty. $G - A$ denotes the subgraph of G obtained by deleting A.

Given a graph $G = (V, E)$ and a graph class $\mathcal{H}$, a set $X \subseteq V$ is called a *modulator* to $\mathcal{H}$ if $G - X \in \mathcal{H}$. A graph class is called *hereditary* if it is closed under vertex deletion. A graph H is an *induced subgraph* of G if H can be obtained by deleting vertices (along with all of their incident edges) from G. For $A \subseteq V(G)$ we use $G[A]$ to denote the subgraph of G obtained by deleting $V(G) \setminus A$. Let $\mathcal{F}$ be a finite set of graphs; then the class of $\mathcal{F}$-*free* graphs is the class of all graphs which do not contain any graph in $\mathcal{F}$ as an induced subgraph. We will often refer to elements of $\mathcal{F}$ as *obstructions*, and we say that the class of $\mathcal{F}$-free graphs is *characterized by* $\mathcal{F}$.

Fixed-Parameter Tractability. We refer the reader to [9, 22] for an introduction to parameterized complexity. A *parameterized problem* $\mathcal{P}$ is a subset of $\Sigma^* \times \mathbb{N}$ for some finite alphabet Σ. For a problem instance $(x, k) \in \Sigma^* \times \mathbb{N}$ we call x the main part and k the parameter. A parameterized problem $\mathcal{P}$ is *fixed-parameter tractable* (FPT in short) if a given instance (x, k) can be solved in time $O(f(k) \cdot p(|x|))$ where f is an arbitrary computable function of k and p is a polynomial function.

Splits. A *split* of a connected graph $G = (V, E)$ is a vertex bipartition $\{A, B\}$ of V such that every vertex of $A' = N(B)$ has the same neighborhood in $B' = N(A)$. The sets A' and B' are called *frontiers* of the split.

Let $G = (V, E)$ be a graph. To simplify our exposition, we will use the notion of *split-modules* instead of splits where suitable. A set $A \subseteq V$ is called a *split-module* of G if there exists a connected component $G' = (V', E')$ of G such that $\{A, V' \setminus A\}$ forms a split of G'. Notice that if A is a split-module then A can be partitioned into A_1 and A_2 such that $N(A_2) \subseteq A$ and for each $v_1, v_2 \in A_1$ it holds that $N(v_1) \cap (V' \setminus A) = N(v_2) \cap (V' \setminus A)$. For technical reasons, V and $\emptyset$ are also considered split-modules. We say that two disjoint split-modules $X, Y \subseteq V$ are *adjacent* if there exist $x \in X$ and $y \in Y$ such that x and y are adjacent.

Rank-width For a graph G and $U, W \subseteq V(G)$, let $\boldsymbol{A}_G[U, W]$ denote the $U \times W$-submatrix of the adjacency matrix over the two-element field GF(2), i.e., the entry $a_{u,w}$, $u \in U$ and $w \in W$, of $\boldsymbol{A}_G[U, W]$ is 1 if and only if $\{u, w\}$ is an edge of G. The *cut-rank* function ρ_G of a graph G is defined as follows: For a bipartition (U, W) of the vertex set $V(G)$, $\rho_G(U) = \rho_G(W)$ equals the rank of $\boldsymbol{A}_G[U, W]$ over GF(2).

A *rank-decomposition* of a graph G is a pair (T, μ) where T is a tree of maximum degree 3 and $\mu : V(G) \to \{t : t \text{ is a leaf of } T\}$ is a bijective function. For an edge e of T, the connected components of $T - e$ induce a bipartition (X, Y) of the set of leaves of T. The *width* of an edge e of a rank-decomposition (T, μ) is $\rho_G(\mu^{-1}(X))$. The *width* of (T, μ) is the maximum width over all edges of T. The *rank-width* of G, rw(G) in short, is the minimum width over all rank-decompositions of G. We denote by $\mathcal{R}_i$ the class of all graphs of rank-width at most i, and say that a graph class $\mathcal{H}$ is of *unbounded rank-width* if $\mathcal{H} \not\subseteq \mathcal{R}_i$ for any $i \in \mathbb{N}$.

Theorem 1 ([18]). *Let $k \in \mathbb{N}$ be a constant and $n \geq 2$. For an n-vertex graph G, we can output a rank-decomposition of width at most k or confirm that the rank-width of G is larger than k in time $f(k) \cdot n^3$, where f is a computable function.*

Monadic Second Order Logic on Graphs. We assume that we have an infinite supply of individual variables, denoted by lowercase letters x, y, z, and an infinite supply of set variables, denoted by uppercase letters X, Y, Z. *Formulas* of *monadic second-order logic* (MSO) are constructed from atomic formulas $E(x, y)$, $X(x)$, and $x = y$ using the connectives $\neg$ (negation), $\wedge$ (conjunction)

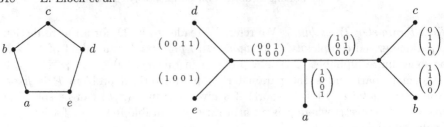

Fig. 1. A rank-decomposition of the cycle C_5

and existential quantification $\exists x$ over individual variables as well as existential quantification $\exists X$ over set variables. Individual variables range over vertices, and set variables range over sets of vertices. The atomic formula $E(x, y)$ expresses adjacency, $x = y$ expresses equality, and $X(x)$ expresses that vertex x in the set X. From this, we define the semantics of monadic second-order logic in the standard way (this logic is sometimes called MSO$_1$).

Free and bound variables of a formula are defined in the usual way. A *sentence* is a formula without free variables. We write $\varphi(X_1, \ldots, X_n)$ to indicate that the set of free variables of formula φ is $\{X_1, \ldots, X_n\}$. If $G = (V, E)$ is a graph and $S_1, \ldots, S_n \subseteq V$ we write $G \models \varphi(S_1, \ldots, S_n)$ to denote that φ holds in G if the variables X_i are interpreted by the sets S_i, for $i \in [n]$. For a fixed MSO sentence φ, the MSO Model Checking problem (MSO-MC$_\varphi$) asks whether an input graph G satisfies $G \models \varphi$.

It is known that MSO formulas can be checked efficiently as long as the graph has bounded rank-width.

Theorem 2 ([11]). *Let φ and $\psi = \psi(X)$ be fixed MSO formulas. Given an n-vertex graph G and a set $S \subseteq V(G)$, there exists a computable function f such that we can decide whether $G \models \varphi$ and whether $G \models \psi(S)$ in time $f(\mathrm{rw}(G)) \cdot n^3$.*

We review MSO *types* roughly following the presentation in [20]. The *quantifier rank* of an MSO formula φ is defined as the nesting depth of quantifiers in φ. For non-negative integers q and l, let MSO$_{q,l}$ consist of all MSO formulas of quantifier rank at most q with free set variables in $\{X_1, \ldots, X_l\}$.

Let $\varphi = \varphi(X_1, \ldots, X_l)$ and $\psi = \psi(X_1, \ldots, X_l)$ be MSO formulas. We say φ and ψ are *equivalent*, written $\varphi \equiv \psi$, if for all graphs G and $U_1, \ldots, U_l \subseteq V(G)$, $G \models \varphi(U_1, \ldots, U_l)$ if and only if $G \models \psi(U_1, \ldots, U_l)$. Given a set F of formulas, let $F/\equiv$ denote the set of equivalence classes of F with respect to $\equiv$. A system of representatives of $F/\equiv$ is a set $R \subseteq F$ such that $R \cap C \neq \emptyset$ for each equivalence class $C \in F/\equiv$. The following statement has a straightforward proof using normal forms (see [20, Proposition 7.5] for details).

Fact 1. *Let q and l be fixed non-negative integers. The set MSO$_{q,l}/\equiv$ is finite, and one can compute a system of representatives of MSO$_{q,l}/\equiv$.*

We will assume that for any pair of non-negative integers q and l the system of representatives of MSO$_{q,l}/\equiv$ given by Fact 1 is fixed.

Definition 1 (MSO Type). *Let q, l be non-negative integers. For a graph G and an l-tuple U of sets of vertices of G, we define $type_q(G, U)$ as the set of formulas $\varphi \in MSO_{q,l}$ such that $G \models \varphi(U)$. We call $type_q(G, U)$ the MSO q-type of U in G.*

It follows from Fact 1 that up to logical equivalence, every type contains only finitely many formulas.

3   Well-Structured Modulators

Definition 2. *Let $\mathcal{H}$ be a hereditary graph class and let G be a graph. A set X of pairwise-disjoint split-modules of G is called a k-well-structured modulator to $\mathcal{H}$ if*

1. *$|X| \le k$, and*
2. *$\bigcup_{X_i \in X} X_i$ is a modulator to $\mathcal{H}$, and*
3. *$rw(G[X_i]) \le k$ for each $X_i \in X$.*

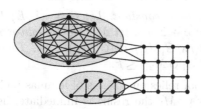

Fig. 2. A graph with a 2-well-structured modulator to K_3-free graphs (in the two shaded areas)

For the sake of brevity and when clear from context, we will sometimes identify X with $\bigcup_{X_i \in X} X_i$ (for instance $G - X$ is shorthand for $G - \bigcup_{X_i \in X} X_i$). To allow a concise description of our parameters, for any hereditary graph class $\mathcal{H}$ we let the *well-structure number* (wsn$^{\mathcal{H}}$ in short) denote the minimum k such that G has a k-well-structured modulator to $\mathcal{H}$. Similarly, we let $mod^{\mathcal{H}}(G)$ denote the minimum k such that G has a modulator of cardinality k to $\mathcal{H}$.

Proposition 1. *Let $\mathcal{H}$ be any hereditary graph class of unbounded rank-width.*

1. *$rw(G) \ge wsn^{\mathcal{H}}(G)$ for any graph G. Furthermore, for every $i \in \mathbb{N}$ there exists a graph G_i such that $rw(G_i) \ge wsn^{\mathcal{H}}(G_i) + i$, and*
2. *$mod^{\mathcal{H}}(G) \ge wsn^{\mathcal{H}}(G)$ for any graph G. Furthermore, for every $i \in \mathbb{N}$ there exists a graph G_i such that $mod^{\mathcal{H}}(G_i) \ge wsn^{\mathcal{H}}(G_i) + i$.*

4   Finding Well-Structured Modulators

The objective of this subsection is to prove the following theorem. Interestingly, our approach only allows us to find well-structured modulators if the rank-width of the graph is sufficiently large. This never becomes a problem though, since on graphs with small rank-width we can always directly use rank-width as our parameter.

Theorem 3. *Let $\mathcal{H}$ be a graph class characterized by a finite obstruction set. There exists an* FPT *algorithm parameterized by k which for any graph G of rank-width at least $k + 2$ either finds a k-well-structured modulator to $\mathcal{H}$ or correctly detects that it does not exist.*

Our starting point on the path to a proof of Theorem 3 is a theorem by Cunningham.

Theorem 4 ([7]). *Let $\{A, C\}$, $\{B, D\}$ be splits of a connected graph G such that $|A \cap B| \geq 2$ and $A \cup B \neq V(G)$. Then $\{A \cap B, C \cup D\}$ is a split of G.*

The following lemma in essence shows that the relation of being in a split-module of small rank-width is transitive (assuming sufficiently high rank-width). The significance of this will become clear later on.

Lemma 1. *Let $k \in \mathbb{N}$ be a constant. Let $G = (V, E)$ be a connected graph with rank-width at least $k + 2$ and let M_1, M_2 be split-modules of G such that $M_1 \cap M_2 \neq \emptyset$ and $\max(\mathrm{rw}(G[M_1]), \mathrm{rw}(G[M_2])) \leq k$. Then $M_1 \cup M_2$ is a split-module of G and $\mathrm{rw}(G[M_1 \cup M_2]) \leq k$.*

Proof (Sketch). The proof relies on a series of lemmas building on Theorem 4.

If $M_1 \subseteq M_2$ or $M_2 \subseteq M_1$ the result is immediate, hence we may assume that they are overlapping. $\mathrm{rw}(G) \geq k + 2$ implies that $M_1 \cup M_2 \neq V$. The fact that $M_1 \cup M_2$ is a split-module of G then follows from Theorem 4. Let $M_{11} = M_1 \setminus M_2, M_{22} = M_2 \setminus M_1$, and $M_{12} = M_1 \cap M_2$. These sets can be shown to be split-modules of G. Let $v_{11} \in N(V \setminus M_{11}), v_{22} \in N(V \setminus M_{22})$, and $v_{12} \in N(V \setminus M_{12})$. We show that $\mathrm{rw}(G[M_1 \cup M_2]) \leq k$. By assumption, both $G[M_1]$ and $G[M_2]$ have rank-width at most k. Since rank-width is preserved by taking induced subgraphs, the graphs $G_{11} = G[M_{11} \cup \{v_{12}\}]$, $G_{12} = G[M_{12} \cup \{v_{22}\}]$, and $G_{22} = G[M_{22} \cup \{v_{12}\}]$ also have rank-width at most k. The proof can be completed by showing how the rank-decompositions of these three graphs can be combined into a rank-decomposition for $G[M_1 \cup M_2]$. $\square$

Definition 3. *Let G be a graph and $k \in \mathbb{N}$. We define a relation $\sim_k^G$ on $V(G)$ by letting $v \sim_k^G w$ if and only if there is a split-module M of G with $v, w \in M$ and $\mathrm{rw}(G[M]) \leq k$. We drop the superscript from $\sim_k^G$ if the graph G is clear from context.*

Using Lemma 1 to deal with transitivity, we prove the following.

Proposition 2. *For every $k \in \mathbb{N}$ and graph $G = (V, E)$ with rank-width at least $k + 2$, the relation $\sim_k$ is an equivalence relation, and each equivalence class U of $\sim_k$ is a split-module of G with $\mathrm{rw}(G[U]) \leq k$.*

Corollary 1. *Any graph G of rank-width at least $k+2$ has its vertex set uniquely partitioned by the equivalence classes of $\sim_k$ into inclusion-maximal split-modules of rank-width at most k.*

Now that we know $\sim_k$ is an equivalence, we show how to compute it in FPT time.

Proposition 3. *Let $k \in \mathbb{N}$ be a constant. Given an n-vertex graph G of rank-width at least $k+2$ and two vertices v, w, we can decide whether $v \sim_k w$ in time $\mathcal{O}(n^3)$.*

Proof (Sketch). The definition of split-modules allows us to consider each connected component of a graph separately. We then compute the so-called *split-tree* [7,14–16] of G and use it to list all minimal split-modules containing v and w. Finally, we check whether any of these split-modules has rank-width at most k by using Theorem 1. □

We are now ready to present an algorithm for finding a k-well-structured modulator to any graph class $\mathcal{H}$ characterized by a finite obstruction set $\mathcal{F}$.

Algorithm 1. FindWSM$_\mathcal{F}$

> **Input**   : $k \in \mathbb{N}_0$, n-vertex graph G, equivalence $\sim$ over a superset of $V(G)$
>
> **Output**   : A k-cardinality set $\boldsymbol{X}$ of subsets of $V(G)$, or *False*

1 **if** G *does not contain any* $D \in \mathcal{F}$ *as an induced subgraph* **then**
2 | **return** $\emptyset$
3 **else**
4 | $D' :=$ an induced subgraph of G isomorphic to an arbitrary $D \in \mathcal{F}$;
5 **end**
6 **if** $k = 0$ **then return** *False*;
7 **foreach** $[a]_\sim$ *of* G *which intersects with* $V(D')$ **do**
8 | $\boldsymbol{X} = $ FindWSM$_\mathcal{F}(k-1, G - [a]_\sim, \sim)$;
9 | **if** $\boldsymbol{X} \neq$ False **then**
10 | | **return** $\boldsymbol{X} \cup \{[a]_\sim\}$
11 | **end**
12 **end**
13 **return** *False*

We will use $\sim_k$ as the input for *FindWSM$_\mathcal{F}$*, however considering general equivalences as inputs is useful for proving correctness.

Lemma 2. *There exists a constant c such that* FindWSM$_\mathcal{F}$ *runs in time $c^k \cdot n^{\mathcal{O}(1)}$. Furthermore, if G is a graph of rank-width at least $k+2$ and $\sim_k$ is the equivalence computed by Proposition 3, then* FindWSM$_\mathcal{F}(k, G, \sim_k)$ *outputs a k-wsm to $\mathcal{H}$ or correctly detects that no such k-wsm exists in G.*

Proof (of Theorem 3). The theorem follows by using Proposition 3 and then Algorithm 1 in conjunction with Lemma 2. □

5   Examples of Algorithmic Applications

In this section, we show how to use the notion of k-well-structured modulators to design efficient parameterized algorithms for two classical NP-hard graph problems, specifically MINIMUM VERTEX COVER (MINVC) and MAXIMUM CLIQUE (MAXCLQ). Given a graph G, we call a set $X \subseteq V(G)$ a *vertex cover* if every edge is incident to at least one $v \in X$ and a *clique* if $G[X]$ is a complete graph.

MINVC, MAXCLQ

Instance: A graph G and an integer m.

Task (MINVC): Find a vertex cover in G of cardinality at most m, or determine that it does not exist.

Task (MAXCLQ): Find a clique in G of cardinality at least m, or determine that it does not exist.

Establishing the following theorem is the main objective of this section.

Theorem 5. *Let* $\mathcal{P} \in \{\text{MINVC}, \text{MAXCLQ}\}$ *and* $\mathcal{H}$ *be a graph class characterized by a finite obstruction set. Then* $\mathcal{P}$ *is FPT parameterized by* $\text{wsn}^{\mathcal{H}}$ *if and only if* $\mathcal{P}$ *is polynomial-time tractable on* $\mathcal{H}$.

Since $\text{wsn}^{\mathcal{H}}(G) = 0$ for any $\mathcal{F}$-free graph G, the "only if" direction is immediate; in other words, being polynomial-time tractable on $\mathcal{H}$ is clearly a necessary condition for being fixed parameter tractable when parameterized by $\text{wsn}^{\mathcal{H}}(G)$. Below we prove that for the selected problems this condition is also sufficient.

Lemma 3. *If* MINVC *is polynomial-time tractable on a graph class* $\mathcal{H}$ *characterized by a finite obstruction set, then* MINVC$[\text{wsn}^{\mathcal{H}}]$ *is FPT.*

Proof (Sketch). We compute a k-well-structured modulator $\boldsymbol{X}$ to $\mathcal{H}$ in G by Theorem 3. For each element $X_i \in \boldsymbol{X}$, it holds that either the frontier of X_i or its neighborhood in $G - X_i$ must be in any vertex cover of G. Branching on these at most 2^k options allows us to reduce the instance to at most 2^k disconnected instances such that each connected component has either rank-width bounded by k or is in $\mathcal{H}$; these connected components can then be solved independently. □

Lemma 4. *If* MAXCLQ *is polynomial-time tractable on a graph class* $\mathcal{H}$ *characterized by a finite obstruction set, then* MAXCLQ$[\text{wsn}^{\mathcal{H}}]$ *is FPT.*

Finally, let us review some concrete graph classes for use in Theorem 5.

Fact 2. MINVC *is polynomial-time tractable on the following graph classes:*

1. $(2K_2, C_4, C_5)$-*free graphs (split graphs);*
2. P_5-*free graphs* [21];
3. *fork-free graphs* [1];
4. *(banner,* $T_{2,2,2}$*)-free graphs and (banner,* $K_{3,3}$-e, *twin-house)-free graphs* [3, 13].

Fact 3. MAXCLQ *is polynomial-time tractable on the following graph classes:*

1. *Any complementary graph class to the classes listed in Fact 2 (such as cofork-free graphs and split graphs);*
2. *Graphs of bounded degree.*

6   MSO Model Checking with Well-Structured Modulators

Here we show how well-structured modulators can be used to solve the MSO Model Checking problem, as formalized in Theorem 6 below. Note that our meta-theorem captures not only the generality of MSO model checking problems, but also applies to a potentially unbounded number of choices of the graph class $\mathcal{H}$. Thus, the meta-theorem supports two dimensions of generality.

Theorem 6. *For every MSO sentence ϕ and every graph class $\mathcal{H}$ characterized by a finite obstruction set such that MSO-MC$_\phi$ is FPT parameterized by $mod^{\mathcal{H}}(G)$, the problem MSO-MC$_\phi$ is FPT parameterized by $wsn^{\mathcal{H}}(G)$.*

The condition that MSO-MC$_\phi$ is FPT parameterized by $mod^{\mathcal{H}}(G)$ is a necessary condition for the theorem to hold by Proposition 1. However, it is natural to ask whether it is possible to use a weaker necessary condition instead, specifically that MSO-MC$_\phi$ is polynomial-time tractable in the class of $\mathcal{F}$-free graphs (as was done for specific problems in Section 5). Before proceeding towards a proof of Theorem 6, we make a digression and show that the weaker condition used in Theorem 5 is in fact not sufficient for the general case of MSO model checking.

Lemma 5. *There exists an MSO sentence ϕ and a graph class $\mathcal{H}$ characterized by a finite obstruction set such that MSO-MC$_\phi$ is polynomial-time tractable on $\mathcal{H}$ but NP-hard on the class of graphs with $wsn^{\mathcal{H}}(G) \leq 2$ or even $mod^{\mathcal{H}}(G) \leq 2$.*

Proof (Sketch). Let ϕ describe vertex 5-colorability and let $\mathcal{H}$ be the class of graphs of degree at most 4. Now consider the class of graphs obtained from $\mathcal{H}$ by adding two adjacent vertices y, z which are adjacent to every other vertex. Hardness follows from hardness of 3-colorability on graphs of degree at most 4 [19]. □

Our strategy for proving Theorem 6 relies on a replacement technique, where each split-module in the well-structured modulator is replaced by a small representative. We use the notion of *similarity* defined below to prove that this procedure does not change the outcome of MSO-MC$_\varphi$.

Definition 4 (Similarity). *Let q and k be non-negative integers, $\mathcal{H}$ be a graph class, and let G and G' be graphs with k-well-structured modulators $\boldsymbol{X} = \{X_1, \ldots, X_k\}$ and $\boldsymbol{X'} = \{X'_1, \ldots, X'_k\}$ to $\mathcal{H}$, respectively. For $1 \leq i \leq k$, let S_i contain the frontier of split module X_i and similarly let S'_i contain the frontier of split module X'_i. We say that $(G, \boldsymbol{X})$ and $(G', \boldsymbol{X'})$ are q-similar if all of the following conditions are met:*

1. *There exists an isomorphism τ between $G - X$ and $G' - X'$.*
2. *For every $v \in V(G) \setminus X$ and $i \in [k]$, it holds that v is adjacent to S_i if and only if $\tau(v)$ is adjacent to S'_i.*
3. *if $k \geq 2$, then for every $1 \leq i < j \leq k$ it holds that S_i and S_j are adjacent if and only if S'_i and S'_j are adjacent.*
4. *For each $i \in [k]$, it holds that $type_q(G[X_i], S_i) = type_q(G'[X'_i], S'_i)$.*

Lemma 6. *Let q and k be non-negative integers, $\mathcal{H}$ be a graph class, and let G and G' be graphs with k-well-structured modulators $X = \{X_1, \ldots, X_k\}$ and $X' = \{X'_1, \ldots, X'_k\}$ to $\mathcal{H}$, respectively. If (G, X) and (G', X') are q-similar, then $type_q(G, \emptyset) = type_q(G', \emptyset)$.*

Proof (Sketch). The proof argument uses the q-round MSO game defined, e.g., in [20]. The notion of q-similarity ensures that the Duplicator has a winning strategy on G', which translates to G and G' having the same $type_q$. If the Spoiler moves in X, then the Duplicator can follow the winning strategies for each $(G[X_i], S_i)$. On the other hand, if the Spoiler moves in $G - X$, then the Duplicator can copy this move in G'. □

The next lemma deals with actually computing small q-similar "representatives" for our split-modules.

Lemma 7. *Let q be a non-negative integer constant and $\mathcal{H}$ be a graph class. Then given a graph G and a k-well-structured modulator $X = \{X_1, \ldots X_k\}$ of G into $\mathcal{H}$, there exists a function f such that one can in time $f(k) \cdot |V(G)|^{\mathcal{O}(1)}$ compute a graph G' with a k-well-structured modulator $X' = \{X'_1, \ldots X'_k\}$ into $\mathcal{H}$ such that (G, X) and (G', X') are q-similar and for each $i \in [k]$ it holds that $|X'_i|$ is bounded by a constant.*

Proof (Sketch). The idea here is to exploit the fact that each split-module has bounded rank-width. In particular, this allows us to determine the MSO type of each $G[X_i]$ and its frontier S_i in the specified time. The size of a minimum representative for each type does not depend on the actual size of G or k. □

Proof (of Theorem 6). Let G be a graph, $k = \text{wsn}^{\mathcal{H}}(G)$ and q be the nesting depth of quantifiers in ϕ. By Theorem 3 it is possible to find a k-well-structured modulator to $\mathcal{H}$ in time $f(k) \cdot |V|^{\mathcal{O}(1)}$. We proceed by constructing (G', X') by Lemma 7. Since each $X'_i \in X'$ has size bounded by a constant and $|X'| \leq k$, it follows that $\bigcup X'$ is a modulator to the class of $\mathcal{F}$-free graphs of cardinality $\mathcal{O}(k)$. Hence MSO-MC$_\phi$ can be decided in FPT time on G'. Finally, since G and G' are q-similar, it follows from Lemma 6 that $G \models \phi$ if and only if $G' \models \phi$. □

We conclude the section by showcasing an example application of Theorem 6. c-COLORING asks whether the vertices of an input graph G can be colored by c colors so that each pair of neighbors have distinct colors. From the connection between c-COLORING, its generalization LIST c-COLORING and modulators [4, Theorem3.3] and tractability results for LIST-c-COLORING [17, Page5], we obtain the following.

Corollary 2. *c-COLORING parameterized by $\text{wsn}^{P_5\text{-}free}$ is FPT for each $c \in \mathbb{N}$.*

7   Conclusion

We have introduced a family of structural parameters which push the frontiers of fixed parameter tractability beyond rank-width and modulator size for a wide range of problems. In particular, the well-structure number can be computed efficiently (Theorem 3) and used to design FPT algorithms for MINIMUM VERTEX COVER, MAXIMUM CLIQUE (Theorem 5) as well any problem which can be described by a sentence in MSO logic (Theorem 6).

In the wake of Theorem 6 and the positive results for the two problems in Section 5, one would expect that it should be possible to strengthen Theorem 6 to also cover LinEMSO problems [6,11] (which extend MSO Model Checking by allowing the minimization/maximization of linear expressions over free set variables). Surprisingly, as our last result we will show that this is in fact not possible if we wish to retain the same conditions. For our hardness proof, it suffices to consider a simplified variant of LinEMSO, defined below. Let φ be an MSO formula with one free set variable.

MSO-OPT$_{\varphi}^{\leq}$
Instance: A graph G and an integer $r \in \mathbb{N}$.
Question: Is there a set $S \subseteq V(G)$ such that $G \models \varphi(S)$ and $|S| \leq r$?

Theorem 7. *There exists an MSO formula φ and a graph class $\mathcal{H}$ characterized by a finite obstruction set such that* MSO-OPT$_{\varphi}^{\leq}$ *is FPT parameterized by* $\mathrm{mod}^{\mathcal{H}}$ *but paraNP-hard parameterized by* $\mathrm{wsn}^{\mathcal{H}}$.

To prove Theorem 7, we let $dom(S)$ express that S is a dominating set in G, and let $cyc(S)$ express that S intersects every C_4 (cycle of length 4). Then we set $\varphi(S) = dom(S) \vee cyc(S)$ and let $\mathcal{H}$ be the class of C_4-free graphs of degree at most 3 (obtained by letting the obstrucion set $\mathcal{F}$ contain C_4 and all 5-vertex supergraphs of $K_{1,4}$).

We conclude with two remarks on Theorem 7. On one hand, the fixed parameter tractability of LinEMSO traditionally follows from the methods used for FPT MSO model checking, and in this respect the theorem is surprising. But on the other hand, our parameters are strictly more general than rank-width and hence one should expect that some results simply cannot be lifted to this more general setting.

References

1. Alekseev, V.E.: Polynomial algorithm for finding the largest independent sets in graphs without forks. Discr. Appl. Math. **135**(1–3), 3–16 (2004)
2. Bodlaender, H.L., Jansen, B.M.P., Kratsch, S.: Kernel bounds for path and cycle problems. Theor. Comput. Sci. **511**, 117–136 (2013)
3. Brandstädt, A., Lozin, V.V.: A note on alpha-redundant vertices in graphs. Discr. Appl. Math. **108**(3), 301–308 (2001)
4. Cai, L.: Parameterized complexity of vertex colouring. Discr. Appl. Math. **127**(3), 415–429 (2003)

5. Corneil, D.G., Lerchs, H., Burlingham, L.S.: Complement reducible graphs. Discr. Appl. Math. **3**, 163–174 (1981)
6. Courcelle, B., Makowsky, J.A., Rotics, U.: Linear time solvable optimization problems on graphs of bounded clique-width. Theory Comput. Syst. **33**(2), 125–150 (2000)
7. Cunningham, W.H.: Decomposition of directed graphs. SIAM J. Algebraic Discrete Methods **3**(2), 214–228 (1982)
8. Diestel, R.: Graph Theory. GTM, vol. 173, 2nd edn. Springer Verlag, New York (2000)
9. Downey, R.G., Fellows, M.R.: Fundamentals of Parameterized Complexity. Texts in Computer Science. Springer Verlag (2013)
10. Gajarský, J., Hliněný, P., Obdržálek, J., Ordyniak, S., Reidl, F., Rossmanith, P., Villaamil, F.S., Sikdar, S.: Kernelization using structural parameters on sparse graph classes. In: Bodlaender, H.L., Italiano, G.F. (eds.) ESA 2013. LNCS, vol. 8125, pp. 529–540. Springer, Heidelberg (2013)
11. Ganian, R., Hliněný, P.: On parse trees and Myhill-Nerode-type tools for handling graphs of bounded rank-width. Discr. Appl. Math. **158**(7), 851–867 (2010)
12. Gaspers, S., Misra, N., Ordyniak, S., Szeider, S., Živný, S.: Backdoors into heterogeneous classes of SAT and CSP. In: Brodley, C.E., Stone, P.(eds.), Proceedings of the Twenty-Eighth AAAI Conference on Artificial Intelligence, pp. 2652–2658. AAAI Press (2014)
13. Gerber, M.U., Lozin, V.V.: Robust algorithms for the stable set problem. Graphs and Combinatorics **19**(3), 347–356 (2003)
14. Gioan, E., Paul, C.: Dynamic distance hereditary graphs using split decomposition. In: Tokuyama, T. (ed.) ISAAC 2007. LNCS, vol. 4835, pp. 41–51. Springer, Heidelberg (2007)
15. Gioan, E., Paul, C.: Split decomposition and graph-labelled trees: characterizations and fully dynamic algorithms for totally decomposable graphs. Discr. Appl. Math. **160**(6), 708–733 (2012)
16. Gioan, E., Paul, C., Tedder, M., Corneil, D.: Practical and efficient split decomposition via graph-labelled trees. Algorithmica **69**(4), 789–843 (2014)
17. Golovach, P.A., Paulusma, D., Song, J.: Closing complexity gaps for coloring problems on h-free graphs. Inf. Comput. **237**, 204–214 (2014)
18. Hliněný, P., Oum, S.I.: Finding branch-decompositions and rank-decompositions. SIAM J. Comput. **38**(3), 1012–1032 (2008)
19. Kochol, M., Lozin, V.V., Randerath, B.: The 3-colorability problem on graphs with maximum degree four. SIAM J. Comput. **32**(5), 1128–1139 (2003)
20. Libkin, L.: Elements of Finite Model Theory. Springer (2004)
21. Lokshantov, D., Vatshelle, M., Villanger, Y.: Independent set in p5-free graphs in polynomial time. In: Proceedings of the Twenty-Fifth Annual ACM-SIAM Symposium on Discrete Algorithms, SODA 2014, pp. 570–581. SIAM (2014)
22. Niedermeier, R.: Invitation to Fixed-Parameter Algorithms. Oxford Lecture Series in Mathematics and its Applications. Oxford University Press, Oxford (2006)
23. Oum, S., Seymour, P.: Approximating clique-width and branch-width. J. Combin. Theory Ser. B **96**(4), 514–528 (2006)

The Parametric Closure Problem

David Eppstein[(✉)]

Computer Science Department, University of California, Irvine, USA
david.eppstein@gmail.com

Abstract. We define the *parametric closure problem*, in which the
input is a partially ordered set whose elements have linearly varying
weights and the goal is to compute the sequence of minimum-weight
lower sets of the partial order as the weights vary. We give polynomial
time solutions to many important special cases of this problem including semiorders, reachability orders of bounded-treewidth graphs, partial
orders of bounded width, and series-parallel partial orders. Our result for
series-parallel orders provides a significant generalization of a previous
result of Carlson and Eppstein on bicriterion subtree problems.

1 Introduction

Parametric optimization problems are a variation on classical combinatorial optimization problems such as shortest paths or minimum spanning trees, in which
the input weights are not fixed numbers, but vary as functions of a parameter. Different parameter settings will give different weights and different optimal
solutions; the goal is to list these solutions and the intervals of parameter values within which they are optimal. As a simple example, consider maintaining
the minimum of n input values as a parameter controlling these values changes.
This parametric minimum problem asks for the *lower envelope* of a collection of
input functions; for linear functions this is equivalent by projective duality to a
planar convex hull [1], and can be constructed in time $O(n \log n)$; more general
function classes such as piecewise-polynomial functions also have efficient lower-
envelope algorithms [2]. The parametric minimum spanning tree problem (with
linear edge weights) has polynomially many solutions that can be constructed
in polynomial time [3–5]; the parametric shortest path problem has a number of
solutions and running time that are exponential in $\log^2 n$ on n-vertex graphs [6].

As well as the obvious applications of this formulation to real-world problems
with time-varying but predictable data (such as rush-hour route planning), parametric optimization problems have another class of applications, to *bicriterion
optimization*. In bicriterion problems, each input has two numbers associated
with it, that can be summed over the elements of a candidate solution. For
instance, these two numbers might be the x and y coordinates of points in the
plane, the mean and variance of a normal distribution, an initial investment cost
and expected profit of a business opportunity, or the cost and log-likelihood of
failure of a communications link. The goal is to find a solution that optimizes a
nonlinear combination of these two sums of values, such as the distance from the

© Springer International Publishing Switzerland 2015
F. Dehne et al. (Eds.): WADS 2015, LNCS 9214, pp. 327–338, 2015.
DOI: 10.1007/978-3-319-21840-3_27

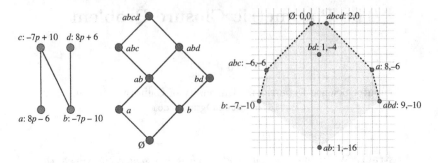

Fig. 1. An instance of the parametric closure problem. Left: The Hasse diagram of a partially ordered set N of four elements, each with a weight that varies linearly with a parameter p. Center: The distributive lattice lower(N) of lower sets of N. Right: The point set project(lower(N)) and its convex hull. The upper hull (dashed) gives in left-to-right order the sequence of six distinct maximum-weight closures as the parameter p varies continuously from $-\infty$ to $+\infty$.

origin, probability of exceeding a given threshold, percentage return on investment (the ratio of total profit to total investment cost), or cost-reliability ratio. Many natural bicriterion problems can be expressed as finding the maximum of a quasiconvex function of the two sums (a function whose lower level sets are convex sets) or equivalently as finding the minimum of a quasiconcave function of the two sums. When this is the case, the optimal solution can always be obtained as one of the solutions of a parametric problem, defined by re-interpreting the two numbers associated with each input element as the slope and y-intercept of a linear parametric function [7,8]. In this way, any algorithm for solving a parametric optimization problem can also be used to solve bicriterion versions of the same type of optimization problem.

In this paper we formulate and provide the first study of the *parametric closure problem*, the natural parametric variant of a classical optimization problem, the *maximum closure problem* [9,10]. A closure in a directed graph is a subset of vertices such that all edges from a vertex in the subset go to another vertex in the subset; the maximum closure problem is the problem of finding the highest-weight closure in a vertex-weighted graph. Equivalently we seek the highest-weight *lower set* of a weighted partial order, where a lower set is a subset of the elements of a partial order such that if $x < y$ in the order, and y belongs to the subset, then x also belongs to the subset. Applications of this problem include open pit mining [11], military attack planning [12], freight depot placement [13,14], scheduling with precedence constraints [15,16], image segmentation [17,18], stable marriage with maximum satisfaction [19], and treemap construction in information visualization [20]. Maximum closures can be found in polynomial time by a reduction to maximum flow [9,21] or by direct algorithms [22].

In the parametric closure problem, we assign weights to the vertices of a directed graph (or the elements of a partial order) that vary linearly as functions of a parameter, and we seek the closures (or lower sets) that have maximum weight as the parameter varies. As described above, an algorithm for this problem can also solve bicriterion closure problems of maximizing a quasiconvex function (or minimizing a quasiconcave function) of two sums of values. Although we have not been able to resolve the complexity of this problem in the general case, we prove near-linear or polynomial complexity for several important special cases of the parametric closure problem.

We do not know of previous work on the general parametric closure problem, but two previous papers can be seen in retrospect as solving special cases. Lawler [15] studied scheduling to minimize weighted completion time with precedence constraints. He used the closure that maximizes the ratio x/y of the priority x and processing time y of a job or set of jobs to decompose instances of this problem into smaller subproblems. As Lawler showed, the optimal closure can be found in polynomial time by a binary search where each step involves the solution of a weighted closure problem. Replacing the binary search by parametric search [23] would make this algorithm strongly polynomial; however, both search methods depend on the specific properties of the ratio function and would not work for other bicriterion problems. A second paper, by Carlson and Eppstein [8], considers bicriterion versions of the problem of finding the best subtree (containing the root) of a given rooted tree with weighted edges. These subtrees can be modeled as lower sets for a partial order on the tree edges in which two edges are comparable when they both belong to a path from the root; this partial order is series-parallel, and we greatly generalize the results of Carlson and Eppstein in our new results on parametric closures for series-parallel partial orders.

Parametric Optimization as an Implicit Convex Hull Problem. Parametric optimization problems can be formulated dually, as problems of computing convex hulls of implicitly defined two-dimensional point sets. Suppose we are given a parametric optimization problem in which weight of element i is a linear function $a_i\lambda + b_i$ of a parameter λ, and in which the weight of a candidate solution S (a subset of elements, constrained by the specific optimization problem in question) is the sum of these functions. Then the solution value is also a linear function, whose coefficients are the sums of the element coefficients:

$$\sum_{i \in S} a_i\lambda + b_i = \left(\sum_{i \in S} a_i\right)\lambda + \left(\sum_{i \in S} b_i\right).$$

Instead of interpreting the numbers a_i and b_i as coefficients of linear functions, we may re-interpret the same two numbers as the x and y coordinates (respectively) of points in the Euclidean plane. In this way any family $\mathcal{F}$ of candidate solutions determines a planar point set, in which each set in $\mathcal{F}$ corresponds to the point given by the sum of its elements' coefficients. We call this point set project($\mathcal{F}$), because the sets in $\mathcal{F}$ can be thought of as vertices of a hypercube $Q_n = \{0,1\}^n$ whose dimension is the number of input elements, and project determines a linear projection from these vertices to the Euclidean plane.

Let hull(project($\mathcal{F}$)) denote the convex hull of this projected planar point set. Then for each parameter value the set in $\mathcal{F}$ minimizing or maximizing the parameterized weight corresponds by projective duality to a vertex of the hull, and the same is true for the maximizer of any quasiconvex function of the two sums of coefficients a_i and b_i. Thus, parametric optimization can be reformulated as the problem of constructing this convex hull, and bicriterion optimization can be solved by choosing the best hull vertex.

New Results. For an arbitrary partially ordered set P, define lower(P) to be the family of lower sets of P. As a convenient abbreviation, we define polygon(P) = hull(project(lower(P))). We consider the following classes of partially ordered set. For each partial order P in one of these classes, we prove polynomial bounds on the complexity of polygon(P) and on the time for constructing the hull. These results imply the same time bounds for parametric optimization over P and for maximizing a quasiconvex function over P.

Semiorders. This class of partial orders was introduced to model human preferences [24] in which each element can be associated with a numerical value, pairs of elements whose values are within a fixed margin of error are incomparable, and farther-apart pairs are ordered by their numerical values. For such orderings, we give a bound of $O(n \log n)$ on the complexity of polygon(P) and we show that it can be constructed in time $O(n \log^2 n)$ using an algorithm based on the quadtree data structure.

Series-parallel partial orders. These are orders formed recursively from smaller orders of the same type by two operations: series compositions (in which all elements from one order are placed earlier in the combined ordering than all elements of the other order) and parallel compositions (in which pairs of one element from each ordering are incomparable). These orderings have been applied for instance in scheduling applications by Lawler [15]. For such orderings, the sets of the form polygon(P) have a corresponding recursive construction by two operations: the convex hull of the union of two convex polygons, and the Minkowski sum of two convex polygons. It follows that polygon(P) has complexity $O(n)$. This construction does not immediately lead to a fast construction algorithm, but we adapt a splay tree data structure to construct polygon(P) in time $O(n \log n)$. Our previous results for optimal subtrees [8] follow as a special case of this result.

Bounded treewidth. Suppose that partial order P has n elements and its transitive reduction forms a directed acyclic graph whose underlying undirected graph has treewidth w. (For prior work on treewidth of partial orders, see [25].) Then we show that polygon(P) has polynomially many vertices, with exponent $O(w)$, and that it can be constructed in polynomial time.

Incidence posets. The incidence poset of a graph G has the vertices and edges as elements, with an order relation $x \leq y$ whenever x is an endpoint of y. One of the initial applications for the closure problem concerned the design of freight delivery systems in which a certain profit could be expected from each of a set of point-to-point routes in the system, but at the cost of setting up depots at each endpoint of the routes [13,14]; this can be modeled with an incidence poset for a graph with a vertex at each depot location and an edge for each potential route. Since the profits and costs have different timeframes, it is reasonable to combine them in a nonlinear way, giving a bicriterion closure problem. The transitive reduction of an incidence poset is a subdivision of G with the same treewidth, so our technique for partial orders of bounded treewidth also applies to incidence posets of graphs of bounded treewidth.

Bounded width. The *width* of a partial order is the maximum number of elements in an *antichain*, a set of mutually-incomparable elements. Low-width partial orders arise, for instance, in the edit histories of version control repositories [26]. The treewidth of a partial order is at most equal to its width, but partial orders of width w have $O(n^w)$ lower sets, tighter than the bound that would be obtained by using treewidth. We show more strongly using quadtrees that in this case $\mathsf{polygon}(P)$ has $O(n^{w-1} + n \log n)$ vertices and can be constructed in time within a logarithmic factor of this bound.

We have been unable to obtain an example of a family of partial orders with a nonlinear lower bound on the complexity of $\mathsf{polygon}(P)$, nor have we been able to obtain a nontrivial upper bound on the hull complexity for unrestricted partial orders. Additionally, we have been unable to obtain polynomial bounds on the hull complexity of the above types of partial orders for dimensions higher than two. We also do not know of any computational complexity bounds (such as NP-hardness) for the parametric closure problem for any class of partial orders in any finite dimension. We leave these problems open for future research.

For space reasons we describe only the semiorder and series-parallel results in the main text of our paper, deferring the remaining results to appendices.

2 Minkowski Sums and Hulls of Unions

Our results on the complexity of the convex polygons $\mathsf{polygon}(P)$ associated with a partial order hinge on decomposing these polygons recursively into combinations of simpler polygons. To do this, we use two natural geometric operations that combine pairs of convex polygons to produce more complex convex polygons.

Definition 1. *For any two convex polygons P and Q, let $P \oplus Q$ denote the Minkowski sum of P and Q (the set of points that are the vector sum of a point in P and a point in Q), and let $P \uplus Q$ denote the convex hull of the union of P and Q.*

Lemma 1 (folklore). *If convex polygons P and Q have p and q vertices respectively, then $P \oplus Q$ and $P \uplus Q$ have at most $p + q$ vertices, and can be constructed from P and Q in time $O(p + q)$.*

We omit the (easy) proof for space reasons.

Corollary 1. *Suppose that P is a convex polygon, described as a formula that combines a set of n points in the plane into a single polygon using a sequence of ⊕ and ⊎ operations. Suppose in addition that, when written as an expression tree, this formula has height h. Then P has at most n vertices and it may be constructed from the formula in time $O(nh)$.*

More complex data structures can reduce this time to $O(n \log n)$; see Section 4.

In higher dimensions, the convex hull of n points and Minkowski sum of n line segments both have polynomial complexity with an exponent that depends linearly on the dimension. However, we do not know of an analogous bound on the complexity of convex sets formed by combining Minkowski sum and hull-of-union operations. If such a bound held, we could extend our results on parametric closures to the corresponding higher dimensional problems.

3   Semiorders

A *semiorder* is a type of partial order defined by Luce [24] to model human preferences. Each element of the order has an associated numerical value (its *utility* to the person whose preferences are being modeled). For items whose utilities are sufficiently far from each other, the ordering of the two items in the semiorder is the same as the numerical ordering of their utilities. However, items whose utilities are within some (global) margin of error of each other are incomparable in the semiorder. Similar concepts of comparisons of numerical values with margins of error give rise to semiorders in many other areas of science and statistics [27]. For efficient computations on semiorders we will assume that the utility values of each element are part of the input to an algorithm, and that the margin of error has been normalized to one. For instance, the semiorder N of Figure 1 can be represented as a semiorder with utilities 2/3, 0, 2, and 4/3 for a, b, c, and d respectively. With this information in hand, the comparison between any two elements can be determined in constant time.

The concept of a lower set is particularly natural for a semiorder: it is a set of elements whose utility values could lie below a sharp numerical threshold, after perturbing each utility value by at most half the margin of error. In this way, the closure problem (the problem of finding a maximum weight lower set) can alternatively be interpreted as the problem of finding the maximum possible discrepancy of a one-dimensional weighted point set in which the location of each point is known imprecisely. Semiorders may have exponentially many lower sets; for instance, if all items have utilities that are within one unit of each other, all sets are lower sets. Nevertheless, as we show in this section, if S is a semiorder, then the complexity of polygon(S) is near-linear.

If S is any parametrically weighted semi-order, we may write the sorted order of the utility values of elements of S as $u_0, u_1, \ldots, u_{n-1}$ where $n = |S|$, and the elements themselves (in the same order) as $x_0, x_1, \ldots, x_{n-1}$. By padding S with items that have a fixed zero weight and a utility that is smaller than that of

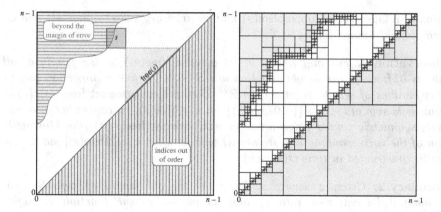

Fig. 2. The grid $[0, n-1]^2$, with the two regions that cannot be part of the image of extremes. The left image shows a square subproblem s and free(s); the right image shows the quadtree decomposition of the grid used to prove Theorem 1.

the elements by more than the margin of error, we may assume without loss of generality that n is a power of two without changing the values of the parametric closure problem on S.

Definition 2. *Let L be an arbitrary lower set in* lower(S). *Let j be the largest index of an element x_i of L. Let i be the smallest index of an element x_i such that x_i does not belong to L and $i < j$, or -1 if no such element exists. Define* extremes(L) *to be the pair of integers $(i + 1, j)$.*

Thus, extremes maps the family lower(S) to the integer grid $[0, n-1]^2$, with potentially many lower sets mapped to each grid point. However, not every grid point is in the image of lower(S): a point (i, j) with $i > j$ cannot be the image of a lower set, because the element defining the first coordinate of extremes must have an index smaller than the element defining the second coordinate. And when $i > 0$, a point (i, j) with $u_{i-1} < u_j - 1$ (i.e. with utility values that are beyond the margin of error for the semiorder) also cannot be the image of a lower set, because in this case $x_{i-1} \leq x_j$ in the semiorder, so every lower set that includes x_j also includes x_{i-1}. Thus, the image of extremes lies in an orthogonally convex subset of the grid, bounded below by its main diagonal and above by a monotone curve (Figure 2).

Definition 3. *Let s be any square subset of the integer grid $[0, n-1]^2$, and define* subproblem(s) *to be the partially-ordered subset of the semiorder S consisting of the elements whose indices are among the rows and columns of s. Define* free(s) *to be the (unordered) set of elements of S that do not belong to* subproblem(s), *but whose indices are between pairs of indices that belong to* subproblem(s). *(See Figure 2, left, for an example.)*

Observation 1. *Given a square s, suppose that the subfamily $\mathcal{F}$ of* lower(S) *that is mapped by* extremes *to s is nonempty. Then each set in $\mathcal{F}$ is the disjoint* *union of a lower set of* subproblem(s) *and an arbitrary subset of* free(s), *and all such disjoint unions belong to* $\mathcal{F}$.

Observation 2. *For any square* s, *let* powerset(free(s)) *be the family of all subsets of* free(s). *Let weight function* $w : S \to \mathbb{R}^2$ *define a projection* project *from families of sets to point sets in* $\mathbb{R}^2$. *Then* project(powerset(free(s))) *is the Minkowski sum of the sets* $\{(0,0), w(x_i)\}$ *for* $x_i \in$ free(s). *Its convex hull is a centrally symmetric convex polygon* hull(project(powerset(free(s)))) *(the Minkowski sum of the corresponding line segments) with at most* $k = 2|$free(s)$|$ *sides, and can be constructed in time* $O(k \log k)$.

Corollary 2. *Given a square* s, *let* $\mathcal{F}$ *be the sub-family of* lower(S) *that is mapped by* extremes *into* s, *and assume a weight function* w *defining a projection* project. *Then* hull(project($\mathcal{F}$)) $=$ polygon(subproblem(s)) $\oplus$ hull(project(powerset(free(s)))).

Lemma 2. *Let* s *be a square in the grid* $[0, n-1]^2$, *and subdivide* s *into four congruent smaller squares* s_i $(0 \leq i < 4)$. *Let* polygon(subproblem(s)) *have* c *vertices, define* c_i *in the same way for each* s_i, *and let* ℓ *be the side length of* s. *Then* $c \leq \sum c_i + O(\ell)$, *and* polygon(subproblem(s)) *can be constructed from the corresponding hulls for the smaller squares in time* $O(\sum c_i + \ell \log \ell)$.

Proof. For each smaller square s_i, define $\mathcal{F}_i$ to be the subset of subproblem(s) mapped to s_i, and define $H_i =$ hull(project($\mathcal{F}_i$)). Then by Corollary 2 (viewing s_i as a subproblem of subproblem(s)),

$$H_i = \text{polygon(subproblem}(s)) \oplus \text{hull(project(powerset(free}(s) \setminus \text{free}(s_i)))).$$

The set free(s) $\setminus$ free(s_i) has cardinality $O(\ell)$, so by Lemma 1 and Observation 2, H_i has complexity $c_i + O(\ell)$ and can be constructed in time $O(c_i + \ell \log \ell)$. Applying Lemma 1 again, polygon(subproblem(s)) $= H_0 \uplus H_1 \uplus H_2 \uplus H_2$ has complexity at most $\sum c_i + O(\ell)$ and can be constructed from the polygons H_i in time $O(\sum c_i + \ell \log \ell)$. $\square$

Theorem 1. *If* S *is a semiorder with* n *elements* x_i, *specified with their utility values* u_i *and a system of two-dimensional weights* $w(x_i)$, *then* polygon(S) *has complexity* $O(n \log n)$ *and can be constructed in time* $O(n \log^2 n)$.

Proof. We sort the utility values, pad n to the next larger power of two if necessary and form a quadtree decomposition of the grid $[0, n-1]^2$ (as shown in Figure 2, right). For each square s of this quadtree, we associate a convex polygon (or empty set) polygon(subproblem(s)) computed according to the following cases:

- If s is a subset of the grid points for which $i > j$, or for which $u_{i-1} < u_j - 1$, then no lower sets are mapped into s by extremes. We associate square s with the empty set.

- If s is a subset of the grid points for which $i \leq j$ and $u_{i-1} \geq u_j - 1$, then every two elements of subproblem(s) are incomparable. In this case, we associate square s with the polygon hull(project(powerset(subproblem(s)))) computed according to Observation 2.
- Otherwise, we split s into four smaller squares. We construct the polygon associated with s by using Lemma 2 to combine the polygons associated with its children.

It follows by induction that the total complexity of the polygon constructed at any square s of the quadtree is $O(\sum \ell_i)$, and the total time for constructing it is $O(\sum \ell_i \log \ell_i)$, where ℓ_i is the side length of the ith square of the quadtree and the sum ranges over all descendants of s. As a base case for the induction, a square containing only a single grid point is associated with a subproblem with one element, with only one lower set that maps to that grid point, and a degenerate convex polygon with a single vertex. The polygon constructed at the root of the quadtree is the desired output, and it follows that it has combinatorial complexity and time complexity of the same form, with a sum ranging over all quadtree squares.

The conditions $i > j$ and $u_{i-1} < u_j - 1$ define two monotone curves through the grid, and we split a quadtree square only when it is crossed by one of these two curves. It follows that the squares of side length ℓ that are subdivided as part of this algorithm themselves form two monotone chains, and that the number of all squares of side length ℓ is $O(n/\ell)$. The results of the theorem follow by summing up the contributions to the polygon complexity and time complexity for the $O(\log n)$ different possible values of ℓ. $\qquad\square$

4   Series-Parallel Partial Orders

Series-parallel partial orders were considered in the context of a scheduling problem by Lawler [15], and include as a special case the tree orderings previously studied in our work on bicriterion optimization [8]. They are the partial orders that can be constructed from single-element partial orders by repeatedly applying the following two operations:

Series composition. Given two series-parallel partial orders P_1 and P_2, form an order from their disjoint union in which every element of P_1 is less than every element of P_2.

Parallel composition. Given two series-parallel partial orders P_1 and P_2, form an order from their disjoint union in which there are no order relations between P_1 and P_2.

Observation 3. *If P is the series composition of P_1 and P_2, then* polygon(P) *is the convex hull of the union of* polygon(P_1) *and a translate (by the sum of the weights of the elements of P_1) of* polygon(P_2). *If P is the parallel composition of P_1 and P_2, then* polygon(P) *is the Minkowski sum of* polygon(P_1) *and* polygon(P_2).

Recursively continuing this decomposition gives us a formula for polygon(P) in terms of the ⊎ and *oplus* operations. By Lemma 1 we immediately obtain:

Corollary 3. *If P is a series-parallel partial order with n elements, then* polygon(P) *has at most $2n$ vertices.*

However, the depth of the formula for polygon(P) may be linear, so using Lemma 1 to construct polygon(P) could be inefficient. We now describe a faster algorithm. The key idea is to follow the same formula to build polygon(P), but to represent each intermediate result (a convex polygon) by a data structure that allows the ⊎ and ⊕ operations to be performed more quickly for pairs of polygons of unbalanced sizes. Note that a Minkowski sum operation between a polygon of high complexity and a polygon of bounded complexity can change a constant fraction of the vertex coordinates, so to allow fast Minkowski sums our representation cannot store these coordinates explicitly.

Lemma 3. *It is possible to store convex polygons in a data structure such that destructively merging the representations of two polygons of m and n vertices respectively by a ⊎ or ⊕ operation (with $m < n$) can be performed in time $O(m \log((m + n)/m))$.*

Proof. We store the lower and upper hulls separately in a binary search tree data structure, in which each node represents a vertex of the polygon, and the inorder traversal of the tree gives the left-to-right order of the vertices. The node at the root of the tree stores the Cartesian coordinates of its vertex; each non-root node stores the vector difference between its coordinates and its parents' coordinates. Additionally, each node stores the vector difference to its clockwise neighbor around the polygon boundary. In this way, we can traverse any path in this tree and (by adding the stored vector difference) determine the coordinates of any vertex encountered along the path. We may also perform a rotation in the tree, and update the stored vector differences, in constant time per rotation.

We will keep this tree balanced (in an amortized sense) by using the splay tree balancing strategy [28]: whenever we follow a search path in the tree, we will immediately perform a splay operation that through a sequence of double rotations moves the endpoint of the path to the root of the tree. By the dynamic finger property for splay trees [29,30], a sequence of m accesses in sequential order into a splay tree of size n will take time $O(m \log((m + n)/m))$.

To compute the hull of the union (the ⊎ operation) we insert each vertex of the smaller polygon (by number of vertices), in left-to-right order, into the larger polygon. To insert a vertex v, we search the larger polygon to find the edges with the same x-coordinate as v and use these edges to check whether v belongs to the lower hull, the upper hull, or neither. If it belongs to one of the two hulls, we search the larger polygon again to find its two neighbors on the hull. By performing a splay so that these neighbors are rotated to the root of the binary tree, and then cutting the tree at these points, we may remove the vertices between v and its new neighbors from the tree without having to consider those vertices one-by-one. We then create a new node for v and add its two neighbors as the left and right child.

To compute the Minkowski sum (the $\oplus$ operation) we must simply merge the two sequences of edges of the two polygons by their slopes. We search for each edge slope in the smaller polygon. When its position is found, we splay the vertex node at the split position to the root of the tree, and then split the tree into its left and right subtrees, each with a copy of the root node. We translate all vertices on one side of the split by the vector difference for the inserted edge (by adding that vector only to the root of its tree), and rejoin the trees.   □

Theorem 2. *If P is a series-parallel partial order, represented by its series-parallel decomposition tree, then* $\mathsf{polygon}(P)$ *has complexity $O(n)$ and may be constructed in time $O(n \log n)$.*

Proof. We follow the formula for constructing $\mathsf{polygon}(P)$ by $\uplus$ and $\oplus$ operations, using the data structure of Lemma 3. We charge each merge operation to the partial order elements on the smaller side of each merge. If a partial order element belongs to subproblems of sizes $n_0 = 1$, n_1, ..., $n_h = n$ where h is the height of the element, then the time charged to it is $\sum_i O(\log(n_i/n_{i-1})) = O(\log \prod_i (n_i/n_{i-1})) = O(\log n)$.   □

Acknowledgments. This research was supported in part by NSF grant 1228639 and ONR grant N00014-08-1-1015.

References

1. Matoušek, J.: Lower envelopes. In: Lectures on Discrete Geometry. Graduate Texts in Mathematics, vol. 212, pp. 165–194. Springer (2002)
2. Hershberger, J.: Finding the upper envelope of n line segments in $O(n \log n)$ time. Information Processing Letters **33**(4), 169–174 (1989)
3. Eppstein, D.: Geometric lower bounds for parametric matroid optimization. Discrete Comput. Geom. **20**(4), 463–476 (1998)
4. Fernández-Baca, D., Slutzki, G., Eppstein, D.: Using sparsification for parametric minimum spanning tree problems. Nordic J. Comput. **3**(4), 352–366 (1996)
5. Agarwal, P.K., Eppstein, D., Guibas, L.J., Henzinger, M.R.: Parametric and kinetic minimum spanning trees. In: Proc. 39th IEEE Symp. Foundations of Computer Science (FOCS 1998), pp. 596–605 (1998)
6. Carstensen, P.J.: Parametric cost shortest path problems. Unpublished memo, Bellcore (1984)
7. Katoh, N.: Bicriteria network optimization problems. IEICE Trans. Fundamentals of Electronics, Communications and Computer Sciences **E75–A**, 321–329 (1992)
8. Carlson, J., Eppstein, D.: The weighted maximum-mean subtree and other bicriterion subtree problems. In: Arge, L., Freivalds, R. (eds.) SWAT 2006. LNCS, vol. 4059, pp. 400–410. Springer, Heidelberg (2006)
9. Picard, J.C.: Maximal closure of a graph and applications to combinatorial problems. Manag. Sci. **22**(11), 1268–1272 (1976)
10. Hochbaum, D.: 50th anniversary article: Selection, provisioning, shared fixed costs, maximum closure, and implications on algorithmic methods today. Manag. Sci. **50**(6), 709–723 (2004)

11. Lerchs, H., Grossmann, I.F.: Optimum design of open-pit mines. Trans. Canad. Inst. Mining and Metallurgy **68**, 17–24 (1965)

12. Orlin, D.: Optimal weapons allocation against layered defenses. Nav. Res. Logist. Q. **34**, 605–617 (1987)

13. Balinski, M.L.: On a selection problem. Manag. Sci. **17**(3), 230–231 (1970)

14. Rhys, J.M.W.: A selection problem of shared fixed costs and network flows. Manag. Sci. **17**(3), 200–207 (1970)

15. Lawler, E.L.: Sequencing jobs to minimize total weighted completion time subject to precedence constraints. Ann. Discrete Math. **2**, 75–90 (1978)

16. Chang, G.J., Edmonds, J.: The poset scheduling problem. Order **2**(2), 113–118 (1985)

17. Gibson, M., Han, D., Sonka, M., Wu, X.: Maximum weight digital regions decomposable into digital star-shaped regions. In: Asano, T., Nakano, S., Okamoto, Y., Watanabe, O. (eds.) ISAAC 2011. LNCS, vol. 7074, pp. 724–733. Springer, Heidelberg (2011)

18. Ahmed, M., Chowdhury, I., Gibson, M., Islam, M.S., Sherrette, J.: On maximum weight objects decomposable into based rectilinear convex objects. In: Dehne, F., Solis-Oba, R., Sack, J.-R. (eds.) WADS 2013. LNCS, vol. 8037, pp. 1–12. Springer, Heidelberg (2013)

19. Irving, R.W., Leather, P., Gusfield, D.: An efficient algorithm for the "optimal" stable marriage. J. ACM **34**(3), 532–543 (1987)

20. Buchin, K., Eppstein, D., Löffler, M., Nöllenburg, M., Silveira, R.I.: Adjacency-preserving spatial treemaps. In: Dehne, F., Iacono, J., Sack, J.-R. (eds.) WADS 2011. LNCS, vol. 6844, pp. 159–170. Springer, Heidelberg (2011)

21. Hochbaum, D.: A new-old algorithm for minimum-cut and maximum-flow in closure graphs. Networks **37**(4), 171–193 (2001)

22. Faaland, B., Kim, K., Schmitt, T.: A new algorithm for computing the maximal closure of a graph. Manag. Sci. **36**(3), 315–33 (1990)

23. Megiddo, N.: Applying parallel computation algorithms in the design of serial algorithms. J. ACM **30**(4), 852–865 (1983)

24. Luce, R.D.: Semiorders and a theory of utility discrimination. Econometrica **24**, 178–191 (1956)

25. Joret, G., Micek, P., Milans, K.G., Trotter, W.T., Walczak, B., Wang, R.: Tree-width and dimension. Unpublished manuscript (2013)

26. Bannister, M.J., Devanny, W.E., Eppstein, D.: Small superpatterns for dominance drawing. In: Proc. Analytic Algorithmics and Combinatorics (ANALCO 2014), pp. 92–103 (2014)

27. Pirlot, M., Vincke, P.: Semiorders: Properties, Representations, Applications. Springer (1997)

28. Sleator, D.D., Tarjan, R.E.: Self-adjusting binary search trees. J. ACM **32**(3), 652–686 (1985)

29. Cole, R., Mishra, B., Schmidt, J., Siegel, A.: On the dynamic finger conjecture for splay trees. I. Splay sorting $\log n$-block sequences. SIAM J. Comput. **30**(1), 1–43 (2000)

30. Cole, R.: On the dynamic finger conjecture for splay trees. II. The proof. SIAM J. Comput. **30**(1), 44–85 (2000)

Rooted Cycle Bases

David Eppstein[1]($\boxtimes$), J. Michael McCarthy[2], and Brian E. Parrish[2]

[1] Computer Science Department, University of California, Irvine, USA
david.eppstein@gmail.com
[2] Department of Mechanical and Aerospace Engineering,
University of California, Irvine, USA

Abstract. A cycle basis in an undirected graph is a minimal set of
simple cycles whose symmetric differences include all Eulerian subgraphs
of the given graph. We define a rooted cycle basis to be a cycle basis in
which all cycles contain a specified root edge, and we investigate the
algorithmic problem of constructing rooted cycle bases. We show that a
given graph has a rooted cycle basis if and only if the root edge belongs
to its 2-core and the 2-core is 2-vertex-connected, and that constructing
such a basis can be performed efficiently. We show that in an unweighted
or positively weighted graph, it is possible to find the minimum weight
rooted cycle basis in polynomial time. Additionally, we show that it is NP-
complete to find a fundamental rooted cycle basis (a rooted cycle basis
in which each cycle is formed by combining paths in a fixed spanning
tree with a single additional edge) but that the problem can be solved
by a fixed-parameter-tractable algorithm when parameterized by clique-
width.

1 Introduction

A cycle basis of an undirected graph is a set of cycles such that all cycles in the
graph have a unique representation as an algebraic sum of basis cycles. In this
paper we study algorithms for finding a special type of cycle basis which we call
a *rooted cycle basis*, in which all cycles in the basis contain a specified root edge.

Cycle bases have diverse applications including subway system scheduling [1],
the analysis of distributed algorithms [2], and bioinformatics [3,4]. The specific
motivation for our rooted variant of the problem comes from mechanical engineering, where cycle bases have long been used in static analysis of structures
such as truss bridges [5] and in the kinematics of moving bodies [6]. We recently
used this method as part of a system for constructing the configuration space
of moving linkages [7], systems that include automobile suspensions, fold-out
sofa-beds, and legs for walking robots.

In this configuration space construction problem, systems of rigid two-dimensional *links* are connected at *joints* where one link can rotate around a point
of another with one degree of freedom. A system of links and joints is called a
kinematic chain; fixing the position of one *ground* link results in a system called
a *mechanism* or *inversion*, and distinguishing a second *input link* (connected to

© Springer International Publishing Switzerland 2015
F. Dehne et al. (Eds.): WADS 2015, LNCS 9214, pp. 339–350, 2015.
DOI: 10.1007/978-3-319-21840-3_28

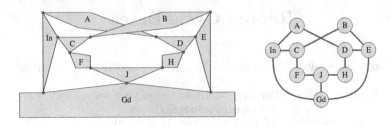

Fig. 1. A linkage and its linkage graph (a subdivision of $K_{3,3}$)

the ground by a joint and to which force is applied to control the rest of the system) results in a system called a *linkage* [7]. The structure of a linkage can be expressed combinatorially by a *linkage graph*, an undirected graph with a vertex for each link and an edge for each joint, including a distinguished ground-input edge. The requirement that the combined motion of the linkage have one degree of freedom can be expressed combinatorially by the property that the linkage graph is $(\frac{3}{2}, 2)$-*tight* [8]: every k-vertex induced subgraph must have at most $\frac{3}{2}k - 2$ edges, and the whole graph must have exactly $\frac{3}{2}n - 2$ edges, where n is the number of vertices in the graph and links in the linkage. Links may cross each other in the plane, resulting in a non-planar linkage graph (Figure 1).

Given a linkage with its linkage graph, each input-to-ground path has an associated equation representing the requirement that the joints along the path have angles consistent with the fixed ground position at both ends of the path. Our system for constructing the configuration space of a linkage chooses a complete and non-redundant subset of these path equations and uses Dixon determinants to solve this system of equations [7]. Each path determining one of these equations can be turned into a cycle by adding the input-ground edge, and a set of equations chosen in this way is complete and non-redundant if and only if the corresponding set of cycles forms a cycle basis of the linkage graph. However, all of these cycles contain the input-ground edge, so the system of equations that we seek comes from a rooted cycle basis. Additionally, because the equation solver forms the computational bottleneck of our system, we would like the system of equations that we construct to be as simple as possible, corresponding to the problem of finding a *minimum rooted cycle basis*.

New Results. We provide the first algorithmic study of the problem of constructing rooted cycle bases. We have the following new results:

- As a warm-up to our main result, we show that an arbitrary graph G with designated root edge e has a rooted cycle basis in which all cycles contain e if and only if the 2-core of G is 2-vertex-connected and contains e. When a rooted cycle basis exists, it can be constructed in time $O(mn)$. This is tight: there exist graphs for which every rooted cycle basis has total size $\Theta(mn)$.
- Our main result is that, in an unweighted or positively weighted graph with a designated root edge, we can find the minimum weight rooted cycle basis by a randomized algorithm with nearly-optimal $O(mn + n^2 \log n)$ expected

time or by a polylogarithmically slower deterministic algorithm. This basis is always *weakly fundamental*: its cycles can be ordered so that each cycle contains an edge that is not in any earlier cycle. Our algorithm uses a greedy method for finding each cycle, with a tie-breaking rule that avoids greedy choices that do not lead to a valid cycle basis.

- In the full version of this paper we show that it is NP-complete to determine whether a graph G with root edge e has a *fundamental* rooted cycle basis, a rooted cycle basis determined from a spanning tree T by choosing all cycles formed by an edge not in T and a path in T. It remains NP-complete even when G is planar. Our proof is based on the observation that, in planar graphs, fundamental rooted cycle bases are dual to a form of Hamiltonian cycle. Additionally, we use Courcelle's theorem to show that finding a fundamental rooted cycle basis is fixed-parameter-tractable in the clique-width of the input.

In comparison, for arbitrary cycle bases, every graph has a fundamental cycle basis, which may be constructed using any spanning tree algorithm. Finding unrestricted minimum weight cycle bases takes polynomial time [9–12]. However, finding an unrestricted minimum weight weakly fundamental cycle basis is NP-hard [13], and cannot be solved by the same greedy strategy that we use for rooted cycle bases, of choosing the shortest cycle that includes a new edge.

2 Preliminaries

By $\mathbb{F}_2$ we mean the field with two elements 0 and 1 under mod-2 arithmetic. If U is an arbitrary finite set, the subsets of U form a vector space $\mathbb{F}_2^U$ over $\mathbb{F}_2$ with the empty set as origin and the symmetric difference of sets as addition.

We define a *rooted graph* to be an undirected graph $G = (V, E)$ with a designated root edge e. A *cycle* is a connected 2-regular subgraph; a cycle is *rooted* if it contains e, and *Hamiltonian* if it contains every vertex of G. The *edge space* of G is the vector space $\mathbb{F}_2^E$. The *cycle space* of G is the subspace of the edge space generated by edge sets of cycles; its elements are subgraphs of G with even degree at every vertex [14]. A *cycle basis* of G is a set of cycles that forms a basis of the cycle space [12]. A cycle basis is *rooted* if all its cycles are rooted.

A *spanning tree* of an undirected graph G is a subgraph that includes all vertices of G, and is connected with no cycles. Any edge f that does not belong to a spanning tree T gives rise to a *fundamental cycle* for T consisting of f plus the unique path in T connecting the endpoints of f. The fundamental cycles for T form a cycle basis; a basis formed in this way is called *fundamental*.

A *matroid* [15] may be defined as a family of subsets of a finite set, called the *independent sets* of the matroid, with two properties:

- Every subset of an independent set is independent.
- If I_1 and I_2 are independent sets and $|I_1| < |I_2|$, then there exists an element x belonging to $I_2 \setminus I_1$ such that $I_1 \cup \{x\}$ is independent.

The linearly independent subsets of a finite family of vectors in any vector space form a *linear matroid*. In a matroid, a *basis* is an independent set all of whose supersets are dependent; for linear matroids, this notion coincides with the standard definition of a basis of a vector space.

If the elements of a matroid are given real-valued weights, then the basis with minimum total weight can be constructed by a greedy algorithm, generalizing Kruskal's algorithm for minimum spanning trees: initialize a set $\mathcal{B}$ to be the empty set, and consider the elements in sorted order by their weights, adding each element to $\mathcal{B}$ if the result would remain independent. In particular, if the edges of an undirected graph G are given weights, the weight of a cycle may be defined as the sum of the weights of its edges, and the weight of a cycle basis may be defined as the sum of the weights of its cycles. Then the minimum weight cycle basis may be found by considering all of the cycles of the graph in sorted order by weight, adding each one to the basis if the result would remain independent. This algorithm may be sped up by considering only a special set of polynomially-many candidate cycles, leading to polynomial-time construction of the minimum weight cycle basis in any graph [9–12].

A *simple path* in a graph G is a connected subgraph with two degree-one vertices (its endpoints) and with all remaining vertices (its interior vertices) having degree exactly two. An *open ear decomposition* of G is a collection of simple paths P_i for $i = 0, 1, 2, \ldots$ (called ears) with the following properties:

- The first ear P_0 is a single edge.
- The two endpoints of each ear P_i with $i > 0$ appear in earlier-numbered ears.
- No interior vertex of an ear appears in any earlier ear.

A graph has an open ear decomposition if and only if it is 2-vertex-connected (no vertex deletion can disconnect the remaining graph) [16]. This decomposition can be constructed in linear time, with any edge as the first ear [17–19]. The number of ears equals one plus the dimension of the cycle space.

A vertex of G belongs to at least one cycle of G if and only if it belongs to the 2-*core* of G, the subgraph formed by removing isolated vertices and degree-one vertices until all remaining vertices have degree ≥ 2. Therefore, the cycle bases of G are the same as the cycle bases of its 2-core.

3   Existence and Construction of Rooted Cycle Bases

The following lemma is a special case of Menger's theorem, but we give a proof as we use the proof construction in our algorithms.

Lemma 1. *Let e be an edge of a 2-vertex-connected graph G. Then for every two distinct vertices u and v of G there exist two vertex-disjoint paths (possibly of length zero) from u and v respectively to the two endpoints of e.*

Proof. Let $P_0 = e, P_1, \ldots, P_k$ be an open ear decomposition of G. We apply induction on k, with the following cases:

- As a base case, if $k = 0$, we have two length-zero paths, one for each endpoint.
- If $k > 0$ and neither u nor v is an interior vertex of P_k, the result follows by induction on the union of the ears up to P_{k-1}.
- If $k > 0$ and exactly one of u or v is an interior vertex of P_k, without loss of generality (by swapping u and v if necessary) we may assume that u is the interior vertex. At least one endpoint of P_k is a vertex w distinct from v. By induction, v and w can be connected by vertex-disjoint paths to e, using only vertices in ears $P_0, \ldots, P_{k-1}$. The result follows by augmenting the path from w with the part of path P_k from u to w.
- If $k > 0$ and both u and v are interior vertices of P_k, then u and v have two disjoint paths within P_k to the endpoints of P_k. By induction, the endpoints of P_k can be connected by paths to e, using only vertices in ears $P_0, \ldots, P_{k-1}$. The result follows by concatenating these paths with the paths within P_k.

Thus, in all cases, the desired two paths exist.   □

An ear with one edge cannot be part of a path constructed by this proof. So for a graph G with n vertices and m edges and a known ear decomposition, we can discard the one-edge ears and transform the case analysis of the proof into an algorithm that constructs the two desired paths in time $O(n)$.

Theorem 1. *An undirected graph G rooted at edge e has a cycle basis that is rooted at e if and only if e belongs to the 2-core of G and the 2-core is 2-vertex-connected. When a rooted cycle basis exists, it can be constructed in time $O(mn)$ and the total length of the cycles in the basis is $O(mn)$.*

Proof. If G has a rooted cycle basis, its 2-core must be 2-vertex-connected. For, suppose that a vertex v is deleted from the 2-core. Every remaining vertex u belongs to a basis cycle from which only v can have been deleted, leaving a path connecting u to the remaining endpoints of e. In this way any two remaining vertices can be connected to each other via e, so the remaining vertices are not disconnected.

In the other direction, suppose that the 2-core of G contains e and is 2-connected. Then it has an open ear decomposition $P_0 = e, P_1, \ldots, P_k$. We may form a set of cycles $C_1, C_2, \ldots, C_k$ in which each cycle C_i consists of e, the edges in P_i, and two paths through the union of ears $P_1, P_2, \ldots P_{i-1}$ connecting the endpoints of P_i to the endpoints of e. These cycles are independent because each one contains at least one edge in P_i that does not belong to any previous cycle. As an independent set of cycles of the correct cardinality to be a basis, they must be a basis.

After computing the ear decomposition, each cycle takes time $O(n)$ to construct (by the remarks following lemma 1) and has length $O(n)$, giving the stated time and length bounds.   □

The length and time bounds of the theorem are tight in the worst case: for a graph consisting of two $\Theta(m)$-vertex cliques connected by two $\Theta(n)$-vertex paths (Figure 2, left), every cycle through e and an edge in the farthest clique from

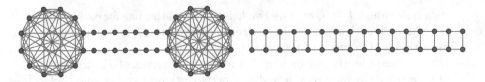

Fig. 2. Two graphs whose rooted cycle bases all have large total length: two cliques connected by two long paths (left), and a ladder graph (right)

e has length n, so every rooted cycle basis has total length $\Theta(mn)$. For linkage graphs with $m = \frac{3}{2}n - 2$, the time and length bounds become $O(n^2)$, which is again tight: every rooted cycle basis of an n-vertex ladder graph (Figure 2, right) has total length $\Theta(n^2)$.

In contrast, unrooted cycle bases may be significantly smaller. Every graph with m vertices and n edges has an (unrooted) cycle basis of total length $O(\min(n^2, m \log n))$, a bound that is close to tight because of the existence of sparse graphs of high girth for which every cycle basis has total length $\Omega(n \log n)$ [12,20].

4   Finding the Minimum Weight Rooted Cycle Basis

In this section we show how to find a rooted cycle basis of minimum total length in biconnected graphs with positive edge weights, in polynomial time. We use a greedy algorithm that chooses one cycle at a time, and prove it correct by showing that the sequence of cycles selected by this algorithm correspond to an ear decomposition. Our strategy is to show that an optimal basis can be derived from an ear decomposition: the cycles of the basis can be sorted from shorter to longer cycles in such a way that, in each successive cycle, the edges that do not belong to earlier cycles form an ear. Our algorithm performs the following steps:

1. Initialize what will eventually become a cycle basis to the empty set.
2. Use Suurballe's algorithm to compute, for each edge, the shortest rooted cycle through that edge.
3. While there exists an edge that is not included in any of the cycles chosen so far, select an edge that has not yet been included and whose computed shortest-cycle length is as small as possible, and add its cycle to the basis.

We will prove this algorithm correct under the additional assumption that no two paths, and no two cycles, have the same weight as each other. We say that a graph is *unambiguously weighted* when this is the case. When paths and cycles can have equal weights, this algorithm can fail by choosing a set of cycles that together cover all edges but do not generate the whole cycle space (Figure 3), so we need a consistent tie-breaking rule in this case, which we describe in the full version of this paper.

4.1   Greedy Cycle Sequences

We define a *greedy cycle sequence* to be a sequence of cycles that could be produced by the algorithm described at the beginning of this section. That is, it is a sequence of rooted cycles $C_1, C_2, \ldots$ in which

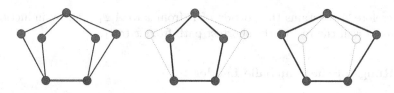

Fig. 3. An unweighted rooted graph (left) with two shortest rooted cycles that together cover the whole graph but do not generate its cycle space (center and right). Our algorithm requires that no two cycles have equal weight, to prevent bad sets of cycles such as these from being chosen.

1. Each cycle includes an edge that is not in any earlier cycle in the sequence, and
2. Subject to constraint (1), each cycle is as short as possible.

We will prove a sequence of lemmas about greedy cycle sequences, with the goal of showing that the set of new edges added by each cycle forms an ear and therefore that our greedy algorithm for rooted cycle bases is correct. To do so, it is helpful to have a notation for the subgraph of G formed by the vertices and edges in the first i cycles in the sequence. We call this subgraph the ith *ambit* of the cycle basis, denoted A_i.

Lemma 2. *Let G be an unambiguously-weighted rooted graph, A_i be the ith ambit of a greedy cycle sequence $C_1, C_2, \ldots$ for G, and x be a point in A_i (a vertex or a point interior to an edge). Then A_i contains the shortest path in G from x to each endpoint of the root edge of G.*

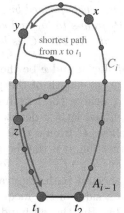

Fig. 4. Notation for Lemma 2. A shortest path is shown diverging from C_i before reaching A_{i-1}, proven impossible by the lemma.

Proof. Let t_1 and t_2 be the endpoints of the root edge. We will show by induction on i that A_i contains the shortest path from x to t_1; by symmetry it also contains a path to t_2. We may assume that x does not belong to A_{i-1}, for otherwise the shortest path is already contained in A_{i-1} by induction.

Then let P be the shortest path in G from x to t_1; we claim that P must remain within C_i until it reaches a vertex of A_{i-1}. For, if P deviated from C_i at some vertex y outside of A_{i-1}, let z be the first point at which P returns to a vertex of C_i; z must exist, because P eventually reaches t_1, which belongs to C_i. In this case the rooted cycle formed from C_i by removing the path in C_i from y to z and replacing it with the part of P from y to z would be strictly shorter than C_i (because the part of P from y to z is a shortest path and no two paths have equal length) and would contain the vertex y outside A_{i-1}, contradicting the greedy choice of C_i as the shortest rooted cycle not contained in A_{i-1}.

Therefore A_i contains the portion of P from x to A_{i-1}, and by induction it contains as well the rest of the shortest path from x to t_1.   □

4.2 Rungs of the Suurballe Ladder

Let P_1 and P_2 be two disjoint paths from s (an arbitrary vertex in the given graph G) to t_1 and t_2 (the endpoints of the root edge of G), as constructed by Suurballe's algorithm. Recall that this algorithm constructs two different paths from s to t_1 and t_2, the first of which is a shortest path in G and the second of which is the shortest path in a derived graph H. The union of these two paths differs from P_1 and P_2 by a collection of paths that we call *rungs*. Each rung is traversed in one direction by the shortest path in G and in the opposite direction by the shortest path in H. The endpoints of the rungs lie on P_1 and P_2 (in the same order on both paths) and each of the two shortest paths is formed by following one of P_1 or P_2 until reaching the endpoint of a rung, then traversing that rung and continuing in the same way along P_2 or P_1 until the next rung, etc.

Lemma 3. *With s, P_1, and P_2 as above, let C be the cycle formed by P_1, P_2, and the root edge, and let R be the rung of P_1 and P_2 closest to s. Let D be the cycle formed by removing the parts of P_1 and P_2 from s to the endpoints of R, and replacing them with R. Then C is longer than D.*

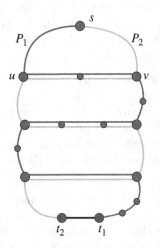

Proof. Let u and v be the endpoints of R, let ℓ_{su} denote the length of the path in P_1 from s to u, let ℓ_{sv} denote the length of the path in P_2 from s to v, and let ℓ_{uv} denote the length of the rung. P_1 follows C from s to u then crosses rung R; by construction, it is the shortest path in G from s to t_1. Thus, $\ell_{su}+\ell_{uv} \leq \ell_{sv}$, for if not then P_1 couldn't be a shortest path. Equivalently, adding $\ell_{su} - \ell_{uv}$ to both sides of the inequality gives $2\ell_{su} \leq \ell_{su} + \ell_{sv} - \ell_{uv}$. But the left hand side of this inequality is positive (by the assumption that the input graph has positive edge weights) and the right hand side is the difference in weights between C and D.   □

Fig. 5. Notation for Lemma 3. P_1 and P_2 are shown in contrasting colors; the horizontal segments with both colors are the rungs.

Corollary 1. *Let C_i be a cycle in the greedy cycle sequence, and let A_{i-1} be the ambit of the previous cycle. Suppose that C_i is constructed by applying Suurballe's algorithm from a starting vertex s, and suppose that the two disjoint paths P_1 and P_2 comprising C_i have a nonempty sequence of rungs. Then these rungs, and all parts of P_1 and P_2 from the first rung endpoint to t_1 and t_2, belong to A_{i-1}.*

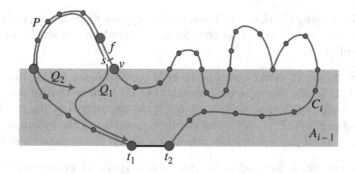

Fig. 6. Notation for Lemma 4. The figure shows a cycle C_i such that $C_i \setminus A_{i-1}$ forms more than one path, a configuration that is proven impossible by the lemma.

Proof. Otherwise the cycle D in the statement of Lemma 3, or one of the other cycles constructed in the same way from one of the other rungs, would be a shorter cycle containing at least one edge that is not in A_{i-1}, and would have been selected in place of C_i in the greedy cycle sequence.   □

4.3 From Cycle Sequences to Ear Decompositions

As we now show, a greedy cycle sequence with cycles $C_1, C_2, \ldots$ and ambits A_1, $A_2, \ldots$ may be used to derive an ear decomposition, in which the first ear P_0 is the root edge and each subsequent ear P_i is the subgraph $A_i \setminus A_{i-1}$. That is, for each i, this subgraph is a single path.

Lemma 4. *Let C_i and A_i be a greedy cycle sequence and the corresponding sequence of ambits in an unambiguously-weighted graph. Then $A_i \setminus A_{i-1}$ forms a single connected path in the given graph G.*

Proof. Let t_1 and t_2 be the endpoints of the root edge. As with any rooted cycle not contained in A_{i-1}, C_i forms one or more connected paths in $A_i \setminus A_{i-1}$, separated by vertices or edges of A_{i-1}. Let P be the closest to t_1 of these paths according to their ordering along C_i, and let f and v be the farthest edge and vertex from t_1 in the path ordering of P. Then v is an endpoint of f and of path P, and belongs to A_{i-1}. By Lemma 2 the shortest path from v to t_1 stays within A_{i-1} and therefore does not use edge f.

Let s be a point on edge f, sufficiently close to v that the shortest path from s to t_1 passes through v. (In other words, subdivide the edge, and place a vertex at this point.) If we apply Suurballe's algorithm starting from the point s, it will find two paths Q_1 (the shortest path from s to t_1) and Q_2. The symmetric difference of these two paths is C_i again, because C_i is the unique shortest rooted cycle containing s. Q_1 passes from s through v and then stays within A_{i-1} by Lemma 2. Q_2 follows the rest of P, then stays within A_{i-1} until it reaches the endpoint of the first rung of $Q_1 \cap Q_2$ (by the choice of P as the first of the paths in $A_i \setminus A_{i-1}$ in cycle order). After reaching this rung endpoint, Q_2 continues to stay within A_{i-1} by Corollary 1. Thus, neither Q_1 nor Q_2 can escape A_{i-1} once they enter it, so the cycle C_i that they form can only have the single component P outside of A_{i-1}.   □

Corollary 2. *Let C_i be the cycles of a greedy cycle sequence for the rooted graph G, and let A_i be the corresponding ambits. Let P_0 be the one-edge path formed by the root edge of G, and for $i > 0$ let P_i be the path $A_i \setminus A_{i-1}$. Then the sequence of paths $P_0, P_1, \ldots P_i$ is an ear decomposition of the subgraph A_i, and the sequence of all paths formed in this way is an ear decomposition of G.*

Proof. By Lemma 4, each of these graphs is a path; its endpoints belong to earlier paths and its edges and interior vertices do not. Thus, this sequence of paths satisfies all the requirements of an ear decomposition.   □

4.4 Greed Is Good

We have nearly completed the proof of correctness of our greedy algorithm for constructing minimum weight rooted cycle bases.

Lemma 5. *Let C_i be the cycles of a greedy cycle sequence for an unambiguously-weighted rooted graph G, and let A_i be the corresponding ambits. Then the set of cycles $C_1, C_2, \ldots C_i$ is a minimum weight rooted cycle basis for A_i.*

Proof. We use induction on i. For $i = 1$, the graph $C_1 = A_1$ has only the one cycle. For $i > 1$, this set of cycles is linearly independent because each contains at least one edge not found in earlier cycles. The number of cycles is the same as the number of ears (after the root edge) in an ear decomposition, by Corollary 2, so it equals the dimension of the cycle space. As an independent set of the correct number of cycles, these cycles must form a cycle basis for A_i.

Because the cycle space of any graph forms a matroid, the minimum weight basis of any subset S of cycles can be found by a greedy algorithm that at each step selects the minimum-weight member of S that is independent of previous selections. By the induction hypothesis, any cycle that is independent of the previous $i - 1$ selections must use at least one edge outside of A_{i-1}, and C_i is the minimum-weight rooted cycle with this property. Therefore, the cycles form a minimum-weight rooted cycle basis.   □

Theorem 2. *The minimum weight rooted cycle basis of a biconnected rooted graph G with positive edge weights can be constructed in polynomial time.*

Proof. As outlined at the beginning of this section, we use Suurballe's algorithm to order the edges of G by the lengths of their shortest cycles through the base edge. Then, using this order as a guide, we construct a greedy cycle sequence by repeatedly choosing an edge f that is not part of the already-chosen cycles and using another instance of Suurballe's algorithm to find the shortest cycle through f and the root edge, breaking ties in favor of cycles that use as few new edges as possible. By Lemma 5, the resulting set of cycles will form a minimum weight rooted cycle basis.

In a graph with n vertices and m edges, Suurballe's algorithm can be implemented in time $O(m + n \log n)$. The first stage of the algorithm may be implemented using Dijkstra's algorithm in this time bound. The second stage involves shortest paths in a graph H with negative edge weights, to which Dijkstra's algorithm does not directly apply. However, in this second stage, we may re-weight each directed edge in H from u to v with length ℓ, giving it the new weight $\ell + d(s, v) - d(s, u)$, where s is the starting vertex of the first path and d is the shortest-path distance between two vertices in the input graph. This reweighting does not modify the comparison between any two path lengths, so the shortest paths in the reweighted version of H remain unchanged. With these weights, the edges whose weights were negative become zero-weight, and all other edge weights remain non-negative, so Dijkstra's algorithm may again be applied.

A naive implementation of the algorithm applies Suurballe's algorithm $O(m)$ times so its total time is $O(m^2 + mn \log n)$. However, this can be improved by observing that there are only $O(n)$ choices for the first path in Suurballe's algorithm, and that for each first path it is possible to handle all starting vertices of the second path, simultaneously, by using Dijkstra's algorithm to perform a single-destination shortest path computation. With this improvement the total runtime is $O(mn + n^2 \log n)$. The algorithm as described so far applies only to unambiguously-weighted graphs but in the full version of this paper we describe how to reduce the general problem to this case in polynomial time.   □

Acknowledgments. The work of the first author was supported by the National Science Foundation under Grant CCF-1228639 and by the Office of Naval Research under Grant No. N00014-08-1-1015.

References

1. Liebchen, C.: Periodic timetable optimization in public transport. Operations Research Proceedings **2006**, 29–36 (2007)
2. Boulinier, C., Petit, F., Villain, V.: When graph theory helps self-stabilization. In: Proc. 23rd ACM Symp. on Principles of Distributed Computing (PODC 2004), pp. 150–159 (2004)
3. Aguiar, D., Istrail, S.: HapCompass: A fast cycle basis algorithm for accurate haplotype assembly of sequence data. J. Computational Biology **19**(6), 577–590 (2012)
4. Lemieux, S., Major, F.: Automated extraction and classification of RNA tertiary structure cyclic motifs. Nucleic Acids Research **34**(8), 2340–2346 (2006)
5. Kaveh, A.: Improved cycle bases for the flexibility analysis of structures. Comput. Methods Appl. Mech. Engrg. **9**(3), 267–272 (1976)
6. Kecskeméthy, A., Krupp, T., Hiller, M.: Symbolic processing of multiloop mechanism dynamics using closed-form kinematics solutions. Multibody System Dynamics **1**(1), 23–45 (1997)
7. Parrish, B.E., McCarty, J.M., Eppstein, D.: Automated generation of linkage loop equations for planar one degree-of-freedom linkages, demonstrated up to 8-bar. J. Mechanisms and Robotics **7**(1), 011006 (2015)
8. Lee, A., Streinu, I.: Pebble game algorithms and sparse graphs. Discrete Math. **308**(8), 1425–1437 (2008)

9. Amaldi, E., Iuliano, C., Rizzi, R.: Efficient deterministic algorithms for finding a minimum cycle basis in undirected graphs. In: Eisenbrand, F., Shepherd, F.B. (eds.) IPCO 2010. LNCS, vol. 6080, pp. 397–410. Springer, Heidelberg (2010)

10. Horton, J.D.: A polynomial-time algorithm to find the shortest cycle basis of a graph. SIAM J. Comput. 16(2), 358–366 (1987)

11. Mehlhorn, K., Michail, D.: Implementing minimum cycle basis algorithms. ACM J. Exp. Algorithmics 11 (2006)

12. Kavitha, T., Liebchen, C., Mehlhorn, K., Michail, D., Rizzi, R., Ueckerdt, T., Zweig, K.A.: Cycle bases in graphs: Characterization, algorithms, complexity, and applications. Comput. Sci. Rev. 3(4), 199–243 (2009)

13. Rizzi, R.: Minimum weakly fundamental cycle bases are hard to find. Algorithmica 53(3), 402–424 (2009)

14. Tutte, W.T.: On the 2-factors of bicubic graphs. Discrete Math. 1(2), 203–208 (1971)

15. Welsh, D.J.A.: Matroid Theory. Dover (2010)

16. Whitney, H.: Non-separable and planar graphs. Trans. Amer. Math. Soc. 34(2), 339–362 (1932)

17. Lovász, L.: Computing ears and branchings in parallel. In: Proc. 26th Symp. Foundations of Computer Science (FOCS 1985), pp. 464–467 (1985)

18. Tarjan, R.: Depth-first search and linear graph algorithms. SIAM J. Comput. 1(2), 146–160 (1972)

19. Ramachandran, V.: Parallel open ear decomposition with applications to graph biconnectivity and triconnectivity. In: Reif, J.H. (ed.) Synthesis of Parallel Algorithms. Morgan Kaufmann, pp. 276–340 (1993)

20. Elkin, M., Liebchen, C., Rizzi, R.: New length bounds for cycle bases. Inform. Process. Lett. 104(5), 186–193 (2007)

On the Chain Pair Simplification Problem

Chenglin Fan[1], Omrit Filtser[2], Matthew J. Katz[2], Tim Wylie[3],
and Binhai Zhu[1(✉)]

[1] Montana State University, Bozeman, MT 59717-3880, USA
chenglin.fan@msu.montana.edu, bhz@cs.montana.edu
[2] Ben-Gurion University of the Negev, 84105 Beer-Sheva, Israel
{omritna,matya}@cs.bgu.ac.il
[3] The University of Texas-Pan American, Edinburg, TX 78539, USA
wylietr@utpa.edu

Abstract. The problem of efficiently computing and visualizing the
structural resemblance between a pair of protein backbones in 3D has
led Bereg et al. [4] to pose the Chain Pair Simplification problem (CPS).
In this problem, given two polygonal chains A and B of lengths m and n,
respectively, one needs to simplify them simultaneously, such that each
of the resulting simplified chains, A' and B', is of length at most k and
the discrete Fréchet distance between A' and B' is at most δ, where
k and δ are given parameters. In this paper we study the complexity of
CPS under the discrete Fréchet distance (CPS-3F), i.e., where the quality
of the simplifications is also measured by the discrete Fréchet distance.
Since CPS-3F was posed in 2008, its complexity has remained open. In
this paper, we prove that CPS-3F is actually polynomially solvable, by
presenting an $O(m^2 n^2 \min\{m,n\})$ time algorithm for the corresponding
minimization problem. On the other hand, we prove that if the vertices of
the chains have integral weights then the problem is weakly NP-complete.

1 Introduction

Polygonal curves play an important role in many applied areas, such as 3D modeling in computer vision, map matching in GIS, and protein backbone structural
alignment and comparison in computational biology. Many different methods
exist to compare curves in these (and in many other) applications, where one of
the more prevalent methods is the Fréchet distance [8].

The *Fréchet distance* is often described by an analogy of a man and a dog
connected by a leash, each walking along a curve from its starting point to
its end point. Both the man and the dog can control their speed but they are
not allowed to backtrack. The Fréchet distance between the two curves is the

A complete version including the one-sided cases and empirical results can be found
at http://arxiv.org/abs/1409.2457.
Work by O. Filtser has been partially supported by the Lynn and William Frankel
Center for Computer Sciences. Work by M. Katz and O. Filtser has been partially
supported by grant 1045/10 from the Israel Science Foundation.

© Springer International Publishing Switzerland 2015
F. Dehne et al. (Eds.): WADS 2015, LNCS 9214, pp. 351–362, 2015.
DOI: 10.1007/978-3-319-21840-3_29

minimum length of a leash that is sufficient for traversing both curves in this manner.

The *discrete Fréchet distance* is a simpler version, where, instead of continuous curves, we are given finite sequences of points, obtained, e.g., by sampling the continuous curves, or corresponding to the vertices of polygonal chains. Now, the man and the dog only hop monotonically along the sequences of points. The discrete Fréchet distance is considered a good approximation of the continuous distance.

One promising application of the discrete Fréchet distance has been protein backbone comparison. Within structural biology, polygonal curve alignment and comparison is a central problem in relation to proteins. Proteins are usually studied using RMSD (Root Mean Square Deviation), but recently the discrete Fréchet distance was used to align and compare protein backbones, which yielded favourable results in many instances [9,10]. In this application, the discrete version of the Fréchet distance makes more sense, because by using it the alignment is done with respect to the vertices of the chains, which represent α-carbon atoms. Applying the continuous Fréchet distance will result in mapping of arbitrary points, which is not meaningful biologically.

There may be as many as 500~600 α-carbon atoms along a protein backbone, which are the nodes (i.e., points) of our chain. This makes efficient computation essential, and is one of the reasons for considering simplification. In general, given a chain A of n vertices, a simplification of A is a chain A' such that A' is "close" to A and the number of vertices in A' is significantly less than n. The problem of simplifying a 3D polygonal chains under the discrete Fréchet distance was first addressed by Bereg et al. [4].

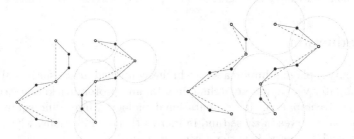

(a) Simplifying the chains independently does not necessarily preserve the resemblance between them.

(b) A simplification of both chains that preserves the resemblance between them.

Fig. 1. Independent simplification vs. simultaneous simplification. Each chain simplification consists of 4 vertices (marked by empty circles) chosen from the corresponding chain. The unit disks illustrate the Fréchet distance between the right chain in each of the figures and its corresponding simplification; their radius in (b) is larger

Simplifying two aligned chains independently does not necessarily preserve the resemblance between the chains; see Figure 1. Thus, the following question arises: Is it possible to simplify both chains in a way that will retain the resemblance between them? This question has led Bereg et al. [4] to pose the Chain Pair Simplification problem (CPS). In this problem, the goal is to simplify both chains simultaneously, so that the discrete Fréchet distance between the resulting simplifications is bounded. More precisely, given two chains A and B of lengths m and n, respectively, an integer k and three real numbers $\delta_1, \delta_2, \delta_3$, one needs to find two chains A', B' with vertices from A, B, respectively, each of length at most k, such that $d_1(A, A') \leq \delta_1$, $d_2(B, B') \leq \delta_2$, $d_{dF}(A', B') \leq \delta_3$ (d_1 and d_2 can be any similarity measures and d_{dF} is the discrete Fréchet distance). When the chains are simplified using the Hausdorff distance, i.e., d_1, d_2 is the Hausdorff distance (CPS-2H), the problem becomes **NP**-complete [4]. However, the complexity of the version in which d_1, d_2 is the discrete Fréchet distance (CPS-3F) has been open since 2008.

Related Work. The Fréchet distance and its variants have been studied extensively in the past two decades. Alt and Godau [2] gave an $O(mn \log mn)$-time algorithm for computing the Fréchet distance between two polygonal curves of lengths m and n. This result in the plane was recently improved by Buchin et al [5]. The discrete Fréchet distance was originally defined by Eiter and Mannila [7], who also presented an $O(mn)$-time algorithm for computing it. A slightly sub-quadratic algorithm was given recently by Agarwal et al. [1].

As mentioned earlier, Bereg et al. [4] were the first to study simplification problems under the discrete Fréchet distance. They considered two such problems. In the first, the goal is to minimize the number of vertices in the simplification, given a bound on the distance between the original chain and its simplification, and, in the second problem, the goal is to minimize this distance, given a bound k on the number of vertices in the simplification. They presented an $O(n^2)$-time algorithm for the former problem and an $O(n^3)$-time algorithm for the latter problem, both using dynamic programming, for the case where the vertices of the simplification are from the original chain. (For the arbitrary vertices case, they solve the problems in $O(n \log n)$ time and in $O(kn \log n \log(n/k))$ time, respectively.) Driemel and Har-Peled [6] showed how to preprocess a polygonal curve in near-linear time and space, such that, given an integer $k > 0$, one can compute a simplification in $O(k)$ time which has $2k - 1$ vertices of the original curve and is optimal up to a constant factor (w.r.t. the continuous Fréchet distance), compared to any curve consisting of k arbitrary vertices.

For the chain pair simplification problem (CPS), Bereg et al. [4] proved that CPS-2H is **NP**-complete, and conjectured that so is CPS-3F. Wylie et al. [10] gave a heuristic algorithm for CPS-3F, using a greedy method with backtracking, and based on the assumption that the (Euclidean) distance between adjacent α-carbon atoms in a protein backbone is almost fixed. More recently, Wylie and Zhu [11] presented an approximation algorithm with approximation ratio 2 for the optimization version of CPS-3F. Their algorithm actually solves the optimization version of a related problem called CPS-$3F^+$, it uses dynamic programming and its running time is between $O(mn)$ and $O(m^2n^2)$ depending on the input simplification parameters.

Some special cases of CPS-3F have recently been studied. Motivated by the need to reduce sensitivity to outliers when comparing curves, Ben Avraham et al. [3] studied the discrete Fréchet distance with shortcuts problem. Both variants of the shortcuts problem can be solved in subquadratic time.

Our Results. In Section 3, we resolve the question concerning the complexity of CPS-3F by proving that it is polynomially solvable, contrary to what was believed. We do this by presenting a polynomial-time algorithm for the corresponding optimization problem. In Section 4 we devise a sophisticated $O(m^2n^2 \min\{m,n\})$-time dynamic programming algorithm for the minimization problem of CPS-3F. Besides being interesting from a theoretical point of view, only after developing (and implementing) this algorithm, were we able to apply the CPS-3F minimization problem to datasets from the Protein Data Bank (PDB), see the full version for the actual empirical results. Finally, in Section 5 we prove that the problem is weakly NP-complete if the vertices of the chains carry integral weights.

2 Preliminaries

Let $A = (a_1 \ldots, a_m)$ and $B = (b_1, \ldots, b_n)$ be two sequences of m and n points, respectively, in $\mathbb{R}^k$. The discrete Fréchet distance $d_{dF}(A, B)$ between A and B is defined as follows. Fix a distance $\delta > 0$ and consider the Cartesian product $A \times B$ as the vertex set of a directed graph G_δ whose edge set is

$$
\begin{aligned}
E_\delta = &\{((a_i, b_j), (a_{i+1}, b_j)) \mid d(a_i, b_j), d(a_{i+1}, b_j) \le \delta\} \cup \\
&\{((a_i, b_j), (a_i, b_{j+1})) \mid d(a_i, b_j), d(a_i, b_{j+1}) \le \delta\} \cup \\
&\{((a_i, b_j), (a_{i+1}, b_{j+1})) \mid d(a_i, b_j), d(a_{i+1}, b_{j+1}) \le \delta\}.
\end{aligned}
$$

Then $d_{dF}(A, B)$ is the smallest $\delta > 0$ for which (a_m, b_n) is reachable from (a_1, b_1) in G_δ.

The chain pair simplification problem (CPS) is formally defined as follows.

Problem 1 (Chain Pair Simplification).
Instance: Given a pair of polygonal chains A and B of lengths m and n, respectively, an integer k, and three real numbers $\delta_1, \delta_2, \delta_3 > 0$.
Problem: Does there exist a pair of chains A', B' each of at most k vertices, such that the vertices of A', B' are from A, B, respectively, and $d_1(A, A') \le \delta_1$, $d_2(B, B') \le \delta_2$, and $d_{dF}(A', B') \le \delta_3$?

When $d_1 = d_2 = d_H$, the problem is **NP**-complete and is called CPS-2H, and when $d_1 = d_2 = d_{dF}$, the problem is called CPS-3F.

3 Chain Pair Simplification (CPS-3F)

We now turn our attention to CPS-3F, which we show to be polynomially solvable in this section. We comment that the running time and space for this solution is $O(m^3 n^3 \min\{m, n\})$ and $O(m^3 n^3)$ respectively, hence this solution is impractical for most of the real protein chains (with m, n as large as 500-600). Nonetheless, this first solution is easier to understand and can be considered as a warm-up. We will present a much better (but more sophisticated) solution in the next section.

We present an algorithm for the minimization version of CPS-3F. That is, we compute the minimum integer k^*, such that there exists a "walk", as above, in which each of the dogs makes at most k^* hops. The answer to the decision problem is "yes" if and only if $k^* < k$.

Returning to the analogy of the man and the dog, we can extend it as follows. Consider a man and his dog connected by a leash of length δ_1, and a woman and her dog connected by a leash of length δ_2. The two dogs are also connected to each other by a leash of length δ_3. The man and his dog are walking on the points of a chain A and the woman and her dog are walking on the points of a chain B. The dogs may skip points. The problem is to determine whether there exists a "walk" of the man and his dog on A and the woman and her dog on B, such that each of the dogs steps on at most k points.

Overview of the Algorithm. We say that (a_i, a_p, b_j, b_q) is a *possible* configuration of the man, woman and the two dogs on the paths A and B, if $d(a_i, a_p) \leq \delta_1$, $d(b_j, b_q) \leq \delta_2$ and $d(a_p, b_q) \leq \delta_3$. Notice that there are at most $m^2 n^2$ such configurations. Now, let G be the DAG whose vertices are the possible configurations, such that there exists a (directed) edge from vertex $u = (a_i, a_p, b_j, b_q)$ to vertex $v = (a_{i'}, a_{p'}, b_{j'}, b_{q'})$ if and only if our gang can move from configuration u to configuration v. That is, if and only if $i \leq i' \leq i + 1$, $p \leq p'$, $j \leq j' \leq j + 1$, and $q \leq q'$. Notice that there are no cycles in G because backtracking is forbidden. For simplicity, we assume that the first and last points of A' (resp., of B') are a_1 and a_m (resp., b_1 and b_n), so the initial and final configurations are $s = (a_1, a_1, b_1, b_1)$ and $t = (a_m, a_m, b_n, b_n)$, respectively. (It is easy, however, to adapt the algorithm below to the case where the initial and final points of A' and B' are not specified, see remark below.) Our goal is to find a path from s to t in G. However, we want each of our dogs to step on at most k points, so, instead of searching for any path from s to t, we search for a path that minimizes the value $max\{|A'|, |B'|\}$, and then check if this value is at most k.

For each edge $e = (u, v)$, we assign two weights, $w_A(e), w_B(e) \in \{0, 1\}$, in order to compute the number of hops in A' and in B', respectively. $w_A(u, v) = 1$ if and only if the first dog jumps to a new point between configurations u and v (i.e., $p < p'$), and, similarly, $w_B(u, v) = 1$ if and only if the second dog jumps to a new point between u and v (i.e., $q < q'$). Thus, our goal is to find a path P from s to t in G, such that $max\{\sum_{e \in P} w_A(e), \sum_{e \in P} w_B(e)\}$ is minimized.

Assume w.l.o.g. that $m \leq n$. Since $|A'| \leq m$ and $|B'| \leq n$, we maintain, for each vertex v of G, an array $X(v)$ of size m, where $X(v)[r]$ is the minimum number z such that v can be reached from s with (at most) r hops of the first dog and z hops of the second dog. We can construct these arrays by processing the vertices of G in topological order (i.e., a vertex is processed only after all its predecessors have been processed). This yields an algorithm of running time $O(m^3 n^3 \min\{m, n\})$, as described in Algorithm 1.

Algorithm 1. CPS-3F

1. Create a directed graph $G = (V, E)$ with two weight functions w_A, w_B, such that:
   - V is the set of all configurations (a_i, a_p, b_j, b_q) with $d(a_i, a_p) \leq \delta_1$, $d(b_j, b_q) \leq \delta_2$, and $d(a_p, b_q) \leq \delta_3$.
   - $E = \{((a_i, a_p, b_j, b_q), (a_{i'}, a_{p'}, b_{j'}, b_{q'})) \mid i \leq i' \leq i+1, p \leq p', j \leq j' \leq j+1, q \leq q'\}$.
   - For each $((a_i, a_p, b_j, b_q), (a_{i'}, a_{p'}, b_{j'}, b_{q'})) \in E$, set
     - $w_A((a_i, a_p, b_j, b_q), (a_{i'}, a_{p'}, b_{j'}, b_{q'})) = \begin{cases} 1, & p < p' \\ 0, & \text{otherwise} \end{cases}$
     - $w_B((a_i, a_p, b_j, b_q), (a_{i'}, a_{p'}, b_{j'}, b_{q'})) = \begin{cases} 1, & q < q' \\ 0, & \text{otherwise} \end{cases}$
2. Sort V topologically.
3. Initialize the array $X(s)$ (i.e., set $X(s)[r] = 0$, for $r = 0, \ldots, m-1$).
4. For each $v \in V \setminus \{s\}$ (advancing from left to right in the sorted sequence) do:
   (a) Initialize the array $X(v)$ (i.e., set $X(v)[r] = \infty$, for $r = 0, \ldots, m-1$).
   (b) For each r between 0 and $m-1$, compute $X(v)[r]$:
   $$X(v)[r] = \min_{(u, v) \in E} \begin{cases} X(u)[r] + w_B(u, v), & w_A(u, v) = 0 \\ X(u)[r-1] + w_B(u, v), & w_A(u, v) = 1 \end{cases}$$
5. Return $k^* = \min_r \max\{r, X(t)[r]\}$.

Running Time. The number of vertices in G is $|V| = O(m^2 n^2)$. By the construction of the graph, for any vertex (a_i, a_p, b_j, b_q) the maximum number of outgoing edges is $O(mn)$. So we have $|E| = O(|V|mn) = O(m^3 n^3)$. Thus, constructing the graph G in Step 1 takes $O(n^3 m^3)$ time. Step 2 takes $O(|E|)$ time, while Step 3 takes $O(m)$ time. In Step 4, for each vertex v and for each index r, we consider all configurations that can directly precede v. So each edge of G participates in exactly m minimum computations, implying that Step 4 takes $O(|E|m)$ time. Step 5 takes $O(m)$ time. Thus, the total running time of the algorithm is $O(m^4 n^3)$.

Theorem 1. *The chain pair simplification problem under the discrete Fréchet distance (CPS-3F) is polynomial, i.e., CPS-3F $\in$ **P**.*

Remark 1. As mentioned, we have assumed that the first and last points of A' (resp., B') are a_1 and a_m (resp., b_1 and b_n), so we have a single initial configuration (i.e., $s = (a_1, a_1, b_1, b_1)$) and a single final configuration (i.e., $t = (a_m, a_m, b_n, b_n)$). However, it is easy to adapt our algorithm to the case where the first and last points of the chains A' and B' are not specified. In this case, any possible configuration of the form (a_1, a_p, b_1, b_q) is considered a potential initial configuration, and any possible configuration of the form (a_m, a_p, b_n, b_q) is considered a potential final configuration, where $1 \leq p \leq m$ and $1 \leq q \leq n$. Let S and T be the sets of potential initial and final configurations, respectively. (Then, $|S| = O(mn)$ and $|T| = O(mn)$.) We thus remove from G all edges entering a potential initial configuration, so that each such configuration becomes a "root" in the (topologically) sorted sequence. Now, in Step 3 we initialize the arrays of each $s \in S$ in total time $O(m^2n)$, and in Step 4 we only process the vertices that are not in S. The value $X(v)[r]$ for such a vertex v is now the minimum number z such that v can be reached from s with r hops of the first dog and z hops of the second dog, over *all* potential initial configurations $s \in S$. In the final step of the algorithm, we calculate the value k^* in $O(m)$ time, for each potential final configuration $t \in T$. The smallest value obtained is then the desired value. Since the number of potential final configurations is only $O(mn)$, the total running time of the final step of the algorithm is only $O(m^2n)$, and the running time of the entire algorithm remains $O(m^4n^3)$.

4   An Efficient Implementation

The time and space complexity of Algorithm 1 (which is $O(m^3n^3 \min\{m,n\})$ and $O(m^3n^3)$, respectively) makes it impractical for our motivating biological application (as m, n could be 500~600). In fact, when m, n are around 200 we already had memory overflows in the implemented Algorithm 1. In this section, we show how to reduce the time and space bounds by a factor of mn, using dynamic programming.

We generate all configurations of the form (a_i, a_p, b_j, b_q), where the outermost for-loop is governed by i, the next level loop by j, then p, and finally q. When a new configuration $v = (a_i, a_p, b_j, b_q)$ is generated, we first check whether it is *possible*. If it is not possible, we set $X(v)[r] = \infty$, for $1 \leq r \leq m$, and if it is, we compute $X(v)[r]$, for $1 \leq r \leq m$.

We also maintain for each pair of indices i and j, three tables $C_{i,j}$, $R_{i,j}$, $T_{i,j}$ that assist us in the computation of the values $X(v)[r]$:

$$C_{i,j}[p,q,r] = \min_{1 \leq p' \leq p} X(a_i, a_{p'}, b_j, b_q)[r]$$

$$R_{i,j}[p,q,r] = \min_{1 \leq q' \leq q} X(a_i, a_p, b_j, b_{q'})[r]$$

$$T_{i,j}[p,q,r] = \min_{\substack{1 \leq p' \leq p \\ 1 \leq q' \leq q}} X(a_i, a_{p'}, b_j, b_{q'})[r]$$

Notice that the value of cell $[p, q, r]$ is determined by the value of one or two previously-determined cells and $X(a_i, a_p, b_j, b_q)[r]$ as follows:

$$C_{i,j}[p,q,r] = \min\{C_{i,j}[p-1,q,r], X(a_i, a_p, b_j, b_q)[r]\}$$
$$R_{i,j}[p,q,r] = \min\{R_{i,j}[p,q-1,r], X(a_i, a_p, b_j, b_q)[r]\}$$
$$T_{i,j}[p,q,r] = \min\{T_{i,j}[p-1,q,r], T_{i,j}[p,q-1,r], X(a_i, a_p, b_j, b_q)[r]\}$$

Observe that in any configuration that can immediately precede the current configuration (a_i, a_p, b_j, b_q), the man is either at a_{i-1} or at a_i and the woman is either at b_{j-1} or at b_j (and the dogs are at $a_{p'}$, $p' \leq p$, and $b_{q'}, q' \leq q$, respectively). The "saving" is achieved, since now we only need to access a constant number of table entries in order to compute the value $X(a_i, a_p, b_j, b_q)[r]$.

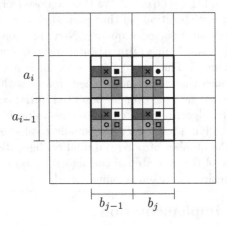

Fig. 2. Illustration of Algorithm 2

One can illustrate the algorithm using the matrix in Figure 2. There are mn large cells, each of them containing a matrix of size mn. The large cells correspond to the positions of the man and the woman. The inner matrices correspond to the positions of the two dogs (for given positions of the man and woman). Consider an optimal "walk" of the gang that ends at cell (a_i, a_p, b_j, b_q) (marked by a full circle), such that the first dog has visited r points. The previous cell in this "walk" must be in one of the 4 large cells $(a_i, b_j), (a_{i-1}, b_j), (a_i, b_{j-1}), (a_{i-1}, b_{j-1})$. Assume, for example, that it is in (a_{i-1}, b_j). Then, if it is in the blue area, then $X(a_i, a_p, b_j, b_q)[r] = C_{i-1,j}[p-1, q, r-1]$ (marked by an empty square), since only the position of the first dog has changed when the gang moved to (a_i, a_p, b_j, b_q). If it is in the purple area, then $X(a_i, a_p, b_j, b_q)[r] = R_{i-1,j}[p, q-1, r]+1$ (marked by a x), since only the position of the second dog has changed. If it is in the orange area, then $X(a_i, a_p, b_j, b_q)[r] = T_{i-1,j}[p-1, q-1, r-1]+1$ (marked by an empty circle), since the positions of both dogs have changed. Finally, if it is the cell marked by the full square, then simply $X(a_i, a_p, b_j, b_q)[r] = X(a_{i-1}, a_p, b_j, b_q)[r]$, since both dogs have not moved. The other three cases, in which the previous cell is in one of the 3 large cells $(a_i, b_j), (a_i, b_{j-1}), (a_{i-1}, b_{j-1})$, are handled similarly.

We are ready to present the dynamic programming algorithm. The initial configurations correspond to cells in the large cell (a_1, b_1). For each initial configuration (a_1, a_p, b_1, b_q), we set $X(a_1, a_p, b_1, b_q)[1] = 1$.

Theorem 2. *The minimization version of the chain pair simplification problem under the discrete Fréchet distance (CPS-3F) can be solved in $O(m^2 n^2 \min\{m, n\})$ time.*

Algorithm 2. CPS-3F using dynamic programming

---

for $i = 1$ to m
   for $j = 1$ to n
      for $p = 1$ to m
         for $q = 1$ to n
            for $r = 1$ to m

$$X_{(-1,0)} = \min \begin{cases} C_{i-1,j}[p-1,q,r-1] \\ R_{i-1,j}[p,q-1,r]+1 \\ T_{i-1,j}[p-1,q-1,r-1]+1 \\ X(a_{i-1},a_p,b_j,b_q)[r] \end{cases}$$

$$X_{(0,-1)} = \min \begin{cases} C_{i,j-1}[p-1,q,r-1] \\ R_{i,j-1}[p,q-1,r]+1 \\ T_{i,j-1}[p-1,q-1,r-1]+1 \\ X(a_i,a_p,b_{j-1},b_q)[r] \end{cases}$$

$$X_{(-1,-1)} = \min \begin{cases} C_{i-1,j-1}[p-1,q,r-1] \\ R_{i-1,j-1}[p,q-1,r]+1 \\ T_{i-1,j-1}[p-1,q-1,r-1]+1 \\ X(a_{i-1},a_p,b_{j-1},b_q)[r] \end{cases}$$

$$X_{(0,0)} = \min \begin{cases} C_{i,j}[p-1,q,r-1] \\ R_{i,j}[p,q-1,r]+1 \\ T_{i,j}[p-1,q-1,r-1]+1 \end{cases}$$

$$X(a_i,a_p,b_j,b_q)[r] = \min\{X_{(-1,0)}, X_{(0,-1)}, X_{(-1,-1)}, X_{(0,0)}\}$$

$$C_{i,j}[p,q,r] = \min\{C_{i,j}[p-1,q,r], X(a_i,a_p,b_j,b_q)[r]\}$$
$$R_{i,j}[p,q,r] = \min\{R_{i,j}[p,q-1,r], X(a_i,a_p,b_j,b_q)[r]\}$$
$$T_{i,j}[p,q,r] = \min\{T_{i,j}[p-1,q,r], T_{i,j}[p,q-1,r], X(a_i,a_p,b_j,b_q)[r]\}$$

return $\min_{r,p,q} \max\{r, X(a_m,a_p,b_n,b_q)[r]\}$

We comment that this algorithm has been implemented with C++ and tested with real datasets from the PDB. Compared with Algorithm FIND-CPS3F$^+$, i.e., the algorithm (mentioned in the introduction) for the optimization version of CPS-3F$^+$, proposed by Wylie and Zhu [11], the improvement is huge. Due to space constraints, we leave the empirical results out and interested readers are referred to the full version for the details.

5 Weighted Chain Pair Simplification

In this section, we consider a more general version of CPS-3F, namely, **Weighted CPS-3F**. In the weighted version of the chain pair simplification problem, the vertices of the chains A and B are assigned arbitrary weights, and, instead of limiting the length of the simplifications, one limits their weights. That is, the total weight of each simplification must not exceed a given value. The problem is formally defined as follows.

Problem 2 (Weighted Chain Pair Simplification).
Instance: Given a pair of 3D chains A and B, with lengths m and n, respectively, an integer k, three real numbers $\delta_1, \delta_2, \delta_3 > 0$, and a weight function $C : \{a_1, \ldots, a_m, b_1, \ldots, b_n\} \to \mathbb{R}^+$.
Problem: Does there exist a pair of chains A',B' with $C(A'), C(B') \leq k$, such that the vertices of A',B' are from A, B respectively, $d_1(A, A') \leq \delta_1$, $d_2(B, B') \leq \delta_2$, and $d_{dF}(A', B') \leq \delta_3$?

When $d_1 = d_2 = d_{dF}$, the problem is called WCPS-3F. When $d_1 = d_2 = d_H$, the problem is **NP**-complete, since the non-weighted version (i.e., CPS-2H) is already **NP**-complete [4].

We prove that WCPS-3F is weakly **NP**-complete via a reduction from the *set partition* problem: Given a set of positive integers $S = \{s_1, \ldots, s_n\}$, find two sets $P_1, P_2 \subset S$ such that $P_1 \cap P_2 = \emptyset$, $P_1 \cup P_2 = S$, and the sum of the numbers in P_1 equals the sum of the numbers in P_2. This is a weakly **NP**-complete special case of the classic subset-sum problem.

Our reduction builds two curves with weights reflecting the values in S. We think of the two curves as the subsets of the partition of S. Although our problem requires positive weights, we also allow zero weights in our reduction for clarity. Later, we show how to remove these weights by slightly modifying the construction.

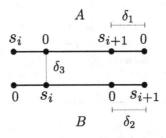

Fig. 3. The reduction for the weighted chain pair simplification problem under the discrete Fréchet distance

Theorem 3. *The weighted chain pair simplification problem under the discrete Fréchet distance is weakly **NP**-complete.*

Proof. Given the set of positive integers $S = \{s_1, \ldots, s_n\}$, we construct two curves A and B in the plane, each of length $2n$. We denote the weight of a vertex x_i by $w(x_i)$. A is constructed as follows. The i'th odd vertex of A has weight s_i, i.e. $w(a_{2i-1}) = s_i$, and coordinates $a_{2i-1} = (i, 1)$. The i'th even vertex of A has coordinates $a_{2i} = (i + 0.2, 1)$ and weight zero. Similarly, the i'th odd vertex of B has weight zero and coordinates $b_{2i-1} = (i, 0)$, and the i'th even vertex of B has coordinates $b_{2i} = (i + 0.2, 0)$ and weight s_i, i.e. $w(b_{2i}) = s_i$. Figure 3 depicts the vertices $a_{2i-1}, a_{2i}, a_{2(i+1)-1}, a_{2(i+1)}$ of A and $b_{2i-1}, b_{2i}, b_{2(i+1)-1}, b_{2(i+1)}$ of B. Finally, we set $\delta_1 = \delta_2 = 0.2$, $\delta_3 = 1$, and $k = \mathfrak{S}$, where $\mathfrak{S}$ denotes the sum of the elements of S (i.e., $\mathfrak{S} = \sum_{j=1}^n s_j$).

We claim that S can be partitioned into two subsets, each of sum $\mathfrak{S}/2$, if and only if A and B can be simplified with the constraints $\delta_1 = \delta_2 = 0.2$, $\delta_3 = 1$ and $k = \mathfrak{S}/2$, i.e., $C(A'), C(B') \leq \mathfrak{S}/2$.

First, assume that S can be partitioned into sets S_A and S_B, such that $\sum_{s \in S_A} s = \sum_{s \in S_B} s = \mathfrak{S}/2$. We construct simplifications of A and of B as follows.

$$A' = \{a_{2i-1} \mid s_i \in S_A\} \cup \{a_{2i} | s_i \notin S_A\} \text{ and } B' = \{b_{2i} \mid s_i \in S_B\} \cup \{b_{2i-1} | s_i \notin S_B\}.$$

It is easy to see that $C(A'), C(B') \leq \mathfrak{S}/2$. Also, since $\{S_A, S_B\}$ is a partition of S, exactly one of the following holds, for any $1 \leq i \leq n$:

1. $a_{2i-1} \in A', b_{2i-1} \in B'$ and $a_{2i} \notin A', b_{2i} \notin B'$.
2. $a_{2i-1} \notin A', b_{2i-1} \notin B'$ and $a_{2i} \in A', b_{2i} \in B'$.

This implies that $d_{dF}(A, A') \leq 0.2 = \delta_1$, $d_{dF}(B, B') \leq 0.2 = \delta_2$ and $d_{dF}(A', B') \leq 1 = \delta_3$.

Now, assume there exist simplifications A', B' of A, B, such that $d_{dF}(A, A') \leq \delta_1 = 0.2$, $d_{dF}(B, B') \leq \delta_2 = 0.2$, $d_{dF}(A', B') \leq \delta_3 = 1$, and $C(A'), C(B') \leq k = \mathfrak{S}/2$. Since $\delta_1 = \delta_2 = 0.2$, for any $1 \leq i \leq n$, the simplification A' must contain one of a_{2i-1}, a_{2i}, and the simplification B' must contain one of b_{2i-1}, b_{2i}. Since $\delta_3 = 1$, for any i, at least one of the following two conditions holds: $a_{2i-1} \in A'$ and $b_{2i-1} \in B'$ or $a_{2i} \in A'$ and $b_{2i} \in B'$. Therefore, for any i, either $a_{2i-1} \in A$ or $b_{2i} \in B$, implying that s_i participates in either $C(A')$ or $C(B')$. However, since $C(A'), C(B') \leq \mathfrak{S}/2$, s_i cannot participate in both $C(A')$ and $C(B')$. It follows that $C(A') = C(B') = \mathfrak{S}/2$, and we get a partition of S into two sets, each of sum $\mathfrak{S}/2$.

Finally, we note that WCPS-3F is in **NP**. For an instance I with chains A, B, given simplifications A', B', we can verify in polynomial time that $d_{dF}(A, A') \leq \delta_1$, $d_{dF}(B, B') \leq \delta_2$, $d_{dF}(A', B') \leq \delta_3$, and $C(A'), C(B') \leq k$. $\square$

Although our construction of A' and B' uses zero weights, a simple modification enables us to prove that the problem is weakly **NP**-complete also when only positive integral weights are allowed. Increase all the weights by 1, that is, $w(a_{2i-1}) = w(b_{2i}) = s_i + 1$ and $w(a_{2i}) = w(b_{2i-1}) = 1$, for $1 \leq i \leq n$, and set $k = \mathfrak{S}/2 + n$. It is easy to verify that our reduction still works. Finally, notice that we could overlay the two curves choosing $\delta_3 = 0$ and prove that the problem is still weakly **NP**-complete in one dimension.

6   Concluding Remarks

In this paper we showed that CPS-3F, which has been an open problem since 2008, is polynomially solvable. We also proved that the weighted version of the problem is weakly NP-complete. In the full version, we include a summary of empirical results that show that Algorithm 2 can handle real datasets, while the $O(m^3 n^3)$ space requirement of Algorithm 1 causes memory overflow for most pairs of protein backbones. Still, it would be interesting and desirable to further reduce the running time of CPS-3F, as some cases take 20 hours to compute.

References

1. Agarwal, P.K., Avraham, R.B., Kaplan, H., Sharir, M.: Computing the discrete Fréchet distance in subquadratic time. SIAM J. Comput. **43**(2), 429–449 (2014)
2. Alt, H., Godau, M.: Computing the Fréchet distance between two polygonal curves. Internat. J. Comput. Geometry Appl. **5**, 75–91 (1995)
3. Avraham, R.B., Filtser, O., Kaplan, H., Katz, M.J., Sharir, M.: The discrete Fréchet distance with shortcuts via approximate distance counting and selection. In: Proc. 30th Annual ACM Sympos. on Computational Geometry, SOCG 2014, p. 377 (2014)
4. Bereg, S., Jiang, M., Wang, W., Yang, B., Zhu, B.: Simplifying 3d polygonal chains under the discrete Fréchet distance. In: Laber, E.S., Bornstein, C., Nogueira, L.T., Faria, L. (eds.) LATIN 2008. LNCS, vol. 4957, pp. 630–641. Springer, Heidelberg (2008)
5. Buchin, K., Buchin, M., Meulemans, W., Mulzer, W.: Four soviets walk the dog – with an application to alt's conjecture. In: Proc. 25th Annual ACM-SIAM Sympos. on Discrete Algorithms, SODA 2014, pp. 1399–1413 (2014)
6. Driemel, A., Har-Peled, S.: Jaywalking your dog: Computing the Fréchet distance with shortcuts. SIAM J. Comput. **42**(5), 1830–1866 (2013)
7. Eiter, T., Mannila, H.: Computing discrete Fréchet distance. Technical Report CD-TR 94/64, Information Systems Dept., Technical University of Vienna (1994)
8. Fréchet, M.: Sur quelques points du calcul fonctionnel. Rendiconti del Circolo Matematico di Palermo **22**(1), 1–72 (1906)
9. Jiang, M., Xu, Y., Zhu, B.: Protein structure-structure alignment with discrete Fréchet distance. J. Bioinformatics and Computational Biology **6**(1), 51–64 (2008)
10. Wylie, T., Luo, J., Zhu, B.: A Practical solution for aligning and simplifying pairs of protein backbones under the discrete Fréchet distance. In: Murgante, B., Gervasi, O., Iglesias, A., Taniar, D., Apduhan, B.O. (eds.) ICCSA 2011, Part III. LNCS, vol. 6784, pp. 74–83. Springer, Heidelberg (2011)
11. Wylie, T., Zhu, B.: Protein chain pair simplification under the discrete Fréchet distance. IEEE/ACM Trans. Comput. Biology Bioinform. **10**(6), 1372–1383 (2013)

Finding Articulation Points of Large Graphs in Linear Time

Martín Farach-Colton[1], Tsan-sheng Hsu[2], Meng Li[1],
and Meng-Tsung Tsai[1 (✉)]

[1] Rutgers University, New Brunswick, NJ 08901, USA
{farach,ml910,mtsung.tsai}@cs.rutgers.edu
[2] Academia Sinica, Taipei 115, Taiwan
tshsu@iis.sinica.edu.tw

Abstract. Given an n-node m-edge graph G, the articulation points of graph G can be found in $\mathcal{O}(m + n)$ time in the RAM model, through a DFS-based algorithm. In the semi-streaming model for large graphs, where memory is limited to $\mathcal{O}(n \, \mathrm{polylog}\, n)$ and edges may only be accessed in one or more sequential passes, no efficient DFS algorithm is known, so another approach is needed.

We show that the articulation points can be found in $\mathcal{O}(m + n)$ time using $\mathcal{O}(n)$ space and one sequential pass of the graph. The previous best algorithm in the semi-streaming model also uses $\mathcal{O}(n)$ space and one pass, but has running time $\mathcal{O}(m\alpha(n) + n \log n)$, where α denotes the inverse of Ackermann function.

Keywords: Articulation points · Semi-streaming algorithm · Linear-time algorithm · Space lower bound

1 Introduction

An **articulation point** is a node whose removal increases the number of connected components of a graph. There are efficient algorithms in various models for finding all articulation points in an n-node m-edge graph G. For example, in the RAM model, Hopcroft and Tarjan [10] give a DFS-based algorithm that runs in $\mathcal{O}(m + n)$ time.

This classical algorithm does not scale to graphs that are larger than memory. We consider algorithms in the semi-streaming model [11–13], in which we are allowed $\mathcal{O}(n \, \mathrm{polylog}\, n)$ working space and edges may be accessed in sequential read-only passes through the graph. The goal is then to minimize the number of passes and the time complexity of the algorithm.

Some graph problems, e.g. connectivity or minimum spanning tree, can be solved optimally [7]. Other graph problems, e.g. counting the number of 3-cycles, maximum matching and graph degeneracy, can be approximated [1,3,6]. Some

This research was supported in part by NSF grants CNS-1408782, IIS-1247750 and by Ministry of Science and Technology, Taiwan, Grant MOST 103-2221-E-001-033.

F. Dehne et al. (Eds.): WADS 2015, LNCS 9214, pp. 363–372, 2015.
DOI: 10.1007/978-3-319-21840-3_30

fundamental problems, such as breath-first search, depth-first search, topological sorting, and directed connectivity, are believed to be difficult to solve in a small number of passes [9,12,13]. Hence, the known algorithms [2,7] for finding articulation points take approaches other than computing a DFS tree.

Feigenbaum et al. [7] gave a first semi-streaming algorithm for finding articulation points. Their algorithm, which we refer to as the FKMSZ algorithm, has quadratic run time $\mathcal{O}(mn\alpha(n))$, where α denotes the inverse Ackermann function. Ausiello et al. [2] later gave an algorithm with run time $\mathcal{O}(m\alpha(n)+n\log n)$. Both these algorithms use $\mathcal{O}(n)$ space and perform one pass. Here, we present the first linear-time algorithm for this problem. It also uses $\mathcal{O}(n)$ space and performs one pass.

Instead of maintaining a structure that processes each incoming edge as it is scanned, we achieve optimality by buffering incoming edges and processing them in batches of size $\mathcal{O}(n)$. We extend this approach to the problems of computing spanning trees and of finding all **bridges**, where a bridge is an edge whose removal increases the number of connected components. Our algorithm has run time $\mathcal{O}(m+n)$, which improves the run time $\mathcal{O}(m\alpha(n))$ that comes from directly using the disjoint union-find set data structure [14].

The proposed algorithm not only has an optimal time complexity but has an optimal space complexity. A lower bound for space complexity can be obtained by noting that biconnectivity[1] is a **balanced property** [8]. For any balanced property $\mathcal{P}$, testing property $\mathcal{P}$ with probability at least 3/4 has a space lower bound of $\Omega(n)$ bits. Since finding articulation points is no easier than biconnectivity, it has a space lower bound of $\Omega(n)$ bits. In Section 6, we give a tighter analysis that finding articulation points in one sequential pass requires $\Omega(n\log n)$ bits. Hence, the space complexity of the proposed algorithm is optimal.

Organizations. In Section 2, we illustrate the idea of batches on two simpler problems. In Section 3, we revisit the FKMSZ algorithm. We explain a simple version of the proposed algorithm in Section 4 and defer the discussion of the full version to Section 5. In Section 6, we prove the space lower bound, $\Omega(n\log n)$ bits.

2   Preliminaries

We begin by showing how to reduce the running time for two simpler problems: finding a spanning tree and all bridges in a given graph G. We illustrate the idea of buffering scanned edges and processing them in a batch. This is the main idea used in our articulation-point algorithm.

Consider a spanning-tree algorithm in the semi-streaming model, and let F be a spanning forest of G, given the edges seen so far. As each edges e gets scanned, it can be added to F if it does not form a cycle. Testing cyclicity can be accomplished via a disjoint union-find data structure, which takes $\mathcal{O}(m\alpha(n))$ in total.

---

[1] A graph is biconnected iff it has no articulation point.

In order to reduce the total running time, process n edges for inclusion into the tree, instead of one at a time. Let B be the set of the next n edges to process, and let F be the current spanning forest. Compute a spanning forest of $B \cup F$ in $\mathcal{O}(n)$ time by an in-memory DFS. After all $\mathcal{O}(m/n)$ batches of edges have been processed in a single pass, the final F is a spanning forest of the original graph and the total computation time is $\mathcal{O}(m + n)$.

We apply the same idea to finding all bridges. Let F denote the spanning forest produced by the above algorithm. Note that if an edge $e \notin F$, then the edge e is on some cycle and thus cannot be a bridge. In addition to computing F, compute F_D, a spanning forest of $G \setminus F$, the discarded edges. This can be computed during the same pass where F is computed. Together they take $\mathcal{O}(n)$ space, one pass and $\mathcal{O}(m + n)$ time to compute. Once F and F_D are computed, the bridges in G can be reduced to find bridges in $F \cup F_D$ due to Lemma 1, thus in $\mathcal{O}(n)$ time by a DFS.

This approach improves the previously best $\mathcal{O}(m\alpha(n) + n \log n)$-time algorithm for finding bridges [2] to linear time.

Lemma 1. *An edge $(u, v) \in bridge(G)$ if and only if $(u, v) \in F \setminus F_D$ and $(u, v) \in bridge(F \cup F_D)$, where $bridge(H)$ denotes the set of bridges in graph H.*

Proof. Let $F_D = T_1 \cup T_2 \cup \cdots \cup T_k$, where each T_i is a maximal tree in F_D.

($\Rightarrow$) If $(u, v) \in bridge(G)$, then $(u, v) \in F$, $(u, v) \notin F_D$. Assume that $(u, v) \notin bridge(F \cup F_D)$, then there is a path P connecting nodes u, v in $F \cup F_D$ without passing through (u, v). The path P is also in G because $F \cup F_D \subseteq G$, a contradiction.

($\Leftarrow$) If $(u, v) \in F \setminus F_D$ and $(u, v) \in bridge(F \cup F_D)$, then $u \in T_a, v \in T_b$ for some $a \neq b$. Assume that $(u, v) \notin bridge(G)$, then there is a path P connecting nodes u, v in G without passing through (u, v). Since $(u, v) \in bridge(F \cup F_D)$, there are some edges $(x_1, y_1), (x_2, y_2), \ldots$ in P are discarded. Note that for any discarded edge (x_i, y_i) the nodes x_i, y_i are both contained in some T_j, implying that a path P_i in T_j connects nodes x_i, y_i. Since $u \in T_a, v \in T_b$ for some $a \neq b$, then $(u, v) \notin P_i$ for all i. Therefore, the closed loop formed by bridge (u, v) and path P with replacing the discarded edges with P_i's (note that $(u, v) \notin P_i$) implies a simple cycle passing through (u, v) in $F \cup F_D$, a contradiction. $\square$

3   The FKMSZ Algorithm

The classical algorithm for finding articulation points in the RAM model generates a DFS tree T and detects articulation points by identifying backedges. However, in the semi-streaming model, no efficient algorithm is known for generating a DFS tree. The FKMSZ algorithm replaces the DFS tree with an arbitrary spanning tree, implicitly relying on Lemmas 2 and 4. Since these lemmas were not stated as such in [7], we provide a statement and proof for each here for completeness.

We define some notions before proceeding to the lemmas. Given a spanning tree T of graph G, if nodes u, v are both tree neighbors of some node x, then we say nodes u, v are **co-paired at node** x or that they are a **co-pair** for short, since x is uniquely defined as the only shared neighbor of u and v.

We say that nodes u and v are **tree-biconnected** if there exists an edge $e \in G \setminus T$ such that u and v are biconnected in $T \cup \{e\}$. Note that if two nodes are tree-biconnected, they are biconnected, but the converse is not true. Tree-biconnectivity is easier to test for than biconnectivity.

Lemma 2. *Given a spanning tree T of graph G, a node x is an articulation point if and only if some co-pair at x is not biconnected.*

Proof. ($\Rightarrow$) By definition, if x is an articulation point in graph G, then, for some nodes $a, b \in G$, $a, b \neq x$, every path connecting a, b passes through x. This implies that, for some neighbors $u, v \in G$ of node x, every path connecting u, v passes through node x.

We divide the x's neighbors into two classes w.r.t. T: tree neighbors and non-tree neighbors. Suppose that node u is a non-tree neighbor of node x, then u, x are connected by a non-tree edge and therefore u and some x's tree neighbor are connected in $G \setminus \{x\}$. Therefore, no matter whether u, v are x's tree neighbors or non-tree neighbors, if nodes u, v are disconnected in $G \setminus \{x\}$, then some pair of x's tree neighbors are also disconnected in $G \setminus \{x\}$. Hence, some co-pair at x is not biconnected.

($\Leftarrow$) Suppose that x is not an articulation point, and let y be an articulation point that separates u and v. Such a y must exist because u and v are not biconnected. But removing $y \neq x$ leaves the u, x, v path intact, contradicting that y separates u and v. □

Corollary 3. *If x is a leaf node in any spanning tree T of graph G, then x is not an articulation point of graph G.*

Lemma 4. *Given a spanning tree T of graph G, a co-pair (u, v) at node x is biconnected if and only if there exist nodes $u = w_0, w_1, \ldots, w_t = v$ such that (w_{i-1}, w_i) is a tree-biconnected co-pair at node x for all $i \in [t]$.*

Proof. ($\Leftarrow$) If (w_{i-1}, w_i) is a tree-biconnected co-pair at node x, then nodes w_{i-1}, w_i are contained in some cycle of $T \cup \{e\}$ for some non-tree edge e. Therefore, nodes w_{i-1}, w_i are connected in $G \setminus \{x\}$. Since connectivity is transitive, u, v are connected in $G \setminus \{x\}$.

($\Rightarrow$) Observe that $T \setminus \{x\}$ is a set of subtrees. Each of x's tree neighbors belongs to an unique subtree and each subtree contains an unique tree neighbor of x. Observe further that $G \setminus \{x\}$ is a set of connected components. The connected components induced by the forest is a refinement of the connected components of the graph. That is, each connected component of the graph is spanned by one or more trees in the forest.

Since (u, v) is a co-pair at x, nodes u, v belong to different subtrees T_u, T_v. Since nodes u, v are biconnected, T_u, T_v are subgraphs of the same connected component $\mathcal{C}$. Suppose $\mathcal{C}$ contains k subtrees, then $k - 1$ non-tree edges suffice to connect the subtrees. Each of the $k-1$ non-tree edges indicates that a co-pair at x is tree-biconnected, implying that there exist nodes $u = w_0, w_1, \ldots, w_t = v$ such that (w_{i-1}, w_i) is a tree-biconnected co-pair at node x.   □

To realize the procedure in Lemma 4, we need an Union-Find data structure. In Section 4, we will introduce an Union-Find data structure that improves the run time of FKMSZ, but for now we will use a standard solution [14]. Let $S(x)$ be such a data structure for x, and initialize $S(x)$ with x's tree neighbors. The main idea of the algorithm is, for each tree-biconnected co-pair (u, v) at node x, to union u and v in $S(x)$. Thus, by Lemmas 2 and 4, we know that when we are done processing all edges, x is an articulation point iff $S(x)$ contains multiple sets, which we can check by performing a find on each element in $S(x)$. Putting this together gives the FKMSZ Algorithm:

---

1 Find a spanning tree T of graph G;
2 Prepare a union-find data structure $S(x)$ for each node x and make an element
    in $S(x)$ for each of x's tree neighbors ;
3 **foreach** *incoming non-tree edge* (u, v) **do**
4   Find the path $P_T(u, v), a_1 = u, a_2, \ldots, a_t = v$ in tree T;
5   For each co-pair (a_{i-1}, a_{i+1}), union a_{i-1} and a_{i+1} in $S(a_i)$;

6 **foreach** *node* x **do**
7   Let r_x be the find of any element in $S(x)$.;
8   **foreach** *element* y *in* $S(x)$ **do**
9   if find$(y) \neq r_x$, report x as an articulation point & break;

Algorithm 1. Pseudo-code of FKMSZ algorithm.

4   A Two Pass Algorithm for Articulation Points

We explain a simple, two-pass version of our algorithm in this section and defer the full one-pass version to Section 5. The simplified algorithm finds all articulation points of an n-node m-edge graph G in $\mathcal{O}(m+n)$ time after two sequential passes on the entire graph. We assume that graph G is connected; otherwise, one can adapt our algorithm to the unconnected cases in a straightforward way.

Our algorithm proceeds as follows. In the first pass, we find a spanning tree T of graph G and preprocess T. In the second pass, we execute Algorithm 1, achieving linear time by exploiting our preprocessing.

4.1 First Pass

We find a spanning tree T of graph G in $\mathcal{O}(m+n)$ time. Before the second pass, we root T at an arbitrary node and preprocess T in $\mathcal{O}(n)$ time to answer the following queries in $\mathcal{O}(1)$ time:

(1) $\text{DEG}_T(x)$: the degree of node x in tree T,
(2) $\text{DEPTH}_T(x)$: the depth of node x in tree T,
(3) $\text{LCA}_T(u,v)$: the lowest common ancestor of nodes u and v in rooted tree T [4],
(4) $\text{LA}_T(u,d)$: the ancestor of node u that has depth d in rooted tree T [5].

In addition, we need to build, for each node x, an union-find data structure, $\text{UF}(x)$. We initialize $\text{UF}(x)$ with all of its neighbors. We specify an union in the typical manner: $\text{UF}(x).\text{union}(u,v)$ performs an union in $\text{UF}(x)$ between the set that contains u and the set that contains v.

In order to beat the bound for union find, we do two things. First, rather than allow arbitrary find queries, we only allow queries $\text{UF}(x).\text{one}()$, which returns TRUE if $\text{UF}(x)$ contains only one set, that is, if all sets have been merged into one. Second, we favor unions over queries. As we will see in our analysis, unions are much more common than queries, so this tradeoff will give us a better total run time than using an off-the-shelf union-find algorithm would.

Lemma 5. *The union-find data structure* $\text{UF}(x)$ *can be implemented using* $\mathcal{O}(\text{DEG}_T(x))$ *space such that* $\text{UF}(x).\text{union}(u,v)$ *takes amortized constant time and* $\text{UF}(x).\text{one}()$ *takes* $\mathcal{O}(\text{DEG}_T(x))$ *time.*

Proof. Let each set in $\text{UF}(x)$ be a node, and let $d = \text{DEG}_T(x)$. We maintain a forest F of all nodes, where two nodes are in the same tree iff they are in the same set. This takes space $O(d)$. We use a buffer of size d. During $\text{UF}(x).\text{union}(u,v)$, an edge (u,v) is placed in the buffer. If the buffer is not full, then $\text{UF}(x).\text{union}(u,v)$ takes constant time. If the buffer is full, let B be the set of edges in the buffer. We compute a new spanning forest of $F \cup B$ in time $\mathcal{O}(d)$. The new spanning forest takes space $\mathcal{O}(d)$, and the buffer is now empty. Since this flushing step happens after every d edge insertions, the amortized edge insertion cost is $\mathcal{O}(1)$.

The query returns true iff $F \cup B$ has a single connected component, which can be checked in $\mathcal{O}(d)$ time. $\qquad\square$

4.2   Second Pass

We need to apply the unions specified by Algorithm 1 for each tree-biconnected co-pair found. However, if we do this, then each of the m non-tree edges found during the second pass would take time equal to the length of the cycle induced by adding the edge to T. In the worst case, we would end up with $\mathcal{O}(mn)$ time.

The problem is that this approach unions the same sets many times. To improve this, instead of enumerating the co-pairs on path $P_T(u,v)$ for each non-tree edge (u,v) individually, we defer the enumeration until there are n such paths waiting for enumeration. Then, we enumerate the co-pairs on n paths in a batch. In this way, we can avoid much of the work of finding the same co-pair many times, as follows.

Decompose each path $P_T(u,v)$ into paths $P_T(u,w)$ and $P_T(v,w)$, where $w = \text{LCA}_T(u,v)$, the lowest common ancestor of node u and node v in tree T. Then, the set of co-pairs on path $P_T(u, v)$ is the union of co-pairs on path $P_T(u, w)$, those on path $P_T(v, w)$, and the co-pair (w_u, w_v) if w_u, w_v exist, where by w_u we denote the child of node w that is an ancestor of node u in tree T and likewise for node w_v. Since there are at most n co-pairs of this last form, the enumeration of such co-pairs takes $\mathcal{O}(n)$ time. Hence, the only difficulty lies in how to union the short paths to reduce the repeated enumeration.

Note that all such paths go from a descendant to an ancestor. We partition the paths by their deepest node. Now, for each u, we union all the paths in u's partition. Notice that if $P_T(u, a)$ and $P_T(u, b)$ are in u's partition, then a and b are both ancestors of u, so one is an ancestor of the other. Furthermore, $P_T(u, a) \subseteq P_T(u, b)$ if b is an ancestor of a, a condition we can check in $\mathcal{O}(1)$ time since we have precomputed the depth of every node. Thus, all we need to do is find the shallowest node in u's partition, and we can discard all other paths. There are at most $2n$ paths total, so these steps take $\mathcal{O}(n)$ time for all paths and all nodes.

This is not enough, however, because the paths we have remaining can still add to length $O(n^2)$. In order to compute all co-pairs specified by these paths, we need to compute, for each node, if it and its grandparent is in one of the specified paths. But we can test this by a single DFS of the tree as follows. Mark every node u with path $P_T(u, w)$ with $\text{DEPTH}_T(w)$. Now by DFS, we can compute for every node v the depth of the shallowest endpoint of every path that goes through v. If this depth is $\text{DEPTH}_T(v) - 2$ or less, then v and its grandparent form a tree-biconnected co-pair. Thus, we can find all tree-biconnected co-pairs specified by n non-tree edges in $\mathcal{O}(n)$ time. We summarize the result in Lemma 6.

Lemma 6. *Given n paths on a tree of n nodes, the (multi-) set of co-pairs on these n paths can be enumerated in $\mathcal{O}(n)$ time.*

We are ready to prove the claimed time complexity. In the second pass, for each n non-tree edges, we enumerate $\mathcal{O}(n)$ tree-biconnected co-pairs in $\mathcal{O}(n)$ time due to Lemma 6. We perform all the unions specified by those co-pairs, that is, if (x, z) is a co-pair at y, we call $\text{UF}(y).\text{union}(x, z)$, and repeat for each such triple. This part also takes $\mathcal{O}(n)$ due to Lemma 5. Therefore, after processing m edges, the running time so far is $\mathcal{O}(m + n)$.

Since a node x is an articulation point if and only if $\text{UF}(x).\text{one}()$ returns FALSE, due to Lemmas 2 and 4, one can find all articulation points in

$$\mathcal{O}\left(\sum_{x \in T} \text{DEG}_T(x)\right) = \mathcal{O}(n)$$

time.

Theorem 7. *Given an n-node m-edge graph G, all articulation points of G can be found in $\mathcal{O}(m + n)$ time using $\mathcal{O}(n)$ space and two sequential passes on the entire graph.*

5   A One Pass Algorithm for Articulation Points

In this section, we modify the above two-pass algorithm into a one-pass algorithm. We do so by bypassing the first pass of the two-pass algorithm and directly moving into the second pass as if the spanning tree T were given. We are able to do this because, for every step of pass two, we don't actually need all of T, but only the parts of T that have some intersection with edges seen so far during the second phase. Thus T can be built incrementally, and the first-pass preprocessing can be computed incrementally, as we encounter edges in the "second" pass.

We first make one modification to the two-pass algorithm. Note that we did not specify which spanning tree T was needed for the two-pass algorithm. Any spanning tree suffices. Thus we have the flexibility to pick one that is suitable for our one-pass algorithm. In Section 2, we present a procedure for finding a spanning tree T of graph G in linear time. In the procedure, we use a buffer of size n to accommodate incoming edges and trim the edges to obtain an intermediate spanning forest every time the buffer is full. We denote those intermediate spanning forests by $F_0 = \phi, F_1, \ldots, F_{m/n} = T$. We say that a such procedure is *stable* if F_i is a subgraph of F_j for all $i < j$. In Lemma 8, we prove that one can generate a spanning tree with a stable procedure in linear time.

Lemma 8. *There is a stable procedure for finding a spanning tree T of an n-node m-edge graph G that runs in $\mathcal{O}(m + n)$ time using $\mathcal{O}(n)$ space and one sequential pass on the entire graph.*

Proof. To make the procedure stable, one need to assert that the newly generated spanning forest F_{i+1} is a supergraph of F_i. In other words, one needs to keep the newer edges with a lower priority than the older ones. To achieve this, one can contract the connected component in the spanning forest F_i and conduct a DFS on the contracted graph F_i union newly added edges. Both the contraction and DFS both takes linear time.   □

To mimic the two-pass algorithm, consider the ith batch of n edges. At this stage, we have spanning forest F_i, which is a subgraph of the spanning tree T. Then, for each non-tree edge (u, v) in the current batch, we need to find the path $P_T(u, v)$ given the subgraph F_i. Node u and node v cannot be contained in two different trees of forest F_i. Otherwise, we would have added edge (u, v) to F_i. We conclude that $P_{F_i}(u, v) = P_T(u, v)$.

The last problem is how to deal with the co-pairs on these paths in the claimed bound. First, we do not know $\text{DEG}_T(x)$ without the entire tree T. However, we only use $\text{DEG}_T(x)$ to allocate space for the union-find data structure $\text{UF}(x)$. One can achieve the same effect without knowing $\text{DEG}_T(x)$ by allocating $2s = \mathcal{O}(1)$ space for $\text{UF}(x)$ and iteratively doubling s whenever a new forest is computed and the degree of a node exceeds it's $s - 1$. In this way, each $\text{UF}(x)$ grows to the size $\mathcal{O}(\text{DEG}_T(x))$ and each $\text{UF}(x).\text{union}(u, v)$ still takes $\mathcal{O}(1)$ amortized time.

Second, for each F_i we preprocess the data structures to answer the queries used in the two-pass algorithm in constant time. Since the preprocessing can be done in time linear to the size of F_i, the total preprocessing time is thus $\mathcal{O}\left(\sum_i |F_i|\right) = \mathcal{O}(m+n)$. Therefore, this variation of the two-pass algorithm can be simulated by one-pass.

Theorem 9. *All articulation point of an n-node m-edge graph G can be done in $\mathcal{O}(m+n)$ time using $\mathcal{O}(n)$ space and one sequential pass on the entire graph.*

6   Space Lower Bound

In this section, we prove the following theorem.

Theorem 10. *Any semi-streaming algorithm that can output all articulation points of an n-node m-edge graph after one sequential pass requires $\Omega(n \log n)$ bits of space.*

Proof. Let function h be a bijection function from $[n]$ to $[n]$. Function h can be encoded with the graph G_h in Figure 1 without the dashed edge $e = (0, n+k)$ where $k \in [n]$ and there are n possible choices for e.

Then, we construct a stream for all edges in $G \cup \{e\}$, where the dashed edge e is placed last. The articulation points of graph $G \cup \{e\}$ are node 0 and every node $h(i)$ for $i \neq k$. Therefore, an algorithm that can output all articulation points of the graph $G \cup \{e\}$ also answers what $h(k)$ is, by computing the sum S_{AP} of the node labels of articulation points

$$S_{AP} = n(n+1)/2 - h(k).$$

At the time that a semi-streaming algorithm processes the last edge $e = (0, n+k)$, the state of memory must include an encoding of the bijection function $h : [n] \rightarrow [n]$ because based on the state of memory and the last edge $e = (0, n+k)$, one has to answer what $h(k)$ is, for any possible k. Since the number of possibilities of such a bijection function $h : [n] \rightarrow [n]$ is $n!$, the memory must have size at least $\Omega(n \log n)$ bits.   □

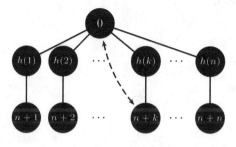

Fig. 1. Graph encoding of the bijection function $h : [n] \rightarrow [n]$

M. Farach-Colton et al.

References

1. Ahn, K.J., Guha, S.: Linear programming in the semi-streaming model with application to the maximum matching problem. In: Aceto, L., Henzinger, M., Sgall, J. (eds.) ICALP 2011, Part II. LNCS, vol. 6756, pp. 526–538. Springer, Heidelberg (2011)
2. Ausiello, G., Firmani, D., Laura, L.: Real-time monitoring of undirected networks: Articulation points, bridges, and connected and biconnected components. Network 59(3), 275–288 (2012)
3. Bar-Yossef, Z., Kumar, R., Sivakumar, D.: Reductions in streaming algorithms, with an application to counting triangles in graphs. In: 13th Annual ACM-SIAM Symposium on Discrete algorithms (SODA), pp. 623–632. SIAM (2002)
4. Bender, M.A., Farach-Colton, M.: The LCA problem revisited. In: Gonnet, G.H., Viola, A. (eds.) LATIN 2000. LNCS, vol. 1776, pp. 88–94. Springer, Heidelberg (2000)
5. Bender, M.A., Farach-Colton, M.: The level ancestor problem simplified. Theoretical Computer Science 321(1), 5–12 (2004)
6. Farach-Colton, M., Tsai, M.-T.: Computing the degeneracy of large graphs. In: Pardo, A., Viola, A. (eds.) LATIN 2014. LNCS, vol. 8392, pp. 250–260. Springer, Heidelberg (2014)
7. Feigenbaum, J., Kannan, S., McGregor, A., Suri, S., Zhang, J.: On graph problems in a semi-streaming model. Theoretical Computer Science 348(2), 207–216 (2005)
8. Feigenbaum, J., Kannan, S., McGregor, A., Suri, S., Zhang, J.: Graph distances in the data-stream model. SIAM Journal on Computing 38(5), 1709–1727 (2008)
9. Guruswami, V., Onak, K.: Superlinear lower bounds for multipass graph processing. In: 28th Conference on Computational Complexity (CCC), pp. 287–298. IEEE (2013)
10. Hopcroft, J., Tarjan, R.: Efficient algorithms for graph manipulation. Commun. ACM 16(6), 372–378 (1973)
11. Muthukrishnan, S.: Data streams: Algorithms and applications. Tech. rep. (2003)
12. O'Connell, T.C.: A survey of graph algorithms under extended streaming models of computation. In: Fundamental Problems in Computing, pp. 455–476. Springer (2009)
13. Ruhl, J.M.: Efficient Algorithms for New Computational Models. Ph.D. thesis, Massachusetts Institute of Technology, September 2003
14. Tarjan, R.E.: Efficiency of a good but not linear set union algorithm. J. ACM 22(2), 215–225 (1975)

LP-Based Approximation Algorithms for Facility Location in Buy-at-Bulk Network Design

Zachary Friggstad[1], Mohsen Rezapour[2]($^{(\boxtimes)}$), Mohammad R. Salavatipour[1], and José A. Soto[3]

[1] Department of Computing Science, University of Alberta, Edmonton, AB, Canada
{zacharyf,mreza}@cs.ualberta.ca
[2] Institute for Mathematics, TU Berlin, Berlin, Germany
rezapour@math.tu-berlin.de
[3] DIM and CMM, Universidad de Chile, Santiago, Chile
jsoto@dim.uchile.cl

Abstract. We study problems that integrate buy-at-bulk network design into the classical (connected) facility location problem. In such problems, we need to open facilities, build a routing network, and route every client demand to an open facility. Furthermore, capacities of the edges can be purchased in discrete units from K different cable types with costs that satisfy economies of scale. We extend the linear programming framework of Talwar [IPCO 2002] for the single-source buy-at-bulk problem to these variants and prove integrality gap upper bounds for both facility location and connected facility location buy-at-bulk problems. For the unconnected variant we prove an integrality gap bound of $O(K)$, and for the connected version, we get an improved bound of $O(1)$.

1 Introduction

We study problems that integrate buy-at-bulk network design into the classical (connected) facility location problem. We are interested in applications with trade-offs between facility opening and network design costs. Problems of this type arise in the planning of optical access networks in telecommunications, for example. An operator must decide on which nodes to install routing and switching devices (these are called central offices, and represented by *facilities*) and on which edges to install transmission technologies (represented by so-called *cable types*) to route traffic demands. In these networks, the traffic originating from each client is sent via tree-like *access networks*, to its respective facility. A combination of different cable types may be installed on the edges of these access trees to support the traffic flow. This allows for multiple fibers emanating

M. Rezapour—Supported by the Deutsche Forschungsgemeinschaft within the research training group 'Methods for Discrete Structures' (GRK 1408).
M.R. Salavatipour—Supported by NSERC.
J.A. Soto—Partially supported by FONDECYT grant 11130266 (Chile) and by Núcleo Milenio Información y Coordinación en Redes ICM/FIC RC130003.

F. Dehne et al. (Eds.): WADS 2015, LNCS 9214, pp. 373–385, 2015.
DOI: 10.1007/978-3-319-21840-3_31

from different clients to share a single, larger cable and the same trunk on their common path towards their common central office. The facilities are connected amongst each other or to some higher network level via a *core network* of (almost) unlimited capacity, which is required to route the traffic further towards its destination. Designing such a network involves selecting the facilities, connecting them via high-bandwidth links, and dimensioning the access links that are used to route the traffic from the clients to facilities.

This planning problem can be modeled as a *connected facility location with buy-at-bulk edge costs problem*, denoted by **BBCFL**. We are given a complete graph $G = (V, E)$ with nonnegative edge lengths $c_e \in \mathbb{Z}_{\geq 0}$, $e \in E$ satisfying triangle inequality; a set $F \subseteq V$ of facilities with opening costs $\mu_i \in \mathbb{Z}_{\geq 0}$, $i \in F$; and a set of clients $D \subseteq V$ with demands $d_j \in \mathbb{Z}_{>0}$, $j \in D$. We are also given K *types of access cables* that may be used to connect clients to open facilities. A cable of type i has capacity $u_i \in \mathbb{Z}_{>0}$ and cost (per unit length) $\sigma_i \in \mathbb{Z}_{\geq 0}$. Furthermore, we are given an extra type of cable, called *core cable*, having a cost (per unit length) of $M > \sigma_K$ and infinite capacity, which may be used to connect the open facilities with each other. We assume that access cable types obey economies of scale. That is, $\sigma_1 < \sigma_2 < \cdots < \sigma_K$ and $\frac{\sigma_1}{u_1} > \frac{\sigma_2}{u_2} > \cdots > \frac{\sigma_K}{u_K}$. A feasible solution or BBCFL consists of (1) A subset $F_0 \subseteq F$ of facilities to open; (2) a Steiner tree of G (core network) connecting all open facilities via core cables; and (3) a forest (access network) connecting all clients to the open facilities. Furthermore, on each edge of this forest we have to specify a list of possibly multiple copies and types of access cables to install, in such a way that the entire demand of each client can be routed along a single path to an open facility. The objective of BBCFL is to minimize the total cost of opening facilities, and constructing core and access networks; where the cost for using edge e in the core network is Mc_e, and the cost for installing a single copy of access cable of type i on an edge e is $\sigma_i c_e$. It is worth noting that we are allowed to install core cables on edges incident to closed facilities, to clients, or even to nodes in $V \setminus (F \cup D)$. Nevertheless, the demand from a client to its facility *is not allowed to* use core cables. The rationality for this constraint is that in real-life situations core and access networks are run independently. The only way to access from the access network to the core network is via an open facility.

There are various interesting variants of BBCFL that differ with respect to the structure of the access or core network. For example, the planning of water and energy supply networks occur in settings where different connection types on the edges of the access network is not motivated by the different capacities but by the different per unit shipping cost of alternative technologies or operational modes. This naturally leads to the *connected facility location with deep-discount edge costs problem*, denoted by **DDCFL**. In this problem, instead of capacitated access cables, we are given K *discount cable types*, where cable type i has a fixed (setup) cost of σ_i, a flow dependent incremental cost of δ_i, and unbounded capacity. We assume that $\delta_1 > \delta_2 > \cdots > \delta_k$ (i.e., discount cables obey economies of scale). The cost for installing one copy of discount type i on edge e and transporting R flow units on e is $(\sigma_i + R\delta_i)c_e$. Yet another variant occurs in logistic networks where the connectivity among facilities is not required, see [12] for more details. This is called *facility location with buy-at-bulk edge costs* problem, denoted by **BBFL**.

Previous Work. The BBFL problem was first considered by Meyerson et al. [11]. They show that BBFL can be seen as a special case of the *Cost-Distance* problem, and thereby provide the first randomized approximation algorithm with approximation guarantee $O(\log(|D|))$ for this problem. Their algorithm works for the more general version of non-uniform buy-at-bulk where one has a different set of cable types for each edge. The algorithm was then derandomized by Chekuri et al. [2], who also show that the integrality gap of the cost-distance problem is $O(\log(|D|))$. Ravi and Sinha [12] later developed an $O(K)$ approximation for this problem extending a combinatorial algorithm for the buy-at-bulk problem presented by Guha et al. [7]. The BBCFL problem was recently considered by Bley and Rezapour [1] who designed an approximation algorithm based on the random sampling techniques, achieving a 192-approximation.

The *Connected facility location problem* (**ConFL**) is the special case of the BBCFL problem with only one access cable type of unit capacity. Gupta et al. [8] obtained a 10.66-approximation for this problem, based on LP rounding. Swamy and Kumar [14] improved the approximation ratio to 8.55, using a primal-dual algorithm. Using sampling techniques, the guarantee was later reduced to 4 by Eisenbrand et al. [3], and to 3.19 by Grandoni et al. [6].

The *unsplittable Single-Sink Buy-at-Bulk problem* (**uSSBB**) can be seen as a further simplification of BBFCL in which the set of interconnected open facilities are given in advance. Several approximation algorithms for uSSBB have been proposed in the literature. Using LP rounding techniques, Garg et al. [4] developed an $O(K)$ approximation, where K is the number of cable types. The first constant factor approximation for this problem is due to Guha et al. [7]. Talwar [15] showed that an LP formulation of this problem has a constant integrality gap and provided a 216 approximation. Using sampling techniques, this factor was reduced to 145.6 by Jothi et al. [9], and later to 40.82 by Grandoni et al. [5].

Our Results. We extend the LP-based approximation for uSSBB in [15] to BBCFL and BBFL, thereby establishing an LP rounding framework for buy-at-bulk variants. Similar to previous work, one can show that the BBCFL and DDCFL problems are closely related, so that a ρ-approximation algorithm for one problem gives a 2ρ-approximation algorithm for the other. Since the integrality gap of the natural flow-based formulation for BBCFL can be arbitrarily large, we focus on the DDCFL problem. In Section 2, we present a strong flow-based IP (IP-1) model for this problem. Our main result is the following.

Theorem 1. *The integrality gap of (IP-1) is at most 234.*

As a consequence, we get an improved constant factor approximation for DDCFL, beating the 384-approximation one can obtain from doubling the 192-approximation guarantee in [1] for BBCFL. We also obtain the first LP based (deterministic) algorithm for the BBCFL problem whose factor is comparable with the (expected) approximation factor of the one in [1]. Finally, using similar techniques, we obtain an integrality gap of $O(K)$ for BBFL in Section 4. This matches the approximation guarantee of the combinatorial algorithm [12].

The reason why we get a better guarantee for BBCFL, even though it may seem more difficult than BBFL, is that the extra constraints in (IP-1) that ensure connectivity among open facilities are helpful in bounding the integrality gap.

2   IP Modeling of DDCFL

We write a flow-based IP formulation for DDCFL. We assume w.l.o.g. that a particular facility r is open and thus it belongs to the core network in the optimal solution and that $D \cap F = \emptyset$. Also, to simplify the description of our algorithm it will be useful to add an artificial *root client* r^* with unit demand, connected to r by an edge of 0 length. For each edge we create a pair of anti-parallel directed arcs, with same length as the original one. Let E be the set of these arcs. The undirected version of an arc $e \in E$ is denoted by $\bar{e}$. For every $e \in E$, cable type $k \in [K] = \{1, \ldots, K\}$ and client $j \in D$, the variable $f_{e;k}^j$ indicates if flow from client j uses cable type k on arc e; for $\bar{e} \in E$ and $k \in [K]$, $x_{\bar{e}}^k$ indicates if cable type k is installed on edge $\bar{e}$; $z_{\bar{e}}$ indicates if the core cable is installed on edge $\bar{e}$; and y_i indicates if facility i is opened. The opening cost C^{fac}, the core cost C^{core}, the fixed cost C^{fixed} and the routing cost C^{route} of a solution are defined as

$$C^{\text{fac}} = \sum_{i \in F} \mu_i y_i; \quad C^{\text{core}} = M \sum_{\bar{e} \in E} c_{\bar{e}} z_{\bar{e}}; \quad C^{\text{fixed}} = \sum_{k=1}^{K} C_k^{\text{fixed}}; \quad C^{\text{route}} = \sum_{k=1}^{K} C_k^{\text{route}},$$

$$\text{where} \quad C_k^{\text{fixed}} = \sigma_k \sum_{\bar{e} \in E} c_{\bar{e}} x_{\bar{e}}^k, \quad \text{and} \quad C_k^{\text{route}} = \delta_k \sum_{j \in D} d_j \sum_{e \in E} c_{\bar{e}} f_{e;k}^j, \tag{1}$$

represent the fixed cost and routing cost of the cables of type k, respectively. We use the notation $\delta^+(S) = \{(u, v) \in E: u \in S, v \notin S\}$, $\delta^-(S) = \delta^+(V \setminus S)$, $\delta(S) = \{uv \in E: u \in S, v \notin S\}$ for each $S \subseteq V$ and $\delta^+(v) = \delta^+(\{v\})$ for each $v \in V$. Given a set of cables $I \subseteq [K]$ and a client $j \in D$, we define the *access flow* on $e \in E$ with respect to I and j as $f_{e;I}^j = \sum_{k \in I} f_{e;k}^j$; and the *net in-flow* on a vertex $v \in V$ with respect to I and j, as $g_I^j(v) = \sum_{e \in \delta^-(v)} f_{e;I}^j - \sum_{e \in \delta^+(v)} f_{e;I}^j$. We also define $h_i^j = \max\{g_{[K]}^j(i), 0\}$ for $j \in D$ and $i \in F$. Formally, this quantity indicates whether facility i is serving client j. With all the notation above, our integer program formulation is as follows. Constraints (2) impose that at least one unit of flow leaves the clients. Constraints (3) are flow conservation constraints at non-facility nodes. Constraints (4) and (5) state that the flow only terminates at open facilities. Constraints (6) ensure that we install access links to support the flow. Finally, Constraints (7) state that if i is the facility serving demand j (the only i for which $h_i^j = 1$) then for each set S containing i and not containing the root there is a core link connecting S with its complement. In other words, all open facilities are connected to the root via core links, where Constraint (8) defines the root facility. Constraints (9) and (10), called *path monotonicity* constraints, strengthen the linear relaxation of (IP-1) – they ensure that the cable types along any path used to connect clients to facilities are nondecreasing from each client to its facility. The validity of these constraints follows from the fact that we have economy of scale, and hence that the flow aggregated on an edge (in the optimum fractional solution) never splits; see [4] for more details.

$$\text{(IP-1)} \quad \min C^{\text{fac}} + C^{\text{core}} + C^{\text{fixed}} + C^{\text{route}}$$

$$g^j_{[K]}(j) \leq -1 \qquad\qquad\qquad \forall j \in D \qquad\qquad\qquad (2)$$

$$g^j_{[K]}(v) = 0 \qquad\qquad\qquad \forall j \in D, v \in V \setminus (F \cup \{j\}) \qquad (3)$$

$$g^j_{[K]}(i) \leq h^j_i \qquad\qquad\qquad \forall j \in D, i \in F \qquad\qquad\quad (4)$$

$$h^j_i \leq y_i \qquad\qquad\qquad\qquad \forall j \in D, i \in F \qquad\qquad\quad (5)$$

$$f^j_{(u,v);k} + f^j_{(v,u);k} \leq x^k_{uv} \qquad\qquad \forall j \in D, k \in [K], uv \in E \qquad (6)$$

$$\sum_{i \in S \cap F} h^j_i - \sum_{\bar{e} \in \delta(S)} z_{\bar{e}} \leq 0 \qquad\qquad \forall j \in D, S \subseteq V \setminus \{r\} \colon S \cap F \neq \emptyset \ (7)$$

$$y_r = 1 \qquad\qquad\qquad\qquad\qquad\qquad\qquad\qquad\qquad (8)$$

$$g^j_{[q,K]}(v) \leq 0 \qquad\qquad\qquad \forall j \in D, v \in V \setminus F, 1 \leq q \leq K \qquad (9)$$

$$g^j_{[q,K]}(i) - \sum_{\bar{e} \in \delta(i)} z_{\bar{e}} \leq 0 \qquad\qquad \forall j \in D, i \in F \setminus \{r\}, 1 \leq q \leq K \ (10)$$

$$x^k_{\bar{e}}, f^j_{e;k}, y_i, z_{\bar{e}}, h^j_i \in \{0,1\} \qquad\qquad\qquad\qquad\qquad\qquad (11)$$

3 Proof of Theorem 1

Let (LP-1) be the linear program relaxation of (IP-1) and (f, x, y, z) be an optimal solution to (LP-1). It is not hard to show that (LP-1) can be solved in polynomial time using, for example, the ellipsoid method. We show how to round this LP solution to an integer one at constant factor loss.

3.1 Rounding Algorithm

We extend the rounding approach of [15] for the single-source buy-at-bulk problem to devise a rounding algorithm for DDCFL. Our algorithm has four phases.

Preprocessing Phase:

Pruning: We prune the set of access cable types such that all cables are considerably different. Similar to [15], this can be done without increasing the cost of the optimal solution too much.

Theorem 2. *Given $\epsilon_1, \epsilon_2 \in (0,1)$, we can prune the set of access cables so that for any i, $\sigma_{i+1} > \sigma_i/\epsilon_1$ and $\delta_{i+1} < \epsilon_2 \cdot \delta_i$ hold, increasing the installation and routing costs of the optimal fractional solution by a factor of at most $1/\epsilon_1$ and $1/\epsilon_2$, resp.*

For the sake of notation, let $[K]$ be the set of cables left and let (f, x, y, z) be the new solution of (LP-1) after the pruning stage. For each client j and positive radius R, define $B(j, R) = \{v \in V : c_{jv} \leq R\}$ to be the *moat* centered at j. We say that two moats $B_1 = B(j_1, R_1)$ and $B_2 = B(j_2, R_2)$ *overlap* if $c_{j_1 j_2} \leq R_1 + R_2$. Define also $L^j_k = \sum_{e \in E} f^j_{e;k} c_{\bar{e}}$ which represents the estimated distance that the flow of client j travels on cables of type k. Note that $C^{\text{route}}_k = \delta_k \sum_{j \in D} d_j L^j_k$.

Flow path decomposition: Every client j sends (at least) one unit of flow from itself to open facilities, specified by the $f^j_{e,[K]}$ variables. We decompose this fractional flow into a set of paths P_j, with path $p \in P_j$ starting from j and ending at some facility. Let $\phi(p)$ denote the amount of flow of path p.

Filtering: For a predefined constant $\theta \in (0,1)$ and for all $j \in D$, choose a subset of paths $\bar{P}_j \subseteq P_j$ such that $\phi_j := \sum_{p \in \bar{P}_j} \phi(p) \geq \theta$, by selecting paths in increasing order of their lengths until their total $\phi(p)$-value is at least θ. For each $j \in D$, let β_j be the length of the longest path in $\bar{P}_j$. Define a new solution $(\bar{f}, \bar{x}, \bar{y}, \bar{z})$ as follows. For each client $j \in D$, scale the amount of flow sent across each $P \in \bar{P}_j$ by $1/\phi_j$ and set the flow sent across each $P \in P_j - \bar{P}_j$ to 0. The new flow $\bar{f}$ is derived naturally from this new path decomposition. For each cable $k \in [K]$ and edge $\bar{e} \in E$, define $\bar{x}^k_{\bar{e}}$ as $x^k_{\bar{e}}/\theta$ if there exists some j with $\bar{f}^j_{e';k} > 0$, where $e' \in E$ is one of the two arcs associated to $\bar{e}$; and 0 otherwise. For each i, set $\bar{y}_i = \min\{y_i/\theta, 1\}$. And finally for each $\bar{e} \in E$, set $\bar{z}_{\bar{e}} = \min\{z_{\bar{e}}/\theta, 1\}$. It is easy to show that this solution is feasible for (LP-1).

Two important points: first, the solution $(\bar{f}, \bar{x}, \bar{y}, \bar{z})$ is such that the entire demand of client j is satisfied by open facilities on the moat $B(j, \beta_j)$. The second property is the following bound which is useful for the analysis. Let $\widetilde{P}_j \subseteq P_j$ be the set of paths with lengths at least β_j. Then, $\widetilde{P}_j$ includes all paths in $P_j \setminus \bar{P}_j$ and at least one path, say p^* (the longest) of $\bar{P}_j$. We conclude that $\sum_{p \in \widetilde{P}_j} \phi(p) \geq \sum_{p \in P_j} \phi(p) - \sum_{p \in \bar{P}_j \setminus \{p^*\}} \phi(p) \geq 1 - \theta$, and so

$$\sum_{k=1}^{K} L^j_k = \sum_{p \in P_j} \sum_{\bar{e} \in p} \phi(p) c_{\bar{e}} \geq \sum_{p \in \widetilde{P}_j} \phi(p) \sum_{\bar{e} \in p} c_{\bar{e}} \geq \beta_j (1 - \theta). \tag{12}$$

Facility Selection Phase:

Moat Selection: For a predefined constant $\eta > 1$, we consider the set of moats $\mathcal{B}_\eta = \{B(j, \eta\beta_j) : j \in D\}$ around clients. We choose a maximal set $\mathcal{B}' \subseteq \mathcal{B}_\eta$ of moats which do not overlap. We do this by processing the moats in $\mathcal{B}_\eta$ in increasing order of their radii, and greedily adding them to $\mathcal{B}'$ so that for each pair of selected moats in $\mathcal{B}'$ with centers $j, j' \in D$, $B(j, \eta\beta_j)$ and $B(j', \eta\beta_{j'})$ do not overlap. Let S_{core} be the set of clients with moats in $\mathcal{B}'$. Observe that for the artificial root client r^*, we have $\beta_{r^*} = 0$ and so $r^* \in S_{\text{core}}$.

Facility Opening: For each $j \in S_{\text{core}}$, let $F_j = \{i : \bar{h}^j_i > 0\}$ be the facilities fractionally serving demand from j with respect to solution $(\bar{f}, \bar{x}, \bar{y}, \bar{z})$. By the first property noted at the end of the preprocessing phase, $F_j \subseteq B(j, \eta\beta_j)$, hence $\{F_j : j \in S_{\text{core}}\}$ consist of disjoint sets. On each F_j we open the facility i^*_j with lowest opening cost. In particular, the root r is opened since $F_{r^*} = \{r\}$. Let I be the set of facilities opened on this stage. The basic idea of this part of the algorithm is inspired by [13]. For the purpose of analysis, associate each client with a special facility denoted as its $(K+1)$-st *proxy*. Formally, for each $j \in S_{\text{core}}$ we set $\text{proxy}_{K+1}(j) = i^*_j$. For the remaining clients $j \in D \setminus S_{\text{core}}$, we set $\text{proxy}_{K+1}(j) = \text{proxy}_{K+1}(j')$, where $j' \in S_{\text{core}}$ is the center of the smallest moat in $\mathcal{B}'$ that overlapped with $B(j, \eta\beta_j)$. Since the moats in $\mathcal{B}'$ were added in increasing radii and (12), we get

$$c(j, \text{proxy}_{K+1}(j)) \leq (1 + 2\eta)\beta_j \leq \frac{(1 + 2\eta)}{(1 - \theta)} \sum_{q=1}^{K} L_q^j \qquad \forall j \in D. \qquad (13)$$

Core Network Phase: Consider the graph G^{K+1} obtained from G by contracting the nodes of each F_j into single nodes, for $j \in S_{\text{core}}$. We construct an approximately optimal Steiner tree T' in G^{K+1} having the contracted nodes as terminals. To do this, we find an approximate Steiner tree whose cost is within a factor 2 of the cut-based relaxation. The edges of T' form a forest in G which touches a subset of the facilities in F_j, called $\bar{F}_j$ which may not include the open facility i_j^*. In order to connect all the open facilities together, we augment T' with the stars $Q_j = \{ji \colon i \in \bar{F}_j \cup \{i_j^*\}\}$, $j \in S_{\text{core}}$. Let T^{core} be the resulting tree, after possibly canceling some cycles. To conclude this stage, we install core cables on T^{core}.

Access Network Phase: We construct the access network in a top-down manner, installing cables progressively in stages numbered from $i = K$ to 1. Let T_{K+1} be a *minimum spanning tree* on the graph induced by the set I of open facilities, and connect them using an artificial cable type $K + 1$. This tree won't appear in the end, as it will be replaced by the core network. In stage i, we augment the current tree T_{i+1}, which uses only cables of type $i + 1$ or higher, by installing cables of type i. Define $\bar{L}_k^j$ to be $\sum_{e \in E} \bar{f}_{e;k}^j \cdot c_e$. This estimates the distance that flow from j goes on cable type k. Let $\bar{R}_l^j = \sum_{k=1}^{l-1} \bar{L}_k^j$ be the estimated distance beyond that flow from j uses cable type l or higher in the new fractional solution. Intuitively, $\bar{R}_l^j$ tells us how far from j to go before the LP solution installs access cable types l or higher. Stage i consists of two steps:

   Step 1. Moat Selection: For predefined $\gamma > \zeta > 1$, we construct the set of moats $\mathcal{B}_\gamma^i = \{B(j, \gamma \bar{R}_i^j) \colon j \in D\}$ around all clients. We define $\hat{S}_i$ to be the set of clients whose moats intersect T_{i+1}. For each $j \in \hat{S}_i$ remove moat $B(j, \gamma \bar{R}_i^j)$ from $\mathcal{B}_\gamma^i$. Similar to what we did for the core network, we choose a maximal set $\mathcal{B}^i \subseteq \mathcal{B}_\gamma^i$ of moats which do not overlap by selecting moats from $\mathcal{B}_\gamma^i$ in increasing order of their radii. Let S_i be the set of clients whose moats are selected in round i.

   Step 2. Cable type i installation: We construct the set $\mathcal{B}_\zeta^i = \{B(j, \zeta \bar{R}_i^j) \colon j \in S_i\}$ of moats around clients in S_i. We obtain a graph G^i from G by contracting each moat in $\mathcal{B}_\zeta^i$ into a super-node, and the current tree T_{i+1} into a super-node called r_{i+1}. We then construct an approximately optimal Steiner tree in G^i (with integrality gap bound 2), where the terminals are all the super-nodes. By uncontracting, we get a forest in G touching at least one node in T_{i+1} and one node from each moat. To get a tree, called $\bar{T}_i$, from the resulting forest, we add direct edges from each client $j \in S_i$ to each node of $B(j, \zeta \bar{R}_i^j)$ that is incident on the forest[1] and then we cancel cycles.

---

[1] This crucial step of adding direct edges is missing from the **uSSBB**-approximation in [15], even though it seems necessary for both that algorithm and ours to work.

Using Khuller et al.'s technique [10], we then convert tree $\bar{T}_i$ rooted at r_{i+1}, into an (α, β)-Light Approximate Shortest-path Tree (LAST), for parameters $\beta = \frac{\alpha+1}{\alpha-1}$ and $\alpha > 1$ to be chosen later. Let LAST_i be the resulting tree. The LAST algorithm [10] transforms tree $\bar{T}_i$ into LAST_i with $c(\text{LAST}_i) \leq \beta c(\bar{T}_i)$ such that the path length of any vertex v to root r_{i+1} in LAST_i is at most α times the length of a shortest v-r_{i+1} path in G^i. We un-contract the moats and install cables of type i on the edges of LAST_i. Let $T_i = T_{i+1} \cup \text{LAST}_i$.

For the purpose of analysis, for each $j \in S_i$, we call an arbitrary node in its moat which is connected to LAST_i as *the proxy*, denoted by $\text{proxy}_i(j)$. For the clients $j \in \hat{S}_i$, we define $\text{proxy}_i(j)$ to be an arbitrary node in $B(j, \gamma \bar{R}_i^j) \cap T_{i+1}$. For the remaining clients $j' \in D \setminus S_i \cup \hat{S}_i$, we define $\text{proxy}_i(j')$ to be $\text{proxy}_i(j)$, where $j \in S_i$ is the center of the smallest moat in $\mathcal{B}^i$ that overlapped with $B(j', \gamma \bar{R}_i^{j'})$. It is easy to verify that $c(j, \text{proxy}_i(j)) \leq 3\gamma \bar{R}_i^j \leq \frac{3\gamma}{\theta} \sum_{k=1}^{i-1} L_k^j$. If we set $\Delta = \max\{\frac{1+2\eta}{1-\theta}, \frac{3\gamma}{\theta}\}$, then by the previous inequality and (13), we get

$$c(j, \text{proxy}_{i+1}(j)) \leq \Delta \cdot \sum_{q=1}^{i} L_q^j \qquad \forall j \in D, 1 \leq i \leq K \qquad (14)$$

which will be useful in bounding the routing cost.

Finally, note that $R_1^j = 0$ for all j. This means that in the first step of the last stage, S_1 consists of all clients that have not been connected to the current tree. Therefore, at the end of the last stage, T_1 is a tree spanning all clients and open facilities. The access network we return consists of the forest obtained by removing the artificial tree T_{K+1} from T_1.

3.2 Analysis

Let $C^{*\text{fac}}$, $C^{*\text{core}}$, $C^{*\text{fixed}}$ and $C^{*\text{route}}$ be the opening cost, core installation cost, fixed installation cost and routing cost paid by the LP optimum (see (1)). And let C^{fac}, C^{core}, C^{fixed} and C^{route} the ones paid by our algorithm. Let gap_{ST} denote the upper bound on the integrality gap of the cut based formulation of Steiner tree problem, which is 2. Let OPT be the cost of LP optimum. The following lemma bounds the opening cost; the proof is omitted as it is similar to that for the facility location problem [13].

Lemma 3. *The opening cost of the returned solution is at most $\frac{1}{\theta} C^{*\text{fac}}$.*

Lemma 4. *The cost of core link installation is at most $\frac{\eta+1}{\theta(\eta-1)} \cdot \text{gap}_{ST} \cdot C^{*\text{core}}$.*

Proof. By (7), one can verify that $\sum_{\bar{e} \in \delta^+(S)} \bar{z}_{\bar{e}} \geq 1$ holds for any arbitrary set $S \subset V$ that contains all facilities in F_j (for some j) and it does not contain r. This means that $\bar{z}$ is a feasible fractional solution to the cut based LP relaxation of the Steiner tree problem on the graph G^{K+1} (see the core network phase) whose terminals are all the contracted sets F_j (recall that $F_{r^*} = \{r\}$). In particular, the Steiner tree T' found in the core network phase has cost at most $\text{gap}_{ST} \cdot \sum_{\bar{e} \in E} c_{\bar{e}} \bar{z}_{\bar{e}}$. The cost of the extra edges included in the final tree T^{core} (i.e., the union of all stars Q_j) can be charged to the cost of T' as follows.

For each facility b_j in $\bar{F}_j$ let $e(b_j) = b_j v \in T'$ be any edge incident to it. Since b_j is in $B(j, \beta_j)$ and v is outside $B(j, \eta\beta_j)$, we conclude that the cost of $e(b_j)$ is at least $(\eta - 1)\beta_j$. By a similar argument, if $e = e(b_j) = e(b_k)$ where $b_j \in \bar{F}_j$ and $b_k \in \bar{F}_k$, then we can use the fact that $B(j, \eta\beta_j)$ and $B(k, \eta\beta_k)$ do not overlap to conclude that the length of e is at least $(\eta - 1)(\beta_j + \beta_k)$. Therefore, the total cost of the union of all Q_j is at most

$$\sum_{j \in S_{core}} \left(c(ji_j^*) + \sum_{b \in \bar{F}_j} c(jb) \right) \leq 2 \sum_{j \in S_{core}} \sum_{b \in \bar{F}_j} \beta_j \leq 2 \sum_{e \in T'} \frac{c(e)}{\eta - 1}.$$

Summing up, the cost of T^{core} is at most $1 + \frac{2}{(\eta - 1)}$ times the cost of T', and therefore it is at most $\frac{\eta + 1}{\theta(\eta - 1)} \cdot \text{gap}_{ST} \cdot \sum_{e \in E} c_{\bar{e}} z_{\bar{e}}$. $\qquad \square$

In the following, we bound the fixed cost and routing cost of the cables installed at stage i of the access network phase, denoted by C_i^{fixed} and C_i^{route}, respectively.

Lemma 5. $C_i^{fixed} \leq \sigma_i \cdot \text{gap}_{ST} \cdot \dfrac{\gamma\beta\zeta}{(\gamma - \zeta)(\zeta - 1)\theta} \left(\sum_{q=i}^{K} \dfrac{1}{\sigma_q} C_q^{*fixed} + \dfrac{1}{M} C^{*core} \right)$

Proof. Let S be an arbitrary subset of $V \setminus \{r\}$ that contains $B(j, \zeta\bar{R}_i^j)$. We first show that $\sum_{q=1}^{i-1} \bar{b}_{q;S}^j \leq \frac{1}{\zeta}$, where $\bar{b}_{q;S}^j := \sum_{e \in \delta^+(S)} \bar{f}_{e,q}^j$ indicates the amount of flow from j crossing the boundary of S thorough cables of type q. The flow we are considering has to travel from j to the boundary of S using only use cables of type q or thinner. So, as $\bar{b}_{q;S}^j$ travels a distance of at least $\zeta\bar{R}_i^j$, it contributes at least $\bar{b}_{q;S}^j\zeta\bar{R}_i^j$ units to $\bar{R}_i^j = \sum_{k=1}^{i-1} \bar{L}_k^j$. As the contributions from each q are disjoint, we have $\bar{R}_i^j \geq \sum_{q=1}^{i-1} \bar{b}_{q;S}^j\zeta\bar{R}_i^j$, which implies that $\sum_{q=1}^{i-1} \bar{b}_{q;S}^j \leq \frac{1}{\zeta}$. This together with the LP constraints guarantee that $\sum_{q=i}^{K} \bar{b}_{q;S}^j + \sum_{e \in \delta^+(S)} \bar{z}_e \geq 1 - \frac{1}{\zeta}$ and hence $\sum_{e \in \delta^+(S)} \left(\sum_{q=i}^{K} \bar{x}_e^q + \bar{z}_e \right) \geq 1 - \frac{1}{\zeta}$. This means that the vector $\bar{z} + \sum_{q=i}^{K} \bar{x}^q$, scaled by a factor $\frac{\zeta}{\zeta - 1}$, is a feasible fractional solution to the LP relaxation of the Steiner tree connecting balls $B(j, \zeta\bar{R}_i^j)$ to T_{i+1}. Therefore, the cost of the Steiner tree computed in step 2 of the access network phase can be bounded by

$$\frac{\text{gap}_{ST}\zeta}{\zeta - 1} \left(\sum_{e \in E} \sum_{q=i}^{K} c_e \bar{x}_e^q + \sum_{e \in E} c_e \bar{z}_e \right) \leq \frac{\text{gap}_{ST}\zeta}{(\zeta - 1)\theta} \left(\sum_{q=i}^{K} \frac{1}{\sigma_q} C_q^{*fixed} + \frac{1}{M} C^{*core} \right).$$

Similar to Lemma 4, one can show that the cost of extra edges of $\bar{T}_i$, added after un-contracting the moats, is at most $\frac{\zeta}{\gamma - \zeta}$ times the cost of the current forest. Altogether, the cost of the $LAST_i$ tree is at most

$$c(LAST_i) \leq \text{gap}_{ST} \cdot \frac{\gamma}{\gamma - \zeta} \cdot \frac{\beta\zeta}{(\zeta - 1)\theta} \left(\sum_{q=i}^{K} \frac{1}{\sigma_q} C_q^{*fixed} + \frac{1}{M} C^{*core} \right). \qquad \square$$

The proof of the next lemma is omitted due to page limitations.

Lemma 6. $C_i^{route} \leq \Delta\delta_i\alpha \sum_{q=1}^{i} (1 + \alpha)^{i-q} \dfrac{1}{\delta_q} C_q^{*route}$.

By Lemma 5, Theorem 2, and by summing over all cable types, the fixed cost paid by the algorithm can be bounded as follows.

$$C^{\text{fixed}} \leq \text{gap}_{\text{ST}} \cdot \frac{\gamma\beta\zeta}{(\gamma-\zeta)(\zeta-1)\theta} \Big[\sum_{s=1}^{K} C_s^{*\text{fixed}} \Big(\sum_{i \leq s} \frac{\sigma_i}{\sigma_s} \Big) + C^{*\text{core}} \Big(\sum_{i=1}^{K} \frac{\sigma_i}{M} \Big) \Big]$$

$$\leq \text{gap}_{\text{ST}} \cdot \frac{\gamma\beta\zeta}{(\gamma-\zeta)(\zeta-1)\theta(1-\epsilon_1)} \big[C^{*\text{fixed}} + C^{*\text{core}} \big]. \tag{15}$$

Similarly, by using Lemma (6), we bound the routing cost as follows.

$$C^{\text{route}} \leq \Delta\alpha \sum_{i=1}^{K} \sum_{s=1}^{i} (1+\alpha)^{i-s} \frac{\delta_i}{\delta_s} C_s^{*\text{route}} \leq \Delta\alpha \sum_{i=1}^{K} \sum_{s=1}^{i} \big((1+\alpha) \cdot \epsilon_2 \big)^{i-s} C_s^{*\text{route}}$$

$$\leq \Delta\alpha \sum_{s=1}^{K} C_s^{*\text{route}} \sum_{i \geq s} \big((1+\alpha) \cdot \epsilon_2 \big)^{i-s} \leq \frac{\Delta\alpha}{1-\epsilon_2(1+\alpha)} \cdot C^{*\text{route}}. \tag{16}$$

Using (15), (16), Lemmas 3 and 4, the total cost of our solution is at most

$$\frac{1}{\theta} C^{*\text{fac}} + \frac{(\eta+1)\text{gap}_{\text{ST}}}{\theta(\eta-1)} C^{*\text{core}} + \frac{\gamma\beta\zeta \cdot \text{gap}_{\text{ST}} (C^{*\text{fixed}} + C^{*\text{core}})}{(\gamma-\zeta)(\zeta-1)\theta(1-\epsilon_1)} + \frac{\Delta\alpha}{1-\epsilon_2(1+\alpha)} C^{*\text{route}}.$$

Finally, using Theorem 2, we can bound the cost of our solution by

$$\leq \max\Big(\frac{1}{\epsilon_1} \cdot \frac{\gamma\beta\zeta \cdot \text{gap}_{\text{ST}}}{(\gamma-\zeta)(\zeta-1)\theta(1-\epsilon_1)} + \frac{(\eta+1)\text{gap}_{\text{ST}}}{\theta(\eta-1)}, \frac{1}{\epsilon_2} \frac{\Delta\alpha}{1-\epsilon_2(1+\alpha)} \Big) OPT. \tag{17}$$

This completes the proof of Theorem 1. Setting $\alpha = 1.47$, $\gamma = 4.10$, $\epsilon_1 = 0.50$, $\epsilon_2 = 0.20$, $\theta = 0.78$, $\eta = 1.27$ and $\zeta = 2$ and recalling $\text{gap}_{\text{ST}} = 2$, inequality (17) implies that the integrality gap of (IP-1) is no more than 234. Thus, we obtain the first LP based (deterministic) algorithm for DDCFL and thereby for BBCFL.

4 On the Integrality Gap of the BBFL Problem

Recall that if we omit the requirement to connect the open facilities, the BBCFL becomes the BBFL problem. In this section we study the integrality gap of an LP formulation for the problem. As with the BBCFL problem, we consider a variant of BBFL, called DDFL, in which we replace the capacitated access cables by discount cable types. Note that similar to the relation between BBCFL and DDCFL, one can transform between BBFL and DDFL with a factor 2 loss.

IP Formulation. Similar to Section 2, DDFL can be formulated as follows:

$$\text{(IP-2)} \quad \min C^{\text{fac}} + C^{\text{fixed}} + C^{\text{route}} \qquad \text{s.t.} \quad (2), (3), (6), (9)$$

$$g_{[K]}^j(i) \leq y_i \qquad \forall j \in D, i \in F \tag{18}$$

$$g_{[q,K]}^j(i) - y_i \leq 0 \qquad \forall j \in D, i \in F, 1 \leq q \leq K \tag{19}$$

$$x_e^k, f_{e;k}^j, y_i \in \{0,1\} \tag{20}$$

We do not need the z and h_i^j variables anymore, as they were used to model facility connectivity. Constr. (18) state that the flow only ends at open facilities, and Constr. (9) and (19) force the path monotonicity discussed in Section 2.

Algorithm. We follow the same general ideas of the rounding algorithm for DDCFL, but we replace the core network and access network phases by a single one denoted network phase. Another key difference is that we may open facilities at any stage of the network phase. Ultimately, this is why our integrality gap bound is $O(K)$ as we have to overestimate and bound the opening cost in each of the K stages by the total opening cost paid by the LP.

Preprocessing Phase. Apply the preprocessing phase (pruning, flow path decomposition and filtering) of Section 3.1, disregarding variables z. Let $(\bar{f}, \bar{x}, \bar{y})$ be the solution after this phase.

Initial Facility Selection Phase. Perform the facility selection phase of Section 3.1 but fixing $\eta = 1$. Let I' be the set of facilities opened in this phase.

Network Phase. We construct a solution in a top-down manner, installing cables *and* possibly opening more facilities in stages, which we number from $i = K$ to 1. We start with solution $(I_{K+1}, T_{K+1}) = (I', \emptyset)$. At stage i we augment the current solution by (1) opening some extra facilities and (2) installing cables of type i. We do this while keeping the invariant that T_i is a forest in G such that each connected component contains an open facility of I_i. Stage i is similar to the i-th stage of the access network phase in Section 3.1.

1. For a predefined constant $\gamma > \zeta > 1$, construct the set of moats $B(j, \gamma \bar{R}_j^i)$ around clients $j \in D$. Remove the moats which intersect T_{i+1} and select from the rest a maximal subset $\mathcal{B}^i$ of non-overlapping moats in increasing order of their radii. Let S_i be the set of selected clients associated to $\mathcal{B}^i$ and construct the set $\mathcal{B}_\zeta^i = \{B(j, \zeta \bar{R}_i^j) : j \in S_i\}$ of moats around clients in S_i.
2. Add a dummy node $\tilde{r}$ and connect it to every facility v fractionally opened by the LP (with $\bar{y}_v > 0$). Set the cost of each dummy edge $\tilde{e} = \tilde{r}v$ to be zero if facility $v \in I_{i+1}$; otherwise set it to be f_v. To simplify the analysis, associate each edge $\tilde{e} = \tilde{r}v$ with a variable $\tilde{x}_{\tilde{e}}$ equal to $\bar{y}_v$.
3. Contract each moat in $\mathcal{B}_\zeta^i$, and each component of T_{i+1} into super-nodes. Call the contracted graph $\widetilde{G}$.
4. Construct an approximately optimal Steiner tree $\hat{T}$ on $\widetilde{G}$, where the terminals are $\tilde{r}$ and all the super-nodes. Without loss of generality we assume that $\hat{T}$ includes a dummy edge of cost 0 from $\tilde{r}$ to every super-node associated to a component of T_{i+1} (or, more precisely, to each facility $v \in I_{i+1}$).
5. For each $v \in F \setminus I_{i+1}$, if edge $\tilde{r}v$ is in $\hat{T}$ then open facility v and put it in I_i.
6. Set $I_i = I_i \cup I_{i+1}$.
7. Contract all the dummy edges that are contained in $\hat{T}$, and uncontract the super-nodes associated to the moats. The edges from $\hat{T}$ form a forest in the resulting graph. To get a tree, add for each moat direct edges from its center to all nodes in the moat that are incident to $\hat{T}$. Let $\widetilde{T}$ be the resulting tree.
8. Using the LAST algorithm for appropriate parameters, transform $\widetilde{T}$ rooted at the contracted node containing $\tilde{r}$ into a tree called LAST_i.
9. Install cables of type i along LAST_i and let $T_i = T_{i+1} \cup \text{LAST}_i$.

Theorem 7. *The integrality gap of (IP-2) is at most $O(K)$.*
The proof is omitted due to page limitations.

384 Z. Friggstad et al.

5 Conclusion

We have shown that the LP rounding framework for **uSSBB** [15] extends to facility location buy-at-bulk problems. Our integrality gap analysis roughly matches the known approximation ratios of combinatorial algorithms for BBCFL [1] and BBFL [12], so the obvious open problem is to improve this analysis to derive better approximation algorithms. In particular, can we get an $O(1)$-approximation for BBFL? We were able to bound the gap by $O(1)$ for BBCFL by exploiting the fact that the facility core network is fractionally connected by the LP. However, in BBFL we do not have this property so we have to pay for the facility opening costs with a copy of the y-values in each stage. A potentially easier problem is to get an α-approximation for BBFL with running time $n^{f(k)}$ for some function f where α is a constant that does not depend on k.

Acknowledgments. A special thank to Babak Behsaz for helpful discussions.

References

1. Bley, A., Rezapour, M.: Approximating connected facility location with buy-at-bulk edge costs via random sampling. In: Proc. of LAGOS 2013, pp. 313–319 (2013)
2. Chekuri, C., Khanna, S., Naor, J.S.: A deterministic algorithm for the cost-distance problem. In: Proc. of SODA 2001, pp. 232–233 (2001)
3. Eisenbrand, F., Grandoni, F., Rothvoß, T., Schäfer, G.: Connected facility location via random facility sampling and core detouring. Journal of Computer and System Sciences **76**(8), 709–726 (2010)
4. Garg, N., Khandekar, R., Konjevod, G., Ravi, R., Salman, F.S., Sinha, A.: On the integrality gap of a natural formulation of the single-sink buy-at-bulk network design formulation. In: Proc. of IPCO 2001, pp. 170–184 (2001)
5. Grandoni, F., Rothvoß, T.: Network design via core detouring for problems without a core. In: Abramsky, S., Gavoille, C., Kirchner, C., Meyer auf der Heide, F., Spirakis, P.G. (eds.) ICALP 2010. LNCS, vol. 6198, pp. 490–502. Springer, Heidelberg (2010)
6. Grandoni, F., Rothvoß, T.: Approximation algorithms for single and multi-commodity connected facility location. In: Günlük, O., Woeginger, G.J. (eds.) IPCO 2011. LNCS, vol. 6655, pp. 248–260. Springer, Heidelberg (2011)
7. Guha, S., Meyerson, A., Munagala, K.: A constant factor approximation for the single sink edge installation problems. In: Proc. of STOC 2001, pp. 383–388 (2001)
8. Gupta, A., Kleinberg, J., Kumar, A., Rastogi, R., Yener, B.: Provisioning a virtual private network: a network design problem for multicommodity flow. In: Proc. of STOC 2001, pp. 389–398 (2001)
9. Jothi, R., Raghavachari, B.: Improved approximation algorithms for the single-sink buy-at-bulk network design problems. In: Hagerup, T., Katajainen, J. (eds.) SWAT 2004. LNCS, vol. 3111, pp. 336–348. Springer, Heidelberg (2004)
10. Khuller, S., Raghavachari, B., Young, N.: Balancing minimum spanning and shortest path trees. Algorithmica **14**(4), 305–321 (1994)
11. Meyerson, A., Munagala, K., Plotkin, S.: Cost-distance: two metric network design. In: Proc. of FOCS 2000, pp. 624–630 (2000)

12. Ravi, R., Sinha, A.: Integrated logistics: approximation algorithms combining facility location and network design. In: Cook, W.J., Schulz, A.S. (eds.) IPCO 2002. LNCS, vol. 2337, pp. 212–229. Springer, Heidelberg (2002)
13. Shmoys, D.B., Tardos, E., Aardal, K.I.: Approximation algorithms for facility location problems. In: Proc. of STOC 1997, pp. 265–274 (1997)
14. Swamy, C., Kumar, A.: Primal-dual algorithms for connected facility location problems. Algorithmica 40(4), 245–269 (2004)
15. Talwar, K.: The single-sink buy-at-bulk LP has constant integrality gap. In: Cook, W.J., Schulz, A.S. (eds.) IPCO 2002. LNCS, vol. 2337, pp. 475–480. Springer, Heidelberg (2002)

Universal Reconstruction of a String

Paweł Gawrychowski[1], Tomasz Kociumaka[1(✉)], Jakub Radoszewski[1],
Wojciech Rytter[1,2], and Tomasz Waleń[1]

[1] Faculty of Mathematics, Informatics and Mechanics,
University of Warsaw, Warsaw, Poland
{gawry,kociumaka,jrad,rytter,walen}@mimuw.edu.pl
[2] Faculty of Mathematics and Computer Science,
Copernicus University, Toruń, Poland

Abstract. Many properties of a string can be viewed as sets of dependencies between substrings of the string expressed in terms of substring equality. We design a linear-time algorithm which finds a solution to an arbitrary system of such constraints: a generic string satisfying a system of substring equations. This provides a general tool for reconstructing a string from different kinds of repetitions or symmetries present in the string, in particular, from runs or from maximal palindromes. The recursive structure of our algorithm in some aspects resembles the suffix array construction by Kärkkäinen and Sanders (J. ACM, 2006).

1 Introduction

Let s be a string of length n, $s = s_0 \ldots s_{n-1}$. For $0 \le p \le q < n$, we denote a substring $s_p \ldots s_q$ by $s[p..q]$. A *substring equation* is a constraint of the form "$s[p..q] = s[p'..q']$". We say that a string s of length n is a *solution* to a system of substring equations E (*satisfies E*) if it satisfies each equation of the system.

Clearly, every system has a solution which is a string over a unary alphabet, i.e., a^n. Thus, we focus on *generic* solutions containing the largest number of different characters. The important feature of any such solution is that it can be used to describe all solutions of a system of substring equations:

Observation 1. *If s is a generic solution of length n to a system E, then for each string s' of length n that satisfies E there exists a letter-to-letter morphism (a coding) μ such that $\mu(s) = s'$.*

In particular, every two generic solutions of the same length are equivalent up to renaming letters. We denote one of the generic solutions by $\Phi(E)$.

P. Gawrychowski—Work done while the author held a post-doctoral position at Warsaw Center of Mathematics and Computer Science.

T. Kociumaka—Supported by Polish budget funds for science in 2013–2017 as a research project under the 'Diamond Grant' program.

T. Kociumaka, J. Radoszewski, W. Rytter, and T. Waleń—Supported by the grant NCN2014/13/B/ST6/00770 of the Polish Science Center.

J. Radoszewski—Supported by the Polish Ministry of Science and Higher Education under the 'Iuventus Plus' program in 2015–2016 grant no 0392/IP3/2015/73. The author also receives financial support of Foundation for Polish Science.

© Springer International Publishing Switzerland 2015
F. Dehne et al. (Eds.): WADS 2015, LNCS 9214, pp. 386–397, 2015.
DOI: 10.1007/978-3-319-21840-3_32

Our Main Result. We design a linear-time algorithm which computes a generic solution $\Phi(E)$ for a system E of substring equations.

The fact that our solution is generic lets us solve in $\mathcal{O}(n)$ time many classic string recovery problems.

Algorithms recovering (reverse engineering, inferring) a string from many internal structures are known. This includes recovery from border array [11–13], strong border array [15], prefix array [7], the set of maximal palindromes [17], minimum and maximum cover array [9,26], Lyndon factorization [27], suffix array [3], directed acyclic word graph (DAWG) [3], suffix tree [5,19,28], and parameterized border array [18]. For all but the last one of the aforementioned problems, there are algorithms running in linear time and constructing a string over the smallest possible alphabet. For parameterized border array reconstruction, the fastest algorithm works in $\mathcal{O}(n^{1.5})$ time. A more difficult task is reconstruction of a string from the set of all runs. In [25] an $\mathcal{O}(n^2)$-time algorithm for this problem is presented and, moreover, it is shown that recovering a string over the smallest alphabet is NP-hard. Another hard problem is string reconstruction from the longest previous factor (LPF) array, which is NP-complete even without restrictions on the alphabet size [16]. Recovery problems have also been investigated for indeterminate strings [1,4].

A solution to our general problem provides a single tool for several existing recovery problems. This mainly includes problems which can be expressed as finding a string satisfying a conjunction of certain explicit substring equality constraints and implicit substring inequality constraints.

Our Further Results. We obtain linear-time algorithms for inferring a string from its border array, prefix array, set of maximal palindromes or set of runs. In particular, we improve the quadratic reconstruction algorithm from runs of [25]. In all cases the algorithms compute a generic solution.

Overview of the Paper. In Section 2, we present a naive algorithm and name certain properties of generic solutions $\Phi(E)$. Also, we provide a way to remove all redundant equations in the system. In Section 3, we present a simple $\mathcal{O}(|E| + n \log n)$-time algorithm, which is based on the doubling technique. It shares some features with the construction algorithm of the KMR identifiers [21], but it processes equations in decreasing lengths, as opposed to the increasing order in KMR. Then, in Section 4, we design a linear-time solution. It is a recursive algorithm which to some extent resembles the suffix array construction algorithm by Kärkkäinen & Sanders [20]. In order to achieve $\mathcal{O}(n)$ running time, in this version of the algorithm we apply the maximum spanning tree construction algorithm by Fredman and Willard [14], which is not feasible in practice. To overcome this issue, in Section 5 we change some details of the algorithm, so that no heavy word-RAM machinery is required. We conclude with Section 6, where we present applications to several string recovery problems.

2 Basic Observations

A naive approach to finding a generic n-character solution to a system of equations E is to transform substring equations into equations of individual letters. If "$s[p..q] = s[p'..q']$" belongs to E, then for every $i \in \{0, \ldots, q-p\}$, the equation on letters $s_{p+i} = s_{p'+i}$ must be satisfied. The latter can be represented as edges in a *positions graph* whose vertices $\{0, \ldots, n-1\}$ correspond to positions in s. This graph lets us easily characterize generic solutions. The idea of using a graph to represent constraints on letters already appeared in the context of prefix array reconstruction for indeterminate strings [4,6].

Observation 2. *For $s = \Phi(E)$, we have $s[i] = s[j]$ if and only if i, j are in the same connected component of the positions graph.*

Example 3. If the equations are:

$$E: \quad s[0..2] = s[3..5], \quad s[2..2] = s[3..3], \quad s[3..5] = s[5..7],$$

then we obtain the following positions graph:

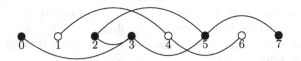

Its connected components are $\{0, 2, 3, 5, 7\}$ and $\{1, 4, 6\}$. The generic string that satisfies E is a string $\Phi(E) = \text{abaababa}$.

The approach described above in general works in $\mathcal{O}(|E|n)$ time, where $|E|$ is the number of equations in E. However, it is much more efficient for short equations.

Observation 4. *If all equations of E are of length 1, then the size of the positions graph is $\mathcal{O}(|E| + n)$. Consequently, $\Phi(E)$ can be computed in linear time.*

We call two systems E and E' *equivalent*, denoted $E \equiv E'$, if both have exactly the same solutions, i.e., $\Phi(E) = \Phi(E')$. Observe that this relation is hereditary in some sense: if $E \equiv E'$, then $E \cup F \equiv E' \cup F$ for any system F.

In the remainder of the paper, we represent each equation "$s[p..q] = s[p'..q']$" as a triple $(p, p', q-p+1)$. We refer to p, p' as the *starting positions* and to $q-p+1$ as the *length* of the equation.

Our algorithms follow the lines of Theorem 4, converting the input system E to an equivalent system composed of equations of length 1. As opposed to the naive algorithm, we control the number of the equations. This is achieved using two kinds of transformations, which we refer to as Split and Reduce.

A Split operation transforms a single equation into an equivalent system of shorter equations. More formally, to split an equation (i, j, ℓ), we choose a collection $\mathcal{I}$ of integer intervals $\{b, \ldots, e-1\}$ such that $\bigcup \mathcal{I} = \{0, \ldots, \ell-1\}$ and replace $\{(i, j, \ell)\}$ with $\{(i+b, j+b, e-b) : \{b, \ldots, e-1\} \in \mathcal{I}\}$. It is easy to verify that this indeed produces an equivalent system of equations.

While a `Split` transformation lets us decrease the lengths of equations, it increases the number of equations. To control the latter, we apply a `Reduce` operation, which finds $E' \subseteq E$ such that $E' \equiv E$. Such an operation can be also seen as a sequence of removals of a *redundant* equation: we find an equation $(i, j, \ell) \in E$ such that $E \setminus \{(i, j, \ell)\} \equiv E$, and remove it from E.

For a system of equations E represented as triples, we define its *equations graph* $G(E)$ as a weighted graph $(V(E), E)$, where $V(E) = \bigcup_{(i,j,\ell) \in E} \{i, j\}$, and triple (i, j, ℓ) represents edge (i, j) with weight ℓ. Note that the equations graph coincides with the positions graph for every system of equations of length 1. The equations graph lets us conveniently describe some properties of the system E using notions of graph theory. We say that a system of equations E is *acyclic* if the underlying graph $G(E)$ is acyclic, i.e., if $G(E)$ is a forest. For a weighted graph G, by $\mathrm{MST}(G)$ we denote the edge-set of a maximum-weight spanning forest of G.

Lemma 5. *Let $F = \mathrm{MST}(G(E))$ be a maximum-weight spanning forest of $G(E)$. Then $F \equiv E$.*

Proof. It is well-known that a maximum spanning forest can be constructed by iteratively removing the lightest edge on a cycle (this is the so-called *red rule*, upon which Kruskal's algorithm is based). Suppose that $G(E)$ contains a cycle C. By removing the lightest edge of C, denoted (i, j, ℓ), we obtain a set of edges E'.

Note that $C' = C \setminus \{(i, j, \ell)\}$ is a sequence of edges $(i, i_1, \ell_1), \ldots, (i_{k-1}, j, \ell_k)$ such that $\ell \leq \min(\ell_1, \ldots, \ell_k)$. By transitivity, the equation (i, j, ℓ) is implied by the equations from C'. Hence, $C \equiv C'$, and consequently, $E \equiv E'$.

Applying this argument inductively, we obtain that $\mathrm{MST}(G(E)) \equiv E$.   □

3 $\mathcal{O}(|E| + n \log n)$-Time Algorithm

We start with an $\mathcal{O}(|E| + n \log n)$-time algorithm, which uses simple split and reduction rules. We say that a system of equations E is *p-uniform* if all equations in E have length p. For uniform systems, it is easy to design an efficient implementation of the reduction rule obtained through Theorem 5. We denote the underlying procedure as `UniformReduce(E)`.

Lemma 6. *Given a p-uniform system E, an equivalent acyclic subsystem $F \subseteq E$ can be constructed in $\mathcal{O}(|E|)$ time.*

Proof. By Lemma 5, it suffices to take $F = \mathrm{MST}(G(E))$. For a uniform system of equations, $G(E)$ has uniform weights, so any spanning forest is maximal. Such a forest can be constructed using a textbook graph search algorithm.   □

A complementary split rule `UniformSplit(E, p)` is used to transform each equation (i, j, ℓ) from a system E into a pair of equations of a specified length p:

$$\{(i, j, \ell)\} \equiv \{(i, j, p), (i + \ell - p, j + \ell - p, p)\}.$$

It is applicable whenever all equations satisfy $p \leq \ell \leq 2p$.

The pseudocode of an algorithm using these two rules is provided below. In each step, it processes a system of equations of length between 2^k and $2^{k+1} - 1$, which consists both of equations obtained from the preceding step and equations from the input system. First, these equations are transformed into a 2^k-uniform system using the `UniformSplit` operation. Then the resulting system is reduced into an acyclic system using Theorem 6.

Algorithm 1. $\mathcal{O}(|E| + n \log n)$-time solution

Input: A system of equations E
Output: A generic solution $\Phi(E)$

$F_{\lfloor \log n \rfloor + 1} := \emptyset$
for $k := \lfloor \log n \rfloor$ **downto** 0 **do**
   $E_k := \{(i, j, \ell) \in E : 2^k \leq \ell < 2^{k+1}\}$
   $F_k := \texttt{UniformSplit}(F_{k+1} \cup E_k,\ 2^k)$ {now F_k is 2^k-uniform}
   $F_k := \texttt{UniformReduce}(F_k,\ 2^k)$ {using Theorem 6}
return $\Phi(F_0)$ {using Theorem 4}

Proposition 7. *Let E be a system of m equations over n positions. Algorithm 1 computes the universal solution $\Phi(E)$ of E in $\mathcal{O}(m + n \log n)$ time.*

Proof. The iteration of the **for**-loop indexed with k runs in $\mathcal{O}(|E_k| + |F_k|)$ time. Note that $\sum_k |E_k| = m$ and $|F_k| < n$, since each F_k is acyclic. $\square$

4   Linear-Time Algorithm

There are two main ideas behind the improvement from $\mathcal{O}(n \log n)$ to $\mathcal{O}(n)$ in the running time of the algorithm. First, we apply advanced machinery to efficiently implement for an arbitrary system the reduction rule following from Lemma 5. The other idea relies on a novel application of splitting. Previously, we used it only to manipulate the lengths of equations: to make them smaller and uniform. Now, we also apply split operations to restrict the starting positions of long equations. This is useful since the $|E| < n$ bound on the size of an acyclic system is actually $|E| < |V(E)|$. Thus, we introduce *special* positions.

4.1   Properties of k-Special Integers

Definition 8. *We say that a non-negative integer i is k-special if none of the k least significant digits of the quaternary representation of i is zero, i.e., if the suffix of length k of $(i)_4$ does not contain a zero.*

We denote the set of k-special integers by $\mathcal{S}_k$.

Observation 9. *Let i, j be non-negative integers such that $i \equiv j \pmod{4^k}$. Then $i \in \mathcal{S}_k$ if and only if $j \in \mathcal{S}_k$.*

Example 10. An integer i is 2-special unless its remainder modulo 16 is one of the following: $\{(00)_4, (01)_4, (02)_4, (03)_4, (10)_4, (20)_4, (30)_4\}$. We have $\mathcal{S}_2 = \{5, 6, 7, 9, 10, 11, 13, 14, 15, 21, 22, 23, 25, 26, 27, 29, 30, 31, \ldots\}$.

Fact 11. *(a) For every positive integer n we have $|\mathcal{S}_k \cap \{0, \ldots, n-1\}| \leq \left(\frac{3}{4}\right)^k n$.*
(b) If $i, j \in \mathcal{S}_k$, then there exists an integer $r \in \{0, 1, 2\}$ such that $i + r4^k, j + r4^k \in \mathcal{S}_{k+1}$. Moreover, such r can be found in constant time.

Proof. (a) Observe that $|\mathcal{S}_k \cap \{0, \ldots, 4^k - 1\}| = 3^k$. Hence, due to Theorem 9, the claim is valid whenever n is a multiple of 4^k. The technical generalization of the proof for arbitrary n is omitted in this version.

(b) Let c and d be the $(k+1)$-th least significant digits of $(i)_4$ and $(j)_4$, respectively. If both c and d are non-zero, we take $r = 0$. If both are equal to 0, we take $r = 1$. Otherwise, exactly one of c, d is 0. If the other is equal to 1 or 2, we choose $r = 1$. In the remaining cases, we take $r = 2$. □

4.2 Split Rules

An equation (i, j, ℓ) is called k-*special* if the positions i, j are both k-special integers. We say that an equation is r-*short* if its length does not exceed r.

Lemma 12. *Every k-special equation can be split in $\mathcal{O}(1)$ time into a constant number of k-special equations each of which is $(k+1)$-special or 4^{k+1}-short.*

Proof. If the input equation (i, j, ℓ) is already 4^{k+1}-short, there is nothing to do. Otherwise, we apply Theorem 11(b) to find $r \in \{0, 1, 2\}$ such that $(i + r4^k, j + r4^k, \ell - r4^k)$ is $(k+1)$-special, and we use it in the decomposition along with $(i, j, 4^{k+1})$. □

By $\texttt{SpecialSplit}(E, k)$ we denote a procedure which applies the lemma above for every equation in E and returns a pair of systems (E_1, E_2) such that equations in E_1 are 4^{k+1}-short, equations in E_2 are $(k+1)$-special, and $E \equiv E_1 \cup E_2$.

Lemma 13. *Every 4^{k+1}-short k-special equation can be split in $\mathcal{O}(1)$ time into a constant number of 4^k-short k-special equations.*

Proof. If the input equation (i, j, ℓ) is already 4^k-short, there is nothing to do. Otherwise, we split it into at most four 4^k-short k-special equations. We replace the input equation with $(i, j, 4^k)$ and $(i + 4^k, j + 4^k, \ell - 4^k)$. By Theorem 9, the latter is also k-special, and we might need to further split it into shorter equations. This step needs to be performed at most 4 times since $\ell \leq 4^{k+1}$. □

By $\texttt{SimpleSplit}(E, k)$ we denote a procedure which applies the lemma above for every equation in E and returns a system E' equivalent with E.

4.3   Reduction Rule

The general reduction procedure Reduce(E) is based on implementing Theorem 5 using a celebrated result by Fredman and Willard:

Theorem 14 ([14]). *In the standard word-RAM model of computation with word size $w = \Omega(\log n)$, maximum-weight spanning forest of a graph with w-bit integer weights can be computed in linear time.*

Corollary 15. *After $\mathcal{O}(n)$-time preprocessing, given an arbitrary equation system E, an equivalent acyclic system $F \subseteq E$ can be constructed in $\mathcal{O}(|E|)$ time.*

Proof. We follow the approach given by Theorem 5: we build the equations graph $G(E)$ and apply Theorem 14 to compute its maximum-weight spanning forest $F \subseteq E$. To avoid renaming vertices, we store them in an array of size n. During the preprocessing phase, we initialize this representation to store an empty graph, and then add an edge for each equation in E. Once we are done computing $\mathrm{MST}(G(E))$, we clean up iterating through E once again.   □

4.4   Algorithm

In this section we apply the split and reduction rules developed above to obtain a linear-time algorithm. The main idea is to use SpecialSplit for subsequent values $k = 0, 1, \dots$ in order to further and further restrict the starting positions of long equations. This is interleaved with reductions which bound the number of such equations to $|\mathcal{S}_k \cap \{0, \dots, n-1\}| \leq (\frac{3}{4})^k n$, which is $\mathcal{O}(n)$ in total. For large enough k there are no k-special equations. In the end we are left with the remaining products of SpecialSplit, i.e., for every k with $\mathcal{O}((\frac{3}{4})^k n)$ equations which are both 4^{k+1}-short and k-special. They are processed in the order of decreasing k using alternating calls of SimpleSplit and Reduce, similarly as in Algorithm 1. The following recursive procedure implements this approach.

Procedure Shorten(E, k)

   Input: An acyclic system E of k-special equations
   Output: An equivalent acyclic system of k-special 4^k-short equations

   if $E = \emptyset$ **then return** $\emptyset$
   $(E_1, E_2) := \texttt{SpecialSplit}(E, k)$ {using Theorem 12}
   $F := \texttt{Shorten}(\texttt{Reduce}(E_2), k+1)$ {recursive call, Theorem 15}
   $F' := \texttt{SimpleSplit}(E_1 \cup F, k)$ {using Theorem 13}
   return $\texttt{Reduce}(F')$ {using Theorem 15}

Let us analyze its running time. By Theorem 11(a), the size of any acyclic system of k-special equations does not exceed $(\frac{3}{4})^k n$. Consequently, we have $|E| \leq (\frac{3}{4})^k n$ and $|F| \leq (\frac{3}{4})^{k+1} n$. Theorems 12 and 13 imply that split operations work in $\mathcal{O}((\frac{3}{4})^k n)$ time and, in particular, return systems of this size. Thus, by

Theorem 15, the reduction also works in $\mathcal{O}((\frac{3}{4})^k n)$ time. Consequently, the total time to compute $\mathtt{Shorten}(E, 0)$ is $\mathcal{O}(\sum_{k \geq 0}(\frac{3}{4})^k n) = \mathcal{O}(n)$.

Before we apply the $\mathtt{Shorten}$ procedure, we need to make sure the input system is acyclic and consists of 0-special equations. The latter condition is void, so we just perform a reduction.

Algorithm 2. MAIN. An $\mathcal{O}(|E| + n)$-time solution

Input: A system of equations E
Output: A generic solution $\Phi(E)$

$E' := \mathtt{Reduce}(E)$
$E'' := \mathtt{Shorten}(E', 0)$ $\{E''$ is 1-short and $E'' \equiv E\}$
return $\Phi(E'')$ $\{$using Theorem 4$\}$

This way, we complete the proof of our main result.

Theorem 16. *Let E be a system of m equations between substrings of a string of length n. There exists an $\mathcal{O}(n+m)$-time algorithm that finds the generic string $\Phi(E)$ that satisfies E.*

5 Practical Implementation

The disadvantage of the algorithm presented in the previous section is that it uses the algorithm of Fredman and Willard (Theorem 14), which is very efficient in asymptotic terms but complicated and thus impractical. However, without much effort we are able to restrict the weights to powers of 2 not exceeding n. In this case, the MST can be computed using a simple solution based on the the Dijkstra-Jarník-Prim algorithm, which uses a priority queue as the underlying data structure (rather than union-find in Kruskal's algorithm). We maintain a single instance of the queue designed to efficiently handle a small universe.

Lemma 17. *In the standard word-RAM model of computation with word size $w = \Omega(\log N)$ (where N is a power of 2), one can implement a priority queue for keys within $\{2^0, 2^1, \ldots, 2^{\log N - 1}\}$ supporting standard operations (insert, find maximum, delete maximum) in $\mathcal{O}(1)$ time after $\mathcal{O}(N)$-time initialization.*

Proof sketch. The idea behind the priority queue is to store an integer whose bits represent keys present in the queue. Elements in the queue are stored in lists with each list responsible for a single key. To implement this queue, we still require word-RAM model, but we just use two standard bit operations. First, given a non-negative integer $x < N$, we want to locate position of the highest bit set to 1 in its binary representation, denoted $\mathrm{msb}(x)$. Second, given $x < N$ and $k < \log N$, we want to flip the k-th bit of x. Details are left for the full version.

Corollary 18. *After $\mathcal{O}(n)$-time preprocessing, given an arbitrary system E of equations whose lengths are powers of two, an equivalent acyclic subsystem $F \subseteq E$ can be constructed in $\mathcal{O}(|E|)$ time.*

5.1   Adjusted Algorithm

We modify the definition of a k-special equation as follows. An equation (i, j, ℓ) is *strongly k-special* if $i, j \in \mathcal{S}_k$ and additionally $\ell = 2^p$ for some integer $p \geq 2k$. In particular, we may use Theorem 18 for any system of strongly special equations. The lower bound $\ell \geq 4^k$ is useful to implement the split operations, both adjusted below for strongly k-special equations. Due to space constraints, we omit the proof of Lemma 19.

Lemma 19. *Every strongly k-special equation can be split in $\mathcal{O}(1)$ time into a constant number of strongly k-special equations each of which is strongly $(k+1)$-special or 4^{k+1}-short.*

Lemma 20. *Every strongly k-special equation of length up to 4^{k+1} can be split in $\mathcal{O}(1)$ time into a constant number of 4^k-short strongly k-special equations.*

Proof. It suffices to split each equation equally into equations of length 4^k.   □

Apart from slightly different implementation of subroutines, the procedure Shorten remains unchanged. However, before we apply it for $k = 0$, we need to take into account that while every equation is 0-special, it does not need to by strongly 0-special. However, it is easy to split any equation into two strongly 0-special ones. This is exactly the UniformSplit operation of Section 3.

6   Applications

In this section we apply Theorem 16 for several string recovery problems. In this class of problems, we are supposed to find an example string of a given length n which satisfies certain properties, or state that no such string exists. We assume an unbounded alphabet $\Sigma = \mathbb{N}$. In each of the problems, the property is expressible as a system of equations on substrings $E_=$ and a system of inequalities (more precisely, non-equalities) of substrings $E_{\neq}$ that the string should satisfy.

Theorem 21 provides a verification tool for the question whether there are strings consistent with $E_=$ and $E_{\neq}$. This is because either $\Phi(E_=)$ is valid or there are no solutions possible.

Lemma 21. *Let $E_=$ and $E_{\neq}$ be sets of equations and inequalities over n positions. If $\Phi(E_=)$ does not satisfy $E_{\neq}$ then there exists no string of length n that satisfies both $E_=$ and $E_{\neq}$. Moreover, one can check in $\mathcal{O}(n+|E_{\neq}|)$ time if $\Phi(E_=)$ satisfies all inequalities $E_{\neq}$.*

Proof. By Observation 1, any string s satisfying $E_=$ must be an image of $t = \Phi(E_=)$ through a letter-to-letter morphism. Therefore, every substring inequality satisfied by s is also satisfied by t. To check if t satisfies $E_{\neq}$, one can use longest common extension queries (i.e., longest common prefix queries); see [8].   □

We start the presentation of applications with two simpler examples. The *prefix array* PREF$[1..n - 1]$ stores in PREF$[i]$ the length of the longest common prefix of s and $s[i..n - 1]$.

Lemma 22. *For every array $A[1..n-1]$ with values in $\{0,\ldots,n-1\}$ there exists a set of equations $E_=$ and a set of inequalities $E_{\neq}$ such that A is the prefix array of a string s if and only if s satisfies both $E_=$ and $E_{\neq}$. Moreover, $|E_=| \leq n$, $|E_{\neq}| \leq n$ and both sets can be constructed in $\mathcal{O}(n)$ time.*

Proof. We take the following equations: $E_= = \{s[0..A[i]-1] = s[i..i+A[i]-1] : i = 1,\ldots,n-1\}$ and inequalities: $E_{\neq} = \{s[A[i]] \neq s[i+A[i]] : i+A[i] < n; i = 1,\ldots,n-1\}$. □

We say that a string u is a *border* of a string v, if u occurs both as a proper prefix and as a proper suffix of v. The *border array* $B[1..n-1]$ is an integer array such that $B[i]$ is the length of the longest border of $s[0..i]$.

Lemma 23. *For every array $A[1..n-1]$ with values in $\{0,\ldots,n-1\}$ there exists a set of equations $E_=$ and a set of inequalities $E_{\neq}$ such that A is the border array of a string s if and only if s satisfies both $E_=$ and $E_{\neq}$. Moreover, $|E_=| \leq n$ and this set can be constructed in $\mathcal{O}(n)$ time.*

Proof. We take the following equations: $E_= = \{s[0..A[i]-1] = s[i-A[i]+1..i] : i = 1,\ldots,n-1\}$. The inequalities state in an analogous way that $s[0..i]$ have no border of length exceeding $A[i]$: $E_{\neq} = \{s[0..j-1] \neq s[i-j+1..i] : i = 1,\ldots,n-1; A[i] < j \leq i\}$. □

A *period* of a string v is such a positive integer $p \leq |v|$ that $v[i] = v[i+p]$ for $0 \leq i < |v|-p$. A *run* is a triple (i,j,p) such that p is the smallest period of $s[i..j]$, $|s[i..j]| \geq 2p$ and neither $s[i-1..j]$ nor $s[i..j+1]$ have period p (possibly because $i = 0$ or $j = |s|-1$). By $Runs(s)$ we denote the set of all runs in s. Every string of length n has less than n runs [2].

Lemma 24. *For a set R ($|R| < n$) of integer triples (i,j,p), $0 \leq i < j \leq n-1$, $p \in \{1,\ldots,\lfloor n/2 \rfloor\}$, there exists a set of equations $E_=$ and a set of inequalities $E_{\neq}$ such that R is the set of all runs of a string s of length n if and only if s satisfies both $E_=$ and $E_{\neq}$. Moreover, $|E_=| \leq n$ and this set can be constructed in $\mathcal{O}(n)$ time.*

Proof. The fact that R is a set of runs of a string s can be described using equations $E_= = \{s[i..j-p] = s[i+p..j] : (i,j,p) \in R\}$. Inequalities say that no run is extendible or has a smaller period, and that no other runs exist in s. □

A string v is called a *palindrome* if v is equal to its reverse $v^R = v_{|v|-1}\ldots v_1 v_0$. A substring $s[i..j]$ is called a *maximal palindrome* if it is a palindrome, but $s[i-1..j+1]$ is not a palindrome. The set of maximal palindromes in s, denoted as $MaxPal(s)$, determines the structure of all palindromic substrings of s. Due to space constraints we omit the proof of the following lemma.

Lemma 25. *For a set P ($|P| < 2n$) of integer pairs (i,j), $0 \leq i \leq j \leq n-1$, there exists a set of equations $E_=$ and a set of inequalities $E_{\neq}$ over $2n$ positions such that P is the set of maximal palindromes of a string t of length n if and only if there exists a string s of length $2n$ that satisfies both $E_=$ and $E_{\neq}$ (then $s = tt^R$). Moreover, $|E_=| \leq 3n$, $|E_{\neq}| = \mathcal{O}(n)$ and both sets can be constructed in $\mathcal{O}(n)$ time.*

Theorem 26. *In $\mathcal{O}(n)$ time one can recover a string of length n from its prefix array, its border array, its runs structure, or its maximal palindromes.*

Proof. First, we apply one of Lemmas 22–25 to generate in $\mathcal{O}(n)$ time a system of equations $E_=$ for the particular recovery problem. Using Theorem 16, we compute a generic string $s = \Phi(E_=)$. Theorem 21 guarantees that if a solution to the recovery problem exists, then s is such a solution. Finally, we use a linear-time algorithm to compute its prefix array/border array/runs/maximal palindromes (see [2,10,24]), and check if the result matches the input. Note that for the maximal palindromes recovery problem, this way we obtain $s = tt^R$, so we need to take the first half of $s = \Phi(E_=)$ as the final solution t.

In the case of prefix array and maximal palindromes we have $|E_{\neq}| = \mathcal{O}(n)$. Therefore in these cases we obtain a simpler algorithm to check if s is a solution using Theorem 21. Either approach gives $\mathcal{O}(n)$-time recovery. $\square$

Our linear-time algorithm to recover a string from its runs improves upon an $\mathcal{O}(n^2)$-time algorithm by Matsubara et al. [25]. Finally, recovery from runs can be extended to gapped repeats and subrepetitions [23] with running time equal to the running time of the respective construction algorithms [22,23].

References

1. Alatabbi, A., Rahman, M.S., Smyth, W.: Inferring an indeterminate string from a prefix graph. J. of Discrete Algorithms (2014)
2. Bannai, H., Tomohiro, I., Inenaga, S., Nakashima, Y., Takeda, M., Tsuruta, K.: The "runs" theorem. ArXiv e-prints 1406.0263v6 (2015)
3. Bannai, H., Inenaga, S., Shinohara, A., Takeda, M.: Inferring strings from graphs and arrays. In: Rovan, B., Vojtáš, P. (eds.) MFCS 2003. LNCS, vol. 2747, pp. 208–217. Springer, Heidelberg (2003)
4. Blanchet-Sadri, F., Bodnar, M., Winkle, B.D.: New bounds and extended relations between prefix arrays, border arrays, undirected graphs, and indeterminate strings. In: Mayr, E.W., Portier, N. (eds.) Symposium on Theoretical Aspects of Computer Science. LIPIcs, vol. 25, pp. 162–173. Schloss Dagstuhl-Leibniz-Zentrum für Informatik (2014)
5. Cazaux, B., Rivals, E.: Reverse engineering of compact suffix trees and links: A novel algorithm. J. Discrete Algorithms **28**, 9–22 (2014)
6. Christodoulakis, M., Ryan, P.J., Smyth, W.F., Wang, S.: Indeterminate strings, prefix arrays & undirected graphs. ArXiv e-prints 1406.3289 (2014)
7. Clément, J., Crochemore, M., Rindone, G.: Reverse engineering prefix tables. In: Albers, S., Marion, J.Y. (eds.) Symposium on Theoretical Aspects of Computer Science. LIPIcs, vol. 3, pp. 289–300. Schloss Dagstuhl-Leibniz-Zentrum für Informatik (2009)
8. Crochemore, M., Hancart, C., Lecroq, T.: Algorithms on Strings. Cambridge University Press, Cambridge (2007)
9. Crochemore, M., Iliopoulos, C.S., Pissis, S.P., Tischler, G.: Cover array string reconstruction. In: Amir, A., Parida, L. (eds.) CPM 2010. LNCS, vol. 6129, pp. 251–259. Springer, Heidelberg (2010)
10. Crochemore, M., Rytter, W.: Jewels of Stringology. World Scientific (2003)

11. Duval, J., Lecroq, T., Lefebvre, A.: Border array on bounded alphabet. Journal of Automata, Languages and Combinatorics **10**(1), 51–60 (2005)
12. Duval, J., Lecroq, T., Lefebvre, A.: Efficient validation and construction of border arrays and validation of string matching automata. RAIRO-Theor. Inf. Appl. **43**(02), 281–297 (2009)
13. Franek, F., Gao, S., Lu, W., Ryan, P., Smyth, W., Sun, Y., Yang, L.: Verifying a border array in linear time. J. on Combinatorial Mathematics and Combinatorial Computing **42**, 223–236 (2002)
14. Fredman, M.L., Willard, D.E.: Trans-dichotomous algorithms for minimum spanning trees and shortest paths. J. Comput. Syst. Sci. **48**(3), 533–551 (1994)
15. Gawrychowski, P., Jeż, A., Jeż, P.: Validating the Knuth-Morris-Pratt failure function, fast and online. Theor. Comput. Syst. **54**(2), 337–372 (2014)
16. He, J., Liang, H., Yang, G.: Reversing longest previous factor tables is hard. In: Dehne, F., Iacono, J., Sack, J.-R. (eds.) WADS 2011. LNCS, vol. 6844, pp. 488–499. Springer, Heidelberg (2011)
17. Tomohiro, I., Inenaga, S., Bannai, H., Takeda, M.: Counting and verifying maximal palindromes. In: Chavez, E., Lonardi, S. (eds.) SPIRE 2010. LNCS, vol. 6393, pp. 135–146. Springer, Heidelberg (2010)
18. Tomohiro, I., Inenaga, S., Bannai, H., Takeda, M.: Verifying and enumerating parameterized border arrays. Theor. Comput. Sci. **412**(50), 6959–6981 (2011)
19. Tomohiro, I., Inenaga, S., Bannai, H., Takeda, M.: Inferring strings from suffix trees and links on a binary alphabet. Discrete Appl. Math. **163**, 316–325 (2014)
20. Kärkkäinen, J., Sanders, P., Burkhardt, S.: Linear work suffix array construction. J. ACM **53**(6), 918–936 (2006)
21. Karp, R.M., Miller, R.E., Rosenberg, A.L.: Rapid identification of repeated patterns in strings, trees and arrays. In: Fischer, P.C., Zeiger, H.P., Ullman, J.D., Rosenberg, A.L. (eds.) 4th Annual ACM Symposium on Theory of Computing, pp. 125–136. ACM (1972)
22. Kociumaka, T., Radoszewski, J., Rytter, W., Waleń, T.: Internal pattern matching queries in a text and applications. In: Indyk, P. (ed.) Twenty-Sixth Annual ACM-SIAM Symposium on Discrete Algorithms, pp. 532–551. SIAM (2015)
23. Kolpakov, R., Podolskiy, M., Posypkin, M., Khrapov, N.: Searching of gapped repeats and subrepetitions in a word. In: Kulikov, A.S., Kuznetsov, S.O., Pevzner, P. (eds.) CPM 2014. LNCS, vol. 8486, pp. 212–221. Springer, Heidelberg (2014)
24. Kolpakov, R.M., Kucherov, G.: Finding maximal repetitions in a word in linear time. In: 40th Annual Symposium on Foundations of Computer Science, FOCS 1999, pp. 596–604. IEEE Computer Society (1999)
25. Matsubara, W., Ishino, A., Shinohara, A.: Inferring strings from runs. In: Holub, J., Ždárek, J. (eds.) Prague Stringology Conference, pp. 150–160. Czech Technical University (2010)
26. Moosa, T.M., Nazeen, S., Rahman, M.S., Reaz, R.: Linear time inference of strings from cover arrays using a binary alphabet. In: Rahman, M.S., Nakano, S. (eds.) WALCOM 2012. LNCS, vol. 7157, pp. 160–172. Springer, Heidelberg (2012)
27. Nakashima, Y., Okabe, T., Tomohiro, I., Inenaga, S., Bannai, H., Takeda, M.: Inferring strings from lyndon factorization. In: Csuhaj-Varjú, E., Dietzfelbinger, M., Ésik, Z. (eds.) MFCS 2014, Part II. LNCS, vol. 8635, pp. 565–576. Springer, Heidelberg (2014)
28. Starikovskaya, T., Vildhøj, H.W.: A suffix tree or not a suffix tree? J. Discrete Algorithms **32**, 14–14 (2015)

The Complexity of Dominating Set Reconfiguration

Arash Haddadan[1], Takehiro Ito[2,3], Amer E. Mouawad[1], Naomi Nishimura[1], Hirotaka Ono[4], Akira Suzuki[2,3]([✉]), and Youcef Tebbal[1]

[1] University of Waterloo, 200 University Ave. West, Waterloo, Ontario N2L 3G1, Canada
{ahaddada,aabdomou,nishi,ytebbal}@uwaterloo.ca
[2] Graduate School of Information Sciences, Tohoku University, Aoba-yama 6-6-05, Sendai 980-8579, Japan
{takehiro,a.suzuki}@ecei.tohoku.ac.jp
[3] CREST, JST, 4-1-8 Honcho, Kawaguchi, Saitama 332-0012, Japan
[4] Faculty of Economics, Kyushu University, Hakozaki 6-19-1, Higashi-ku, Fukuoka 812-8581, Japan
hirotaka@econ.kyushu-u.ac.jp

Abstract. Suppose that we are given two dominating sets D_s and D_t of a graph G whose cardinalities are at most a given threshold k. Then, we are asked whether there exists a sequence of dominating sets of G between D_s and D_t such that each dominating set in the sequence is of cardinality at most k and can be obtained from the previous one by either adding or deleting exactly one vertex. This decision problem is known to be PSPACE-complete in general. In this paper, we study the complexity of this problem from the viewpoint of graph classes. We first prove that the problem remains PSPACE-complete even for planar graphs, bounded bandwidth graphs, split graphs, and bipartite graphs. We then give a general scheme to construct linear-time algorithms and show that the problem can be solved in linear time for cographs, trees, and interval graphs. Furthermore, for these tractable cases, we can obtain a desired sequence if it exists such that the number of additions and deletions is bounded by $O(n)$, where n is the number of vertices in the input graph.

1 Introduction

Consider the art gallery problem modeled on graphs: Each vertex corresponds to a room which has a monitoring camera and each edge represents the adjacency of two rooms. Assume that each camera in a room can monitor the room itself and its adjacent rooms. Then, we wish to find a subset of cameras that can monitor all rooms; the corresponding vertex subset D of the graph G is called a *dominating set*, that is, every vertex in G is either in D or adjacent to a vertex in D. For example, Fig. 1 shows six different dominating sets of the same graph. Given a graph G and a positive integer k, the problem of determining whether G has a dominating set of cardinality at most k is a classical NP-complete problem [4].

© Springer International Publishing Switzerland 2015
F. Dehne et al. (Eds.): WADS 2015, LNCS 9214, pp. 398–409, 2015.
DOI: 10.1007/978-3-319-21840-3_33

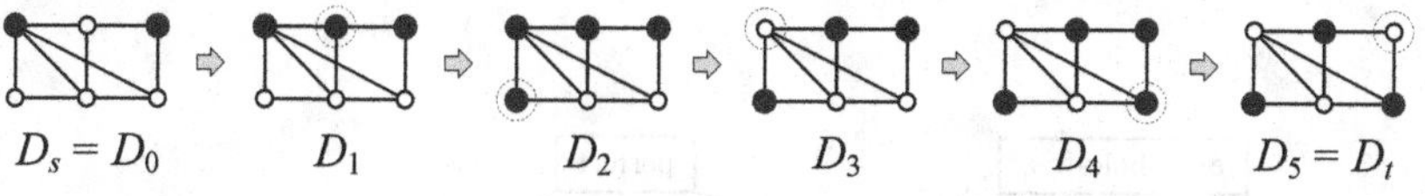

Fig. 1. A sequence $\langle D_0, D_1, \ldots, D_5 \rangle$ of dominating sets in the same graph, where $k = 4$ and the vertices in dominating sets are depicted by large (blue) circles

1.1 Our Problem

However, the art gallery problem could be considered in more "dynamic" situations: In order to maintain the cameras, we sometimes need to change the current dominating set into another one. This transformation needs to be done by switching the cameras individually and we certainly need to keep monitoring all rooms, even during the transformation.

In this paper, we thus study the following problem: Suppose that we are given two dominating sets of a graph G whose cardinalities are at most a given threshold $k > 0$ (e.g., the leftmost and rightmost ones in Fig. 1, where $k = 4$), and we are asked whether we can transform one into the other via dominating sets of G such that each intermediate dominating set is of cardinality at most k and can be obtained from the previous one by either adding or deleting a single vertex. We call this decision problem the DOMINATING SET RECONFIGURATION (DSR) problem. For the particular instance of Fig. 1, the answer is yes as illustrated in Fig. 1.

1.2 Known and Related Results

Recently, similar problems have been extensively studied under the reconfiguration framework [8], which arises when we wish to find a step-by-step transformation between two feasible solutions of a combinatorial problem such that all intermediate solutions are also feasible. The reconfiguration framework has been applied to several well-studied problems, including SATISFIABILITY [5], INDEPENDENT SET [7,8,10,12,15], VERTEX COVER [8,9,11,12], CLIQUE, MATCHING [8], VERTEX-COLORING [2], and so on. (See also a survey [14].)

Mouawad et al. [12] proved that DOMINATING SET RECONFIGURATION is $W[2]$-hard when parameterized by $k + \ell$, where k is the cardinality threshold of dominating sets and ℓ is the length of a sequence of dominating sets.

Haas and Seyffarth [6] gave sufficient conditions for the cardinality threshold k for which any two dominating sets can be transformed into one another. They proved that the answer to DOMINATING SET RECONFIGURATION is yes for a graph G with n vertices if $k = n - 1$ and G has a matching of cardinality at least two; they also gave a better sufficient condition when restricted to bipartite and chordal graphs. Recently, Suzuki et al. [13] improved the former condition and showed that the answer is yes if $k = n - \mu$ and G has a matching of cardinality at least $\mu + 1$, for any nonnegative integer μ.

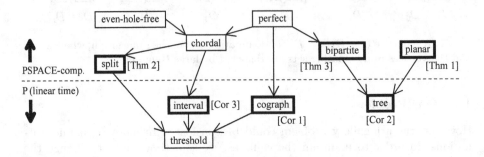

Fig. 2. Our results, where each arrow represents the inclusion relationship between graph classes: $A \rightarrow B$ represents that B is properly included in A [3]. We also show PSPACE-completeness on graphs of bounded bandwidth (Theorem 1).

1.3 Our Contribution

To the best of our knowledge, no algorithmic results are known for the DOMI-NATING SET RECONFIGURATION problem, and it is therefore desirable to obtain a better understanding of what separates "hard" from "easy" instances. To that end, we study the problem from the viewpoint of graph classes and paint an interesting picture of the boundary between intractability and polynomial-time solvability. (See also Fig. 2.)

We first prove that the problem is PSPACE-complete even on planar graphs, bounded bandwidth graphs, split graphs, and bipartite graphs. Our reductions for PSPACE-hardness follow from the classical reductions for proving the NP-hardness of DOMINATING SET [1,4]. However, the reductions should be constructed carefully so that they preserve not only the existence of dominating sets but also the reconfigurability.

We then give a general scheme to construct linear-time algorithms for the problem. As examples of its application, we demonstrate that the problem can be solved in linear time on cographs (also known as P_4-free graphs), trees, and interval graphs. Furthermore, for these tractable cases, we can obtain a desired sequence if it exists such that the number of additions and deletions (i.e., the length of the sequence) can be bounded by $O(n)$, where n is the number of vertices in the input graph.

Due to the page limitation, we omit proofs of lemmas and theorems marked with a star from this extended abstract.

2   Preliminaries

In this section, we define some basic terms and notation which will be used throughout the paper.

2.1   Graph Notation and Dominating Set

We assume that each input graph G is a simple undirected graph with vertex set $V(G)$ and edge set $E(G)$, where $|V(G)| = n$ and $|E(G)| = m$. For a vertex v in G, we let $N_G(v) = \{u \in V(G) \mid vu \in E(G)\}$ and $N_G[v] = N_G(v) \cup \{v\}$. For a set $S \subseteq V(G)$ of vertices, we define $N_G[S] = \bigcup_{v \in S} N_G[v]$ and $N_G(S) = N_G[S] \setminus S$.

For a graph G, a set $D \subseteq V(G)$ is a *dominating set* of G if $N_G[D] = V(G)$. Note that $V(G)$ always forms a dominating set of G. For a vertex $u \in V(G)$ and a dominating set D of G, we say that u is *dominated* by $v \in D$ if $u \notin D$ and $u \in N_G(v)$. A vertex w in a dominating set D is *deletable* if $D \setminus \{w\}$ is also a dominating set of G. A dominating set D of G is *minimal* if there is no deletable vertex in D.

2.2   Dominating Set Reconfiguration

We say that two dominating sets D and D' of the same graph G are *adjacent* if there exists a vertex $u \in V(G)$ such that $D \bigtriangleup D' = (D \setminus D') \cup (D' \setminus D) = \{u\}$, i.e., u is the only vertex in the *symmetric difference* of D and D'. For two dominating sets D_p and D_q of G, a sequence $\langle D_0, D_1, \ldots, D_\ell \rangle$ of dominating sets of G is called a *reconfiguration sequence* between D_p and D_q if it has the following properties:

(a) $D_0 = D_p$ and $D_\ell = D_q$; and
(b) D_{i-1} and D_i are adjacent for each $i \in \{1, 2, \ldots, \ell\}$.

Note that any reconfiguration sequence is *reversible*, that is, $\langle D_\ell, D_{\ell-1}, \ldots, D_0 \rangle$ is also a reconfiguration sequence between D_q and D_p. We say a vertex $v \in V(G)$ is *touched* in a reconfiguration sequence $\sigma = \langle D_0, D_1, \ldots, D_\ell \rangle$ if v is either added or deleted at least once in σ.

For two dominating sets D_p and D_q of a graph G and an integer $k > 0$, we write $D_p \overset{k}{\leftrightsquigarrow} D_q$ if there exists a reconfiguration sequence $\langle D_0, D_1, \ldots, D_\ell \rangle$ between D_p and D_q in G such that $|D_i| \leq k$ holds for every $i \in \{0, 1, \ldots, \ell\}$, for some $\ell \geq 0$. Note that $k \geq \max\{|D_p|, |D_q|\}$ clearly holds if $D_p \overset{k}{\leftrightsquigarrow} D_q$. Then, the DOMINATING SET RECONFIGURATION (DSR) problem is defined as follows:

> **Input:** A graph G, two dominating sets D_s and D_t of G, and an integer threshold $k \geq \max\{|D_s|, |D_t|\}$
> **Question:** Determine whether $D_s \overset{k}{\leftrightsquigarrow} D_t$ or not.

We denote by a 4-tuple (G, D_s, D_t, k) an instance of DOMINATING SET RECONFIGURATION. Note that DSR is a decision problem and hence it does not ask for an actual reconfiguration sequence. We always denote by D_s and D_t the *source* and *target* dominating sets of G, respectively.

3   PSPACE-Completeness

In this section, we prove that DOMINATING SET RECONFIGURATION remains PSPACE-complete even for restricted classes of graphs; some of these classes show nice contrasts to our algorithmic results in Section 4. (See also Fig. 2.)

Theorem 1. DSR *is PSPACE-complete on planar graphs of maximum degree six and on graphs of bounded bandwidth.*

Proof. One can observe that the problem is in PSPACE [8, Theorem 1]. We thus show that it is PSPACE-hard for those graph classes by a polynomial-time reduction from VERTEX COVER RECONFIGURATION [8,9,11]. In VERTEX COVER RECONFIGURATION, we are given two vertex covers C_s and C_t of a graph G' such that $|C_s| \leq k$ and $|C_t| \leq k$, for some integer k, and asked whether there exists a reconfiguration sequence of vertex covers $C_0, C_1, \ldots, C_\ell$ of G such that $C_0 = C_s$, $C_\ell = C_t$, $|C_i| \leq k$, and $|C_{i-1} \triangle C_i| = 1$ for each $i \in \{1, 2, \ldots, \ell\}$.

Our reduction follows from the classical reduction from VERTEX COVER to DOMINATING SET [4]. Specifically, for every edge uw in $E(G')$, we add a new vertex v_{uw} and join it with each of u and w by two new edges uv_{uw} and $v_{uw}w$; let G be the resulting graph. Then, let $(G, D_s = C_s, D_t = C_t, k)$ be the corresponding instance of DOMINATING SET RECONFIGURATION. Clearly, this instance can be constructed in polynomial time.

We now prove that $D_s \overset{k}{\leftrightsquigarrow} D_t$ holds if and only if there is a reconfiguration sequence of vertex covers in G' between C_s and C_t. However, the if direction is trivial, because any vertex cover of G' forms a dominating set of G and both problems employ the same reconfiguration rule (i.e., the symmetric difference is of size one). Therefore, suppose that $D_s \overset{k}{\leftrightsquigarrow} D_t$ holds, and hence there exists a reconfiguration sequence of dominating sets in G between D_s and D_t. Recall that neither D_s nor D_t contain a newly added vertex in $V(G) \setminus V(G')$. Thus, if a vertex v_{uw} in $V(G) \setminus V(G')$ is touched, then v_{uw} must be added first. By the construction of G, both $N_G[v_{uw}] \subseteq N_G[u]$ and $N_G[v_{uw}] \subseteq N_G[w]$ hold. Therefore, we can replace the addition of v_{uw} by that of either u or w and obtain a (possibly shorter) reconfiguration sequence of dominating sets in G between D_s and D_t which touches vertices only in G'. Then, it is a reconfiguration sequence of vertex covers in G' between C_s and C_t, as needed.

VERTEX COVER RECONFIGURATION is known to be PSPACE-complete on planar graphs of maximum degree three [9,11] and on graphs of bounded bandwidth [15]. Thus, the reduction above implies PSPACE-hardness on planar graphs of maximum degree six and on graphs of bounded bandwidth; note that, since the number of edges in G is only the triple of that in G', the bandwidth increases only by a constant multiplicative factor.   □

We note that both pathwidth and treewidth of a graph G are bounded by the bandwidth of G. Thus, Theorem 1 yields that DOMINATING SET RECONFIGURATION is PSPACE-complete on graphs of bounded pathwidth and treewidth.

Adapting known techniques from NP-hardness proofs for the DOMINATING SET problem [1], we also show PSPACE-completeness of DOMINATING SET RECONFIGURATION on split graphs and on bipartite graphs; a graph is *split* if its vertex set can be partitioned into a clique and an independent set [3].

Theorem 2 (*). DSR *is PSPACE-complete on split graphs.*

Theorem 3 (*). DSR *is PSPACE-complete on bipartite graphs.*

4   General Scheme for Linear-Time Algorithms

In this section, we show that DOMINATING SET RECONFIGURATION is solvable in linear time on cographs, trees, and interval graphs. Interestingly, these results can be obtained by the application of the same strategy; we first describe the general scheme in Section 4.1. We then show in Sections 4.2–4.4 that the problem can be solved in linear time on those graph classes.

4.1   General Scheme

The general idea is to introduce the concept of a "canonical" dominating set for a graph G. We say that a minimum dominating set C of G is *canonical* if for every dominating set D of G, it holds that $D \overset{k}{\longleftrightarrow} C$, where $k = |D| + 1$. Note that $|C| \leq |D|$ always holds, since C is a minimum dominating set of G. Then, we have the following theorem.

Theorem 4. *For every graph G admitting a canonical dominating set,* DOMINATING SET RECONFIGURATION *can be solved in linear time.*

We note that proving the existence of a canonical dominating set is sufficient for solving the decision problem (DOMINATING SET RECONFIGURATION). Therefore, we do not need to find an actual canonical dominating set in linear time. In Sections 4.2–4.4, we will show that cographs, trees, and interval graphs admit canonical dominating sets, and hence the problem can be solved in linear time on those graph classes. Note that, however, Theorem 4 can be applied to any graph which has a canonical dominating set. In the remainder of this subsection, we prove Theorem 4 starting with the following lemma.

Lemma 1. *Suppose that a graph G has a canonical dominating set. Then, an instance (G, D_s, D_t, k) of* DOMINATING SET RECONFIGURATION *is a yes-instance if $k \geq \max\{|D_s|, |D_t|\} + 1$.*

Proof. Let C be a canonical dominating set of G. Then, $D_s \overset{k'}{\longleftrightarrow} C$ holds for $k' = |D_s| + 1$. Suppose that $k \geq \max\{|D_s|, |D_t|\} + 1$. Since $k \geq |D_s| + 1 = k'$, we clearly have $D_s \overset{k}{\longleftrightarrow} C$. Similarly, we have $D_t \overset{k}{\longleftrightarrow} C$. Since any reconfiguration sequence is reversible, we have $D_s \overset{k}{\longleftrightarrow} C \overset{k}{\longleftrightarrow} D_t$, as needed. $\square$

Lemma 1 implies that if a graph G has a canonical dominating set C, then it suffices to consider the case where $k = \max\{|D_s|, |D_t|\}$. Note that there exist no-instances of DOMINATING SET RECONFIGURATION in such a case, but we show that they can be easily identified in linear time, as implied by the following lemma.

Lemma 2. *Let (G, D_s, D_t, k) be an instance of* DOMINATING SET RECONFIGURATION, *where G is a graph admitting a canonical dominating set, $D_s \neq D_t$, and $k = \max\{|D_s|, |D_t|\}$. Then, (G, D_s, D_t, k) is a yes-instance if and only if D_i is not minimal for every $i \in \{s, t\}$ such that $|D_i| = k$.*

Lemma 2 can be immediately obtained from the following lemma.

Lemma 3. *Suppose that a graph G has a canonical dominating set C. Let D be an arbitrary dominating set of G such that $D \neq C$, and let $k = |D|$. Then, $D \overset{k}{\rightsquigarrow} C$ holds if and only if D is not a minimal dominating set.*

Proof. We first prove the if direction. Suppose that D is not minimal. Then, D contains at least one vertex x which is deletable from D, that is, $D \setminus \{x\}$ forms a dominating set of G. Since $k = |D| = |D \setminus \{x\}| + 1$, we have $D \setminus \{x\} \overset{k}{\rightsquigarrow} C$. Therefore, $D \overset{k}{\rightsquigarrow} D \setminus \{x\} \overset{k}{\rightsquigarrow} C$ holds.

We then prove the only-if direction by taking the contrapositive. Suppose that D is minimal. Then, no vertex in D is deletable and hence any dominating set D' which is adjacent to D must be obtained by adding a vertex to D. Therefore, $|D'| = k + 1$ for any dominating set D' which is adjacent to D. Since $D \neq C$, $D \overset{k}{\rightsquigarrow} C$ does not hold. □

We note again that Lemmas 1 and 2 imply that an actual canonical dominating set is not required to solve the problem. Furthermore, it can be easily determined in linear time whether a dominating set of a graph G is minimal or not. Thus, Theorem 4 follows from Lemmas 1 and 2.

Before constructing canonical dominating sets in Sections 4.2–4.4, we give the following lemma showing that it suffices to construct a canonical dominating set for a connected graph.

Lemma 4 (*). *Let G be a graph consisting of p connected components $G_1, G_2, \ldots, G_p$. For each $i \in \{1, 2, \ldots, p\}$, suppose that C_i is a canonical dominating set for G_i. Then, $C = C_1 \cup C_2 \cup \cdots \cup C_p$ is a canonical dominating set for G.*

4.2   Cographs

We first define the class of cographs (also known as P_4-free graphs) [3]. For two graphs G_1 and G_2, their *union* $G_1 \cup G_2$ is the graph such that $V(G_1 \cup G_2) = V(G_1) \cup V(G_2)$ and $E(G_1 \cup G_2) = E(G_1) \cup E(G_2)$, while their *join* $G_1 \vee G_2$ is the graph such that $V(G_1 \vee G_2) = V(G_1) \cup V(G_2)$ and $E(G_1 \vee G_2) = E(G_1) \cup E(G_2) \cup \{vw \mid v \in V(G_1), w \in V(G_2)\}$. Then, a *cograph* can be recursively defined as follows:

(1) a graph consisting of a single vertex is a cograph;
(2) if G_1 and G_2 are cographs, then the union $G_1 \cup G_2$ is a cograph; and
(3) if G_1 and G_2 are cographs, then the join $G_1 \vee G_2$ is a cograph.

In this subsection, we show that DOMINATING SET RECONFIGURATION is solvable in linear time on cographs. By Theorem 4, it suffices to prove the following lemma.

Lemma 5. *Any cograph admits a canonical dominating set.*

As a proof of Lemma 5, we will construct a canonical dominating set for any cograph G. By Lemma 4, it suffices to consider the case where G is connected and we may assume that G has at least two vertices, because otherwise the problem is trivial. Then, from the definition of cographs, G must be obtained by the join operation applied to two cographs G_a and G_b, that is, $G = G_a \vee G_b$. Notice that any pair $\{w_a, w_b\}$ of vertices $w_a \in V(G_a)$ and $w_b \in V(G_b)$ forms a dominating set of G. Let C be a dominating set of G, defined as follows:
  - If there exists a vertex $w \in V(G)$ such that $N_G[w] = V(G)$, then let $C = \{w\}$.
  - Otherwise choose an arbitrary pair of vertices $w_a \in V(G_a)$ and $w_b \in V(G_b)$ and let $C = \{w_a, w_b\}$.
Clearly, C is a minimum dominating set of G. We thus give the following lemma, which completes the proof of Lemma 5.

Lemma 6 (*). *For every dominating set D of G, $D \overset{k}{\rightsquigarrow} C$ holds, where $k = |D| + 1$.*

We have thus proved that any cograph has a canonical dominating set. Then, Theorem 4 gives the following corollary.

Corollary 1. DSR *can be solved in linear time on cographs.*

4.3   Trees

In this subsection, we show that DOMINATING SET RECONFIGURATION is solvable in linear time on trees. As for cographs, it suffices to prove the following lemma.

Lemma 7. *Any tree admits a canonical dominating set.*

As a proof of Lemma 7, we will construct a canonical dominating set for a tree T. We choose an arbitrary vertex r of degree one in T and regard T as a rooted tree with root r.

We first label each vertex in T either 1, 2, or 3, starting from the leaves of T up to the root r of T, as in the following steps (1)–(3); intuitively, the vertices labeled 2 will form a dominating set of T, each vertex labeled 1 will be dominated by its parent, and each vertex labeled 3 will be dominated by at least one of its children (see also Fig. 3(a)):
  (1) All leaves in T are labeled 1.
  (2) Pick an internal vertex v of T, which is not the root, such that all children of v have already been labeled. Then,
      - assign v label 1 if all children of v are labeled 3;
      - assign v label 2 if at least one child of v is labeled 1; and
      - otherwise assign v label 3.
  (3) Assign the root r (of degree one) label 3 if its child is labeled 2, otherwise assign r label 2.
For each $i \in \{1, 2, 3\}$, we denote by V_i the set of all vertices in T that are assigned label i. Then, $\{V_1, V_2, V_3\}$ forms a partition of $V(T)$.

We will prove that V_2 forms a canonical dominating set of T. We first prove, in Lemmas 8 and 9, that V_2 is a minimum dominating set of T and then prove, in Lemma 10, that $D \overset{k}{\rightsquigarrow} V_2$ holds for every dominating set D of T and $k = |D| + 1$.

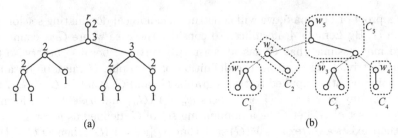

(a)　　　　　　　　　　　　　　　　　　(b)

Fig. 3. (a) The labeling of a tree T, and (b) the partition of $V(T)$ into $C_1, C_2, \ldots, C_5$.

Lemma 8. V_2 *is a dominating set of* T.

Proof. It suffices to show that both $V_1 \subseteq N_T(V_2)$ and $V_3 \subseteq N_T(V_2)$ hold.

Let v be any vertex in V_1, and hence v is labeled 1. Then, by the construction above, v is not the root of T and the parent of v must be labeled 2. Therefore, $v \in N_T(V_2)$ holds, as claimed.

Let u be any vertex in V_3, and hence u is labeled 3. Then, u is not a leaf of T. Notice that label 3 is assigned to a vertex only when at least one of its children is labeled 2. Thus, $u \in N_T(V_2)$ holds. □

We now prove that V_2 is a minimum dominating set of T. To do so, we introduce some notation. Suppose that the vertices in V_2 are ordered as $w_1, w_2, \ldots, w_{|V_2|}$ by a post-order depth-first traversal of the tree starting from the root r of T. For each $i \in \{1, 2, \ldots, |V_2|\}$, we denote by T_i the subtree of T which is induced by w_i and all its descendants in T. Then, for each $i \in \{1, 2, \ldots, |V_2|\}$, we define a vertex subset C_i of $V(T)$ as follows (see also Fig. 3(b)):

$$C_i = \begin{cases} V(T_i) \setminus \bigcup_{j<i} V(T_j) & \text{if } i \neq |V_2|; \\ V(T) \setminus \bigcup_{j<i} V(T_j) & \text{if } i = |V_2|. \end{cases}$$

Note that $\{C_1, C_2, \ldots, C_{|V_2|}\}$ forms a partition of $V(T)$. Furthermore, notice that

$$V_2 \cap C_i = \{w_i\} \tag{1}$$

holds for every $i \in \{1, 2, \ldots, |V_2|\}$. Then, Eq. (1) and the following lemma imply that V_2 is a minimum dominating set of T.

Lemma 9 (*). *Let D be an arbitrary dominating set of T. Then, $|D \cap C_i| \geq 1$ holds for every $i \in \{1, 2, \ldots, |V_2|\}$.*

We finally claim the following lemma, which completes the proof of Lemma 7.

Lemma 10 (*). *For every dominating set D of T, $D \overset{k}{\rightsquigarrow} V_2$ holds, where $k = |D| + 1$.*

We have thus proved that V_2 forms a canonical dominating set for any tree T. Then, Theorem 4 gives the following corollary.

Corollary 2. DSR *can be solved in linear time on trees.*

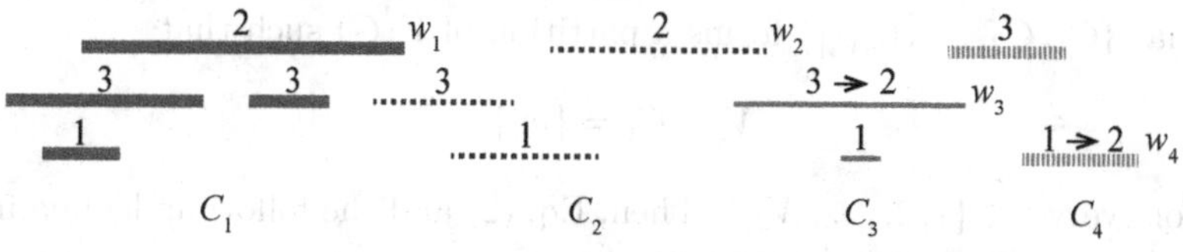

Fig. 4. The labeling of an interval graph in the interval representation

4.4   Interval Graphs

A graph G with $V(G) = \{v_1, v_2, \ldots, v_n\}$ is an *interval graph* if there exists a set $\mathcal{I}$ of (closed) intervals $I_1, I_2, \ldots, I_n$ such that $v_i v_j \in E(G)$ if and only if $I_i \cap I_j \neq \emptyset$ for each $i, j \in \{1, 2, \ldots, n\}$. We call the set $\mathcal{I}$ of intervals an *interval representation* of the graph. In this subsection, we show that DOMINATING SET RECONFIGURATION is solvable in linear time on interval graphs. As for cographs, it suffices to prove the following lemma.

Lemma 11. *Any interval graph admits a canonical dominating set.*

As a proof of Lemma 11, we will construct a canonical dominating set for any interval graph G. By Lemma 4 it suffices to consider the case where G is connected. Let $\mathcal{I}$ be an interval representation of G. For an interval $I \in \mathcal{I}$, we denote by $l(I)$ and $r(I)$ the left and right endpoints of I, respectively; we sometimes call the values $l(I)$ and $r(I)$ the *l-value* and *r-value* of I, respectively. As for trees, we first label each vertex in G either 1, 2, or 3, from left to right; the vertices labeled 2 will form a dominating set of G (see Fig. 4 as an example):

(1) Pick an unlabeled vertex v_i which has the minimum r-value among all unlabeled vertices and assign v_i label 1.
(2) Let v_j be the vertex in $N_G[v_i]$ which has the maximum r-value among all vertices in $N_G[v_i]$. Note that v_j may have been already labeled, and $v_j = v_i$ may hold. We (re)label v_j to 2.
(3) For each unlabeled vertex in $N_G(v_j)$, we assign it label 3.

We execute Steps (1)–(3) above until all vertices are labeled. For each $i \in \{1, 2, 3\}$, we denote by V_i the set of all vertices in G that are assigned label i. Then, $\{V_1, V_2, V_3\}$ forms a partition of $V(G)$.

By the construction above, it is easy to see that V_2 forms a dominating set of G. We thus prove that V_2 is canonical in Lemmas 12 and 13, that is, V_2 is a minimum dominating set of G (in Lemma 12) and $D \overset{k}{\leadsto} V_2$ holds for every dominating set D of G and $k = |D| + 1$ (in Lemma 13).

We now prove that the dominating set V_2 of G is minimum. To do so, we introduce some notation. Assume that the vertices in V_2 are ordered as $w_1, w_2, \ldots, w_{|V_2|}$ such that $r(w_1) < r(w_2) < \cdots < r(w_{|V_2|})$. For each $i \in \{1, 2, \ldots, |V_2|\}$, we define the vertex subset C_i of $V(G)$ as follows (see Fig. 4 as an example):

$$C_i = \begin{cases} \{v \mid \qquad\qquad r(v) \leq r(w_1) \} & \text{if } i = 1; \\ \{v \mid \ r(w_{i-1}) < r(v) \leq r(w_i) \} & \text{if } 2 \leq i \leq |V_2| - 1; \\ \{v \mid r(w_{|V_2|-1}) < r(v) \} & \text{if } i = |V_2|. \end{cases}$$

Note that $\{C_1, C_2, \ldots, C_{|V_2|}\}$ forms a partition of $V(G)$ such that

$$V_2 \cap C_i = \{w_i\} \tag{2}$$

holds for every $i \in \{1, 2, \ldots, |V_2|\}$. Then, Eq. (2) and the following lemma imply that V_2 is a minimum dominating set of G.

Lemma 12 (*). *Let D be an arbitrary dominating set of G. Then, $|D \cap C_i| \geq 1$ holds for every $i \in \{1, 2, \ldots, |V_2|\}$.*

We finally claim the following lemma, which completes the proof of Lemma 11.

Lemma 13 (*). *For every dominating set D of G, $D \overset{k}{\leadsto} V_2$ holds, where $k = |D| + 1$.*

Combining Lemma 11 and Theorem 4 yields the following corollary.

Corollary 3. DSR *can be solved in linear time on interval graphs.*

5   Concluding Remarks

In this paper, we delineated the complexity of the DOMINATING SET RECONFIG-URATION problem restricted to various graph classes. As shown in Fig. 2, our results clarify some interesting boundaries on the graph classes lying between tractability and PSPACE-completeness: For example, the structure of interval graphs can be seen as a path-like structure of cliques. As a super-class of interval graphs, the well-known class of chordal graphs has a tree-like structure of cliques. We have proved that DOMINATING SET RECONFIGURATION is solvable in linear time on interval graphs, while it is PSPACE-complete on chordal graphs.

We note again that our linear-time algorithms for cographs, trees, and interval graphs employ the same strategy. We also emphasize that this general scheme can be applied to any graph which admits a canonical dominating set. It is easy to modify our algorithms so that they actually find a reconfiguration sequence for a yes-instance (G, D_s, D_t, k) on cographs, trees, or interval graphs. Observe that each vertex is touched at most once in the reconfiguration sequence from D_s (or D_t) to the canonical dominating set. Therefore, for a yes-instance on an n-vertex graph belonging to one of those classes, there exists a reconfiguration sequence between D_s and D_t which touches vertices only $O(n)$ times. In other words, the length of a shortest reconfiguration sequence between D_s and D_t can be bounded by $O(n)$.

Acknowledgments. We thank anonymous referees for their helpful suggestions. This work is partially supported by the Natural Science and Engineering Research Council of Canada (A. Mouawad, N. Nishimura and Y. Tebbal) and by MEXT/JSPS KAKENHI 25330003 and 15H00849 (T. Ito), 25104521 and 26540005 (H. Ono), and 26730001 (A. Suzuki).

References

1. Bertossi, A.A.: Dominating sets for split and bipartite graphs. Information Processing Letters **19**, 37–40 (1984)
2. Bonsma, P., Cereceda, L.: Finding paths between graph colourings: PSPACE-completeness and superpolynomial distances. Theoretical Computer Science **410**, 5215–5226 (2009)
3. Brandstädt, A., Le, V.B., Spinrad, J.P.: Graph Classes: A Survey. SIAM (1999)
4. Garey, M.R., Johnson, D.S.: Computers and Intractability: A Guide to the Theory of NP-Completeness. Freeman, San Francisco (1979)
5. Gopalan, P., Kolaitis, P.G., Maneva, E.N., Papadimitriou, C.H.: The connectivity of Boolean satisfiability: computational and structural dichotomies. SIAM J. Computing **38**, 2330–2355 (2009)
6. Haas, R., Seyffarth, K.: The k-dominating graph. Graphs and Combinatorics **30**, 609–617 (2014)
7. Hearn, R.A., Demaine, E.D.: PSPACE-completeness of sliding-block puzzles and other problems through the nondeterministic constraint logic model of computation. Theoretical Computer Science **343**, 72–96 (2005)
8. Ito, T., Demaine, E.D., Harvey, N.J.A., Papadimitriou, C.H., Sideri, M., Uehara, R., Uno, Y.: On the complexity of reconfiguration problems. Theoretical Computer Science **412**, 1054–1065 (2011)
9. Ito, T., Nooka, H., Zhou, X.: Reconfiguration of vertex covers in a graph. In: Jan, K., Miller, M., Froncek, D. (eds.) IWOCA 2014. LNCS, vol. 8986, pp. 164–175. Springer, Heidelberg (2015)
10. Kamiński, M., Medvedev, P., Milanič, M.: Complexity of independent set reconfigurability problems. Theoretical Computer Science **439**, 9–15 (2012)
11. Mouawad, A.E., Nishimura, N., Raman, V.: Vertex cover reconfiguration and beyond. In: Ahn, H.-K., Shin, C.-S. (eds.) ISAAC 2014. LNCS, vol. 8889, pp. 452–463. Springer, Heidelberg (2014)
12. Mouawad, A.E., Nishimura, N., Raman, V., Simjour, N., Suzuki, A.: On the parameterized complexity of reconfiguration problems. In: Gutin, G., Szeider, S. (eds.) IPEC 2013. LNCS, vol. 8246, pp. 281–294. Springer, Heidelberg (2013)
13. Suzuki, A., Mouawad, A.E., Nishimura, N.: Reconfiguration of dominating sets. In: Cai, Z., Zelikovsky, A., Bourgeois, A. (eds.) COCOON 2014. LNCS, vol. 8591, pp. 405–416. Springer, Heidelberg (2014)
14. van den Heuvel, J.: The complexity of change. Surveys in Combinatorics (2013)
15. Wrochna, M.: Reconfiguration in bounded bandwidth and treedepth. arXiv:1405.0847 (2014)

Editing Graphs Into Few Cliques: Complexity, Approximation, and Kernelization Schemes

Falk Hüffner, Christian Komusiewicz, and André Nichterlein[(✉)]

Institut für Softwaretechnik und Theoretische Informatik,
TU Berlin, Berlin, Germany
{falk.hueffner,christian.komusiewicz,andre.nichterlein}@tu-berlin.de

Abstract. Given an undirected graph G and a positive integer k, the NP-hard SPARSE SPLIT GRAPH EDITING problem asks to transform G into a graph that consists of a clique plus isolated vertices by performing at most k edge insertions and deletions; similarly, the P_3-BAG EDITING problem asks to transform G into a graph which is the union of two possibly overlapping cliques. We give a simple linear-time 3-approximation algorithm for SPARSE SPLIT GRAPH EDITING, an improvement over a more involved known factor-3.525 approximation. Further, we show that P_3-BAG EDITING is NP-complete. Finally, we present a kernelization scheme for both problems and additionally for the 2-CLUSTER EDITING problem. This scheme produces for each fixed ε in polynomial time a kernel of order εk. This is, to the best of our knowledge, the first example of a kernelization scheme that converges to a known lower bound.

1 Introduction

The study of graph modification problems is a classic topic in theoretical computer science. The typical task in this context is, given a graph class Π and a graph G, to modify G by a minimum number of operations such that the resulting graph is contained in Π. By a general result, graph modification is NP-hard if the operation is vertex deletion and Π is hereditary [18]. In contrast, for *edge* modification problems where one may insert or delete edges, no such general hardness result is possible. One nontrivially tractable example is the case when Π is the class of split graphs, that is, graphs whose vertex set can be partitioned into a clique and an independent set (edges between the independent set and the clique are allowed). The problem of modifying a graph into a split graph by a minimum number of edge modifications (insertions or deletions) is polynomial-time solvable [14]. This result relies on the fact that a split graph can be recognized by its degree sequence. In contrast, Natanzon et al.[19] showed that the problem becomes NP-hard when allowing either only edge deletions or only edge insertions.

Damaschke and Mogren [5,6] considered several graph modification problems for very restricted graph classes where, informally, the number of different neighborhoods is constant. In this paper, we study two problems of this kind.

F. Hüffner—Supported by DFG project ALEPH (HU 2139/1).

F. Dehne et al. (Eds.): WADS 2015, LNCS 9214, pp. 410–421, 2015.
DOI: 10.1007/978-3-319-21840-3_34

First, we consider a very restricted subclass of split graphs where no edges between the independent set and the clique are allowed. More specifically, we call a graph G a *sparse split graph* if G consists of a clique and isolated vertices. The corresponding graph modification problem is defined as follows.

SPARSE SPLIT GRAPH EDITING
Input: A graph $G = (V, E)$ and an integer $k \in \mathbb{N}$.
Question: Can G be transformed into a sparse split graph by at most k
          edge insertions and deletions?

SPARSE SPLIT GRAPH EDITING was studied by Damaschke and Mogren [5] under the names $K_1[0]$-BAG EDITING and CLIQUE EDITING. For example, it was shown that SPARSE SPLIT GRAPH EDITING can be solved in $2^{O(\sqrt{k} \log k)} \cdot n^{O(1)}$ time whereas the NP-hardness of SPARSE SPLIT GRAPH EDITING was initially left open [5]; it was later shown to be NP-hard by Kovác et al. [17]. We also consider the following further problem, as introduced by Damaschke and Mogren [5].

P_3-BAG EDITING
Input: A graph $G = (V, E)$ and an integer $k \in \mathbb{N}$.
Question: Can G be transformed into two possibly overlapping cliques
          by at most k edge insertions and deletions?

We call such graphs P_3-*bag graphs*. The term refers to the fact that in such a graph merging all vertices with the same closed neighborhood results in a P_3 or an induced subgraph of a P_3. An equivalent definition is as follows: the graph class is the set of all graphs with edge clique cover number at most two.

Further Related Work. To obtain a sparse split graph by a minimum number of edge insertions is trivially solvable in polynomial time. If one allows only edge deletions, the problem is NP-hard [6]. SPARSE SPLIT GRAPH EDITING has applications in the identification of core–periphery structures in social networks [3]. Other models considered in this context include SPLIT EDITING and DENSE SPLIT GRAPH EDITING which asks to transform the input graph into a dense split graph, that is, a graph which consists of a clique and an independent set and in which all edges are present between the clique and the independent set [3].

Many graph classes defined by existence of a certain vertex partitioning can be captured with the notion of a *pattern* [15]. A pattern for a partition into d parts is a symmetric $d \times d$ matrix M with entries from $\{0, 1, *\}$. Then, an M-*partition* of a graph $G = (V, E)$ is a partition $V_1, \ldots, V_d$ of V such that two distinct vertices in (possibly equal) parts V_i and V_j are adjacent if $M(i, j) = 1$ and nonadjacent if $M(i, j) = 0$ (the entry $M(i, j) = *$ signifies no restriction). Thus, sparse split graphs are the graphs with a $\left(\begin{smallmatrix} 1 & 0 \\ 0 & 0 \end{smallmatrix}\right)$-partition and P_3-bag graphs are the graphs with a $\left(\begin{smallmatrix} 1 & 1 & 1 \\ 1 & 1 & 0 \\ 1 & 0 & 1 \end{smallmatrix}\right)$-partition. Expressed with these definitions, Damaschke and Mogren [6] consider editing problems for the case where the diagonal is 1 and off-diagonal elements are 0 or 1.

2-CLUSTER EDITING (also known as 2-CORRELATION CLUSTERING on complete graphs) is to find a minimum number of edge modifications to convert a graph into two disjoint cliques (that is, into a graph with a $\left(\begin{smallmatrix} 1 & 0 \\ 0 & 1 \end{smallmatrix}\right)$-partition). It is NP-hard [20], but has a kernel with at most $4k + 2$ vertices [13]. It can be solved in subexponential time $2^{O(\sqrt{k})} + n^{O(1)}$ [10]; a subexponential running time follows also from the more general result of Damaschke and Mogren [6]. Wu and Chen [21] give a different subexponential algorithm.

Our Results. First, we complement and improve on results for Sparse Split Graph Editing and P_3-Bag Editing. In particular, we show a factor-3 approximation for Sparse Split Graph Editing in Section 2, and prove NP-hardness of P_3-Bag Editing in Section 3. The former result improves a factor-3.524 approximation from Kovác et al. [17] and the latter answers an open question of Damaschke and Mogren [5].

Second, we provide kernelization schemes for Sparse Split Graph Editing, P_3-Bag Editing, and 2-Cluster Editing in Section 4. Analogous to a polynomial-time approximation scheme (PTAS), a kernelization scheme provides increasingly good bounds on the kernel size, at the cost of an increasing running time bound. Only few kernelization schemes are known (e.g. [1,2,9]), and they provide kernel size bounds of the form $(1+\varepsilon)k$, where the limit bound k is not known to be sharp (unlike for a PTAS). Abu-Khzam and Fernau [1] ask whether there are kernelization schemes that converge to a provable lower bound. We answer this question positively by providing, for the three above-mentioned problems, such schemes where the size bound converges, in fact, to 0. We formalize this by introducing the notion of strict kernelization schemes.

Definition 1. *A* strict kernelization scheme *is an algorithm $\mathcal{A}$ which takes as input an instance (I, k) of a parameterized problem and a constant $\varepsilon > 0$ and produces in $(|I| + k)^{f(1/\varepsilon)}$ time an instance (I', k') such that $(I, k) \in L \iff (I', k') \in L$, $|I'| \leq \varepsilon \cdot g(k)$, and $k' \leq k$ for some functions f and g.*

Note that, by first kernelizing with $\varepsilon = 1$ and then in a second step kernelizing the resulting instance with the intended value of ε, the running time of a strict kernelization scheme can always be improved to $g(k)^{f(1/\varepsilon)} + |I|^{O(1)}$.

Preliminaries. For a graph $G = (V, E)$ we set $n := |V|$ and $m := |E|$. The *open neighborhood* of a vertex u is $N_G(u) := \{v \mid \{u, v\} \in E\}$. The *closed neighborhood* of a vertex u is $N_G[u] := \{u\} \cup N_G(u)$. For a vertex subset $V' \subseteq V$, the *subgraph induced by V'* is denoted by $G[V']$. For two disjoint vertex subsets $V_1, V_2 \subseteq V$, the set of edges with one endpoint in V_1 and one endpoint in V_2 is denoted by $E_G(V_1, V_2)$. We omit the subscript if the graph G is clear from the context. A clique on $k \in \mathbb{N}$ vertices is denoted by K_k, and a complete bipartite graph with $k_1 \in \mathbb{N}$ vertices in one part and $k_2 \in \mathbb{N}$ vertices in the other part is denoted by K_{k_1, k_2}. The "$\triangle$" operator denotes the symmetric difference with $A \triangle B := (A \cup B) \setminus (A \cap B)$.

For the relevant notions of parameterized complexity, such as kernelization, we refer to the monograph by Downey and Fellows [7]. Due to space constraints, several proofs are deferred to a full version.

2   Sparse Split Graph Editing

We first make several simple observations on the structure of sparse split graphs and on dense split graphs. These observations can be useful in applications of SPARSE SPLIT GRAPH EDITING.

Characterizations. The class of sparse split graphs is hereditary, that is, it is closed under vertex deletions. Hence, sparse split graphs can be characterized by a set of forbidden induced subgraphs. In general, such characterizations can be useful for example for obtaining recognition algorithms for a graph class Π or for obtaining fixed-parameter algorithms for hard graph modifications problems for Π [4]. For sparse split graphs, the following simple characterization is known.

Theorem 1 ([22, Theorem 5.2.7]). *A graph G is a sparse split graph if and only if it does not contain a $2K_2$ or a P_3 as an induced subgraph.*

Like split graphs, sparse split graphs can be characterized by their degree sequence, that is, the list of degrees of their vertices sorted in descending order.

Theorem 2. *A graph is a sparse split graph if and only if its degree sequence is* $\underbrace{c, c, \ldots, c}_{c+1}, \underbrace{0, 0, \ldots, 0}_{n-c-1}$ *for some $c \geq 1$.*

Sparse split graphs are closely related to dense split graphs.

Lemma 1. *A graph G is a dense split graph if and only if its complement is a sparse split graph.*

By building the complement of the forbidden induced subgraphs for sparse split graphs, we can thus obtain the following forbidden subgraph characterization for dense split graphs.

Corollary 1. *A graph G is a dense split graph if and only if it does not contain a C_4 or a $K_2 + K_1$ as an induced subgraph.*

Similarly, we obtain the following corollary to Theorem 2.

Corollary 2. *A graph is a dense split graph if and only if its degree sequence is* $\underbrace{n-1, n-1, \ldots, n-1}_{c}, \underbrace{c, c, \ldots, c}_{n-c}$ *for some $c \geq 1$.*

Approximation. Kovác et al [17] present an approximation algorithm for SPARSE SPLIT GRAPH EDITING and prove an approximation factor of 3.524; they conjecture that the algorithm is a 3.383-approximation. We give a simpler 3-approximation, inspired by the polynomial-time algorithm for SPLIT GRAPH EDITING [14] that is based on a characterization by the degree sequence. This algorithm sorts the vertices by degree, and chooses the vertices up to a certain point in the sequence for the independent set and the remaining ones for the clique of the resulting split graph. Since sparse split graphs have a similar characterization by degree sequence (Theorem 2), we use the same algorithm to get an approximation for SPARSE SPLIT GRAPH EDITING. For separating the clique from the independent set, however, we do not calculate the threshold but try all of them. More precisely, for each $0 \leq x \leq n$, choose the x vertices with the highest degree as clique (resolving ties arbitrarily), and retain the best of these $n+1$ solutions. Here, "choosing as clique" means to add all missing edges within the vertex set and delete all other edges, yielding a sparse split graph.

Theorem 3. SPARSE SPLIT GRAPH EDITING *can be approximated in linear time within a factor of* 3.

Proof. A linear running time can be achieved by processing the vertices in order of decreasing degree, where the clique would be formed by all vertices processed so far. We maintain m_c, the number of edges within the clique; updating m_c can be done in $O(m)$ time total. The number of modifications for clique size c can then be calculated as $(\binom{c}{2} - m_c) + (m - m_c)$.

We now analyze the approximation factor. The analysis is based on the proof of Hammer and Simeone [14] showing that SPLIT GRAPH EDITING is polynomial time solvable. Let C_{opt} be the clique of an optimal solution S_{opt} of cost k_{opt} and C the clique of the solution S calculated by the approximation algorithm for $c = |C_{\text{opt}}|$, with cost k. For $S_1, S_2 \subseteq V$, we denote by $E(S_1)$ the edges that have both endpoints in $S_1 \subseteq V$ and by $E(S_1, S_2)$ the edges with one endpoint in S_1 and the other endpoint in S_2. With this, the cost k_{opt} can be decomposed:

$$k_{\text{opt}} = \underbrace{\frac{c(c-1)}{2} - |E(C_{\text{opt}})|}_{\text{edges added in } C_{\text{opt}}} + \underbrace{|E(V \setminus C_{\text{opt}})| + |E(C_{\text{opt}}, V \setminus C_{\text{opt}})|}_{\text{deleted edges with endpoint(s) in } V \setminus C_{\text{opt}}}. \quad (1)$$

Observe that for any set $S \subseteq V$ it holds that:

$$\sum_{v \in S} \deg(v) = 2|E(S)| + |E(S, V \setminus S)|, \quad (2)$$

where $\deg(v)$ denotes the degree of v. Rearranging Equality (2) to have $|E(S)|$ on the left-hand side and inserting the right-hand side in Equality (1) for $|E(C_{\text{opt}})|$ and $|E(V \setminus C_{\text{opt}})|$ yields:

$$k_{\text{opt}} = \frac{1}{2}\left(c(c-1) - \sum_{v \in C_{\text{opt}}} \deg(v) + \sum_{v \in V \setminus C_{\text{opt}}} \deg(v)\right) + |E(C_{\text{opt}}, V \setminus C_{\text{opt}})|. \quad (3)$$

Let $d_1 \geq d_2 \geq \ldots \geq d_n$ be the degrees of the vertices in descending order. It follows that:

$$\sum_{v \in C_{\text{opt}}} \deg(v) \leq \sum_{i=1}^{c} d_i \quad \text{and} \quad \sum_{v \in V \setminus C_{\text{opt}}} \deg(v) \geq \sum_{i=c+1}^{n} d_i. \quad (4)$$

Inserting this into Equality (3) yields:

$$k_{\text{opt}} \geq \frac{1}{2}\left(c(c-1) - \sum_{i=1}^{c} d_i + \sum_{i=c+1}^{n} d_i\right) + |E(C_{\text{opt}}, V \setminus C_{\text{opt}})| \quad (5)$$

Observe that if C_{opt} contains the vertices with the highest degree in G, then Inequality (5) becomes an equality. Furthermore, our approximation algorithm for $x = c$ actually contains the c vertices with highest degree in C. Thus, using the same analysis as above for k and C instead of k_{opt} and C_{opt}, we obtain

$$k = \frac{1}{2}\left(c(c-1) - \sum_{i=1}^{c} d_i + \sum_{i=c+1}^{n} d_i\right) + |E(C, V \setminus C)|. \quad (6)$$

It remains to bound the size of $E(C, V \setminus C)$. To this end, observe that

$$|E(C, V \setminus C)| \leq \sum_{v \in V \setminus C} \deg(v) = \sum_{i=c+1}^{n} d_i \overset{(4)}{\leq} \sum_{v \in V \setminus C_{\text{opt}}} \deg(v)$$

$$\overset{(2)}{\leq} 2|E(V \setminus C_{\text{opt}})| + 2|E(V \setminus C_{\text{opt}}, C_{\text{opt}})| \overset{(1)}{\leq} 2k_{\text{opt}}$$

Putting this together yields $k \leq 3k_{\text{opt}}$. □

Using a computer program, we determined the worst-case approximation factor (i.e., with unlucky tie resolving) for all graphs up to 11 vertices. The worst case is a factor of 2.5, and only one graph with this factor was found (up to adding singletons): a disjoint union of a triangle and a P_3.

3 P_3-Bag Editing

We now turn to P_3-BAG EDITING. Recall that a P_3-bag graph is a graph that consists of exactly two possibly overlapping cliques.

Characterizations. We first give a forbidden subgraph characterization of P_3-bag graphs. Note that P_3-bag graphs cannot be characterized by their degree sequence: Two disjoint triangles and a cycle on six vertices have both the degree sequence $2, 2, 2, 2, 2, 2$. However, only the former is a P_3-bag graph.

Theorem 4. *A graph G is a P_3-bag graph if and only if it does not contain a $3K_1$, P_4, or C_4 as an induced subgraph.*

Proof. It is easy to see that P_4 and C_4 are not P_3-bag graphs. From a more general result on forbidden subgraphs for graphs with certain M-partitions [8, Corollary 3.3], it follows that a minimal forbidden subgraph for P_3-bag graphs can have at most four vertices, and there can be at most two minimal forbidden subgraphs with four vertices. Finally, it is easy to verify that all graphs with three or fewer vertices except for $3K_1$ are P_3-bag graphs. □

For P_3-BAG EDITING, we can also consider the complement problem. The following characterizations follow from our characterizations of P_3-bag graphs.

Lemma 2. *For a graph G, the following are equivalent.*

1. *G is a complement of a P_3-bag graph.*
2. *G consists of a complete bipartite graph (biclique) plus isolated vertices.*
3. *G does not contain a K_3, P_4, or $2K_2$ as an induced subgraph.*

Thus, while SPARSE SPLIT GRAPH EDITING is the problem of editing a graph into a clique plus isolated vertices, P_3-BAG EDITING is the problem of editing the complement of a graph into a biclique plus isolated vertices. Note that the problem of editing a graph into a biclique (without isolated vertices) is the complement problem of 2-CLUSTER EDITING.

NP-hardness. We now show that P_3-BAG EDITING is NP-complete. This demonstrates the value of the subexponential fixed-parameter algorithm that solves P_3-BAG EDITING in $O(2^{\sqrt{k}\log k})$ time [6].

Theorem 5. *P_3-BAG EDITING is NP-complete.*

Proof (sketch). Containment in NP is obvious. To prove NP-hardness, we provide a polynomial-time reduction from the BISECTION problem, which was shown to be NP-hard by Garey et al. [11].

> BISECTION
> **Input:**   A graph $G = (V, E)$ and an integer $k \in \mathbb{N}$.
> **Question:** Does G have a bisection with cut size at most k, that is, a partition of V into two sets V_1 and V_2 such that $|V_1| = |V_2|$ and $|E(V_1, V_2)| \leq k$?

Given a BISECTION instance $(G = (V, E), k)$ with $m > k$ we construct an equivalent P_3-BAG EDITING instance $(G' = (V', E'), k')$ as follows. First, copy G into G'. Next, add for each vertex $v \in V$ a clique with n^2 vertices to G' and make all vertices in this clique adjacent to v in G'. Denote the vertices in this clique by $C(v)$ (with $v \notin C(v)$). We call these cliques *pendant cliques* to distinguish them from the at most two maximal cliques in the P_3-bag graph. We first explain the intuition behind the construction. The pendant cliques are pairwise non-adjacent. This forces a balanced "distribution" of the pendant cliques to the two maximal cliques of the P_3-bag graph as any non-balanced distribution exceeds the budget (which we define below). Then the balanced distribution of the pendant cliques forces the original vertex set V to be also split into two equal size sets. Hence, choosing the budget k' appropriately ensures a cut size of at most k between these two sets. To define k', we use $t := n/2$ to denote the size of the two parts in a bisection of G and set

$$k' := \underbrace{n^4 \cdot 2\binom{t}{2}}_{\substack{\text{edges added} \\ \text{between cliques}}} + \underbrace{n^2 \cdot n(t-1)}_{\substack{\text{edges added} \\ \text{between cliques and} \\ \text{original vertices}}} + \underbrace{k}_{\substack{\text{edges removed} \\ \text{in cut of} \\ \text{bisection}}} + \underbrace{2\binom{t}{2} - (m-k)}_{\substack{\text{edges added between} \\ \text{original vertices inside the} \\ \text{two parts of the partition}}}.$$

It now holds that (G, k) is a yes-instance of BISECTION $\iff$ (G', k') is a yes-instance of P_3-BAG EDITING; we omit the proof. $\square$

In the proof above, the intersection of the maximal cliques in the optimal solution for the constructed instance is empty. Thus, the reduction also provides an alternative NP-hardness proof for 2-CLUSTER EDITING.

4   Kernelization Schemes

We now give strict kernelization schemes (see Definition 1 in Section 1) for 2-CLUSTER EDITING, SPARSE SPLIT GRAPH EDITING, and P_3-BAG EDITING. Since complementing the graph does not affect k, they also apply to BICLIQUE EDITING, DENSE SPLIT GRAPH EDITING, and BICLIQUE+SINGLETONS EDITING.

The idea of all three schemes is to apply data reduction that ensures that the number of edge modifications incident on each vertex is at least some constant c. Then, if we can solve the instance with k modifications, the number of vertices remaining is at most $2k/c$, and by setting $c := 2/\varepsilon$, we can achieve any kernel of order εk. The critical property that allows the data reduction is that from knowing the neighborhood of just one vertex in an optimal solution, we can easily construct a complete optimal solution graph. Here, for simplicity, we use *solution* to refer either to the set of editing operations or to the graph from the target class of the editing problem that is obtained by applying the editing operations.

We formulate the data reduction for any graph modification problem that is "neighborhood-reconstructible" and "allows isolation", and prove that our problems have these properties. For convenience, instead of P_3-BAG EDITING we consider the complement problem BICLIQUE+SINGLETONS EDITING, that is, the problem of editing into a biclique plus isolated vertices.

Definition 2. *A graph modification problem is* neighborhood-reconstructible *in $p(n)$ time for some polynomial p when given the nonempty neighborhood of a vertex in a solution G', one can in $p(n)$ time either find a solution with at most k edge modifications or determine that the solution G' incurs more than k edge modifications. This method is called* neighborhood reconstruction.

Observe that we demand the reconstructibility only for nonempty neighborhoods. This is done to cope with vertices that can become singletons in the solution.

Lemma 3. *2-CLUSTER EDITING, SPARSE SPLIT GRAPH EDITING, and BICLIQUE+SINGLETONS EDITING are neighborhood-reconstructible in linear time.*

Proof. For 2-CLUSTER EDITING, neighborhood reconstruction is possible even given an empty neighborhood of a vertex. Assume we know the neighborhood $N(u)$ of any vertex u in a solution. Then we can reconstruct the solution in linear time: one clique is $C_1 := N[u]$ and the other is $C_2 := V \setminus N[u]$. We can in linear time determine $m_{1,2}$, the number of edges between C_1 and C_2. Then the number of modifications k can be calculated as $\binom{|C_1|}{2} + \binom{|C_2|}{2} - m + 2m_{1,2}$.

For SPARSE SPLIT GRAPH EDITING, let C and I be the clique and the isolated vertices of a solution, respectively. Only vertices in C have nonempty neighborhoods in a solution. Assume we know the neighborhood $N(u)$ of a vertex $u \in C$ in a solution. Then $C = N[u]$ and $I = V \setminus N[u]$. We can count in linear time the number m_C of edges within C, and the number of edge modifications is $m + \binom{|C|}{2} - 2m_C$.

For BICLIQUE+SINGLETONS EDITING, let B_1 and B_2 be the two parts of the biclique, and I the isolated vertices. Assume that for a vertex $u \in B_1 \cup B_2$ (without loss of generality $u \in B_1$), we know the neighborhood $N(u)$ of u in a solution. Then we have $B_2 = N(u)$ and $B_1 \cup I = V \setminus N(u)$. It remains to allocate the vertices in $B_1 \cup I$ to B_1 or I. Each decision for a vertex $v \in B_1 \cup I$ can be made independently: if there are at least $|B_2|/2$ edges from v to B_2, we place v in B_1, and otherwise we place it in I. Since each edge will be considered at most once, this can be done in $O(m)$ time. We can then count in linear time the number m_B of edges between B_1 and B_2, and the number of edge modifications is $m + |B_1||B_2| - 2m_B$. □

The first rule directly exploits neighborhood reconstructibility: Assume that there is a vertex with nonempty neighborhood in the solution and that the difference between the input neighborhood and solution neighborhood is small. Then, we can find an optimal solution by guessing this small difference and then using neighborhood reconstruction. When this fails for all vertices, we know that each vertex has many incident edge modifications or is isolated in the solution.

Rule 1. Consider a constant c and a graph modification problem that is neighborhood-reconstructible in $p(n)$ time. For each vertex u, try all ways of changing up to $c - 1$ incidences with the other vertices, that is, consider the neighborhoods $\{N(u) \triangle T \mid T \subseteq V \setminus \{u\}, |T| \leq c - 1\}$. If for some u and some T, neighborhood reconstruction finds a solution with at most k edge modifications, then replace the instance by a trivial "yes"-instance.

Lemma 4. *Rule 1 is sound and can be executed in $O(n^c \cdot p(n))$ time.*

Proof. It is clear that the rule is sound, that is, it produces a "yes"-instance if and only if the original instance is a "yes"-instance. The running time can be seen as follows. There are n vertices and $O(n^{c-1})$ vertex sets to try, and each choice can be checked in $p(n)$ time. □

Observation 1. Exhaustively applying Rule 1 yields an instance in which it holds for every solution with at most k edge modifications that each vertex is incident with at least c edge modifications or isolated in the solution.

For 2-CLUSTER EDITING, at most two vertices have empty neighborhood in the solution. Hence, Rule 1 is already sufficient to bound the number of incident edge modifications for all except two vertices. This is not sufficient for SPARSE SPLIT GRAPH EDITING and BICLIQUE+SINGLETONS EDITING where a solution may contain many singletons. Here, we exploit another problem property.

Definition 3. *A graph modification problem allows isolation if the property of being a solution is hereditary, and adding an isolated vertex to a solution yields another solution.*

Thus, any solution can be transformed into a new one by picking an arbitrary vertex and removing all incident edges, hence the name. Observe that SPARSE SPLIT GRAPH EDITING and BICLIQUE+SINGLETONS EDITING allow isolation.

Rule 2. For a graph modification problem that allows isolation, assume it is known that in every solution, each vertex has at least c incident edge modifications or degree 0 in the solution (or both). If G contains a vertex u with $\deg(u) \leq c$, then remove u from G and reduce k by $\deg(u)$.

Lemma 5. *Rule 2 is sound and can be performed exhaustively in $O(nm)$ time.*

Observation 2. Exhaustively applying Rules 1 and 2 yields an instance in which it holds for any solution with at most k editing operation each vertex has at least c incident edge modifications.

Rule 3. For a graph modification problem, assume we can for some constant $c \geq 1$ in polynomial time reduce to an instance where the number of edge modifications incident on each vertex is at least c. If the graph contains more than $2k/c$ vertices, then return a trivial no-instance.

The above observation together with Rule 3 yields the problem kernel of oder $2k/c$ for graph modification problems that are neighborhood-reconstructible and allow isolation. For the running time bound of the kernelization, we make use of the fact that neighborhood reconstruction runs in linear time for all three problems.

Theorem 6. *For any $c \geq 1$, 2-CLUSTER EDITING, SPARSE SPLIT GRAPH EDITING, and P_3-BAG EDITING have a kernel with at most $2k/c$ vertices that can be computed in $O\left(nm + ck^2 \cdot \left(\frac{2k}{c-1}\right)^c\right)$ time.*

Proof. Let $\delta(u)$ denote the number of edge modifications incident on a vertex u. By Observation 2, we have $2k = \sum_{u \in V} \delta(u) \geq cn$, implying $n \leq 2k/c$. The straightforward running time of $O(n^c m)$ caused by Rule 1 can be improved by applying data reduction in rounds for $c' = 1$ to $c' = c$. The first round with $c' = 1$ takes $O(nm)$ time and produces an instance with at most $2k$ vertices. Before a round with $c' \geq 2$, there are at most $2k/(c'-1)$ vertices left, and the number of edges can be bounded by $O(k^2)$. Thus, the remaining rounds run in time

$$\sum_{c'=2}^{c} O\left(\left(\frac{2k}{c'-1}\right)^{c'} k^2\right) = O\left(ck^2 \cdot \left(\frac{2k}{c-1}\right)^c\right). \qquad \square$$

Note that for 2-CLUSTER EDITING, already for $c = 1$, we obtain a kernel with $2k$ vertices, improving the $4k + 2$-vertex kernel derived from a more general result for d-CLUSTER EDITING [13].

Subexponential-time Algorithms. We can use our strict kernelization schemes to obtain subexponential-time algorithms.

Theorem 7. *If a graph problem can be solved in $2^{O(n)}$ time and it has a strict kernelization scheme that produces for any $c > 0$ in $n^{O(c)}$ time a kernel of at most $O(k/c)$ vertices, then it can be solved in $2^{O(\sqrt{k \log k})} + n^{O(1)}$ time.*

Proof. We kernelize for increasing c up to $c = \sqrt{k/\log k}$ and then solve the instance. This requires $k^{O(\sqrt{k/\log k})} + n^{O(1)} + 2^{O(\sqrt{k \log k})} = 2^{O(\sqrt{k \log k})} + n^{O(1)}$ time.   □

For SPARSE SPLIT GRAPH EDITING and P_3-BAG EDITING, this running time is similar to the known subexponential-time algorithms [6], for 2-CLUSTER EDITING this running time almost meets the best known running time [10]. We can also use Theorem 7 to rule out strict kernelization schemes for certain problems with known lower bounds on their running time.

Theorem 8. CLUSTER EDITING *does not not have a kernelization scheme that produces for any $c > 0$ in $n^{O(c)}$ time a kernel of at most $O(k/c)$ vertices, unless the exponential time hypothesis (ETH) is false.*

Proof. CLUSTER EDITING is easily solved in $2^{O(n)}$ time by standard dynamic programming over vertex subsets. Moreover, assuming ETH, there is no $2^{o(k)} \cdot n^{O(1)}$ time algorithm for CLUSTER EDITING [16].   □

5   Outlook

Several open questions remain. For example, is SPARSE SPLIT GRAPH EDITING APX-hard, or does it have a PTAS? Possibly a PTAS for the 2-CORRELATION CLUSTERING problem [12] can be adapted. Furthermore, it seems worthwhile to explore the relation between kernelization schemes, subexponential-time solvability and polynomial-time approximation schemes more closely. For example, it would be interesting to investigate whether, similar to efficient polynomial-time approximation schemes (EPTAS) there are *efficient* strict kernelization schemes with running time $f(1/\varepsilon) \cdot (|I| + k)^{O(1)}$. Finally, it is open to find further applications of strict kernelization schemes. We would like to remark that the schemes also apply to the edge deletion variants of the considered problems. Preliminary considerations indicate that some of these problems admit efficient strict kernelization schemes.

Acknowledgments. We are grateful to Henning Fernau for fruitful discussions about the problems considered in this work.

References

1. Abu-Khzam, F.N., Fernau, H.: Kernels: annotated, proper and induced. In: Bodlaender, H.L., Langston, M.A. (eds.) IWPEC 2006. LNCS, vol. 4169, pp. 264–275. Springer, Heidelberg (2006)

2. Bessy, S., Fomin, F.V., Gaspers, S., Paul, C., Perez, A., Saurabh, S., Thomassé, S.: Kernels for feedback arc set in tournaments. J. Comput. System Sci. **77**(6), 1071–1078 (2011)
3. Borgatti, S.P., Everett, M.G.: Models of core/periphery structures. Soc. Networks **21**(4), 375–395 (1999)
4. Cai, L.: Fixed-parameter tractability of graph modification problems for hereditary properties. Inf. Process. Lett. **58**(4), 171–176 (1996)
5. Damaschke, P., Mogren, O.: Editing the simplest graphs. In: Pal, S.P., Sadakane, K. (eds.) WALCOM 2014. LNCS, vol. 8344, pp. 249–260. Springer, Heidelberg (2014)
6. Damaschke, P., Mogren, O.: Editing simple graphs. J. Graph Algorithms Appl. **18**(4), 557–576 (2014). doi:10.7155/jgaa.00337
7. Downey, R.G., Fellows, M.R.: Fundamentals of Parameterized Complexity. Texts in Computer Science. Springer (2013)
8. Feder, T., Hell, P.: On realizations of point determining graphs, and obstructions to full homomorphisms. Discrete Math. **308**(9), 1639–1652 (2008)
9. Fernau, H.: Parameterized algorithmics: A graph-theoretic approach. Wilhelm-Schickard-Institut für Informatik. Universität Tübingen, Habilitationsschrift (2005)
10. Fomin, F.V., Kratsch, S., Pilipczuk, M., Pilipczuk, M., Villanger, Y.: Tight bounds for parameterized complexity of cluster editing with a small number of clusters. J. Comput. System Sci. **80**(7), 1430–1447 (2014)
11. Garey, M.R., Johnson, D.S., Stockmeyer, L.J.: Some simplified NP-complete graph problems. Theor. Comput. Sci. **1**(3), 237–267 (1976)
12. Giotis, I., Guruswami, V.: Correlation clustering with a fixed number of clusters. Theory of Computing **2**(1), 249–266 (2006)
13. Guo, J.: A more effective linear kernelization for cluster editing. Theor. Comput. Sci. **410**(8–10), 718–726 (2009)
14. Hammer, P.L., Simeone, B.: The splittance of a graph. Combinatorica **1**(3), 275–284 (1981)
15. Hell, P.: Graph partitions with prescribed patterns. Eur. J. Combin. **35**, 335–353 (2014)
16. Komusiewicz, C., Uhlmann, J.: Cluster editing with locally bounded modifications. Discrete Appl. Math. **160**(15), 2259–2270 (2012)
17. Kováč, I., Selečéniová, I., Steinová, M.: On the clique editing problem. In: Csuhaj-Varjú, E., Dietzfelbinger, M., Ésik, Z. (eds.) MFCS 2014, Part II. LNCS, vol. 8635, pp. 469–480. Springer, Heidelberg (2014)
18. Lewis, J.M., Yannakakis, M.: The node-deletion problem for hereditary properties is NP-complete. J. Comput. System Sci. **20**(2), 219–230 (1980)
19. Natanzon, A., Shamir, R., Sharan, R.: Complexity classification of some edge modification problems. Discrete Appl. Math. **113**, 109–128 (2001)
20. Shamir, R., Sharan, R., Tsur, D.: Cluster graph modification problems. Discrete Appl. Math. **144**(1–2), 173–182 (2004)
21. Wu, B.Y., Chen, L.-H.: Parameterized algorithms for the 2-clustering problem with minimum sum and minimum sum of squares objective functions. Algorithmica (2014, to appear). doi:10.1007/s00453-014-9874-8
22. Xie, W.: Obstructions to trigraph homomorphisms. Master's thesis, School of Computing Science. Simon Fraser University, British Columbia, Canada (2006)

Competitive Diffusion on Weighted Graphs

Takehiro Ito[1,2], Yota Otachi[3], Toshiki Saitoh[2,4], Hisayuki Satoh[1],
Akira Suzuki[1,2], Kei Uchizawa[5(✉)], Ryuhei Uehara[3],
Katsuhisa Yamanaka[6], and Xiao Zhou[1]

[1] Tohoku University, Sendai, Japan
{takehiro,h.satoh,a.suzuki,zhou}@ecei.tohoku.ac.jp
[2] CREST, JST, Saitama, Japan
saitoh@eedept.kobe-u.ac.jp
[3] Japan Advanced Institute of Science and Technology, Nomi, Japan
{otachi,uehara}@jaist.ac.jp
[4] Kobe University, Kobe, Japan
[5] Yamagata University, Yonezawa, Japan
uchizawa@yz.yamagata-u.ac.jp
[6] Iwate University, Morioka, Japan
yamanaka@cis.iwate-u.ac.jp

Abstract. Consider an undirected and vertex-weighted graph modeling
a social network, where the vertices represent individuals, the edges do
connections among them, and weights do levels of importance of individuals. In the competitive diffusion game, each of a number of players
chooses a vertex as a seed to propagate his/her idea which spreads along
the edges in the graph. The objective of every player is to maximize the
sum of weights of vertices infected by his/her idea. In this paper, we
study a computational problem of asking whether a pure Nash equilibrium exists in a given graph, and present several negative and positive
results with regard to graph classes. We first prove that the problem
is W[1]-hard when parameterized by the number of players even for
unweighted graphs. We also show that the problem is NP-hard even
for series-parallel graphs with positive integer weights, and is NP-hard
even for forests with arbitrary integer weights. Furthermore, we show
that the problem for forests of paths with arbitrary weights is solvable
in pseudo-polynomial time; and it is solvable in quadratic time if a given
graph is unweighted. We also prove that the problem is solvable in polynomial time for chain graphs, cochain graphs, and threshold graphs with
arbitrary integer weights.

1 Introduction

Ideas, innovations or trends spread by interactions between individuals. Social
networks such as Facebook and Twitter facilitate their diffusion; an idea of an
influential individual spreads along the connections over a network, and a small
number of initial seeds can yield widespread infection. Since we can employ
the so-called word-of-mouth effect as a tool for viral marketing, analysis of the

© Springer International Publishing Switzerland 2015
F. Dehne et al. (Eds.): WADS 2015, LNCS 9214, pp. 422–433, 2015.
DOI: 10.1007/978-3-319-21840-3_35

dynamics and process of the diffusion receive increasing attention in computer science. A number of papers focus on a task for a single company that wishes to advertise their product through a network; they investigate a problem of finding key individuals for maximizing the largest expected infection based on a given stochastic model of diffusion process [9,19,20,22]. Another active line of research stems from a task for multiple competing companies which try to advertise their products through a network, where the diffusion process is set in a game-theoretic formulation [1–7,10,15,16,23,24].

In this paper, we focus on the latter setting, and consider the one introduced by Alon *et al.* [1]. In their setting, a network is modeled by an unweighted graph, and each of a given number of competing companies chooses a vertex in the graph as a seed of their advertisement. Then their advertisements deterministically spread along the edges of a graph so that every infected vertex adopts its neighbors in a discrete time step. The objective of every player is to maximize the number of infected vertices. (The precise definition of the game is given in Section 2.) Alon *et al.* call the game *competitive diffusion game*, and show that there exists an unweighted graph of diameter three that does not admit a Nash equilibrium for two players. Following the paper [1], several results are known for the competitive diffusion game. Takehara *et al.* provided an unweighted graph G of diameter two that does not admit a Nash equilibrium for two players [24]. Small and Mason considered the case where a social network has a tree structure, and show that any tree admits a Nash equilibrium for two players [23]. More recently, Bulteau *et al.* consider certain graph classes including paths, cycles and grid graphs; in particular, they prove that there is no Nash equilibrium for three players on $m \times n$ grids with $\min\{m, n\} \geq 5$ [6].

We generalize the game to weighted graphs, where a weight on a vertex represents a level of importance of an individual; negative weights are admitted to express very demanding customers. We then focus on a problem COMPETITIVE DIFFUSION of deciding whether, given the number k, a graph G and weight function w, the competitive diffusion game on G with w for k players has a Nash equilibrium.

We establish solid complexity foundation of COMPETITIVE DIFFUSION with regard to graph classes. Since there are a number of theoretical models of social networks, and some of them are directly related to restricted graph classes (such as random trees with scale free properties [8]), our results give useful tools for obtaining algorithmic results on such models.

Our contributions are twofold. On the one hand, we provide the following three hardness results:

(i) COMPETITIVE DIFFUSION is W[1]-hard when parameterized by the number of players even for unweighted graphs;

(ii) COMPETITIVE DIFFUSION is NP-complete even for series-parallel graphs with positive integer weights;

(iii) COMPETITIVE DIFFUSION is NP-complete even for forests with arbitrary integer weights.

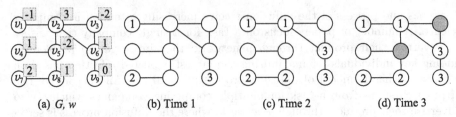

(a) G, w (b) Time 1   (c) Time 2   (d) Time 3

Fig. 1. Example of competitive diffusion with $k = 3$ players. (a) The graph G and weight w; numbers in the gray squares are weights. (b) p_1, p_2 and p_3 choose v_1, v_7 and v_9 in G, respectively; thus a strategy profile $s = (v_1, v_7, v_9)$. (c) Each player dominates the neighbor. (d) The game ends; the two gray vertices are neutral. Consequently, $U_1(s) = 2, U_2(s) = 3$ and $U_3(s) = 1$.

Very recently, Etesami and Basar studied unweighted version of the problem, and showed that COMPETITIVE DIFFUSION is a NP-complete problem [12], but their result does not imply ours. On the other hand, we obtain the following two algorithmic results.

(iv) For forests of paths, we prove that COMPETITIVE DIFFUSION is solvable in pseudo-polynomial time. In particular, we give a quadratic-time algorithm for forests of unweighted paths;

(v) For chain graphs, cochain graphs, and threshold graphs with arbitrary integer weights, we show that COMPETITIVE DIFFUSION is solvable in polynomial time.

Note that, while four years past after Alon *et al.* introduced the competitive diffusion game, no nontrivial algorithm for the k-player game is known, even for unweighted trees with $k \geq 3$. Our research breaks this situation, and provides a new landscape of the computational aspect of the game.

The rest of the paper is organized as follows. In Section 2, we formally define the competitive diffusion game and the problem COMPETITIVE DIFFUSION. In Section 3, we present our hardness results for COMPETITIVE DIFFUSION. In Section 4, we give algorithms for forests of paths. In Section 5, we provide an algorithm for chain, cochain, and threshold graphs.

2 Preliminaries

We model a network as an undirected graph $G = (V, E)$, where the vertex set V represents individuals in the network, and the edge set E does the connections among them. The weight function $w : V \to \mathbb{Z}$ represents a level of importance of each individual. For a positive integer k, we define $[k] = \{1, 2, \ldots, k\}$, and call the k players $p_1, p_2, \ldots, p_k$.

The competitive diffusion game (k, G, w) proceeds as follows (see Fig. 1(a)–(d) for an explicit example). At time one, each player chooses a vertex in V; suppose a player p_i, $i \in [k]$, chooses a vertex $v \in V$. If any other player p_j,

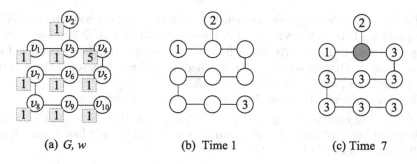

| (a) G, w | (b) Time 1 | (c) Time 7 |

Fig. 2. The vertex v_3 becomes neutral at time 2, and consequently, p_3 dominates v_4 at time 7

$i \neq j$, does not choose the vertex v, then p_i *dominates* v; and otherwise (that is, if there exists a player p_j, $i \neq j$, who chooses v), v becomes a *neutral* vertex. In the subsequent time steps, no player can dominate the neutral vertex. For each time t, $t \geq 2$, a vertex $v \in V$ is dominated by a player p_i at time t if (i) v is neither neutral nor dominated by any player by time $(t-1)$, and (ii) v has a neighbor dominated by p_i, but does not have a neighbor dominated by any player p_j, $i \neq j$. If v satisfies (i) and there are two or more players who dominate neighbors of v, then v becomes a neutral vertex at time t. The game ends when no player can dominate a vertex any more.

We note that the notion of a neutral vertex plays important role in the game; it sometimes gives critical effect on the result. (See Fig 2.) This contrasts to a similar game, called Voronoi game, where a player can dominate all the nearest vertices; if there is a vertex whose distances to seeds of two or more players tie, then they do not dominate but share the vertex [11,13,21,25].

Let $s = (s^{(1)}, s^{(2)}, \ldots, s^{(k)}) \in V^k$ be the vector of vertices which the players choose at the beginning of the game. We call s a *strategy profile*. For every $i \in [k]$, we define a *utility* $U_i(s)$ of p_i for s as the sum of the weights of the vertices which p_i dominates at the end. (See Fig. 1(d).)

For an index $i \in [k]$, we define (s_{-i}, v') as a strategy profile such that p_i chooses v' instead of $s^{(i)}$, but any other player p_j, $i \neq j$, chooses $s^{(j)}$: $(s_{-i}, v') = (s^{(1)}, s^{(2)}, \ldots, s^{(i-1)}, v', s^{(i+1)}, \ldots, s^{(k)})$. For simplicity, we write $U_i(s_{-i}, v')$ for $U_i((s_{-i}, v'))$. Then, if s satisfies $U_i(s_{-i}, v') \leq U_i(s)$ for every $i \in [k]$ and every $v' \in V$, we say that s is a *(pure) Nash equilibrium*. The strategy profile given in Fig. 1(b) is, in fact, a Nash equilibrium. We define COMPETITIVE DIFFUSION as the problem of deciding whether (k, G, w) has a Nash equilibrium.

3   Hardness Results on COMPETITIVE DIFFUSION

In this section, we observe computational complexity of COMPETITIVE DIFFUSION. Our first hardness result is the following theorem.

Theorem 1. COMPETITIVE DIFFUSION *is* $W[1]$-*hard even for unweighted graphs when parameterized by the number of players.*

To prove the theorem, we construct a reduction from a well-known $W[1]$-hard problem, INDEPENDENT SET [14]. Given a graph $G = (V, E)$ and a positive integer k, INDEPENDENT SET asks whether there exists an independent set I of size at least k, where a set I ($\subseteq V$) is called an independent set if there is no pair of vertices $u, v \in I$ such that $(u, v) \in E$.

Below we provide the desired reduction and a proof overview.

Proof idea. We construct a graph $G' = (V', E')$ such that $G = (V, E)$ has an independent set I of size $|I| \geq k$ if and only if $(k + 3, G', w')$ has a pure Nash equilibrium, where $w' : V' \to \{1\}$.

Construction of G'

Let $n = |V|$, and d_v be the degree of v for every $v \in V$. The graph G' consists of two connected components $A = (V_A, E_A)$ and $B = (V_B, E_B)$.

We obtain the component A as follows. We construct a path of four vertices a_1, a_2, a_3, a_4; and make $2n$ vertices $a'_1, a'_2, \ldots, a'_n$ and $a''_1, a''_2, \ldots, a''_n$. Then we connect the terminal a_1 to $a'_1, a'_2, \ldots, a'_n$, and connect the terminal a_4 to $a''_1, a''_2, \ldots, a''_n$. We obtain the component B from the original graph G as follows. For every edge $e = (u, v) \in E$, we add a vertex b_e subdividing e. Then, for each $v \in V$, we introduce a set D_v of $n - d_v$ vertices, and connect v to every $u \in D_v$. Lastly we make a vertex b and λ vertices $b_1, b_2, \ldots, b_\lambda$, where λ is a sufficiently large number satisfying $\lambda = \Theta(n^3)$, and connect b to every $v \in V$, and connect b to $b_1, b_2, \ldots, b_\lambda$. Thus, we have $V' = V_A \cup V_B$ and $E' = E_A \cup E_B$.

Consider the game $(k + 3, G', w')$. We can easily observe that any Nash equilibrium includes a strategy of a single player choosing the vertex b, since the strategy always give the maximum utility. Consequently, we can show that exactly two players can choose vertices other than the ones in the original graph G to hold a Nash equilibrium; otherwise, some player has extremely low utility (that is, below two) due to the player choosing b. In fact, we can show that any Nash equilibrium includes strategies of the two players choosing the vertex a_2 and a_3. Then the existence of a Nash equilibrium depends on whether there exists a strategy profile such that the other k players choose vertices composing an independent set: If the strategy profile of the other k players does not compose an independent set, then one of the k player obtains the utility less than $n + 1$; but the player can obtain the utility exactly $n + 1$ by changing its strategy to a_1 or a_4. □

For the cases where weights can be nonnegative or arbitrary integers, we can obtain the following stronger hardness results.

Theorem 2. COMPETITIVE DIFFUSION *is NP-complete even for series-parallel graphs with nonnegative integer weights.*

Theorem 3. COMPETITIVE DIFFUSION *is NP-complete even for forests of two components with integer weights.*

The proofs for Theorems 2 and 3 are similar to the one for Theorem 1, but we use other tricks by means of a neutral vertex together with positive and negative weights; we omit them due to the page limitation.

4   Algorithms for Forests of Paths

In the last section, we have shown that COMPETITIVE DIFFUSION is basically a computational hard problem. However, we can solve the problem for some particular graph classes. In Section 4.1, we give a pseudo-polynomial-time algorithm to solve COMPETITIVE DIFFUSION for forests of weighted paths; as its consequence, we show that the problem is solvable in polynomial time for forest of unweighted paths. In Section 4.2, we improve the running time of our algorithm to quadratic for the unweighted case.

4.1   Forests of Weighted Paths

Let F be a forest consisting of weighted m paths $P_1, P_2, \ldots, P_m$, and let W_j be the sum of the positive weights in a path P_j, $j \in [m]$. Then, we define $W = \max_{j \in [m]} W_j$ as the *upper bound on utility* for F, that is, any player can obtain at most W in F. In this subsection, we prove the following theorem.

Theorem 4. *Let F be a forest of weighted paths. Let n and W be the number of vertices in F and the upper bound on utility for F, respectively. Then, we can solve* COMPETITIVE DIFFUSION, *and find a Nash equilibrium, if any, in $O(Wn^9)$ time.*

We note that $W = O(n)$ if F is an unweighted graph. Therefore, by Theorem 4, COMPETITIVE DIFFUSION is solvable in $O(n^{10})$ time for an unweighted graph F; this running time will be improved to $O(n^2)$ in Section 4.2.

Idea and Definitions

Let F be a given forest consisting of weighted m paths $P_1, P_2, \ldots, P_m$. Let w be a given weight function; we sometimes denote by w_j the weight function restricted to the path P_j, $j \in [m]$. Suppose that, for an integer k, there exists a strategy profile s for the game (k, F, w) that is a Nash equilibrium. Then, the strategy profile restricted to each path P_j, $j \in [m]$, forms a Nash equilibrium for (k_j, P_j, w_j), where k_j is the number of players who chose vertices in P_j. However, the other direction does not always hold: A Nash equilibrium s_j for (k_j, P_j, w_j) is not always extended to a Nash equilibrium for the whole forest F, because some player may increase its utility by moving to another path in F. To capture such a situation, we classify a Nash equilibrium for a (single) path P_j more precisely.

Consider the game (κ_j, P_j, w_j) for an integer $\kappa_j \geq 0$. For a strategy profile s_j for (κ_j, P_j, w_j), we define $\mu_{P_j}(s_j)$ as the minimum utility over all the κ_j players: $\mu_{P_j}(s_j) = \min_{i \in [\kappa_j]} U_i(s_j)$. In other words, any player in P_j obtains the utility at least $\mu_{P_j}(s_j)$. For the case where $\kappa_j = 0$, we define $s_j = \emptyset$ as the unique strategy profile for (κ_j, P_j, w_j); then, s_j is a Nash equilibrium and we define $\mu_{P_j}(s_j) = +\infty$.

For a strategy profile $s_j = \left(s_j^{(1)}, s_j^{(2)}, \ldots, s_j^{(\kappa_j)}\right)$ for (κ_j, P_j, w_j), we then define the "potential" of the maximum utility under s_j that can be expected to gain by an extra player other than the κ_j players. More formally, for a vertex v in P_j, we denote by $s_j + v$ the strategy profile $\left(s_j^{(1)}, s_j^{(2)}, \ldots, s_j^{(\kappa_j)}, s_j^{(\kappa_j+1)}\right)$

for $(\kappa_j + 1, P_j, w_j)$ such that $s_j^{(\kappa_j+1)} = v$. Then, we define $\nu_{P_j}(s_j) = \max_{v \in V(P_j)} U_{\kappa_j+1}(s_j + v)$.

For two nonnegative integers κ_j and t, we say that P_j *admits* κ_j *players with a boundary* t if there exists a strategy profile s_j such that s_j is a Nash equilibrium for (κ_j, P_j, w_j) and $\nu_{P_j}(s_j) \le t \le \mu_{P_j}(s_j)$ holds. Then, the following lemma characterizes a Nash equilibrium of the game (k, F, w) in terms of the components of F; we omit the proof.

Lemma 1. *The game (k, F, w) has a Nash equilibrium if and only if there exist nonnegative integers $\kappa_1, \kappa_2, \ldots, \kappa_m$ and t such that $k = \sum_{j=1}^{m} \kappa_j$ and P_j admits κ_j players with the common boundary t for every $j \in [m]$.*

Algorithm
We first focus on a weighted single path.

Lemma 2. *Let P be a weighted path of n vertices, and t be a nonnegative integer. Then, one can find in $O(n^9)$ time the set $K \subseteq \{0, 1, \ldots, 2n\}$ of all the integers κ such that P admits κ players with boundary t.*

Based on Lemma 2, we can obtain the m sets $K_1, K_2, \ldots, K_m$, where $K_j \subseteq \{0, 1, \ldots, 2n\}$, $j \in [m]$, is the set of all the integers κ such that P admits κ players with boundary t. This can be done in $O(n^9)$ time, where n is the number of vertices in the whole forest F.

We now claim that, for a given integer t, it can be decided in $O(n^3)$ time whether there exist nonnegative integers $\kappa_1, \kappa_2, \ldots, \kappa_m$ such that $k = \sum_{j=1}^{m} \kappa_j$ and P_j admits κ_j players with the common boundary t for every $j \in [m]$; later we will apply this procedure to all possible values of t, $0 \le t \le W$. To show this, observe that finding desired m integers $\kappa_1, \kappa_2, \ldots, \kappa_m$ from the m sets $K_1, K_2, \ldots, K_m$ can be regarded as solving an instance of the multiple-choice knapsack problem [18]: The capacity c of the knapsack is equal to k; each integer κ' in K_j, $j \in [m]$, corresponds to an item with profit κ' and cost κ'; the items from the same set K_j form one class, from which at most one item can be packed into the knapsack. The multiple-choice knapsack problem can be solved in $O(cN)$ time [18], where N is the number of all items. Since $c = k$ and $N = O(mn)$, we can solve the corresponding instance in time $O(kmn) = O(n^3)$.

We finally apply the procedure above to all possible values of boundaries t. Since any player can obtain at most the upper bound W on utility for F, it suffices to consider $t \in [W]$. Therefore, our algorithm runs in $O(Wn^9)$ time in total.

4.2   Forests of Unweighted Paths

In this subsection, we improve the running time of our algorithm in Section 4.1 to quadratic when restricted to the unweighted case.

Theorem 5. *Let F be a forest of unweighted paths, and n be the number of vertices in F. Then, we can solve COMPETITIVE DIFFUSION, and find a Nash equilibrium, if any, in $O(n^2)$ time.*

In the rest of this subsection, we consider unweighted graphs, and thus define $w : V \to \{1\}$ for the vertex set V of a given forest. We assume that the number k of players is less than n; otherwise, a Nash equilibrium always exists. Note that, in this case, every player has utility at least one for any Nash equilibrium.

We first show that the set K_j of Lemma 2 can be obtained in $O(1)$ time, instead of $O(n^9)$ time, by characterizing Nash equilibriums for (κ, P, w) in terms of κ, t and n.

Lemma 3. *Let P be a single unweighted path of n vertices, and let κ and t be nonnegative and positive integers, respectively.*

(1) *P admits $\kappa = 0$ player with boundary t if and only if $n \le t$.*
(2) *P admits $\kappa = 1$ player with t if and only if $t \le n \le 2t + 1$.*
(3) *P admits $\kappa = 2$ players with t if and only if $2t \le n \le 2t + 2$.*
(4) *P admits $\kappa = 3$ players with t if and only if $t = 1$ and $n = 3, 4$ or 5.*
(5) *For any integer $\kappa \ge 4$, P admits κ players with t if and only if*

$$(\kappa + 1)t - 1 \le n \le (2\kappa - 4)t + \kappa \quad \text{if } \kappa \text{ is odd};$$
$$\kappa t \le n \le (2\kappa - 4)t + \kappa \qquad \qquad \text{if } \kappa \text{ is even}.$$

By Lemme 3, we can immediately obtain the number of players which P admits with a given boundary t:

Corollary 1. *Consider a fixed boundary t. If P is a path of n vertices, the numbers of players which P admits with a boundary t is given as follows.*

(1) *If $n \le t - 1$, the number is only 0.*
(2) *If $n = t$, the numbers are 0 and 1.*
(3) *If $t + 1 \le n \le 2t - 1$, the number is only 1.*
(4) *If $2t \le n \le 2t + 1$ and $n = 3$, the numbers are $1, 2$ and 3; and if $2t \le n \le 2t + 1$ and $n \ne 3$, the numbers are 1 and 2.*
(5) *If $n = 2t + 2$ and $n = 4$, the numbers are $2, 3$ and 4; and if $n = 2t + 2$ and $n \ne 4$, the number is only 2.*
(6) *If $2t + 3 \le n \le 4t - 1$, P has no desired Nash equilibrium.*
(7) *If $4t \le n$ and $5 \le n$, the numbers are integers κ such that*

$$\left\lceil \frac{n + 4t}{2t + 1} \right\rceil \le \kappa \le \max(k_{odd}, k_{even}),$$

where k_{odd} is the maximum odd integer satisfying $k_{odd} \le (n - t + 1)/t$, and k_{even} is the maximum even integer satisfying $k_{even} \le n/t$.

We use Corollary 1 to design our algorithm for forests of paths.

Without loss of generality, we assume that P_1 is a longest path among the m paths, and has n_1 vertices. For each t, $1 \le t \le n_1$, we repeat the following procedure: For every j, $1 \le j \le m$, we obtain, using Corollary 1, the minimum number $k_j^{\min}$ and the maximum number $k_j^{\max}$ of players which P_j admits with the boundary t. Corollary 1 implies that, for every j, $1 \le j \le m$, P_j admits κ

players with t for any κ between $k_j^{\min}$ and $k_j^{\max}$, and hence (k, F, w) has a Nash equilibrium with the common boundary t if and only if

$$\sum_{j=1}^{m} k_j^{\min} \leq k \leq \sum_{j=1}^{m} k_j^{\max}. \tag{1}$$

We thus complete the procedure by checking if the two inequalities in (1) both hold. Since Corollary 1 implies that we can obtain $k_j^{\min}$ and $k_j^{\max}$ in constant time for every j, the running time of the procedure above for single t is $O(m)$, and hence that of our entire algorithm is $O(n_1 m) = O(n^2)$, as desired.

5   Algorithms for Chain, Cochain, and Threshold Graphs

A bipartite graph $B = (X, Y; E)$ with $|X| = p$ and $|Y| = q$ is a *chain graph* if there is an ordering $(x_1, x_2, \ldots, x_p)$ on X such that $N(x_1) \subseteq N(x_2) \subseteq \cdots \subseteq N(x_p)$, where $N(u)$ denote a set of neighbors of a vertex u. If there is such an ordering on X, then there also exists an ordering $(y_1, y_2, \ldots, y_q)$ on Y such that $N(y_1) \subseteq N(y_2) \subseteq \cdots \subseteq N(y_q)$. We call such orderings *inclusion orderings*. A graph B' is a *cochain graph* if it can be obtained from a chain graph $B = (X, Y; E)$ by making the independent sets X and Y into cliques. A graph B'' is a *threshold graph* if it can be obtained from a chain graph $B = (X, Y; E)$ by making one of the independent sets X and Y into a clique. Observe that inclusion orderings on X and Y in B can be seen as *inclusion orderings* in B' and B'' if we use closed neighborhoods in cliques. Such inclusion orderings can be found in linear time [17]. Because the algorithm for chain graphs we will describe in this section depends only on its property of having inclusion orderings, we can apply the exactly same algorithm for cochain graphs and threshold graphs.

The following lemma follows directly from the definitions. Note that we denote $N[u] = N(u) \cup \{u\}$.

Lemma 4. *If $N(u) \subseteq N(v)$ or $N[u] \subseteq N[v]$ holds for $u = s^{(i)} \neq v = s^{(j)}$, then*

$$U_i(s) = \begin{cases} 0 & \text{if there is } h \neq i \text{ such that } s^{(h)} = u, \\ w(u) & \text{otherwise.} \end{cases}$$

In what follows, let $B = (X, Y; E)$ be a chain graph with inclusion orderings $(x_1, \ldots, x_p)$ and $(y_1, \ldots, y_q)$ on X and Y, respectively. We define $\eta(s, X) = \max(\{0\} \cup \{i \mid x_i \in V(s)\})$ and $\eta(s, Y) = \max(\{0\} \cup \{i \mid y_i \in V(s)\})$.

Lemma 5. *Let s be a Nash equilibrium of B. If $s^{(i)} \notin \{x_{\eta(s,X)}, y_{\eta(s,Y)}\}$, then*

$$w(s^{(i)}) \geq \max\left\{w(u) \mid u \in (\{x_j \mid j \leq \eta(s, X)\} \cup \{y_j \mid j \leq \eta(s, Y)\}) \setminus V(s)\right\}. \tag{2}$$

Proof. Since $N(s^{(i)}) \subseteq N(x_{\eta(s,X)})$ or $N(s^{(i)}) \subseteq N(y_{\eta(s,Y)})$, it follows that $U_i(s) \leq w(s^{(i)})$ by Lemma 4. Suppose for the contrary that there exists $u \in (\{x_j \mid j \leq \eta(s, X)\} \cup \{y_j \mid j \leq \eta(s, Y)\}) \setminus V(s)$ such that $w(s^{(i)}) < w(u)$.

Algorithm 1. Find a Nash equilibrium $s \in V^k$ of a chain graph $B = (X, Y; E)$

1: Let $(x_1, \ldots, x_p)$ on X and $(y_1, \ldots, y_q)$ on Y be inclusion orderings.
2: // The following is for the case where $\eta(s, X) \neq 0$.
3: **for all** guesses $(\eta(s, X), \eta(s, Y)) \in \{1, \ldots, p\} \times \{0, \ldots, q\}$ **do**
4:     $s^{(1)} := x_{\eta(s,X)}$. $s^{(2)} := y_{\eta(s,Y)}$ if $\eta(s, Y) \neq 0$.
5:     $R := \{x_i \mid i < \eta(s, X)\} \cup \{y_i \mid i < \eta(s, Y)\}$.
6:     **while** there is a player i not assigned to a vertex **do**
7:         $v := \arg\max_{u \in R} w(u)$.
8:         **if** $w(v) \geq 0$ **then**
9:             $s^{(i)} := v$. $R := R \setminus \{v\}$.
10:        **else**
11:            $s^{(i)} := x_{\eta(s,X)}$.
12:        **end if**
13:    **end while**
14:    **return** s if it is a Nash equilibrium.
15: **end for**
16: **return** "no Nash equilibrium"

Now it holds that $N(u) \subseteq N(x_{\eta(s,X)})$ or $N(u) \subseteq N(y_{\eta(s,Y)})$. Thus, by Lemma 4, we have $U_i(s_{-i}, u) = w(u) > w(s^{(i)}) \geq U_i(s)$. This contradicts the assumption that s is a Nash equilibrium. $\qquad\square$

Thus, it suffices to check the strategy profiles satisfying Eq. (2) for our purpose.

Theorem 6. *Let G be a chain, cochain, or threshold graph of n vertices and m edges. Then, we can solve* COMPETITIVE DIFFUSION *for G, and find a Nash equilibrium, if any, in $O(n^4(m + n))$ time.*

Proof. We present an algorithm for chain graphs only. As previously described, we can apply the same algorithm for cochain and threshold graph.

We first guess $\eta(s, X)$ and $\eta(s, Y)$. Here we assume $\eta(s, X) \neq 0$. The other case can be treated in the same way by swapping X and Y. We assign $x_{\eta(s,X)}$ to the first player. If $\eta(s, Y) \neq 0$, then we assign $y_{\eta(s,Y)}$ to the second player. By Lemma 5, if s is a Nash equilibrium, then the other players have to select the heaviest vertices in $\{x_i \mid i < \eta(s, X)\} \cup \{y_i \mid i < \eta(s, Y)\}$. For each of the remaining players, we assign a vacant vertex with the maximum non-negative weight. If there is no such a vertex, we assign $x_{\eta(s,X)}$. We then test whether the strategy profile is a Nash equilibrium. See Algorithm 1.

Lemma 5 implies that if the algorithm assigns at most one player to $x_{\eta(s,X)}$, then the algorithm is correct. If two or more players are assigned to $x_{\eta(s,X)}$, then these players have utility 0. In such a case, there are not enough number of vertices of non-negative weights in $\{x_i \mid i < \eta(s, X)\} \cup \{y_i \mid i < \eta(s, Y)\}$. Thus every s with the guesses $\eta(s, X)$ and $\eta(s, Y)$ has a player with non-positive utility. If such a player, say p_i, has negative utility, then s is clearly not a Nash equilibrium. If p_i has utility 0, then it may improve its utility only if there is a vertex $v \in \{x_{\eta(s,X)+1}, \ldots, x_p\} \cup \{y_{\eta(s,Y)+1}, \ldots, y_q\}$ such that $U_i(s_{-i}, v) > 0$. However, in this case, there is no Nash equilibrium with the guesses $\eta(s, X)$ and $\eta(s, Y)$. Therefore, the algorithm is correct.

We now analyze the running time of the algorithm. We have $O(n^2)$ options for guessing $x_{\eta(s,X)}$ and $y_{\eta(s,Y)}$. For each guess, the bottle-neck of the running time is to test whether the strategy profile is a Nash equilibrium or not. It takes $O(n^2(m + n))$ time as follows: we have $O(n^2)$ candidates of moves of players; for each candidate, we can compute the utility of the player moved by running a breadth-first search once in $O(m + n)$ time by adding a virtual root connecting to all the vertices occupied by the players. In total, the algorithm runs in $O(n^4(m + n))$ time.   □

Acknowledgment. This work is partially supported by MEXT/JSPS KAKENHI, including the ELC project. (Grant Numbers: 24106004, 24106007 24106010, 24700130, 25330001, 25330003, 25730003, 26730001.)

References

1. Alon, N., Feldman, M., Procaccia, A.D., Tennenholtz, M.: A note on competitive diffusion through social networks. Information Processing Letters **110**(6), 221–225 (2010)
2. Apt, K.R., Markakis, E.: Social networks with competing products. Fundamenta Informaticae **129**(3), 225–250 (2014)
3. Bharathi, S., Kempe, D., Salek, M.: Competitive influence maximization in social networks. In: Deng, X., Graham, F.C. (eds.) WINE 2007. LNCS, vol. 4858, pp. 306–311. Springer, Heidelberg (2007)
4. Borodin, A., Braverman, M., Lucier, B., Oren, J.: Strategyproof mechanisms for competitive influence in networks. In: Proc. of the 22nd International Conference on World Wide Web, pp. 141–151 (2013)
5. Borodin, A., Filmus, Y., Oren, J.: Threshold models for competitive influence in social networks. In: Saberi, A. (ed.) WINE 2010. LNCS, vol. 6484, pp. 539–550. Springer, Heidelberg (2010)
6. Bulteau, L., Froese, V., Talmon, N.: Multi-Player diffusion games on graph classes. In: Proc. of the 12th Annual Conference on Theory and Applications of Models of Computation (to appear)
7. Clark, A., Poovendran, R.: Maximizing influence in competitive environments: a game-theoretic approach. In: Proc. of the Second International Conference on Decision and Game Theory for Security, pp. 151–162
8. Cooper, C., Uehara, R.: Scale Free Properties of Random k-Trees. Mathematics in Computer Science **3**(4), 489–496 (2010)
9. Domingos, P., Richardson, M.: Mining the network value of customers. In: Proc. of the 7th ACM SIGKDD International Conference on Knowledge Discovery and Data Mining, pp. 57–66 (2001)
10. Draief, M., Heidari, H., Kearns, M.: New models for competitive contagion. In: Proc. of the 28th AAAI Conference on Artificial Intelligence, pp. 637–644 (2014)
11. Dürr, C., Thang, N.K.: Nash equilibria in Voronoi games on graphs. In: Proc. of the 15th Annual European Symposium on Algorithms, pp. 17–28 (2007)

12. Etesami, S.R., Basar, T.: Complexity of equilibrium in diffusion games on social networks. 1403.3881
13. Feldmann, R., Mavronicolas, M., Monien, B.: Nash equilibria for voronoi games on transitive graphs. In: Leonardi, S. (ed.) WINE 2009. LNCS, vol. 5929, pp. 280–291. Springer, Heidelberg (2009)
14. Flum, J., Grohe, M.: Parameterized Complexity Theory. Texts in Theoretical Computer Science. An EATCS Series (2006)
15. Goyal, S., Heidari, H., Kearns, M.: Competitive contagion in networks. Games and Economic Behavior (to appear)
16. He, X., Kempe, D.: Price of anarchy for the N-player competitive cascade game with submodular activation functions. In: Proc. of the 9th International Workshop on Internet and Network Economics, pp. 232–248 (2013)
17. Heggernes, P., Kratsch, D.: Linear-time certifying recognition algorithms and forbidden induced subgraphs. Nordic Journal of Computing 14, 87–108 (2008)
18. Kellerer, H., Pferschy, U., Pisinger, D.: Knapsack Problems. Springer-Verlag (2004)
19. Kempe, D., Kleinberg, J., Tardos, E.: Maximizing the spread of influence through a social network. In: Proc. of the 9th ACM SIGKDD International Conference on Knowledge Discovery and Data Mining, pp. 137–146 (2003)
20. Kempe, D., Kleinberg, J.M., Tardos, É.: Influential nodes in a diffusion model for social networks. In: Caires, L., Italiano, G.F., Monteiro, L., Palamidessi, C., Yung, M. (eds.) ICALP 2005. LNCS, vol. 3580, pp. 1127–1138. Springer, Heidelberg (2005)
21. Mavronicolas, M., Monien, B., Papadopoulou, V.G., Schoppmann, F.: Voronoi games on cycle graphs. In: Proc. of the 33rd International Symposium on Mathematical Foundations of Computer Science, pp. 503–514 (2008)
22. Richardson, M., Domingos, P.: Mining knowledge-sharing sites for viral marketing. In: Proc. of the 8th ACM SIGKDD International Conference on Knowledge Discovery and Data Mining, pp. 61–70 (2002)
23. Small, L., Mason, O.: Nash Equilibria for competitive information diffusion on trees. Information Processing Letters 113(7), 217–219 (2013)
24. Takehara, R., Hachimori, M., Shigeno, M.: A comment on pure-strategy Nash equilibria in competitive diffusion games. Information Processing Letters 112(3), 59–60 (2012)
25. Teramoto, S., Demaine, E.D., Uehara, R.: The Voronoi game on graphs and its complexity. Journal of Graph Algorithms and Applications 15(4), 485–501 (2011)

Sorting and Selection with Equality Comparisons

Varunkumar Jayapaul[1]([✉]), J. Ian Munro[2],
Venkatesh Raman[3], and Srinivasa Rao Satti[4]

[1] Chennai Mathematical Institute, H1, SIPCOT IT Park,
Siruseri, Chennai 603 103, India
varunkumarj@cmi.ac.in
[2] Cheriton School of Computer Science, University of Waterloo,
Waterloo, ON N2L 3G1, Canada
imunro@uwaterloo.ca
[3] The Institute of Mathematical Sciences, CIT Campus,
Taramani, Chennai 600 113, India
vraman@imsc.res.in
[4] Seoul National University, 1 Gwanak-ro, Gwanak-gu, Seoul 151-744, Korea
ssrao@cse.snu.ac.kr

Abstract. We consider the fundamental sorting and selection problems
on a list of elements that are not necessarily from a totally ordered set.
Here relation between elements are determined by 'equality' comparisons whose outcome is $=$ when the two elements being compared are
equal and $\neq$ otherwise. We determine the complexity of sorting (finding the frequency of every element), finding mode and other frequently
occurring elements using only $=, \neq$ comparisons. We show that $\Omega(n^2/m)$
comparisons are necessary and this many comparisons are sufficient to
find an element that appears at least m times. This is in sharp contrast
to the bound of $\Theta(n \log(n/m))$ bound in the model where comparisons
are $<, =, >$ or $\leq, >$.

1 Introduction

Sorting and selection are fundamental well studied problems in computer science.
We consider these problems when the input sequence is not necessarily from a
totally ordered set. Interestingly, this corresponds to the dictionary meaning of
the word 'sort'. The definition of "sort" in Oxford English Dictionary [13] goes on
for about 4 pages. The relevant sense is indeed the earliest as a verb: "to arrange
(things, etc.) according to kind or quality ..." (from the mid 14th century, and
from the Old French and Latin "sors" (lot, share, fortune)). Note that, there
is no notion of an inherent linear order. The word's association with putting
things in numerical or alphabetical order, came later with the development of
computing.

The only way the relation between a pair of elements is determined in this
scenario is by making equality comparisons. While this is a natural variant that
occurs when dealing with heterogenous sets of elements, to the best of our knowledge the only problem studied in this model is the problem of determining the

F. Dehne et al. (Eds.): WADS 2015, LNCS 9214, pp. 434–445, 2015.
DOI: 10.1007/978-3-319-21840-3_36

majority element (an element that appears at least $\lceil n/2 \rceil$ times) if exists, and there is a classical linear time algorithm for this [6]. Exact comparison complexity including upper and lower bounds, and average case complexity of this problem have been studied [2–4,8,12].

Starting with the (natural) problem of determining the mode, the most frequently occurring element, we study the comparison complexity of sorting (determining the frequency of every element) and to determine the least frequent element. We show that $\Omega(n^2/m)$ (equality) comparisons are necessary and this many comparisons are sufficient to find an element that appears at least m times. This is in sharp contrast to the bound of $\Theta(n \log(n/m))$ [7] bound in the traditional comparison model. The lack of transitivity of the 'not equal' operation throws interesting challenges. We develop a simple mode finding algorithm (which is then developed to a sorting algorithm), we believe that its analysis is quite subtle and interesting. Our lower bounds are through adversary arguments.

The next section discusses algorithms to find the mode or an element with a specific frequency in a list of n elements. In Section 3 we discuss bounds for sorting - i.e. to determine the frequency of all elements in the list. In Section 4, we discuss lower bounds for finding the mode and the least frequent element. Finally in Section 5, we conclude with some remarks.

1.1   Related Work

As referred earlier, we know of only the majority problem [6] studied with $=$, $\neq$ comparisons. In one of the earliest papers studying optimal algorithms on sets, Reingold [11] proved lower bounds for determining the intersection/union of two sets if only $=$, $\neq$ comparisons are allowed. Munro and Spira [9] considered optimal algorithms and lower bounds to find the mode and the spectrum (the frequencies of all elements), albeit in the three way comparison model. Misra and Gries [10] give algorithms to determine an element that appears at least n/k times for various values of k, in the three way comparison model.

2   Finding Mode (Or Elements with Specific Frequency)

A natural randomized algorithm to find the mode (given its frequency m) in this model is to pick a random element and find its frequency by comparing it with all other elements. If m is the frequency of the mode, then the probability that this algorithm picks the mode in any given round is m/n, and hence in about n/m rounds, the algorithm finds the mode with high probability. As it makes $n - 1$ comparisons in each step, the expected number of comparisons is around n^2/m. A high confidence bound can then be shown for this randomized approach. We show that this bound of $O(n^2/m)$ is achievable by a deterministic algorithm even without the knowledge of m. In addition, we give an adversary argument to show that $\Omega(n^2/m)$ comparisons are necessary. We then extend these results for sorting and finding the least frequent element.

Consider the naive algorithm that repeatedly finds the frequency of every element by a scan of the list. I.e. it picks an element, scans the list to find all elements equal to it and removes them in $n-1$ comparisons, and continues until all the frequencies are found. If, suppose, all elements appear m times, then the algorithm performs $(n-1)+(n-m-1)+(n-2m-1)+\ldots 1 = n^2/2m-n/m+n/2$ comparisons. But if the frequencies of the elements are different, in particular, if only the last m elements are the same and the rest are distinct, then this algorithm performs roughly $(n^2 - m^2)/2$ comparisons which is far from optimal (as we show in Section 4). But in this case, the following simple algorithm finds the mode in at most $n^2/2(m-1) + n$ comparisons.

1. Divide the given sequence into $m - 1$ blocks of size $\lceil n/(m - 1) \rceil$ each.
2. Compare every pair of elements within each block to find the frequency within the block of each element.
3. At least one of the blocks will have two copies of the mode and we can declare the mode as that element after confirming by comparing it with elements of the other blocks.

The no. of comparisons made by the algorithm is atmost $(m-1)\binom{(n/(m-1)+1)}{2}+n$ which is at most $n^2/2(m-1) + n$.

Now we give our mode finding algorithm that finds all modes in roughly n^2/m comparisons. (We subsequently generalize this for sorting.)

Theorem 1. *There exists an algorithm that performs at most $n^2/m+n$ (equality) comparisons to find a mode or even all modes and their frequency m, in a given list of n elements.*

Proof. Let $a_1, a_2, \ldots a_n$ be the given sequence of n elements, and consider them arranged clockwise in a circular list. The algorithm repeatedly compares, in sequence, every element with the first element in the clockwise order with which it has not determined its (equal/notequal) relation, until an element with frequency m is found. If m is not known to the algorithm, then the algorithm performs a sequence of rounds of comparisons until it finds an element that appears at least $\lceil (n-1)/k \rceil$ times at the end of k rounds. We show that the (circular order) sequence in which the comparisons are made achieves the desired upper bound.

The following pseudocode describes the algorithm.

0. Initialize $r = 0$; for $i = 1$ to n $eq(a_i) = \{i\}$; $neq(a_i) = \emptyset$;
1. Repeat
2. $r = r + 1$
3.   for $i = 1$ to n
4.   find the next j if any, starting from $i + 1$, wrapped around after n
5.   if necessary, such that $j \notin eq(a_i) \cup neq(a_i)$.
6.   if such a j is found then
7.   if $a_i = a_j$ then
8.   for all $x \in eq(a_i) \cup eq(a_j)$,
9. $eq(a_x) \leftarrow eq(a_i) \cup eq(a_j)$ and

10. $neq(a_x) \leftarrow neq(a_i) \cup neq(a_j)$
11.   else if $a_i \neq a_j$ then
12.   for all $x \in eq(a_i), neq(a_x) \leftarrow neq(a_x) \cup eq(a_j)$ and
13.   for all $y \in eq(a_j), neq(a_y) \leftarrow neq(a_y) \cup eq(a_i)$
14.   endfor
15. until there exists an element i such that $|eq(a_i)| \geq (n-1)/r$

In the algorithm $eq(a_i)$ corresponds to the set of indices of all elements that are known to the algorithm to be equal to a_i, and similarly $neq(a_i)$ corresponds to the set of elements known to be not equal to a_i. We refer to the comparison made in Step 7 as the one *initiated* by a_i (in round r) and associate such a comparison with a_i (note that a_j will also, by the same token, make one such comparison which will be associated with a_j). Though we count only comparisons between elements, the book keeping required to keep track of the sets $eq()$ and $neq()$ are not hard. It is clear that the algorithm maintains the invariant that if the algorithm knows that $a_i = a_j$, then $eq(a_i) = eq(a_j)$ and $neq(a_i) = neq(a_j)$. Hence we could keep these two lists for each group of elements that are known to be equal as one pair of lists instead of keeping them with each element. In what follows, we will continue to assume that every element has these two lists available.

As the algorithm performs at most n comparisons in each round r, and stops in $\lceil (n-1)/m \rceil$ rounds, the algorithm performs at most $n\lceil (n-1)/m \rceil \leq n(n+m-2)/m = (n^2-2n)/m + n < n^2/m + n$ comparisons.

The rest of the proof gives the correctness of the algorithm. We show that the first time an element that appears at least $(n-1)/r$ times at the end of r rounds, it is the mode. First we show that if an element appears m times, then the element will be discovered (to have exactly m copies) in at most $\lceil (n-1)/m \rceil$ rounds. We deal with the case $m = 1$ first where we can show a slightly better bound.

Lemma 1. *When all elements are distinct (i.e. when $m = 1$), the algorithm determines this in $\lfloor n/2 \rfloor$ rounds and the number of comparisons made by the algorithm is $n(n-1)/2$.*

Proof. When n is even, each element gains information regarding two new elements in each round (one due to the comparison initiated by the element, and another due to the comparison initiated on this element). So by round $n/2 - 1$, all elements have discovered their relation with all other but one element of the input. So in one more round of $n/2$ comparisons, all relationships will be found.

Thus the total number of comparisons made in this case is $(n/2-1)n+n/2 = n(n-1)/2$.

Similarly when n is odd, the algorithm takes $(n-1)/2$ rounds for each element to find its relation with the rest of the elements. Thus, total number of comparisons made in this case is also $n(n-1)/2$.   □

We prove the correctness for $m \geq 2$ through a series of lemmas. The first lemma follows from the fact that we maintain the 'invariant' for the sets $eq(a_i)$ and $neq(a_i)$ for each i, throughout the algorithm.

Lemma 2. *At any point in the algorithm, if $a_i = a_j$ has been discovered by the algorithm, then $eq(a_i) = eq(a_j)$ and $neq(a_i) = neq(a_j)$.*

To understand the next lemma (and hence the total runtime of the algorithm), consider the (lucky) situation where all initial comparisons were equality comparisons and so all groups have found (a lower bound for) their frequencies. Now to determine their exact frequency, all we need to do is to make *one* comparison between each pair of groups. But we may not be that lucky, as several wasteful 'not equal' comparisons may have been made between groups of (equal) elements before we even discover that a pair of elements in a group are equal. Lemma 3 says that because of the *order* in which we make the comparisons, the algorithm will not do too many wasteful comparisons.

Lemma 3. *Let a_i, a_j be two elements such that $a_i = a_j$, and let $k \notin eq(i)$. If a_i initiates a comparison with a_k in a round, then a_j will subsequently not initiate a comparison with a_k (even if $a_i = a_j$ was not determined when a_i initiated a comparison with a_k).*

Proof. If $a_i = a_j$ has already been discovered by the algorithm when a_k was directly compared with a_i, then clearly the outcome of the comparison between a_k and a_i also gives the relation between a_k and a_j (as updated by the eq and neq sets) and so the algorithm will not compare a_j with a_k.

Suppose $a_i = a_j$ has not been discovered by the algorithm when a_k was compared with a_i. Suppose $k > i$. If $j < i$, then the way the algorithm makes the comparisons, a_j would be compared with a_i (and be found equal and hence will learn its unequal relation with a_k from a_i's neq set) before comparing with a_k and hence will not initiate a comparison with a_k thereafter. Hence assume that $j > i$. Now, if $j < k$, then a_i would have initiated a comparison with a_j before initiating with a_k – a contradiction to the fact that $a_i = a_j$ has not been discovered when a_i was initiating a comparison with a_k. So $j > k$. Then again a_j would initiate a comparison (in the wrap around) with a_i before initiating a comparison with a_k. Hence a_j will not initiate a comparison with a_k.

A similar argument proves the claim if $k < i$.   □

A similar mirroring lemma also holds.

Lemma 4. *Let a_i, a_j be two elements such that $a_i = a_j$, and let $k \notin eq(i)$. If a_k initiates a comparison with a_i in a round, then a_k will subsequently not initiate a comparison with a_j (even if $a_i = a_j$ was not determined when a_k initiated a comparison with a_i).*

Proof. Similar to proof of Lemma 3.

Lemma 5. *Let X be the set of a_is whose value is x, for some value x, and let $|X| = m \geq 2$. Then all elements of X together initiate at most $n-1$ comparisons in $\lceil (n-1)/m \rceil$ rounds and know their relation with every other element.*

Proof. From Lemma 3, all elements of X initiate together at most $n - m$ comparisons with elements not in X. As the equality relation is transitive, at most $m - 1$ equality comparisons will be made among themselves to determine that they are all equal. Thus together elements of X initiate at most $n - 1$ comparisons. Thus at least one element of X initiates at most $(n - 1)/m$ comparisons and hence knows its relation to others in this many rounds (as otherwise it would have initiated more comparisons). Note that if any element of X has determined its relationship to all other elements, then all elements of X immediately know their relationship to all other elements. So in at most $\lceil (n - 1)/m \rceil$ rounds, each element of X will determine its relationship with all other elements.   □

Hence it is clear that the first r for which an element appears at least $\lceil (n - 1)/r \rceil$ times is the frequency of the mode. For, if there is an element whose frequency is more, then from Lemma 5, that element would have been discovered in the previous rounds.   □ (of Theorem 1)

In fact, it follows from the proof of the above theorem that

Theorem 2. *Given a list of n elements and an integer k, all elements (if any) with frequency at least k can be found using at most $O(n^2/k)$ comparisons.*

3   Sorting

Recall the naive algorithm that repeatedly picks an element a_i, which is known to be different from the elements whose exact frequencies have been discovered, and finds its frequency by comparing it with all elements not in $eq(a_i)$ and $neq(a_i)$ updating these two sets after every comparison. Let c be the number of distinct elements in the sequence, and let $x_1, x_2, \ldots x_c$ be the values of the elements, and let m_i, $i = 1$ to c be the number of times x_i occurs, where $m_1 \geq m_2 \geq m_3 \ldots \geq m_c$. In the worst case this algorithm will take at most $(n - 1) + (n - 1 - (m_c)) + (n - 1 - (m_c + m_{c-1})) + \ldots + (n - 1 - (m_c + m_{c-1} - \ldots + m_2)))$ comparisons. This is at most $nc - c - \sum_{i=2}^{c} (i - 1)m_i = \sum_{i=1}^{c}(c - i + 1)m_i - c = \sum_{i=1}^{c}(c - i)m_i + n - c$ which is at most $c(n - 1) + n$ comparisons. Thus we have

Theorem 3. *Let c be the number of distinct values in the sequence, and let $x_1, x_2, \ldots x_c$ be the values of the elements, and let $m_1, m_2, \ldots m_c$ be the number of times x_i occurs, where $m_1 \geq m_2 \geq m_3 \ldots \geq m_c$. Then there exists an algorithm to sort the list using at most $\sum_{i=1}^{c}(c - i)m_i + n - c$ comparisons.*

In what follows, we analyze the number of comparisons made by our mode algorithm (Theorem 1) if we run it until all frequencies are determined. Let c be the number of distinct elements in the list. Then, it follows from Lemma 4 that an element a_k never initiates a comparison with two elements a_i and a_j that are equal to each other, but are not equal to a_k. Thus every element initiates at most $c - 1$ comparisons with elements not equal to it, and hence the total number of comparisons resulting in 'not equal' answer, made by the algorithm is at most $n(c - 1)$. Along with the $n - c$ equality comparisons, the total number of comparisons is at most $c(n-1)$. In the following, we give a tighter upper bound for the number of comparisons made by the algorithm.

We begin with the following corollary that follows from Lemmas 3 and 4.

Corollary 1. *Let x and y be two distinct values that occur m_1 and m_2 times respectively in the sequence, then the number of comparisons made by the algorithm between elements $a_i = x$ and $a_j = y$ is at most $2\min\{m_1, m_2\}$.*

Proof. Let A be the set of indices i such that $a_i = x$ and B be the set of indices j such that $a_j = y$. Draw a bipartite graph with two parts as A and B and orient an edge from i to j if a_i initiates a comparison with a_j. Lemma 3 says that the indegree of any vertex in this bipartite graph is at most one and Lemma 4 says that the outdegree of any vertex is at most one. Thus the number of edges, that corresponds to the number of comparisons made between A and B is at most $2\min\{m_1, m_2\}$. □

Theorem 4. *Let c be the number of distinct elements in the sequence, and let $x_1, x_2, \ldots x_c$ be the values of the elements, and let $m_1, m_2, \ldots m_c$ be the number of times x_i occurs, where $m_1 \geq m_2 \geq m_3 \ldots \geq m_c$. Then the number of comparisons made by the mode algorithm to identify the frequency of every element is at most $2(\sum_{i=1}^{c} im_i) - n - c \leq c(n-1)$.*

Proof. Consider the number of comparisons made together by the set of all elements that are equal to x_c with all the elements outside this set. By Corollary 1, this number is at most $2m_c(c-1)$. In general, the number of comparisons made together by the set of elements that equal x_i with elements that equal x_j for all $j < i$, is at most $2m_i(i-1)$. Thus the total number of comparisons made by the algorithm is at most $2\sum_{i=1}^{c} m_i(i-1) = 2\sum_{i=1}^{c}(m_i i - m_i) = 2\sum_{i=1}^{c} im_i - 2n$. As the equality comparisons are transitive, the number of equality comparisons (made within each group) is at most $\sum_{i=1}^{c}(m_i - 1)$ which is at most $n - c$.

Thus the total number of comparisons made by the algorithm is at most $2\sum_{i=1}^{c} im_i - 2n + n - c = 2(\sum_{i=1}^{c} im_i) - n - c$

To show that $2(\sum_{i=1}^{c} im_i) - n - c \leq c(n-1)$, it suffices to show $2\sum_{i=1}^{c} im_i - 2n \leq n(c-1)$ or $\sum_{i=1}^{c} im_i \leq n(c+1)/2$.

Suppose c is odd. Then

$$n(\frac{c+1}{2}) - \sum_{i=1}^{c} im_i = \sum_{i=1}^{c}(\frac{c+1}{2} - i)(m_i)$$

$$= \sum_{i=1}^{(c+1)/2-1} m_i(\frac{c+1}{2} - i) - \sum_{i=(c+1)/2+1}^{c} m_i(i - \frac{c+1}{2})$$

$$= \sum_{i=1}^{(c-1)/2} m_i(\frac{c+1}{2} - i) - \sum_{j=1}^{(c-1)/2} j(m_{j+(c+1)/2})$$

$$= \sum_{i=1}^{(c-1)/2} i(m_{(c+1)/2-i}) - \sum_{i=1}^{(c-1)/2} i(m_{i+(c+1)/2})$$

$$= \sum_{i=1}^{(c-1)/2} i[m_{(c+1)/2-i} - m_{(c+1)/2+i}] \geq 0$$

The last inequality is true since every term in the summand is nonnegative as m_i's are in nondecreasing order. This proves the claim.

Suppose c is even. Then

$$n(c+1)/2 - \sum_{i=1}^{c} im_i = \sum_{i=1}^{c}((c+1)/2) - i)(m_i)$$

$$= \sum_{i=1}^{c/2} m_i(\frac{c+1}{2} - i) - \sum_{i=c/2+1}^{c} m_i(i - \frac{c+1}{2})$$

$$= \sum_{i=1}^{c/2} m_i(\frac{c+1}{2} - i) - \sum_{i=1}^{c/2} m_{i+c/2}(i - 1/2)$$

$$= \sum_{i=1}^{c/2} m_{c/2-i+1}(i - 1/2) - \sum_{i=1}^{c/2} m_{i+c/2}(i - 1/2)$$

$$= \sum_{i=1}^{c/2}(i - 1/2)(m_{c/2-i+1} - m_{c/2+i}) \geq 0$$

This proves the theorem. □

Corollary 2. *The least frequent element in a sequence of n elements can be found in at most n^2/ℓ equality comparisons where ℓ is the frequency of the least frequent element.*

Proof. If ℓ is the frequency of the least frequent element, then every element appears at least ℓ times and hence the number c of distinct elements is at most n/ℓ. So if we apply our sorting algorithm (Theorem 4), we can sort, and hence find the least frequent element in at most $n(n - 1)/\ell$ comparisons. □

4 Lower Bound

In this section, we give lower bounds for finding the mode, the least frequent element and sorting with equality comparisons. The lower bounds are proved by the adversary first modeling the input elements as vertices of a graph. Then for every comparison made by the algorithm, the adversary answers equal/not equal and constructs edges (based on the structure of the graph it has constructed till then) appropriately. The adversary answers in a way that it can instantiate the input elements, consistent with its answers.

4.1 Lower Bound for Finding the Mode

For giving a lower bound for finding the mode, we use Turán's theorem stated below.

Theorem 5. *[14] (see Theorem 1.1 in Chapter VI in [5]) Let G be any graph with n vertices such that G has no K_{m+1}, the complete graph on $m+1$ vertices. Then the number of edges in G is at most $(1 - 1/m)n^2/2 = n^2/2 - n^2/2m$.*

Theorem 6. *At least $n^2/2m - n/2$ equality comparisons are necessary for any algorithm to determine a mode with frequency m from a given list of n elements, even if the algorithm knows m.*

Proof. The proof is by an adversary argument. The adversary models the n elements as n vertices of an undirected graph G. Whenever the algorithm makes a comparison between a pair of elements, the adversary will answer 'not equal' and draws an edge between the pair of vertices, if after the addition of this edge, there is still an independent set in the graph of size $m + 1$ or more. Otherwise, the adversary will answer 'equal'.

As long as the modeled graph has an independent set of size $m + 1$, the algorithm cannot determine the mode. For, the adversary can make any subset of the elements of the independent set of size $m + 1$ as being equal to the mode.

Now when the algorithm gets the first 'equality' answer, the graph G has no independent set of size $m+1$. Hence $\bar{G}$, the complement of G has no clique of size $m + 1$. Hence by Theorem 5, the number of edges in $\bar{G}$ is at most $n^2/2 - n^2/2m$. Hence the number of edges in G, that corresponds to the number of comparisons made by the algorithm, is at least $\binom{n}{2} - n^2/2 + n^2/2m = n^2/2m - n/2$.   □

Recall the classical (textbook) algorithm [6] to determine the majority, if exists, of a list of n elements using at most $2n$ equality comparisons. Suppose the list has no majority element, one could ask for a pair of elements whose combined frequency is at least $\lceil n/2 \rceil$. We show, by an adversary argument, using Theorem 6 that finding such a pair requires $\Omega(n^2)$ comparisons.

The adversary sets up the input in such a way that one element appears exactly $n/2-2$ times (and hence not a majority) and every other element appears once or twice. The adversary gives away the element that appears exactly $n/2-2$ times. So the algorithm's task is to determine whether there is an element that appears at least twice among the remaining elements. The adversary answers the comparisons between the remaining elements as in the proof of Theorem 6 (with $m = 2$) forcing the algorithm to perform $\Omega(n^2)$ comparisons. Thus we have

Theorem 7. *Given a list of n elements, $\Omega(n^2)$ comparisons are necessary to determine whether there exists a pair of elements that together appear at least $n/2$ times.*

4.2   Lower Bound for Finding the Least Frequent Element

Note that the adversary in the proof of Theorem 6 does not work well to prove a lower bound for finding the least frequent element, as an algorithm can pick an element and compare it with every other element receiving 'not equal' answers and can declare it to be the least frequent element using only $n - 1$ comparisons.

To obtain a better lower bound for the least frequent element, we resort to a different adversary.

It is easy to see that any algorithm to determine whether the given n elements are distinct, using equality comparisons, requires $\Omega(n^2)$ comparisons. We convert that into a bound for finding the least frequent element.

Lemma 6. *In an n element list, $\Omega(n^2)$ comparisons are needed to find the least frequent value.*

Proof. The adversary answers a comparison between a_i and a_j 'not equal' as long as there is some other element with which a_i and a_j have not found their relation; i.e. the adversary makes them equal if this is the 'last' comparison for a_i or a_j. The adversary follows this strategy except for the last pair of elements for which it makes them not equal even on their last comparison. Thus it makes the last pair of elements as the least frequent elements appearing only once.

More precisely the adversary's answer for a comparison between a pair of elements a_i and a_j is as follows:

- If there exists an index $k(\neq j) \notin eq(a_i) \cup neq(a_i)$, then answer 'not equal'
- else if there exists an index $k(\neq i) \notin eq(a_j) \cup neq(a_j)$, then answer 'not equal'
- else if there exists an index $k \neq i, k \neq j$ such that $|eq(a_k) \cup neq(a_k)| < n-1$, then answer 'equal'
- else answer 'not equal'.

Thus the adversary makes all but two elements with frequency two and the last pair of elements to determine their frequencies to be of frequency 1. The adversary can declare the least frequent element to have frequency 1. Thus it makes the algorithm to perform at least $n - 1$ comparisons to eliminate every two elements out of contention for the least frequent element, thus forcing the algorithm to make $\Omega(n^2)$ comparisons. $\square$

We strengthen the bound to show the following.

Theorem 8. *In an n elements list, $\Omega(n^2/l^2)$ equality comparisons are necessary for any algorithm to determine a least frequent element with frequency l even if the algorithm knows l.*

Proof. Let n be a multiple of l. The adversary gives away n/l sets of size l each and reveals that all the elements in each set are equal to each other. Now the algorithm's task is to determine whether any element of any of the sets has copies elsewhere. Now using the adversary as in the proof of Lemma 6, it follows that $\Omega(n^2/l^2)$ comparisons are necessary to find the least frequent element. $\square$

We also show that

Theorem 9. *In an n elements list, $\Omega(n^2/m^2)$ equality comparisons are necessary for any algorithm to find all modes or to sort if every element is a mode appearing m times even if the algorithm knows m.*

Proof. The adversary is similar to that in the proof of Theorem 8, but if the algorithm is told that all elements in each set (of size m) are the same, then the algorithm has nothing to do. So the adversary partitions the set into two, one of size n/m containing all distinct elements, which is revealed to the algorithm, and the other of size $n - n/m$ grouped into n/m groups of size $m - 1$ each having all equal elements. Now the algorithm's task is for each element in the first set to find its group in the second set. The adversary can answer in a way that the algorithm is forced to make roughly $n^2/2m^2$ comparisons. □

We note that this bound falls short of our upper bound of $O(n^2/m)$ proved in Theorem 1 to find all modes or to sort when all elements appear m times.

5   Conclusions

We have determined (up to constant factors) the comparison complexity of finding a mode in a given list of n elements using only equality comparisons. There is a gap of a factor of 2 between upper and lower bounds (Theorem 1 and Theorem 6). In the special case when all elements other than the mode (that appears $m \geq 2$ times) are distinct, we did discuss an algorithm that achieves the lower bound. So we conjecture that the lower bound is correct, and if so improving our algorithm to match the lower bound is an interesting open problem.

Here are other specific interesting questions.

- Can we find a least frequent element that is known to appear ℓ times, in $O(n^2/\ell^2)$ comparisons, as claimed in our lower bound (Theorem 8), or can the lower bound be improved? Our algorithm (Theorem 2) finds *all* least frequent elements in $O(n^2/\ell)$ time.
- If we are told that every element appears $m \geq 2$ times, can we identify the groups, i.e. sort or equivalently find all the modes in better than $O(n^2/m)$ time? We could only prove a lower bound of $\Omega(n^2/m^2)$ (Theorem 9) and we conjecture that the lower bound can be strengthened to n^2/m even in this case.

References

1. Ajtai, M., Feldman, V., Hassidim, A., Nelson, J.: Sorting and selection with imprecise comparisons. In: Albers, S., Marchetti-Spaccamela, A., Matias, Y., Nikoletseas, S., Thomas, W. (eds.) ICALP 2009, Part I. LNCS, vol. 5555, pp. 37–48. Springer, Heidelberg (2009)
2. Alonso, L., Reingold, E.M., Schott, R.: The Average-Case Complexity of Determining the Majority. SIAM Journal on Computing **26**, 1–14 (1997)
3. Alonso, L., Reingold, E.M., Schott, R.: Determining the majority. Information Processing Letters **47**, 253–255 (1993)
4. Alonso, L., Reingold, E.M., Schott, R.: Analysis of Boyer and Moore's MJRTY. Information Processing Letters **113**, 495–497 (2013)
5. Bollobás, B.: Extremal Graph Theory. Academic Press (1978)

6. Boyer, R.S., Moore, J.S.: MJRTY - A fast majority vote algorithm. In: Boyer, R.S. (ed.) Automated Reasoning: Essays in Honor of Woody Bledsoe. Automated Reasoning Series, pp. 105–117. Kluwer (1991)
7. Dobkin, D.P., Munro, J.I.: Determining the Mode. Theoretical Computer Science **12**, 255–263 (1980)
8. Fischer, M.J., Salzberg, S.L.: Solution to problem 81-5. Journal of Algorithms **3**, 376–379 (1982)
9. Munro, J.I., Spira, P.M.: Sorting and Searching in Multisets. SIAM Journal on Computing **5**(1), 1–8 (1976)
10. Misra, J., Gries, D.: Finding Repeated Elements. Science of Computing Programming **2**(2), 143–152 (1982)
11. Reingold, E.M.: On the optimality of some set algorithms. Journal of the ACM **19**(4), 649–659 (1972)
12. Saks, M.E., Werman, M.: On computing majority by comparisons. Combinatorica **11**(4), 383–387 (1991)
13. Simpson, J.A., Weiner, E.S.C.: The Oxford English Dictionary, vol. XVI. Clarendon Press (1989)
14. Turań, P.: On an extremal problem in graph theory (in Hungarian). Mat. Fiz. Lapk **48**, 436–452 (1941)

Polynomial Delay Algorithm for Listing Minimal Edge Dominating Sets in Graphs

Mamadou Moustapha Kanté[1]([⊠]), Vincent Limouzy[1], Arnaud Mary[2],
Lhouari Nourine[1], and Takeaki Uno[3]

[1] Clermont-Université, Université Blaise Pascal, LIMOS, CNRS,
Aubiere Cedex, France
kante@isima.fr
[2] Université de Lyon, Université Lyon 1, UMR CNRS 5558,
LBBE, INRIA Erable, Lyon, France
[3] National Institute of Informatics, Tokyo, Japan

Abstract. It was proved independently and with different techniques in [Golovach et al. - ICALP 2013] and [Kanté et al. - ISAAC 2012] that there exists an incremental output polynomial algorithm for the enumeration of the minimal edge dominating sets in graphs, *i.e.*, minimal dominating sets in line graphs. We provide the first polynomial delay and polynomial space algorithm for the problem. We propose a new technique to enlarge the applicability of Berge's algorithm that is based on skipping hard parts of the enumeration by introducing a new search strategy. The new search strategy is given by a strong use of the structure of line graphs.

1 Introduction

The Minimum Dominating Set problem is a classic and well-studied graph optimization problem. A *dominating set* in a graph G is a subset D of its vertex set such that each vertex is either in D or has a neighbor in D. Computing a minimum dominating set has numerous applications in many areas, *e.g.*, networks, graph theory (see for instance the book [9]). The Minimum Edge Dominating Set problem is a classic well-studied variant of the Minimum Dominating Set problem [9]. An edge dominating set is a subset F of the edge set such that each edge is in F or is adjacent to an edge in F. In this paper, we are interested in an output polynomial algorithm for listing without duplications the (inclusionwise) minimal edge dominating sets of a graph. An output polynomial algorithm is an algorithm whose running time is bounded by a polynomial depending on the sum of the sizes of the input and of the output. The enumeration of minimal or maximal subsets of vertices satisfying some property in a (hyper)graph is a central area in graph algorithms and for several properties output polynomial algorithms have been proposed, e.g., [1,3,5,6,15,18], while for others it was proven that no output polynomial time algorithm exists unless P=NP [14,15].

The existence of an output polynomial algorithm for the enumeration of minimal dominating sets of graphs (Dom-Enum problem) is a widely open question and is closely related to the well-known Trans-Enum problem in hypergraphs

F. Dehne et al. (Eds.): WADS 2015, LNCS 9214, pp. 446–457, 2015.
DOI: 10.1007/978-3-319-21840-3_37

which asks for an output polynomial algorithm for the enumeration of all minimal transversals in hypergraphs. A *transversal* (or a *hitting-set*) in a hypergraph is a subset of its vertex set that intersects all its hyper-edges. This is a long-standing open problem (see for instance [4]) and is well-studied due to its applications in several areas [4,5,8]. Up to now only few tractable cases are known (see [11] for some examples). It is easy to see that the minimal dominating sets of a graph are the minimal transversals of its *closed neighbourhoods*[1], and then, as a particular case, it seems that the DOM-ENUM problem is more tractable than the TRANS-ENUM problem, but some of the authors have very recently proved in [11] that the TRANS-ENUM problem can be polynomially reduced to the DOM-ENUM problem. They also investigate the enumeration of minimal dominating sets in the perspective of graph theory and exhibit several new tractable cases, split graphs [11], undirected path graphs [10], interval and permutation graphs [12] and chordal P_6-free graphs [11]. In particular, they prove that the enumeration of minimal edge dominating sets can be done in (incremental) output polynomial time (a result obtained independently by Golovach et al. [7]). Many enumeration problems admit an output polynomial algorithm but no polynomial delay algorithm [20, Section 2.3], and so a natural question is whether one can enumerate with polynomial delay all the minimal edge dominating sets, and we answer this question positively in this paper.

Further, some of the aforementioned tractable cases of the TRANS-ENUM problem are based on Berge's algorithm [2]. Berge's algorithm consists in ordering the hyper-edges $E_1, \ldots, E_m$ of a given hypergraph $\mathcal{H}$ and computes incrementally the minimal transversals of $\{E_1, \ldots, E_i\}$ from the minimal transversals of $\{E_1, \ldots, E_{i-1}\}$. The algorithm is not output polynomial when there is a possibility that the intermediate steps have huge output sizes compared to the output solution. Indeed, it is proved in [21] that there exist hypergraphs for which Berge's algorithm is not output polynomial for any ordering. But, even though the applicability of Berge's algorithm seems to be limited, for several cases, *e.g.*, bounded tree-width graphs, planar graphs and more generally k-degenerate graphs, one can prove that Berge's algorithm is output polynomial. On the other hand, in the presence of some instances Berge's algorithm has a great potential for being turned into a polynomial delay polynomial space algorithm, compared to the other algorithms (for instance Khachiyan's algorithm). Indeed, Berge's algorithm admits a depth-first search on the solution space and then there is no need to store the solutions already found. In this sense, expanding the applicability of Berge's algorithm is quite important for more understanding the TRANS-ENUM problem.

In this paper, we propose a new way of expanding the applicability of Berge's algorithm. One of the disadvantages of Berge's algorithm is a huge computation time on the intermediate steps. Our first idea is to identify intermediate steps which produce intermediate solutions that can be extended to an output solution, and "skip" the other costly intermediate steps. But the search cost for finding

---

[1] The *closed neighborhood* of a vertex v in a graph is the set containing v and all its neighbors.

neighboring solutions becomes hard; we have to spend much time and use much space for finding each solution in the next intermediate step, from the solution of the current step. Our idea is to introduce another enumeration scheme to cope with this difficulty, to enable polynomial delay enumeration of the solutions to the next intermediate step. We apply this idea to the minimal edge dominating set enumeration problem, and obtain the first polynomial delay polynomial space algorithm for this problem.

2 Preliminaries

The power-set of a set V is denoted by 2^V and its size is denoted by $|V|$. Our graph terminology is standard and we deal only with finite and simple undirected graphs. A graph G is denoted by the pair (V, E) with vertex set V and edge set E. An edge between x and y in a graph is denoted by xy (equivalently yx) and sometimes it will be convenient to consider an edge xy as the set $\{x, y\}$, but this will be clear from the context. A hypergraph is a pair $(V, \mathcal{F} \subseteq 2^V)$ with V called its vertex set and $\mathcal{F}$ its set of hyper-edges.

Let $G := (V, E)$ be a graph. For a vertex x, we denote by $\widetilde{N}(x)$ the set of edges incident to x, and $N(x)$ denotes the set of vertices adjacent to x. For every edge xy, we denote by $N[xy]$ the set $\widetilde{N}(x) \cup \widetilde{N}(y)$. A subset D of E is called an *edge dominating set* if for every edge e of G, we have $N[e] \cap D \neq \emptyset$, and D is *minimal* if no proper subset of it is an edge dominating set.

An *enumeration* algorithm for a search problem consists in listing completely all the solutions without duplications. When an enumeration algorithm always terminates in time polynomial in n and N where n is the input size and N is the output size, the algorithm is called *output polynomial*, and it is called *polynomial space* if it uses space bounded by a polynomial in n. The *delay* of an enumeration algorithm is the maximum computation time from the time that a solution is output to the time the next solution is output, or the termination of the algorithm. The algorithm is called *polynomial delay* if its preprocessing time and the delay are both polynomial in the input size. It is worth noticing that such an algorithm has a running time bounded by the sum of the preprocessing time, and the delay multiplied by the number of solutions.

Let $\mathcal{H} := (V, \mathcal{F})$ be a hypergraph. The set of *private neighbors of a vertex x* w.r.t. $T \subseteq V$, denoted by $P_{\mathcal{H}}(x, T)$, is $\{E \in \mathcal{F} \mid E \cap T = \{x\}\}$. A subset T of V is called an *irredundant set* if $P_{\mathcal{H}}(x, T) \neq \emptyset$ for all $x \in T$. A *transversal* (or *hitting set*) of $\mathcal{H}$ is a subset of V that has a non-empty intersection with every hyper-edge of $\mathcal{H}$; it is *minimal* if it does not contain any other transversal as a proper subset. It is known that a transversal is minimal if and only if $P_{\mathcal{H}}(x, T) \neq \emptyset$ for all $x \in T$. The set of all minimal transversals of $\mathcal{H}$ is denoted by $tr(\mathcal{H})$.

For a graph $G := (V, E)$ and $E' \subseteq E$, we denote by $\mathcal{H}(E')$ the hypergraph $(E, \{N[e] \mid e \in E'\})$, and the *edge neighborhood hypergraph* of G is the hypergraph $\mathcal{H}(E)$. The following proposition is easy to obtain.

Proposition 1. *For any graph $G := (V, E)$, $T \subseteq E$ is an edge dominating set of G if and only if T is a transversal of $\mathcal{H}(E)$. Therefore, $T \subseteq E(G)$ is a minimal edge dominating set of G if and only if T is a minimal transversal of $\mathcal{H}(E)$.*

For a better readability we say that an edge f is a *private neighbor of an edge e* w.r.t. T in $\mathcal{H}(E')$, for $E' \subseteq E$, if $N[f] \in P_{\mathcal{H}(E')}(e, T)$, and by abuse of notation we will write $f \in P_{\mathcal{H}(E')}(e, T)$ instead of $N[f] \in P_{\mathcal{H}(E')}(e, T)$.

3 Berge's Algorithm and Basic Strategy

Our strategy for the enumeration is based on Berge's algorithm [2]. For a given hypergraph $\mathcal{H} := (V, \mathcal{F})$ with hyper-edges enumerated as $F_1, \ldots, F_m$, let $\mathcal{F}_j$ be $\{F_1, \ldots, F_j\}$ for each $1 \leq j \leq m$. Roughly, Berge's algorithm computes, for each $1 < j \leq m$, $tr(\mathcal{F}_j)$ from $tr(\mathcal{F}_{j-1})$. Although the algorithm is not polynomial space, there is a way to reduce the space complexity to polynomial. The algorithm follows a tree of height m rooted at $\emptyset$ and such that the nodes located on the i-th level correspond to the minimal transversals of $\mathcal{F}_i$. Thus, the leaves at level m correspond to the minimal transversals of the hypergraph. The tree can be described by the following parent-child relation. For $j \geq 1$ and $T \in tr(\mathcal{F}_j)$, we define the parent $Q'(T, j)$ of T as follows

$$Q'(T, j) := \begin{cases} T & \text{if } T \in tr(\mathcal{F}_{j-1}), \\ T \setminus \{v\} & \text{if } v \text{ is such that } P_{\mathcal{F}_j}(v, T) = \{F_j\}. \end{cases}$$

We can observe that $T \notin tr(\mathcal{F}_{j-1})$ if and only if $P_{\mathcal{F}_j}(v, T) = \{F_j\}$ holds for some $v \in T$, thus the parent is well defined and always in $tr(\mathcal{F}_{j-1})$ [13,17]. One can moreover compute the parent of any $T \in tr(\mathcal{F}_j)$ in time polynomial in $|V| + \sum_{F \in \mathcal{F}} |F|$. The tree induced by the parent-child relation spans all members of $\bigcup_{1 \leq j \leq m} tr(\mathcal{F}_j)$. We can traverse this tree in a depth-first search manner from the root by recursively generating the children of the current visited minimal transversal. Any child is obtained by adding at most one vertex, then the children can be listed in polynomial time. In this way, we can enumerate all the minimal transversals of a hypergraph with polynomial space.

Formally and generally, we consider the problem of enumerating all elements of a set $\mathcal{Z}$ that is a subset of an implicitly given set $\mathcal{X}$. Assume that we have a polynomial time computable parent function $P : \mathcal{X} \to \mathcal{X} \cup \{nil\}$. For each $X \in \mathcal{X}$, $P(X)$ is called the *parent* of X, and the elements Y such that $P(Y) = X$ are called *children* of X. The parent-child relation of P is *acyclic* if any $X \in \mathcal{X}$ is not a proper ancestor of itself, that is, it always holds that $X \neq P(P(\cdots P(X)) \cdots)$. We say that an acyclic parent-child relation is *irredundant* when any $X \in \mathcal{X}$ has a descendant in $\mathcal{Z}$, in the parent-child relation. The *depth* of an acyclic parent-child relation P is the size of the longest chain between nil and an element of $\mathcal{X}$. The following statements are well-known in the literature [1,13,16,17,19].

Proposition 2. *All elements in $\mathcal{Z}$ can be enumerated with polynomial space if there is a polynomial depth acyclic parent-child relation $P : \mathcal{X} \to \mathcal{X} \cup \{nil\}$ such that there is a polynomial space algorithm for enumerating all the children of each $X \in \mathcal{X} \cup \{nil\}$.*

Proposition 3. *All elements in $\mathcal{Z}$ can be enumerated with polynomial delay and polynomial space if there is a polynomial depth irredundant parent-child relation $P : \mathcal{X} \to \mathcal{X} \cup \{nil\}$ such that there is a polynomial delay polynomial space algorithm for enumerating all the children of each $X \in \mathcal{X} \cup \{nil\}$.*

With acyclic (resp., irredundant) parent-child relation $P : \mathcal{X} \to \mathcal{X} \cup \{nil\}$, the following algorithm enumerates all elements in $\mathcal{Z}$, with polynomial space (resp., with polynomial delay and polynomial space).

Algorithm ReverseSearch(X)
 1. **if** $X \in \mathcal{Z}$ **then output** X
 2. **for each** child Y of X **call** ReverseSearch(Y)

The call ReverseSearch(nil) enumerates all elements in $\mathcal{Z}$. Since the above parent-child relation for transversals Q' is acyclic, the algorithms proposed in [13, 17] use polynomial space. However, the parent-child relation Q' is not irredundant and hence ReverseSearch(nil) does not guarantee a polynomial delay neither an output polynomiality. Indeed, we can expect that the size of $tr(\mathcal{F}_j)$ increases as the increase of j, and it can be observed in practice. However, $tr(\mathcal{F}_j)$ can be exponentially larger than $tr(\mathcal{F}_m)$, thus Berge's algorithm is not output polynomial [21]. Examples of irredundant parent-child relations can be found in the literature [1, 16, 19].

One idea to avoid the lack of irredundancy is to certify the existence of minimal transversals in the descendants. Suppose that we choose some levels $1 = l_1, \ldots, l_k = m$ of Berge's algorithm, and state that for any $T \in tr(\mathcal{F}_{l_j})$, we have at least one descendant in $tr(\mathcal{F}_{l_{j+1}})$. This implies that any transversal in $tr(\mathcal{F}_{l_j})$ has a descendant in $tr(\mathcal{F}_m)$, thus we can have an irredundant parent-child relation by looking only at these levels, and the enumeration can be polynomial delay and polynomial space.

We will use this idea to obtain a polynomial delay polynomial space algorithm to enumerate the minimal edge dominating sets, the levels are determined with respect to a *maximal matching*. From now we assume that we have a fixed graph $G := (V, E)$ and we will show how to enumerate all its minimal edge dominating sets. A subset of E is a *matching* if every two of its edges e and f are not adjacent. A matching is *maximal* if it is not included in any other matching. Let $\{b_1, \ldots, b_k\}$ be a maximal matching of G, and let $b_i = x_i y_i$. For each $0 \le i \le k$, let $V_i := V \setminus \left(\bigcup_{i' > i} b_{i'} \right)$, and let $E_i := \{e \mid e \subseteq V_i\}$). Let $B_i := E_i \setminus E_{i-1}$ for $i > 1$. Note that any edge in E_1 is adjacent to b_1 and by definition B_i never includes an edge $b_j \ne b_i$. Without loss of generality, we here assume that we have taken a linear ordering $\le$ on the edges of G so that: (1) for each $e \in B_i$ and each $f \in E_{i-1}$ we have $f < e$, (2) for each $e \in \widetilde{N}(x_i) \cap B_i$, each $f \in \widetilde{N}(y_i) \cap B_i$ we have $b_i < e < f$. Observe that with that ordering we have $e < f$ whenever $e \in B_i$ and $f \in B_j$ with $i < j$. We consider that Berge's algorithm on $\mathcal{H}(E)$ follows that ordering. In fact we will prove using Berge's algorithm that we can define an irredundant parent-child relation to enumerate $tr(\mathcal{H}(E_i))$ from $tr(\mathcal{H}(E_{i-1}))$.

Lemma 1. *Let $1 \leq i < k$. Any $T \in tr(\mathcal{H}(E_{i-1}))$ has at least one descendant in $tr(\mathcal{H}(E_i))$.*

Proof. If $T' \in tr(\mathcal{H}(E_i))$ satisfies $T' = T$, then T' is a descendant of T since the parent is never greater than the child. If $T \notin tr(\mathcal{H}(E_i))$, some edges X of B_i are not dominated by T, and consider $T' := T \cup \{b_i\}$. We observe that b_i is adjacent to all edges of B_i and the edges in X are private neighbors of b_i in T', thus T' is included in $tr(\mathcal{H}(E_i))$. Let us compute the ancestor of T' in $tr(\mathcal{H}(E_{i-1}))$ as follows: set $T" := T'$ and repeatedly compute the parent of $T"$ and set $T"$ to its parent, until reaching a minimal transversal in $tr(\mathcal{H}(E_{i-1}))$. In this process no vertex of T is removed since each vertex in T has a private neighbor in E_{i-1}. But, at some point b_i is removed from T' since it is the only one in T' which has a private neighbor in B_i. This means that T is an ancestor of T', and thus T always has a descendant in $tr(\mathcal{H}(E_i))$. $\square$

For conciseness, we introduce a new parent-child relation for edge dominating set enumeration. For $T \in tr(\mathcal{H}(E_i))$, let $Q'_j(T, |E_i|)$ be the ancestor of T located on the j-th level of Berge's algorithm, i.e., $Q'_j(T, |E_i|) = Q'(Q'(\cdots(T, |E_i|), |E_i| - 1), \cdots, j+1)$. Then, we define the *skip parent* $Q(T, i)$ of T by $Q'_{|E_{i-1}|}(T, |E_i|)$. T' is a *skip-child* of $T \in tr(\mathcal{H}(E_{i-1}))$ if and only if $T' \in tr(\mathcal{H}(E_i))$ and $Q(T', i) = T$. The set of skip-children of $T \in tr(\mathcal{H}(E_i))$ is denoted by $\mathcal{C}(T, i)$. From Propositions 2 and 3, and Lemma 1 we have the following proposition.

Proposition 4. *If we can list all skip-children of $T \in tr(\mathcal{H}(E_i))$, for each $1 \leq i \leq k$, with polynomial delay and polynomial space, then we can enumerate all minimal edge dominating sets with polynomial delay and polynomial space.*

But, as we will show in the next section, for a transversal T in $tr(\mathcal{H}(E_{i-1}))$, the problem of finding a transversal of $tr(\mathcal{H}(E_i))$ including T is NP-complete in general. In order to overcome this difficulty, we will identify a pattern, that we call an *H-pattern*, that makes the problem difficult. We will first show that one can enumerate with polynomial delay and polynomial space all the skip-children that include no edges from H-patterns, and then define a new parent-child relation that will allow to enumerate also with polynomial delay and polynomial space the other skip-children in a different way. In the following sections, we explain the methods for the enumeration.

4   Computing Skip-Children

Let T be in $tr(\mathcal{H}(E_{i-1}))$ and $T' \in tr(\mathcal{H}(E_i))$ a skip-child of T. First notice that every edge in $T' \setminus T$ can have a private neighbor only in B_i. Indeed every edge in E_{i-1} is already dominated by T and an edge in $T' \setminus T$ is only used to dominate an edge in B_i. Moreover, an edge $e \neq b_i$ in $\widetilde{N}(x_i) \cap (T' \setminus T)$ (resp. in $\widetilde{N}(y_i) \cap (T' \setminus T)$) can only have private neighbors in $\widetilde{N}(x_i) \cap B_i$ (resp. $\widetilde{N}(y_i) \cap B_i$). And from the proof of Lemma 1 if $b_i \in T' \setminus T$ then $T' \setminus T = \{b_i\}$.

Let us first consider the case that every edge in $T' \setminus T$ is adjacent to b_i. From our discussion above, when two edges in $T' \setminus T$ are incident to x_i (resp. y_i), they cannot have both private neighbors. Thus $T' \setminus T$ can include at most two such edges. Therefore, by choosing all combinations of one or two edges adjacent to b_i, adding them to T and then checking if the skip-parent of the resulting set is T, we can enumerate all the skip-children T' of T such that $T' \setminus T \subseteq B_i$ with polynomial delay and polynomial space.

We now consider the remaining case that an edge in $T' \setminus T$ is not adjacent to b_i. We call such a skip-child *extra*. We can see that at least one edge $f \neq b_i$ adjacent to b_i has to be included in T' to dominate b_i. Actually, since $b_i < e$ for any $e \in B_i \setminus \{b_i\}$, any extra skip-child of T is a descendant of some $T \cup \{f\}$ with $f \neq b_i$ incident to x_i or y_i in the original parent-child relation. So, without loss of generality, we will assume that such an edge $f \neq b_i$ is incident to x_i and is included in T. Hereafter, we suppose that $N(y_i) := \{z_1, \ldots, z_k\}$ and assume T' is an extra skip-child of T.

A vertex $z_h \in N(y_i) \cap V_i$ is *free* if it is not incident to an edge in T, and is *non-free* otherwise. A free vertex is said to be *isolated* if it is not incident to an edge in E_{i-1}. Clearly, if there is an isolated free vertex, then T has no extra skip-child. Thus, we assume that there is no isolated free vertex. Edges in $E_i \setminus B_i$ that are incident to some free vertices are called *border edges*. Observe that any border edge vz_h incident to a free vertex z_h is adjacent to an edge $vw \in T$ if $v \in V_{i-1}$. The set of border edges is denoted by $Bd(T, i)$. Note that no edge in $Bd(T, i)$ is incident to two free vertices, otherwise the edge is in E_{i-1} but not dominated by T, and then any border edge is incident to exactly one free vertex. We can see that an edge of B_i incident to y_i is not dominated by T if and only if it is incident to a free vertex, and any edge in $T' \setminus T$ that is not incident to x_i is a border edge. Then, for any border edge set $Z \subseteq Bd(T, i)$, $T \cup Z \in tr(\mathcal{H}(E_i))$ only if each free vertex has a border edge $e \in Z$ incident to it. Since any border edge is incident to exactly one free vertex, for any $Z \subseteq Bd(T, i)$ such that $T \cup Z$ is irredundant and for any edge $vz_h \in Z$ with free vertex z_h, $P_{\mathcal{H}(E_i)}(e, T \cup Z)$ is always $\{vz_h\}$. This implies that $T \cup Z$ is in $tr(\mathcal{H}(E_i))$ only if $Z \subseteq Bd(T, i)$ includes exactly one edge incident to each free vertex. We call such an edge set Z a *selection*. We observe that all border edges are dominated by Z. We have the following lemma which is straightforward to prove.

Lemma 2. *For any edge subset Z with $Z \cap T = \emptyset$, there holds $T \cup Z \in tr(\mathcal{H}(E_i))$ only if Z is a selection.*

An edge $e \in T$ is called *redundant* if all edges in $P_{\mathcal{H}(E_{i-1})}(e, T)$ are border edges and no edge $y_i z_h$ is in $P_{\mathcal{H}(E_i)}(e, T)$.

Lemma 3. *If T has a redundant edge, then any selection Z does not satisfy $T \cup Z \in tr(\mathcal{H}(E_i))$.*

Proof. Let e be a redundant edge of T. Since any border edge f is incident to a free vertex z_h, any selection Z should contain one edge incident to z_h and then if f is incident to e, we have $f \notin P_{\mathcal{H}(E_i)}(e, T \cup Z)$. Since no edge $y_i z_h$ is in $P_{\mathcal{H}(E_i)}(e, T)$, there holds that $P_{\mathcal{H}(E_i)}(e, T \cup Z) = \emptyset$ for any selection Z. $\square$

Let $X_T := \{e \in Bd(T, i) \mid \exists e' \in T \text{ and } P_{\mathcal{H}(E_i)}(e', T \cup \{e\}) \subseteq Bd(T, i)\}$. The addition of any edge $e \in X_T$ to T transforms an edge e' of T into a redundant one with respect to $T \cup \{e\}$, and thus by Lemma 3 for any $Z \subseteq Bd(T, i)$, $T \cup Z \in tr(\mathcal{H}(E_i))$ holds only if $Z \cap X_T = \emptyset$. Therefore, the following follows.

Lemma 4. *If a free vertex is not incident to an edge in* $Bd(T, i) \setminus X_T$, *then any* $Z \subseteq Bd(T, i)$ *does not satisfy* $T \cup Z \in tr(\mathcal{H}(E_i))$.

One can hope that we can characterize the selections Z not intersecting X_T such that $T \cup Z \in tr(\mathcal{H}(E_i))$ and be able to use it for listing the extra skip-children. Unfortunately, checking whether there is such a selection Z is NP-complete.

Theorem 1. *Given* $T \in tr(\mathcal{H}(E_{i-1}))$, *it is np-complete to check whether there is a selection* Z *such that* $Z \cap X_T = \emptyset$ *and* $T \cup Z \in tr(\mathcal{H}(E_i))$.

In order to overcome this difficulty, we identify a pattern, that we call an H-pattern, that makes the problem difficult.

Definition 1 (H-Pattern). *A vertex set* $\{z_\ell, v_\ell, z_j, v_j\}$ *is an* H-*pattern if* z_ℓ *and* z_j *are free vertices,* $v_\ell v_j$ *is in* T, *and* $v_\ell v_j$ *has two non-border private neighbors in* $E_{i-1} \setminus T$: *one is adjacent to* v_ℓ *and the other to* v_j. *We also say that the edges* $z_\ell v_\ell$, $z_j v_j$ *and* $v_\ell v_j$ *induces an* H-*pattern.*

We will see that the np-completeness comes from the presence of H-patterns. Indeed, for an H-pattern $\{z_\ell, v_\ell, z_j, v_j\}$, any private neighbor of $v_\ell v_j$ is adjacent to either $z_\ell v_\ell$ or to $z_j v_j$, thus we cannot add both to a selection Z since in that case $P_{\mathcal{H}(E_i)}(v_\ell v_j, T \cup Z)$ will be empty. Let H_T be the set of border edges included in an H-pattern. In the next two subsections we will see how to list selections including no edge from H_T, and those that do.

Lemma 5. *If* T *has no redundant edge, then* $T \cup Z \in tr(\mathcal{H}(E_i))$ *holds for any selection* $Z \subseteq Bd(T, i) \setminus (X_T \cup H_T)$.

Proof. From the definition, $T \cup Z$ dominates all the edges in E_i and for each $e \in Z$ it holds that $P_{\mathcal{H}(E_i)}(e, T \cup Z) \neq \emptyset$. Since Z includes no edge from $H_T \cup X_T$, and T has no redundant edge, one easily checks from Lemmas 2, 3 and 4 by case analysis that any edge $e \in T$ has a private neighbor f that is adjacent to no border edge, or an edge $y_i z_h$ is adjacent to e and not to edges in $T \setminus \{e\}$. Thus, either $f \in P_{\mathcal{H}(E_i)}(e, T \cup Z)$ or $y_i z_h \in P_{\mathcal{H}(E_i)}(e, T \cup Z)$. These imply that $T \cup Z$ is in $tr(\mathcal{H}(E_i))$.   □

4.1   Dealing with Redundancies

The lemmas above demonstrate how to construct transversals $T' \in tr(\mathcal{H}(E_i))$ from T, but some generated transversals may not be extra skip-children of T. This is because such T' can be also generated from other transversals in $tr(\mathcal{H}(E_{i-1}))$. Such redundancies happen for example when two edges f_1 and

f_2 in T' have private neighbors only in B_i, but after the removal of either one from T', the other will have a private neighbor outside B_i. Assuming in this case that $f_1 \in T$ and $f_2 \in T' \setminus T$, it holds that T' can be generated from T or from $(T \setminus \{f_1\}) \cup \{f_2\}$. And since the number of selections Z such that $T \cup Z \in tr(\mathcal{H}(E_i))$ can be arbitrarily large, we need to avoid such redundancies.

To address this issue, we state the following lemmas to characterize the edges not to be added to selections Z such that $T \cup Z$ is an extra skip-child of T. We say that a border edge vz_ℓ is *preceding* if there is an edge vz_h in T satisfying $P_{\mathcal{H}(E_{i-1})}(vz_h, T) \subseteq N[vz_\ell]$ and $y_i z_\ell < y_i z_h$, and denote the set of preceding edges by X'_T. We also say that an edge $vz_h \in T$ is *fail* if $P_{\mathcal{H}(E_{i-1})}(vz_h, T) \subseteq Bd(T, i)$, $y_i z_h$ is in $P_{\mathcal{H}(E_i)}(vz_h, T)$, and no edge $wz_\ell \in P_{\mathcal{H}(E_{i-1})}(vz_h, T)$ satisfies $y_i z_h < y_i z_\ell$.

Lemma 6. *For any selection Z including a preceding edge, $T \cup Z$ is not an extra skip-child of T.*

Lemma 7. *If T has a fail edge, then $T \cup Z$ is not an extra skip-child of T for any selection Z.*

We are now able to characterize exactly those selections Z not intersecting H_T and such that $T \cup Z$ is an extra skip-child of T.

Lemma 8. *Suppose that T has neither redundant edges nor fail edges and any free vertex is incident to an edge in $Bd(T, i)$. Then, $T \cup Z$ with $T \cap Z = \emptyset$ is an extra skip-child of T including no edge of H_T if and only if Z is a selection including no edge of $X_T \cup X'_T \cup H_T$.*

As a corollary we have the following.

Proposition 1. *One can enumerate with polynomial delay and space all the extra skip-children of T that do not contain edges of H_T.*

Proof. If T has redundant edges or fail edges or has a free vertex not incident to an edge in $Bd(T, i) \setminus X_T$, then by Lemmas 3, 4 and 7 we can conclude that T has no extra skip-child. Since we can compute X_T in polynomial time and check in polynomial time whether an edge is redundant or is a fail edge, this step can be done in polynomial time. So, assume T has no redundant edges, no fail edges and every free vertex is incident to an edge in $Bd(T, i) \setminus X_T$. By Lemma 8 by removing all edges in $H_T \cup X_T \cup X'_T$, any selection Z is such that $T \cup Z$ is a skip-child of T. One easily checks that the enumeration of these selections can be performed by picking exactly one edge in each incident star.   □

4.2   Dealing with the Presence of H-patterns

As we saw in Theorem 1, it is hard to enumerate all extra skip-children having some edges in H-patterns from a given transversal $T \in tr(\mathcal{H}(E_{i-1}))$. Let us call these children *slide-children*. We approach this difficulty by introducing a new parent-child relation among slide-children, and enumerate them by traversing the forest induced by the new relation. In this way, we now do not follow the skip-parent skip-child relation for slide-children. However, the root of each tree in the induced forest is a transversal obtained with the skip-child skip-parent relation. Let us be more precise now. For two sets S and S' of edges we write $S <_{lex} S'$ if $\min(S \Delta S') \in S$, called *lexicographical ordering*.

Hereafter, we consider an extra skip-child $T' = T \cup Z$ of $T \in tr(\mathcal{H}(E_{i-1}))$ such that $T' \cap H_T \neq \emptyset$. Let $H^*(T') := \{v_h z_h, v_\ell z_\ell, v_h v_\ell\}$ be the lexicographically minimum H-pattern among all H-patterns of T that includes an edge of Z. Without loss of generality, we assume that $v_\ell z_\ell$ is in Z. Let $u z_h$ be the edge in Z incident to z_h. Notice that such an edge exists because z_h is a free vertex. Then, we define the *slide-parent* $Q^*(T', i)$ of T' by $T' \cup \{v_h z_h\} \setminus \{u z_h, v_h v_\ell\}$.

Lemma 9. *The slide-parent of T' is well-defined and is a member of $tr(\mathcal{H}(E_i))$.*

Proof. Since z_h is a free vertex for T, Z includes exactly one edge incident to z_h, thus $u z_h$ is uniquely determined, and thus the slide-parent is uniquely defined. Since $u z_h$ is a border edge, either $u \notin V_{i-1}$ or u is incident to an edge of T. This together with that $v_h z_h$ and $v_\ell z_\ell$ dominate all edges in $N[v_h v_\ell]$ leads that $Q^*(T', i)$ dominates all edges in E_i.

By adding $v_h z_h$ to T', no edge in $T' \setminus \{u z_h, v_h v_\ell\}$ loses its private neighbor. The edge $v_h z_h$ is adjacent to no edge in $T' \setminus \{u z_h, v_h v_\ell\}$, and $v_h z_h \in P_{\mathcal{H}(E_i)}(v_h z_h, Q^*(T', i))$. These imply that $Q^*(T', i)$ is a member of $tr(\mathcal{H}(E_i))$. $\square$

The slide-parent of T has less edges than T, thus the (slide-parent)-(slide-child) relationship is acyclic, and for each $T' \in tr(\mathcal{H}(E_i))$, there is an ancestor $T'' \in tr(\mathcal{H}(E_i))$ in the (slide-parent)-(slide-child) relation such that the skip-parent of T'' has no H-pattern. Similar to the depth-first search versions of Berge's algorithm [13,17], we will traverse the (slide-parent)-(slide-child) relation to enumerate all transversals including H-pattern edges. The following follows from the definition of slide-parent.

Proposition 5. *Any slide-child T' of T'' is obtained from T'' by adding two edges and remove one edge.*

The computation of the slide-parent of any $T' \in tr(\mathcal{H}(E_i))$ including edges of H-patterns can be easily done in polynomial time: compute its skip-parent T in polynomial time, choose $H^*(T)$ and then compute its slide-parent in polynomial time as described above. Proposition 5 shows that there are at most n^3 candidates for slide-children, thus the enumeration of slide-children can be done with polynomial delay and polynomial space.

Lemma 10. *For any $T' \in tr(\mathcal{H}(E_i))$, all its slide-children can be enumerated with polynomial delay and polynomial space.*

We can now summarize the steps of the algorithm.

1. All transversals in $tr(\mathcal{H}(E_1))$ can be enumerated with polynomial delay and polynomial space, since they include at most two edges from $N[b_1]$.

2. In Section 4 (second paragraph), we have explained how to enumerate all non-extra skip-children with polynomial delay and polynomial space.
3. By Proposition 1 all the extra skip-children not including any edges of H-patterns can be enumerated with polynomial delay and polynomial space.
4. By Lemma 10 all the extra skip-children including some edges from H-patterns can be enumerated with polynomial delay and space.
5. Therefore, by executing these three enumeration algorithms for each minimal transversal $T \in tr(\mathcal{H}(E_{i-1}))$, we can generate all the members in $tr(\mathcal{H}(E_i))$ with polynomial delay and polynomial space.

All these show that the conditions of Proposition 4 are satisfied. And thus we can state our main result.

Theorem 2. *All edge minimal dominating sets in a graph G can be enumerated with polynomial delay and polynomial space.*

The greatest delay is reached by the computation of the slide-children of a given $T \in tr(\mathcal{H}(E_i))$. We have $O(n^3)$ candidates and for each one we compute its slide-parent in time $O(m)$. Then the slide-children can be enumerated with delay $O(mn^3)$. Since the depth of the (skip-parent)-(skip-child) relation is the size of the maximal matching, it is bounded by n, and then the total delay is bounded by $O(n^6)$.

5   Conclusion

In this paper, we propose a polynomial delay polynomial space algorithm for listing all minimal edge dominating sets in a given graph. This improves drastically the previously known algorithms which were incremental output polynomial and use exponential space. We state furthermore that usual approaches with Berge's algorithm involves an NP-complete problem, and thus it is difficult with usual approaches of Berge's algorithm to produce an efficient algorithm. To cope with this difficulty, we introduce a new idea of "changing the traversal routes in the area of difficult solutions" (the notion of skip-children and the removal of edges involved in H-patterns). Based on this idea, we give a new traversal route on these difficult solutions, that is totally independent from Berge's traversal route (the (slide-parent)-(slide-child) relation). As a result, we are able to construct a polynomial delay polynomial space algorithm.

The idea of changing the traversal routes seems to be new and to be able to apply to many other kind of algorithms in enumeration area. Interesting future works are applications of this idea to other kind of enumeration algorithms, *e.g.* the one used by Lawler et al. for enumerating maximal subsets [15] or other algorithms for enumerating minimal transversals (see for instance [5]).

References

1. Avis, D., Fukuda, K.: Reverse search for enumeration. Discrete Applied Mathematics **65**(1–3), 21–46 (1996)
2. Berge, C.: Hypergraphs: Combinatorics of Finite Sets. North-Holland (1989)

3. Boros, E., Elbassioni, K.M., Gurvich, V.: Transversal hypergraphs to perfect matchings in bipartite graphs: Characterization and generation algorithms. Journal of Graph Theory **53**(3), 209–232 (2006)
4. Eiter, T., Gottlob, G.: Identifying the minimal transversals of a hypergraph and related problems. SIAM J. Comput. **24**(6), 1278–1304 (1995)
5. Eiter, T., Gottlob, G., Makino, K.: New results on monotone dualization and generating hypergraph transversals. SIAM J. Comput. **32**(2), 514–537 (2003)
6. Fomin, F.V., Heggernes, P., Kratsch, D., Papadopoulos, C., Villanger, Y.: Enumerating minimal subset feedback vertex sets. Algorithmica **69**(1), 216–231 (2014)
7. Golovach, P.A., Heggernes, P., Kratsch, D., Villanger, Y.: An incremental polynomial time algorithm to enumerate all minimal edge dominating sets. In: Fomin, F.V., Freivalds, R., Kwiatkowska, M., Peleg, D. (eds.) ICALP 2013, Part I. LNCS, vol. 7965, pp. 485–496. Springer, Heidelberg (2013)
8. Gunopulos, D., Khardon, R., Mannila, H., Toivonen, H.: Data mining, hypergraph transversals, and machine learning. In: PODS, pp. 209–216 (1997)
9. Haynes, T.W., Hedetniemi, S.T., Slater, P.J.: Fundamentals of Domination in Graphs. Pure and Applied Mathematics, vol. 208. M. Dekker (1998)
10. Kanté, M.M., Limouzy, V., Mary, A., Nourine, L.: On the neighbourhood helly of some graph classes and applications to the enumeration of minimal dominating sets. In: Chao, K.-M., Hsu, T., Lee, D.-T. (eds.) ISAAC 2012. LNCS, vol. 7676, pp. 289–298. Springer, Heidelberg (2012)
11. Kanté, M.M., Limouzy, V., Mary, A., Nourine, L.: On the enumeration of minimal dominating sets and related notions. SIAM J. Discrete Math. **28**(4), 1916–1929 (2014)
12. Kanté, M.M., Limouzy, V., Mary, A., Nourine, L., Uno, T.: On the enumeration and counting of minimal dominating sets in interval and permutation graphs. In: Cai, L., Cheng, S.-W., Lam, T.-W. (eds.) Algorithms and Computation. LNCS, vol. 8283, pp. 339–349. Springer, Heidelberg (2013)
13. Kavvadias, D.J., Stavropoulos, E.C.: An efficient algorithm for the transversal hypergraph generation. J. Graph Algorithms Appl. **9**(2), 239–264 (2005)
14. Khachiyan, L., Boros, E., Borys, K., Elbassioni, K.M., Gurvich, V., Makino, K.: Generating cut conjunctions in graphs and related problems. Algorithmica **51**(3), 239–263 (2008)
15. Lawler, E.L., Lenstra, J.K., Rinnooy Kan, A.H.G.: Generating all maximal independent sets: np-hardness and polynomial-time algorithms. SIAM J. Comput. **9**(3), 558–565 (1980)
16. Makino, K., Uno, T.: New algorithms for enumerating all maximal cliques. In: Hagerup, T., Katajainen, J. (eds.) SWAT 2004. LNCS, vol. 3111, pp. 260–272. Springer, Heidelberg (2004)
17. Murakami, K., Uno, T.: Efficient algorithms for dualizing large-scale hypergraphs. Discrete Applied Mathematics **170**, 83–94 (2014)
18. Schwikowski, B., Speckenmeyer, E.: On enumerating all minimal solutions of feedback problems. Discrete Applied Mathematics **117**(1–3), 253–265 (2002)
19. Shioura, A., Tamura, A., Uno, T.: An optimal algorithm for scanning all spanning trees of undirected graphs. SIAM J. Comput. **26**(3), 678–692 (1997)
20. Strozecki, Y.: Enumeration Complexity and Matroid Decomposition. PhD thesis, Université Paris Diderot - Paris 7 (2010)
21. Takata, K.: A worst-case analysis of the sequential method to list the minimal hitting sets of a hypergraph. SIAM J. Discrete Math. **21**(4), 936–946 (2007)

Fast and Simple Connectivity
in Graph Timelines

Adam Karczmarz[✉] and Jakub Łącki

University of Warsaw, Warszawa, Poland
{a.karczmarz,j.lacki}@mimuw.edu.pl

Abstract. In this paper we study the problem of answering connectivity queries about a *graph timeline*. A graph timeline is a sequence of undirected graphs $G_1, \ldots, G_t$ on a common set of vertices of size n such that each graph is obtained from the previous one by an addition or a deletion of a single edge. We present data structures, which preprocess the timeline and can answer the following queries:

- `forall`(u, v, a, b) – does the path $u \to v$ exist in *each* of $G_a, \ldots, G_b$?
- `exists`(u, v, a, b) – does the path $u \to v$ exist in *any* of $G_a, \ldots, G_b$?
- `forall2`(u, v, a, b) – do there exist two edge-disjoint paths connecting u and v in *each* of $G_a, \ldots, G_b$?

We show data structures that can answer `forall` and `forall2` queries in $O(\log n)$ time after preprocessing in $O(m + t \log n)$ time. Here by m we denote the number of edges that remain unchanged in each graph of the timeline. For the case of `exists` queries, we show how to extend an existing data structure to obtain a preprocessing/query trade-off of $\langle O(m + \min(nt, t^{2-\alpha})), O(t^\alpha) \rangle$ and show a matching conditional lower bound.

1   Introduction

In this paper we revisit the problem of maintaining the connectivity information in a *graph timeline*. The problem was formulated and solved in a recent paper by Łącki and Sankowski [9]. They define a graph timeline to be a sequence of graphs $G_1, G_2, \ldots, G_t$ on a common set of vertices V of size n such that the graph G_i is obtained from G_{i-1} by adding or deleting a single edge. Their goal was to preprocess the graph timeline to build a data structure that may answer connectivity queries regarding a contiguous fragment of the timeline:

- `forall`(u, v, a, b) — are vertices u and v connected by a path in *each* of $G_a, G_{a+1}, \ldots, G_b$?
- `exists`(u, v, a, b) — are vertices u and v connected by a path in *any* of $G_a, G_{a+1}, \ldots, G_b$?

A. Karczmarz—Supported by the grant NCN2014/13/B/ST6/01811 of the Polish Science Center. Partially supported by FET IP project MULTIPLEX 317532.

J. Łącki —Jakub Łącki is a recipient of the Google Europe Fellowship in Graph Algorithms, and this research is supported in part by this Google Fellowship.

© Springer International Publishing Switzerland 2015
F. Dehne et al. (Eds.): WADS 2015, LNCS 9214, pp. 458–469, 2015.
DOI: 10.1007/978-3-319-21840-3_38

We stress that the entire timeline is revealed in the very beginning for preprocessing, and after that the queries may arrive in an online fashion.

Throughout this paper, we write $\langle f(n, m, t), g(n, m, t) \rangle$ to denote a data structure, whose preprocessing time is $f(n, m, t)$ and the query time is $g(n, m, t)$.

In the case of forall queries, Łącki and Sankowski presented an $\langle O(m + t \log t \log \log t \log n), O(\log n \log \log t) \rangle$ data structure. Here by m we denote the number of edges that remain unchanged in each of $G_1, \ldots, G_t$. Their data structure is Monte Carlo randomized and the query time is amortized. For exists queries they give an $\langle O(m + nt), O(1) \rangle$ data structure.

We improve the results of [9] and show new algorithms, which are more efficient, simpler and deterministic. In addition, we also develop an extended data structure that may efficiently answer an even more complex query regarding 2-edge-connectivity:

- forall2(u, w, a, b) — are vertices u and v connected by two edge-disjoint paths in *each* of $G_a, G_{a+1}, \ldots, G_b$?

Moreover, we give new conditional lower bounds for the problem of answering exists queries, which also improves the results of [9].

1.1   Related Work

A rich body of connectivity-related dynamic problems has been studied in the area of networks and distributed computing. A number of such problems has been surveyed in [2]. In a typical scenario, we work with a sequence of graphs $G^t = G_1, \ldots, G_t$ that represent the states of an evolving network at different points in time. However, the properties of these graphs, which are of interest, such as *T-interval connectivity* [8] or *time-respecting paths* [7] are usually much more complex than what can be studied with ordinary connectivity queries, that is queries about the existence of a path connecting two given vertices in a particular graph. For example, the problem of T-interval connectivity consists of deciding if for every subsequence $G_a, \ldots, G_{a+T-1}$ of T consecutive graphs in G^t, the intersection $G_a \cap \ldots \cap G_{a+T-1}$ of these graphs contains a connected component spanning all vertices. Here we define the intersection of two graphs to be the graph obtained by intersecting their edge sets.

We believe that the queries we consider in this paper are powerful enough to study interesting properties of evolving networks. A forall query checks if two vertices are connected with a path in every graph among $G_a, \ldots, G_b$, but the path can be different in each of the graphs and may not even exist in the intersection of these graphs. Even stronger is a forall2 query, checking whether two vertices are connected with two edge-disjoint paths in each graph of the given fragment. This may serve as a measure of robustness of connection between two nodes of a network.

The algorithms that process graph timelines can also be considered *semi-offline* counterparts of dynamic graph algorithms. The updates are given upfront, but the queries may arrive in an online fashion, i.e. they are issued one by one, only after the preprocessing is finished. A possible scenario for the semi-offline model would be to collect and index the history of evolving network up to some point of time and then use the queries to analyze various properties of the network efficiently.

It is worth noting that the knowledge of the entire history of changes in most cases leads to data structures faster and simpler than the best online ones. However, this property has rarely been exploited to design efficient algorithms. Eppstein [4] has shown an algorithm, which, given a weighted graph G and a sequence of k edge weight updates, computes the weight of the minimum spanning tree after each update in $O((m + k) \log n)$ time.

1.2   Our Results

We show $\langle O(m + t \log n), O(\log n) \rangle$ data structures for answering `forall` and `forall2` queries. The data structures use $O(t \log n)$ space. This improves the results of [9] in a number of ways: our algorithms are faster and deterministic, use less space, the time bounds are worst-case and the query time is independent of the length of the timeline. We also introduce `forall2` queries, which were not considered before. On top of that, our algorithms are arguably simpler.

What is interesting, we obtain a solution for the 2-edge-connectivity problem, which is much more efficient than what has been achieved in the dynamic case. The best known algorithm for 2-edge-connectivity is due to Holm et al. [5]. It processes t updates in $O((t + m) \log^4 n)$ time, where m is the initial number of edges, and answers queries in $O(\log n)$ time. Our algorithm may preprocess the timeline in only $O(m + t \log n)$ time to answer queries in $O(\log n)$ time.

In the construction of the algorithm for answering `forall` queries we use the following two observations. Consider a timeline $G_1, \ldots, G_t$. If there is an edge uw present in every graph among $G_1, \ldots, G_t$, vertices u and w are equivalent from the point of view of any query, so the edge uw can be contracted in each graph. Once we do that, we are left with $O(t)$ edges in total, each being added or deleted at some point of time. Thus, if there are much more than t vertices, some vertices are isolated in every $G_1, \ldots, G_t$, and can be safely treated separately in the beginning and removed. These ideas are then used recursively in a divide-and-conquer algorithm, which at each step halves the length of the timeline to compute a segment tree over the sequence $G_1, \ldots, G_t$. This segment tree stores connectivity information about every individual graph in the timeline. Here we adapt the ideas of Eppstein's reduction and contraction scheme used for offline computation of minimum spanning trees [4].

Next, we use a fingerprinting scheme to identify vertices belonging to the same connected components in multiple consecutive graphs, which allows us to answer `forall` queries. Additionally, our fast algorithm for answering queries uses a data structure for efficient testing of equality of contiguous subsequences of a given sequence. This is then extended to handle `forall2` queries.

For `exists` queries, we show how to leverage the $\langle O(m + nt), O(1) \rangle$ data structure from [9] to build an $\langle O(m + \min(nt, t^{2-\alpha})), O(t^\alpha) \rangle$ data structure, where α is a parameter from the range $[0, 1)$, which can be chosen arbitrarily. All of the presented algorithms are simple and can easily be implemented.

Moreover, we develop a conditional lower bound for the problem of answering `exists` queries. We show that answering t `exists` queries on a timeline of length t, consisting of graphs with $O(t)$ edges, can be used to detect triangles in a graph with $O(t)$ edges. This implies a conditional lower bound of $\Omega(t^{1.41})$ and improves the result of [9], where a weaker lower bound was shown. We also show that an $O(t^{1.5-\epsilon})$ *combinatorial* algorithm for the aforementioned problem would imply a subcubic *combinatorial* algorithm for the Boolean matrix multiplication problem, which would be a major breakthrough. At the same time, our improved data structure for `exists` queries may solve this problem in $O(t^{1.5})$ time, which means that it is, in some sense, optimal.

1.3   Organization of This Paper

In Section 2 we introduce notation and give a few simple properties of segment trees, which we later use. Section 3 describes the basic version of our data structure, which is then extended to handle `forall` and `forall2` queries. Then, in Section 4 we present an algorithm for answering `forall` queries. Next, in Section 5 we develop improved lower bounds for the problem of answering `exists` queries, as well as show that a trade-off between query and preprocessing time is possible. Finally, in Section 6 we discuss the possible directions of future research. Due to space constraints, some parts of the description of our algorithms have not been included in this extended abstract. In particular, the omitted proofs of some lemmas can be found in the full version of this paper.

2   Preliminaries

A *graph timeline* is a sequence G^t of graphs $G_1, G_2, \ldots, G_t$, where $G_i = (V, E_i)$. We call each individual graph in G^t a *version*. For each $i \in [1, t)$ we have $|E_i \oplus E_{i+1}| = 1$, i.e. E_{i+1} is obtained from E_i by adding or deleting a single edge. We assume that the input is given as the set E_1 and a list of $t - 1$ operations that describe, for each $i \in [1, t - 1]$, how to obtain E_{i+1} from E_i.

Throughout this paper we work with intervals of integers, that is $[a, b]$ denotes $\{a, a+1, \ldots, b\}$. We say that edge (u, v) is *alive* in the interval $[x, y]$ iff $(u, v) \in E_j$ for each $j \in [x, y]$. For each edge $e \in E_1 \cup \ldots \cup E_t$ we define $L(e)$ to be the set of maximal intervals such that e is alive in each of them. An edge e is called *permanent* iff $L(e) = \{[1, t]\}$, that is, it is present in every version. Otherwise, we say that e is a *temporary* edge. We denote by m the number of permanent edges. The number of temporary edges is at most t. We begin the initialization of our data structures by finding the sets $L(e)$ in $O(|E_1| + t) = O(m + t)$ time.

We denote by Δ_a^+ the set of edges e such that $[a, x] \in L(e)$ for some $x \in [a, t]$, i.e., edges present in G_a, but not in G_{a-1}. Similarly, let Δ_b^- be the set of edges e such that $[x, b] \in L(e)$ for some $x \in [1, b]$. It is easy to verify that $\sum_{i=1}^{t} |\Delta_i^+| +$

$\sum_{i=1}^{t} |\Delta_i^-| = O(m + t)$. Moreover, for $a \in (1, t]$, we have $|\Delta_a^+| \leq 1$, while for $b \in [1, t)$ we have $|\Delta_b^-| \leq 1$.

Throughout the paper, we assume that $t \geq n$ and $t = 2^B$ for some integer $B \geq 0$. The latter assumption can be achieved by adding dummy graphs to the timeline.

2.1 Elementary Intervals and The Segment Tree

Given $t = 2^B$, the set of *elementary intervals* is defined inductively:

1. $[1, t]$ is an elementary interval,
2. if $[a, b]$ is an elementary interval, and $a < b$ we let mid $= \lfloor \frac{a+b}{2} \rfloor$, and define $[a, \text{mid}]$ and $[\text{mid} + 1, b]$ to be elementary intervals as well.

The set of elementary intervals can be naturally organized into a complete binary tree, which we call a *segment tree*. Assuming the above notation, we call left($[a, b]$) = $[a, \text{mid}]$ the left child of interval $[a, b]$. Similarly, right($[a, b]$) = $[\text{mid} + 1, b]$. The parent interval of P is denoted by par(P).

Lemma 1. *Every interval* $[c, d] \subseteq [1, t]$ *can be partitioned into no more than* $2\log_2(d - c + 1) + 2$ *disjoint elementary intervals such that no two intervals from the partition can be merged into a bigger elementary interval. The partition can be computed in time* $O(\log(d - c + 1))$.

As it is much easier to work with elementary intervals, for each edge e we partition all intervals from $L(e)$ into elementary intervals.

Lemma 2. *For each* e, *all intervals in* $L(e)$ *can be partitioned into* $O(m + t \log n)$ *elementary intervals. The partition can be performed in time* $O(m + t \log n)$.

For an elementary interval $[a, b]$, we set $E_{[a,b]}$ to be the set of edges that contain $[a, b]$ in their partition. From Lemmas 1 and 2 it follows that each edge is contained in $O(\log t)$ sets $E_{[a,b]}$ and the sum over elementary intervals $\sum_{[a,b]} E_{[a,b]}$ is of order $O(m + t \log n)$.

3 The Data Structure

We now describe a tree-like data structure T, which is a crucial part of all our algorithms. In the following we reserve the name T for this particular data structure. The data structure T is based on the set of all elementary intervals organized into a complete binary tree. This tree has a single node $T_{[a,b]}$ for each elementary interval $[a, b]$. Denote by $G_{[a,b]}$ the graph $(V, E_a \cap \ldots \cap E_b)$. Roughly speaking, our goal is to associate with $T_{[a,b]}$ the information about the connected components of $G_{[a,b]}$. We first give a simple approach for constructing the data structure T, and then show how to speed it up. We use the following fact.

Lemma 3. *Let* $[a, b]$ *be an elementary interval such that* $[a, b] \neq [1, t]$. *Then* $E(G_{[a,b]}) = E(G_{par([a,b])}) \cup E_{[a,b]}$.

In the simple approach, we associate with $T_{[a,b]}$ a graph $S_{[a,b]}$, which has a single vertex for each connected component of $G_{[a,b]}$, and does not contain any edges. By Lemma 3, $G_{[a,b]}$ is obtained from $G_{\mathrm{par}([a,b])}$ by adding some edges. This implies that each component of $G_{[a,b]}$ is a sum of some components of $G_{\mathrm{par}([a,b])}$. To compute $S_{[a,b]}$ we build a graph H on a vertex set $V(S_{\mathrm{par}([a,b])})$ and add to it edges of $E_{[a,b]}$ (each edge endpoint has to be mapped to its connected component in $G_{\mathrm{par}([a,b])}$) and then find its connected components. These components are exactly the components of $G_{[a,b]}$. Observe that during this computation we may also compute a mapping between the vertices of $S_{\mathrm{par}([a,b])}$ and $S_{[a,b]}$. In the case of $S_{[1,t]}$ we compute a mapping between individual vertices and connected components of $G_{[1,t]}$.

T represents the connected components of every graph in the timeline. Consider a graph G_c. In order to find a connected component of a vertex v in G_c, we traverse the path in T from $T_{[1,t]}$ to $T_{[c,c]}$. We compute the connected component of vertex v in every graph $G_{[a,b]}$ on the path. Observe that if we know the connected component of v in $G_{\mathrm{par}([a,b])}$, we may compute the connected component of v in $G_{[a,b]}$ by following the mapping between the components of $G_{\mathrm{par}([a,b])}$ and $G_{[a,b]}$. At the end of the traversal, we find the component of v in $G_{[c,c]} = G_c$.

3.1 An Efficient Construction

In order to compute the data structure T efficiently, we need to make an additional optimization, which is crucial for obtaining good running time.

Consider an elementary interval $[a,b]$ and a connected component C of $G_{[a,b]}$. Assume that within the graphs $G_a, \ldots, G_b$ no edge incident to a vertex of C is ever added or deleted. In other words, the edges incident to vertices of C are the same in each of $G_a, \ldots, G_b$. This means that in each of $G_a, \ldots, G_b$ vertices of C are connected to each other, but not connected to *any* vertex outside C. Hence, C is also a connected component in each of $G_a, \ldots, G_b$.

As a result, there is no need to store C in the descendants of $T_{[a,b]}$. When searching for a connected component of a vertex $v \in C$ in G_c, where $c \in [a,b]$, we may simply stop the search in the representation of C in $T_{[a,b]}$. This observation will be used in the reduction phase of the construction of the tree T.

We now describe the efficient construction of the tree T. For each node $T_{[a,b]}$ of T, where $[a,b]$ is an elementary interval, we compute a graph $S_{[a,b]}$. The vertices of $S_{[a,b]}$ correspond to *some* of the components of $G_{[a,b]}$. We say that $v \in V$ is *represented* in $S_{[a,b]}$ if there is a vertex $s \in V(S_{[a,b]})$ that corresponds to a component containing v. The graphs $S_{[a,b]}$ have no edges.[1]

Let $[a,b]$ be an elementary interval. $S_{[a,b]}$ is computed based on $S_{\mathrm{par}([a,b])}$ (or $(V, \emptyset)$, if $[a,b] = [1,t]$) in two phases called *reduction* and *contraction*.

In the reduction phase some vertices of $H = S_{\mathrm{par}([a,b])}$ are removed, as they are not affected by any edge addition or deletion that is carried out among $G_a, \ldots, G_b$. Namely, we mark endpoints of edges in $F = E_{[a,b]} \cup \bigcup_{i=a+1}^{b} \Delta_i^+ \cup$

---

[1] Defining a graph with no edges may look confusing. However, we define $S_{[a,b]}$ to be a graph, as we add edges to $S_{[a,b]}$ in our data structure for 2-edge-connectivity.

$\bigcup_{i=a}^{b-1} \Delta_i^-$ and then remove the unmarked vertices. Note that the sets $E_{[a,b]}$, Δ_i^+ and Δ_i^- contain edges of the original graph, so their endpoints have to be mapped to the corresponding vertices of H. The reduction phase is performed only when $b - a + 1 < n$. It is done by a call REDUCE(H, F), which produces a pair (S', M), where S' is the reduced graph and M is a mapping between $V(S_{\text{par}([a,b])})$ and $V(S') \cup \{\bot\}$. The value of $\bot$ means that a vertex has been removed and does not have a corresponding vertex in S'. The procedure can be implemented with a simple graph search to work in $O(|H| + |F|)$ time.

In the second phase, called the contraction phase, some of the remaining vertices of $H = S'$ are merged to form $S_{[a,b]}$. Specifically, the components formed in S' after adding edges $F = E_{[a,b]}$ are contracted. Again, we use a function CONTRACT(H, F), which produces a pair (S', M) consisting of the contracted graph S' and the mapping between H and S'. This function can also be easily implemented to work in linear time.

Consider an elementary interval P. Together with S_P, the node T_P stores two tables 1_P and r_P mapping vertices of S_P to $V(S_{\text{left}(P)}) \cup \{\bot\}$ and $V(S_{\text{right}(P)}) \cup \{\bot\}$ respectively. If $1_P[k] \neq \bot$, $1_P[k]$ is the vertex of $S_{\text{left}(P)}$ that corresponds to $k \in V(S_P)$. $1_P[k] = \bot$ means that P is a leaf, or there is no vertex corresponding to k in $S_{\text{left}(P)}$. The table r_P is defined analogously. For simplicity, we also assume that $T_{[1,t]}$ is a left child of a special node $T_{[0,\infty]}$ and $S_{[0,\infty]} = (V, \emptyset)$, so that for each $v \in V$, $1_{[0,\infty]}[v]$ points to the vertex of $S_{[1,t]}$ representing the original vertex v.

The graphs S_P along with 1 and r pointers are sufficient to find the component of any vertex v in any of $G_1, \ldots, G_t$. To access the component of vertex v in G_c we start at vertex v in $S_{[0,\infty]}$ and follow 1 or r pointers in order to reach the leaf $T_{[c,c]}$. The traversal stops once we reach $T_{[c,c]}$ or the pointer we want to use ($1[k]$ or $r[k]$) is equal to $\bot$. Let P be the elementary interval, where the traversal finishes and k be the vertex in S_P, which we reached. Then, as we later show, (k, P) uniquely identifies the component of vertex v in G_c. The above process can be seen as a function COMP-ID(w, a, b, c) that follows the path to $T_{[c,c]}$ starting at vertex $w \in V(S_{[a,b]})$. The pair (k, P), defined as above, is what the call COMP-ID$(1_{[0,\infty]}[v], 1, t, c)$ returns. The full text of the COMP-ID function can be found in the full version of this paper.

Lemma 4. *Let $1 \leq c \leq t$. For any $u \in V$, denote by (k_u, P_u) the value returned by* COMP-ID$(1_{[0,\infty]}[v], 1, t, c)$. *Then, two vertices $v, w \in V$ are connected by a path in G_c iff $k_v = k_w$ and $P_v = P_w$.*

Let us bound the time needed to build T. We begin with an auxiliary lemma, whose proof is based on the fact that we perform the reduction.

Lemma 5. *Let $[a, b]$ be an elementary interval. Then $|V(S_{[a,b]})| \leq \min(8(b - a + 1), n)$.*

To build T we use a recursive procedure COMPUTE-TREE(a, b), which computes the subtree rooted at $T_{[a,b]}$. It produces each graph $S_{[a,b]}$ based on $S_{\text{par}([a,b])}$ by applying reduction and contraction. After the graph $S_{[a,b]}$ is computed, the tables $l_{\text{par}([a,b])}$ and $r_{\text{par}([a,b])}$ are filled. Finally, the subtrees $T_{\text{left}([a,b])}$ and $T_{\text{right}([a,b])}$ are computed by calling COMPUTE-TREE recursively.

The details of this procedure, including the pseudocode, can be found in the full version.

Lemma 6. *The total running time of* COMPUTE-TREE$(1,t)$ *is* $O(m + t \log n)$.

Proof. We first analyze the time spent in the call COMPUTE-TREE(a,b), excluding the work in recursive calls. Let C be $\bigcup_{i=a+1}^{b} \Delta_i^+ \cup \bigcup_{i=a}^{b-1} \Delta_i^-$. Thus $O(|C|) = O(b-a)$. Recall that the functions CONTRACT and REDUCE run in linear time. For $b - a + 1 \geq n$, we only perform contraction of $E_{[a,b]}$ in a graph of size $O(n)$, which requires $O(n + |E_{[a,b]}|)$ time. The amount of work for $b - a + 1 < n$ can be bounded by $O(|V(S_{\text{par}([a,b])})| + |C| + |E_{[a,b]}|)$, as REDUCE is passed the edges $C \cup E_{[a,b]}$.

To complete the proof, we sum these running times over all elementary intervals. The term $|E_{a,b}|$ appears in both cases and, by Lemma 2, we have $\sum_P E_P = O(m + t \log n)$, thus we can focus on the other summands. For the case $b - a + 1 \geq n$, the remaining work is $O(n)$, but there are only $O(\frac{t}{n})$ such intervals, so the total work is $O(t)$. On the other hand, if $b - a + 1 < n$, by Lemma 5, $O(|V(S_{\text{par}([a,b])})|) = O(b-a)$, so the total work is $O(b-a)$. Hence, the total work on each level of the tree such that its elementary intervals are shorter than n, is $O(t)$. The number of such levels is $O(\log n)$, which gives $O(t \log n)$ total time. The lemma follows. $\square$

Having computed T, the function COMP-ID allows us to access the component of some vertex v in G_c in time $O(\log t)$. However, as we now show, this can be speeded up to $O(\log n)$ time. Recall that $t = 2^B$. Let 2^D be the smallest power of 2 such that $2^D \geq n$ and fix some $k \in [0, 2^{B-D})$. Then, for each $c \in [k \cdot 2^D + 1, (k+1) \cdot 2^D]$, the call COMP-ID$(1_{[0,\infty]}[v], 1, t, c)$ descends down T through the first $B - D$ levels in the same way, independent of c. We can thus add another preprocessing phase, building the table shortcut. For a vertex v and $0 \leq k < 2^{B-D}$, shortcut$[v][k]$ is defined to be a pair (s, P) such that for $c \in [k \cdot 2^D + 1, (k+1) \cdot 2^D]$, COMP-ID$(1_{[0,\infty]}[v], 1, t, c)$, after going through at most $B - D$ levels of T, ends up in the interval P and $s \in V(S_P)$ represents v. There are only $O(t/n)$ allowed values of k, so the table shortcut has size $O(t)$.

The table can be computed by finding the components of each vertex v in all the graphs S_P from the first $B - D$ levels of the tree. As the component of v in S_P can be computed in constant time based on the component of v in $S_{\text{par}(P)}$, we spend $O(t/n)$ time for each v, and thus $O(t)$ time in total.

The optimized procedure COMP-ID starts by looking up the shortcut through the first $B - D$ levels of T and then calls the original COMP-ID, starting at an elementary interval of length $O(n)$. Thus, its running time is $O(\log n)$.

We may build a data structure similar to T that represents information about 2-edge-connectivity in individual versions. In this case, the graphs S_P are forests, whose vertices represent (some) 2-edge-connected components of S_P. The details of this construction are deferred to the full version of this paper.

4   Answering forall Queries

In this section we show how to extend the data structure T, so that it can be used for answering forall queries. The preprocessing for forall queries constitutes another phase, that we apply only after we computed the data structure T.

Let us begin with a simple observation. Assume that we want to answer a forall(u, w, a, b) query, where $[a, b]$ is an elementary interval. Then, if the same vertex of $S_{[a,b]}$ represents both u and w, then there is actually a path between u and w in $G_{[a,b]}$ and we can immediately give a positive answer. However, the reverse relation is not true. It may happen that u and w are represented by distinct vertices in $S_{[a,b]}$, but are connected in each of $G_a, \ldots, G_b$. Thus, our first goal in this section is to compute, for each two vertices in each of $S_{[a,b]}$, whether the vertices represented by them are connected in each of $G_a, \ldots, G_b$.

For an elementary interval $[a, b]$, let $c_{[a,b]}(s, x)$, where $s \in V(S_{[a,b]})$, $x \in [a, b]$, be the result of the call COMP-ID(s, a, b, x). Our goal is to compute for each vertex $s \in S_{[a,b]}$ a *fingerprint*, that is, an integer $H_{[a,b]}(s) \in [1, |V(S_{[a,b]})|]$ with the following property: the sequences $c_{[a,b]}(s, a) c_{[a,b]}(s, a + 1) \ldots c_{[a,b]}(s, b)$ and $c_{[a,b]}(s', a) c_{[a,b]}(s', a + 1) \ldots c_{[a,b]}(s', b)$ are equal iff $H_{[a,b]}(s) = H_{[a,b]}(s')$.

To answer a forall(u, v, a, b) query, where $[a, b]$ is an elementary interval, we first map u and v into vertices u' and v' of $S_{[a,b]}$ and then report a positive answer iff $H_{[a,b]}(u') = H_{[a,b]}(v')$. In order to handle arbitrary intervals, we decompose the query interval into $O(\log t)$ elementary intervals. The decomposition as well as the mapping can be implemented as a function FORALL-AUX(s_1, s_2, x, y, a, b), whose pseudocode can be found in the full version of the paper. To answer a forall(u, v, x, y) query we execute FORALL-AUX$(1_{[0,\infty]}[v], 1_{[0,\infty]}[w], x, y, 1, t)$.

Let us now describe the computation of fingerprints. They are computed in a bottom-up fashion, starting from the leaves of T.

Lemma 7. *Let $P = [a, b]$ be an elementary interval and $s \in V(S_P)$. Define:*

$$\tilde{H}_P(s) = \begin{cases} (s, 0) & \text{if } 1_P(s) = \perp \text{ or } r_P(s) = \perp \\ (H_{left(P)}(1_P[s]), H_{right(P)}(r_P[s])) & \text{otherwise.} \end{cases}$$

Then $c_P(s_1, a) \ldots c_P(s_1, b) = c_P(s_2, a) \ldots c_P(s_2, b)$ iff $\tilde{H}_P(s_1) = \tilde{H}_P(s_2)$.

Observe that the pairs $\tilde{H}_P(s)$ from the above lemma satisfy the desired properties of fingerprints, with the exception that they are pairs of integers, not integers. Thus, in order to compute the values $H_P(s)$, it suffices to map the values of $\tilde{H}_P(s)$ into distinct positive integers (two pairs are assigned the same integer iff they are equal). As both numbers in each pair $\tilde{H}_P(s)$ are at most $O(|V(S_P)|)$ we may compute the mapping in linear time by using radix-sort algorithm. Note that this resembles the Karp-Miller-Rosenberg [6] algorithm. The total additional time and space used is $O(\sum_P |V(S_P)|) = O(t \log n)$. Thus, we obtain an $\langle O(m + t \log n), O(\log t) \rangle$ data structure for answering forall queries.

However, the query time can be made independent of the length of the timeline and speeded up to $O(\log n)$. In order to do that, we employ a shortcutting technique similar to the one used for finding connected components of vertices in individual graphs combined with an optimal data structure for comparing the subwords of a given word [3]. For details, refer to the full version of this paper.

The above construction can be generalized to work with 2-edge-connectivity within the same time and space bounds, as shown in the full version.

Theorem 1. *There exists an* $\langle O(m + t \log n), O(\log n) \rangle$ *data structure for answering* forall *and* forall2 *queries. The data structure uses* $O(t \log n)$ *space.*

5 Improved Lower and Upper Bounds for exists Queries

In this section we focus on exists queries. We first give improved conditional lower bounds for answering these queries, and then show an algorithm, whose running time matches one of the new bounds. As shown in [9], the problem of multiplying two Boolean $n \times n$ matrices can be reduced to the problem of answering $\Theta(n^2)$ exists queries about a graph timeline G^t, where $t = \Theta(n^2)$. Denote by $O(n^{\omega'})$ the time required to perform $n \times n$ Boolean matrix multiplication (BMM). Thus, unless $\omega' = 2$, it is not possible to develop a data structure, which after almost linear preprocessing answers exists queries in polylogarithmic time. In this section we give several new lower bounds.

Throughout this section, we repeatedly use ϵ to denote an arbitrarily small, positive number. The exact value of ϵ may vary and depend on the context. We also denote by $\delta(\epsilon)$ some other small positive number, dependent on ϵ.

Let us recall the somewhat informal, yet important, partition of algorithms into *algebraic* and *combinatorial*. The combinatorial algorithms do not make use of the fact that the matrices are defined over a ring, i.e., they do not use subtraction. No $O(n^{3-\epsilon})$ combinatorial algorithm is known for BMM.

We show a connection between the exists data structure and algorithmic problems related to detecting triangles in graphs. In the *triangle detection* problem we are given a graph $G = (V, E)$, where $|E| = m$, and the goal is to find three vertices $a, b, c \in V$ such that $(a, b), (a, c), (b, c) \in E$. The best known known algorithm for triangle detection was given by Alon et al. [1] and works in $O(m^{1.41})$ time. The best combinatorial algorithm is folklore and runs in $O(m\sqrt{m})$ time. The following relation between triangle detection and BMM was shown in [11]:

Lemma 8. *An* $O(m^{1.5-\epsilon})$ *combinatorial algorithm for triangle detection implies an* $O(n^{3-\delta(\epsilon)})$ *combinatorial algorithm for BMM.*

The related problem is *triangle listing*, where we are asked to find c triangles in a graph with m edges. Pătraşcu [10] proved the following lemma.

Lemma 9. *If one can list* m *triangles from a graph with* m *edges in* $O(m^{4/3-\epsilon})$ *time, then there exists an* $O(n^{2-\delta(\epsilon)})$ *algorithm for 3-SUM.*

We now show a relation between triangle listing and exists queries.

Lemma 10. *The problem of listing* c *triangles in a graph with* m *edges can be reduced to answering* $O(m + c \log n)$ exists *queries in a timeline* G^t *of length* $t = O(m)$ *and no permanent edges.*

Proof. Let H be the input graph, in which we are supposed to list triangles. Moreover, let $V(H) = \{v_1, \ldots, v_n\}$. We build a timeline G^t of graphs on vertex set $V(H)$ by processing vertices $v_1, \ldots, v_n$ one by one. First, we add an empty graph to G^t. Then, for a vertex v_i, we append $2 \deg_H(v_i)$ new versions to G^t ($\deg_H(v)$ denotes the degree of vertex v in H), which we call a *block* of vertex v_i. Within each block, we first create $\deg_H(v_i)$ new versions, at each step adding one more edge incident to v_i. The edges are added in arbitrary order. Then, we create $\deg_H(v_i)$ more versions by removing the edges incident to v_i. Note that the last graph in every block is empty, and in the middle graph the vertex v_i has degree $\deg_H(v_i)$. Let the the block of a vertex v_i start at G_{a_i} and end at G_{b_i}.

Observe that we obtain a timeline G^t, where $t = 4m + 1$, as each edge of H is added and removed exactly twice. For each edge $(v_i, v_j) \in E(G)$, $i < j$, we can test if there is a triangle (v_i, v_j, v_k), where $j < k$, with a single query $\mathtt{exists}(v_i, v_j, b_j + 1, t)$. Indeed, the answer to such a query is positive iff there exists v_k such that there is a path from v_i to v_j in $G_{a_k + \deg_H(v_k) - 1}$. The path, along with the edge (v_i, v_j), forms a triangle.

Note that the query $\mathtt{exists}(v_i, v_j, a_p, b_q)$, for $j < p \leq q$, tells us if there is any triangle (v_i, v_j, v_k) such that $k \in [p, q]$. Thus, we may use a divide-and-conquer approach for listing triangles, which is based on the following observation. If we are looking for triangles such that $k \in [p, q]$, a negative answer to an $\mathtt{exists}(v_i, v_j, a_p, b_{(p+q)/2})$ query allows us to halve the search interval. Hence, we can find all l vertices v_k such that (v_i, v_j, v_k) is a triangle in time $O(l \log n)$. The detailed procedure REPORT-TRIANGLES is given in the full version.   □

By combining Lemmas 8, 9 and 10, we obtain the following.

Theorem 2. *Let Ψ be a problem of answering $\Theta(t)$ $\mathtt{exists}$ queries about an arbitrary graph timeline G^t with no permanent edges.*

- *An $O(t^{1.4})$ algorithm for Ψ implies an $O(t^{1.4})$ algorithm for triangle finding.*
- *An $O(t^{1.5-\epsilon})$ combinatorial algorithm for Ψ implies an $O(n^{3-\delta(\epsilon)})$ combinatorial algorithm for BMM.*
- *An $O(t^{4/3-\epsilon})$ algorithm for Ψ implies an $O(n^{2-\delta(\epsilon)})$ algorithm for 3-SUM.*

In addition, we show that an $\mathtt{exists}$ data structure with preprocessing/query time product of $O(t^{2-\epsilon})$ and queries substantially faster than $O(\sqrt{t})$ implies a faster BMM algorithm.

Lemma 11. *Suppose there exists an $\langle O(t^{2-q-\epsilon}), O(t^q) \rangle$ combinatorial data structure for answering $\mathtt{exists}$ queries, where $q \in [0, \frac{1}{2})$ is a parameter. Then there exists an $O(n^{3-\delta(\epsilon)})$ combinatorial algorithm for BMM.*

What is interesting, we can give a combinatorial data structure, whose running time matches the above lower bound.

Theorem 3. *For every $0 \leq \alpha < 1$ there exists an $\langle O(m + \min(nt, t^{2-\alpha})), O(t^\alpha) \rangle$ data structure for answering $\mathtt{exists}$ queries. It uses $O(\min(nt, t^{2-\alpha}))$ space.*

The main idea is to split the timeline into blocks of size $O(t^\alpha)$ and use the $\langle O(m + nt), O(1) \rangle$ $\mathtt{exists}$ data structure of Łącki and Sankowski [9].

6 Open Problems

For `forall` and `forall2` queries, we gave an $\langle O(m + t \log n), O(\log n) \rangle$ data structure. What about the biconnectivity? Although it is possible to propose a similar tree-like structure that represents biconnectivity in individual versions, it seems hard to extend it to `forall2`-like queries. The main obstacle is biconnectivity relation on vertices not being an equivalence relation.

It would be also interesting to know whether even faster query (without sacrificing $O(t \log n)$ initialization time) is possible for `forall` queries.

Concerning `exists` queries, we proved that beating our trade-off structure in the domain of combinatorial algorithms implies a faster combinatorial matrix multiplication algorithm. However, is there a way to employ fast matrix multiplication to obtain a data structure for `exists` queries with preprocessing/query time product of $O(t^{2-\epsilon})$?

References

1. Alon, N., Yuster, R., Zwick, U.: Finding and counting given length cycles. Algorithmica **17**(3), 209–223 (1997)
2. Casteigts, A., Flocchini, P., Quattrociocchi, W., Santoro, N.: Time-varying graphs and dynamic networks. IJPEDS **27**(5), 387–408 (2012)
3. Crochemore, M., Hancart, C., Lecroq, T.: Algorithms on Strings. Cambridge University Press, New York (2007)
4. Eppstein, D.: Offline algorithms for dynamic minimum spanning tree problems. J. Algorithms **17**(2), 237–250 (1994)
5. Holm, J., de Lichtenberg, K., Thorup, M.: Poly-logarithmic deterministic fully-dynamic algorithms for connectivity, minimum spanning tree, 2-edge, and biconnectivity. J. ACM **48**(4), 723–760 (2001)
6. Karp, R.M., Miller, R.E., Rosenberg, A.L.: Rapid identification of repeated patterns in strings, trees and arrays. In: Proceedings of the 4th Annual ACM Symposium on Theory of Computing, May 1–3, 1972, Denver, Colorado, USA, pp. 125–136 (1972)
7. Kempe, D., Kleinberg, J.M., Kumar, A.: Connectivity and inference problems for temporal networks. J. Comput. Syst. Sci. **64**(4), 820–842 (2002)
8. Kuhn, F., Lynch, N.A., Oshman, R.: Distributed computation in dynamic networks. In: Schulman, L.J. (ed.) Proceedings of the 42nd ACM Symposium on Theory of Computing, STOC 2010, Cambridge, Massachusetts, USA, 5–8 June 2010, pp. 513–522. ACM (2010)
9. Łącki, J., Sankowski, P.: Reachability in graph timelines. In: Proceedings of the 4th Conference on Innovations in Theoretical Computer Science, ITCS 2013, pp. 257–268. ACM, New York (2013)
10. Pătraşcu, M.: Towards polynomial lower bounds for dynamic problems. In: Proceedings of the Forty-second ACM Symposium on Theory of Computing, STOC 2010, pp. 603–610. ACM, New York (2010)
11. Williams, V.V., Williams, R.: Subcubic equivalences between path, matrix and triangle problems. In: Proceedings of the 2010 IEEE 51st Annual Symposium on Foundations of Computer Science, FOCS 2010, pp. 645–654. IEEE Computer Society, Washington, DC (2010)

Dynamic Set Intersection

Tsvi Kopelowitz[1][(✉)], Seth Pettie[1], and Ely Porat[2]

[1] University of Michigan, Ann Arbor, USA
kopelot@gmail.com
[2] Bar-Ilan University, Ramat Gan, Israel

Abstract. Consider the problem of maintaining a family F of dynamic sets subject to insertions, deletions, and set-intersection reporting queries: given $S, S' \in F$, report every member of $S \cap S'$ in any order. We show that in the word RAM model, where w is the word size, given a cap d on the maximum size of any set, we can support set intersection queries in $O(\frac{d}{w/\log^2 w})$ expected time, and updates in $O(1)$ expected time. Using this algorithm we can list all t triangles of a graph $G = (V, E)$ in $O(m + \frac{m\alpha}{w/\log^2 w} + t)$ expected time, where $m = |E|$ and α is the arboricity of G. This improves a 30-year old triangle enumeration algorithm of Chiba and Nishizeki running in $O(m\alpha)$ time.

We provide an incremental data structure on F that supports intersection *witness* queries, where we only need to find *one* $e \in S \cap S'$. Both queries and insertions take $O\left(\sqrt{\frac{N}{w/\log^2 w}}\right)$ expected time, where $N = \sum_{S \in F} |S|$. Finally, we provide time/space tradeoffs for the fully dynamic set intersection reporting problem. Using M words of space, each update costs $O(\sqrt{M \log N})$ expected time, each reporting query costs $O(\frac{N\sqrt{\log N}}{\sqrt{M}} \sqrt{op+1})$ expected time where op is the size of the output, and each witness query costs $O(\frac{N\sqrt{\log N}}{\sqrt{M}} + \log N)$ expected time.

1 Introduction

In this paper we explore the power of *word level parallelism* to speed up algorithms for dynamic set intersection and triangle enumeration. We assume a w-bit word-RAM model, $w > \log n$, with the standard repertoire of unit-time operations on w-bit words: bitwise Boolean operations, left/right shifts, addition, multiplication, comparison, and dereferencing. Using the modest parallelism intrinsic in this model (sometimes in conjunction with tabulation) it is often possible to obtain a nearly factor-w (or factor-$\log n$) speedup over traditional algorithms. The *Four Russians* algorithm for boolean matrix multiplication is perhaps the oldest algorithm to use this technique. Since then it has been applied to computing edit distance [2], regular expression pattern matching [3], APSP in dense

A more detailed version of this paper appears in [1]. Supported by NSF grants CCF-1217338 and CNS-1318294 and a grant from the US-Israel Binational Science Foundation. This research was performed in part at the Center for Massive Data Algorithmics (MADALGO) at Aarhus University, which is supported by the Danish National Research Foundation grant DNRF84.

© Springer International Publishing Switzerland 2015
F. Dehne et al. (Eds.): WADS 2015, LNCS 9214, pp. 470–481, 2015.
DOI: 10.1007/978-3-319-21840-3_39

weighted graphs [4], APSP and transitive closure in sparse graphs [5,6], and more recently, to computing the Fréchet distance [7] and solving 3SUM in sub-quadratic time [8,9]. Refer to [10] for more examples.

Set Intersection. The problem is to represent a (possibly dynamic) family of sets F with total size $N = \sum_{S \in F} |S|$ so that given $S, S' \in F$, one can quickly determine if $S \cap S' = \emptyset$ (emptiness query) or report some $x \in S \cap S'$ (witness query) or report all members of $S \cap S'$. Let d be an *a priori* bound on the size of any set. We give a randomized algorithm to preprocess F in $O(N)$ time such that reporting queries can be answered in $O(d/\frac{w}{\log^2 w} + |S \cap S'|)$ *expected* time. Subsequent insertion and deletion of elements can be handled in $O(1)$ expected time.

We give $O(N)$-space structures for the three types of queries when there is no restriction on the size of sets. For emptiness queries the expected update and query times are $O(\sqrt{N})$; for witness queries the expected update and query times are $O(\sqrt{N \log N})$; for reporting queries the expected update time is $O(\sqrt{N \log N})$ and the expected query time is $O(\sqrt{N \log N}(1 + |S \cap S'|))$. These fully dynamic structures do not benefit from word-level parallelism. When only insertions are allowed we give another structure that handles both insertions and emptiness/witness queries in $O(\sqrt{N/\frac{w}{\log^2 w}})$ expected time.[1]

3SUM Hardness. Data structure lower bounds can be proved unconditionally, or conditionally, based on the *conjectured* hardness of some problem. One of the most popular conjectures for conditional lower bounds is that the 3SUM problem (given n real numbers, determine if any three sum to zero) cannot be solved in truly subquadratic (expected) time, i.e. $O(n^{2-\Omega(1)})$ time. Even if the inputs are integers in the range $[-n^3, n^3]$ (the Integer3SUM problem), the problem is still conjectured to be insoluble in truly subquadratic (expected) time. See [9,11,12] and the references therein.

Pătraşcu in [11] showed that the Integer3SUM problem can be reduced to offline set-intersection, thereby obtaining conditional lower bounds for offline data structures for set-intersection. The parameters of this reduction were tightened by us in [12]. Converting a conditional lower bound for the offline version of a problem to a conditional lower bound for the incremental (and hence dynamic) version of the same problem is straightforward, and thus we can prove conditional lower bounds for the incremental (and hence dynamic) set intersection problems. In particular, we are able to show that conditioned on the Integer3SUM conjecture, for the incremental emptiness version either the update or query time must be at least $\Omega(N^{1/2-o(1)})$ time. This is discussed in more detail, including lower bounds for the reporting version, in the full version of this paper (see [1]).

Related Work. Most existing set intersection data structures, e.g., [13–15], work in the comparison model, where sets are represented as sorted lists or

---

[1] These data structures offer a tradeoff between space M, query time, and update time. We restricted our attention to $M = O(N)$ here for simplicity.

arrays. In these data structures the main benchmark is the minimum number of comparisons needed to certify the answer. Bille, Pagh, and Pagh [16] also used similar word-packing techniques to evaluate expressions of set intersections and unions. Their query algorithm finds the intersection of m sets with a total of n elements in $O(n/\frac{w}{\log^2 w} + m \cdot op)$ time, where op is the size of the output. Cohen and Porat [17] designed a *static* $O(N)$-space data structure for answering reporting queries in $O(\sqrt{N(1 + |S \cap S'|)})$ time, which is only $O(\sqrt{\log N})$ faster than the data structure presented here.

Triangle Enumeration. Itai and Rodeh [18] showed that all t triangles in a graph could be enumerated in $O(m^{3/2})$ time. Thirty years ago Chiba and Nishizeki [19] generalized [18] to show that $O(m\alpha)$ time suffices, where α is the *arboricity* of the graph. This algorithm has only been improved for dense graphs using fast matrix multiplication. The recent algorithm of Björklund, Pagh, Williams, and Zwick [20] shows that when the matrix multiplication exponent $\omega = 2$, triangle enumeration takes $\tilde{O}(\min\{n^2 + nt^{2/3}, m^{4/3} + mt^{1/3}\})$ time. (The actual running time is expressed in terms of ω.) We give the first asymptotic improvement to Chiba and Nishizeki's algorithm for graphs that are too sparse to benefit from fast matrix multiplication. Using our set intersection data structure, we can enumerate t triangles in $O(m + m\alpha/\frac{w}{\log^2 w} + t)$ expected time.

For simplicity we have stated all bounds in terms of an arbitrary word size w. When $w = O(\log n)$ the $w/\log^2 w$ factor becomes $\log n/\log \log n$.

Overview of the Paper. The paper is structured as follows. In Section 2 we discuss a packing algorithm for (dynamic) set intersection, and in Section 3 we show how the packing algorithm for set intersection can be used to speed up triangle listing. In Section 4 we present our data structure for emptiness queries on a fully dynamic family of sets, with time/space tradeoffs. In Section 5 we combine the packing algorithm for set intersection with the emptiness query data structure to obtain a packed data structure for set intersection witness queries on an incremental family of sets. In Section 6 we present non-packed data structures for emptiness, witness, and reporting set intersection queries on a fully dynamic family of sets, with time/space tradeoffs. Finally, the discussion of conditional lower bounds based on the 3SUM conjecture for dynamic versions of the set intersection problem appears in the full version of the paper [1].

2   Packing Sets

Theorem 1. *A family of sets* $F = \{S_1, \cdots, S_t\}$ *with* $d > \max_{S \in F} |S|$ *can be preprocessed in linear time to facilitate the following set intersection queries. Given two* $S, S' \in F$, *one can find a witness in* $S \cap S'$ *in* $O(\frac{d \log^2 w}{w})$ *expected time and list all of the elements of* $S \cap S'$ *in* $O(|S \cap S'|)$ *additional expected time. If* $w = O(\log n)$ *then the query time is reduced to* $O(\frac{d \log \log n}{\log n})$. *Furthermore, updates (insertions/deletions of elements) to sets in* F *can be performed* $O(1)$ *expected time, subject to the constraint that* $d > \max_{S \in F} |S|$.

Proof. Every set $S \in F$ is split into ℓ buckets $B_1^S, \ldots, B_\ell^S$ where $\ell = \frac{d \log w}{w}$. We pick a function h from a pairwise independent family of hash functions and assign each element $e \in S$ into a bucket $B_{h(e)}^S$. The expected number of elements from a set S in each bucket is $\frac{w}{\log w}$. We use a second hash function h' from another family of pairwise independent hash functions which reduces the universe size to w^2. An $h'(e)$ value is represented with $2 \log w + 1$ bits, the extra *control bit* being necessary for certain manipulations described below. For each S and i we represent $h'(B_i^S)$ as a packed, sorted sequence of h'-values. In expectation each $h'(B_i^S)$ occupies $O(1)$ words, though some buckets may be significantly larger. Finally, for each bucket B_i^S we maintain a lookup table that translates from $h'(e)$ to e. If there is more than one element that is hashed to $h'(e)$ then all such elements are maintained in the lookup table via a linked list.

Notice that $S \cap S' = \bigcup_{i=1}^\ell B_i^S \cap B_i^{S'}$. Thus, we can enumerate $S \cap S'$ by enumerating the intersections of all $B_i^S \cap B_i^{S'}$. Fix one such i. We first merge the packed sorted lists $h'(B_i^S)$ and $h'(B_i^{S'})$. Albers and Hagerup [21] showed that two words of sorted numbers (separated by control bits) can be merged using Batcher's algorithm in $O(\log w)$ time. Using this as a primitive we can merge the sorted lists $h'(B_i^S)$ and $h'(B_i^{S'})$ in time $O(|B_i^S| + |B_i^{S'}|/(w/\log^2 w))$. Let C be the resulting list, with control bits set to 0. Our task is now to enumerate all numbers that appear twice (necessarily consecutively) in C. Let C' be C with control bits set to 1. We shift C one field to the right ($2 \log w + 1$ bit positions) and subtract it from C'.[2] Let C'' be the resulting list, with all control bits reset to 0. A field is zero in C'' iff it and its predecessor were identical, so the problem now is to enumerate zero fields. By repeated halving, we can distill each field to a single bit (0 for zero, 1 for non-zero) in $O(\log \log w)$ time and then take the complement of these bits (1 for zero, 0 for non-zero). We have now reduced the problem to reading off all the 1s in a w-bit word, which can be done in $O(1)$ time per 1 using the most-significant-bit algorithm of [22].[3] For each repeated h'-value we lookup all elements in B_i^S and $B_i^{S'}$ with that value and report any occurring in both sets. Every unit of time spent in this step corresponds to an element in the intersection or a false positive.

The cost of intersecting buckets B_i^S and $B_i^{S'}$ is

$$O\left(1 + \left(\lceil \frac{|B_i^S|}{w/\log w} \rceil + \lceil \frac{|B_i^{S'}|}{w/\log w} \rceil\right) \log w + |B_i^S \cap B_i^{S'}| + f_i\right),$$

where f_i is the number of false positives. The expected value of f_i is $o(1)$ since the expected sizes of B_i^S and $B_i^{S'}$ are $w/\log w$ and for $e \in B_i^S, e' \in B_i^{S'}$, $\Pr(h'(e) = $

---

[2] The control bits stop carries from crossing field boundaries.

[3] This algorithm uses multiplication. Without unit-time multiplication [23] one can read off the 1s in $O(\log \log w)$ time per 1. If $w = O(\log n)$ then the instruction set is not as relevant since we can build $o(n)$-size tables to calculate most significant bits and other useful functions.

$h'(e')) = 1/w^2$. Thus, the expected runtime for a query is

$$\sum_{i=1}^{\ell} O\left(1 + \left(\lceil \frac{|B_i^S|}{w/\log w}\rceil + \lceil \frac{|B_i^{S'}|}{w/\log w}\rceil\right)\log w + |B_i^S \cap B_i^{S'}| + f_i\right)$$

$$= O(\ell \log w + |S \cap S'|) = O\left(\frac{d\log^2 w}{w} + |S \cap S'|\right).$$

It is straightforward to implement insertions and deletions in $O(1)$ time in expectation. Suppose we must insert e into S. Once we calculate $i = h(e)$ and $h'(e)$ we need to insert $h'(e)$ into the packed sorted list representing $h'(B_i^S)$. Suppose that $h'(B_i^S)$ fits in one word; let it be D, with all control bits set to 1.[4] With a single multiplication we form a word D' whose fields each contain $h'(e)$ and whose control bits are zero. If we subtract D' from D and mask everything but the control bits, the most significant bit identifies the location of the successor of $h'(e)$ in $h'(B_i^S)$. We can then insert $h'(e)$ into the sorted list in D with $O(1)$ masks and shifts. The procedure for deleting an element in $O(1)$ time follows the same lines.   □

3   A Faster Triangle Enumeration Algorithm

Theorem 2. *Given an undirected graph $G = (V, E)$ with $m = |E|$ edges and arboricity α, all t triangles can be enumerated in $O(m + \frac{m\alpha}{w/\log^2 w} + t)$ expected time or in $O\left(m + \frac{m\alpha}{\log n/\log\log n} + t\right)$ expected time if $w = O(\log n)$.*

Proof. We will make use of the data structure in Theorem 1. To do this we first find an acyclic orientation of E in which the out-degree of any vertex is $O(\alpha)$. Such an orientation can be found in linear time using the peeling algorithm of Chiba and Nishizeki [19]. Define $\Gamma^+(u) = \{v \mid (u, v)\}$ to be the set of out-neighbors of u according to this orientation. Begin by preprocessing the family $F = \{\Gamma^+(u) \mid u \in V\}$, where all sets have size $O(\alpha)$. For each edge (u, v), enumerate all elements in the intersection $\Gamma^+(u) \cap \Gamma^+(v)$. For each vertex w in the intersection output the triangle $\{u, v, w\}$. Since the orientation is acyclic, every triangle is output exactly once. There are m set intersection queries, each taking $O(1 + \alpha/\max\{\frac{w}{\log^2 w}, \frac{\log n}{\log\log n}\})$ time, aside from the cost of reporting the output, which is $O(1)$ per triangle.   □

4   Dynamic Emptiness Queries with Time/Space Tradeoff

Theorem 3. *There exists an algorithm that maintains a family F of dynamic sets using $O(M)$ space where each update costs $O(\sqrt{M})$ expected time, and each emptiness query costs $O(\frac{N}{\sqrt{M}})$ expected time.*

---

[4] If $h'(B_i^S)$ is larger we apply this procedure to each word of the list $h'(B_i^S)$. It occupies $O(1)$ words in expectation.

Proof. Each set $S \in F$ maintains its elements in a lookup table using a perfect dynamic hash function. So the cost of inserting a new element into S, deleting an element from S, or determining whether some element x is in S is expected $O(1)$ time. Let $N = \sum_{S \in F} |S|$. We make the standard assumption that N is always at least $N'/2$ and at most $2N'$ for some natural number N'. Standard rebuilding de-amortization techniques are used if this is not the case.

The Structure. We say a set S is *large* if at some point $|S| > 2N'/\sqrt{M}$, and since the last time S was at least that large, its size was never less than $N'/\sqrt{M}$. If S is not large, and its size is at least $N'/\sqrt{M}$ then we say it is *medium*. If S is neither large nor medium then it is *small*. Notice that the size of a small set is less than $N'/\sqrt{M} = O(N/\sqrt{M})$. Let $L \subseteq F$ be the sub-family of large and medium sets, and let $\ell = |L|$. Notice that $\ell \leq \sqrt{M}$. For each set $S \in L$ we maintain a unique integer $1 \leq i_S \leq \ell$, and an *intersection-size* dynamic lookup table T_S of size ℓ such that for a large set S' we have $T_S[i_{S'}] = |S \cap S'|$. Adding and deleting entries from the table takes expected constant time using hashing. Due to the nature of our algorithm we cannot guarantee that all of the intersection-size tables will always be fully updated. However, we will guarantee the following invariant.

Invariant 1. *For every two large sets S and S', $T_S[i_{S'}]$ and $T_{S'}[i_S]$ are correctly maintained.*

Query. For two sets $S, S' \in F$ where either S or S' is not large, say S, we determine if they intersect by scanning the elements in S and using the lookup table for S'. The time cost is $O(|S|) = O(N'/\sqrt{M})$. If both sets are large, then we examine $T_S[i_{S'}]$ which determines the size of the intersection (by Invariant 1) and decide accordingly if it is empty or not. This takes $O(1)$ time.

Insertions. When inserting a new element x into S, we first update the lookup table of S to include x. Next, if S was small and remained small then no additional work is done. Otherwise, for each $S' \in L$ we must update the size of $S \cap S'$ in the appropriate intersection-size tables. This is done directly in $O(\sqrt{M})$ time by determining whether x is in S', for each S', via the lookup tables. We briefly recall, as mentioned above, that it is possible that some of the intersection-size tables will not be fully updated, and so incrementing the size of an intersection is only helpful if the intersection size was correctly maintained before. Nevertheless, as explained soon, Invariant 1 will be guaranteed to hold, which suffices for the correctness of the algorithm since the intersection-size tables are only used when intersecting two large sets.

   The more challenging case is when S becomes medium. If this happens we would like to increase ℓ by 1, assign i_S to be the new ℓ, allocate and initialize T_S in $O(\sqrt{M})$ time, and for each $S' \in L$ we compute $|S \cap S'|$ and insert the answer into $T_S[i_{S'}]$ and $T_{S'}[i_S]$. This entire process is dominated by the the task of computing $|S \cap S'|$ for each $S' \in L$, taking a total of $O(\sum_{S' \in L} |S|)$ time, which could be as large as $O(N)$ and is too costly. However, this work can be spread over the next $N'/\sqrt{M}$ insertions made into S until S becomes large. This is done as follows. When S becomes medium we create a list L_S of all of the large and medium sets at this time (without their elements). This takes $O(\sqrt{M})$ time. Next, for every insertion into S we compute the values of $O(M/N')$ locations in T_S by computing the intersection size of S and each of $O(M/N')$ sets from L_S in $O(\frac{M}{N'} \cdot \frac{N}{\sqrt{M}}) = O(\sqrt{M})$ time. For each such set S' we also update $T_{S'}[i_S]$. By the time S becomes large we will have correctly computed the values in T_S for all $O(\sqrt{M})$ of the sets in L_S, and for every set $S' \in L_S$ we will have correctly computed $T_{S'}[i_S]$. It is possible that between the time S became medium to the time S became large, there were other sets such as S' which became medium and perhaps even large, but $S' \notin L_S$. Notice that in such a case $S \in L_{S'}$ and so it is guaranteed that by the time both S and S' are large, the indicators $T_S[i_{S'}]$ and $T_{S'}[i_S]$ are correctly updated, thereby guaranteeing that Invariant 1 holds. Thus the total cost of performing an insertion is $O(\sqrt{M})$ expected time.

Deletions. When deleting an element x from S, we first update the lookup table of S to remove x in $O(1)$ expected time. If S was small and remained small then no additional work is done. If S was in L then we scan all of the $S' \in L$ and check if x is in S' in order to update the appropriate locations in the intersection-size tables. This takes $O(\sqrt{M})$ time.

If S was medium and now became small, we need to decrease ℓ by 1, remove the assignment to i_S to be the new ℓ, delete T_S, and for each $S' \in L$ we need to remove $T_{S'}[i_S]$. In addition, in order to accommodate the update process of medium sized sets, for each medium set S' we must remove S from $L_{S'}$ if it was in there. □

Corollary 2. *There exists an algorithm that maintains a family F of dynamic sets using $O(N)$ space where each update costs $O(\sqrt{N})$ expected time, and each emptiness query costs $O(\sqrt{N})$ expected time.*

5   Incremental Witness Queries

Theorem 4. *Suppose there exists an algorithm A that maintains a family F of incremental sets, each of size at most d, such that set intersection witness queries can be answered in $O(\frac{d}{\tau_q})$ expected time and inserts can be performed in $O(\tau_u)$ expected time. Then there exists an algorithm to maintain a family F of incremental sets—with no upper bound on set sizes—that uses $O(N)$ space and performs insertions and witness queries in $O(\sqrt{N'/\tau_q})$ expected time, where $N = \sum_{S \in F} |S|$.*

Proof. We make the standard assumption that N is always at least $N'/2$ and at most $2N'$ for some natural number N'. Standard rebuilding de-amortization techniques are used if this is not the case. In our context, we say that a set is large if its size is at least $\sqrt{N'\tau_q}$, and is medium if its size is between $\sqrt{N'/\tau_q}$ and $\sqrt{N'\tau_q}$. Each medium and large set S maintains a *stash* of the at most $\sqrt{N'\tau_q}$

last elements that were inserted into S (these elements are part of S). This stash is the entire set S if S is medium. If S is large then the rest of S (the elements not in the stash) is called the *primary* set of S. Stashes are maintained using algorithm A with $d = \sqrt{N'\tau_q}$. Thus, answering intersection queries between two medium sets takes $O(\sqrt{N'/\tau_q})$ expected time.

We maintain for each medium and large set S a witness table P_S such that for any large set S' we have that $P_S[i_{S'}]$ is either an element (witness) in the intersection of S and the primary set of S', or null if no such element exists. This works in the incremental setting as once a witness is established it never changes. Since there are at most $\sqrt{N'/\tau_q}$ large sets and at most $\sqrt{N'/\tau_q}$ medium sets, the space usage is $O(N')$. If a query is between S_1 and S_2 and S_1 is large, then: (1) if S_2 is small we lookup each element in S_2 to see if it is in S_1, (2) if S_2 is medium or large then we use the witness tables to see if there is a witness of an intersection between S_2 and the primary set of S_1 or between S_1 and the primary set of S_2, and if there is no such witness then we use algorithm A to intersect the stashes of S_2 and S_1. In any case, the cost of a query is $O(\sqrt{N'/\tau_q})$ expected time. The details for maintaining these tables are similar to the details of maintaining the intersection-size array tables from Section 4.

Insertion. When inserting an element x into S, if S is small then we do nothing. If S is medium then we add x to the stash of S in algorithm A. If S is large then we add x to the stash of S and verify for every other large set if x is in that set, updating the witness table accordingly. If S became medium then we add it to the structure of algorithm A. Since the size of S is $O(\sqrt{N'/\tau_q})$ this takes $O(\sqrt{N'/\tau_q})$ expected time. Furthermore, when S becomes medium the table P_S needs to be prepared. To do this, between the time S is of size $\sqrt{N'/2\tau_q}$ and the time S is of size $\sqrt{N'/\tau_q}$, the table P_S is inclemently constructed. If S became large then we now allow its primary set to be nonempty, and must also update the witness tables. The changes to witness tables in this case is treated using the same techniques as in Theorem 3, and so we omit their description. This will cost $O(\sqrt{N'/\tau_q} + \tau_u)$ expected time.

Finally, for a large set S, once its stash reaches size $\sqrt{N'\tau_q}$ we dump the stash into the primary set of S, thereby emptying the stash. We describe an amortized algorithm for this process, which is deamortized using a standard lazy approach. To combine the primary set and the stash we only need to update the witness tables for set intersection witnesses between medium sets and the new primary set of S as it is possible that a witness was only in the stash. To do this, we directly scan all of the medium sets and check if a new witness can be obtained from the stash. The number of medium sets is $O(\sqrt{N'\tau_q})$ and the cost of each intersection will be $O(\sqrt{N'/\tau_q})$ for a total of $O(N')$ time. Since this operation only happens after $\Omega(\sqrt{N'\tau_q})$ insertions into S the amortized cost is $O(\sqrt{N'/\tau_q})$ time.

$\square$

Combining Theorem 1 with Theorem 4 we obtain the following.

Corollary 1. *There exists an algorithm in the word-RAM model that maintains a family F of incremental sets using $O(N)$ space where each insertion costs $O(\sqrt{\frac{N}{w/\log^2 w}})$ expected time and a witness query costs $O(\sqrt{\frac{N}{w/\log^2 w}})$ expected time.*

6 Fully Dynamic Set Intersection with Witness and Reporting Queries

Each element in $\bigcup_{S \in F} S$ is assigned an integer from the range of $[2N']$. When a new element not appearing in $\bigcup_{S \in F} S$ arrives, it is assigned to the smallest available integer, and that integer is used as its key. When keys are deleted (no longer in use), we do not remove their assignment, and instead, we conduct a standard rebuilding technique in order to reassign the elements. Finally, we use a second assignment via a random permutation of the integers in order to uniformly spread the assignments within the range.

The structure. Consider the following binary tree T of height $\log N' + 1$ where each vertex v covers some range from U, denoted by $[\alpha_v, \beta_v]$, such that the range of the root covers all of U, and the left (right) child of v covers the first (second) half of $[\alpha_v, \beta_v]$. A vertex at depth i covers $\frac{2N'}{2^i}$ elements of U. For a vertex v let $S^v = S \cap [\alpha_v, \beta_v]$. Let $N_v = \sum_{S \in F} |S^v|$. Let $M_v = \frac{N_v \cdot M}{N'}$. We say a set S is v-*large* if at some point $|S^v| > \frac{2N_v}{\sqrt{M_v}}$, and since the last time S^v was at least that large, its size was never less than $\frac{N_v}{\sqrt{M_v}}$.

Each vertex $v \in T$ with children v_0 and v_1 maintains a structure for emptiness queries as in Theorem 3, using M_v space, on the family $F^v = \{S^v : S \in F\}$. In addition, we add auxiliary data to the intersection-size tables as follows. For sets $S_1, S_2 \in F$ the set of all vertices in which S_1 and S_2 intersect under them defines a connected tree T'. This tree has some branching vertices which have 2 children, some non-branching internal vertices with only 1 child, and some leaves. Consider the vertices v in T for which S_1 and S_2 are v-large and define $\hat{T}$ to be the connected component of these vertices that includes the root r. (It may be that $\hat{T}$ does not exist.) To facilitate a fast traversal of $\hat{T}$ during a query we maintain *shortcut* pointers for every two sets $S_1, S_2 \in F$ and for every vertex $v \in T$ such that both S_1 and S_2 are v-large. To this end, we say v is a *branching-(S_1, S_2)-vertex* if both $S_1^{v_0} \cap S_2^{v_0} \neq \emptyset$ and $S_1^{v_1} \cap S_2^{v_1} \neq \emptyset$. Consider the path starting from the left (right) child of v and ending at the first descendent v' of v such that:(1) S_1 and S_2 are relatively large for all of the vertices on the path, (2) $S_1^{v'} \cap S_2^{v'} \neq \emptyset$, and (3) either v' is a *branching-(S_1, S_2)-vertex* or one of the sets S_1 and S_2 is not v'-large. The left (right) shortcut pointer of v will point to v'. Notice that the shortcut pointers are maintained for every vertex v even if on the path from r to v there are some vertices for which either S_1 or S_2 are not relatively large, which helps to reduce the update time during insertions/deletions. Also notice that using these pointers it is straightforward to check in $O(1)$ time if $S_1^{v_0} \cap S_2^{v_0}$ and $S_1^{v_1} \cap S_2^{v_1}$ are empty or not.

The space complexity of the structure is as follows. Each vertex v uses $O(M_v)$ words of space which is $O(MN_v/N')$. So the space usage is $\sum_v M_v = O(M \log N)$ words, since in each level of T the sum of all M_v for the vertices in that level is $O(M)$, and there are $O(\log N)$ levels.

Reporting Queries. For a reporting query on S_1 and S_2, if $op = 0$ then either the emptiness test at the root will conclude in $O(1)$ time, or we spend $O(\frac{N_r}{\sqrt{M_r}}) = O(\frac{N}{\sqrt{M}})$ time. Otherwise, we recursively examine vertices v in T starting with the root r. If both S_1 and S_2 are v-large and $S_1^v \cap S_2^v \neq \emptyset$, then we continue recursively to the vertices pointed to by the appropriate shortcut pointers. If either S_1 or S_2 is not v-large then we wish to output all of the elements in the intersection of S_1^v and S_2^v. To do this, we check for each element in the smaller set if it is contained within the larger set using the lookup table which takes $O(\frac{N_v}{\sqrt{M_v}})$ time.

For the runtime, as we traverse down T from r using appropriate shortcut pointers, we encounter only two types of vertices. The first type are vertices v for which both S_1 and S_2 are v-large, and the second type are vertices v for which either S_1 or S_2 is not v-large. Each vertex of the first type performs $O(1)$ work, and the number of such vertices is at most the number of vertices of the second type, due to the branching nature of the shortcut pointers. For vertices of the second type, the intersection of S_1 and S_2 must both be non-empty relative to such vertices and so the $O(\frac{N_v}{\sqrt{M_v}})$ time cost can be charged to at least one element in the output. Denote the vertices of the second type by $v_1, v_2, \ldots, v_t$. Notice that $t \leq op$ as each v_i contains at least one element from the intersection, and that $\sum_i N_{v_i} < 2N'$ since the vertices are not ancestors of each other. We will make use of the following Lemma.

Lemma 3. *If* $\sum_{i=1}^t x_i \leq k$ *then* $\sum_{i=1}^t \sqrt{x_i} \leq \sqrt{k \cdot t}$.

Proof. Since $\sum_{i=1}^t \sqrt{x_i}$ is maximized whenever all the x_i are equal, we have that $\sum_{i=1}^t \sqrt{x_i} \leq t\sqrt{\frac{k}{t}} = \sqrt{kt}$. $\square$

Therefore, the total time cost is

$$\sum_i \frac{N_{v_i}}{\sqrt{M_{v_i}}} = \sum_i \frac{N_{v_i}\sqrt{N'}}{\sqrt{MN_{v_i}}} = \sqrt{\frac{N'}{M}} \sum_i \sqrt{N_{v_i}} \leq \sqrt{\frac{N'}{M}}\sqrt{2N'}\sqrt{t} \leq O\left(\frac{N\sqrt{op}}{\sqrt{M}}\right).$$

Witness Queries. A witness query is answered by traversing down T using shortcut pointers, but instead of recursively looking at both shortcut pointers for each vertex, we only consider one. Thus the total time it takes until we reach a vertex v for which either S_1 or S_2 is not v-large is $O(\log N)$. Next, we use the hash function to find an element in the intersection in $O(\frac{N}{\sqrt{M}})$ time, for a total of $O(\log N + \frac{N}{\sqrt{M}})$ time to answer a witness query.

Insertions and Deletions. When inserting a new element x into S_1, we first locate the leaf ℓ of T which covers x. Next, we update our structure on the path from ℓ to r as follows. Starting from ℓ, for each vertex v on the path we insert x into S_1^v. This incurs a cost of $\sqrt{M_v}$ for updating the emptiness query structure at v. If there exists some set S_2 such that $|S_1^v \cap S_2^v|$ becomes non-zero, then we may need to update some shortcut pointers on the path from ℓ to r relative to S_1 and S_2. Being that such a set S_2 must be large, the number of such sets is at most $\frac{N_v}{\sqrt{M_v}}$.

To analyze the expected running time of an insertion notice that since the elements in the universe are randomly distributed, the expected value of N_v and M_v for a vertex v at depth i are $\frac{N}{2^i}$ and $\frac{M}{2^i}$ respectively. So the number of v-large sets is at most $\frac{N_v}{\sqrt{M_v}} = \frac{N}{\sqrt{2^i M}}$. The expected time costs of updating the emptiness structure is at most $\sum_{i=0}^{\log N'} \frac{N}{\sqrt{2^i M}} = O(\frac{N}{\sqrt{M}})$. The same analysis holds for the shortcut pointer. The deletion process is exactly the reverse of the insertions process, and also costs $O(\frac{N}{\sqrt{M}})$ expected time.

The total space usage is $O(M \log N)$. With a change of variable (substituting $M/\log N$ for M in the construction above), we can make the space $O(M)$ and obtain the following result.

Theorem 5. *There exists an algorithm that maintains a family F of dynamic sets using $O(M)$ space where each update costs $O(\sqrt{M \log N})$ expected time, each reporting query costs $O(\frac{N \sqrt{\log N}}{\sqrt{M}} \sqrt{op+1})$ time, and each witness query costs $O(\frac{N \sqrt{\log N}}{\sqrt{M}} + \log N)$ expected time.*

References

1. Kopelowitz, T., Pettie, S., Porat, E.: Dynamic set intersection (2014). CoRR abs/1407.6755v2
2. Masek, W.J., Paterson, M.: A faster algorithm computing string edit distances. J. Comput. Syst. Sci. **20**(1), 18–31 (1980)
3. Myers, G.: A Four Russians algorithm for regular expression pattern matching. J. ACM **39**(2), 432–448 (1992)
4. Chan, T.M.: More algorithms for all-pairs shortest paths in weighted graphs. SIAM J. Comput. **39**(5), 2075–2089 (2010)
5. Chan, T.M.: All-pairs shortest paths for unweighted undirected graphs in $o(mn)$ time. ACM Transactions on Algorithms **8**(4), 34 (2012)
6. Chan, T.M.: All-pairs shortest paths with real weights in $o(n^3/\log n)$ time. Algorithmica **50**(2), 236–243 (2008)
7. Buchin, K., Buchin, M., Meulemans, W., Mulzer, W.: Four soviets walk the dog - with an application to Alt's conjecture. In: Proceedings 25th Annual ACM-SIAM Symposium on Discrete Algorithms (SODA), pp. 1399–1413 (2014)
8. Baran, I., Demaine, E.D., Pătraşcu, M.: Subquadratic algorithms for 3SUM. Algorithmica **50**(4), 584–596 (2008)
9. Grønlund, A., Pettie, S.: Threesomes, degenerates, and love triangles. In: Proceedings 55th IEEE Symposium on Foundations of Computer Science (FOCS) (2014). Full manuscript available as arXiv:1404.0799

10. Chan, T.M.: The art of shaving logs. In: Dehne, F., Solis-Oba, R., Sack, J.-R. (eds.) WADS 2013. LNCS, vol. 8037, p. 231. Springer, Heidelberg (2013)

11. Pătraşcu, M.: Towards polynomial lower bounds for dynamic problems. In: Proceedings 42nd ACM Symposium on Theory of Computing (STOC), pp. 603–610 (2010)

12. Kopelowitz, T., Pettie, S., Porat, E.: 3sum hardness in (dynamic) data structures (2014). CoRR abs/1407.6756

13. Demaine, E.D., López-Ortiz, A., Munro, J.I.: Adaptive set intersections, unions, and differences. In: Proceedings of the Eleventh Annual ACM-SIAM Symposium on Discrete Algorithms, pp. 743–752 (2000)

14. Barbay, J., Kenyon, C.: Adaptive intersection and t-threshold problems. In: Proceedings 13th Annual ACM-SIAM Symposium on Discrete Algorithms (SODA), pp. 390–399 (2002)

15. Baeza-Yates, R.: A fast set intersection algorithm for sorted sequences. In: Sahinalp, S.C., Muthukrishnan, S., Dogrusoz, U. (eds.) CPM 2004. LNCS, vol. 3109, pp. 400–408. Springer, Heidelberg (2004)

16. Bille, P., Pagh, A., Pagh, R.: Fast evaluation of union-intersection expressions. In: Tokuyama, T. (ed.) ISAAC 2007. LNCS, vol. 4835, pp. 739–750. Springer, Heidelberg (2007)

17. Cohen, H., Porat, E.: Fast set intersection and two-patterns matching. Theor. Comput. Sci. 411(40–42), 3795–3800 (2010)

18. Itai, A., Rodeh, M.: Finding a minimum circuit in a graph. SIAM J. Comput. 7(4), 413–423 (1978)

19. Chiba, N., Nishizeki, T.: Arboricity and subgraph listing algorithms. SIAM J. Comput. 14(1), 210–223 (1985)

20. Björklund, A., Pagh, R., Williams, V.V., Zwick, U.: Listing triangles. In: Esparza, J., Fraigniaud, P., Husfeldt, T., Koutsoupias, E. (eds.) ICALP 2014. LNCS, vol. 8572, pp. 223–234. Springer, Heidelberg (2014)

21. Albers, S., Hagerup, T.: Improved parallel integer sorting without concurrent writing. Inf. Comput. 136(1), 25–51 (1997)

22. Fredman, M.L., Willard, D.E.: Surpassing the information theoretic bound with fusion trees. J. Comput. Syst. Sci. 47(3), 424–436 (1993)

23. Brodnik, A., Miltersen, P.B., Munro, J.I.: Trans-dichotomous algorithms without multiplication—some upper and lower bounds. In: Rau-Chaplin, A., Dehne, F., Sack, J.-R., Tamassia, R. (eds.) WADS 1997. LNCS, vol. 1272, pp. 426–439. Springer, Heidelberg (1997)

Time-Space Trade-offs
for Triangulations and Voronoi Diagrams

Matias Korman[1], Wolfgang Mulzer[2]($\boxtimes$), André van Renssen[1],
Marcel Roeloffzen[3], Paul Seiferth[2], and Yannik Stein[2]

[1] JST, ERATO, Kawarabayashi Large Graph Project,
National Institute of Informatics (NII), Tokyo, Japan
{korman,andre}@nii.ac.jp
[2] Institut für Informatik, Freie Universität Berlin, Berlin, Germany
{mulzer,pseiferth,yannikstein}@inf.fu-berlin.de
[3] Tohoku University, Sendai, Japan
marcel@dais.is.tohoku.ac.jp

Abstract. Let S be a planar n-point set. A *triangulation* for S is a maximal plane straight-line graph with vertex set S. The *Voronoi diagram* for S is the subdivision of the plane into cells such that each cell has the same nearest neighbors in S. Classically, both structures can be computed in $O(n \log n)$ time and $O(n)$ space. We study the situation when the available workspace is limited: given a parameter $s \in \{1, \dots, n\}$, an s-workspace algorithm has read-only access to an input array with the points from S in arbitrary order, and it may use only $O(s)$ additional words of $\Theta(\log n)$ bits for reading and writing intermediate data. The output should then be written to a write-only structure. We describe a deterministic s-workspace algorithm for computing a triangulation of S in time $O(n^2/s + n \log n \log s)$ and a randomized s-workspace algorithm for finding the Voronoi diagram of S in expected time $O((n^2/s) \log s + n \log s \log^* s)$.

1 Introduction

Since the early days of computer science, a major concern has been to cope with strong memory constraints. This started in the '70s [21] when memory was expensive. Nowadays, the motivation comes from a proliferation of small embedded devices where large memory is neither feasible nor desirable (e.g., due to constraints on budget, power, size, or simply to discourage potential thievery).

Even when memory size is not an issue, we might want to limit the number of write operations: one can read flash memory quickly, but writing (or even reordering) data is slow and may reduce the lifetime of the storage system; write-access to removable memory may be limited for technical or security reasons, (e.g., when using read-only media such as DVDs or to prevent leaking

W. Mulzer, P. Seiferth, and Y. Stein—WS and PS were supported in part by DFG Grants MU 3501/1 and MU 3501/2. YS was supported by the DFG within the research training group MDS (GRK 1408).

© Springer International Publishing Switzerland 2015
F. Dehne et al. (Eds.): WADS 2015, LNCS 9214, pp. 482–494, 2015.
DOI: 10.1007/978-3-319-21840-3_40

information about the algorithm). Similar problems occur when concurrent algorithms access data simultaneously. A natural way to address this is to consider algorithms that do not modify the input.

The exact setting may vary, but there is a common theme: the input resides in read-only memory, the output must be written to a write-only structure, and we can use $O(s)$ additional variables to find the solution (for a parameter s). The goal is to design algorithms whose running time decreases as s grows, giving a *time-space trade-off* [22]. One of the first problems considered in this model is *sorting* [18,19]. Here, the time-space product is known to be $\Omega(n^2)$ [8], and matching upper bounds for the case $b \in \Omega(\log n) \cap O(n/\log n)$ were obtained by Pagter and Rauhe [20] (b denotes the available workspace in *bits*).

Our current notion of memory constrained algorithms was introduced to computational geometry by Asano *et al.* [4], who show how to compute many classic geometric structures with $O(1)$ workspace (related models were studied before [9]). Later, time-space trade-offs were given for problems on simple polygons, e.g., shortest paths [1], visibility [6], or the convex hull of the vertices [5].

In our model, we are given an array S of n points in the plane such that random access to each input point is possible, but we may not change or even reorder the input. Additionally, we have $O(s)$ variables (for a parameter $s \in \{1, \ldots, n\}$). We assume that each variable or pointer contains a data word of $\Theta(\log n)$ bits. Other than this, the model allows the usual word RAM operations. We consider two problems: computing an arbitrary triangulation for S and computing the Voronoi diagram VD(S) for S. Since the output cannot be stored explicitly, the goal is to report the edges of the triangulation or the vertices of VD(S) successively, in no particular order. Dually, the latter goal may be phrased in terms of Delaunay triangulations. We focus on Voronoi diagrams, as they lead to a more natural presentation.

Both problems can be solved in $O(n^2)$ time with $O(1)$ workspace [4] or in $O(n \log n)$ time with $O(n)$ workspace [7]. However, to the best of our knowledge, no trade-offs were known before. Our triangulation algorithm achieves a running time of $O(n^2/s + n \log n \log s)$ using $O(s)$ variables. A key ingredient is the recent time-space trade-off by Asano and Kirkpatrick for a special type of simple polygons [3]. This also lets us obtain significantly better running times for the case that the input is sorted in x-order; see Section 2. For Voronoi diagrams, we use random sampling to find the result in expected time $O((n^2 \log s)/s + n \log s \log^* s))$; see Section 3. Together with recent work of Har-Peled [15], this appears to be one of the first uses of random sampling to obtain space-time trade-offs for geometric algorithms. The sorting lower bounds also apply to triangulations (since we can reduce the former to the latter). By duality between Voronoi diagrams and Delaunay triangulations this implies that our second algorithm for computing the Voronoi diagram is almost optimal.

2   Triangulating a Sorted Point Set

We first describe our *s*-workspace algorithm for triangulating a planar point set that is given in sorted x-order. The input points $S = \{q_1, \ldots, q_n\}$ are stored by increasing x-coordinate, and we assume that all x-coordinates are distinct, i.e., $x_i < x_{i+1}$ for $1 \leq i < n$, where x_i denotes the x-coordinate of q_i, for $1 \leq i \leq n$.

A crucial ingredient in our algorithms is a recent result by Asano and Kirkpatrick for triangulating *monotone mountains*[1] (or *mountains* for short). A mountain is a simple polygon with vertex sequence $v_1, v_2, \ldots, v_k$ such that the x-coordinates of the vertices increase monotonically. The edge $v_1 v_k$ is called the *base*. Mountains can be triangulated very efficiently with bounded workspace.

Theorem 2.1 (Lemma 3 in [3], rephrased). *Let H be a mountain with n vertices, stored in sorted x-order in read-only memory. Let $s \in \{2, \ldots, n\}$. We can report the edges of a triangulation of H in $O(n \log_s n)$ time and $O(s)$ space.*

Since S is given in x-order, the edges $q_i q_{i+1}$, for $1 \leq i < n$, form a monotone simple polygonal chain. Let Part(S) be the subdivision obtained by the union of this chain with the edges of the convex hull of S. We say that a convex hull edge is *long* if the difference between its indices is at least two (i.e., the endpoints are not consecutive). The following lemma lets us decompose the problem into smaller pieces. The proof can be found in the full version.

Lemma 2.2. *Any bounded face of* Part(S) *is a mountain whose base is a long convex hull edge. Moreover, no point of S lies in more than four faces of* Part(S).

Let $e_1, \ldots, e_k$ be the long edges of the convex hull of S, let F_i be the unique mountain of Part(S) whose base is e_i, and let n_i be the number of vertices of F_i.

With the above definitions we can give an overview of our algorithm. We start by computing the edges of the upper convex hull, from left to right. Each time an edge e_i of the convex hull is found, we check if it is long. If so, we triangulate the corresponding mountain F_i (otherwise we do nothing), and we proceed with the computation of the convex hull. The algorithm finishes once all convex hull edges are computed (and their corresponding faces have been triangulated).

Theorem 2.3. *Let S be a set of n points, sorted in x-order. We can report the edges of a triangulation of S in $O(n^2)$ time using $O(1)$ variables, $O(n^2 \log n / 2^s)$ time using $O(s)$ additional variables (for any $s \in \Omega(\log \log n) \cap o(\log n)$), or $O(n \log_p n)$ time using $O(p \log_p n)$ additional variables (for any $2 \leq p \leq n$).*

Proof. Correctness follows directly from the first claim of Lemma 2.2. Thus, it suffices to show the performance bounds. The main steps are: (i) computing the convex hull of a point set given in x-order; (ii) determining if an edge is long; and (iii) triangulating a mountain. We can identify long edges in constant time, by comparing the endpoint indices. Moreover, by Theorem 2.1, we can triangulate the polygon F_i of n_i vertices in time $O(n_i \log_s n_i)$ with $O(s)$ variables.

Further note that we never need to store the polygons to triangulate explicitly: once a long edge e_i of the convex hull is found, we can implicitly give it as input for the triangulation algorithm by specifying the indices of e_i: since the points are sorted by x-coordinate, the other vertices of F_i are exactly the input

---

[1] Also known as *unimonotone polygon* [14].

points between the endpoints of e_i. We then pause our convex-hull computation, keeping its current state in memory, and reserve $O(s)$ memory for triangulating F_i. Once this is done, we can discard all used memory, and we reuse that space for the next mountain. We then continue the convex hull algorithm to find the next long convex-hull edge. By the second claim of Lemma 2.2, no vertex appears in more than four mountains, and thus the total time for triangulating all mountains is bounded by $\sum_i O(n_i \log_s n_i) = O(n \log_s n)$.

Finally, we need to bound the time for computing the convex hull of S. Recall that we may temporarily pause the algorithm to triangulate a generated polygon, but overall it is executed only once over the input. There exist several algorithms for computing the convex hull of a set of points sorted by x-coordinate under memory constraints. When $s \in \Theta(1)$, we can use the gift-wrapping algorithm (Jarvis march [16]) which runs in $O(n^2)$ time. Barba et al. [5] provided a different algorithm that runs in $O(n^2 \log n / 2^s)$ time using $O(s)$ variables (for any $s \in o(\log n)$).[2] This approach is desirable for $s \in \Omega(\log \log n) \cap o(\log n)$. As soon as the workspace can fit $\Omega(\log n)$ variables, we can use the approach of Chan and Chen [10]. This algorithm runs in $O(n \log_p n)$ time and uses $O(p \log_p n)$ variables, for any $2 \le p \le n$. In any case, the time for the convex hull dominates the time for triangulating the mountain mountains.   □

General Input. The previous algorithm uses the sorted input order in two ways. Firstly, the algorithms of Barba et al. [5] and of Chan and Chen [10] work only for simple polygons (e.g., for sorted input). Instead, we may use the algorithm by Darwish and Elmasry [13] that gives the upper (or lower) convex hull of any sequence of n points in $O(n^2/(s \log n) + n \log n)$ time with $O(s)$ variables[3], matching known lower bounds. Secondly, and more importantly, the Asano-Kirkpatrick (AK) algorithm requires the input to be sorted. To address this issue, we simulate sorted input using multiple heap structures. For this, we require some technical details on how the AK-algorithm accesses its input.

Let F be a mountain with n vertices. Let $F^\uparrow$ denote the vertices of F in ascending x-order, and $F^\downarrow$ denote F in descending x-order. The AK-algorithm makes one pass over $F^\uparrow$ and one pass over $F^\downarrow$.[4] Each pass computes half of the triangulation, uses $O(s)$ variables and has $\Theta(\log_s n)$ rounds. In round i, it partitions $F^\uparrow$ ($F^\downarrow$) into blocks of $O(|F|/s^i)$ consecutive points that are processed from left to right. Each block is further subdivided into $O(s)$ sub-blocks $b_1, \ldots, b_k$ of size $O(|F|/s^{i+1})$. The algorithm does two scans over the sub-blocks. The first scan processes the elements in x-order. Whenever the first scan finishes reading a sub-block b_i, the algorithm makes b_i active and creates a pointer l_i to the

---

[2] In fact, Barba et al. show how to compute the convex hull of a simple polygon, but also show that both problems are equivalent. The monotone chain can be completed to a polygon by adding a vertex with a very high or low y-coordinate.

[3] Darwish and Elmasry [13] state a running time of $O(n^2/s + n \log n)$, but they measure workspace in bits while we use words.

[4] AK reduce triangulation to the *all next smaller right neighbor* (NSR) and the *all next smaller left neighbor* (NSL) problem and present an algorithm for NSR if the input is in x-order. This implies an NSL-algorithm by reading the input in reverse.

rightmost element of b_i. The second scan goes from right to left and is concurrent to the first scan. In each step, it reads the element at l_i in the rightmost active sub-block b_i, and it decreases l_i by one. If l_i leaves b_i, then b_i becomes inactive. As the first scan creates new active sub-blocks as it proceeds, the second scan may jump between sub-blocks.

We use the heap by Asano et al. [2] to provide the input for the AK-algorithm. We shortly restate its properties.

Lemma 2.4 ([2]). *Let S be a set of n points. There is a heap that supports insert and extract-min (resp. extract-max) in $O\big((n/(s\log n) + \log s)D(n)\big)$ time using $O(s)$ variables, where $D(n)$ is the time to decide whether a given element currently resides in the heap (is alive).*[5]

Lemma 2.5 ([2]). *Let S be a set of n points. We can build a heap with all elements in S in $O(n)$ time that supports extract-min in $O\big(n/(s\log n) + \log(n)\big)$ time using $O(s)$ variables.*

Proof. The construction time is given in [2]. To decide in $O(1)$ time if some $x \in S$ is alive, we store the last extracted minimum m and test whether $x > m$.   □

We now present the complete algorithm. We first show how to subdivide S into mountains F_i and how to run the AK-algorithm on each $F_i^\uparrow$. Finally, we discuss the changes to run the AK-algorithm on each $F_i^\downarrow$. Sorted input is emulated by constructing two heaps H_1, H_2 for S according to x-order. By Lemma 2.5, each heap uses $O(s)$ space, can be constructed in $O(n)$ time, and supports extract-min in $O(n/(s\log n) + \log n)$ worst-case time. With H_1 we determine the size of the next mountain F_i, with H_2 we process the points of F_i.

We execute the convex hull algorithm with $\Theta(s)$ space until it reports the next convex hull edge pq. Throughout, the heaps H_1 and H_2 contain exactly the points to the right of p. To determine if pq is long (i.e., if there are points between p and q), we repeatedly extract the minimum of H_1 until q becomes the minimum element. Let k be the number of removed points.

If $k = 1$, then pq is short. We extract the minimum of H_2, and we continue with the convex hull algorithm. If $k \geq 2$, Lemma 2.2 shows that pq is the base of a mountain F that consists of all points between p and q. These are exactly the $k+1$ smallest elements in H_2 (including p and q). If $k \leq s$, we extract them from H_2, and we triangulate F in memory. If $k > s$, we execute the AK-algorithm on F using $O(s)$ variables. At the beginning of the ith round, we create a copy $H_{(i)}$ of H_2, i.e., we duplicate the $O(s)$ variables that determine the state of H_2. Further, we create an empty max-heap $H_{(ii)}$ using $O(s)$ variables to provide input for the second scan. To be able to reread a sub-block, we create a further copy $H'_{(i)}$ of H_2. Whenever the AK-algorithm requests the next point p in the first scan, we simply extract the minimum of $H_{(i)}$. When a sub-block is fully read, we use $H'_{(i)}$ to reread the elements and insert them into $H_{(ii)}$. Now, the

---

[5] The bounds in [2] do not include the factor $D(n)$ since the authors studied a setting similar to Lemma 2.5 where it takes $O(1)$ time to decide whether an element is alive.

rightmost element of all active sub-blocks corresponds exactly to the maximum of $H_{(ii)}$. One step in the second scan is equivalent to an extract-max on $H_{(ii)}$.

At the end of one round, we delete $H_{(i)}$, $H'_{(i)}$, and $H_{(ii)}$. The space can be reused in the next round. Once the AK-algorithm finishes, we repeatedly extract the minimum of H_2 until the minimum becomes q.

For each mountain F with $|F| > s$, the algorithm runs the AK-algorithm on $F^\uparrow$. To output a complete triangulation of F, we repeat the whole algorithm on S in reverse order, so that the AK-algorithm is run on each $F^\downarrow$.

Theorem 2.6. *We can report the edges of a triangulation of a set S of n points in time $O(n^2/s + n \log n \log s)$ using $O(s)$ additional variables.*

Proof. As before, correctness directly follows from Lemma 2.2 and the correctness of the AK-algorithm. The bound on the space usage is immediate.

Computing the convex hull now needs $O(n^2/(s \log n) + n \log n)$ time [13]. By Lemma 2.5, the heaps H_1 and H_2 can be constructed in $O(n)$ time. During execution, we perform n extract-min operations on each heap, requiring $O(n^2/(s \log n) + n \log n)$ time in total.

Let F_j be a mountain with n_j vertices that is discovered by the convex hull algorithm. If $n_j \leq s$, then F_j is triangulated in memory in $O(n_j)$ time, and the total time for such mountains is $O(n)$. If $n_j > s$, then the AK-algorithm runs in $O(n_j \log_s n_j)$ time. We must also account for providing the input for the algorithm. For this, consider some round $i \geq 1$. We copy H_2 to $H_{(i)}$ in $O(s)$ time. This time can be charged to the first scan, since $n_j > s$. Furthermore, we perform n_j extract-min operations on $H_{(i)}$. Hence the total time to provide input for the first scan is $O(n_j n/(s \log n) + n_j \log n)$.

For the second scan, we create another copy $H'_{(i)}$ of H_2. Again, the time for this can be charged to the scan. Also, we perform n_j extract-min operations on $H'_{(i)}$ which takes $O(n_j n/(s \log n) + n_j \log n)$ time. Additionally, we insert each fully-read block into $H_{(ii)}$. The main problem is to determine if an element in $H_{(ii)}$ is alive: there are at most $O(s)$ active sub-blocks. For each active sub-block b_i, we know the first element y_i and the element z_i that l_i points to. An element is alive if and only if it is in the interval $[y_i, z_i]$ for some active b_i. This can be checked in $O(\log s)$ time. Thus, by Lemma 2.4, each insert and extract-max on $H_{(ii)}$ takes $O((n/(s \log n) + \log s) \log s)$ time. Since each element is inserted once, the total time to provide input to the second scan is $O(n_j \log(s)(n/(s \log n) + \log s))$. This dominates the time for the first scan. There are $O(\log_s n_j)$ rounds, so we can triangulate F_j in time $O(n_j \log_s n_j + n_j \log(n_j)(n/(s \log n) + \log s))$. Summing over all F_j, the total time is $O(n^2/s + n \log n \log s)$. $\qquad\square$

3   Voronoi Diagrams

Given a planar n-point set S, we would like to find the vertices of VD(S). Let $K = \{p_1, p_2, p_3\}$ be a triangle with $S \subseteq \text{conv}(K)$ so that all vertices of VD(S) are vertices of VD($S \cup K$). We use random sampling to divide the problem of computing VD($S \cup K$) into $O(s)$ subproblems of size $O(n/s)$. First, we show how to take a random sample from S with small workspace. One of many possible approaches is the following deterministic one that ensures a worst-case guarantee:

Lemma 3.1. *We can sample a uniform random subset $R \subseteq S$ of size s in time $O(n + s \log s)$ and space $O(s)$.*

Proof. We sample a random sequence I of s distinct numbers from $\{1, \ldots, n\}$. This is done in s *rounds*. At the beginning of round k, for $k = 1, \ldots, s$, we have a sequence I of $k - 1$ numbers from $\{1, \ldots, n\}$. We store I in a binary search tree T. We maintain the invariant that each node in T with value in $\{1, \ldots, n-k+1\}$ stores a pointer to a unique number in $\{n - k + 2, \ldots, n\}$ that is not in I. In round k, we sample a random number x from $\{1, \ldots, n - k + 1\}$, and we check in T whether $x \in I$. If not, we add x to I. Otherwise, we add to I the number that x points to. Let y be the new element. We add y to T. Then we update the pointers: if $x = n - k + 1$, we do nothing. Now suppose $x < n - k + 1$. Then, if $n - k + 1 \notin I$, we put a pointer from x to $n - k + 1$. Otherwise, if $n - k + 1 \in I$, we let x point to the element that $n - k + 1$ points to. This keeps the invariant and takes $O(\log s)$ time and $O(s)$ space. We continue for s rounds. Any sequence of s distinct numbers in $\{1, \ldots, n\}$ is sampled with equal probability.

Finally, we scan through S to obtain the elements whose positions correspond to the numbers in I. This requires $O(n)$ time and $O(s)$ space. $\qquad \square$

We use Lemma 3.1 to find a random sample $R \subseteq S$ of size s. We compute $\mathrm{VD}(R \cup K)$, triangulate the bounded cells and construct a planar point location structure for the triangulation. This takes $O(s \log s)$ time and $O(s)$ space [17]. Given a vertex $v \in \mathrm{VD}(R \cup K)$, the *conflict circle* of v is the largest circle with center v and no point from $R \cup K$ in its interior. The *conflict set* B_v of v contains all points from S that lie in the conflict circle of v, and the *conflict size* b_v of v is $|B_v|$. We scan through S to find the conflict size b_v for each vertex $v \in \mathrm{VD}(R \cup K)$: every Voronoi vertex has a counter that is initially 0. For each $p \in S \setminus (R \cup K)$, we use the point location structure to find the triangle Δ of $\mathrm{VD}(R \cup K)$ that contains it. At least one vertex v of Δ is in conflict with p. Starting from v, we walk along the edges of $\mathrm{VD}(R \cup K)$ to find all Voronoi vertices in conflict with p. We increment the counters of all these vertices. This may take a long time in the worst case, so we impose an upper bound on the total work. For this, we choose a *threshold* M. When the sum of the conflict counters exceeds M, we start over with a new sample R. The total time for one attempt is $O(n \log s + M)$, and below we prove that for $M = \Theta(n)$ the success probability is at least $3/4$. Next, we pick another threshold T, and we compute for each vertex v of $\mathrm{VD}(R \cup K)$ the *excess* $t_v = b_v s/n$. The excess measures how far the vertex deviates from the desired conflict size n/s. We check if $\sum_{v \in \mathrm{VD}(R \cup K)} t_v \log t_v \leq T$. If not, we start over with a new sample. Below, we prove that for $T = \Theta(s)$, the success probability is at least $3/4$. The total success probability is $1/2$, and the expected number of attempts is 2. Thus, in expected time $O(n \log s + s \log s)$, we can find a sample $R \subseteq S$ with $\sum_{v \in \mathrm{VD}(R \cup K)} b_v = O(n)$ and $\sum_{v \in \mathrm{VD}(R \cup K)} t_v \log t_v = O(s)$.

We now analyze the success probabilities, using the classic Clarkson-Shor method [12]. We begin with a variant of the Chazelle-Friedman bound [11].

Lemma 3.2. *Let X be a planar point set of size o, and let $Y \subset \mathbb{R}^2$ with $|Y| \leq 3$. For fixed $p \in (0,1]$, let $R \subseteq X$ be a random subset of size po and let $R' \subseteq X$ be a random subset of size $p'o$, for $p' = p/2$. Suppose that $p'o \geq 4$. Fix $\mathbf{u} \in X^3$, and let $v_\mathbf{u}$ be the Voronoi vertex defined by $\mathbf{u}$. Let $b_\mathbf{u}$ be the number of points from X in the largest circle with center $v_\mathbf{u}$ and with no points from R in its interior. Then,*

$$\Pr[v_\mathbf{u} \in \mathrm{VD}(R \cup Y)] \leq 64 e^{-p b_\mathbf{u}/2} \Pr[v_\mathbf{u} \in \mathrm{VD}(R' \cup Y)].$$

Proof. Let $\sigma = \Pr[v_\mathbf{u} \in \mathrm{VD}(R \cup Y)]$ and $\sigma' = \Pr[v_\mathbf{u} \in \mathrm{DT}(R' \cup Y)]$. The vertex $v_\mathbf{u}$ is in $\mathrm{VD}(R \cup Y)$ precisely if $\mathbf{u} \subseteq R \cup Y$ and $B_\mathbf{u} \cap (R \cup Y) = \emptyset$, where $B_\mathbf{u}$ are the points from X in the conflict circle of $v_\mathbf{u}$. If $Y \cap B_\mathbf{u} \neq \emptyset$, then $\sigma = \sigma' = 0$, and the lemma holds. Thus, assume that $Y \cap B_\mathbf{u} = \emptyset$. Let $d_\mathbf{u} = |\mathbf{u} \setminus Y|$, the number of points in $\mathbf{u}$ not in Y. There are $\binom{o-b_\mathbf{u}-d_\mathbf{u}}{po-d_\mathbf{u}}$ ways to choose a po-subset from X that avoids all points in $B_\mathbf{u}$ and contains all points of $\mathbf{u} \cap X$, so

$$\sigma = \binom{o - b_\mathbf{u} - d_\mathbf{u}}{po - d_\mathbf{u}} \bigg/ \binom{o}{po} = \frac{\prod_{j=0}^{po-d_\mathbf{u}-1}(o - b_\mathbf{u} - d_\mathbf{u} - j)}{\prod_{j=0}^{po-d_\mathbf{u}-1}(po - d_\mathbf{u} - j)} \bigg/ \frac{\prod_{j=0}^{po-1}(o - j)}{\prod_{j=0}^{po-1}(po - j)}$$

$$= \prod_{j=0}^{d_\mathbf{u}-1} \frac{po - j}{o - j} \cdot \prod_{j=0}^{po-d_\mathbf{u}-1} \frac{o - b_\mathbf{u} - d_\mathbf{u} - j}{o - d_\mathbf{u} - j} \leq p^{d_\mathbf{u}} \prod_{j=0}^{po-d_\mathbf{u}-1} \left(1 - \frac{b_\mathbf{u}}{o - d_\mathbf{u} - j}\right).$$

Similarly, we get

$$\sigma' = \prod_{i=0}^{d_\mathbf{u}-1} \frac{p'o - i}{o - i} \prod_{j=0}^{p'o-d_\mathbf{u}-1} \left(1 - \frac{b_\mathbf{u}}{o - d_\mathbf{u} - j}\right),$$

and since $p'o \geq 4$ and $i \leq 2$, it follows that

$$\sigma' \geq \left(\frac{p'}{2}\right)^{d_\mathbf{u}} \prod_{j=0}^{p'o-d_\mathbf{u}-1} \left(1 - \frac{b_\mathbf{u}}{o - d_\mathbf{u} - j}\right).$$

Therefore, since $p' = p/2$,

$$\frac{\sigma}{\sigma'} \leq \left(\frac{2p}{p'}\right)^{d_\mathbf{u}} \prod_{j=p'o-d_\mathbf{u}}^{po-d_\mathbf{u}-1} \left(1 - \frac{b_\mathbf{u}}{o - d_\mathbf{u} - j}\right) \leq 64 \left(1 - \frac{b_\mathbf{u}}{o}\right)^{po/2} \leq 64 \, e^{pb_\mathbf{u}/2}.$$

$\square$

We can now bound the total expected conflict size.

Lemma 3.3. *We have* $\mathbf{E}\left[\sum_{v \in \mathrm{VD}(R \cup K)} b_v\right] = O(n).$

Proof. By expanding the expectation, we get

$$\mathbf{E}\left[\sum_{v \in \mathrm{VD}(R \cup K)} b_v\right] = \sum_{\mathbf{u} \in S^3} \Pr[v_{\mathbf{u}} \in \mathrm{VD}(R \cup K)] b_{\mathbf{u}},$$

$v_{\mathbf{u}}$ being the Voronoi vertex of $\mathbf{u}$ and $b_{\mathbf{u}}$ its conflict size. By Lemma 3.2 with $X = S$, $Y = K$ and $p = s/n$,

$$\leq \sum_{\mathbf{u} \in S^3} 64 e^{-p b_{\mathbf{u}}/2} \Pr[v_{\mathbf{u}} \in \mathrm{VD}(R' \cup K)] b_{\mathbf{u}},$$

where $R' \subseteq S$ is a sample of size $s/2$. We estimate

$$\leq \sum_{t=0}^{\infty} \sum_{\substack{\mathbf{u} \in S^3 \\ b_{\mathbf{u}} \in [\frac{t}{p}, \frac{t+1}{p})}} \frac{64 e^{-t/2}(t+1)}{p} \Pr[v_{\mathbf{u}} \in \mathrm{VD}(R' \cup K)]$$

$$\leq \frac{1}{p} \sum_{\mathbf{u} \in S^3} \Pr[v_{\mathbf{u}} \in \mathrm{VD}(R' \cup K)] \sum_{t=0}^{\infty} 64 e^{-t/2}(t+1)$$

$$= O(s/p) = O(n),$$

since $\sum_{\mathbf{u} \in S^3} \Pr[v_{\mathbf{u}} \in \mathrm{VD}(R' \cup K)] = O(s)$ is the size of $\mathrm{VD}(R' \cup K)$ and $\sum_{t=0}^{\infty} e^{-t/2}(t+1) = O(1)$. $\square$

By Lemma 3.3 and Markov's inequality, it follows that there is an $M = \Theta(n)$ with $\Pr[\sum_{v \in \mathrm{VD}(R \cup K)} b_v > M] \leq 1/4$. The proof for the excess is very similar to the previous calculation and can be found in the full version.

Lemma 3.4. $\mathbf{E}\left[\sum_{v \in \mathrm{VD}(R \cup K)} t_v \log t_v\right] = O(s)$.

By Markov's inequality and Lemma 3.4, we can conclude that there is a $T = \Theta(s)$ with $\Pr[\sum_{v \in \mathrm{VD}(R \cup K)} t_v \log t_v \geq T] \leq 1/4$. This finishes the first sampling phase. The next goal is to sample for each vertex v with $t_v \geq 2$ a random subset $R_v \subseteq B_v$ of size $\alpha t_v \log t_v$ for large enough $\alpha > 0$ (recall that B_v is the conflict set of v).

Lemma 3.5. *In total time $O(n \log s)$, we can sample for each vertex $v \in \mathrm{VD}(R \cup K)$ with $t_v \geq 2$ a random subset $R_v \subseteq B_v$ of size $\alpha t_v \log t_v$.*

Proof. First, we perform $O(s)$ rounds to sample for each vertex v with $t_v \geq 2$ a sequence I_v of $\alpha t_v \log t_v$ distinct numbers from $\{1, \ldots, b_v\}$. For this, we use the algorithm from Lemma 3.1 in parallel for each relevant vertex from $\mathrm{VD}(R \cup K)$. Since $\sum_v t_v \log t_v = O(s)$, this takes total time $O(s \log s)$ and total space $O(s)$.

After that, we scan through S. For each vertex v, we have a counter c_v, initialized to 0. For each $p \in S$, we find the conflict vertices of p, and for each conflict vertex v, we increment c_v. If c_v appears in the corresponding set I_v, we add p to R_v. The total running time is $O(n \log s)$, as we do one point location for each input point and the total conflict size is $O(n)$.   □

We next show that for a *fixed* vertex $v \in \mathrm{VD}(R \cup K)$, with constant probability, all vertices in $\mathrm{VD}(R_v)$ have conflict size n/s with respect to B_v.

Lemma 3.6. *Let $v \in \mathrm{VD}(R \cup K)$ with $t_v \geq 2$, and let $R_v \subseteq B_v$ be the sample for v. The expected number of vertices v' in $\mathrm{VD}(R_v)$ with at least n/s points from B_v in their conflict circle is at most $1/4$.*

Proof. Recall that $t_v = b_v s/n$. We have

$$\mathbf{E}\left[\sum_{\substack{v' \in \mathrm{VD}(R_v) \, b'_{v'} \geq n/s}} 1\right] = \sum_{\substack{u \in B_v^3 \\ b'_u \geq n/s}} \Pr[v'_{\mathbf{u}} \in \mathrm{VD}(R_v)],$$

where $b'_{\mathbf{u}}$ is the conflict size of $v'_{\mathbf{u}}$ with respect to B_v. Using Lemma 3.2 with $X = B_v$, $Y = \emptyset$, and $p = (\alpha t_v \log t_v)/b_v = \alpha(s/n) \log t_v$, this is $O(t_v^{-\alpha/2} t_v \log t_v) \leq 1/4$, for α large enough (remember that $t_v \geq 2$).   □

By Lemma 3.6 and Markov's inequality, the probability that all vertices from $\mathrm{VD}(R_v)$ have at most n/s points from B_v in their conflict circles is at least $3/4$. If so, we call v *good*. Scanning through S, we can identify the good vertices in time $O(n \log s)$ and space $O(s)$. Let s' be the size of $\mathrm{VD}(R \cup K)$. If we have less than $s'/2$ good vertices, we repeat the process. Since the expected number of good vertices is $3s'/4$, the probability that there are at least $s'/2$ good vertices is at least $1/2$ by Markov's inequality. Thus, in expectation, we need to perform the sampling twice. For the remaining vertices, we repeat the process, but now we take two samples per vertex, decreasing the failure probability to $1/4$. We repeat the process, taking in each round the maximum number of samples that fit into the work space. In general, if we have s'/a_i active vertices in round i, we can take a_i samples per vertex, resulting in a failure probability of 2^{-a_i}. Thus, the expected number of active vertices in round $i + 1$ is $s'/a_{i+1} = s'/(a_i 2^{a_i})$. After $O(\log^* s)$ rounds, all vertices are good. To summarize:

Lemma 3.7. *In total expected time $O(n \log s \log^* s)$ and space $O(s)$, we can find sets $R \subseteq S$ and $R_v \subset B_v$ for each vertex $v \in \mathrm{VD}(R' \cup K)$ such that (i) $|R| = s$: (ii) $\sum_{v \in \mathrm{VD}(R \cup K)} |R_v| = O(s)$; and (iii) for every R_v, all vertices of $\mathrm{VD}(R_v)$ have at most n/s points from B_v in their conflict circle.*

We set $R_2 = R \cup \bigcup_{v \in \mathrm{VD}(R \cup K)} R_v$. By Lemma 3.7, $|R_2| = O(s)$. We compute $\mathrm{VD}(R_2 \cup K)$ and triangulate its bounded cells. For a triangle Δ of the triangulation, let $r \in R_2 \cup K$ be the site whose cell contains Δ, and v_1, v_2, v_3 the vertices of Δ. We set $B_\Delta = \{r\} \cup \bigcup_{i=1}^{3} B_{v_i}$. Using the next lemma, we show that $|B_\Delta| = O(n/s)$. The proof is in the full version.

Lemma 3.8. *Let $S \subset \mathbb{R}^2$ and $\Delta = \{v_1, v_2, v_3\}$ a triangle in the triangulation of* VD(S). *Let $x \in \Delta$. Then any circle C with center x that contains no points from S is covered by the conflict circles of v_1, v_2 and v_3.*

Lemma 3.9. *Any triangle Δ in the triangulation of* VD($R_2 \cup K$) *has $|B_\Delta| = O(n/s)$.*

Proof. Let v be a vertex of Δ. We show that $b_v = O(n/s)$. Let $\Delta_R = \{v_1, v_2, v_3\}$ be the triangle in the triangulation of VD(R) that contains v. By Lemma 3.8, we have $B_v \subseteq \bigcup_{i=1}^{3} B_{v_i}$. We consider the intersections $B_v \cap B_{v_i}$, for $i = 1, 2, 3$. If $t_{v_i} < 2$, then $b_{v_i} = O(n/s)$ and $|B_v \cap B_{v_i}| = O(n/s)$. Otherwise, we have sampled a set R_{v_i} for v_i. Let $\Delta_i = \{w_1, w_2, w_3\}$ be the triangle in the triangulation of VD(R_{v_i}) that contains v. Again, by Lemma 3.8, we have $B_v \subseteq \bigcup_{j=1}^{3} B_{w_j}$ and thus also $B_v \cap B_{v_i} \subseteq \bigcup_{j=1}^{3} B_{w_j} \cap B_{v_i}$. However, by construction of R_{v_i}, $|B_{w_j} \cap B_{v_i}|$ is at most n/s for $j = 1, 2, 3$. Hence, $|B_v \cap B_{v_i}| = O(n/s)$ and $b_v = O(n/s)$. $\quad\square$

The following lemma enables us to compute the Voronoi diagram of $R_2 \cup K$ locally for each triangle Δ in the triangulation of VD($R_2 \cup K$) by only considering sites in B_Δ. It is a direct consequence of Lemma 3.8.

Lemma 3.10. *For every triangle Δ in the triangulation of* VD($R_2 \cup K$), *we have* VD($S \cup K$) $\cap \Delta =$ VD(B_Δ) $\cap \Delta$.

Theorem 3.11. *Let S be a planar n-point set. In expected time $O((n^2/s) \log s + n \log s \log^* s)$ and space $O(s)$, we can compute all Voronoi vertices of S.*

Proof. We compute a set R_2 as above. This takes $O(n \log s \log^* s)$ time and space $O(s)$. We triangulate the bounded cells of VD($R_2 \cup K$) and compute a point location structure for the result. Since there are $O(s)$ triangles, we can store the resulting triangulation in the workspace. Now, the goal is to compute simultaneously for all triangles Δ the Voronoi diagram VD(B_Δ) and to output all Voronoi vertices that lie in Δ and are defined by points from S. By Lemma 3.10, this gives all Voronoi vertices of VD(S).

Given a planar m-point set X, the algorithm by Asano et al. finds all vertices of VD(X) in $O(m)$ scans over the input, with constant workspace [4]. We can perform a simultaneous scan for all sets B_Δ by determining for each point in S all sets B_Δ that contain it. This takes total time $O(n \log s)$, since we need one point location for each $p \in S$ and since the total size of the B_Δ's is $O(n)$. We need $O(\max_\Delta |B_\Delta|) = O(n/s)$ such scans, so the second part of the algorithm needs $O((n^2/s) \log s)$ time. $\quad\square$

As mentioned in the introduction, Theorem 3.11 also lets us report all edges of the Delaunay triangulation of S in the same time bound: by duality, the three sites that define a vertex of VD(S) also define a triangle for the Delaunay triangulation. Thus, whenever we discover a vertex of VD(S), we can instead output the corresponding Delaunay edges, while using a consistent tie-breaking rule to make sure that every edge is reported only once.

Acknowledgments. This work began while W. Mulzer, P. Seiferth, and Y. Stein visited the Tokuyama Laboratory at Tohoku University. We would like to thank Takeshi Tokuyama and all members of the lab for their hospitality and for creating a conducive and stimulating research environment.

References

1. Asano, T., Buchin, K., Buchin, M., Korman, M., Mulzer, W., Rote, G., Schulz, A.: Memory-constrained algorithms for simple polygons. Comput. Geom. **46**(8), 959–969 (2013)
2. Asano, T., Elmasry, A., Katajainen, J.: Priority queues and sorting for read-only data. In: Chan, T.-H.H., Lau, L.C., Trevisan, L. (eds.) TAMC 2013. LNCS, vol. 7876, pp. 32–41. Springer, Heidelberg (2013)
3. Asano, T., Kirkpatrick, D.: Time-space tradeoffs for all-nearest-larger-neighbors problems. In: Dehne, F., Solis-Oba, R., Sack, J.-R. (eds.) WADS 2013. LNCS, vol. 8037, pp. 61–72. Springer, Heidelberg (2013)
4. Asano, T., Mulzer, W., Rote, G., Wang, Y.: Constant-work-space algorithms for geometric problems. J. of Comput. Geom. **2**(1), 46–68 (2011)
5. Barba, L., Korman, M., Langerman, S., Sadakane, K., Silveira, R.: Space-time trade-offs for stack-based algorithms. Algorithmica 1–33 (2014)
6. Barba, L., Korman, M., Langerman, S., Silveira, R.I.: Computing the visibility polygon using few variables. Comput. Geom. **47**(9), 918–926 (2013)
7. de Berg, M., Cheong, O., van Kreveld, M., Overmars, M.: Computational geometry: Algorithms and applications, third edition. Springer (2008)
8. Borodin, A., Cook, S.: A time-space tradeoff for sorting on a general sequential model of computation. SIAM J. Comput. **11**, 287–297 (1982)
9. Brönnimann, H., Chan, T.M., Chen, E.Y.: Towards in-place geometric algorithms and data structures. In: Proc. 20th Annu. ACM Sympos. Comput. Geom. (SoCG), pp. 239–246 (2004)
10. Chan, T.M., Chen, E.Y.: Multi-pass geometric algorithms. Discrete Comput. Geom. **37**(1), 79–102 (2007)
11. Chazelle, B., Friedman, J.: A deterministic view of random sampling and its use in geometry. Combinatorica **10**(3), 229–249 (1990)
12. Clarkson, K.L., Shor, P.W.: Applications of random sampling in computational geometry, II. Discrete Comput. Geom. **4**, 387–421 (1989)
13. Darwish, O., Elmasry, A.: Optimal time-space tradeoff for the 2D convex-hull problem. In: Schulz, A.S., Wagner, D. (eds.) ESA 2014. LNCS, vol. 8737, pp. 284–295. Springer, Heidelberg (2014)
14. Fournier, A., Montuno, D.Y.: Triangulating simple polygons and equivalent problems. ACM Transactions on Graphics **3**, 153–174 (1984)
15. Har-Peled, S.: Shortest path in a polygon using sublinear space. To appear in: SoCG 2015
16. Jarvis, R.: On the identification of the convex hull of a finite set of points in the plane. Inform. Process. Lett. **2**(1), 18–21 (1973)
17. Kirkpatrick, D.: Optimal search in planar subdivisions. SIAM J. Comput. **12**(1), 28–35 (1983)
18. Munro, J.I., Paterson, M.: Selection and sorting with limited storage. Theoret. Comput. Sci. **12**, 315–323 (1980)
19. Munro, J.I., Raman, V.: Selection from read-only memory and sorting with minimum data movement. Theoret. Comput. Sci. **165**(2), 311–323 (1996)

20. Pagter, J., Rauhe, T.: Optimal time-space trade-offs for sorting. In: Proc. 39th Annu. IEEE Sympos. Found. Comput. Sci. (FOCS), pp. 264–268 (1998)
21. Pohl, I.: A minimum storage algorithm for computing the median. Technical Report RC2701, IBM (1969)
22. Savage, J.E.: Models of computation–exploring the power of computing. Addison-Wesley (1998)

A $2k$-vertex Kernel
for Maximum Internal Spanning Tree

Wenjun Li[1], Jianxin Wang[1], Jianer Chen[1,2], and Yixin Cao[3]([⊠])

[1] School of Information Science and Engineering,
Central South University, Changsha, China
[2] Department of Computer Science and Engineering,
Texas A&M University, College Station, TX, USA
[3] Department of Computing, Hong Kong Polytechnic University, Hong Kong, China
yixin.cao@polyu.edu.hk

Abstract. We consider the parameterized version of the maximum internal spanning tree problem: given an n-vertex graph and a parameter k, does the graph have a spanning tree with at least k internal vertices? Fomin et al. [J. Comput. System Sci., 79:1–6] crafted a very ingenious reduction rule, and showed that a simple application of this rule is sufficient to yield a $3k$-vertex kernel for this problem. Here we propose a novel way to use the same reduction rule, resulting in an improved $2k$-vertex kernel. Our algorithm applies first a greedy procedure consisting of a sequence of local exchange operations, which ends with a local-optimal spanning tree, and then uses this special tree to find a reducible structure. As a corollary of our kernel, we obtain a $4^k \cdot n^{O(1)}$-time deterministic algorithm, improving all previous algorithms for the problem.

Keywords: Parameterized computation · Kernelization algorithms · Local-search

1 Introduction

A *spanning tree* of a connected graph G is a minimal connected subgraph of G including all its vertices. Spanning tree is a fundamental concept in graph theory, and finding a spanning tree of the input graph is a routine step of most graph algorithms, though it usually induces no extra cost: most algorithms start from exploring the input graph anyway, and both breadth- and depth-first-search procedures produce a spanning tree as a byproduct. However, a graph can have an exponential number of spanning trees, of which some might suit a specific application better than others. We are hence asked to find spanning trees that minimize or maximize certain objective functions. The most classic example is the minimum-weight spanning tree problem (in weighted graphs), which has an equivalent but less known formulation, i.e., maximum-weight spanning tree.

Supported by the National Natural Science Foundation of China under grants 61232001, 61472449, and 61420106009.

© Springer International Publishing Switzerland 2015
F. Dehne et al. (Eds.): WADS 2015, LNCS 9214, pp. 495–505, 2015.
DOI: 10.1007/978-3-319-21840-3_41

Other spanning tree problems that have received wide attention include minimum diameter spanning tree [7], degree constrained spanning tree [10, 11], maximum leaf spanning tree [14], and maximum internal spanning tree [20]. Unlike the minimum-weight spanning tree problem [8], most of these constrained versions are NP-hard [16].

The optimization objective we consider here is to maximize the number of internal vertices (i.e., non-leaf vertices) of the spanning tree, or equivalently, to minimize the number of its leaves. More formally, the *maximum internal spanning tree* problem asks whether a given graph G has a spanning tree with at least k internal vertices. Containing the Hamiltonian path problem as a special case $(k = n - 2)$, it is clearly NP-hard. This paper is focused on its parameterized version; here the parameter is k, and hence we use the name k-*internal spanning tree*. Given an instance (G, k) of the k-internal spanning tree problem, a *kernelization algorithm* produces in polynomial time an "equivalent" instance (G', k') such that $k' \leq k$ and that the *kernel size* (i.e., the number of vertices in G') is upper bounded by some function of k'. Prieto and Sloper [17] presented an $O(k^3)$-vertex kernel for the problem, and improved it to $O(k^2)$ in its journal version [18]. Fomin et al. [4] crafted a very ingenious reduction rule, and showed that a simple application of this rule is sufficient to yield a $3k$-vertex kernel. Answering a question asked by Fomin et al. [4], we further improve the kernel size to $2k$.

Theorem 1. *The k-internal spanning tree problem has a $2k$-vertex kernel.*

We obtain this improved result by revisiting the reduction rule proposed by Fomin et al. [4]. A nonempty independent set X (i.e., a subset of vertices that are pairwise nonadjacent in G) as well as its neighborhood are called a *reducible structure* if $|X|$ is at least twice as the cardinality of its neighborhood. The observation in [4] is that the leaves of a depth-first-search tree T are independent. Therefore, if the graph has more than $3k - 3$ vertices, then either the problem has been solved (when T has k or more internal vertices), or the set of (at least $2k - 2$) leaves of T will be the required independent set. It is, however, very nontrivial to find a reducible structure when $2k < n < 3k - 3$, and this will be the focus of this paper. We first preprocess the tree T using a greedy procedure that applies a sequence of local-exchange operations to increase the number of its internal vertices. After a local-optimal spanning tree is obtained, we show that if it has more leaves than internal vertices, then a subset of its leaves and its neighborhood make the reducible structure. We apply the reduction rule of [4] to reduce it and then repeat the process, which terminates on either a $2k$-vertex kernel or a solution. Indeed, we are proving a stronger statement that implies Theorem 1 as a corollary.

Theorem 2. *Given an n-vertex graph G, we can find in polynomial time either a spanning tree of G with at least $n/2$ internal vertices, or a reducible structure.*

It is interesting to point out that our kernelization algorithm never directly ends with a NO situation, which is common in kernelization algorithms in literature. Our algorithm either returns a trivial YES instance, or continuously reduces the graph until it has a spanning tree with at least half its vertices are internal. This also means that our kernelization algorithm does not rely on the parameter k.

Priesto and Sloper [18] also initiated the study of parameterized algorithms (i.e., algorithms running in time $O^*(f(k))$ for some function f independent of n)[1] for k-internal spanning tree, which have undergone a sequence of improvement. Closely related here is the k-*internal out-branching* problem, which, given a *directed* graph G and an integer k, asks if G has an out-branching (i.e., a spanning tree having exactly one vertex of in-degree 0) with at least k vertices of positive out-degrees. Any $O^*(f(k))$-time algorithm for k-internal out-branching can solve k-internal spanning tree in the same time—replacing every edge by two arcs of opposite directions, calling the algorithm for k-internal out-branching, and then dropping the directions from the obtained out-branching,—but not necessarily the other way. After a successive sequence of studies [2,3,5,6,22], the current best deterministic and randomized parameterized algorithms for k-internal out-branching run in time $O^*(6.86^k)$ and $O^*(4^k)$ respectively, which are also the best known for k-internal spanning tree. Table 1 summarizes the history of this line of research.

Table 1. Known parameterized algorithms for k-internal out-branching and k-internal spanning tree (note that an algorithm for the former applies to the later as well)

Problem	Running time	Reference	Remark
k-internal out-branching	$O^*(k^{O(k)})$	Gutin et al. [6]	
	$O^*(55.8^k)$	Cohen et al. [2]	
	$O^*(16^{k+o(k)})$	Fomin et al. [5]	
	$O^*(6.86^k)$	Shachnai and Zehavi [22]	
	$O^*(4^k)$	Daligault and Kim [3]	*randomized*
k-internal spanning tree	$O^*(k^{2.5k})$	Priesto and Sloper [18]	
	$O^*(2.14^k)$	Binkele-Raible et al. [1]	*cubic graphs*
	$O^*(8^k)$	Fomin et al. [4]	
	$O^*(4^k)$	This paper	

The $O^*(4^k)$-time *randomized* algorithm for k-internal out-branching [3, Theorem 180] was obtained using an algebraic technique developed by Koutis and Williams [12], which, however, is very unlikely to be derandomized. As a corollary of Theorem 1, we obtain an $O^*(4^k)$-time deterministic algorithm for k-internal spanning tree,—it suffices to apply the $O^*(2^n)$-time algorithm of Nederlof [15] to the $2k$-vertex kernel produced by Theorem 1,—matching the running time of the best-known randomized algorithm.

---

[1] Following convention, we use the $O^*(f(k))$ notation to suppress the polynomial factor in the running time $O(f(k) \cdot n^{O(1)})$.

Theorem 3. *The k-internal spanning tree problem can be solved in time $O^*(4^k)$.*

It remains an open problem to develop a deterministic $O^*(4^k)$-time algorithm for the k-internal out-branching problem. Note that the minimum spanning tree problem has been long known to be solvable in randomized linear time [8], while a deterministic linear-time algorithm is still elusive. As a final remark, there is also a line of research devoted to developing approximation algorithms for maximum internal spanning tree [9,18,19,21]. In a companion paper [13], we have used a similar but deeper (it needs depth-5 in contrast to depth-3 used in the present paper) local-search procedure to improve the approximation ratio to 1.5.

2   Local-Optimal Spanning Trees

All graphs discussed in this paper shall always be undirected and simple, and the input graph is assumed to be connected and contain at least two vertices. The vertex set and edge set of a graph G are denoted by $V(G)$ and $E(G)$ respectively. For a vertex $v \in V(G)$, let $N_G(v)$ denote the neighborhood of v in G, and let $d_G(v) := |N_G(v)|$ be its degree in G. The neighborhood of a subset $U \subseteq V(G)$ of vertices is defined to be $N_G(U) = \bigcup_{v \in U} N_G(v) \backslash U$. A tree T is a *spanning tree* of a graph G if $V(T) = V(G)$ and $E(T) \subseteq E(G)$. A vertex $u \in V(T)$ is a *leaf* of T if $d_T(u) = 1$, and an *internal vertex* of T otherwise. Let $L(T)$ and $I(T)$ denoted the set of leaves and the set of internal vertices of T respectively. We further divide $I(T)$ into two subsets $I_2(T) := \{v \in V(T) : d_T(v) = 2\}$ and $I_3(T) := \{v \in V(T) : d_T(v) \geq 3\}$. Hence, the three vertex sets $L(T), I_2(T)$, and $I_3(T)$ partition $V(T)$. For any pair of vertices $u, v \in V(T)$, denote by $P_T(u, v)$ the unique path in T from u to v.

Since $|I(T)| = |V(T)| - |L(T)|$, to maximize it is equivalent to minimizing the number of leaves. Also connecting leaves and internal vertices, especially $I_3(T)$, of a tree T is the following elementary fact:

$$|L(T)| - 2 = \sum_{v \in I(T)} (d_T(v) - 2) = \sum_{v \in I_3(T)} (d_T(v) - 2).$$

Therefore, informally speaking, we need to decrease the number and degrees of vertices in $I_3(T)$. This is achieved by swapping edges in T and out of T. As a matter of fact, we can transform a spanning tree T to any other spanning tree T' of the same graph by exchanging $|V(T)| - 1 - |E(T) \cap E(T')|$ $(< n)$ edges.

Let T be a spanning tree of graph G. An edge $uv \in E(G) \backslash E(T)$ is called a *cotree edge* of T. Note that for a cotree edge uv, the length of $P_T(u, v)$ is at least two. We try to turn leaves of T into internal vertices by swapping a cotree edge and a tree edge, and hence we will be only concerned with cotree edges that are incident to leaves of T. On the other hand, the tree edge to be removed must be selected in a way that the resulting subgraph remains a spanning tree of G. Let ℓw be a cotree edge with $\ell \in L(T)$, and let uv be an edge on the path $P_T(\ell, w)$. An *edge swapping* with ℓw and uv on T is to add ℓw to T and remove

uv, denoted by $uv \to \ell w$. Edge swappings can be applied successively. A set of edge swapping(s) is *improving*, *weakening*, and *holding*, respectively, if their applications in turn increases, decreases, and maintains the number of internal vertices in the spanning tree. Note that the cotree edge used in the second or later swappings may or may not be incident to a leaf of the original tree.

A spanning tree T of a graph G is *local-optimal* if every set of three or less edge swappings is not improving. We can in polynomial time check whether a spanning tree is local-optimal and find an improving set of edge swappings if it is not. Since a spanning tree of G has at most $|V(G)| - 2$ internal vertices, it follows that a local-optimal spanning tree of a graph can be constructed in polynomial time.

We now study the structural properties of a local-optimal spanning tree T. The problem has been solved if T is already a path. Instead of polluting every statement to follow, we will tacitly assume that T is *not* a path throughout. As a result, for any pair of leaves ℓ_1, ℓ_2 of T, the path $P_T(\ell_1, \ell_2)$ necessarily visits some vertex $v \in I_3(T)$; let u be a neighbor of v in the path (chosen arbitrarily). Suppose $\ell_1 \ell_2 \in E(G)$, then the edge swapping $uv \to \ell_1 \ell_2$ will increase the internal vertices by either one or two (depending on $d_T(u)$). This is not possible as T is local-optimal.

Proposition 1. *Let ℓ_1, ℓ_2 be any pair of leaves of a local-optimal spanning tree T. The path $P_T(\ell_1, \ell_2)$ visits some vertex in $I_3(T)$, and $\ell_1 \ell_2 \notin E(G)$.*

Proposition 1 is straightforward: indeed, a depth-first-search tree always has this property. An edge swapping is *trivial* if both ends of the cotree edge are leaves. What concern us more are the other edge swappings, where the cotree edge has one end in $L(T)$ and the other in $I(T)$.

Definition 1. *A cotree edge of T is* good *if it connects a leaf ℓ and an internal vertex w of T. We say that ℓw crosses every edge in the path $P_T(\ell, w)$.*

For notational convenience, when referring to a good cotree edge ℓw, we always put the leaf ℓ first, and when referring to an edge uv crossed by ℓw, we always put the vertex closer to ℓ in T first; hence, $P_T(\ell, w)$ can be written as $\ell \cdots uv \cdots w$. We would like to point out that the same edge uv can be crossed by two different good cotree edges and they may be referred to by different orders.

Now consider the edge swapping $uv \to \ell w$. The following three types are the most fundamental ones and hence called *basic*; here possibly $v = w$ but u must be different from ℓ. Three or four vertices are involved in this swapping: ℓ and w are always internal vertices of the new tree (independent of whether $v = w$ or not), and hence our focus is on u and v (when $v \neq w$), whose degrees decrease by one.

(A) The simplest case is when the edge swapping $uv \to \ell w$ itself is improving. It turns the only leaf ℓ into an internal vertex while keeping u, v internal. In particular, $u \in I_3(T)$ and $v \in I_3(T) \cup \{w\}$.

(B) The edge swapping $uv \to \ell w$ does not change the number of internal vertices if only one of the two conditions in (A) holds true. In this case, either (a)

$u \in I_3(T)$ and $v \in I_2(T) \setminus \{w\}$; or (b) $u \in I_2(T)$ and $v \in I_3(T) \cup \{w\}$. The vertex v in case (a) or u in case (b) is in $I_2(T)$, and was turned into a leaf by $uv \rightarrow \ell w$. If v is adjacent to some leaf $\ell' \in L(T) \setminus \{\ell\}$ in T, then a trivial edge swapping can be applied subsequently. Therefore, we have an improving set of two edge swappings.

(C) With little effort, we can push this one step further. Now that neither of two conditions in (A) holds true, both u and v must be in $I_2(T) \setminus \{w\}$. They are both turned into leaves by the edge swapping $uv \rightarrow \ell w$, which decreases the number of internal vertices. However, if there are two different leaves ℓ_u and ℓ_v such that $\ell_u u, \ell_v v \in E(G)$ and $\ell_u \neq \ell$,—we may assume that $\ell_v \neq \ell$, as otherwise the edge swapping $\ell v \rightarrow uv$ is already in type (B),—then we can subsequently apply trivial edge swappings to $\ell_u u$ and/or[2] $\ell_v v$ to increase the number of internal vertices by at least two. We have thus an improving set of two or three edge swappings.

By definition, none of such structures can be found in a local-optimal spanning tree T. To characterize vertices participating in these basic edge swappings, we need the following technical definition.

Definition 2. *All vertices in $I_3(T)$ are* detachable. *A vertex $w \in I_2(T)$ is* detachable *if there exists a good cotree edge ℓw of T such that the path $P_T(\ell, w)$ contains an inner vertex v that is*

(1) in $I_3(T)$, or
(2) incident to a good cotree edge $\ell' v$ of T with $\ell' \neq \ell$.

A tree edge connecting a detachable vertex and an internal vertex of T is critical *(in T) if it is not crossed by any good cotree edge in either direction.*

Let $D(T)$ denote the set of detachable vertices of T, and let $C(T)$ denote the (possibly empty) set of critical edges in T. We remark that in Definition 2, if condition (2) is satisfied, then the vertex v is detachable as well: indeed, if v itself is not in $I_3(T)$, then the path $P_T(\ell', v)$ must visit $I_3(T)$. We can find in polynomial time all detachable vertices and all critical edges.

In all the three kinds of basic edge swappings, the tree edge uv is not critical and both u, v are detachable. But these two conditions are not sufficient to ensure the application of a basic edge swapping: the definition of a detachable vertex does not specify anything about its neighbors in T, which, however, is crucial. In other words, a local-optimal spanning tree might still contain non-critical edges connecting detachable vertices, of which at least one is in $I_2(T)$. This situation is characterize by the following lemma. Recall that the removal of $uv \in E(T)$ from a tree T breaks it into two components, one containing u and the other containing v. In general, the removal of all edges of an edge subset $E' \subseteq E(T)$ from T breaks it into $|E'| + 1$ components, each being a subtree of T.

---

[2] One should be noted that in the very special case when $|L(T)| = 3$, which becomes 4 after $uv \rightarrow \ell w$, the first trivial edge swapping (on $\ell_u u$) might decrease the number of leaves by two, thereby leaving a Hamiltonian path and the second trivial edge swapping is no longer applicable. But this is irrelevant for our discussion.

Lemma 1. *Let u, v be two detachable vertices of a local-optimal spanning tree T. If $uv \in E(T) \setminus C(T)$, then there exist a leaf ℓ, an internal vertex w, and $x \in \{u, v\}$ such that*

(1) $x \in I_2(T)$;
(2) $\ell w \in E(G) \setminus E(T)$, and ℓ is in the component of $T - uv$ with x but w is not; and
(3) ℓx is the only good cotree edge of T incident to x.

Proof. Since u, v are detachable and uv is not critical, there must be a good cotree edge e of T connecting the two components of $T - uv$. At most one of u, v can be in $I_3(T)$, as otherwise $uv \to e$ is an improving edge swapping. Assume first $|\{u, v\} \cap I_3(T)| = 1$; we consider $v \in I_3(T)$ and the case $u \in I_3(T)$ follows by symmetry. Then $u \in I_2(T)$, and by definition, there is a good cotree edge ℓu of T. The leaf ℓ must be in the component of $T - uv$ with u; otherwise, $vu \to \ell u$ is a basic edge swapping (A). Since $vu \to e$ is not a basic edge swapping (A, B), e crosses uv. On the other hand, since $uv \to e$ is not a basic edge swapping (B), e has to be incident to ℓ and ℓu is the only cotree edge incident to u. In other words, $\ell w = e$ and $x = u$ verify the claim.

In the following both $u, v \in I_2(T)$, each incident to some good cotree edge of T. If there is a unique leaf ℓ such that ℓu and ℓv are the only good cotree edges of T incident to u and v respectively, then $\{u, v\} = \{x, w\}$ depending on which component of $T - uv$ the leaf ℓ is in. Otherwise, let $\ell_u u, \ell_v v \in E(G) \setminus E(T)$ with $\ell_u \neq \ell_v$. The following hold true because T is local-optimal.

- The leaves ℓ_u and ℓ_v must be in the component of $T - uv$ with u and v respectively. Suppose for contradiction that ℓ_u is in the component of $T - uv$ with v, then the edge swapping $vu \to \ell_u u$ is basic (B). A symmetric arguments applies when ℓ_v is in the component of $T - uv$ with u.
- Every good cotree edge of T crossing uv is incident to ℓ_u and every good cotree edge of T crossing vu is incident to ℓ_v. Suppose that there is an edge $e' \in E(G) \setminus E(T)$ crossing uv and non-incident to ℓ_u, then $uv \to e'$ is a basic edge swapping (C).

If uv is crossed by a good cotree edge $\ell_u w$, then $\ell = \ell_u$ and $x = u$; otherwise $\ell = \ell_v$ and $x = v$. The proof is now complete. $\qquad\square$

We use $D_B(T)$ to denote this set of vertices x stipulated in Lemma 1, and define $D_G(T) := D(T) \setminus D_B(T)$. Note that $D_B(T) \subseteq I_2(T)$ and $I_3(T) \subseteq D_G(T)$.

Proposition 2. *Let T be a local-optimal spanning tree. If $u \in D_B(T)$, then $|L(T) \cap N_G(u)| = 1$.*

Proof. Let ℓw be the good cotree edge specified in Lemma 1, then ℓ is the component of $T - uv$ containing u and ℓu is a good cotree edge of T. Noting that $d_T(u) = 2$, let u' be the other neighbor of u (different from v). Since $u'u \in E(T)$ and $\ell u \notin E(T)$, it follows $\ell \neq u'$ and $N_T(u) \cap L(T) = \emptyset$. Therefore, the statement follows from Lemma 1(3). $\qquad\square$

Note that in a local-optimal spanning tree T, vertices of $D_G(T)$ cannot be adjacent in $T - C(T)$. We now strengthen this result by characterizing paths connecting vertices of $D_G(T)$ in $T - C(T)$.

Lemma 2. *Let T be a local-optimal spanning tree. For every pair of vertices $u, w \in D_G(T)$ that are in the same component of $T - C(T)$, the path $P_T(u, w)$ visits at least one vertex $v \in I(T) \setminus D(T)$. Moreover, for any $\ell \in L(T)$, the path $P_T(\ell, v)$ visits $D(T)$.*

Proof. Let $P_T(u, w) = uv_1 \cdots v_p w$; noting that $uw \notin C(T)$, by Lemma 1, $uw \notin E(T)$, and hence $p \geq 1$. No generality will be lost by assuming that $P_T(u, w)$ is minimal (in the sense that it visits no other vertex in $D_G(T)$); hence $v_i \in D_B(T)$ and $d_T(v_i) = 2$ for each $1 \leq i \leq p$. Note that $P_T(u, w)$ is retained in $T - C(T)$.

We find first an inner vertex v_i of $P_T(u, w)$ that is not in $D(T)$ as follows. If $v_1 \notin D(T)$, then $i = 1$ and we are done. We proceed only when $v_1 \in D_B(T)$, and then there is a unique good cotree edge $\ell_1 v_1$. We prove by contradiction that ℓ_1 is in the same component of $T - uv_1$ with v_1. Suppose the contrary, then

- if $u \in I_3(T)$, then $uv_1 \rightarrow \ell_1 v_1$ is a basic edge swapping (A); or
- if $u \in I_2(T) \cap D_G(T)$, then there is a good cotree edge $\ell'u$ with $\ell' \neq \ell_1$, and hence $uv_1 \rightarrow \ell_1 v_1$ is a basic edge swapping (B).

Noting that every inner vertex of $P_T(u, w)$ has degree 2, this actually implies that ℓ_1 must be in the same component of $T - v_p w$ with w. With a symmetric argument, we can either find a leaf ℓ_p in the same component of $T - uv_1$ with u such that $\ell_p v_p$ is the only good cotree edge incident to v_p, or conclude that $v_p \notin D(T)$. In either case, $\ell_1 v_p \notin E(G)$ and $p \geq 2$. Let i be the smallest such that $\ell_1 v_i \notin E(G)$; clearly, $2 \leq i \leq p$, and its existence is ensured by $\ell_1 v_p \notin E(G)$. If there exists a good cotree edge $\ell' v_i$ of T, then $\ell' \neq \ell_1$ (possibly $\ell' = \ell_p$) and $v_i v_{i-1} \rightarrow \ell_1 v_{i-1}$ is a basic edge swapping (B). This contradiction implies that there exists no such a good cotree edge of T. Noting that $v_i \in I_2(T)$, we have verified that v_i is an inner vertex of $P_T(u, w)$ not in $D(T)$.

For any $\ell \in L(T)$, the path $P_T(\ell, v)$ necessarily visits either u or w, which is in $D(T)$. This concludes the proof. □

3   The Kernelization Algorithm

We use the reduction rule of Fomin et al. [4], which is recalled below. Let $\mathsf{opt}(G)$ denote the maximum number of internal vertices a spanning tree of G can have.

Lemma 3 ([4]). *Let L' be an independent set of G satisfying $|L'| \geq 2|N_G(L')|$. We can find in polynomial time nonempty subsets $S \subseteq N_G(L')$ and $L \subseteq L'$ such that:*

(1) $N_G(L) = S$, and
(2) the graph $(S \cup L, E(G) \cap (S \times L))$ has a spanning tree such that all vertices of S and $|S| - 1$ vertices of L are internal.

Moreover, let G' be obtained from G by adding a vertex v_S adjacent to every vertex in $N_G(S) \backslash L$, adding a vertex v_L adjacent to v_S, and removing all vertices of $S \cup L$, then $\mathsf{opt}(G') = \mathsf{opt}(G) - 2|S| + 2$.

Reduction Rule ([4]). Find nonempty subsets S and L of vertices as in Lemma 3. Return (G', k') where G' is defined in Lemma 3 and $k' = k - 2|S| + 2$.

The safeness of the reduction rule is ensured by Lemma 3. Note that $|L| \geq 2$ (otherwise the graph mentioned in Lemma 3(2) cannot have internal vertices), and hence each application of the reduction rule decreases the number of vertices by at least 1.

The main technical obstacle is then to identify a vertex set L' with $|L'| \geq 2|N_G(L')|$. This is trivial when $|V(G)| \geq 3k - 3$. In any spanning tree T of G with $|I(T)| < k$ it holds that $|L(T)| \geq 2k - 2 \geq 2|I(T)| \geq 2|N_G(L(T))|$. Hence, we can use $L(T)$ as L' and a $3k$-vertex kernel follows. However, it becomes very nontrivial to find such a set when $2k < |V(G)| < 3k - 3$. Our approach here is to separate a local-optimal spanning tree T into several subtrees (by removing all critical edges) and bound the number of $L(T)$ by the number of $I(T)$ residing in each subtree individually. It is worth mentioning that a leaf of a subtree may not be a leaf of T. We are now ready for proving the main result of the paper. Recall that the two ends of any good cotree edge always reside in the same subtree.

Lemma 4. *Let T be a local-optimal spanning tree of G with $|V(G)| \geq 4$. If $|L(T)| > |I(T)|$, then we can find in polynomial time an independent set L' of G such that $|L'| \geq 2|N_G(L')|$.*

Proof. We find all critical edges $C(T)$, and take the forest $T - C(T)$. By assumption, there must be some component T_0 of $T - C(T)$ of which more than half vertices are from $L(T)$. Let X and Y denote $L(T) \cap V(T_0)$ and $I(T) \cap V(T_0)$ respectively; then $|X| \geq |Y| + 1$. By Proposition 1, X is an independent set. We divide X into the following three subsets:

$$X_1 := X \cap N_G(D_B(T)); \quad X_2 := X \cap N_T(D(T)) \backslash X_1; \quad \text{and } X_3 := X \backslash (X_1 \cup X_2).$$

We will show that $|X_2| \geq 2|N_G(X_2)|$, and hence X_2 satisfies the claimed condition and can be used as L'. We accordingly divide Y into subsets. The detachable vertices are either in $Y_1 := D_B(T) \cap V(T_0)$ or $Y_2 := (D(T) \backslash D_B(T)) \cap V(T_0)$, while a vertex $y \in Y \backslash D(T)$ is in Y_3 if there exists $\ell \in L(T)$ such that the path $P_T(\ell, y)$ does not visit $D(T)$, or in Y_4 otherwise. Note that $|X| = |X_1| + |X_2| + |X_3|$ and $|Y| = |Y_1| + |Y_2| + |Y_3| + |Y_4|$.

We argue first that $N_G(X_2) \subseteq Y_2$. By the definition of critical edges, there is no good cotree edge of T connecting two different components of $T - C(T)$; hence $N_G(X) \subseteq V(T_0) \backslash X = Y$. It suffices to show $N_G(X_2) \subseteq D(T)$ (the definition of X_2 requires that a vertex in it is nonadjacent to $D_B(T)$ in G), which further boils down to showing $N_G(X_2) \cap I_2(T) \subseteq D(T)$: by Proposition 1, X_2 has no neighbor in $L(T)$; on the other hand, $I_3(T) \subseteq D(T)$. Consider a vertex $x \in X_2$, and let y be the unique neighbor of x in T. By assumption, $y \in D_G(T) \cap I_2(T)$, and hence there is a good cotree edge ℓy of T with $\ell \neq x$. For each $y' \in N_G(x) \cap I_2(T)$ different from y, the path $P_T(x, y')$ visits y, using the definition of $D(T)$ we can conclude that $y' \in D(T)$.

Each $x \in X_1$ has a neighbor $y \in Y_1$. By Proposition 2(5), x is the only vertex in $N_G(y) \cap L(T)$. Thus, $|X_1| \leq |Y_1|$. The unique neighbor y of a vertex $x \in X_3$ in T must be in $I_2(T) \setminus D(T)$. Since the trivial path $P_T(x, y)$ (consisting of a single edge xy) does not visit $D(T)$, we have $y \in Y_3$. The other neighbor of y in T cannot be a leaf of T (G has at least four vertices). Thus, $|X_3| \leq |Y_3|$. By Lemma 2, for any two different vertices u and w of Y_2, the path $P_T(u, w)$ visits at least one vertex in Y_4. Since T_0 is a tree, using induction it is easy to show $|Y_4| \geq |Y_2| - 1$.

Summarizing above, we have

$$
\begin{aligned}
|X_2| = |X| - |X_1| - |X_3| \quad &\text{(because } |X| = |X_1| + |X_2| + |X_3|.) \\
\geq |Y| + 1 - |Y_1| - |Y_3| \quad &\text{(because } |X| \geq |Y| + 1; |X_1| \leq |Y_1|; |X_3| \leq |Y_3|.) \\
= |Y_2| + |Y_4| + 1 \quad &\text{(because } |Y| = |Y_1| + |Y_2| + |Y_3| + |Y_4|.) \\
\geq 2|Y_2| \quad &\text{(because } |Y_4| \geq |Y_2| - 1.) \\
\geq 2|N_G(X_2)|. \quad &\text{(because } N_G(X_2) \subseteq Y_2.)
\end{aligned}
$$

Hence X_2 can be used as the independent set L'. This concludes the proof. $\square$

Lemmas 4 and 3 imply Theorem 2.

References

1. Binkele-Raible, D., Fernau, H., Gaspers, S., Liedloff, M.: Exact and parameterized algorithms for max internal spanning tree. Algorithmica 65(1), 95–128 (2013)
2. Cohen, N., Fomin, F.V., Gutin, G., Kim, E.J., Saurabh, S., Yeo, A.: Algorithm for finding k-vertex out-trees and its application to k-internal out-branching problem. Journal of Computer and System Sciences 76(7), 650–662 (2010)
3. Daligault, J.: Combinatorial Techniques for Parameterized Algorithms and Kernels, with Applicationsto Multicut. Ph.D. thesis, Université Montpellier II, Montpellier, Hérault, France (2011)
4. Fomin, F.V., Gaspers, S., Saurabh, S,, Thomassé, S.: A linear vertex kernel for maximum internal spanning tree. Journal of Computer and System Sciences 79(1), 1–6 (2013)
5. Fomin, F.V., Grandoni, F., Lokshtanov, D., Saurabh, S.: Sharp separation and applications to exact and parameterized algorithms. Algorithmica 63(3), 692–706 (2012)
6. Gutin, G., Razgon, I., Kim, E.J.: Minimum leaf out-branching and related problems. Theoretical Computer Science 410(45), 4571–4579 (2009)
7. Hassin, R., Tamir, A.: On the minimum diameter spanning tree problem. Information Processing Letters 53(2), 109–111 (1995)
8. Karger, D.R., Klein, P.N., Tarjan, R.E.: A randomized linear-time algorithm to find minimum spanning trees. Journal of the ACM 42(2), 321–328 (1995). a preliminary version appeared in STOC 1994

9. Knauer, M., Spoerhase, J.: Better approximation algorithms for the maximum internal spanning tree problem. In: Dehne, F., Gavrilova, M., Sack, J.-R., Tóth, C.D. (eds.) WADS 2009. LNCS, vol. 5664, pp. 459–470. Springer, Heidelberg (2009)

10. Könemann, J., Ravi, R.: A matter of degree: Improved approximation algorithms for degree-bounded minimum spanning trees. SIAM Journal on Computing **31**(6), 1783–1793 (2002). a preliminary version appeared in STOC 2000

11. Könemann, J., Ravi, R.: Primal-dual meets local search: Approximating MSTs with nonuniform degree bounds. SIAM Journal on Computing **34**(3), 763–773 (2005). a preliminary version appeared in STOC 2003

12. Koutis, I., Williams, R.: Limits and applications of group algebras for parameterized problems. In: Albers, S., Marchetti-Spaccamela, A., Matias, Y., Nikoletseas, S., Thomas, W. (eds.) ICALP 2009, Part I. LNCS, vol. 5555, pp. 653–664. Springer, Heidelberg (2009)

13. Li, W., Chen, J., Wang, J.: Deeper local search for better approximation on maximum internal spanning trees. In: Schulz, A.S., Wagner, D. (eds.) ESA 2014. LNCS, vol. 8737, pp. 642–653. Springer, Heidelberg (2014)

14. Lu, H., Ravi, R.: Approximating maximum leaf spanning trees in almost linear time. Journal of Algorithms **29**(1), 132–141 (1998)

15. Nederlof, J.: Fast polynomial-space algorithms using inclusion-exclusion. Algorithmica **65**(4), 868–884 (2013)

16. Ozeki, K., Yamashita, T.: Spanning trees: A survey. Graphs and Combinatorics **27**(1), 1–26 (2011)

17. Prieto, E., Sloper, C.: Either/or: using VERTEX COVER structure in designing FPT-algorithms — the case of k-INTERNAL SPANNING TREE. In: Dehne, F., Sack, J.-R., Smid, M. (eds.) WADS 2003. LNCS, vol. 2748, pp. 474–483. Springer, Heidelberg (2003)

18. Prieto, E., Sloper, C.: Reducing to independent set structure–the case of k-internal spanning tree. Nordic Journal of Computing **12**(3), 308–318 (2005)

19. Salamon, G.: Approximating the maximum internal spanning tree problem. Theoretical Computer Science **410**(50), 5273–5284 (2009)

20. Salamon, G.: Degree-Based Spanning Tree Optimization. Ph.D. thesis, Budapest University of Technology and Economics, Budapest, Hungary (2010)

21. Salamon, G., Wiener, G.: On finding spanning trees with few leaves. Information Processing Letters **105**(5), 164–169 (2008)

22. Shachnai, H., Zehavi, M.: Representative families: a unified tradeoff-based approach. In: Schulz, A.S., Wagner, D. (eds.) ESA 2014. LNCS, vol. 8737, pp. 786–797. Springer, Heidelberg (2014)

Reconfiguration on Sparse Graphs

Daniel Lokshtanov[1], Amer E. Mouawad[2]([⊠]), Fahad Panolan[3],
M.S. Ramanujan[1], and Saket Saurabh[3]

[1] University of Bergen, Bergen, Norway
{daniello,ramanujan.sridharan}@ii.uib.no
[2] David R. Cheriton School of Computer Science, University of Waterloo,
Ontario, Canada
aabdomou@uwaterloo.ca
[3] Institute of Mathematical Sciences, Chennai, India
{fahad,saket}@imsc.res.in

Abstract. A vertex-subset graph problem Q defines which subsets of the vertices of an input graph are feasible solutions. A reconfiguration variant of a vertex-subset problem asks, given two feasible solutions S_s and S_t of size k, whether it is possible to transform S_s into S_t by a sequence of vertex additions and deletions such that each intermediate set is also a feasible solution of size bounded by k. We study reconfiguration variants of two classical vertex-subset problems, namely INDEPENDENT SET and DOMINATING SET. We denote the former by ISR and the latter by DSR. Both ISR and DSR are PSPACE-complete on graphs of bounded bandwidth and W[1]-hard parameterized by k on general graphs. We show that ISR is fixed-parameter tractable parameterized by k when the input graph is of bounded degeneracy or nowhere dense. As a corollary, we answer positively an open question concerning the parameterized complexity of the problem on graphs of bounded treewidth. Moreover, our techniques generalize recent results showing that ISR is fixed-parameter tractable on planar graphs and graphs of bounded degree. For DSR, we show the problem fixed-parameter tractable parameterized by k when the input graph does not contain large bicliques, a class of graphs which includes degenerate and nowhere dense graphs.

1 Introduction

Given an n-vertex graph G and two vertices s and t in G, determining whether there exists a path and computing the length of the shortest path between s and t are two of the most fundamental graph problems. In the classical battle of P versus NP or "easy" versus "hard", both of these problems are on the easy side. That is, they can be solved in $poly(n)$ time, where $poly$ is any polynomial function. But what if our input consisted of a 2^n-vertex graph? Of course, we can no longer assume G to be part of the input, as reading the input alone requires

---

The research leading to these results has received funding from the European Research Council under the European Union's Seventh Framework Programme (FP/2007-2013)/ERC Grant Agreement n. 267959.

F. Dehne et al. (Eds.): WADS 2015, LNCS 9214, pp. 506–517, 2015.
DOI: 10.1007/978-3-319-21840-3_42

more than $poly(n)$ time. Instead, we are given an oracle encoded using $poly(n)$ bits and that can, in constant or $poly(n)$ time, answer queries of the form "is u a vertex in G" or "is there an edge between u and v?". Given such an oracle and two vertices of the 2^n-vertex graph, can we still determine if there is a path or compute the length of the shortest path between s and t in $poly(n)$ time?

This seemingly artificial question is in fact quite natural and appears in many practical and theoretical problems. In particular, these are exactly the types of questions asked under the reconfiguration framework, the main subject of this work. Under the reconfiguration framework, instead of finding a feasible solution to some instance $\mathcal{I}$ of a search problem $\mathcal{Q}$, we are interested in structural and algorithmic questions related to the solution space of $\mathcal{Q}$. Naturally, given some adjacency relation $\mathcal{A}$ defined over feasible solutions of $\mathcal{Q}$, the solution space can be represented using a graph $R_\mathcal{Q}(\mathcal{I})$. $R_\mathcal{Q}(\mathcal{I})$ contains one node for each feasible solution of $\mathcal{Q}$ on instance $\mathcal{I}$ and two nodes share an edge whenever their corresponding solutions are adjacent under $\mathcal{A}$. An edge in $R_\mathcal{Q}(\mathcal{I})$ corresponds to a *reconfiguration step*, a walk in $R_\mathcal{Q}(\mathcal{I})$ is a sequence of such steps, a *reconfiguration sequence*, and $R_\mathcal{Q}(\mathcal{I})$ is a *reconfiguration graph*.

Studying problems related to reconfiguration graphs has received considerable attention in the literature [3,12,15,16,20,21], the most popular problem being to determine whether there exists a reconfiguration sequence between two given feasible solutions/configurations. In many cases, this problem was shown PSPACE-hard in general, although some polynomial-time solvable restricted cases have been identified. For PSPACE-hard cases, it is not surprising that shortest paths between solutions can have exponential length. More surprising is that for most known polynomial-time solvable cases the diameter of the reconfiguration graph has been shown to be polynomial. Some of the problems that have been studied under the reconfiguration framework include INDEPENDENT SET [19], SHORTEST PATH [2], COLORING [4], BOOLEAN SATISFIABILITY [12], and FLIP DISTANCE [3,5]. We refer the reader to the recent survey by van den Heuvel [27] for a detailed overview of reconfiguration problems and their applications. Recently, a systematic study of the parameterized complexity [9] of reconfiguration problems was initiated by Mouawad et al. [21]; various problems were identified where the problem was not only NP-hard (or PSPACE-hard), but also W-hard under various parameterizations. The reader is referred to [9] for more on parameterized complexity.

Overview of our Results. In this work, we focus on reconfiguration variants of the INDEPENDENT SET (IS) and DOMINATING SET (DS) problems. Given two independent sets I_s and I_t of a graph G such that $|I_s| = |I_t| = k$, the INDEPENDENT SET RECONFIGURATION (ISR) problem asks whether there exists a sequence of independents sets $\sigma = \langle I_0, I_1, \ldots, I_\ell \rangle$, for some ℓ, such that:

(1) $I_0 = I_s$ and $I_\ell = I_t$,
(2) I_i is an independent set of G for all $0 \leq i \leq \ell$,
(3) $|\{I_i \setminus I_{i+1}\} \cup \{I_{i+1} \setminus I_i\}| = 1$ for all $0 \leq i < \ell$, and
(4) $k - 1 \leq |I_i| \leq k$ for all $0 \leq i \leq \ell$.

Alternatively, given a graph G and integer k, the reconfiguration graph $R_{\mathrm{IS}}(G, k-1, k)$ has a node for each independent set of G of size k or $k-1$ and two nodes are adjacent in $R_{\mathrm{IS}}(G, k-1, k)$ whenever the corresponding independent sets can be obtained from one another by either the addition or the deletion of a single vertex. The reconfiguration graph $R_{\mathrm{DS}}(G, k, k+1)$ is defined similarly for dominating sets. Hence, ISR and DSR can be formally stated as follows:

INDEPENDENT SET RECONFIGURATION (ISR)
Input: Graph G, integer $k > 0$, and two independent sets I_s and I_t of size k
Question: Is there a path from I_s to I_t in $R_{\mathrm{IS}}(G, k-1, k)$?

DOMINATING SET RECONFIGURATION (DSR)
Input: Graph G, integer $k > 0$, and two dominating sets D_s and D_t of size k
Question: Is there a path from D_s to D_t in $R_{\mathrm{DS}}(G, k, k+1)$?

Note that since we only allow independent sets of size k and $k-1$ the ISR problem is equivalent to reconfiguration under the token jumping model considered by Ito et al. [17,18]. ISR is known to be PSPACE-complete on graphs of bounded bandwidth [28] (hence pathwidth and treewidth) and W[1]-hard when parameterized by k on general graphs [18]. On the positive side, the problem was shown fixed-parameter tractable, with parameter k, for graphs of bounded degree, planar graphs, and graphs excluding $K_{3,d}$ as a (not necessarily induced) subgraph, for any constant d [17,18]. We push this boundary further by showing that the problem remains fixed-parameter tractable for graphs of bounded degeneracy and nowhere dense graphs. As a corollary, we answer positively an open question concerning the parameterized complexity of the problem parameterized by k on graphs of bounded treewidth.

For DSR, we show that the problem is fixed-parameter tractable, with parameter k, for graphs excluding $K_{d,d}$ as a (not necessarily induced) subgraph, for any constant d. Note that this class of graphs includes both nowhere dense and bounded degeneracy graphs and is the "largest" class on which the DOMINATING SET problem is known to be in FPT [25,26].

Clearly, our main open question is whether ISR remains fixed-parameter tractable on graphs excluding $K_{d,d}$ as a subgraph. Intuitively, all of the classes we consider fall under the category of "sparse" graph classes. Hence, in some sense, one would not expect a sparse graph to have "too many" dominating sets of fixed small size k as n becomes larger and larger. For independent sets, the situation is reversed. As n grows larger, so does the number of independent sets of fixed size k. So it remains to be seen whether some structural properties of graphs excluding $K_{d,d}$ as a subgraph can be used to settle our open question or whether the problem becomes W[1]-hard. In the latter case, this would be the first example of a W[1]-hard problem (in general), which is in FPT on a class $\mathcal{C}$ of graphs but where the reconfiguration version is not; finding such a problem, we believe, is interesting in its own right. Another open question is whether we can adapt our results for ISR to find shortest reconfiguration sequences. Our algorithm for DSR does in fact guarantee shortest reconfiguration sequences but, as we shall see, the same does not hold for either of the two ISR algorithms.

Due to space limitations, some proofs (marked with a star) have been omitted from the current version of the paper.

2   Preliminaries

For an in-depth review of general graph theoretic definitions we refer the reader to the book of Diestel [8]. Unless otherwise stated, we assume that each graph G is a simple, undirected graph with vertex set $V(G)$ and edge set $E(G)$, where $|V(G)| = n$ and $|E(G)| = m$. The *open neighborhood*, or simply *neighborhood*, of a vertex v is denoted by $N_G(v) = \{u \mid uv \in E(G)\}$, the *closed neighborhood* by $N_G[v] = N_G(v) \cup \{v\}$. Similarly, for a set of vertices $S \subseteq V(G)$, we define $N_G(S) = \{v \mid uv \in E(G), u \in S, v \notin S\}$ and $N_G[S] = N_G(S) \cup S$. The *degree* of a vertex is $|N_G(v)|$. We drop the subscript G when clear from context. A *subgraph* of G is a graph G' such that $V(G') \subseteq V(G)$ and $E(G') \subseteq E(G)$. The *induced subgraph* of G with respect to $S \subseteq V(G)$ is denoted by $G[S]$; $G[S]$ has vertex set S and edge set $E(G[S]) = \{uv \in E(G) \mid u, v \in S\}$. For $r \geq 0$, the *r-neighborhood* of a vertex $v \in V(G)$ is defined as $N_G^r[v] = \{u \mid dist_G(u, v) \leq r\}$, where $dist_G(u, v)$ is the length of a shortest uv-path in G.

Contracting an edge uv of G results in a new graph H in which the vertices u and v are deleted and replaced by a new vertex w that is adjacent to $N_G(u) \cup N_G(v) \setminus \{u, v\}$. If a graph H can be obtained from G by repeatedly contracting edges, H is said to be a *contraction* of G. If H is a subgraph of a contraction of G, then H is said to be a *minor* of G, denoted by $H \preceq_m G$. An equivalent characterization of minors states that H is a minor of G if there is a map that associates to each vertex v of H a non-empty connected subgraph G_v of G such that G_u and G_v are disjoint for $u \neq v$ and whenever there is an edge between u and v in H there is an edge in G between some node in G_u and some node in G_v. The subgraphs G_v are called *branch sets*. H is a *minor at depth r of G*, $H \preceq_m^r G$, if H is a minor of G which is witnessed by a collection of branch sets $\{G_v \mid v \in V(H)\}$, each of which induces a graph of radius at most r. That is, for each $v \in V(H)$, there is a $w \in V(G_v)$ such that $V(G_v) \subseteq N_{G_v}^r[w]$.

Sparse Graph Classes.   We define the three main classes we consider.

Definition 1 ([22,24]). *A class of graphs $\mathcal{C}$ is said to be* nowhere dense *if for every $d \geq 0$ there exists a graph H_d such that $H_d \npreceq_m^d G$ for all $G \in \mathcal{C}$. $\mathcal{C}$ is* effectively nowhere dense *if the map $d \mapsto H_d$ is computable. Otherwise, $\mathcal{C}$ is said to be* somewhere dense.

Nowhere dense classes of graphs were introduced by Nesetril and Ossona de Mendez [22,24] and "nowhere density" turns out to be a very robust concept with several natural characterizations [13]. We use one such characterization in Section 3.2. It follows from the definition that planar graphs, graphs of bounded treewidth, graphs of bounded degree, H-minor-free graphs, and H-topological-minor-free graphs are nowhere dense [22,24]. As in the work of Dawar and Kreutzer [7], we are only interested in effectively nowhere dense classes; all natural nowhere dense classes are effectively nowhere dense, but it is possible to construct artificial classes that are nowhere dense, but not effectively so.

Definition 2. *A class of graphs $\mathcal{C}$ is said to be d-degenerate if every induced subgraph of any graph $G \in \mathcal{C}$ has a vertex of degree at most d.*

Graphs of bounded degeneracy and nowhere dense graphs are incomparable [14]. In other words, graphs of bounded degeneracy are somewhere dense. Degeneracy is a hereditary property, hence any induced subgraph of a d-degenerate graph is also d-degenerate. It is well-known that graphs of treewidth at most d are also d-degenerate. Moreover a d-degenerate graph cannot contain $K_{d+1,d+1}$ as a subgraph, which brings us to the class of biclique-free graphs. The relationship between bounded degeneracy, nowhere dense, and $K_{d,d}$-free graphs was shown by Philip et al. and Telle and Villanger [25,26].

Definition 3. *A class of graphs $\mathcal{C}$ is said to be d-biclique-free, for some $d > 0$, if $K_{d,d}$ is not a subgraph of any $G \in \mathcal{C}$, and it is said to be biclique-free if it is d-biclique-free for some d.*

Proposition 1 ([25,26]). *Any degenerate or nowhere dense class of graphs is biclique-free, but not vice-versa.*

Reconfiguration. For any vertex-subset problem $\mathcal{Q}$, graph G, and positive integer k, we consider the *reconfiguration graph* $R_\mathcal{Q}(G, k, k+1)$ when $\mathcal{Q}$ is a minimization problem (e.g. DOMINATING SET) and the reconfiguration graph $R_\mathcal{Q}(G, k-1, k)$ when $\mathcal{Q}$ is a maximization problem (e.g. INDEPENDENT SET). A set $S \subseteq V(G)$ has a corresponding node in $V(R_\mathcal{Q}(G, r_l, r_u))$, $r_l \in \{k-1, k\}$ and $r_u \in \{k, k+1\}$, if and only if S is a feasible solution for $\mathcal{Q}$ and $r_l \leq |S| \leq r_u$. We refer to *vertices* in G using lower case letters (e.g. u, v) and to the *nodes* in $R_\mathcal{Q}(G, r_l, r_u)$, and by extension their associated feasible solutions, using upper case letters (e.g. A, B). If $A, B \in V(R_\mathcal{Q}(G, r_l, r_u))$ then there exists an edge between A and B in $R_\mathcal{Q}(G, r_l, r_u)$ if and only if there exists a vertex $u \in V(G)$ such that $\{A \setminus B\} \cup \{B \setminus A\} = \{u\}$. Equivalently, for $A \Delta B = \{A \setminus B\} \cup \{B \setminus A\}$ the *symmetric difference* of A and B, A and B share an edge in $R_\mathcal{Q}(G, r_l, r_u)$ if and only if $|A \Delta B| = 1$.

We write $A \leftrightarrow B$ if there exists a path in $R_\mathcal{Q}(G, r_l, r_u)$, a reconfiguration sequence, joining A and B. Any reconfiguration sequence from *source* feasible solution S_s to *target* feasible solution S_t, which we sometimes denote by $\sigma = \langle S_0, S_1, \ldots, S_\ell \rangle$, for some ℓ, has the following properties:

- $S_0 = S_s$ and $S_\ell = S_t$,
- S_i is a feasible solution for $\mathcal{Q}$ for all $0 \leq i \leq \ell$,
- $|S_i \Delta S_{i+1}| = 1$ for all $0 \leq i < \ell$, and
- $r_l \leq |S_i| \leq r_u$ for all $0 \leq i \leq \ell$.

We denote the *length* of σ by $|\sigma|$. For $0 < i \leq |\sigma|$, we say vertex $v \in V(G)$ is *added* at step/index/position/slot i if $v \notin S_{i-1}$ and $v \in S_i$. Similarly, a vertex v is *removed* at step/index/position/slot i if $v \in S_{i-1}$ and $v \notin S_i$. A vertex $v \in V(G)$ is *touched* in the course of a reconfiguration sequence if v is either added or removed at least once; it is *untouched* otherwise. A vertex is *removable* (*addable*) from feasible solution S if $S \setminus \{v\}$ ($S \cup \{v\}$) is also a feasible solution for $\mathcal{Q}$. For any pair of consecutive solutions (S_{i-1}, S_i) in σ, we say S_i (S_{i-1}) is the *successor* (*predecessor*) of S_{i-1} (S_i). A reconfiguration sequence $\sigma' = \langle S_0, S_1, \ldots, S_{\ell'} \rangle$ is a *prefix* of $\sigma = \langle S_0, S_1, \ldots, S_\ell \rangle$ if $\ell' < \ell$.

We adapt the concept of irrelevant vertices from parameterized complexity to introduce the notions of irrelevant and strongly irrelevant vertices for reconfiguration. Since these notions apply to almost any reconfiguration problem, we give general definitions.

Definition 4. *For any vertex-subset problem $\mathcal{Q}$, n-vertex graph G, positive integers r_l and r_u, and $S_s, S_t \in V(R_{\mathcal{Q}}(G, r_l, r_u))$ such that there exists a reconfiguration sequence from S_s to S_t in $R_{\mathcal{Q}}(G, r_l, r_u)$, we say a vertex $v \in V(G)$ is irrelevant (with respect to S_s and S_t) if and only if $v \notin S_s \cup S_t$ and there exists a reconfiguration sequence from S_s to S_t in $R_{\mathcal{Q}}(G, r_l, r_u)$ which does not touch v. We say v is strongly irrelevant (with respect to S_s and S_t) if it is irrelevant and the length of a shortest reconfiguration sequence from S_s to S_t which does not touch v is no greater than the length of a shortest reconfiguration sequence which does (if the latter sequence exists).*

At a high level, it is enough to ignore irrelevant vertices when trying to find *any* reconfiguration sequence between two feasible solutions, but only strongly irrelevant vertices can be ignored if we wish to find a *shortest* reconfiguration sequence. As we shall see, our kernelization algorithm for DSR does in fact find strongly irrelevant vertices and can therefore be used to find shortest reconfiguration sequences. For ISR, we are only able to find irrelevant vertices and reconfiguration sequences are not guaranteed to be of shortest possible length.

3 Independent Set Reconfiguration

3.1 Graphs of Bounded Degeneracy

To show that the ISR problem is fixed-parameter tractable on d-degenerate graphs, for some integer d, we will proceed in two stages. In the first stage, we will show, for an instance (G, I_s, I_t, k), that as long as the number of low-degree vertices in G is "large enough" we can find an irrelevant vertex (Definition 4). Once the number of low-degree vertices is bounded, a simple counting argument (Proposition 2) shows that the size of the remaining graph is also bounded and hence we can solve the instance by exhaustive enumeration.

Proposition 2 $(\star)$. *Let G be an n-vertex d-degenerate graph, $S_1 \subseteq V(G)$ be the set of vertices of degree at most $2d$, and $S_2 = V(G) \setminus S_1$. If $|S_1| < s$, then $|V(G)| \leq (2d + 1)s$.*

To find irrelevant vertices, we make use of the following classical result of Erdős and Rado [11], also known in the literature as the sunflower lemma. We first define the terminology used in the statement of the theorem. A *sunflower* with k *petals* and a *core* Y is a collection of sets $S_1, \ldots, S_k$ such that $S_i \cap S_j = Y$ for all $i \neq j$; the sets $S_i \setminus Y$ are petals and we require all of them to be non-empty. Note that a family of pairwise disjoint sets is a sunflower (with an empty core).

Theorem 1 (Sunflower Lemma [11]). *Let $\mathcal{A}$ be a family of sets (without duplicates) over a universe $\mathcal{U}$, such that each set in $\mathcal{A}$ has cardinality at most d. If $|\mathcal{A}| > d!(k-1)^d$, then $\mathcal{A}$ contains a sunflower with k petals and such a sunflower can be computed in time polynomial in $|\mathcal{A}|$, $|\mathcal{U}|$, and k.*

Lemma 1. *Let (G, I_s, I_t, k) be an instance of* ISR *and let B be the set of vertices in $V(G) \setminus \{I_s \cup I_t\}$ of degree at most $2d$. If $|B| > (2d+1)!(2k-1)^{2d+1}$, then there exists an irrelevant vertex $v \in V(G) \setminus \{I_s \cup I_t\}$ such that (G, I_s, I_t, k) is a yes-instance if and only if (G', I_s, I_t, k) is a yes-instance, where G' is obtained by deleting v and all edges incident on v.*

Proof. Let $b_1, b_2, \ldots, b_{|B|}$ denote the vertices in B and let $\mathcal{A} = \{N_G[b_1], N_G[b_2], \ldots, N_G[b_{|B|}]\}$ denote the family of the closed neighborhoods of each vertex in B and set $\mathcal{U} = \bigcup_{b \in B} N[b]$. Since $|B|$ is greater than $(2d+1)!(2k-1)^{2d+1}$, we know from Theorem 1 that $\mathcal{A}$ contains a sunflower with $2k$ petals and such a sunflower can be computed in time polynomial in $|\mathcal{A}|$ and k. Note that we assume, without loss of generality, that there are no two vertices u and v in $V(G) \setminus \{I_s \cup I_t\}$ such that $N_G[u] = N_G[v]$, as we can safely delete one of them from the input graph otherwise, i.e. $uv \in E(G)$ and one of the two is (strongly) irrelevant. Let v_{ir} be a vertex whose closed neighborhood is one of those $2k$ petals. We claim that v_{ir} is irrelevant and can therefore be deleted from G to obtain G'.

To see why, consider any reconfiguration sequence $\sigma = \langle I_s = I_0, I_1, \ldots, I_t = I_\ell \rangle$ from I_s to I_t in $R_{\mathrm{IS}}(G, k-1, k)$. Since $v_{ir} \notin I_s \cup I_t$, we let p, $0 < p < \ell$, be the first index in σ at which v_{ir} is added, i.e. $v_{ir} \in I_p$ and $v_{ir} \notin I_i$ for all $i < p$. Moreover, we let $q + 1$, $p < q + 1 \leq \ell$ be the first index after p at which v_{ir} is removed, i.e. $v_{ir} \in I_q$ and $v_{ir} \notin I_{q+1}$. We will consider the subsequence $\sigma_s = \langle I_p, \ldots, I_q \rangle$ and show how to modify it so that it does not touch v_{ir}. Applying the same procedure to every such subsequence in σ suffices to prove the lemma.

Since the sunflower constructed to obtain v_{ir} has $2k$ petals and the size of any independent set in σ (or any reconfiguration sequence in general) is at most k, there must exist another *free* vertex v_{fr} whose closed neighborhood corresponds to one of the remaining $2k - 1$ petals which we can add at index p instead of v_{ir}, i.e. $v_{fr} \notin N_G[I_p]$. We say v_{fr} *represents* v_{ir}. Assume that no such vertex exists. Then we know that either some vertex in the core of the sunflower is in I_p contradicting the fact that we are adding v_{ir}, or every petal of the sunflower contains a vertex in I_p, which is not possible since the size of any independent set is at most k and the number of petals is larger. Hence, we first modify the subsequence σ_s by adding v_{fr} instead of v_{ir}. Formally, we have $\sigma'_s = \langle (I_p \setminus \{v_{ir}\}) \cup \{v_{fr}\}, \ldots, (I_q \setminus \{v_{ir}\}) \cup \{v_{fr}\} \rangle$.

To be able to replace σ_s by σ'_s in σ and obtain a reconfiguration sequence from I_s to I_t, then all of the following conditions must hold:

(1) $|(I_q \setminus \{v_{ir}\}) \cup \{v_{fr}\}| = k$.
(2) $(I_i \setminus \{v_{ir}\}) \cup \{v_{fr}\}$ is an independent set of G for all $p \leq i \leq q$,
(3) $|(I_i \setminus \{v_{ir}\}) \cup \{v_{fr}\} \Delta (I_{i+1} \setminus \{v_{ir}\}) \cup \{v_{fr}\}| = 1$ for all $p \leq i < q$, and
(4) $k - 1 \leq |(I_i \setminus \{v_{ir}\}) \cup \{v_{fr}\}| \leq k$ for all $p \leq i \leq q$.

It is not hard to see that if there exists no i, $p < i \leq q$, such that σ'_s adds a vertex in $N[v_{fr}]$ at position i, then all four conditions hold. If there exists such a position, we will modify σ'_s into yet another subsequence σ''_s by finding a new vertex to represent v_{ir}. The length of σ''_s will be two greater than that of σ'_s.

We let i, $p < i \leq q$, be the first position in σ'_s at which a vertex in $u \in N[v_{fr}]$ (possibly equal to v_{fr}) is added (hence $|I_{i-1}| = k-1$). Using the same arguments discussed to find v_{fr}, and since we constructed a sunflower with $2k$ petals, we can find another vertex v'_{fr} such that $N[v_{fr}] \cap ((I_{i-1} \setminus \{v_{ir}\}) \cup \{v_{fr}\}) = \emptyset$. This new vertex will represent v_{ir} instead of v_{fr}. We construct σ''_s from σ'_s as follows: $\sigma''_s = \langle (I_p \setminus \{v_{ir}\}) \cup \{v_{fr}\}, \ldots, (I_{i-1} \setminus \{v_{ir}\}) \cup \{v_{fr}\}, (I_{i-1} \setminus \{v_{ir}\}) \cup \{v_{fr}\} \cup \{v'_{fr}\}, (I_{i-1} \setminus \{v_{ir}\}) \cup \{v'_{fr}\}, (I_i \setminus \{v_{ir}\}) \cup \{v'_{fr}\}, \ldots, (I_q \setminus \{v_{ir}\}) \cup \{v'_{fr}\} \rangle$. If σ''_s now satisfies all four conditions then we are done. Otherwise, we repeat the same process (at most $q - p$ times) until we reach such a subsequence. $\square$

Theorem 2. ISR *on d-degenerate graphs is fixed-parameter tractable parameterized by $k + d$.*

Proof. For an instance (G, I_s, I_t, k) of ISR, we know from Lemma 1 that as long as $V(G) \setminus \{I_s \cup I_t\}$ contains more than $(2d+1)!(2k-1)^{2d+1}$ vertices of degree at most $2d$ we can find an irrelevant vertex and reduce the size of the graph. After exhaustively reducing the graph to obtain G', we known that $G'[V(G') \setminus \{I_s \cup I_t\}]$, which is also d-degenerate, has at most $(2d+1)!(2k-1)^{2d+1}$ vertices of degree at most $2d$. Hence, applying Proposition 2, we know that $|V(G') \setminus \{I_s \cup I_t\}| \leq (2d+1)(2d+1)!(2k-1)^{2d+1}$ and $|V(G')| \leq (2d+1)(2d+1)!(2k-1)^{2d+1} + 2k$. $\square$

3.2 Nowhere Dense Graphs

Nesetril and Ossona de Mendez [23] showed an interesting relationship between nowhere dense classes and a property of classes of structures introduced by Dawar [6] called *quasi-wideness*. We will use quasi-wideness and show a rather interesting relationship between ISR on graphs of bounded degeneracy and nowhere dense graphs. That is, our algorithm for nowhere dense graphs will closely mimic the previous algorithm in the following sense. Instead of using the sunflower lemma to find a large sunflower, we will use quasi-wideness to find a "large enough almost sunflower" with an initially "unknown" core and then use structural properties of the graph to find this core and complete the sunflower. We first state some of the results that we need. Given a graph G, a set $S \subseteq V(G)$ is called *r-scattered* if $N^r_G(u) \cap N^r_G(v) = \emptyset$ for all distinct $u, v \in S$.

Proposition 3. *Let G be a graph and let $S = \{s_1, s_2, ..., s_k\} \subseteq V(G)$ be a 2-scattered set of size k in G. Then the closed neighborhoods of the vertices in S form a sunflower with k petals and an empty core.*

Definition 5 ([7,23]). *A class $\mathcal{C}$ of graphs is* uniformly quasi-wide *with margin $s_{\mathcal{C}} : \mathbb{N} \to \mathbb{N}$ and $N_{\mathcal{C}} : \mathbb{N} \times \mathbb{N} \to \mathbb{N}$ if for all $r, k \in \mathbb{N}$, if $G \in \mathcal{C}$ and $W \subseteq V(G)$ with $|W| > N_{\mathcal{C}}(r, k)$, then there is a set $S \subseteq W$ with $|S| < s_{\mathcal{C}}(r)$, such that W contains an r-scattered set of size at least k in $G[V(G) \setminus S]$. $\mathcal{C}$ is* effectively uniformly quasi-wide *if $s_{\mathcal{C}}(r)$ and $N_{\mathcal{C}}(r, k)$ are computable.*

Examples of effectively uniformly quasi-wide classes include graphs of bounded degree with margin 1 and H-minor-free graphs with margin $|V(H)| - 1$.

Theorem 3 ([7]). *A class $\mathcal{C}$ of graphs is effectively nowhere dense if and only if $\mathcal{C}$ is effectively uniformly quasi-wide.*

Theorem 4 ([7]). *Let $\mathcal{C}$ be an effectively nowhere dense class of graphs and h be the computable function such that $K_{h(r)} \not\preceq_m^r G$ for all $G \in \mathcal{C}$. Let G be an n-vertex graph in $\mathcal{C}$, $r, k \in \mathbb{N}$, and $W \subseteq V(G)$ with $|W| \geq N(h(r), r, k)$, for some computable function N. Then in $\mathcal{O}(n^2)$ time, we can compute a set $B \subseteq V(G)$, $|B| \leq h(r) - 2$, and a set $A \subseteq W$ such that $|A| \geq k$ and A is an r-scattered set in $G[V(G) \setminus B]$.*

Lemma 2. *Let $\mathcal{C}$ be an effectively nowhere dense class of graphs and h be the computable function such that $K_{h(r)} \not\preceq_m^r G$ for all $G \in \mathcal{C}$. Let (G, I_s, I_t, k) be an instance of ISR where $G \in \mathcal{C}$ and let R be the set of vertices in $V(G) \setminus \{I_s \cup I_t\}$. Moreover, let $\mathcal{P} = \{P_1, P_2, ...\}$ be a family of sets which partitions R such that for any two distinct vertices $u, v \in R$, $u, v \in P_i$ if and only if $N_G(u) \cap \{I_s \cup I_t\} = N_G(v) \cap \{I_s \cup I_t\}$. If there exists a set $P_i \in \mathcal{P}$ such that $|P_i| > N(h(2), 2, 2^{h(2)+1}k)$, for some computable function N, then there exists an irrelevant vertex $v \in V(G) \setminus \{I_s \cup I_t\}$ such that (G, I_s, I_t, k) is a yes-instance if and only if (G', I_s, I_t, k) is a yes-instance, where G' is obtained from G by deleting v and all edges incident on v.*

Proof. By construction, we known that the family $\mathcal{P}$ contains at most 4^k sets, as we partition R based on their neighborhoods in $I_s \cup I_t$. Note that some vertices in R have no neighbors in $I_s \cup I_t$ and will therefore belong to the same set in $\mathcal{P}$.

Assume that there exists a $P \in \mathcal{P}$ such that $|P| > N(h(2), 2, 2^{h(2)+1}k)$. Consider the graph $G[R]$. By Theorem 4, we can, in $\mathcal{O}(|R|^2)$ time, compute a set $B \subseteq R$, $|B| \leq h(2) - 2$, and a set $A \subseteq P$ such that $|A| \geq 2^{h(2)+1}k$ and A is a 2-scattered set in $G[R \setminus B]$. Now let $\mathcal{P}' = \{P_1', P_2', ...\}$ be a family of sets which partitions A such that for any two distinct vertices $u, v \in A$, $u, v \in P_i'$ if and only if $N_G(u) \cap B = N_G(v) \cap B$. Since $|A| \geq 2^{h(2)+1}k$ and $|\mathcal{P}'| \leq 2^{h(2)}$, we know that at least one set in $\mathcal{P}'$ will contain at least $2k$ vertices of A. Denote these $2k$ vertices by A'. All vertices in A' have the same neighborhood in B and the same neighborhood in $I_s \cup I_t$ (as all vertices in A' belonged to the same set $P \in \mathcal{P}$). Moreover, A' is a 2-scattered set in $G[R \setminus B]$. Hence, the sets $\{N_G[a_1'], N_G[a_2'], ..., N_G[a_{2k}']\}$, i.e. the closed neighborhoods of the vertices in

A', form a sunflower with $2k$ petals (Proposition 3); the core of this sunflower is contained in $B \cup I_s \cup I_t$. Using the same arguments as we did in the proof of Lemma 1, we can show that there exists at least one irrelevant vertex $v \in V(G) \setminus \{B \cup I_s \cup I_t\}$. □

Theorem 5. ISR *restricted to any effectively nowhere dense class* $\mathcal{C}$ *of graphs is fixed-parameter tractable parameterized by* k.

Proof. If after partitioning $V(G) \setminus \{I_s \cup I_t\}$ into at most 4^k sets the size of every set $P \in \mathcal{P}$ is bounded by $N(h(2), 2, 2^{h(2)+1}k)$, then we can solve the problem by exhaustive enumeration, as $|V(G)| \leq 2k + 4^k N(h(2), 2, 2^{h(2)+1}k)$. Otherwise, we can apply Lemma 2 and reduce the size of the graph in polynomial time. □

4   Dominating Set Reconfiguration

4.1   Graphs Excluding $K_{d,d}$ as a Subgraph

The parameterized complexity of the DOMINATING SET problem (parameterized by k the solution size) on various classes of graphs has been studied extensively in the literature; the main goal has been to push the tractability frontier as far as possible. The problem was shown fixed-parameter tractable on nowhere dense graphs by Dawar and Kreutzer [7], on degenerate graphs by Alon and Gutner [1], and on $K_{d,d}$-free graphs by Philip et al. [25] and Telle and Villanger [26]. Our fixed-parameter tractable algorithm relies on many of these earlier results. Interestingly, and since the class of $K_{d,d}$-free graphs includes all those other graph classes, our algorithm (Theorem 6) implies that the diameter of the reconfiguration graph $R_{\mathrm{DS}}(G, k, k+1)$ (or of its connected components), for G in any of the aforementioned classes, is bounded above by $f(k, c)$, where f is a computable function and c is constant which depends on the graph class at hand. We start with some definitions and needed lemmas.

Definition 6. *A bipartite graph* G *with bipartition* (A, B) *is* B-twinless *if there are no vertices* $u, v \in B$ *such that* $N(u) = N(v)$.

Lemma 3 $(\star)$. *If* G *is a bipartite graph with bipartition* (A, B) *such that* $|A| \geq 2(d-1)$, G *is* B-twinless, *and* G *excludes* $K_{d,d}$ *as a subgraph, then* $|B| \leq 2d|A|^d$.

Definition 7 ([10]). *Given a graph* G, *the* domination core *of* G *is a set* $C \subseteq V(G)$ *such that any set* $D \subseteq V(G)$ *is a dominating set of* G *if and only if* D *dominates* C, *i.e.* D *is a dominating set of* G *if and only if* $C \subseteq N_G[D]$.

Lemma 4 $(\star)$. *If* G *is a graph which excludes* $K_{d,d}$ *as a subgraph and* G *has a dominating set of size at most* k *then the size of the domination core* C *of* G *is at most* $2dk^{2d}$ *and* C *can be computed in* $\mathcal{O}^*(dk^d)$ *time.*

Since Lemma 4 implies a bound on the size of the domination core and allows us to compute it efficiently, our main concern is to deal with vertices outside of the core, i.e. vertices in $V(G) \setminus C$. The next lemma shows that we can in fact find strongly irrelevant vertices outside of the domination core.

Lemma 5 (⋆). *For G an n-vertex graph, C the domination core of G, and D_s and D_t two dominating sets of G, if there exist $u, v \in V(G) \setminus \{C \cup D_s \cup D_t\}$ such that $N_G(u) \cap C = N_G(v) \cap C$ then u (or v) is strongly irrelevant.*

Theorem 6. DSR *parameterized by $k + d$ is fixed-parameter tractable on graphs that exclude $K_{d,d}$ as a subgraph.*

Proof. Given a graph G, integer k, and two dominating sets D_s and D_t of G of size at most k, we first compute the domination core C of G, which by Lemma 4 can be accomplished in $\mathcal{O}^*(dk^d)$ time. Next, and due to Lemma 5, we can delete all strongly irrelevant vertices from $V(G) \setminus \{C \cup D_s \cup D_t\}$. We denote this new graph by G'.

Now consider the bipartite graph G'' with bipartition $(A = C \setminus \{D_s \cup D_t\}, B = V(G') \setminus \{C \cup D_s \cup D_t\})$. This graph is B-twinless, since for every pair of vertices $u, v \in V(G) \setminus \{C \cup D_s \cup D_t\}$ such that $N_G(u) \cap C = N_G(v) \cap C$ either u or v is strongly irrelevant and is therefore not in $V(G')$ nor $V(G'')$. Moreover, since every subgraph of a $K_{d,d}$-free graph is also $K_{d,d}$-free, G'' is $K_{d,d}$-free. Hence, if $|A| < 2(d - 1)$ then $|B| \leq 2^{2(d-1)} = 4^{d-1}$. Otherwise, by Lemmas 3 and 4, we have $|B| \leq 2d|A|^d \leq 2d(2dk^{2d})^d$.

Putting it all together, we know that after deleting all strongly irrelevant vertices, the number of vertices in the resulting graph G' is at most $|V(G')| = |V(C)| + |D_s \cup D_t| + |V(G') \setminus \{C \cup D_s \cup D_t\}| \leq 2dk^{2d} + 2k + 2d(2dk^{2d})^d$.

Hence, we can solve DSR by exhaustively enumerating all $2^{|V(G')|}$ subsets of $V(G')$ and building the reconfiguration graph $R_{DS}(G', k, k + 1)$.   □

References

1. Alon, N., Gutner, S.: Linear time algorithms for finding a dominating set of fixed size in degenerated graphs. Algorithmica **54**(4), 544–556 (2009)
2. Bonsma, P.: Rerouting shortest paths in planar graphs. In: Proceedings of the 32nd Annual Conference on Foundations of Software Technology and Theoretical Computer Science, pp. 337–349 (2012)
3. Bose, P., Hurtado, F.: Flips in planar graphs. Computational Geometry **42**(1), 60–80 (2009)
4. Cereceda, L., van den Heuvel, J., Johnson, M.: Finding paths between 3-colorings. Journal of Graph Theory **67**(1), 69–82 (2011)
5. Cleary, S., John, K.S.: Rotation distance is fixed-parameter tractable. Information Processing Letters **109**(16), 918–922 (2009)
6. Dawar, A.: Homomorphism preservation on quasi-wide classes. Journal of Computer and System Sciences **76**(5), 324–332 (2010). Workshop on Logic, Language, Information and Computation
7. Dawar, A., Kreutzer, S.: Domination problems in nowhere-dense classes of graphs. CoRR (2009). arXiv:0907.42837
8. Diestel, R.: Graph theory, Electronic Edition. Springer (2005)
9. Downey, R.G., Fellows, M.R.: Parameterized complexity. Springer, New York (1997)

10. Drange, P.G., Dregi, M.S., Fomin, F.V., Kreutzer, S., Lokshtanov, D., Pilipczuk, M., Pilipczuk, M., Reidl, F., Saurabh, S., Villaamil, F.S., Sikdar, S.: Kernelization and sparseness: the case of dominating set. CoRR (2014). arXiv:1411.4575
11. Erdős, P., Rado, R.: Intersection theorems for systems of sets. Journal of the London Mathematical Society **35**, 85–90 (1960)
12. Gopalan, P., Kolaitis, P.G., Maneva, E.N., Papadimitriou, C.H.: The connectivity of Boolean satisfiability: computational and structural dichotomies. SIAM Journal on Computing **38**(6), 2330–2355 (2009)
13. Grohe, M., Kreutzer, S., Siebertz, S.: Characterisations of nowhere dense graphs. In: Proceedings of the 33rd Annual Conference on Foundations of Software Technology and Theoretical Computer Science, pp. 21–40 (2013)
14. Grohe, M., Kreutzer, S., Siebertz, S.: Deciding first-order properties of nowhere dense graphs. In: Proceedings of the 46th Annual ACM Symposium on Theory of Computing, pp. 89–98 (2014)
15. Ito, T., Demaine, E.D., Harvey, N.J.A., Papadimitriou, C.H., Sideri, M., Uehara, R., Uno, Y.: On the complexity of reconfiguration problems. Theoretical Computer Science **412**(12–14), 1054–1065 (2011)
16. Ito, T., Kamiński, M., Demaine, E.D.: Reconfiguration of list edge-colorings in a graph. Discrete Applied Mathematics **160**(15), 2199–2207 (2012)
17. Ito, T., Kamiński, M., Ono, H.: Fixed-parameter tractability of token jumping on planar graphs. In: Ahn, H.-K., Shin, C.-S. (eds.) ISAAC 2014. LNCS, vol. 8889, pp. 208–219. Springer, Heidelberg (2014)
18. Ito, T., Kamiński, M., Ono, H., Suzuki, A., Uehara, R., Yamanaka, K.: On the parameterized complexity for token jumping on graphs. In: Gopal, T.V., Agrawal, M., Li, A., Cooper, S.B. (eds.) TAMC 2014. LNCS, vol. 8402, pp. 341–351. Springer, Heidelberg (2014)
19. Kamiński, M., Medvedev, P., Milanič, M.: Complexity of independent set reconfigurability problems. Theoretical Computer Science **439**, 9–15 (2012)
20. Lubiw, A., Pathak, V.: Flip distance between two triangulations of a point-set is NP-complete. CoRR (2012). arXiv:1205.2425
21. Mouawad, A.E., Nishimura, N., Raman, V., Simjour, N., Suzuki, A.: On the parameterized complexity of reconfiguration problems. In: Gutin, G., Szeider, S. (eds.) IPEC 2013. LNCS, vol. 8246, pp. 281–294. Springer, Heidelberg (2013)
22. Nesetril, J., de Mendez, P.O.: Structural properties of sparse graphs. In: Building Bridges. Bolyai Society Mathematical Studies, vol. 19, pp. 369–426. Springer, Heidelberg (2008)
23. Nesetril, J., de Mendez, P.O.: First order properties on nowhere dense structures. Journal of Symbolic Logic **75**(3), 868–887 (2010)
24. Nesetril, J., de Mendez, P.O.: From sparse graphs to nowhere dense structures: Decompositions, independence, dualities and limits. European Congress of Mathematics (2010)
25. Philip, G., Raman, V., Sikdar, S.: Solving dominating set in larger classes of graphs: FPT algorithms and polynomial kernels. In: Fiat, A., Sanders, P. (eds.) ESA 2009. LNCS, vol. 5757, pp. 694–705. Springer, Heidelberg (2009)
26. Telle, J.A., Villanger, Y.: FPT algorithms for domination in biclique-free graphs. In: Epstein, L., Ferragina, P. (eds.) ESA 2012. LNCS, vol. 7501, pp. 802–812. Springer, Heidelberg (2012)
27. van den Heuvel, J.: The complexity of change. Surveys in Combinatorics **2013**(409), 127–160 (2013)
28. Wrochna, M.: Reconfiguration in bounded bandwidth and treedepth. CoRR (2014). arXiv:1405.0847

Smoothed Analysis of Local Search Algorithms

Bodo Manthey[✉]

Department of Applied Mathematics,
University of Twente, Enschede, The Netherlands
b.manthey@utwente.nl

Abstract. Smoothed analysis is a method for analyzing the performance of algorithms for which classical worst-case analysis fails to explain the performance observed in practice. Smoothed analysis has been applied to explain the performance of a variety of algorithms in the last years.

One particular class of algorithms where smoothed analysis has been used successfully are local search algorithms. We give a survey of smoothed analysis, in particular applied to local search algorithms.

1 Smoothed Analysis

1.1 Motivation

The goal of the analysis of algorithms is to provide measures for the performance of algorithms. In this way, it helps to compare algorithms and to understand their behavior. The most commonly used method for the performance of algorithms is *worst-case analysis*. If an algorithm has a good worst-case performance, then this is a very strong statement and, up to constants and lower order terms, the algorithm should also perform well in practice. However, there are many algorithms that work surprisingly well in practice although they have a very poor worst-case performance. The reason for this is that the worst-case performance can be dominated by a few pathological instances that hardly or never occur in practice.

A frequently used alternative to worst-case analysis is *average-case analysis*. In average-case analysis, the expected performance is measured with respect to some fixed probability distribution. Many algorithms with poor worst-case but good practical performance show a good average-case performance. However, the drawback of average-case analysis is that random instances drawn according to some fixed probability distribution often have very special properties with high probability. These properties of random instances distinguish them from typical instances. Thus, a good average-case running-time does not necessarily explain a good practical performance.

In order to get a more realistic measure for the performance of algorithms in cases where worst-case analysis is too pessimistic, Spielman and Teng [56] proposed *smoothed analysis* as a new paradigm to analyze algorithms. The key idea is that practical inputs are often not pathological, but are subject to a

© Springer International Publishing Switzerland 2015
F. Dehne et al. (Eds.): WADS 2015, LNCS 9214, pp. 518–527, 2015.
DOI: 10.1007/978-3-319-21840-3_43

small amount of random noise. This random noise can, for instance, stem from measurement errors. It can also come from numerical imprecision or other circumstances, where we have no reason to believe that these influences change the input in a worst-case manner.

1.2 Definition

In smoothed analysis, we measure the maximum expected running-time, where the maximum is taken over the (adversarial) choices of the adversary, and the expected value is taken over the random perturbation of the input. The random perturbation is controlled by some perturbation parameter.

In almost all cases, this *perturbation parameter* is either the standard deviation σ of the perturbation or an upper bound ϕ on the density of the underlying probability distributions. In the former case, larger σ means more randomness, and the analysis approaches the worst-case analysis for very small σ. This model is also called the *two-step model* of smoothed analysis. The most commonly used type of perturbations are Gaussian distributions of standard deviation σ.

In the latter case, smaller ϕ means more randomness, and the analysis approaches the worst-case analysis for large ϕ. This model is also called the *one-step model* of smoothed analysis.

We restrict ourselves here to the two-step model with Gaussian noise, and we define this model in the following. We assume that our instances $X = \{x_1, \ldots, x_n\}$ of size n consist of n points $x_i \in \mathbb{R}^d$ ($1 \leq i \leq n$). We denote by $\mathcal{N}(\mu, \sigma^2)$ a d-dimensional Gaussian distribution with mean $\mu \in \mathbb{R}^d$ and variance σ^2 (more precisely, its covariance matrix is a diagonal matrix with σ^2 on all diagonal entries).

Assume that we have a performance measure m that maps instances to, e.g., the number of iterations that the algorithm under consideration needs on an instance X or the approximation ratio that the algorithm achieves on X. Then the worst-case performance as a function of the input size is given as

$$M_{\text{worst}}(n) = \max_{\substack{X = \{x_1, \ldots, x_n\} \\ \subseteq [0, 1]^d}} \big(m(X)\big). \tag{1}$$

The average-case performance is given by

$$M_{\text{average}}(n) = \mathop{\mathbb{E}}_{\substack{Y = \{y_1, \ldots, y_n\} \\ y_i \sim \mathcal{N}(0, 1)}} \big(m(Y)\big).$$

Here, the points y_i (for $1 \leq i \leq n$) are drawn according to independent d-dimensional Gaussian distributions with mean 0 and standard deviation 1. Another probability distribution that is frequently used is drawing the points independently and uniformly from the unit hypercube $[0, 1]^d$.

The smoothed performance is a combination of both:

$$M_{\text{smoothed}}(n, \sigma) = \max_{\substack{X = \{x_1, \ldots, x_n\} \\ \subseteq [0, 1]^d}} \mathop{\mathbb{E}}_{\substack{Y = \{y_1, \ldots, y_n\} \\ y_i \sim \mathcal{N}(x_i, \sigma^2)}} \big(m(Y)\big). \tag{2}$$

An adversary specifies the instance X, and then Y is obtained by perturbing the points in X.

Note that M_{smoothed} depends also on the perturbation parameter σ: For very small σ, we have $Y \approx X$ and the smoothed performance approaches the worst-case performance. For large σ, the influence of X is negligible compared to the perturbation, and the smoothed performance approaches the average-case performance.

Note further that we have restricted the choices of the adversary to points in $[0,1]^d$. Assuming scale-invariance of the underlying problem, this is no restriction and makes no difference for worst-case analysis. For smoothed analysis, however, we would have to scale σ in the same way.

Moreover, we observe that the (adversarial) choice of X in (2) can be different from the choice of X in (1). In worst-case analysis, the adversary picks an instance with worst performance. In smoothed analysis, the adversary chooses an instance X that maximizes the expected performance subject to the perturbation.

Finally, we remark that we do not require that a feasible or optimal solution for X remains a feasible or an optimal solution for Y, respectively. Roughly speaking, we are interested in the distribution of difficult instances and if difficult instances are isolated. This does not require that we can obtain a solution for X from a solution for the instance Y obtained by perturbing X.

1.3   Overview of Results Besides Local Search

Since its invention, smoothed analysis has been applied to a variety of algorithms and problems using a variety of perturbation models. We do not discuss the models here, but only give an overview to which algorithms and problems smoothed analysis has been applied. We also refer to two surveys about smoothed analysis that highlight different perspectives of smoothed analysis [46,57].

Linear programming and matrix problems. Smoothed analysis has originally been applied to the simplex method [56]. This analysis has subsequently been improved and simplified significantly [28,59]. Besides this, smoothed analysis has been applied to a variety of related algorithms and problems such as the perceptron algorithm [13], interior point methods [55], and condition numbers [19,20,29,52,58].

Integer programming and multi-criteria optimization. Starting with a smoothed analysis of the knapsack problem [7], a significant amount of research has been dedicated to understanding the solvability of integer programming problems and the size of Pareto curves in multi-criteria optimization problems [6,8,10,16,17, 49–51]. Beier and Vöcking's characterization of integer programming problems that can be solved in smoothed polynomial time [8] inspired an embedding of smoothed analysis into the existing worst-case and average-case complexity theory [11].

Graphs and formulas. Smoothed analysis can also be applied to purely discrete problems such as satisfiability of Boolean formulas [24,34,41] or graph problems [35,41,44,54]. However, it is much less obvious what a meaningful perturbation model is than in problems involving numbers.

Sorting and searching. Smoothed analysis has been applied to analyze problems based on permutations, most notably the quicksort algorithm [4,27,36,45].

Approximation ratios. Smoothed analysis has mostly been applied to analyze the running-time of algorithms, but there are also a few analyses of approximation ratios for Euclidean optimization problems [12,26] and packing problems [26,39].

Other applications. Other applications of smoothed analysis to concrete algorithms include online algorithms [5,53], algorithms for computing minimum cost flows [15,25], computational geometry [9,22], finding Nash equilibria [23], PAC learning [38], computing the edit distance [1], minimizing concave functions [40], balancing networks [37], and belief propagation for discrete optimization problems [14].

2   Local Search Algorithms

Local search algorithms are often very powerful tools to compute near-optimal solutions for hard combinatorial optimization problems. Starting from an initial solution, they iteratively try to improve the solution by small changes, until they terminate in a local optimum. While often showing a surprisingly good performance in practice, the theoretical performance of many local search heuristics is poor.

Smoothed Analysis has successfully been used to bridge the gap between the theoretical prediction of performance and the performance observed in practice and to explain the practical performance of a couple of local search algorithms. In most cases, the number of iterations until a local optimum is reached has been analyzed. Examples of local search algorithms whose running-time has been analyzed in the framework of smoothed analysis include the 2-opt heuristic for the traveling salesman problem (TSP) [31,48], the iterative closest point (ICP) algorithm to match point clouds [3], the k-means method for clustering [2,3,47], and the flip heuristic for the maximum cut problem [30,33].

Only a few results are known about the smoothed approximation ratio of local search algorithms. Examples are the 2-opt heuristic for the TSP [31,42] and the jump and lex-jump heuristic in scheduling [18,32].

In the following, we briefly sketch the main ideas how these results have been obtained.

2.1   Smoothed Analysis of the Running-Time

The key idea of all smoothed analyses of running-times of local search heuristics is the following: we use the objective function to measure progress. Then we show that, after perturbation, the objective function decreases (in case of a minimization problem) significantly with high probability either in every iteration or in every sequence of iterations.

More precisely: assume that the objective value of our initial solution is at most I, and assume further that the objective value decreases by at least δ in every iteration of the local search algorithm. Then (assuming that the objective value cannot become negative) we must reach a local optimum within at most I/δ iterations. An upper bound I for the initial solution is often relatively easy to get, and usually one that holds with high probability suffices. Thus, the main task is to analyze the minimal improvement δ.

The general outline to analyze δ is as follows: often, it is quite straightforward to show that the probability that some fixed iteration yields a small improvement is small. Then a simple union bound over all possible iterations yields a first bound for the probability that δ is small. However, the number of possible iterations can be quite large, which renders this bound useless. Thus, the goal is to analyze similar iterations together to avoid the wasteful and naive union bound. Hence, we want to come up with as few classes as possible such that for every class, we can show that it is unlikely that it contains an iteration that yields only a small improvement.

For the 2-opt heuristic for the TSP, one can get polynomial bounds for the smoothed running-time by considering single iterations, although better bounds can be obtained by considering pairs of iterations that share an edge [31].

For the k-means method for clustering, considering single iterations does not seem to be sufficient. In the case that the clustering does not change much from iteration to iteration, it seems to be possible that very small improvements occur. However, it is unlikely that a short sequence of such iterations yields only very small improvements [2]. Even more iterations have been considered together to analyze the flip heuristic for the maximum cut problem [33].

2.2   Smoothed Analysis of the Approximation Ratio

Much less is known about the smoothed approximation ratios of local search algorithms than about their smoothed running-time. This might be because the approximation ratio depends heavily on the initialization, and the worst local optima are often quite robust against slight perturbations. In light of this, the running-time becomes crucial for the approximation performance: if the local search heuristic terminates very quickly, we can afford to run it many times with different initializations. Hopefully, at least one initialization yields a good solution.

One way to get rid of the dependency of the initialization in the analysis is to compare the worst local optimum to the global optimum [31, 42]. This keeps the analysis tractable, but still often leads to results that are too pessimistic to reflect the performance observed in practice. In the following, we denote by WLO the objective value of the worst local optimum and by OPT the objective value of a global optimum.

A second technical difficulty is that WLO and OPT are not independent, and we would like to analyze their ratio. The simplest approach to circumvent this challenge is to replace WLO by a worst-case upper bound. While this again seems too pessimistic, it simplifies the analysis a lot: we are only left with analyzing $\mathbb{E}(\frac{1}{\text{OPT}})$ instead of $\mathbb{E}(\frac{\text{WLO}}{\text{OPT}})$. This approach has in particular been used for the 2-opt heuristic for the TSP. For the 2-opt heuristic, it is known that WLO $= O(n^{\frac{d-1}{d}})$ for tours of n points in $[0, 1]^d$ [21]. This has been exploited by Englert et al. [31] to prove a bound on the smoothed approximation ratio of the 2-opt heuristic.

However, ignoring the dependency between global and local optimum has significant limitations. What is bad for the approximation ratio is a large WLO together with a small OPT. Intuitively, in terms of the TSP, we get a very short optimal tour if the points are very close. But then also WLO should be small. The other way around, if there is a locally optimal TSP tour that is very long, then the points cannot be too close to each other. Hence, OPT cannot be too small. This information has been exploited to prove that the 2-opt heuristic achieves smoothed approximation ratio of $O(\log(1/\sigma))$ [42].

Still, simple construction heuristics for the TSP achieve approximation ratios of 2. Thus, the obvious open problem concerning smoothed approximation ratios is to analyze hybrid heuristics consisting of a clever initialization together with local search (see also Section 3). (It has been shown that using the nearest-neighbor heuristic to initialize 2-opt does not yield a better bound than $\Omega(\log n/ \log \log n)$ for sufficiently small σ [42].)

3   Open Problems

To conclude, we list three open problems concerning smoothed analysis of local search algorithms.

Lin-Kernighan heuristic for the TSP. The Lin-Kernighan heuristic [43] is an extremely powerful heuristic for finding near-optimal TSP tours quickly in practice. Unfortunately, different to the 2-opt heuristic, it seems to be difficult to describe iterations or sequences of iterations in a compact form in order to avoid a too wasteful union bound.

Flip heuristic for Max-Cut. Etscheid and Röglin [33] have recently shown that the smoothed number of iterations that the flip heuristic needs is bounded by a polynomial in $n^{\log n}$ and the perturbation parameter ϕ, where n is the number of nodes of the graph.

More general, we observe that the running-time of the flip heuristic is pseudo-polynomial. For integer programming problems, it is known that every problem that can be solved in pseudo-polynomial time can also be solved in smoothed polynomial time [8]. It would be interesting to see if something similar holds for local search heuristics, i.e., if every local search algorithm with pseudo-polynomial running-time has also smoothed polynomial running-time.

Approximation ratios with initialization. The existing results about smoothed approximation ratios of local search algorithms compare the worst local optimum to the global optimal solution [31,42]. However, the performance of local search heuristics relies heavily on a good initialization. The 2-opt heuristic is no exception, and the smoothed guarantees for the approximation ratio are easily beaten by choosing the initial tour with a constant-factor approximation algorithm.

Consequently, an obvious open problem is to take into account clever initializations when analyzing the approximation ratios of local search algorithms.

References

1. Andoni, A., Krauthgamer, R.: The smoothed complexity of edit distance. ACM Transactions on Algorithms 8(4), 44:1–44:25 (2012)
2. Arthur, D., Manthey, B., Röglin, H.: Smoothed analysis of the k-means method. Journal of the ACM 58(5) (2011)
3. Arthur, D., Vassilvitskii, S.: Worst-case and smoothed analysis of the ICP algorithm, with an application to the k-means method. SIAM Journal on Computing 39(2), 766–782 (2009)
4. Banderier, C., Beier, R., Mehlhorn, K.: Smoothed analysis of three combinatorial problems. In: Rovan, B., Vojtáš, P. (eds.) MFCS 2003. LNCS, vol. 2747, pp. 198–207. Springer, Heidelberg (2003)
5. Becchetti, L., Leonardi, S., Marchetti-Spaccamela, A., Schäfer, G., Vredeveld, T.: Average case and smoothed competitive analysis of the multilevel feedback algorithm. Mathematics of Operations Research 31(1), 85–108 (2006)
6. Beier, R., Röglin, H., Vöcking, B.: The smoothed number of pareto optimal solutions in bicriteria integer optimization. In: Fischetti, M., Williamson, D.P. (eds.) IPCO 2007. LNCS, vol. 4513, pp. 53–67. Springer, Heidelberg (2007)
7. Beier, R., Vöcking, B.: Random knapsack in expected polynomial time. Journal of Computer and System Sciences 69(3), 306–329 (2004)
8. Beier, R., Vöcking, B.: Typical properties of winners and losers in discrete optimization. SIAM Journal on Computing 35(4), 855–881 (2006)
9. de Berg, M., Haverkort, H.J., Tsirogiannis, C.P.: Visibility maps of realistic terrains have linear smoothed complexity. Journal of Computational Geometry 1(1), 57–71 (2010)
10. Berger, A., Röglin, H., van der Zwaan, R.: Internet routing between autonomous systems: Fast algorithms for path trading. Discrete Applied Mathematics 185, 8–17 (2015)
11. Bläser, M., Manthey, B.: Smoothed complexity theory. ACM Transactions on Computation Theory (to appear)
12. Bläser, M., Manthey, B., Rao, B.V.R.: Smoothed analysis of partitioning algorithms for Euclidean functionals. Algorithmica 66(2), 397–418 (2013)
13. Blum, A.L., Dunagan, J.D.: Smoothed analysis of the perceptron algorithm for linear programming. In: Proc. of the 13th Ann. ACM-SIAM Symp. on Discrete Algorithms (SODA), pp. 905–914. SIAM (2002)
14. Brunsch, T., Cornelissen, K., Manthey, B., Röglin, H.: Smoothed analysis of belief propagation for minimum-cost flow and matching. Journal of Graph Algorithms and Applications 17(6), 647–670 (2013)

15. Brunsch, T., Cornelissen, K., Manthey, B., Röglin, H., Rösner, C.: Smoothed analysis of the successive shortest path algorithm. Computing Research Repository 1501.05493 [cs.DS], arXiv (2015), a preliminary version has been presented at SODA (2013)

16. Brunsch, T., Goyal, N., Rademacher, L., Röglin, H.: Lower bounds for the average and smoothed number of pareto-optima. Theory of Computing 10(10), 237–256 (2014)

17. Brunsch, T., Röglin, H.: Improved smoothed analysis of multiobjective optimization. Journal of the ACM 62(1), 4 (2015)

18. Brunsch, T., Röglin, H., Rutten, C., Vredeveld, T.: Smoothed performance guarantees for local search. Mathematical Programming 146(1–2), 185–218 (2014)

19. Bürgisser, P., Cucker, F., Lotz, M.: Smoothed analysis of complex conic condition numbers. Journal de Mathématiques Pures et Appliquées 86(4), 293–309 (2006)

20. Bürgisser, P., Cucker, F., Lotz, M.: The probability that a slightly perturbed numerical analysis problem is difficult. Mathematics of Computation 77(263), 1559–1583 (2008)

21. Chandra, B., Karloff, H., Tovey, C.: New results on the old k-opt algorithm for the traveling salesman problem. SIAM Journal on Computing 28(6), 1998–2029 (1999)

22. Chaudhuri, S., Koltun, V.: Smoothed analysis of probabilistic roadmaps. Computational Geometry 42(8), 731–747 (2009)

23. Chen, X., Deng, X., Teng, S.H.: Settling the complexity of computing two-player Nash equilibria. Journal of the ACM 56(3) (2009)

24. Coja-Oghlan, A., Feige, U., Frieze, A.M., Krivelevich, M., Vilenchik, D.: On smoothed k-CNF formulas and the Walksat algorithm. In: Proc. of the 20th Ann. ACM-SIAM Symp. on Discrete Algorithms (SODA), pp. 451–460. SIAM (2009)

25. Cornelissen, K., Manthey, B.: Smoothed analysis of the minimum-mean cycle canceling algorithm and the network simplex algorithm. In: Proc. of the 21st Ann. Int. Computing and Combinatorics Conf. (COCOON). LNCS. Springer (to appear, 2015)

26. Curticapean, R., Künnemann, M.: A quantization framework for smoothed analysis of euclidean optimization problems. In: Bodlaender, H.L., Italiano, G.F. (eds.) ESA 2013. LNCS, vol. 8125, pp. 349–360. Springer, Heidelberg (2013)

27. Damerow, V., Manthey, B., Auf der Heide, F.M., Räcke, H., Scheideler, C., Sohler, C., Tantau, T.: Smoothed analysis of left-to-right maxima with applications. ACM Transactions on Algorithms 8(3) (2012)

28. Deshpande, A., Spielman, D.A.: Improved smoothed analysis of the shadow vertex simplex method. In: Proc. of the 46th Ann. IEEE Symp. on Foundations of Computer Science (FOCS), pp. 349–356. IEEE Computer Society (2005)

29. Dunagan, J., Spielman, D.A., Teng, S.H.: Smoothed analysis of condition numbers and complexity implications for linear programming. Mathematical Programming 126(2), 315–350 (2011)

30. Elsässer, R., Tscheuschner, T.: Settling the complexity of local max-cut (almost) completely. In: Aceto, L., Henzinger, M., Sgall, J. (eds.) ICALP 2011, Part I. LNCS, vol. 6755, pp. 171–182. Springer, Heidelberg (2011)

31. Englert, M., Röglin, H., Vöcking, B.: Worst case and probabilistic analysis of the 2-Opt algorithm for the TSP. Algorithmica 68(1), 190–264 (2014)

32. Etscheid, M.: Performance guarantees for scheduling algorithms under perturbed machine speeds. Discrete Applied Mathematics (to appear)

33. Etscheid, M., Röglin, H.: Smoothed analysis of local search for the maximum-cut problem. In: Proc. of the 25th Ann. ACM-SIAM Symp. on Discrete Algorithms (SODA), pp. 882–889. SIAM (2014)

34. Feige, U.: Refuting smoothed 3CNF formulas. In: Proc. of the 48th Ann. IEEE Symp. on Foundations of Computer Science (FOCS), pp. 407–417. IEEE Computer Society (2007)

35. Flaxman, A.D., Frieze, A.M.: The diameter of randomly perturbed digraphs and some applications. Random Structures and Algorithms $30(4)$, 484–504 (2007)

36. Fouz, M., Kufleitner, M., Manthey, B., Zeini Jahromi, N.: On smoothed analysis of quicksort and Hoare's find. Algorithmica $62(3\text{--}4)$, 879–905 (2012)

37. Friedrich, T., Sauerwald, T., Vilenchik, D.: Smoothed analysis of balancing networks. Random Structures and Algorithms $39(1)$, 115–138 (2011)

38. Kalai, A.T., Samorodnitsky, A., Teng, S.H.: Learning and smoothed analysis. In: Proc. of the 50th Ann. IEEE Symp. on Foundations of Computer Science (FOCS), pp. 395–404. IEEE Computer Society (2009)

39. Karger, D., Onak, K.: Polynomial approximation schemes for smoothed and random instances of multidimensional packing problems. In: Proc. of the 18th Ann. ACM-SIAM Symp. on Discrete Algorithms (SODA), pp. 1207–1216. SIAM (2007)

40. Kelner, J.A., Nikolova, E.: On the hardness and smoothed complexity of quasiconcave minimization. In: Proc. of the 48th Ann. IEEE Symp. on Foundations of Computer Science (FOCS), pp. 472–482 (2007)

41. Krivelevich, M., Sudakov, B., Tetali, P.: On smoothed analysis in dense graphs and formulas. Random Structures and Algorithms $29(2)$, 180–193 (2006)

42. Künnemann, M., Manthey, B.: Towards understanding the smoothed approximation ratio of the 2-opt heuristic. In: Proc. of the 42nd Int. Coll. on Automata, Languages and Programming (ICALP). LNCS. Springer (to appear, 2015)

43. Lin, S., Kernighan, B.W.: An effective heuristic for the traveling-salesman problem. Operations Research $21(2)$, 498–516 (1973)

44. Manthey, B., Plociennik, K.: Approximating independent set in perturbed graphs. Discrete Applied Mathematics $161(12)$, 1761–1768 (2013)

45. Manthey, B., Reischuk, R.: Smoothed analysis of binary search trees. Theoretical Computer Science $378(3)$, 292–315 (2007)

46. Manthey, B., Röglin, H.: Smoothed analysis: Analysis of algorithms beyond worst case. it - Information Technology $53(6)$, 280–286 (2011)

47. Manthey, B., Röglin, H.: Worst-case and smoothed analysis of k-means clustering with Bregman divergences. Journal of Computational Geometry $4(1)$, 94–132 (2013)

48. Manthey, B., Veenstra, R.: Smoothed analysis of the 2-Opt heuristic for the TSP: polynomial bounds for gaussian noise. In: Cai, L., Cheng, S.-W., Lam, T.-W. (eds.) ISAAC 2013. LNCS, vol. 8283, pp. 579–589. Springer, Heidelberg (2013)

49. Moitra, A., O'Donnell, R.: Pareto optimal solutions for smoothed analysts. SIAM Journal on Computing $41(5)$, 1266–1284 (2012)

50. Röglin, H., Teng, S.H.: Smoothed analysis of multiobjective optimization. In: Proc. of the 50th Ann. IEEE Symp. on Foundations of Computer Science (FOCS), pp. 681–690. IEEE Computer Society (2009)

51. Röglin, H., Vöcking, B.: Smoothed analysis of integer programming. Mathematical Programming $110(1)$, 21–56 (2007)

52. Sankar, A., Spielman, D.A., Teng, S.H.: Smoothed analysis of the condition numbers and growth factors of matrices. SIAM Journal on Matrix Analysis and Applications $28(2)$, 446–476 (2006)

53. Schäfer, G., Sivadasan, N.: Topology matters: Smoothed competitiveness of metrical task systems. Theoretical Computer Science **241**(1–3), 216–246 (2005)
54. Spielman, D.A., Teng, S.-H.: Smoothed analysis. In: Dehne, F., Sack, J.-R., Smid, M. (eds.) WADS 2003. LNCS, vol. 2748, pp. 256–270. Springer, Heidelberg (2003)
55. Spielman, D.A., Teng, S.H.: Smoothed analysis of termination of linear programming algorithms. Mathematical Programming, Series B **97**(1–2), 375–404 (2003)
56. Spielman, D.A., Teng, S.H.: Smoothed analysis of algorithms: Why the simplex algorithm usually takes polynomial time. Journal of the ACM **51**(3), 385–463 (2004)
57. Spielman, D.A., Teng, S.H.: Smoothed analysis: An attempt to explain the behavior of algorithms in practice. Communications of the ACM **52**(10), 76–84 (2009)
58. Tao, T., Vu, V.H.: Smooth analysis of the condition number and the least singular value. Mathematics of Computation **79**(272), 2333–2352 (2010)
59. Vershynin, R.: Beyond Hirsch conjecture: Walks on random polytopes and smoothed complexity of the simplex method. SIAM Journal on Computing **39**(2), 646–678 (2009)

Optimal Shuffle Code with Permutation Instructions

Sebastian Buchwald, Manuel Mohr$^{(\boxtimes)}$, and Ignaz Rutter

Karlsruhe Institute of Technology, Karlsruhe, Germany
{sebastian.buchwald,manuel.mohr,rutter}@kit.edu

Abstract. During compilation of a program, register allocation is the task of mapping program variables to machine registers. During register allocation, the compiler may introduce *shuffle code*, consisting of copy and swap operations, that transfers data between the registers. Three common sources of shuffle code are conflicting register mappings at joins in the control flow of the program, e.g, due to if-statements or loops; the calling convention for procedures, which often dictates that input arguments or results must be placed in certain registers; and machine instructions that only allow a subset of registers to occur as operands.

Recently, Mohr et al. [9] proposed to speed up shuffle code with special hardware instructions that arbitrarily permute the contents of up to five registers and gave a heuristic for computing such shuffle codes.

In this paper, we give an efficient algorithm for generating optimal shuffle code in the setting of Mohr et al. An interesting special case occurs when no register has to be transferred to more than one destination, i.e., it suffices to permute the contents of the registers. This case is equivalent to factoring a permutation into a minimal product of permutations, each of which permutes up to five elements.

1 Introduction

One of the most important tasks of a compiler during code generation is register allocation, which is the task of mapping program variables to machine registers. During this phase, it is frequently necessary to insert so-called *shuffle code* that transfers values between registers. Common reasons for the insertion of shuffle code are control flow joins, procedure calling conventions and constrained machine instructions.

The specification of a shuffle code, i.e., a description which register contents should be transferred to which registers, can be formulated as a directed graph whose vertices are the registers and an edge (u, v) means that the content of u before the execution of the shuffle code must be in v after the execution. Naturally, every vertex must have at most one incoming edge. Note that vertices may have several outgoing edges, indicating that their contents must be transferred to several destinations, and even loops (u, u), indicating that the content of register u must be preserved. We call such a graph a *Register Transfer Graph* or *RTG*.

© Springer International Publishing Switzerland 2015
F. Dehne et al. (Eds.): WADS 2015, LNCS 9214, pp. 528–541, 2015.
DOI: 10.1007/978-3-319-21840-3_44

Fig. 1. Two example RTGs where the optimal shuffle code is not obvious

Two important special types of RTGs are outdegree-1 RTGs where the maximum out-degree is 1 and PRTGs where $\deg^-(v) = \deg^+(v) = 1$ for all vertices v ($\deg^-$ and $\deg^+$ denote the in- and out-degree of a vertex, respectively).

We say that a shuffle code, consisting of a sequence of copy and swap operations on the registers, *implements* an RTG if after the execution of the shuffle code every register whose corresponding vertex has an incoming edge has the correct content. The *shuffle code generation* problem asks for a shortest shuffle code that implements a given RTG.

The amount of shuffle code directly depends on the quality of copy coalescing, a subtask of register allocation [9]. As copy coalescing is NP-complete [2], reducing the amount of shuffle code is expensive in terms of compilation time, and thus cannot be afforded in all contexts, e.g., just-in-time compilation.

Therefore, it has been suggested to allow more complicated operations than simply copying and swapping to enable more efficient shuffle code. Mohr et al. [9] propose to allow performing permutations on the contents of small sets of up to five registers. The processor they develop offers three instructions to implement shuffle code:

copy: copies the content of one register to another one

permi5: cyclically shifts the contents of up to five registers

permi23: swaps the contents of two registers and performs a cyclic shift of the contents of up to three registers; the two sets of registers must be disjoint.

In fact, the two operations permi5 and permi23 together allow to arbitrarily permute the contents of up to five registers in a single operation. A corresponding hardware and a modified compiler that employs a greedy approach to generate the shuffle code have been shown to improve performance in practice [9]. While the greedy heuristic works well in practice, it does not find an optimal shuffle code in all cases.

It is not obvious how to generate optimal shuffle code using the three instructions copy, permi5 and permi23 even for small RTGs. In the left RTG from Fig. 1, a naive solution would implement edges $(1, 2)$ and $(1, 3)$ using copies and the remaining cycle (4 5 6) using a permi5. However, using one permi23 to implement the cycle (4 5 6) and swap registers 1 and 2, and then copying register 2 to 3 requires only two instructions. This is legal because the contents of register 1 can be overwritten. The same trick is not applicable for the right RTG in Fig. 1 because of the loop $(1, 1)$ and hence three instructions are necessary to implement that RTG.

A maximum permutation size of 5 may seem arbitrary at first but is a consequence of instruction encoding constraints. In each permi instruction, the register numbers and their order must be encoded in the instruction word. Hence,

$\lceil \log_2 \left(\binom{n}{k} k! \right) \rceil$ bits of an instruction word are needed to be able to encode all permutations of k registers out of n total registers. As many machine architectures use a fixed size for instruction words, e.g., 32 or 64 bits, and the operation type must also be encoded in the instruction word, space is very limited. In fact, for a 32 bit instruction word, 34 is the maximum number of registers that leave enough space for the operation type.

Related Work. As long as only copy and swap operations are allowed, finding an optimal shuffle code for a given RTG is a straightforward task [7, p. 56–57]. Therefore work in the area of compiler construction in this context has focused on coalescing techniques that reduce the number and the size of RTGs [1,2,6,8].

From a theoretical point of view, the most closely related work studies the case where the input RTG consists of a union of disjoint directed cycles, which can be interpreted as a permutation π. Then, no copy operations are necessary for an optimal shuffle code and hence the problem of finding an optimal shuffle code using permi23 and permi5 is equivalent to writing π as a shortest product of permutations of maximum size 5, where a permutation of n elements has size k if it fixes $n - k$ elements.

There has been work on writing a permutation as a product of permutations that satisfy certain restrictions. The factorization problem on permutation groups from computational group theory [10] is the task of writing an element g of a permutation group as a product of given generators S. Hence, an algorithm for solving the factorization problem could be applied in our context by using all possible permutations of size 5 or less as the set S. However, the algorithms do not guarantee minimality of the product. For the case that S consists of all permutations that reverse a contiguous subsequence of the elements, known as the pancake sorting problem, it has been shown that computing a factoring of minimum size is NP-complete [4].

Farnoud and Milenkovic [5] consider a weighted version of factoring a permutation into transpositions. They present a polynomial constant-factor approximation algorithm for factoring a given permutation into transpositions where transpositions have arbitrary non-negative costs. In our problem, we cannot assign costs to an individual transposition as its cost is context-dependent, e.g., four transpositions whose product is a cycle require one operation, whereas four arbitrary transpositions may require two.

Contribution and Outline. In this paper, we present an efficient algorithm for generating optimal shuffle code using the operations copy, permi5, and permi23, or equivalently, using copy operations and permutations of size at most 5.

We first prove the existence of a special type of optimal shuffle codes whose copy operations correspond to edges of the input RTG in Section 2. Removing the set of edges implemented by copy operations from an RTG leaves an outdegree-1 RTG.

We show that the greedy algorithm proposed by Mohr et al. [9] finds optimal shuffle codes for outdegree-1 RTGs and that the size of an optimal shuffle code can be expressed as a function that depends only on three characteristic numbers of the outdegree-1 RTG rather than on its structure. Since PRTGs are a special case of outdegree-1 RTGs, this shows that GREEDY is a linear-time algorithm for factoring an arbitrary permutation into a minimum number of permutations of size at most 5.

Finally, in Section 4, we show how to compute an optimal set of RTG edges that will be implemented by copy operations such that the remaining outdegree-1 RTG admits a shortest shuffle code. This is done by several dynamic programs for the cases that the input RTG is disconnected, is a tree, or is connected and contains a (single) cycle. Proofs omitted due to space constraints can be found in the full version of this paper [3].

2   Register Transfer Graphs and Optimal Shuffle Codes

In this section, we rephrase the shuffle code generation problem as a graph problem. An RTG that has only self-loops needs no shuffle-code and is called *trivial*.

It is easy to define the effect of a permutation on an RTG. Let G be an RTG and let π be an arbitrary permutation that is applied to the contents of the registers. We define $\pi G = (V, \pi E)$, where $\pi E = \{(\pi(u), v) \mid (u, v) \in E\}$. This models the fact that if v should receive the data contained in u, then after π moves the data contained in u to some other register $\pi(u)$, the data contained in $\pi(u)$ should end up in v. We observe that for two permutations π_1, π_2 of V, it is $(\pi_2 \circ \pi_1)G = \pi_2(\pi_1(G))$, i.e., we have defined a group action of the symmetric group on RTGs. For PRTGs, the shuffle code generation problem asks for a shortest shuffle code that makes the given PRTG trivial.

Unfortunately, it is not possible to directly express copy operations in RTGs. Instead, we rely on the following observation. Consider an arbitrary shuffle code that contains a copy $a \to b$ with source a and target b that is followed by a transposition τ of the contents of registers c and d. We can replace this sequence by a transposition of the registers $\{c, d\}$ and a copy $\tau(a) \to \tau(b)$. Thus, given a sequence of operations, we can successively move the copy operations to the end of the sequence without increasing its length. Thus, for any RTG there exists a shuffle code that consists of a pair of sequences $((\pi_1, \ldots, \pi_p), (c_1, \ldots, c_t))$, where the π_i are permutation operations and the c_i are copy operations. We now strengthen our assumption on the copy operations.

Lemma 1. *Every instance of the shuffle code generation problem has an optimal shuffle code $((\pi_1, \ldots, \pi_p), (c_1, \ldots, c_t))$ such that*
(i) No register occurs as both a source and a target of copy operations.
(ii) Every register is the target of at most one copy operation.
(iii) There is a bijection between the copy operations c_i and the edges of πG that are not loops, where $\pi = \pi_p \circ \pi_{p-1} \circ \cdots \circ \pi_1$.

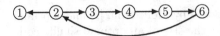

(a) The original RTG G needs one per-   (b) After removing the edge $(2,3)$, the
mutation and one copy operation   RTG needs two permutation operations

Fig. 2. The RTG G obtains the normalized optimal shuffle code (π_1, c_1), where $\pi_1 = (23456)$ and $c_1 = 3 \to 1$. However, after removing the edge $(2,3)$ (instead of $(1,2)$) we cannot achieve an optimal solution anymore.

 (iv) *If u is the source of a copy operation, then u is incident to a loop in πG.*
 (v) *The number of copies is $\sum_{v \in V} \max\{\deg_G^+(v) - 1, 0\}$.*

We call a shuffle code satisfying the conditions of Lemma 1 *normalized*. Observe that the number of copy operations used by a normalized shuffle code is a lower bound on the number of necessary copy operations since permutations cannot copy values.

Consider now an RTG G together with a normalized optimal shuffle code and one of its copy operations $u \to v$. Since the code is normalized, the value transferred to v by this copy operation is the one that stays there after the shuffle code has been executed. If v had no incoming edge in G, then we could shorten the shuffle by omitting the copy operation. Thus, v has an incoming edge (u', v) in G, and we associate the copy $u \to v$ with the edge (u', v) of G. In fact, $u' = \pi^{-1}(u)$, where $\pi = \pi_p \circ \cdots \circ \pi_1$. In this way, we associate every copy operation with an edge of the input RTG. In fact, this is an injective mapping by Lemma 1 (ii).

Lemma 2. *Let $((\pi_1, \ldots, \pi_p), (c_1, \ldots, c_t))$ be an optimal shuffle code S for an RTG $G = (V, E)$ and let $C \subseteq E$ be the edges that are associated with copies in S. Then*
 (i) *Every vertex v has $\max\{\deg_G^+(v) - 1, 0\}$ outgoing edges in C.*
 (ii) *$G - C$ is an outdegree-1 RTG.*
 (iii) *$\pi_1, \ldots, \pi_p$ is an optimal shuffle code for $G - C$.*

Lemma 2 shows that an optimal shuffle code for an RTG G can be found by 1) picking for each vertex one of its outgoing edges (if it has any) and removing the remaining edges from G, 2) finding an optimal shuffle code for the resulting outdegree-1 RTG, and 3) creating one copy operation for each of the previously removed edges. Fig. 2 shows that the choice of the outgoing edges is crucial to obtain an optimal shuffle code.

In the following, we first show how to compute an optimal shuffle code for an outdegree-1 RTG in Section 3. Afterwards, in Section 4, we design an algorithm for efficiently determining a set of edges to be removed such that the resulting outdegree-1 RTG admits a shuffle code with the smallest number of operations.

3 Optimal Shuffle Code for Outdegree-1 RTGs

In this section we prove the optimality of the greedy algorithm proposed by Mohr et al. [9] for outdegree-1 RTGs. Before we formulate the algorithm, let us look at the effect of applying a transposition $\tau = (u\ v)$ to contiguous vertices of a k-cycle $K = (V_K, E_K)$ in a PRTG G, where k-cycle denotes a cycle of size k. Hence, $u, v \in V_K$ and $(u, v) \in E_K$. Then, in τG, the cycle K is replaced by a $(k - 1)$-cycle and a vertex v with a loop. We say that τ has reduced the size of K by 1. If τK is trivial, we say that τ resolves K. It is easy to see that `permi5` reduces the size of a cycle by up to 4 and `permi23` reduces the sizes of two distinct cycles by 1 and up to 2, respectively. We can now formulate GREEDY as follows.

1. Complete each directed path of the input outdegree-1 RTG into a directed cycle, thereby turning the input into a PRTG.
2. While there exists a cycle K of size at least 4, apply a `permi5` operation to reduce the size of K as much as possible.
3. While there exist a 2-cycle and a 3-cycle, resolve them with a `permi23` operation.
4. Resolve pairs of 2-cycles by `permi23` operations.
5. Resolve triples of 3-cycles by pairs of `permi23` operations.

We claim that GREEDY computes an optimal shuffle code. Let G be an outdegree-1 RTG and let Q denote the set of paths and cycles of G. For a path or cycle $\sigma \in Q$, we denote by $\text{size}(\sigma)$ the number of vertices of σ. Define $X = \sum_{\sigma \in Q} \lfloor \text{size}(\sigma)/4 \rfloor$ and $a_i = |\{\sigma \in Q \mid \text{size}(\sigma) = i \mod 4\}|$ for $i = 2, 3$. We call the triple $\text{sig}(G) = (X, a_2, a_3)$ the *signature* of G.

Lemma 3. *Let G be an outdegree-1 RTG with $\text{sig}(G) = (X, a_2, a_3)$. The number GREEDY$(G)$ of operations in the shuffle code produced by the greedy algorithm is GREEDY$(G) = X + \max\{\lceil (a_2 + a_3)/2 \rceil, \lceil (a_2 + 2a_3)/3 \rceil\}$.*

In particular, the length of the shuffle code computed by GREEDY only depends on the signature of the input RTG G. In the following, we prove that GREEDY is optimal for outdegree-1 RTGs and hence GREEDY(G) is the length of an optimal shuffle code.

Lemma 4. *Let G, G' be PRTGs with $\text{sig}(G) = (X, a_2, a_3)$, $\text{sig}(G') = (X', a_2', a_3')$ and GREEDY(G) − GREEDY$(G') \geq c$, and let $(\Delta_X, \Delta_2, \Delta_3) = \text{sig}(G) - \text{sig}(G')$. If $a_2 \geq a_3$, then $2\Delta_X + \Delta_2 + \Delta_3 \leq -2c + 1$. If $a_3 > a_2$, then $3\Delta_X + \Delta_2 + 2\Delta_3 \leq -3c + 2$.*

Proof (sketch). We assume that $a_2 \geq a_3$, the other case is analogous. By Lemma 3 GREEDY$(G) \leq X + (a_2 + a_3 + 1)/2$ and GREEDY$(G') \geq X' + (a_2' + a_3')/2$. Therefore, GREEDY$(G)$ − GREEDY$(G') \leq -\Delta_X - (\Delta_2 + \Delta_3 - 1)/2 = -(2\Delta_X + \Delta_2 + \Delta_3 - 1)/2$. By assumption, $-(2\Delta_X + \Delta_2 + \Delta_3 - 1)/2 \geq c$; this is equivalent to the claim. $\square$

Lemma 4 gives us necessary conditions for when the GREEDY solutions of two RTGs differ by some value c. These necessary conditions depend only on

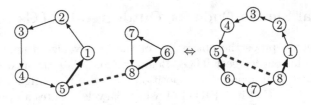

Fig. 3. The transposition $\tau = (5\ 8)$ acting on PRTGs. Affected edges are drawn thick. Read from left to right, the transposition is a merge; read from right to left, it is a split.

Table 1. Signature changes and Ψ values for merges. Row and column are the cycle sizes modulo 4 before the merge.

$\|$	0	1	2	3
0	$(0,0,0)$	$(0,0,0)$	$(0,0,0)$	$(0,0,0)$
1		$(0,1,0)$	$(0,-1,1)$	$(1,0,-1)$
2			$(1,-2,0)$	$(1,-1,-1)$
3				$(1,1,-2)$

$\|$	0	1	2	3
0	0	0	0	0
1		1	0	1
2			0	0
3				1

$\|$	0	1	2	3
0	0	0	0	0
1		1	1	1
2			1	0
3				0

(a) Signature change $(\Delta_X, \Delta_2, \Delta_3)$ (b) Values of Ψ_1 (left) and Ψ_2 (right)

the difference of the two signatures. To study them more precisely, we define $\Psi_1(\Delta_X, \Delta_2, \Delta_3) = 2\Delta_X + \Delta_2 + \Delta_3$ and $\Psi_2(\Delta_X, \Delta_2, \Delta_3) = 3\Delta_X + \Delta_2 + 2\Delta_3$. Next, we study the effect of a single transposition on these two functions.

Let $G = (V, E)$ be a PRTG with $\text{sig}(G) = (X, a_2, a_3)$ and let τ be a transposition of two elements in V. We distinguish cases based on whether the swapped elements are in different connected components or not. In the former case, we say that τ is a *merge*, in the latter we call it a *split*; see Fig. 3 for an illustration.

We start with the merge operations. When merging two cycles of size s_1 and s_2, respectively, they are replaced by a single cycle of size $s_1 + s_2$. Note that removing the two cycles may decrease the values a_2 and a_3 of the signature by at most 2 in total. The new cycle can potentially increase one of these values by 1. The value X never decreases, and it increases by 1 if and only if $s_1 \mod 4 + s_2 \mod 4 \geq 4$. Table 1a shows the possible signature changes $(\Delta_X, \Delta_2, \Delta_3)$ resulting from a merge. The entry in row i and column j shows the result of merging two cycles whose sizes modulo 4 are i and j, respectively. Table 1b shows the corresponding values of Ψ_1 and Ψ_2. Only entries with $i \leq j$ are shown, the remaining cases are symmetric.

Lemma 5. *Let G be a PRTG with $\text{sig}(G) = (X, a_2, a_3)$ and let τ be a merge. Then* $\text{GREEDY}(G) \leq \text{GREEDY}(\tau G)$.

Proof. Suppose $\text{GREEDY}(\tau G) < \text{GREEDY}(G)$. Then $\text{GREEDY}(G) - \text{GREEDY}(\tau G) \geq 1$ and by Lemma 4 either $\Psi_1 \leq -1$ or $\Psi_2 \leq -1$. However, Table 1b shows the values of Ψ_1 and Ψ_2 for all possible merges. In all cases it is $\Psi_1, \Psi_2 \geq 0$. A contradiction.   □

Fig. 4. Transition graphs for Ψ_1 (left) and Ψ_2 (right)

In particular, the lemma shows that merges never decrease the cost of the greedy solution, even if they were for free. We now make a similar analysis for splits. It is, however, obvious that splits indeed may decrease the cost of greedy solutions. In fact, one can always split cycles in a PRTG until it is trivial.

First, we study again the effect of splits on the signature change $(\Delta_X, \Delta_2, \Delta_3)$. Since a split is an inverse of a merge, we can essentially reuse Table 1a. If merging two cycles whose sizes modulo 4 are i and j, respectively, results in a signature change of $(\Delta_X, \Delta_2, \Delta_3)$, then, conversely, we can split a cycle whose size modulo 4 is $i + j$ into two cycles whose sizes modulo 4 are i and j, respectively, such that the signature change is $(-\Delta_X, -\Delta_2, -\Delta_3)$, and vice versa. Note that given a cycle whose size modulo 4 is s one has to look at all cells (i, j) with $i + j \equiv s \pmod 4$ to consider all the possible signature changes. Since Ψ_1, Ψ_2 are linear, negating the signature change also negates the corresponding value. Thus, we can reuse Table 1b for splits by negating each entry.

Lemma 6. *Let $G = (V, E)$ be a PRTG and let π be a cyclic shift of c vertices in V. Let further $(\Delta_X, \Delta_2, \Delta_3)$ be the signature change affected by π. Then $\Psi_1(\Delta_X, \Delta_2, \Delta_3) \geq -\lceil (c-1)/2 \rceil$ and $\Psi_2(\Delta_X, \Delta_2, \Delta_3) \geq -\lceil (3c-3)/4 \rceil$.*

Proof. We can write $\pi = \tau_{c-1} \circ \cdots \circ \tau_1$ as a product of $c - 1$ transpositions such that any two consecutive transpositions τ_i and τ_{i+1} affect a common element for $i = 1, \ldots, c - 1$.

Each transposition decreases Ψ_1 (or Ψ_2) by at most 1, but a decrease happens only for certain split operations. However, it is not possible to reduce Ψ_1 (or Ψ_2) with every single transposition since for two consecutive splits the second has to split one of the connected components resulting from the previous splits. To get an overview of the sequences of splits that reduce the value of Ψ_1 (or of Ψ_2) by 1 for each split, we consider the following transition graphs T_k for Ψ_k $(k = 1, 2)$ on the vertex set $S = \{0, 1, 2, 3\}$. In the graph T_k there is an edge from i to j if there is a split that splits a component of size i mod 4 such that one of the resulting components has size j mod 4 and this split decreases Ψ_k by 1. The transition graphs T_1 and T_2 are shown in Fig. 4.

For Ψ_1 the longest path in the transition graph has length 1. Thus, the value of Ψ_1 can be reduced at most every second transposition and $\Psi_1(\Delta_X, \Delta_2, \Delta_3) \geq -\lceil (c - 1)/2 \rceil$.

For Ψ_2 the longest path has length 3 (vertex 1 has out-degree 0). Therefore, after at most three consecutive steps that decrease Ψ_2, there is one that does not. It follows that at least $\lfloor (c-1)/4 \rfloor$ operations do not decrease Ψ_2, and consequently at most $\lceil (3c-3)/4 \rceil$ operations decrease Ψ_2 by 1. Thus, $\Psi_2(\Delta_X, \Delta_2, \Delta_3) \geq -\lceil (3c-3)/4 \rceil$. $\qquad\qquad\qquad\qquad\qquad\qquad\qquad\qquad\qquad\qquad\qquad\qquad\qquad\quad\square$

Since `permi5` performs a single cyclic shift and `permi23` is the concatenation of two cyclic shifts, Lemmas 6 and 4 can be used to show that no such operation may decrease the number of operations GREEDY has to perform by more than 1.

Corollary 1. *Let G be a PRTG and let π be an operation, i.e., either a* `permi23` *or a* `permi5`*. Then* GREEDY$(G) \leq$ GREEDY$(\pi G) + 1$.

Using this corollary and an induction on the length of an optimal shuffle code, we show that GREEDY is optimal for PRTGs; if no operation reduces the number of operations GREEDY needs by more than 1, why not use the operation suggested by GREEDY?

Theorem 1. *Let G be a PRTG. An optimal shuffle code for G takes* GREEDY(G) *operations. Algorithm* GREEDY *computes an optimal shuffle code in linear time.*

Moreover, since merge operations may not decrease the cost of GREEDY and any PRTG that can be formed from the original outdegree-1 RTG G by inserting edges can be obtained from the PRTG G' formed by GREEDY and a sequence of merge operations, it follows that the length of an optimal shuffle for G is GREEDY(G'). Thus, GREEDY is optimal for outdegree-1 RTGs.

Theorem 2. *Let G be an outdegree-1 RTG. Then an optimal shuffle code for G requires* GREEDY(G) *operations.* GREEDY *computes such a shuffle code in linear time.*

4   The General Case

In this section we study the general case. A *copy set* of an RTG $G = (V, E)$ is a set $C \subseteq E$ such that $G - C = (V, E - C)$ is an outdegree-1 RTG and $|C| = \sum_{v \in V} \max\{\deg^+(v) - 1, 0\}$. We denote by $\mathcal{C}(G)$ the set of all copy sets of G. According to Lemma 2 an optimal shuffle code for G can be found by finding a copy set $C \in \mathcal{C}(G)$ such that the outdegree-1 RTG $G - C$ admits a shortest shuffle code. By Theorem 2 an optimal shuffle code for $G - C$ can be computed with the greedy algorithm and its length can be computed according to Lemma 3. We thus seek a copy set $C \in \mathcal{C}(G)$ that minimizes the cost function GREEDY$(G - C) = X + \max\{\lceil (a_2 + a_3)/2 \rceil, \lceil (a_2 + 2a_3)/3 \rceil\}$, where (X, a_2, a_3) is the signature of $G - C$. Such a copy set is called *optimal*. Clearly, this is equivalent to minimizing the function

$$\text{GREEDY}'(G-C) = X + \max\left\{\frac{a_2 + a_3}{2}, \frac{a_2 + 2a_3}{3}\right\} = \begin{cases} X + \frac{a_2}{2} + \frac{a_3}{2} & \text{if } a_2 \geq a_3 \\ X + \frac{a_2}{3} + \frac{2a_3}{3} & \text{if } a_2 < a_3 \end{cases}$$

To keep track of which case is used for evaluating GREEDY$'$, we define diff$(G - C) = a_2 - a_3$ and compute for each of the two function parts and every possible value d a copy set C_d with diff$(G - C_d) = d$ that minimizes that function. More formally, we define cost$^1(G - C) = X + \frac{1}{2}a_2 + \frac{1}{2}a_3$ and cost$^2(G - C) = X + \frac{1}{3}a_2 + \frac{2}{3}a_3$ and we seek two tables $T_G^1[\cdot], T_G^2[\cdot]$, such that $T_G^i[d]$ is the smallest cost cost$^i(G-C)$ that can be achieved with a copy set $C \in \mathcal{C}(G)$ with diff$(G-C) = d$. We observe that $T_G^i[d] = \infty$ for $d < -n$ and for $d > n$. The following lemma shows that the length of an optimal shuffle code can be computed from these two tables.

Lemma 7. *Let $G = (V, E)$ be an RTG. The length of an optimal shuffle code for G is $\sum_{v \in V} \max\{\deg^+(v) - 1, 0\} + \min\{\min_{d \geq 0} \lceil T_G^1[d] \rceil, \min_{d < 0} \lceil T_G^2[d] \rceil\}$.*

In the following, we show how to compute for an RTG G a table $T_G[\cdot]$ with

$$T_G[d] = \min_{\substack{C \in \mathcal{C}(G) \\ \text{diff}(G-C)=d}} \text{cost}(G - C)$$

for an arbitrary cost function cost$(G - C) = c(\text{sig}(G - C))$, where c is a linear function. This is done in several steps depending on whether G is disconnected, is a tree, or is connected and contains a cycle. Before we continue, we introduce several preliminaries to simplify the following calculations. We denote by P_s a directed path on s vertices.

Definition 1. *A map f that assigns a value to an outdegree-1 RTG is signature-linear if there exists a linear function $g: \mathbb{R}^3 \to \mathbb{R}$ such that $f(G) = g(\text{sig}(G))$ for every outdegree-1 RTG G. For a signature-linear function f, $\Delta_f(s) = f(P_{s+1}) - f(P_s)$ is the correction term.*

Note that both cost $= c \circ \text{sig}$ and diff $= d \circ \text{sig}$ with $d(X, a_2, a_3) = a_2 - a_3$ are signature-linear. The correction term $\Delta_f(s)$ describes the change of f when the size of one connected component is increased from s to $s + 1$.

Lemma 8. *Let f be a signature-linear function. Then the following hold:*
(i) $f(G_1 \cup G_2) = f(G_1) + f(G_2)$ for disjoint outdegree-1 RTGs G_1, G_2,
(ii) Let $G = (V, E)$ be an outdegree-1 RTG and let $v \in V$ with in-degree 0. Denote by s the size of the connected component containing v and let $G^+ = (V \cup \{u\}, E \cup \{(u, v)\})$ where u is a new vertex. Then $f(G^+) = f(G) + \Delta_f(s)$.

Note that $\Delta_f(s) = \Delta_f(s+4)$ for all values of s and hence it suffices to know the size of the enlarged component modulo 4.

The main idea for computing table $T_G[\cdot]$ by dynamic programming is to decompose G into smaller edge-disjoint subgraphs $G = G_1 \cup \cdots \cup G_k$ such that the copy sets of G can be constructed from copy sets for each of the G_i. We call such a decomposition *proper partition* if for every vertex v of G there exists an index i such that G_i contains all outgoing edges of v. Let $G_1, \ldots, G_k$ be a proper partition of G and let $\mathcal{C}_i \subseteq \mathcal{C}(G_i)$ for $i = 1, \ldots, k$. We define $\mathcal{C}_1 \otimes \cdots \otimes \mathcal{C}_k = \{C_1 \cup \cdots \cup C_k \mid C_i \in \mathcal{C}_i, i = 1, \ldots, k\}$. It is not hard to see that $\mathcal{C}(G_1 \cup \cdots \cup G_k) = \mathcal{C}(G_1) \otimes \cdots \otimes \mathcal{C}(G_k)$.

Disconnected RTGs. We start with the case that G is disconnected and consists of connected components $G_1, \ldots, G_k$, which form a proper partition of G. The main issue is to keep track of diff and cost. For an RTG G, we define $\mathcal{C}(G; d) = \{C \in \mathcal{C}(G) \mid \text{diff}(G - C) = d\}$. By Lemma 8(i) and the signature-linearity of diff, if $C_i \in \mathcal{C}(G_i; d_i)$ for $i = 1, 2$, then $C_1 \cup C_2 \in \mathcal{C}(G_1 \cup G_2; d_1 + d_2)$. This leads to the following lemma.

Lemma 9. *Let G be an RTG and let G_1, G_2 be vertex-disjoint RTGs. Then*
(i) $\mathcal{C}(G) = \bigcup_d \mathcal{C}(G; d)$ and (ii) $\mathcal{C}(G_1 \cup G_2; d) = \bigcup_{d'} (\mathcal{C}(G_1; d') \otimes \mathcal{C}(G_2; d - d'))$.

By further exploiting the signature-linearity of cost, we also get $\text{cost}((G_1 \cup G_2) - (C_1 \cup C_2)) = \text{cost}(G_1 - C_1) + \text{cost}(G_2 - C_2)$, allowing us to compute the cost of copy sets formed by the union of copy sets of vertex-disjoint graphs.

Lemma 10. *Let G_1, G_2 be two vertex-disjoint RTGs and let $G = G_1 \cup G_2$. Then $T_G[d] = \min_{d'}\{T_{G_1}[d'] + T_{G_2}[d - d']\}$.*

Proof. Applying the definition of $T_G[\cdot]$ as well as Lemma 9 (ii) and Lemma 8 (i) yields

$$T_G[d] = \min_{C \in \mathcal{C}(G;d)} \text{cost}(G - C) = \min_{C \in \bigcup_{d'}(\mathcal{C}(G_1;d') \otimes \mathcal{C}(G_2;d-d'))} \text{cost}(G - C)$$

$$= \min_{d'} \left\{ \min_{C \in \mathcal{C}(G_1;d') \otimes \mathcal{C}(G_2;d-d')} \text{cost}(G - C) \right\}$$

$$= \min_{d'} \left\{ \min_{C_1 \in \mathcal{C}(G_1;d')} \text{cost}(G_1 - C_1) + \min_{C_2 \in \mathcal{C}(G_2;d-d')} \text{cost}(G_2 - C_2) \right\}$$

$$= \min_{d'}\{T_{G_1}[d'] + T_{G_2}[d - d']\}. \qquad \square$$

By iteratively applying Lemma 10, we compute $T_G[\cdot]$ for a disconnected RTG G with an arbitrary number of connected components.

Lemma 11. *Let G be an RTG with n vertices and connected components $G_1, \ldots, G_k$. Given the tables $T_{G_i}[\cdot]$ for $i = 1, \ldots, k$, the table $T_G[\cdot]$ can be computed in $O(n^2)$ time.*

Tree RTGs. For a tree RTG G, we compute $T_G[\cdot]$ in a bottom-up fashion. The direction of the edges naturally defines a unique root vertex r that has no incoming edges and we consider G as a rooted tree. For a vertex v, we denote by $G(v)$ the subtree of G with root v. Let v be a vertex with children $v_1, \ldots, v_k$. How does a copy set C of $G(v)$ look like? Clearly, $G(v) - C$ contains precisely one of the outgoing edges of v, say (v, v_j). Then $Z_j = \{(v, v_i) \mid i \neq j\} \subseteq C$. Graph $G(v) - Z_j$ has connected components $G(v_i)$ for $i \neq j$, whose union we denote $G_{\neg j}$, and one additional connected component $G^+(v_j)$ that is obtained from $G(v_j)$ by adding the vertex v and the edge (v, v_j). This forms a proper partition of $G(v) - Z_j$. As above, we decompose the copy set $C - Z_j$ further into a union of a copy set $C_{\neg j}$ of $G_{\neg j}$ and a copy set C_j of $G^+(v_j)$. Graph $G_{\neg j}$ is disconnected and can be handled as above. Note that the only child of the root of $G^+(v_j)$ is v_j and hence C_j is a copy set of $G(v_j)$. For expressing the cost and difference measures for copy sets of $G^+(v_j)$ in terms of copy sets of $G(v_j)$, we use the correction terms Δ_{cost} and Δ_{diff}. By Lemma 8 (ii), $\text{diff}(G^+(v_j) - C_j) = \text{diff}(G(v_j) - C_j) + \Delta_{\text{diff}}(s)$, where s is the size of the *root path* $P(v_j, C_j)$ of $G(v_j) - C_j$, i.e., the size of the connected component of $G(v_j) - C_j$ containing v_j. An analogous statement holds for cost. More precisely, it suffices to know s modulo 4. Therefore, we further decompose our copy sets as follows, which allows us to formalize our discussion.

Definition 2. *For a tree RTG G with root v and children $v_1, \ldots, v_k$, we define $\mathcal{C}(G; d, s) = \{C \in \mathcal{C}(G; d) \mid |P(v, C)| \equiv s \pmod 4\}$. We further decompose these by $\mathcal{C}(G; d, s, j) = \{C \in \mathcal{C}(G; d, s) \mid (v, v_j) \notin C\}$, according to which outgoing edge of the root is not in the copy set.*

Lemma 12. *Let G be a tree RTG with root v and children $v_1, \ldots, v_k$ and for a fixed vertex v_j, $1 \le j \le k$, let $G^+(v_j)$ be the subgraph of G induced by the vertices in $G(v_j)$ together with v. Let further $G_{\neg j} = \bigcup_{i=1, i \ne j}^k G(v_i)$ and $Z_j = \{(v, v_i) \mid i \ne j\}$. Then*

(i) $\mathcal{C}(G; d) = \bigcup_{s=0}^3 \mathcal{C}(G; d, s)$ and $\mathcal{C}(G; d, s) = \bigcup_{j=1}^k \mathcal{C}(G; d, s, j)$.

(ii) $\mathcal{C}(G^+(v_j); d, s) = \mathcal{C}(G(v_j); d - \Delta_{\text{diff}}(s), s - 1)$.

(iii) $\mathcal{C}(G; d, s, j) = \bigcup_{d'} (\mathcal{C}(G_{\neg j}; d') \otimes \mathcal{C}(G^+(v_j); d - d', s) \otimes \{Z_j\})$.

To make use of this decomposition of copy sets, we extend our table T with an additional parameter s to keep track of the size of the root path modulo 4. We call the resulting table $\tilde{T}$. More formally, $\tilde{T}_v[d, s] = \min_{C \in \mathcal{C}(G(v); d, s)} \text{cost}(G(v) - C)$. It is not hard to see that $T_G[\cdot]$ can be computed from $\tilde{T}_r[\cdot, \cdot]$ for the root r of a tree RTG G.

Lemma 13. *Let G be a tree RTG with root r. Then $T_G[d] = \min_s \tilde{T}_r[d, s]$.*

To compute $\tilde{T}_v[\cdot, \cdot]$ in a bottom-up fashion, we exploit the decompositions from Lemma 12 and the fact that we can update the cost function from $G(v_j) - C_j$ to $G^+(v_j) - C_j$ using the correction term Δ_{cost}. The proof is similar to that of Lemma 10 but more technical.

Lemma 14. *Let G be a tree RTG, let v be a vertex of G with children $v_1, \ldots, v_k$, and let $G(v_i) = (V_i, E_i)$ for $i = 1, \ldots, k$. Then with $G_{\neg j} = (V_{\neg j}, E_{\neg j}) = \bigcup_{i=1, i \ne j}^k G(v_i)$*

$$\tilde{T}_v[d, s] = \min_{j \in \{1, \ldots, k\}} \min_{d'} T_{G_{\neg j}}[d'] + \tilde{T}_{v_j}[d - d' - \Delta_{\text{diff}}(s), (s - 1) \bmod 4] + \Delta_{\text{cost}}(s).$$

For leaves v of a tree RTG G, $\tilde{T}_v[0, 1] = 0$ and all other entries are ∞. We compute $T_G[\cdot]$ by iteratively applying Lemma 14 in a bottom-up fashion, using Lemma 13 to compute $T[\cdot]$ from $\tilde{T}[\cdot, \cdot]$ in linear time when needed.

Lemma 15. *Let $G = (V, E)$ be a tree RTG with n vertices and root r. The tables $\tilde{T}_r[\cdot, \cdot]$ and $T_G[\cdot]$ can be computed in $O(n^3)$ time.*

Connected RTGs Containing a Cycle. We only give a sketch. The idea is that such an RTG contains a single directed cycle. Every copy set contains either an edge of that cycle or it contains all edges that have their source on the cycle but do not belong to the cycle. This leads to a linear number of tree instances, which we solve using Lemma 15.

Lemma 16. *Let G be a connected RTG containing a directed cycle. The table $T_G[\cdot]$ can be computed in $O(n^4)$ time.*

Putting Things Together. To compute $T_G[\cdot]$ for an arbitrary RTG G, we first compute $T_K[\cdot]$ for each connected component K of G using Lemmas 15 and 16. Then, we compute $T_G[\cdot]$ using Lemma 11 and the length of an optimal shuffle code using Lemma 7. To actually compute the shuffle code, we augment the dynamic program computing $T_G[\cdot]$ such that an optimal copy set C can be found by backtracking in the tables. An optimal shuffle code is then found by applying GREEDY to $G - C$ and adding one copy operation for each edge in C.

Theorem 3. *Given an RTG G, an optimal shuffle code can be computed in $O(n^4)$ time.*

Conclusion. We have presented an efficient algorithm for generating optimal shuffle code using copy instructions and permutation instructions, which allow to arbitrarily permute the contents of up to five registers. As an intermediate result, we have proven the optimality of the greedy algorithm for factoring a permutation into a minimal product of permutations, each of which permutes up to five elements. It would be interesting to allow permutations of larger size.

Acknowledgments. This work was partly supported by the German Research Foundation (DFG) as part of the Transregional Collaborative Research Center "Invasive Computing" (SFB/TR 89).

References

1. Blazy, S., Robillard, B.: Live-range unsplitting for faster optimal coalescing. In: Languages, Compilers, and Tools for Embedded Systems (LCTES 2009), pp. 70–79. ACM (2009)
2. Bouchez, F., Darte, A., Rastello, F.: On the complexity of register coalescing. In: Code Generation and Optimization (CGO 2007), pp. 102–114. IEEE (2007)
3. Buchwald, S., Mohr, M., Rutter, I.: Optimal shuffle code with permutation instructions. CoRR abs/1504.07073 (2015). http://arxiv.org/abs/1504.07073
4. Caprara, A.: Sorting by reversals is difficult. In: Computational Molecular Biology (RECOMB 1997), pp. 75–83. ACM (1997)
5. Farnoud, F., Milenkovic, O.: Sorting of permutations by cost-constrained transpositions. IEEE Transactions on Information Theory 58(1), 3–23 (2012)
6. Grund, D., Hack, S.: A fast cutting-plane algorithm for optimal coalescing. In: Adsul, B., Odersky, M. (eds.) CC 2007. LNCS, vol. 4420, pp. 111–125. Springer, Heidelberg (2007)

7. Hack, S.: Register Allocation for Programs in SSA Form. Ph.D. thesis, Universität Karlsruhe (2007). http://digbib.ubka.uni-karlsruhe.de/volltexte/documents/6532

8. Hack, S., Goos, G.: Copy coalescing by graph recoloring. SIGPLAN Notices **43**(6), 227–237 (2008)

9. Mohr, M., Grudnitsky, A., Modschiedler, T., Bauer, L., Hack, S., Henkel, J.: Hardware acceleration for programs in SSA form. In: Compilers, Architecture and Synthesis for Embedded Systems (CASES 2013). ACM (2013)

10. Seress, Á.: Permutation Group Algorithms, vol. 152. Cambridge University Press (2003)

Non-preemptive Scheduling on Machines
with Setup Times

Alexander Mäcker[⊠], Manuel Malatyali,
Friedhelm Meyer auf der Heide, and Sören Riechers

Heinz Nixdorf Institute and Computer Science Department, University of Paderborn,
Paderborn, Germany
{alexander.maecker,manuel.malatyali,fmadh,soeren.riechers}@upb.de

Abstract. Consider the problem in which n jobs that are classified into k types are to be scheduled on m identical machines without preemption. A machine requires a proper setup taking s time units before processing jobs of a given type. The objective is to minimize the makespan of the resulting schedule. We design and analyze an approximation algorithm that runs in time polynomial in n, m and k and computes a solution with an approximation factor that can be made arbitrarily close to $^3/_2$.

Keywords: Scheduling · Approximation algorithms · Setup times

1 Introduction

In this paper, we consider a scheduling problem where a set of n jobs, each with an individual processing time, that is partitioned into k disjoint classes has to be scheduled on m identical machines. Before a machine is ready to process jobs belonging to a certain class, this machine has to be configured properly. That is, whenever a machine switches from processing a job of one class to a job of another class, a setup taking s time units is required. Meanwhile a machine is not available for processing. The objective is to assign jobs (and the respective setup operations) to machines so as to minimize the makespan of the resulting non-preemptive schedule.

The considered problem models situations where the preparation of machines for processing jobs requires a non-negligible setup time. These setups depend on the classes of jobs to be processed (i.e. they are class-dependent), however, the required setup time is class-independent. Also, jobs might not be preempted, e.g. because of additional high preemption costs. Possible examples of problems for which this model is applicable are (1) the processing of jobs on (re-)configurable machines (e.g. Field Programmable Gate Arrays) which only provide functionalities required for certain operations (or jobs of a certain class) after a suitable setup or (2) a scenario where large tasks (consisting of smaller jobs) have to

---

This work was partially supported by the German Research Foundation (DFG) within the Collaborative Research Centre "On-The-Fly Computing" (SFB 901).

F. Dehne et al. (Eds.): WADS 2015, LNCS 9214, pp. 542–553, 2015.
DOI: 10.1007/978-3-319-21840-3_45

be scheduled on remote machines and it takes a certain (setup) time to make task-dependent data available on these distributed machines.

Surprisingly, although a lot of research has been done on scheduling with setup times, we are not aware of results concerning the considered model. This is due to the fact that the motivation for considering setup times are often related to preemption of jobs, which is not true for our model. We discuss some results on these alternative models in the following section on related work. Thereafter, we settle some preliminaries in Section 3. Section 4 presents the main contribution of this paper which is an algorithm whose approximation factor can be made arbitrarily close to $3/2$ with a runtime that is polynomial in the input quantities n, k and m. For an online version where jobs arrive over time our offline algorithm implies an online strategy with a competitiveness arbitrarily close to 4.

For ommited proofs please refer to the full version of this paper [7].

1.1 Related Work

The scheduling problem considered in this paper is a generalization of the classical problem of scheduling jobs on identical machines without preemption and in which setup times are equal to 0. This problem has been extensively studied in theoretical research and PTASs with runtimes that are linear in the number n of jobs are known for objective functions such as minimizing (maximizing) the maximum (minimum) completion time or sum of completion times [2,5]. If the number m of machines is constant, even FPTASs exist [6].

When setup times are larger than 0, the problem is usually refered to as scheduling with setup times (or setup costs). It has also been studied for quite a long time and there is a rich literature analyzing different models and objective functions. Usually models are distinguished by whether or not setup times are job-, machine- and/or sequence-dependent. For an overview on studied problems and results in this context the reader is refered to detailed surveys on scheduling with setup times [1,9]. We discuss some closely related problems in the following. In [8], Monma and Potts consider a model quite similar to ours but they allow preemption of jobs and setup times may be different for each class. They design two simple algorithms, one with an approximation factor of at most $\max\{3/2 - 1/(4m - 4), 5/3 - 1/m\}$ if each class is small (i.e. setup time plus size of all jobs of a class are not larger than the optimal makespan), and a second one with approximation factor of at most $2 - 1/(\lfloor m/2 \rfloor + 1)$ for the general case. Later, Schuurman and Woeginger [10] improve the result for the case that each class consists of only one job that, together with its setup time, is not larger than the optimal makespan. The authors design a PTAS for the case where all setup times are identical and a polynomial time algorithm with approximation factor arbitrary close to $4/3$ for non-identical setup times.

A closely related problem was also studied in another context by Shachnai and Tamir [11]. They design a dual PTAS for a class-constrained packing problem. In contrast to the basic bin packing problem, in this variant each item belongs to a class and each bin has an upper bound on the number of different classes that might be placed in one bin.

The dual problem of our scheduling problem was studied by Xavier and Miyazawa and is known as class-constrained shelf bin packing. For a constant number of classes, an asymptotic PTAS is known for this problem [13] as well as a dual approximation scheme [14], i.e. a PTAS for our problem if k is constant.

Very recently, Correa et al. [3] studied the problem of scheduling splittable jobs on unrelated machines. Here, unrelated refers to the fact that each job may have a different processing time on each of the machines. In their model, jobs may be split and each part might be assigned to a different machine but each requires a setup before being processed. For this problem and the objective of minimizing the makespan they show their algorithm to have an approximation factor of at most $1 + \phi$, where $\phi \approx 1.618$ is the golden ratio.

In [4], an online variant of scheduling with setup times is considered. The authors propose a $O(1)$-competitive online algorithm for minimizing the maximum flow time if jobs arrive over time at one single machine.

2   Model and Notation

We consider a model in which there is a set $J = \{1, \dots, n\}$ of n independent jobs (i.e. there are no precedence constraints for jobs) that are to be scheduled on m identical machines $M = \{M_1, \dots, M_m\}^1$. Each job i is available at the beginning and comes with a *processing time* (or *size*) $p_i \in \mathbb{N}_{>0}$. Additionally, the job set is partitioned into k disjoint classes $C = \{C_1, \dots, C_k\}$, i.e. $J = \bigcup_{i=0}^{k} C_i$ and $C_i \cap C_j = \emptyset$ for all $i \neq j$. Before a job $j \in C_i$ can be processed on a machine, this machine has to be configured properly and afterward jobs of class C_i can be processed without additional setups until the machine is reconfigured for a class $C_{i'} \neq C_i$. That is, a setup needs to take place before the first job is processed on a machine and whenever the machine switches from processing a job $j \in C_i$ to a job $j' \in C_{i'}$ with $C_i \neq C_{i'}$. Such a setup takes $s \in \mathbb{N}_{>0}$ time units and while setting up a machine, it is blocked and cannot do any processing.

Given this setting, the objective is to find a feasible schedule that minimizes the makespan, i.e. the maximum completion time of a job, and does not preempt any job, i.e. once the processing of a job is started at a machine it finishes at this machine without interruption.

In the following we refer to the overall processing time of all jobs of a class C_i as its *workload* and denote it $w(C_i) := \sum_{j \in C_i} p_j$ and we assume that for all $1 \leq i \leq n$ it holds that $w(C_i) \leq \gamma OPT$ for some constant γ and OPT being the optimal makespan. By abuse of notation, by $w(C_i)$ we sometimes also represent (an arbitrary sequence of) those jobs belonging to class C_i. To refer to the class C_i of a job $j \in C_i$, we use a mapping $c : J \to C$ with $c(j) = C_i$ and we say a job $j \in C_i$ forms an *individual class* if $c^{-1}(C_i) = \{j\}$. The processing time of the largest job in a given instance is denoted by $p_{max} := \max_{1 \leq i \leq n}(p_i)$. We say a machine is an *exclusive machine* (of a class C_i) if it only processes jobs of a single class (class C_i).

---

1 We do not assume m to be a constant. Although still being NP-hard, for constant m there is a simple FPTAS that can be found in the full version.

3   Preliminaries

As a preliminary for our approximation algorithm presented in Section 4, we need to know the optimal makespan before we can actually compute a schedule fulfilling the desired approximation guarantee concerning its makespan. However, this assumption is feasible and justified by the applicability of a common notion known as an α-relaxed decision procedure [5].

Definition 1. *Given an instance I and a candidate makespan T, an α-relaxed decision procedure either outputs no or provides a schedule with makespan at most $\alpha \cdot T$. In case it outputs no, there is no schedule with makespan at most T.*

Using such an α-relaxed decision procedure (that runs in polynomial time) to guide a binary search on an interval $[l, u]$ with $OPT \in [l, u]$, we directly obtain a polynomial time approximation algorithm with approximation factor α. We can find a suitable interval containing the optimal makespan by applying a greedy algorithm that provides an interval of length OPT in time $O(n)$ as described in the full version.

For the sake of simplicity, we assume in the following that by means of this approach we have guessed OPT correctly and show how to obtain an effective approximation algorithm. Particularly, using the presented algorithm within the binary search framework as an α-relaxed decision procedure, provides the final result.

4 $(^3/_2 + \varepsilon)$-Approximation Algorithm

In this section, we present the main algorithm of the paper. The outline of our approach is as follows:

(1) We first identify a class of schedules that features a certain structural property and show that if we narrow our search for a solution to schedules belonging to this class, we will still find a good schedule, i.e. one whose makespan is not too far away from an optimal one.

(2) We then show how to perform a rounding of the involved job sizes and further transformations and thereby significantly decrease the size of the search space.

(3) Finally, given such a (transformed) instance, it will be easy to optimize over the restricted class of schedules studied in (1) to obtain an approximate solution to any given instance.

4.1   Block-Schedules

We start by discussing the question how to narrow our study to a class of schedules that fulfill a certain property and still be able to find a provably good approximate solution. Particularly, we focus on block-schedules, which are schedules satisfying a structural property, and which we define in Definition 2. Intuitively speaking, in a block-schedule jobs of a class are assigned to consecutive machines instead of being widely scattered.

Definition 2. *Given an instance I, we call a schedule for I block-schedule if for all $1 \leq i \leq m$ the following holds: In the (partial) schedule for the machines $M_1, \ldots, M_i$, there is at most one class of which some but not all jobs are processed on $M_1, \ldots, M_i$.*

In order to prove our main theorem about block-schedules, we first have to take care of jobs having a large processing time in terms of the optimal makespan. Let $L_i = \{j \in C_i : \frac{1}{2}OPT - s < p_j < \frac{1}{2}OPT\}$ be the set of *large* jobs of class C_i and $H_i = \{j \in C_i : p_j \geq \frac{1}{2}OPT\}$ be the set of *huge* jobs of class C_i. Based on these definitions we show the following lemma.

Lemma 1. *With an additive loss of s in the makespan we may assume that*

1. *Each huge job forms an individual class,*
2. *There is a schedule with the property that all large jobs of class C_i are processed on exclusive machines, except (possibly) one large job $q_i \in L_i$, for each C_i, and*
3. *$q_i = argmin_{j \in L_i}\{p_j\}$ is the smallest large job in C_i and the machine it is processed on has makespan at most OPT.*

Proof. We prove the lemma by showing how to establish the three properties by transformations of the given instance I and an optimal schedule S for I with makespan OPT. To establish the first property, transform I into I' by putting each job $j \in H_i$ into a new individual class, for each class C_i. Because any machine processing such a huge job j cannot process any other huge or large job due to their definitions, the transformation increases the makespan of any machine by at most s.

Next, we focus on the second property. In S no machine can process two large jobs of different classes. Hence, we distinguish the following two cases: A machine processes one large job or a machine processes at least two large jobs. We start with the latter case and consider any machine that processes at least two large jobs of a class C_i. Because these two jobs already require at least $2\lceil (OPT+1)/2 - s \rceil + s \geq OPT - s + 1$ time units including the setup time, no job of another class can be processed and thus, this machine already is an exclusive machine. On the other hand, if a machine M_p only processes one large job $j \in C_i$, we can argue as follows. The machine M_p works on j for at least $\lceil (OPT+1)/2 \rceil$ time units (including the setup). Thus, the remaining jobs and setups processed by M_p can have a size of at most $\lfloor (OPT-1)/2 \rfloor$. If there is still another machine processing a single large job of C_i, we can exchange these jobs and setups with this large job and both involved machines have a makespan of at most $OPT + s$. Also, the machine from which the large job was removed does not contain any huge or large jobs anymore ensuring there is no machine where this process can happen twice. Therefore, we can repeat this procedure until all (but possibly one) large jobs are paired so that the second property holds.

Finally, to establish the third property, we can argue as follows: If the smallest large job q_i is the only large one on a machine in the schedule S, we can do the grouping just described without shifting q_i to another machine satisfying the desired bound on the makespan. If q_i is already processed on a machine together with another large job, we may pair the remaining jobs but (possibly) one (one that is not processed together with another large job on a machine). In case there is such a remaining unpaired job, we finally exchange q_i with the unpaired job. The resulting schedule fulfills the desired properties.   □

By our next transformation we put the smallest large job q_i of each class C_i into a new individual class. Based on the previous result there is still a schedule with makespan at most $OPT + s$ for the resulting instance.

In the next lemma, we directly deduce that there is a block-schedule with makespan at most $OPT + s$ if we allow some jobs to be split, i.e. some jobs are cut into two parts that are treated as individual jobs and processed on different machines. To this end, fix a schedule S for I fulfilling the properties of Lemma 1. By $\tilde{M}$ denote the exclusive machines according to schedule S and by $\tilde{C}_i$ the class C_i without those jobs processed on machines belonging to $\tilde{M}$.

Lemma 2. *Given the schedule S fulfilling the properties of Lemma 1, there is a schedule S' with makespan at most $OPT + s$ with the following properties:*

1. *A machine is exclusive in S' if and only if it belongs to $\tilde{M}$ and the partial schedule of these machines is unchanged.*
2. *When removing the machines belonging to $\tilde{M}$ and their jobs from S, we can schedule the remaining jobs on the remaining machines such that*
   (a) The block-property holds and
   (b) only jobs with size at most $\frac{1}{2}OPT - s$ are split s.t. one part is processed until the completion time of some M_j and one from time s on by M_{j+1}.

Proof. Remove machines belonging to $\tilde{M}$ and the jobs scheduled on them from the schedule S obtaining $\tilde{S}$. We now show that there is a schedule S' with the desired properties. Similar to [10] consider a graph $G = (V, E)$ in which the nodes correspond to the machines in $\tilde{S}$ and there is an edge between two nodes if and only if in $\tilde{S}$ the respective machines process jobs of the same class. We argue for each connected component of G. Let m' be the number of nodes/machines in this component. Furthermore, let $C' = \{C'_1, \dots C'_l\}$ be the set of classes processed on these machines without those formed by single huge or large jobs and $H = \{h_1, \dots, h_r\}$ be the set of jobs processed on these machines that are either huge jobs or large jobs forming individual classes. Note that $r \leq m'$ since all jobs of H must be processed on different machines in $\tilde{S}$. Note that the number of setups is at least $l + r + m' - 1$ because there are $l + r$ classes and at least $m' - 1$ additional setups by the definition of the edges. By an averaging argument we know $OPT + s \geq \frac{1}{m'} \left(\sum_{i=1}^{l} w(\tilde{C}'_i) + \sum_{i=1}^{r} w(h_i) + (l + r + m' - 1)s \right)$ and hence,

$$\sum_{i=1}^{l} w(\tilde{C}'_i) + (l - 1)s \leq (m' - r)OPT + \sum_{i=1}^{r}(OPT - w(h_i) - s). \quad (1)$$

Consider the sequence $w(\tilde{C}'_1), s, w(\tilde{C}'_2), s, \dots, s, w(\tilde{C}'_l)$ of length $\sum_{i=1}^{l} w(\tilde{C}'_i) + (l - 1)s$ and split it from the left to the right into blocks of length $OPT -$

$w(h_1) - s, \ldots, OPT - w(h_r) - s$, followed by blocks of length OPT. Note that each block has non-negative length. By equation (1) we obtain at most m' blocks and by adding a setup to each block and the jobs h_i plus setup to the first r blocks, we can process each block on one machine.

Consequently, if we apply these arguments to each connected component and add the removed exclusive machines, we have shown that there is a schedule S' with makespan at most $OPT + s$ satisfying the required properties of the lemma. $\qquad \square$

Lemma 2 proves the existence of a schedule that almost fulfills the properties of block-schedules, whose existence is the major concern in this section. However, it remains to show how to handle jobs that are split as we do not allow splitting or preemption of jobs and how to place exclusive machines belonging to $\tilde{M}$, which are not taken care of by the previous lemma, into the obtained schedule in order to yield a block-schedule.

To simplify the description in the following, when we say we place an exclusive machine M_i before machine M_j, we think of a re-indexing of the machines such that the ordering of machines other than M_i and M_j stays untouched but now the new indices of M_i and M_j are consecutive. Also, a job j is *started* at the machine that processes (parts of) j and has the smallest index among all those processing j. A class C_i is processed at the end (beginning) of a machine if there is a job $j \in C_i$ that is processed as the last job (as the first job) on M_j.

Lemma 3. *A schedule fulfilling the properties of Lemma 2 can be transformed into a block-schedule with makespan at most $\frac{3}{2}OPT$.*

Proof. Consider an arbitrary class C_i. We distinguish three cases depending on where the jobs of C_i are placed in the schedule S' according to the proof of the previous lemma.

(1) There is a job in $\tilde{C}_i$ that is split among two machines M_j and M_{j+1}.
(2) There is no job in $\tilde{C}_i$ that is split.
(3) $\tilde{C}_i = \emptyset$.

In case (1) there is a job in $\tilde{C}_i$ that is split. Hence, we can simply place all exclusive machines of C_i between M_j and M_{j+1}. Since jobs that are split have size at most $\frac{1}{2}OPT - s$, we can process any split job completely on the machine on which it was started increasing its makespan to at most $\frac{3}{2}OPT$. We repeat this process as long as there are jobs with property (1) left. Note that for each class C_i, after having finished case (1), there is no split job left.

In case (2), we distinguish two cases. If the jobs in $\tilde{C}_i$ have an overall size of at most $\frac{1}{2}OPT$ (including setup), there either is no exclusive machine of C_i and hence no violation of the block-property, or we can process the jobs on an exclusive machine of C_i increasing its makespan to at most $\frac{3}{2}OPT$. If the jobs have an overall size of more than $\frac{1}{2}OPT$, we distinguish whether $\tilde{C}_i$ is processed at the end or beginning of a machine M_j or not. In the positive case, we can simply place any exclusive machines of C_i behind or before machine M_j. If $\tilde{C}_i$

is not processed at the end or beginning of a machine M_j, there must be a second class $\tilde{C}_{i'}$ that is processed at the beginning and a third class $\tilde{C}_{i''}$ that is processed at the end of machine M_j. Note that consequently the workload of $\tilde{C}_{i'}$ processed on M_j cannot be larger than $\frac{1}{2}OPT - s$. We can perform the following steps on the currently considered machine M_j:

1. Move all jobs from the class $C_{i'}$ that is processed at the beginning of M_j to machine M_{j-1} if $C_{i'}$ is also processed at the end of M_{j-1}, thus only increasing the makespan of M_{j-1} by at most $\frac{1}{2}OPT - s$.
2. Move all other jobs processed before some workload of C_i to one of their exclusive machines, if they exist.
3. Shift all the workload $w(\tilde{C}_i)$ to time 0 on machine M_j and shift other jobs to a later point in time.
4. Place all exclusive machines of C_i in front of M_j.

In case (3), there are only exclusive machines. Such machines can simply be placed behind all other machines.

These steps establish the block-schedule property and no jobs are split anymore. Also note that each machine gets an additional workload of at most $\frac{1}{2}OPT - s$ without requiring additional setups, proving the lemma.   □

Theorem 1. *Given an instance I with optimal makespan OPT, there are transformations to I' such that there is a block-schedule for I' with makespan at most $OPT_{BL} := \min\{OPT + p_{max} - 1, \frac{3}{2}OPT\}$ and it can be turned into a schedule for I with makespan not larger than OPT_{BL}.*

Proof. The bound $OPT_{BL} \leq \frac{3}{2}OPT$ directly follows from Lemma 3 and the fact that there are only transformations performed on instance I by Lemma 1. The second bound (which gives a better result if $p_{max} \leq \frac{1}{2}OPT$) follows by arguments quite similar to those used before: If $p_{max} \leq \frac{1}{2}OPT$ holds, we skip the transformation of Lemma 1. Additionally, in the proof of Lemma 2 we do not remove exclusive machines (thus, considering all machines). Note that, since we skipped the transformation of Lemma 1, the set H is empty. Then, it is straightforward to calculate the second bound of $OPT_{BL} \leq OPT + p_{max} - 1$.

□

4.2 Grouping and Rounding

In this section, we show how we can reduce the search space by rounding the involved processing times to integer multiples of some value depending on the desired precision $\varepsilon > 0$ of the approximation. We assume that the transformations described in previous sections have already been performed. In order to be able to ensure that the rounding of processing times cannot increase the makespan of the resulting schedule too much, we first need to get rid of classes and jobs that have a very small workload in terms of OPT_{BL} and ε. In the following, we use $\lambda > 0$ to represent the desired precision, i.e. λ essentially depends on the reciprocal of ε. We call every job j with $p_j \leq OPT_{BL}/\lambda$ a *tiny job* and every class C_i with $w(C_i) \leq OPT_{BL}/\lambda$ a *tiny class*.

Lemma 4. *Given a block-schedule for an instance I, with an additive loss of at most $4OPT_{BL}/\lambda$ in the makespan we may assume that tiny jobs only occur in tiny classes.*

Proof. We prove the lemma by applying the following transformations to each class C_i: In a first step, we greedily group as many tiny jobs of class C_i as possible to new jobs with sizes in the interval $[OPT_{BL}/\lambda, 2OPT_{BL}/\lambda)$. In a second step, combine the (possibly) remaining tiny grouped job $j \in C_i$ with a size less than OPT_{BL}/λ, with an arbitrary other job $j' \in C_i$. By this transformation we ensure that tiny jobs only occur in tiny classes and it remains to show the claimed bound on the makespan.

First of all, focus on the first step of the transformation and assume that we do not perform the second step. Let S be the given block-schedule for instance I. Lemma 2.3 in the work of Shachnai and Tamir [11] proves (speaking in our terms) that for the transformed instance there is a schedule S' with makespan of at most $OPT_{BL} + 2OPT_{BL}/\lambda$. The proof also implies that S' is still a block-schedule: For each machine M_j it holds that if M_j is configured for class C_i in the new schedule S', it has also been configured for C_i in the original block-schedule S. Thus, if S is a block schedule, so is S' since we do not have any additional setups in S'.

Now assume that also the second step of the transformation is carried out and consider the block-schedule S' we just proved to exist. Distinguish two cases, depending on where the tiny grouped job $j \in C_i$, which was paired in the second step, is processed in schedule S'. If j was paired with a job j' and both j and j' are assigned to the same machine in S', the schedule S' already is feasible for the transformed instance (possibly after shifting j and j' such that they are processed consecutively). If the paired jobs j and j' are processed on different machines in schedule S', there is a schedule whose makespan is by an additive of at most $2\frac{OPT_{BL}}{\lambda}$ larger than that of S'. To see this, note that in S' this case can happen at most twice per machine (for the classes processed at the beginning and end of the machine). Hence, we can place any paired jobs j and j' on the same machine yielding a schedule for the transformed instance with the claimed bound on the makespan. Finally, note that we can easily turn a schedule fulfilling the claimed bound on the makespan into a schedule for the original instance I satisfying the same bound on the makespan. □

Next, we take care of tiny classes that still might occur in a given instance. Again, without losing too much with respect to the optimal makespan we may assume a simplifying property as shown in the next lemma.

Lemma 5. *With an additive loss of at most $4OPT_{BL}/\lambda$ in the makespan we may assume the following properties:*

1. *Each tiny class consists of a single job.*
2. *In case that $OPT_{BL}/\lambda > s$, this job has size $OPT_{BL}/\lambda - s$.*

Proof. At first note that with an additive loss of at most $2OPT_{BL}/\lambda$ in the makespan, we may assume that a tiny class is completely scheduled on one machine in a block-schedule. This is true because of reasons similar to those used in the proof of the previous lemma: For each machine it holds that there are at most two different tiny classes of which some but not all jobs are processed on this machine. Hence, we may shift all jobs of such classes to one machine and thereby increase the makespan by at most $2OPT_{BL}/\lambda$.

Now distinguish two cases depending on whether $OPT_{BL}/\lambda > s$ or not. If this is the case, determine the length L of the sequence of all tiny classes (including setup times), round up L to an integer multiple of OPT_{BL}/λ, remove all tiny classes from the instance and instead, introduce $\lambda L/OPT_{BL}$ new classes each comprised of a single job with workload $OPT_{BL}/\lambda - s$. Observe that, given a block-schedule in which each tiny class is completely scheduled on one machine, we can simply replace tiny classes by these new classes, increasing the makespan by an additive of at most OPT_{BL}/λ. Also, this schedule implies a schedule for the instance in which tiny classes have not been grouped and its makespan is by an additive of at most OPT_{BL}/λ larger. This schedule is simply obtained by again replacing grouped tiny classes by its respective original classes.

In case that $OPT_{BL}/\lambda \leq s$, we simply group all jobs of a tiny class C_i to a new job j of the same size $p_j = w(C_i)$. Due to the fact that we might assume that a tiny class is completely scheduled on one machine, this proves the lemma.   □

From now on, we assume that we have already conducted the grouping from the previous lemmas and we describe how to round job sizes to reduce the search space for later optimization. The approach is quite common for scheduling.

Given an instance I, we compute its rounded version I' by rounding up the size of each job to the next integer multiple of OPT_{BL}/λ^2. We know that there is a block-schedule with makespan at most $OPT_{BL} + 8\frac{OPT_{BL}}{\lambda}$ and we also assume that the properties from Lemma 5 hold.

In case that $OPT_{BL}/\lambda > s$ each job has either a processing time of at least OPT_{BL}/λ or forms a tiny class with workload at least $OPT_{BL}/\lambda - s$. On the other hand, in case that $OPT_{BL}/\lambda \leq s$ and there are tiny classes consisting of a single job, to execute such a job, we need to perform a setup first which yields a processing time of at least OPT_{BL}/λ as well. Hence, we can have at most $\lambda+8$ jobs on one machine in the considered block-schedule, leading to an additive rounding error of at most $(\lambda + 8) \cdot OPT_{BL}/\lambda^2$ in the makespan. Therefore, by choosing λ appropriately, there is a solution to the rounded instance that approximates OPT_{BL} up to any desired precision $\varepsilon > 0$.

4.3   Optimization over Block-Schedules

We are ready to show how to compute a block-schedule for the rounded instance with makespan at most $(1+\varepsilon)OPT_{BL}$ for any $\varepsilon > 0$. The schedule directly implies a schedule for the original instance with the same bound on the makespan.

We say that all classes that can be represented by the same tuple are of the same *class-type* and show that there are not too many different class-types.

Lemma 6. *If all job sizes are a multiple of OPT_{BL}/λ^2 and $\lambda > 0$ is a constant, there is only a constant number c_{cl} of different class-types.*

Proof. We can represent any class C_i by a tuple of length λ^2 describing how many jobs of each size $l \cdot OPT_{BL}/\lambda^2$, $1 \le l \le \lambda^2$, occur in class C_i. Recall that each class has a size of at most $\gamma \cdot OPT$ and hence, each entry of the tuple is limited by $\gamma\lambda^2$. Thus, there is at most a constant number $c_{cl} := (\gamma\lambda^2)^{\lambda^2}$ of different tuples describing the classes and hence, only a constant number of different class-types.   □

We can represent the classes that have to be scheduled as a tuple of size c_{cl} where each entry contains the number of times classes of the respective class-type occur. Given a block-schedule S, we consider machine configurations that describe which classes are finished on the first i machines. We denote the sub-schedule induced by these first i machines by S_i.

Lemma 7. *If all job sizes are a multiple of* OPT_{BL}/λ^2 *and* $\lambda > 0$ *is a constant, the number of machine configurations representing* S_i *for some block-schedule S and some $i > 0$ is bounded by a value c_{conf} that is polynomial in m.*

Proof. First, note that in a block-schedule S, for every S_i, there is at most one class that is split due to the block-schedule property. Now, to uniquely define a candidate configuration, we need to store information about the classes that are finished, and in case a class has been split, the type of this class and which jobs of this class are finished. We reserve c_{cl} entries for the finished classes, where each entry corresponds to the number of classes of the certain type that has been fully finished. Each entry is at most $m \cdot (\lambda + 8)$ with similar arguments as in the proof of Lemma 6 and the reasoning concerning the maximum rounding error. For the class that has been split, we store the type of that class in an extra entry, which gives c_{cl} possible values. If there is no class that has been split, we leave this entry empty adding another possible value to the entry. Finally, we store the number of jobs from the split class that have been finished for each job size as λ^2 additional entries, where each entry does not exceed $c_{cl} \cdot \lambda$ similar to the structure in Lemma 6. Overall, we write a configuation as a tuple $(n_1, \ldots, n_{c_{cl}}, j, u_1, \ldots, u_{\lambda^2})$ and thus there are at most $c_{conf} := (m(\lambda + 8))^{c_{cl}} \cdot (c_{cl} + 1) \cdot (c\lambda)^{\lambda^2}$ possible configurations, which proves the lemma.   □

We now build a graph where we add a node for each machine configuration. We draw a directed edge from node u to v if and only if the machine configuration corresponding to v can be reached from the configuration u by using at most one additional machine with makespan not larger than $(1 + \varepsilon)OPT_{BL}$. That is, assuming u is a possible sub-schedule induced by the first i machines, we verify whether v is a possible sub-schedule induced by the first $i + 1$ machines. We can do so as we assume that we have guessed OPT correctly and we can hence determine $(1 + \varepsilon)OPT_{BL}$ which is the amount of workload we will fit on one machine. Using this idea, we obtain the following lemma and theorem, which proves our main result. The proofs can be found in the full version.

Lemma 8. *We can construct a graph G such that there is a path from the node representing no job at all (source) to the node representing the entire instance I' (target) that has a length of at most m.*

Theorem 2. *Using breadth-first search on G, a schedule with makespan at most*

$$(1 + \varepsilon) \min \left\{ \frac{3}{2} OPT, OPT + p_{max} - 1 \right\}$$

for the original instance I can be determined. It implies an algorithm with exactly this approximation guarantee and runtime polynomial in n, k and m.

For an online variant where jobs arrive over time and are not known to the scheduler before their release times, we have the following result (cf. full version).

Theorem 3. *No online algorithm can be c-competitive for $c \le 2 - \varepsilon$ and arbitrary small $\varepsilon > 0$. Furthermore, our offline algorithm implies an online strategy which is c-competitive for c arbitrary close to 4.*

References

1. Allahverdi, A., Gupta, J.N., Aldowaisan, T.: A review of scheduling research involving setup considerations. Omega **27**(2), 219–239 (1999)
2. Alon, N., Azar, Y., Woeginger, G.J., Yadid, T.: Approximation Schemes for Scheduling on Parallel Machines. Journal of Scheduling **1**(1), 55–66 (1998)
3. Correa, J.R., Marchetti-Spaccamela, A., Matuschke, J., Stougie, L., Svensson, O., Verdugo, V., Verschae, J.: Strong LP Formulations for Scheduling Splittable Jobs on Unrelated Machines. In: Lee, J., Vygen, J. (eds.) IPCO 2014. LNCS, vol. 8494, pp. 249–260. Springer, Heidelberg (2014)
4. Divakaran, S., Saks, M.E.: An Online Algorithm for a Problem in Scheduling with Set-ups and Release Times. Algorithmica **60**(2), 301–315 (2011)
5. Hochbaum, D.S., Shmoys, D.B.: Using Dual Approximation Algorithms for Scheduling Problems. Journal of the ACM **34**(1), 144–162 (1987)
6. Horowitz, E., Sahni, S.: Exact and Approximate Algorithms for Scheduling Non-identical Processors. Journal of the ACM **23**(2), 317–327 (1976)
7. Mäcker, A., Malatyali M., Meyer auf der Heide, F., Riechers, S.: Non-Preemptive Scheduling on Machines with Setup Times. CoRR (2015). 1504.07066
8. Monma, C.L., Potts, C.N.: Analysis of Heuristics for Preemptive Parallel Machine Scheduling with Batch Setup Times. Operations Research **41**(5), 981–993 (1993)
9. Potts, C.N., Kovalyov, M.Y.: Scheduling with batching: A review. European Journal of Operational Research **120**(2), 228–249 (2000)
10. Schuurman, P., Woeginger, G.J.: Preemptive scheduling with job-dependent setup times. In: Proceedings of the 10th Annual ACM-SIAM Symposium on Discrete Algorithms (SODA 1999), pp. 759–767. ACM/SIAM (1999)
11. Shachnai, H., Tamir, T.: Polynomial Time Approximation Schemes for Class-Constrained Packing Problems. In: Jansen, K., Khuller, S. (eds.) APPROX 2000. LNCS, vol. 1913, pp. 238–249. Springer, Heidelberg (2000)
12. Shmoys, D.B., Wein, J., Williamson, D.: Scheduling Parallel Machines On-line. In: Proceedings of the 32nd Annual Symposium on Foundations of Computer Science (FOCS 1991), pp. 131–140. IEEE (1991)
13. Xavier, E.C., Miyazawa, F.K.: A one-dimensional bin packing problem with shelf divisions. Discrete Applied Mathematics **156**(7), 1083–1096 (2008)
14. Xavier, E.C., Miyazawa, F.K.: A Note on Dual Approximation Algorithms for Class Constrained Bin Packing Problems. RAIRO - Theoretical Informatics and Applications **43**(2), 239–248 (2009)

A Moderately Exponential Time Algorithm for k-IBDD Satisfiability

Atsuki Nagao[1,2], Kazuhisa Seto[3], and Junichi Teruyama[4,5($\boxtimes$)]

[1] Kyoto University, Kyoto, Japan
[2] Research Fellow of Japan Society for the Promotion of Science, Tokyo, Japan
a-nagao@kuis.kyoto-u.ac.jp
[3] Seikei University, Musashino, Japan
seto@st.seikei.ac.jp
[4] National Institute of Informatics, Tokyo, Japan
[5] JST, ERATO, Kawarabayashi Large Graph Project, Tokyo, Japan
teruyama@nii.ac.jp

Abstract. A k-indexed Binary Decision Diagram (k-IBDD) is a branching program with k-layers and each layer consists of an Ordered Binary Decision Diagram (OBDD). This paper studies the satisfiability of k-IBDD (k-IBDD SAT). A k-IBDD SAT is, given a k-IBDD, to ask whether there exists a consistent path from the root to the 1-sink. We propose a moderately exponential time algorithm using exponential space for k-IBDD SAT of n variables and cn size. Our algorithm runs in time $O\left(2^{(1-\mu(c))n}\right)$, where $\mu(c) = \Omega\left(\frac{1}{(\log c)^{2^{k-1}-1}}\right)$. As a corollary, we obtain a polynomial space and deterministic algorithm, which solves k-IBDD SAT of size polynomial in n and runs in $O\left(2^{n-n^{1/2^{k-1}}}\right)$ time.

Keywords: Indexed binary decision diagram · Ordered binary decision diagram · Satisfiability · Moderately exponential time

1 Introduction

The satisfiability problem (SAT) is one of central problems in theoretical computer science. There exist many variants of SAT and in many cases, they are known to be NP-complete. One can solve SAT by an exhaustive search which checks all possible assignments to the input variables. Therefore, it is a natural task to design a faster algorithm than an exhaustive search.

CNF SAT is one of well studied problems. CNF SAT is, given a conjunctive normal form, to ask whether there exists an assignment satisfying it. An exhaustive search algorithm solves this problem with n variables and m clauses in time $O(m \cdot 2^n)$. A lot of excellent algorithms for this problem have been developed such as [1,5,7,8,10,13,15]. The current best algorithm for CNF SAT runs in time $O\left(2^{(1-\frac{1}{\log(m/n)})n}\right)$ [5]. This states that if the number of clauses is bounded by cn (c is an arbitrary positive constant), CNF SAT is solvable in time $O\left(2^{(1-\mu(c))n}\right)$,

© Springer International Publishing Switzerland 2015
F. Dehne et al. (Eds.): WADS 2015, LNCS 9214, pp. 554–565, 2015.
DOI: 10.1007/978-3-319-21840-3_46

where $\mu(c)$ is some constant depending on c. Such algorithms are called "Moderately exponential time algorithms." Recently, Circuit SAT has been extensively studied. Circuit SAT is, given a Boolean Circuit C with n variables and m gates, to ask whether there exists an assignment to the input variables such that C outputs 1. CNF SAT is a special case of Circuit SAT. In some class of circuits, moderately exponential time satisfiability algorithms have been designed such as [11,14,16].

However, there are a very few researches on the satisfiability of branching programs. One important result is about k-OBDD SAT. An OBDD is a branching program such that all paths from the root to any sink have the same order of the variables. A k-OBDD is an extension of OBDD such that it can be separated to k layers and all layers are OBDD with the same order of the variables. k-OBDD SAT, given a k-OBDD, asks whether there exists a consistent path from the root to the 1-sink. For any constant k, this problem is solvable in polynomial time [3]. For the satisfiability of general branching programs, a deterministic algorithm running in $O(2^{n-\omega(\log n)})$ time exists [6]. This algorithm requires exponential space and cannot solve BP SAT with $m = \Omega(n^2)$ states, but counts the number of satisfying assignments.

In this paper, we consider the satisfiability of a k-Indexed Binary Decision Diagram (k-IBDD). A k-IBDD is an extension of k-OBDD, which is the same as k-OBDD except that each layer may have a different order of the variables. k-IBDD SAT was shown to be NP-complete for any $k \geq 2$ [3]. Our task is to design an algorithm super-polynomially faster than an exhaustive search: i.e., an $O(m \cdot 2^{n-\omega(\log n)})$ time algorithm. Jain et al. proposed an experimental efficient algorithm [12], but the worst case complexity has not been given.

We design a moderately exponential time algorithm for k-IBDD SAT and analyze its time and space complexity, then we obtain the followings.

Theorem 1. *There exists a deterministic and exponential space algorithm for* k-IBDD SAT *with n variables and cn nodes, which runs in time* $O\left(2^{(1-\mu(c))n}\right)$, *where* $\mu(c) = \Omega\left(\frac{1}{(\log c)^{2^{k-1}-1}}\right)$ *and c is an arbitrary positive constant.*

Corollary 1. *There exists a deterministic and polynomial space algorithm for* k-IBDD SAT *with n variables and poly(n) nodes, which runs in time* $O\left(2^{n-n^\alpha}\right)$, *where* $\alpha = \frac{1}{2^{k-1}}$.

Paper Organization: In Section 2, we give notations and definitions used in this paper. In Section 3, we provide two key transformations of OBDDs to use our proposed algorithm. In Section 4, we give an algorithm for k-IBDD SAT.

2   Preliminaries

Let $X = \{x_1, \ldots, x_n\}$ be a set of variables and $\bar{x}$ is the negation of variable $x \in X$. A *nondeterministic branching program*, denoted by $B = (V, E)$, is a rooted directed acyclic multigraph. Each $v \in V$ is called *node*, and has a label from $X \cup \{0, 1\}$. The root node is denoted by r and there exists exact two sink nodes denoted by t_0 and t_1 with labels **0** and **1**, respectively. Nodes t_0 and t_1 are called the *0-sink* and the *1-sink*, respectively. For all nodes $v \in V$ except for t_0 and t_1, v gets a label from X. A node v is called an x_i-*node* when v's label is x_i. Each edge $e \in E$ has a label 0 or 1. An edge e is called a *0-edge* (resp. *1-edge*) when its label is 0 (resp. 1). For an edge $e = (u, v) \in E$, u is a *parent* of v and the *head* of e, and v is the *tail* of e. The number of parents of v is called the *in-degree* of v. In addition, when e is a 0-edge (resp. 1-edge), we call v as the 0-*successor* (resp. 1-*successor*) of u. For a branching program B on X, each input $a = (a_1, \ldots, a_n) \in \{0, 1\}^n$ activates all a_i-edges leaving x_i-nodes in B, where $1 \le i \le n$. B outputs 0 if there is no path from the root node r to the 1-sink t_1 using only activated edges. In other words, B outputs 1 if there is at least one path from r to t_1 using only activated edges. A *computation path* is a path from r to t_0 or from r to t_1 using only activated edges. Let $f : \{0, 1\}^n \to \{0, 1\}$ be a boolean function. A branching program B *represents* f if $f(a)$ is equal to the output of B for any assignment $a \in \{0, 1\}^n$. If both of branching programs B and B' represent the same function, then B is *equivalent* to B'. The size of B, denoted by $|B|$, is defined as the number of edges in B. A branching program B is *deterministic* if any nodes except for t_0 and t_1 in B have exact two outgoing edges: one is 0-edge and the other is 1-edge. Note that if B is deterministic, B shows exact one computation path for any inputs. In this paper, a branching program is deterministic unless otherwise noted.

A order $\pi = (\pi(1), \pi(2), \ldots, \pi(n))$ is an arbitrary order from 1 to n. For $i \in \{1, \ldots, n\}$, we define that $\pi^{-1}(i) = j$ when $\pi(j) = i$. An *ordered binary decision diagrams (OBDD)* and a *k-indexed binary decision diagrams (k-IBDD)* are defined as below.

Definition 1. *An $OBDD_\pi$ is a branching program with a fixed ordering π. It holds that $\pi^{-1}(i) < \pi^{-1}(j)$ if there exists an edge from an x_i-node to an x_j-node.*

Definition 2. *A $k\text{-}IBDD_{(\pi_1, \ldots, \pi_k)}$ is a branching program as follows: It can be separated into k layers and the i-th layer is an $OBDD_{\pi_i}$. An arbitrary edge from the i-th layer reaches to the j-th layer or a sink node where $i < j$. If all π_i are the same order π, it is called as a $k\text{-}OBDD_\pi$.*

Figure 1 and Figure 2 represent a nondeterministic $OBDD_\pi$ and a deterministic $OBDD_\pi$, where $\pi = (1, 2, 3)$, respectively. Figure 3 represents a (deterministic) $2\text{-}IBDD_{(\pi_1, \pi_2)}$ with $\pi_1 = (1, 2, 3)$ and $\pi_2 = (2, 3, 1)$.

Remark 1. In [17], the definition of OBDD is as follows: A π'-OBDD is a branching program with a fixed π', and $\pi'(i) < \pi'(j)$ holds if there exists an edge from an x_i-node to an x_j-node. An $OBDD_\pi$ is also a π'-OBDD, if π and π' satisfy the following property: For all i if $\pi(i) = j$, then $\pi'(j) = i$. For example, if $\pi = (2, 3, 1)$ and $\pi' = (3, 1, 2)$, then an $OBDD_\pi$ is also a π'-OBDD. For the simple description of our algorithm, in this paper we adopt Definition 1 as the definition of OBDDs.

k-IBDD SAT is, given a $k\text{-}IBDD_{(\pi_1, \ldots, \pi_k)}$ B with n variables and m nodes, to ask whether there exists an input $a \in \{0, 1\}^n$ such that B outputs 1.

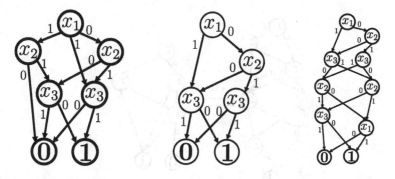

Fig. 1. A nondeterministic OBDD   **Fig. 2.** An OBDD   **Fig. 3.** 2-IBDD

If $k = 1$, a 1-IBDD$_{(\pi)}$ (equal to an OBDD$_\pi$) can be solved in $O(m)$ time by solving reachability from the root node to the 1-sink. k-IBDD SAT is known to be NP-complete when $k \geq 2$ [3].

A *partial assignment* to $x = (x_1, \ldots, x_n)$ is $a = (a_1, \ldots, a_n) \in \{0, 1, *\}^n$. This means that x_i is assigned to 0 or 1 if a_i is 0 or 1, respectively. For any partial assignment $a \in \{0, 1, *\}^n$, a *support* of a is defined as $S(a) := \{x_i \mid a_i \neq *\}$. Let $B|_a$ be a *partial branching program* of B followed by a partial assignment a, and it is constructed as follows:

(1) For all $x_i \in S(a)$, remove all $\overline{a_i}$-edges whose heads are x_i-nodes.
(2) For any node v with in-degree 0 except for the root node, remove all edges whose heads are v and the node v.
(3) For all $x_i \in S(a)$ and any x_i-node v except for the root node, let U be $\{u \mid (u, v) \in E\}$. Note that there exists an a_i-edge (v, w). For all $u \in U$, we add an edge (u, w) labeled the same label of (u, v) and remove an edge (u, v). Remove an edge (v, w).
(4) If the label of the root node r is in $S(a)$, then remove an edge (r, v) and set the node v as a new root node.

Figure 4 is a partial branching program of Figure 3 followed by a partial assignment $a = (1, *, *)$. Note that this partial branching program is a 2-OBDD$_\pi$, where $\pi = (2, 3)$ or $(3, 2)$. For partial assignments a and a' such that $S(a)$ and $S(a')$ are disjoint, $a \circ a'$ denotes a *composition* of a and a': $a \circ a'(i) = a(i)$ if $x_i \in S(a)$, $a \circ a'(i) = a(i)$ if $x_i \in S(a')$, $a \circ a'(i) = *$ otherwise. For instance, when $a = (1, *, *)$ and $a' = (*, *, 0)$, $a \circ a' = (1, *, 0)$.

Recall $\pi = (\pi(1), \pi(2), \ldots, \pi(n))$. A *reverse order* of π is defined as $\pi^R = (\pi^R(1), \pi^R(2), \ldots, \pi^R(n)) = (\pi(n), \pi(n-1), \ldots, \pi(1))$. A *subsequence* of length m of π is $\pi' = (\pi'(1), \pi'(2), \ldots, \pi'(m)) = (\pi(i_1), \pi(i_2), \ldots, \pi(i_m))$, where $1 \leq i_1 < i_2 < \cdots < i_m \leq n$. A *longest increasing subsequence* or *LIS* is a subsequence π' of length m of π with maximum m such that it satisfies $\pi'(i) < \pi'(j)$ for all $1 \leq i < j \leq m$. A *longest decreasing subsequence* or *LDS* is a subsequence π' of length m of π with maximum m such that it satisfies $\pi'(i) > \pi'(j)$ for all

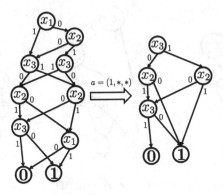

Fig. 4. A partial branching program

$1 \leq i < j \leq m$. Note that LIS and LDS can be found in $O(n \log n)$ time [2] and the following is well known.

Theorem 2 (The Erdős-Szekeres theorem [9]). *A sequence of real number of length n contains a longest increasing subsequence of length $m \geq \lceil \sqrt{n} \rceil$ or a longest decreasing subsequence of length $m \geq \lceil \sqrt{n} \rceil$.*

3   Algorithms of Transformation for OBDDs

In this section, we introduce two key transformations for our algorithm.

First, we can construct a nondeterministic $OBDD_\pi$ B by conjunction of two nondeterministic $OBDD_\pi$s B_1 and B_2. The nondeterministic $OBDD_\pi$ B represents the function $f_1 \wedge f_2$, where f_1 and f_2 are functions represented by B_1 and B_2, respectively. Bryant provided such an algorithm when B_1 and B_2 are deterministic $OBDD_\pi$s [4]. We modify Bryant's algorithm to be applied to nondeterministic $OBDD_\pi$s and get the following lemma. We omit the proof in this paper, see the full version of this paper.

Lemma 1. *Let B_1 and B_2 be nondeterministic $OBDD_\pi$s which represent boolean functions f_1 and f_2, respectively. There exists an algorithm which constructs an $OBDD_\pi$ B which represents $f_1 \wedge f_2$ from B_1 and B_2, and it runs in $O(|B_1| \cdot |B_2|)$ time. Moreover, $|B| \leq |B_1| \cdot |B_2|$ holds.*

Next, we provide an algorithm which constructs a nondeterministic $OBDD_{\pi^R}$ B^R from a deterministic $OBDD_\pi$ B. B^R and B represent the same function.

Lemma 2. *There exists an algorithm which, given an $OBDD_\pi$ B, constructs a nondeterministic $OBDD_{\pi^R}$ B^R which is equivalent to B in $O(|B|)$ time. The size of B^R, denoted by $|B^R|$, is at most $7|B|$.*

Proof. Let B be a given $OBDD_\pi$. Our transformation is divided into two parts. We call the first part **Preprocess** and the second part **Reverse**.

Preprocess: First, we construct B' from B by removing all edges whose tails are the 0-sink. Next, we construct B'' by adding some new nodes into B' to satisfy that for each node v, parents of v have the same label. Now, we describe the detail of these operations. Let $U(v)$ be a set of parents of v, i.e., $U(v) := \{u \mid (u, v) \in E\}$ and $\Phi(v)$ be a set of labels of nodes in $U(v)$, i.e., $\Phi(v) := \{x_i \mid u \text{ is an } x_i\text{-node and } u \in U(v)\}$. If $\Phi(v)$ includes at least two different labels, we choose a label x_ℓ such that $\pi^{-1}(\ell)$ is the maximum in $\Phi(U(v))$. For each b-edge $e = (u, v)$, $b \in \{0, 1\}$, such that $u \in U(v)$ and u is not an x_ℓ-node,

(P1) Add a new x_ℓ-node w.

(P2) Add a new b-edge (u, w).

(P3) Add two new edges, the 0-edge (w, v) and the 1-edge (w, v).

(P4) Remove e.

The second of the left of Figure 5 shows an example of this operation applied to Figure 2. For each node in B'', its parents have the same label. Note that the function represented by B'' is the same function of B. This operation is not applied to any new node created by this operation because such node has exactly one parent. Since (P1)–(P4) are applied to B' at most $|E|$ times, the number of edges of B'' is at most $3|E|$ and the number of nodes in B'' is at most $|V| + |E|$. **Preprocess** takes only $O(|B|)$ time.

Reverse: We construct B^R from B'' as follows:

(R1) Label each node by the following way in bottom up fashion: Label a node v x_i, where x_i is the label of v's parents. Finally, label the root node $\mathbf{1}$, i.e., the root node is replaced with the 1-sink.

(R2) Reverse the direction of all edges.

(R3) For any node v and $b \in \{0, 1\}$, if there exists no a b-successor of v, add a new b-edge (v, t_0), where t_0 is the 0-sink.

For any b-edge $e = (u, v)$ on B'', let us assume that u is an x_i-node. Applying (R1) and (R2), v becomes an x_i-node and B^R has a b-edge $e^R = (v, u)$. Both e and e^R are activated by inputs a such that $a_i = b$. Thus, a path $p = (e_1, \ldots, e_{|p|})$ which is a computation path to the 1-sink on B'' and a path $p^R := (e^R_{|p|}, \ldots, e^R_1)$ which is a computation path to the 1-sink on B^R are in one-to-one correspondence, where $|p|$ is the length of p. In addition, sets of inputs which activate p and p^R are equivalent. Therefore, the satisfying input set for B^R and B'' are equivalent. In this way, we can construct a nondeterministic OBDD$_{\pi^R}$ B^R which is equivalent to B. Note that B^R can be a nondeterministic OBDD$_{\pi^R}$ if there exist at least 2 edges whose labels are the same and tails are the same. The rightmost of Figure 5 shows a nondeterministic OBDD$_{\pi^R}$ constructed by completing this transformation. Since we add at most two edges per one node at (R3), $|B^R| \leq 3|E| + 2(|V| + |E|) \leq 7|E| = 7|B|$. **Reverse** takes only $O(|B|)$ time.

The proof is complete. □

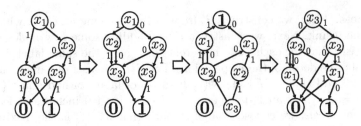

Fig. 5. Preprocess and Reverse

4   Satisfiability Algorithms for k-IBDD

In this section, we explain the proposed algorithm. At first, we provide a polynomial time algorithm for a special k-IBDD SAT satisfying the following property: each layer has either the order π or π^R. Next, we give an algorithm for general case.

4.1   Special Cases

We show that one can solve k-IBDD SAT in polynomial time in the size of input, when a given k-IBDD$_{(\pi_1,\ldots,\pi_k)}$ satisfies that $\pi_i = \pi$ or $\pi_i = \pi^R$ $(1 \leq i \leq k)$ for some order π and k is a constant. We call such a k-IBDD as k-IBDD$_{\{\pi,\pi^R\}}$.

Lemma 3. *Let k be a positive integer. There exists an $O\left((6m/k)^{2k}\right)$ time and $O\left((14m/k)^k\right)$ space algorithm for k-IBDD$_{\{\pi,\pi^R\}}$ SAT with n variables and m nodes.*

Proof. We extend Theorem 2 in [3] to k-IBDD$_{\{\pi,\pi^R\}}$s. Let B be an arbitrary k-IBDD$_{\{\pi,\pi^R\}}$. For $1 \leq h \leq k$, let B_h be the h-th layer of B and V_h be a set of nodes of B_h.

We choose a set $I = \{\ell_i\} \subseteq \{1,\ldots,k\}$ such that $\ell_1 = 1 < \ell_2 < \cdots < \ell_{|I|} \leq k$ holds. Let $L := |I|$. Then, we choose $R = \{r_i\}$ such that $r_i \in V_{\ell_i}$ and there exists some edge (v, r_i), where $v \in V_{\ell_{i-1}}$. For each choice (I, R), we check whether there exists an input $a \in \{0,1\}^n$ which satisfies the following condition: The computation path for a of B reaches to the 1-sink through edges (v_{i-1}, r_i), where $v_{i-1} \in V_{\ell_{i-1}}$ for all $1 < i \leq L$.

For simplicity, let r_1 and r_{L+1} be the root and the 1-sink of B, respectively. From B_{ℓ_i}, for each $1 \leq i \leq L$, we construct an OBDD$_{\pi_i}$ B'_i with a root node r_i as follows. Let V'_i and E'_i be a set of nodes and edges of B'_i, respectively. At first, V'_i consists of nodes such as a 0-sink $t_{i,0}$, a 1-sink $t_{i,1}$ and all nodes $v \in V_{\ell_i}$ which are reachable from r_i. For each edge $e = (u, v) \in E$ such that $u \in V'_i$, E'_i contains an edge e' has the same label as e and satisfies the following property:

– $e' := (u, v)$ if $v \in V'_i$.
– $e' := (u, t_{i,1})$ if $v = r_{i+1}$.
– $e' := (u, t_{i,0})$ if $v \notin V'_i \cup \{r_{i+1}\}$.

From this construction, the following two statements are equivalent:

(1) There is an input $a \in \{0,1\}^n$ such that the computation path on B reaches to the 1-sink through edges (v_{i-1}, r_i), where $v_{i-1} \in V'_{i-1}$ for all $1 < i \leq L$.

(2) There is a common input $a \in \{0,1\}^n$ that satisfies B'_i for all $1 \leq i \leq L$.

If $\pi^R_{\ell_i} = \pi$ for $1 \leq i \leq L$, then we construct a nondeterministic OBDD$_\pi$ B_i^R which is equivalent to the OBDD$_{\pi^R}$ B'_i. By Lemma 2, we can construct $B_i'^R$ in time $O(|B'_i|)$ and $|B_i'^R| \leq 7|B'_i| \leq 7|B_{\ell_i}|$ holds. Then, we rename this $B_i'^R$ by B'_i. Here, it is sufficient to check whether there exists an assignment which satisfies all nondeterministic OBDD$_\pi$s $B'_1, B'_2, \ldots, B'_L$. Applying Lemma 1 to $B'_1, B'_2, \ldots, B'_L$ continuously, we have a nondeterministic OBDD$_\pi$ B^* such that B^* outputs 1 if and only if there exists an assignment that satisfies $B'_1, B'_2, \ldots, B'_L$ simultaneously. The constructing B^* ends in $O(7^k \prod_{i=1}^{k} |B_i|)$ time and the size of B^* is $O(7^k \prod_{i=1}^{k} |B_i|)$. Finally, we check the satisfiability of B^* in $O(|B^*|)$ time and $O(|B^*|)$ space by solving reachability.

The number of pairs (I', R) is at most the number of choices when at most one node must be chosen from each layer. Then, it is at most $\prod_{i=2}^{k} |B_i|$. For each choice, we take $O(7^k \prod_{i=1}^{k} |B_i|)$ time to check the satisfiability. Thus, a satisfiability of B can be solvable in $O(7^k \prod_{i=1}^{k} |B_i|^2)$ time and $O(7^k \prod_{i=1}^{k} |B_i|)$ space. Using the arithmetic-geometric mean inequality and the assumption $\sum_{i=1}^{k} |B_i| = |B| \leq 2m$, we have

$$\prod_{i=1}^{k} |B_i| \leq \left(\sum_{i=1}^{k} \frac{|B_i|}{k} \right)^k \leq \left(\frac{2m}{k} \right)^k$$

holds. Then, the computational time is

$$O\left(7^k \prod_{i=1}^{k} |B_i|^2 \right) = O\left(\left(\frac{2\sqrt{7}m}{k} \right)^{2k} \right) = O\left(\left(\frac{6m}{k} \right)^{2k} \right).$$

In addition, the computational space is $O(7^k \prod_{i=1}^{k} |B_i|) = O((14m/k)^k)$.   □

4.2   Main Algorithm

We provide a moderately exponential time algorithm that solves k-IBDD SAT when k is a constant and the size of input is linear in n. To understand the idea of our proposed algorithm, we describe the outline of our algorithm.

Let B be a given k-IBDD$_{(\pi_1, \pi_2, \ldots, \pi_k)}$. Without loss of generality, we assume $\pi_1 = (1, 2, \ldots, n)$. We focus layers of B from the second layer to the k-th layer. First, we focus on the second layer and divide the order π_2 into L parts $\pi_2^1, \ldots, \pi_2^L$. This means that the second layer is divided into L layers such that each layer respects to orders $\pi_2^1, \ldots, \pi_2^L$, respectively. For each divided layer, we compute the longer sequence σ^ℓ of an LIS or an LDS of π_2^ℓ. Let σ be the ascending

---

Algorithm 1. k-IBDD SAT(B, L)

Require: k-IBDD$_{(\pi_1,\ldots,\pi_k)}$ B, where $\pi_1 = (1,\ldots,n)$, an integer L
Ensure: Output "Yes" if B is satisfiable, otherwise "No".
1: **return Solve**$(B, L, 1, \pi_1, a = (*,\ldots,*))$

Algorithm 2. Solve(B, L, h, σ_h, a)

Require: k-IBDD$_{(\pi_1,\ldots,\pi_k)}$ B, an integer L
Ensure: Output "Yes" if B is satisfiable, otherwise "No".
1: **if** $h = k$ { $B|_a$ is a $((k-1)L+1)$-IBDD$_{\{\sigma_k, \sigma_k^R\}}$ } **then**
2:     Solve $((k-1)L+1)$-IBDD$_{\{\sigma_k, \sigma_k^R\}}$ SAT for $B|_a$
3:     **if** $B|_a$ is satisfiable **then**
4:         **return** "Yes"
5:     **else**
6:         **return** "No"
7:     **end if**
8: **else**
9:     $\pi'_{h+1} :=$ the subsequence of π_{h+1} that consists of all elements of σ_h
10:     Divide π'_{h+1} into L parts such as $\pi^1_{h+1}, \ldots, \pi^L_{h+1}$.
11:     **for** $\ell = 1$ **to** L **do**
12:         $\sigma^\ell_{h+1} :=$ the longer sequence of an LIS or an LDS of π^ℓ_{h+1}
13:     **end for**
14:     $\sigma_{h+1} :=$ the ascending order of all elements of $\{\sigma^\ell_{h+1}\}$
15:     $X_{h+1} := \{x_i \mid i \in I_{h+1}\}$
16:     **for all** partial assignments a_{h+1}, where $S(a_{h+1}) = X \setminus S(a_h) \setminus X_{h+1}$ **do**
17:         **if Solve**$(B, L, h+1, \sigma_{h+1}, a \circ a_{h+1}) =$ "Yes" **then**
18:             **return** "Yes"
19:         **end if**
20:     **end for**
21:     **return** "No"
22: **end if**

---

order of all elements of $\{\sigma^1, \ldots, \sigma^L\}$ and X_2 be a set of variables whose indices are elements of σ. We assign to all variables in $X \setminus X_2$. For any partial assignment a such that $S(a) = X \setminus X_2$, a partial k-IBDD $B|_a$ satisfies that the first layer and all divided L parts respect to orders σ or σ^R. Next, we focus on the third layer and applying the above division and partial assignment for $B|_a$. After operations with regard to the k-th layer, we have a partial $((k-1)L+1)$-IBDD$_{\{\pi,\pi^R\}}$ $B|_{a'}$ for some order π. We check the satisfiability of $B|_{a'}$ by applying Lemma 3. We can solve the satisfiability of B by computing the satisfiability of all partial IBDDs for all partial assignments. Selecting the parameter L optimally, we have a moderately exponential time algorithm.

Here, we provide the moderately exponential time algorithm for k-IBDD SAT with linear size inputs (See Algorithm 1 and its module Algorithm 2). Theorem 1 is directly implied by the following theorem.

Theorem 3. *There exists a deterministic and $O\left(2^{\nu(c)n}\right)$ space algorithm for k-IBDD SAT with n variables and $m = cn$ nodes which runs in $O\left(2^{(1-\mu(c))n}\right)$ time, where $\mu(c) > 0$ for all $c > 0$ and $\mu(c) = \Omega\left(\frac{1}{(\log c)^{2^{k-1}-1}}\right)$, $\nu(c) = O\left(\frac{1}{(\log c)^{2^{k-1}-1}}\right)$ for sufficiently large c.*

Proof. Let B be a k-IBDD$_{(\pi_1,\pi_2,\ldots,\pi_k)}$. Without loss of generality, we assume $\pi_1 = (1, 2, \ldots, n)$. We describe the detail of the proposed algorithm. Let L be a parameter.

First, divide π_2 into L consecutive subsequences $\pi_2^1, \ldots, \pi_2^L$. This means that we divide the second layer into L parts such that each divided layer respects to the order π_2^ℓ. We denote an LIS and an LDS of π_2^ℓ by $\sigma_{2,inc}^\ell$ and $\sigma_{2,dec}^\ell$, respectively. Let a sequence σ_2^ℓ be the longer one of $\sigma_{2,inc}^\ell$ or $\sigma_{2,dec}^\ell$. Let σ_2 be the ascending order of all elements of $\{\sigma_2^1, \ldots, \sigma_2^L\}$. We define $X_2 := \{x_{\sigma_2(i)} \mid 1 \leq i \leq |\sigma_2|\}$, i.e., the set of variables whose indices appear in σ_2. We assign to all variables in $X \setminus X_2$. For any partial assignment a_2 such that $S(a_2) = X \setminus X_2$, the first layer and all divided parts of the second layer of $B|_{a_2}$ respect to orders σ_2 or σ_2^R. Then, we focus on the third layer.

Here, we focus on the $(h + 1)$-th layer, where $2 \leq h \leq k - 1$. Let us assume that we have a partial assignment $a_2 \circ \cdots \circ a_h$ and a sequence σ_h such that all of divided layers from the second to the h-th layer respect to orders σ_h or σ_h^R on $B|_{a_2 \circ \cdots \circ a_h}$. Let $X_h := X \setminus S(a_2 \circ \cdots \circ a_h)$, i.e., the set of unassigned variables. Let π_{h+1}' be a subsequence of π_{h+1} that consists of all elements of σ_h. The $(h + 1)$-th layer respects to the order π_{h+1}' on $B|_{a_2 \circ \cdots \circ a_h}$. We divide π_{h+1}' into L parts $\pi_{h+1}^1, \ldots, \pi_{h+1}^L$. Let σ_{h+1}^ℓ be the longer sequence of an LIS or an LDS of π_{h+1}^ℓ. Let σ_{h+1} be the ascending order of all elements of $\{\sigma_{h+1}^1, \ldots, \sigma_{h+1}^L\}$. We define $X_{h+1} := \{x_{\sigma_{h+1}(i)} \mid 1 \leq i \leq |\sigma_{h+1}|\}$ and assign to all variables in $X_h \setminus X_{h+1}$. For any partial assignment a_{h+1} such that $S(a_{h+1}) = X_h \setminus X_{h+1}$, $S(a_2 \circ \cdots \circ a_h \circ a_{h+1}) = X \setminus X_{h+1}$ holds. A partial branching program $B|_{a_2 \circ \cdots \circ a_h \circ a_{h+1}}$ satisfies that all divided parts of $(h + 1)$-th layer are respect to orders σ_{h+1} or σ_{h+1}^R. In addition, since σ_{h+1} is a subsequence of σ_h, the first layer and all divided parts of the second to h-th layer are also respect to orders σ_{h+1} or σ_{h+1}^R. We focus on the next $(h + 2)$-th layer, if $h + 1 < k$.

Inductively, we have a partial $((k-1)L + 1)$-IBDD$_{\{\sigma_k, \sigma_k^R\}}$ $B|_{a_2 \circ \cdots \circ a_k}$ after the above operations for the k-th layer. Applying Lemma 3, we check the satisfiability of $B|_{a_2 \circ \cdots \circ a_k}$. We can compute the satisfiability of B by checking the satisfiability of $B|_{a_2 \circ \cdots \circ a_k}$ for all partial assignments $a_2 \circ \cdots \circ a_k$.

Here, we give the analysis of the computational time and space. We denote the size of X_h by n_h. Let T_h be the computational time which **Solve**(B, L, h, σ_h, a) requires. Then, T_1 means the computational time for k-IBDD SAT and T_k means the computational time for $((k-1)L + 1)$-IBDD$_{\{\sigma_k, \sigma_k^R\}}$ SAT. By Lemma 3, we have

$$T_k \leq \text{poly}(n) \cdot (6cn/(k-1)L)^{2(k-1)L}. \tag{1}$$

For $1 \leq h \leq k - 1$, when we focus on the $(h + 1)$-th layer, the number of partial assignments is $2^{n_h - n_{h+1}}$. We can compute LISs and LDSs in polynomial time.

Then,

$$T_h \leq \text{poly}(n) \cdot 2^{n_h - n_{h+1}} \cdot T_{h+1}. \tag{2}$$

For each $1 \leq \ell \leq L$, the length of divided order $\pi_{h+1}^{\prime\ell}$ is $\lfloor n_h/L \rfloor$ or $\lceil n_h/L \rceil$. From Theorem 2, $|\sigma_{h+1}^\ell| \geq \left\lceil \sqrt{\lfloor n_h/L \rfloor} \right\rceil$ or $\left\lceil \sqrt{\lceil n_h/L \rceil} \right\rceil$. By calculating, we have

$$n_{h+1} = \sum_{\ell=1}^{L} |\sigma_{h+1}^\ell| \geq \sqrt{L n_h}$$

holds, where $n_1 = n$. Inductively, $n_k \geq L^{1 - \frac{1}{2^{k-1}}} \times n^{\frac{1}{2^{k-1}}}$ holds. Calculating inequalities (1) and (2) inductively,

$$T_1 \leq n^{O(k)} \cdot 2^{(n_1 - n_k)} \cdot T_k$$
$$\leq n^{O(k)} \cdot 2^{\left(n - L^{1 - \frac{1}{2^{k-1}}} n^{\frac{1}{2^{k-1}}} \right)} \cdot \left(\frac{6cn}{(k-1)L} \right)^{2(k-1)L}$$

holds. Using a new parameter $\delta := L/n$ $(0 < \delta \leq 1)$, we have

$$T_1 \leq 2^{\left(1 - \delta^{1 - \frac{1}{2^{k-1}}} + 2(k-1)\delta \log \frac{6c}{(k-1)\delta} \right) n + O(k \log n)}.$$

Let $\mu(c) = \delta^{1 - \frac{1}{2^{k-1}}} - 2(k-1)\delta \log \frac{6c}{(k-1)\delta}$. Setting $\delta = \frac{1}{\beta \cdot (\log c)^{2^{k-1}}}$ and $\beta > (2(k-1))^{2^{k-1}}$, $\mu(c) = \Omega\left(\frac{1}{(\log c)^{2^{k-1}-1}} \right)$ holds for sufficiently large c. The complete analysis of $\mu(c)$ is shown in the full version of this paper.

When we focus on each layer, we need only polynomial space. We use $\text{poly}(n) \cdot (7cn/(k-1)L)^{(k-1)L}$ space for $((k-1)L+1)\text{-IBDD}_{\{\sigma_k, \sigma_k^R\}}$ SAT by Lemma 3. Therefore, the computational space is $\text{poly}(n) \cdot 2^{\nu(c)n}$, where $\nu(c) = O\left(\frac{1}{(\log c)^{2^{k-1}}} \right)$.

Therefore, the proof is complete. $\qquad\square$

We obtain the following corollary by setting the parameter $L = 1$ and $c = \text{poly}(n)$ in the proof of Theorem 3. The precise proof is shown in the full version of this paper.

Corollary 2 (Recall of Corollary 1). *There exists a deterministic and polynomial space algorithm for k-IBDD SAT with n variables and poly(n) nodes, which runs in time $O\left(2^{n - n^\alpha} \right)$, where $\alpha = \frac{1}{2^{k-1}}$.*

Acknowledgments. This work is supported by MEXT KAKENHI 24106003, JSPS KAKENHI 26730007, JST, ERATO, Kawarabayashi Large Graph Project and Japan Society for the Promotion of Science.

References

1. Arvind, V., Schuler, R.: The Quantum Query Complexity of 0-1 Knapsack and Associated Claw Problems. In: Ibaraki, T., Katoh, N., Ono, H. (eds.) ISAAC 2003. LNCS, vol. 2906, pp. 168–177. Springer, Heidelberg (2003)
2. Bergroth, L., Hakonen, H., Raita, T.: A Survey of Longest Common Subsequence Algorithms. In: Proceedings of the 7th International Symposium on String Processing and Information Retrieval (SPIRE), pp. 39–48 (2000)
3. Bollig, B., Sauerhoff, M., Sieling, D., Wegener, I.: On the Power of Different Types of Restricted Branching Programs. In: Electronic Colloquium on Computational Complexity (ECCC), vol. 1(26) (1994)
4. Bryant, R.E.: Graph-based Algorithm for Boolean Function Manipulation. IEEE Transaction on Computers 35(8), 677–691 (1986)
5. Calabro, C., Impagliazzo, R., Paturi, R.: A Duality between Clause Width and Clause Density for SAT. In: Proceedings of the 21st Annual IEEE Conference Computational Complexity (CCC), pp. 252–260 (2006)
6. Chen, R., Kabanets, V., Kolokolova, A., Shaltiel, R., Zuckerman, D.: Mining Circuit Lower Bound Proofs for Meta-Algorithms. In: Proceedings of the 29th Annual IEEE Conference on Computational Complexity (CCC), pp. 262–273 (2014)
7. Dantsin, E., Hirsch, E.A., Wolpert, A.: Algorithms for SAT Based on Search in Hamming Balls. In: Proceedings of the 21st Annual Symposium on Theoretical Aspects of Computer Science (STACS), pp. 141–151 (2004)
8. Dantsin, E., Hirsch, E.A., Wolpert, A.: Clause Shortening Combined with Pruning Yields a New Upper Bound for Deterministic SAT Algorithms. In: Calamoneri, T., Finocchi, I., Italiano, G.F. (eds.) CIAC 2006. LNCS, vol. 3998, pp. 60–68. Springer, Heidelberg (2006)
9. Erdős, P., Szekeres, G.: A Combinatorial Problem in Geometry. Compositio Mathematica 2, 463–470 (1935)
10. Hirsch, E.A.: Exact Algorithm for General CNF SAT, Encyclopedia of Algorithms. Springer (2008)
11. Impagliazzo, R., Matthews, W., Paturi, R.: A satisfiability algorithm for AC^0. In: Proceedings of the 23rd Annual ACM-SIAM Symposium on Discrete Algorithms (SODA), pp. 961–972 (2012)
12. Jain, J., Bitner, J., Abadir, M.S., Abraham, J.A., Fussell, D.S.: Indexed BDDs: Algorithmic Advances in Techniques to Represent and Verify Boolean Functions. IEEE Transaction on Computers 46(11), 1230–1245 (1997)
13. Pudlák, P.: Satisfiability - Algorithms and Logic. In: Brim, L., Gruska, J., Zlatuška, J. (eds.) MFCS 1998. LNCS, vol. 1450, pp. 129–141. Springer, Heidelberg (1998)
14. Santhanam, R.: Fighting Perebor: New and Improved Algorithms for Formula and QBF Satisfiability. In: Proceedings of the 51st International Symposium on Foundations of Computer Science (FOCS), pp. 183–192 (2010)
15. Schuler, R.: An algorithm for the satisfiability problem of formulas in conjunctive normal form. J. Algorithms 54(1), 40–44 (2005)
16. Seto, K., Tamaki, S.: A Satisfiability Algorithm and Average-Case Hardness for Formulas over the Full Binary Basis. Computational Complexity 22(2), 245–274 (2013)
17. Wegener, I.: Branching Programs and Binary Decision Diagrams. SIAM Monographs on Discrete Mathematics and Applications (2000)

On the Parameterized Complexity of GIRTH and CONNECTIVITY Problems on Linear Matroids

Fahad Panolan[1,2]([⊠]), M.S. Ramanujan[2], and Saket Saurabh[1,2]

[1] The Institute of Mathematical Sciences, Chennai, India
{fahad,saket}@imsc.res.in
[2] University of Bergen, Bergen, Norway
Ramanujan.Sridharan@ii.uib.no

Abstract. Computing the minimum distance of a linear code is a fundamental problem in coding theory. This problem is a special case of the MATROID GIRTH problem, where the objective is to compute the length of a shortest circuit in a given matroid. A closely related problem on matroids is the MATROID CONNECTIVITY problem where the objective is to compute the connectivity of a given matroid. Given a matroid $M = (E, \mathcal{I})$, a k-separation of M is a partition (X, Y) of E such that $|X| \geq k$, $|Y| \geq k$ and $r(X) + r(Y) - r(E) \leq k - 1$, where r is the rank function. The connectivity of a matroid M is the smallest k such that M has a k-separation.

In this paper we study the parameterized complexity of MATROID GIRTH and MATROID CONNECTIVITY on linear matroids representable over a field $\mathbb{F}_q$. We consider the parameters–(i) solution size, k, (ii) rank(M), and (iii) rank$(M)+q$, where M is the input matroid.

We prove that MATROID GIRTH and MATROID CONNECTIVITY when parameterized by rank(M), hence by solution size, k, are not expected to have FPT algorithms under standard complexity hypotheses. We then design *fast* FPT algorithms for MATROID GIRTH and MATROID CONNECTIVITY when parameterized by rank$(M)+q$. Finally, since the field size of the linear representation of transversal matroids and gammoids are large we also study MATROID GIRTH on these specific matroids and give algorithms whose running times do not depend exponentially on the field size.

1 Introduction

One of the most fundamental problems in coding theory is the problem of computing the minimum distance of a linear code. The decision version of this problem for binary linear codes was conjectured to be NP-complete by Berlekamp, McEliece, and van Tilborg [1] in 1978 and it was only in 1997 that its NP-completeness was proved by Vardy [13]. In fact, the result of Vardy can be

The research leading to these results has received funding from the European Research Council under the European Union's Seventh Framework Programme (FP/2007-2013) / ERC Grant Agreement n. 267959.

F. Dehne et al. (Eds.): WADS 2015, LNCS 9214, pp. 566–577, 2015.
DOI: 10.1007/978-3-319-21840-3_47

extended to linear codes over all finite fields. The MINIMUM DISTANCE problem is a special case of the MATROID GIRTH problem, where the objective is to compute the length of a shortest circuit in a given matroid. A closely related problem on matroids is the MATROID CONNECTIVITY problem where the objective is to compute the connectivity of a given matroid. This problem generalizes the classical graph problem of computing the connectivity of a given graph. In this paper we study the parameterized complexity of natural decision problems associated with GIRTH and CONNECTIVITY on linear matroids.

Matroids are mathematical objects which have many applications in algorithms. Certain problems on matroids are known to be equivalent to fundamental combinatorial problems like MINIMUM WEIGHT SPANNING TREE or PERFECT MATCHING. Matroids are an exact characterization of structures on which a greedy algorithm produces an optimum solution. This motivates our study on several natural matroid problems. In order to give an unambiguous descriptions of the problems we study, we first recall certain notions from matroid theory.

Matroids. A matroid is a pair $M = (E, \mathcal{I})$, where E is a ground set and $\mathcal{I}$ is a family of subsets (called independent sets) of E, and it satisfies the following conditions: (i) $\emptyset \in \mathcal{I}$, (ii) If $A' \subseteq A$ and $A \in \mathcal{I}$ then $A' \in \mathcal{I}$, and (iii) If $A, B \in \mathcal{I}$ and $|A| < |B|$, then $\exists e \in (B \setminus A)$ such that $A \cup \{e\} \in \mathcal{I}$.

An inclusion-wise maximal set of $\mathcal{I}$ is called a *basis* of the matroid. Using axiom (iii) it is easy to show that all the bases of a matroid have the same size. This size is called the *rank* of the matroid M, and is denoted by $\mathsf{rank}(M)$.

Linear Matroids and Representable Matroids. Let A be a matrix over an arbitrary field $\mathbb{F}$ and let E be the set of columns of A. Given A we define the matroid $M = (E, \mathcal{I})$ as follows. A set $X \subseteq E$ is independent (that is $X \in \mathcal{I}$) if the corresponding columns are linearly independent over $\mathbb{F}$. The matroids that can be defined by such a construction are called *linear matroids*, and if a matroid can be defined by a matrix A over a field $\mathbb{F}$, then we say that the matroid is representable over $\mathbb{F}$. A matroid $M = (E, \mathcal{I})$ is called *representable* or *linear* if it is representable over some field $\mathbb{F}$. For a matrix A, we also use $\mathsf{rank}(A)$ to denote its rank.

Girth and Connectivity. A subset of the ground set E that is not independent is called *dependent*. A *circuit* in a matroid M is an inclusion-wise minimal dependent subset of E. The *girth* of a matroid M, denoted by $g(M)$, is the cardinality of a minimum sized circuit. We now consider the notion of a rank function associated with a matroid: $r : 2^E \rightarrow \mathbb{N}^+ \cup \{0\}$. Here, $r(S)$ is the maximum size of an independent set contained in S. Note that $r(E) = \mathsf{rank}(M)$. It is also well known that r is a submodular function, which is extremely useful in designing algorithms for matroids. Now we define the notions of k-*separation* and connectivity.

Definition 1. *A k-separation of M is a partition (X, Y) of E such that $|X| \geq k$, $|Y| \geq k$ and $r(X) + r(Y) - r(E) \leq k - 1$. The connectivity of a matroid M, denoted by $\kappa(M)$, is the smallest k such that M has a k-separation.*

We first discuss how the notions of girth and connectivity defined for matroids relate with the ones defined for graphs. Given a graph G, one can define a matroid $M = (E, \mathcal{I})$ as follows. We take E as $E(G)$ and a set $F \subseteq E(G)$ is in $\mathcal{I}$ if it forms a forest. This is called a *graphic matroid*. The notion of girth for graphic matroids coincides with the notion of girth defined in the context of graphs. The notion of connectivity defined for matroids is slightly different from the standard one defined for graphs. It would be desirable to have a notion of connectivity from graphs extending to matroids. Unfortunately, the standard notion of edge-connectivity in graphs when extended to matroids does not fit well with the duals of these matroids. With these issues in mind, Tutte [12] proposed the above definition of connectivity for a matroid, which renders it dual-invariant. That is, a matroid is ℓ-connected if and only if its dual is. Finally, we note that Oxley [10] has shown how Tutte's definition of matroid connectivity can be modified to give a matroid concept that directly generalizes the notion of connectivity on graphs.

The Problems and our Results. In this paper we study the following two problems on linear matroids.

MATROID GIRTH
Input: A linear matroid $M = (E, \mathcal{I})$ together with its representation matrix A_M of dimension $\mathrm{rank}(M) \times |E|$ over a field $\mathbb{F}_q$, and a positive integer k.
Parameters: (1) k, (2) $\mathrm{rank}(M)$ and (3) $\mathrm{rank}(M) + q$
Question: Does there exist a circuit of size at most k in M?

MATROID CONNECTIVITY
Input: A linear matroid $M = (E, \mathcal{I})$ together with its representation matrix A_M of dimension $\mathrm{rank}(M) \times |E|$ over a field $\mathbb{F}_q$, and a positive integer k.
Parameters: (1) k, (2) $\mathrm{rank}(M)$ and (3) $\mathrm{rank}(M) + q$
Question: Does M has a k-separation?

It was known in the 80's that MATROID GIRTH is NP-complete [9]. Later Vardi showed that the problem is NP-complete even for binary matroids (matroids over $\mathbb{F}_2$) [13]. Similarly, it is possible to derive that MATROID CONNECTIVITY is not in P unless P=NP (see Section 3). Although it is easy to observe that MATROID GIRTH admits an algorithm with running time $|E|^{\mathcal{O}(k)}$, it is not at all obvious that MATROID CONNECTIVITY admits an algorithm that is polynomial for every fixed integer k. Bixby and Cunningham, using an algorithm for matroid intersection, gave an algorithm for MATROID CONNECTIVITY running in time $|E|^{\mathcal{O}(k)}$ (see [3] for details). Therefore, these problems are natural candidates for a study in the realm of parameterized complexity.

In parameterized complexity each problem instance comes with a parameter k and a central notion in parameterized complexity is *fixed parameter tractability* (FPT). This means, for a given instance (x, k), solvability in time $\tau(k)|x|^{\mathcal{O}(1)}$, where τ is an arbitrary function of k. There is also an accompanying theory of intractability which enables us to show that certain problems are unlikely to be FPT. These problems are called W-hard problems (see [4] for more details).

A first natural parameter for our problems is the solution size, k. We argue that this problem is unlikely to have an FPT algorithm in general. For this, we consider the HALL SET problem. In this problem we are given a bipartite graph G with bipartition into A and B and a positive integer k, and the objective is to find a set $S \subseteq A$ of size at most k such that the number of neighbors of S in B is strictly smaller than $|S|$, that is, $|N(S)| < |S|$. It is known that HALL SET is W-hard [6]. This problem is clearly a special case of MATROID GIRTH. Indeed, if the input to MATROID GIRTH is a transversal matroid then the problem is precisely HALL SET. We would also like to point out that EVEN SET, the parameterized version of the problem of computing the minimum distance of a binary linear code is a long standing open problem in the area and is stated among the most "infamous open problems" in the Research Horizons section of the recent textbook by Downey and Fellows [4, Chapter33.1]. However, the *exact* version of EVEN SET, where we want to check for a circuit of size exactly k is known to be W-hard [5]. In this work, we prove a similar hardness result for MATROID CONNECTIVITY. These intractability results force us to look for alternate parameterizations.

The next natural parameter would be the rank of the input matroid. Since it is a larger parameter than the solution size k, one might hope for tractability results in place of previous intractability results. However, we show (Section 3) that this is also unlikely. Therefore, we choose as our parameter, rank(M)+q, where q is the size of the field in which the matroid is represented. Indeed, q is constant for a fixed finite field such as $\mathbb{F}_2$. Observe that since the number of distinct column vectors in $\mathbb{F}_q^r$ is q^r, MATROID GIRTH can be solved in time $q^{r^2}|E|^{\mathcal{O}(1)}$, where $r = \text{rank}(M)$. Furthermore, an algorithm for MATROID GIRTH with running time $\mathcal{O}(q^{\text{rank}(M)+k}\text{rank}(M)\log k)$ can be found as a byproduct in [2] (see Theorem 14 in [2]). However, it was unknown if the additive dependence on k in the exponent can be avoided. In this paper, we give a faster algorithm for MATROID GIRTH. This algorithm gives an exponential speedup over the previous algorithm when k is close to rank(M).

Theorem 1. MATROID GIRTH *can be solved in time* $\mathcal{O}(q^{\text{rank}(M)}\text{rank}(M) + |E|k^2)$.

In the context of MATROID CONNECTIVITY, we first show that the problem is not FPT unless FPT= $W[1]$, when parameterized by rank(M). Subsequently, we give a branching algorithm for MATROID CONNECTIVITY which has a single-exponential dependence on the rank. The main features of this algorithm are the use of the rank of matroids obtained in the subproblems as a measure to quantify the progress of the algorithm and application of the algorithm given by Theorem 1 to solve the base cases. Thus, one can view this algorithm as a parameterized Turing-reduction from MATROID CONNECTIVITY to MATROID GIRTH. Formally, we obtain the following theorem.

Theorem 2. MATROID CONNECTIVITY *can be solved in time*
$$\mathcal{O}\left(2^{\text{rank}(M)+k} \cdot \text{rank}(M)^2 \cdot |E| \cdot \left(q^{\text{rank}(M)}\text{rank}(M) + |E|k^2\right)\right).$$

Theorems 1 and 2 imply FPT algorithms parameterized by rank(M) for all matroids defined over a constant size field (such as graphic matroids and co-graphic matroids). Our lower bounds rule out having an algorithm without having any dependence on the field size. However, it is possible that for certain matroids, for instance, transversal matroids, gammoids and strict gammoids, which are only representable over fields whose size depends on $|E|$, one can obtain an FPT algorithm parameterized by rank(M) alone. In fact, we give such an algorithm, running in time $2^{\mathsf{rank}(M)}|E|^{\mathcal{O}(1)}$, for transversal matroids. For strict gammoids however, we give a polynomial time algorithm and leave open the same problem for gammoids. Due to lack of space, the proofs of these two results are deferred to the full version of the paper.

2   Preliminaries

In this section we give some basic definitions and terminology we use in the paper.

Matroids. Here we give definitions related to matroids that are not presented in the introduction (Section 1). For a broader overview on matroids we refer to [11]. Let $M = (E, \mathcal{I})$ be a matroid. Recall that $r : 2^E \to \mathbb{N}^+ \cup \{0\}$ is the rank function associated with the matroid $M = (E, \mathcal{I})$. The closure $\mathsf{cl}(A)$ of a subset A of E is the set $\mathsf{cl}(A) = \left\{x \in E \mid r(A) = r(A \cup \{x\})\right\}$. For $X \subseteq E$, the contraction of M by X, written M/X, is the matroid on the underlying set $E \setminus X$ whose rank function $r' : 2^{E \setminus X} \to \mathbb{N}^+ \cup \{0\}$ is defined as follows. For all $A \subseteq E \setminus X$, $r'(A) = r(A \cup X) - r(X)$. We now define the notion of the dual of a matroid.

Definition 2. *For a matroid, $M = (E, \mathcal{I})$, the dual matroid $M^* = (E, \mathcal{I}^*)$ is a matroid such that the bases of M^* are the complements of the bases of M.*

We note that the rank of the dual matroid M^* is $|E| - \mathsf{rank}(M)$. The dual of a linear matroid is also linear. Further, given a representation A of a matroid M, a representation of the dual matroid M^* can be found in polynomial time.

Uniform Matroid. A pair $M = (E, \mathcal{I})$, is called a uniform matroid if the family of independent sets is given by $\mathcal{I} = \{A \subseteq E \mid |A| \leq k\}$, where k is a constant. This matroid is also denoted as $U_{n,k}$. Every uniform matroid is linear and can be represented over a finite field by a $k \times n$ matrix A_M where the $A_M[i,j] = j^{i-1}$.

3 $W[1]$-Hardness When Parameterized by Rank

In this section we show that MATROID GIRTH and MATROID CONNECTIVITY parameterized by rank(M) are not FPT unless FPT=$W[1]$. It is known that MATROID GIRTH is NP-hard [9,13]. In this section we first show that MATROID CONNECTIVITY cannot be solved in polynomial time unless $P = NP$. Then we

explain how the same proof gives $W[1]$-hardness result for MATROID CONNEC-
TIVITY and prove the $W[1]$-hardness result for MATROID GIRTH. Towards the
proof, we need to consider the UNIFORM MATROID ISOMORPHISM problem.

UNIFORM MATROID ISOMORPHISM (UMI)
Input: A $k \times m$ matrix M of rank k.
Question: Is M isomorphic to $U_{m,k}$? I.e., is every k sized subset of columns
of M linearly independent?

Khachiyan et al [7] showed that UMI is NP-hard, by giving a reduction from
SMALL SUBSET SUM, which is defined as follows.

SMALL SUBSET SUM
Input: A set of n positive integers $S = \{\alpha_1, \ldots, \alpha_n\}$ and $k, \beta \in \mathbb{N}^+$.
Question: Does there exist a subset of k integers in S which sum up to β?

The following known lemma is needed to prove the hardness of MATROID
CONNECTIVITY and MATROID GIRTH, and its proof is deferred to the full version
of the paper.

Lemma 1. *Let* $M = (E, \mathcal{I})$ *be a matroid of rank* k *and* $m = |E| > 2k+1$. *Then*

1. M *is isomorphic to* $U_{m,k}$, *if and only if the girth of* M *is* $k+1$.
2. *If* $\kappa(M) = k+1$ *then* $g(M) = k+1$.

Theorem 3. MATROID CONNECTIVITY *is not in* P *unless* P=NP.

Proof. We prove the theorem by designing a polynomial time algorithm for UMI,
which uses an algorithm for MATROID CONNECTIVITY as a subroutine. Now
given a $k \times m$ matrix M of rank k, we want to test whether M is isomorphic to
$U_{m,k}$. We assume that $m > 2k+1$. Note that M is a uniform matroid if and only
if the dual of M is also a uniform matroid. So without loss of generality we can
assume that $k \leq m - k$, otherwise instead of checking whether M is isomorphic
to $U_{m,k}$, it is enough to check whether dual of M is isomorphic to $U_{m,m-k}$.

Now we describe an algorithm to solve UMI using an algorithm for MATROID
CONNECTIVITY. Using an algorithm for MATROID CONNECTIVITY, find the least
integer, $i \leq k + 1$, if it exists, such that M has i-separation. This implies that
$\kappa(M) = i$. If $i = k + 1$ then we output YES, otherwise we output NO. Now we
show that our algorithm is correct. If the algorithm outputs YES, then $\kappa(M) =
k + 1$. Then by Lemma 1, we have that $g(M) = k + 1$ and M is isomorphic to
$U_{m,k}$. Now consider the following claim.

Claim 1. $\kappa(U_{m,k}) = k + 1$

Proof. Let E be the ground set of $U_{m,k}$. Let $C \subseteq E$ be an arbitrary subset of size
$k + 1$. Note that $r(C) + r(E \setminus C) - r(E) \leq r(C) \leq |C| - 1$ and $|C|, |E \setminus C| \geq k + 1$.
Hence, $\kappa(M) \leq k + 1$. Next we show that there does not exists $i < k + 1$ such
that $\kappa(M) = i$. We prove the statement via contradiction. Suppose there exists
$i < k + 1$ such that $\kappa(M) = i$. Let $(A, E \setminus A)$ be an i-separation. Since $m > 2k$,

we have that $|A| > k$ or $|E \setminus A| > k$. Assume without loss of generality that $|E \setminus A| > k$. Then $r(A) + r(E \setminus A) - r(E) = r(A) \geq i$ (since $|A| \geq i$ and $i \leq k$). This leads to a contradiction that $(A, E \setminus A)$ be an i-separation. This completes the proof of the lemma.   □

If M is isomorphic to $U_{m,k}$ then by Claim 1, we have that $\kappa(M) = k + 1$ and thus the algorithm will output YES. This completes the proof of correctness of our algorithm. Hence, if we do have a polynomial time algorithm for MATROID CONNECTIVITY then we do have a polynomial time algorithm for UMI.   □

Now we explain that in fact the proof of Theorem 3 also gives the lower bound for MATROID CONNECTIVITY parameterized by $\mathsf{rank}(M)$. The reduction to prove UMI NP-hard [7, Theorem 1] takes as an input (S, k, β) to SMALL SUBSET SUM and produces an instance $(M, k+3)$ to UMI, where M is a $(k+3) \times (n+2)$ matrix. It is known that SMALL SUBSET SUM parameterized by k is $W[1]$-hard [4]. This implies that UMI parameterized by k is $W[1]$-hard. Thus, combining this fact with Theorem 3 we have that, if MATROID CONNECTIVITY parameterized by $\mathsf{rank}(M)$ is FPT, then UMI parameterized by, $k= \mathsf{rank}(M)$ is FPT. Thus we have the following theorem.

Theorem 4. MATROID CONNECTIVITY *parameterized by* $\mathsf{rank}(M)$ *is not* FPT *unless* FPT=$W[1]$.

As $\kappa(M) \leq \mathsf{rank}(M)+1$, MATROID CONNECTIVITY parameterized by k is also not FPT unless FPT=$W[1]$. Now in the proof of Theorem 3, if we replace the MATROID CONNECTIVITY subroutine with a subroutine of MATROID GIRTH, then by condition 1 in Lemma 1, we get an algorithm for UMI. Again, by the fact that UMI parameterized by k is $W[1]$-hard, we get the following theorem.

Theorem 5. MATROID GIRTH *parameterized by* $\mathsf{rank}(M)$ *is not* FPT *unless* FPT=$W[1]$.

4   Algorithms for Girth of a Matroid

In this section we design a $q^{\mathsf{rank}(M)}|E|^{\mathcal{O}(1)}$ time algorithm for MATROID GIRTH parameterized by $\mathsf{rank}(M)$ using the MacWilliams identity. In what follows we give basics of coding theory and recall the MacWilliams identity.

Coding Theory. A linear code C over a finite field $\mathbb{F}_q$, defined by $n \times m$ matrix A, is the set of m-dimensional vectors $C = \{vA \mid v \in \mathbb{F}_q^n\}$. The matrix A is called the generator matrix of C. The code C is the linear subspace of $\mathbb{F}_q^m$ spanned by the row vectors of A and its dimension is equal to $\mathsf{rank}(A)$. Without loss of generality we can assume $n = \mathsf{rank}(A)$. A (m,n)-*linear code* is one such that the length of codewords is m and its dimension is n.

Let C be a linear code with generator matrix A. Let $\mathbf{0}$ be the zero vector $(0, \ldots, 0)^T$. The length of $\mathbf{0}$ will be clear from the context. The parity check matrix H of C is a $(m - n) \times m$ matrix satisfying $Hw^T = \mathbf{0}$ for any codeword

$w \in C$. It is well-known that there is a duality between generator matrices and parity check matrices: For the code $C^\perp$ with generator matrix H^T, it is easily verified that $Av = \mathbf{0}$ holds for any $v \in C^\perp$. That is, A is the parity check matrix of $C^\perp$. The code $C^\perp$ is called the dual code of C. Given a codeword w, the number of non-zero entries in w is called the weight of w and is denoted by $\mathsf{wt}(w)$. The *weight enumerator* of an (m, n)-linear code C is a polynomial in x, y and is given by,

$$W_C(x, y) = \sum_{c \in C} x^{m - \mathsf{wt}(c)} y^{\mathsf{wt}(c)} = \sum_{i=0}^{m} \xi_i x^{m-i} y^i,$$

where ξ_i is the number of words of weight i in C. The following theorem shows that the *weight enumerator* of $C^\perp$ can be calculated from that of C.

Proposition 1 (MacWilliams identity [8]). $W_{C^\perp}(x, y) = \frac{1}{|C|} W_C(x + (q - 1)y, x - y)$.

Algorithm for MATROID GIRTH. We next prove Theorem 1.

Proof (Proof of Theorem 1). Let (M, k) be an instance to MATROID GIRTH and let A_M be the representation matrix of M of order $r \times m$ over $\mathbb{F}_q$, where $m = |E|$ and $r = \mathrm{rank}(M)$. Consider the system of linear equations $A_M v = \mathbf{0}$, where $v = (v_1, \ldots, v_m)^t$ is a vector of variables. We have the following claim.

Claim 2. $g(M) \leq k$ if and only if there exists a vector $z \in \mathbb{F}_q^m$ with $\mathsf{wt}(z) \leq k$ and $A_M z = \mathbf{0}$.

Proof. Let $C' \subseteq E$ be a circuit of length at most $\ell \leq k$ in M. Let $W \subseteq \{1, \ldots, m\}$ be the set of indices corresponding to the elements of the circuit C'. Since C' is a circuit in M, C' is linearly dependent. Thus, the columns corresponding to indices in W are also linearly dependent. Hence there exist $\lambda_1, \ldots, \lambda_\ell \in \mathbb{F}_q$, not all zeros, such that $\sum_{j \in W} \lambda_j A_j = \mathbf{0}$. Here, A_j denotes the j-th column of A_M. Now consider the vector $z = (z_1, \ldots, z_m)^T$, where $z_j = \lambda_j$ if $j \in W$, else $z_j = 0$. Note that $\mathsf{wt}(z) = \ell \leq k$. Since $\sum_{j \in W} \lambda_j A_j = \mathbf{0}$, we have that $A_M z = \mathbf{0}$.

Suppose there exists a vector z with $\mathsf{wt}(z) \leq k$ such that $A_M z = \mathbf{0}$, then $\sum_{i=1}^{m} z_i A_i = \mathbf{0}$ where A_i is i^{th} column in A_M. Let $W \subseteq \{1, \ldots, m\}$ such that $i \in W$ if and only if $z_i \neq 0$, then $\sum_{i \in W} z_i A_i = \mathbf{0}$. Since $|W| \leq k$, there exist (at most) k columns in A which are linearly dependent. Hence, $g(M) \leq k$. □

By Claim 2, to show that $g(M) \leq k$, it is sufficient to show that there exists a vector $z \in \mathbb{F}_q^m$ with $\mathsf{wt}(z) \leq k$ and $A_M z = \mathbf{0}$. Let C be the code generated by the matrix A_M, i.e., $C = \{v A_M \mid v \in \mathbb{F}_q^r\}$. Let $C^\perp$ be the dual code of C, i.e $C^\perp = \{u \mid A_M u = \mathbf{0}\}$. Using the MacWilliams identity we have,

$$W_{C^\perp}(x, y) = \frac{1}{|C|} W_C(x + (q - 1)y, x - y). \tag{1}$$

Since $|C| \leq q^r$, the polynomial $W_C(x,y)$ can be computed in $\mathcal{O}(q^r r + |E|)$ time. Now we have,

$$W_C(x + (q-1)y, x - y) = \sum_{i=0}^{m} \xi_i (x + (q-1)y)^{m-i}(x-y)^i,$$

where ξ_i is the number of codewords in C of weight i. Claim 2 implies that there exists a circuit of size at most k in M if and only if there exists a codeword z in $C^\perp$ such that $\mathrm{wt}(z) \leq k$, that is, the coefficient of $x^{m-j}y^j$ in $W_{C^\perp}(x,y)$ is non zero for some $j \leq k$ (by the definition of weight enumerator of $C^\perp$). Due to the MacWilliams identity (Equation 1), there exists a codeword z in $C^\perp$ such that $\mathrm{wt}(v) = j$ if and only if coefficient of $x^{m-j}y^j$ in $W_C(x + (q-1)y, x - y)$ is not equal to zero. Using the binomial theorem, we have that the coefficient of $x^{m-j}y^j$ in $\xi_i(x + (q-1)y)^{m-i}(x-y)^i$ is

$$\xi_i \sum_{j'+j''=j} (-1)^{j''} (q-1)^{j'} \binom{m-i}{j'} \binom{i}{j''}.$$

Hence the coefficient of $x^{m-j}y^j$ in $W_C(x + (q-1)y, x - y)$ is

$$\sum_{i=0}^{m} \xi_i \sum_{j'+j''=j} (-1)^{j''} (q-1)^{j'} \binom{m-i}{j'} \binom{i}{j''} \tag{2}$$

Thus we can check whether the coefficient of $x^{m-j}y^j$ in $W_C(x + (q-1)y, x - y)$ is non zero or not in time $\mathcal{O}(q^r + mj)$. We output YES if Equation 2 is non zero for any $j \leq k$. Hence the total running time is $\mathcal{O}(q^r r + |E|k^2)$. □

5   Algorithm for k-Connectivity of a Matroid

In this section we design a fast FPT algorithm for MATROID CONNECTIVITY parameterized by $\mathrm{rank}(M)+q$. Our algorithm is a recursive algorithm and at the leaves of the search tree it runs the algorithm for MATROID GIRTH as a subroutine. In what follows, we say that a partition (X,Y) of the ground set E of a matroid M, obeys the pair (X_1,Y_1) where $X_1, Y_1 \subseteq E$, if $X_1 \subseteq X$ and $Y_1 \subseteq Y$. We start with two lemmas which are useful for our algorithm.

Lemma 2. *Let $M = (E,\mathcal{I})$ be a matroid. Let $X,Y \subseteq E$ such that $X \cap Y = \emptyset$ and $|Y| \geq k$. Let $S_x = \mathrm{cl}(X) \cap (E \setminus (X \cup Y))$. If there exist a k-separation (X',Y') obeying the pair (X,Y), then there exist a k-separation (X'',Y'') obeying the pair $(X \cup S_x, Y)$.*

Proof. Let (X',Y') be a k-separation obeying the pair (X,Y). Then we know that $r(X') + r(Y') - r(E) \leq k - 1$. Now consider the partition $(X' \cup S_x, Y' \setminus S_x)$. We claim that $(X' \cup S_x, Y' \setminus S_x)$ is a k-separation because $|X' \cup S_x| \geq |X'| \geq k$, $|Y' \setminus S_x| \geq |Y| \geq k$ and $r(X' \cup S_x) + r(Y' \setminus S_x) - r(E) \leq r(X') + r(Y') - r(E) \leq k - 1$. The pair $(X' \cup S_x, Y' \setminus S_x)$ obeys $(X \cup S_x, Y)$ because $X \cup S_x \subseteq X' \cup S_x$ and $Y \subseteq Y' \setminus S_x$. □

Lemma 3. *Let $M = (E, \mathcal{I})$ be a matroid. Let $X, Y \subseteq E$ such that $X \cap Y = \emptyset$, $|Y| \geq k, |X| < k, r(Y) = r(E)$, and $r(X) = |X|$. Then there exists a k-separation (X', Y') obeying (X, Y) if and only if there exists a circuit C in the matroid M/X of size at most $k - |X|$ contained in $E \setminus (X \cup Y)$.*

Proof. ($\Rightarrow$) Suppose there exists a k-separation (X', Y') obeying (X, Y). We need to show that there exists a circuit C in the matroid M/X, of size at most $k - |X|$, contained in $E \setminus (X \cup Y)$. Since (X', Y') is a k-separation, we have that $r(X') + r(Y') - r(E) \leq k - 1$. This implies that $r(X') \leq k - 1$, because $r(Y') = r(E)$. Since (X', Y') is a k-separation, $|X'| \geq k$. Consider a minimum sized set $S \subseteq X'$ such that $X \subseteq S$ and $r(S) = |S| - 1$ (such a set S exists because $|X'| \geq k$, $r(X') \leq k - 1$ and $r(X) = |X|$). Also note that $|S| \leq k$. Since $r(S) = |S| - 1$, there exists a circuit C' in S. Now we claim that there exists a circuit in M/X of size at most $k - |X|$ contained in $C' \setminus X$. It is easy to see that $r_{M/X}(C' \setminus X) \leq |C' \setminus X| - 1$. Now consider the size of $C' \setminus X$.

$$\begin{aligned} |C' \setminus X| &= |C'| - |X| + |X \setminus C'| \\ &\leq |S| - |X \setminus C'| - |X| + |X \setminus C'| \quad \text{(Since } C' \subseteq S \text{ and } X \subseteq S) \\ &\leq k - |X| \quad \text{(Since } |S| \leq k) \end{aligned}$$

Hence there exists a circuit in M/X of size at most $k - |X|$ contained in $C' \setminus X \subseteq E \setminus (X \cup Y)$.

($\Leftarrow$) Suppose there exists a circuit C in M/X of size at most $k - |X|$, contained in $E \setminus (X \cup Y)$. Then we claim that $(C \cup X \cup S, E \setminus (C \cup X \cup S))$ is a k-separation obeying (X, Y), where S is an arbitrary $k - |C \cup X|$ sized set in $E \setminus (C \cup X \cup Y)$. Note that $|C \cup X \cup S|, |E \setminus (C \cup X \cup S)| \geq k$ and

$$\begin{aligned} r(C \cup X \cup S) + r(E \setminus (C \cup X \cup S)) - r(E) &\leq r(C \cup X \cup S) \\ &\leq r_{M/X}(C) + r(X) + r(S) \\ &\leq |C| - 1 + |X| + k - |C \cup X| \\ &\leq k - 1. \end{aligned}$$

This completes the proof. $\qquad\square$

Now we give the main proof of this section.

Proof (Proof of Theorem 2). Let $r = \operatorname{rank}(M)$. Since $\kappa(M) \leq \operatorname{rank}(M) + 1$, we can assume that $k \leq r$. We design a branching algorithm which gradually creates a solution (X, Y) starting from the pair $(\emptyset, \emptyset)$. At any point in the branching algorithm, we branch on a carefully chosen element from $E \setminus (X \cup Y)$. Our branching rules are the following, applied in the order in which they are listed.

- **Rule 1:** If there exists an element $e \in E \setminus (X \cup Y)$ such that $e \notin \operatorname{cl}(X) \cup \operatorname{cl}(Y)$, we branch on e by adding e to X or Y.
- **Rule 2:** If $|X|, |Y| < k$, then we branch on an arbitrary element $e \in E \setminus (X \cup Y)$.

In any node of the branching tree of the algorithm we have a potential partial solution (X, Y), and we abort if $r(X) + r(Y) - r(E) \geq k$. Now we claim that there will not be an application of Rule 1 after an application of Rule 2. Consider a node of the branching tree of the algorithm, with a potential partial solution (X, Y). We apply Rule 2, only if Rule 1 is not applicable, that is when for all $e \in E \setminus (X \cup Y)$, $e \in \mathsf{cl}(X) \cup \mathsf{cl}(Y)$. Hence for any $X' \supseteq X, Y' \supseteq Y$, for all $e \in E \setminus (X' \cup Y')$, $e \in \mathsf{cl}(X') \cup \mathsf{cl}(Y')$. This implies Rule 1 is not applicable after an application of Rule 2. Now consider any root to leaf path in the branching tree of the algorithm. If there exists an application of Rule 2 in this path then the length of the path is at most $2k$, because Rule 2 is applicable only if $|X|, |Y| < k$. Otherwise, we claim that the length of the path is at most $r + k$. Suppose not. Consider the leaf node and the potential partial solution (X, Y) associated with it. If the length of the path is more than $r + k$ and if we only used branching Rule 1, then $r(X) + r(Y) - r(E) > r + k - r(E) > k$, which is a contradiction (because we should have aborted this branch). Hence the height of the branching tree is at most $r + k$ (since $k \leq r$).

Now we explain how to compute a solution from a leaf node labeled (X, Y), if there exists a solution obeying (X, Y). Note that for all $e \in E \setminus (X \cup Y)$, $e \in \mathsf{cl}(X) \cup \mathsf{cl}(Y)$ because of Rule 1. Also note that either $|X| \geq k$ or $|Y| \geq k$. Without loss of generality assume that $|Y| \geq k$. Now we can apply Lemma 2 and add S_x to X where $S_x = \mathsf{cl}(X) \cap (E \setminus (X \cup Y))$. Now if $|X \cup S_x| \geq k$, then we can output the partition $(X \cup S_x, E \setminus (X \cup S_x))$ as k-separation, because $r(X \cup S_x) + r(E \setminus (X \cup S_x)) - r(E) = r(X) + r(Y) - r(E) \leq k - 1$ (because we did not abort this branch). Otherwise $|X \cup S_x| < k$. For convenience, now we use (X, Y) to denote the partial solution $(X \cup S_x, Y)$. The properties of (X, Y) are $|X| < k, |Y| \geq k$, for all $e \in E \setminus (X \cup Y)$ $e \in \mathsf{cl}(Y)$ and $e \notin \mathsf{cl}(X)$. If $r(X) < |X|$, then we can output $(X \cup S, E \setminus (X \cup S))$ as a k-partition where S is an arbitrary set of $k - |X|$ elements from $E \setminus (X \cup Y)$, because

$$r(X \cup S) + r(E \setminus (X \cup S)) - r(E) \leq r(X) + r(S) + r(E \setminus (X \cup S)) - r(E)$$
$$\leq r(X) + k - |X| + r(E \setminus (X \cup S)) - r(E)$$
$$\leq k - 1 \qquad \text{(Since } r(X) < |X|)$$

If $r(X) = |X|$ and $r(Y) < r(E)$, then we can output the $(X \cup S, E \setminus (X \cup S))$ as a k-partition where S is an arbitrary set of $k - |X|$ elements from $E \setminus (X \cup Y)$, because

$$r(X \cup S) + r(E \setminus (X \cup S)) - r(E) \leq k + r(E \setminus (X \cup S)) - r(E)$$
$$\leq k + r(Y) - r(E) \leq k - 1.$$

Now if $r(X) = |X|, |X| < k, r(Y) = r(E), |Y| \geq k$, then we apply Lemma 3 and output YES if there exists a circuit of size at most $k - |X|$ in M/X (if there exists one), otherwise abort this particular branch. A linear representation of M/X and different case analysis explained above can be computed in time $\mathcal{O}((\mathrm{rank}(M))^2 |E|)$ field operations using Gaussian elimination. Since the height of the branching tree is at most $r + k$, the algorithm runs in time $\mathcal{O}(2^{\mathrm{rank}(M)+k} \cdot (\mathrm{rank}(M))^2 |E| (q^{\mathrm{rank}(M)} \mathrm{rank}(M) + |E|k^2))$.   □

6   Conclusion

In this paper we proved that MATROID GIRTH and MATROID CONNECTIVITY in a linear matroid when parameterized by $\text{rank}(M)$ are not FPT unless FPT$= W[1]$, but FPT when parameterized by $\text{rank}(M)+q$, where q is the field size. Other than the EVEN SET problem which remains notoriously open, we draw attention to the following interesting open problem arising from our work. Is MATROID GIRTH FPT on gammoids when parameterized by $\text{rank}(M)$?

References

1. Berlekamp, E.R., McEliece, R.J., van Tilborg, H.C.A.: On the inherent intractability of certain coding problems (corresp.). IEEE Transactions on Information Theory **24**(3), 384–386 (1978)
2. Bhattacharyya, A., Indyk, P., Woodruff, D.P., Xie, N.: The complexity of linear dependence problems in vector spaces. In: Proceedings Innovations in Computer Science - ICS 2010, Tsinghua University, Beijing, China, January 7–9, 2011, pp. 496–508 (2011)
3. Bixby, R.E., Cunningham, W.H.: Matroid optimization and algorithms. In: Graham, R., Grötschel, M., Lovász, L. (eds.) Handbook of combinatorics, vol. 1, pp. 550–609. MIT Press, Cambridge (1996)
4. Downey, R.G., Fellows, M.R.: Fundamentals of Parameterized complexity. Springer (2013)
5. Downey, R.G., Fellows, M.R., Vardy, A., Whittle, G.: The parametrized complexity of some fundamental problems in coding theory. SIAM J. Comput. **29**(2), 545–570 (1999)
6. Gaspers, S., Kim, E.J., Ordyniak, S., Saurabh, S., Szeider, S.: Don't be strict in local search! In: AAAI (2012)
7. Khachiyan, L.: On the complexity of approximating extremal determinants in matrices. Journal of Complexity **11**(1), 138–153 (1995)
8. Macwilliams, J.: A theorem on the distribution of weights in a systematic code. Bell System Technical Journal **42**(1), 79–94 (1963)
9. McCormick, S.T.: A Combinatorial Approach to Some Sparse Matrix Problems. PhD thesis, Stanford University, CA Systems Optimization, Lab (1983)
10. Oxley, J.G.: On a matroid generalization of graph connectivity. In: Mathematical Proceedings of the Cambridge Philosophical Society, vol. 90, pp. 207–214. Cambridge Univ Press (1981)
11. Oxley, J.G.: Matroid theory, vol. 3. Oxford University Press (2006)
12. Tutte, W.: Connectivity in matroids. Canad. J. Math **18**, 1301–1324 (1966)
13. Vardy, A.: The intractability of computing the minimum distance of a code. IEEE Transactions on Information Theory **43**(6), 1757–1766 (1997)

Elastic Geometric Shape Matching
for Point Sets under Translations

Christian Knauer and Fabian Stehn[✉]

Institut Für Informatik, Universität Bayreuth, Bayreuth, Germany
{christian.knauer,fabian.stehn}@uni-bayreuth.de

Abstract. In *geometric shape matching problems* one is given a pattern
P, a model Q, a distance measure d (which formalizes the intuitive notion
of similarity of such shapes), and a class of geometric transformations
applicable to the pattern P. The task is to find a transformation t in the
given class that minimizes the distance of the transformed pattern $t(P)$
to the model Q (as measured by d) in order to compute the similarity of
the given shapes.

In many applications, among them medical-image analysis, industrial
design, robotics or computer vision, where local distortions and complex
deformations can occur, this setting is too restrictive, since only the single transformation t is used to align the entire pattern with the model.
Almost all known strategies that deal with non-rigid deformations apply
heuristics (based on local descent, relaxed LP formulations, simulated
annealing, or alike). The quality of the solution found by these heuristics
can usually not be related to the quality of a global optimum.

Elastic geometric shape matching tries to remedy this situation by
computing a whole set of transformations T. Each transformation $t \in T$
is applied to a subpattern of P. The objective of the optimization problem becomes twofold: *Minimize the distance* of the (union of the) transformed subpatterns to the model while also *maximizing the similarity* of
the transformations in the ensemble T. This modeling aims at strategies
that compute provably optimal solutions, or alternatively approximative
results of a guaranteed quality.

We consider variations of a simple elastic geometric shape matching
problem in the plane where each subshape is just a single point. We show
that this problem already is NP-hard for the directed Hausdorff- or bottleneck distance under arbitrary translations. We complement our result
with efficient algorithms to compute transformation ensembles under
both distances for variants of the problem where only translations *in
a prescribed, fixed direction* are allowed.

1 Introduction

Determining the similarity of two geometric shapes and computing a deformation
of a geometric shape to maximize its similarity to another one are two central
problems studied in computational geometry. Due to the vast number of applications (character recognition, logo detection, human-computer-interaction, etc.)

© Springer International Publishing Switzerland 2015
F. Dehne et al. (Eds.): WADS 2015, LNCS 9214, pp. 578–592, 2015.
DOI: 10.1007/978-3-319-21840-3_48

and the plethora of implications for other scientific fields (robotics [14], computer aided medicine [6], drug design [13], etc.) such problems have received a considerable amount of attention. We refer to the survey papers by Alt et al. [2] and Veltkamp et al. [12] for an extensive overview of this field.

1.1 The Standard Approach vs. the Elastic Setting

Most geometric shape matching problems can be stated in the following form: Given two geometric shapes P (the pattern) and Q (the model), both from a class S of shapes at hand (point sets, polygons, etc.), a transformation class T acting on S (e.g., translations, rigid motions, affine transformations, etc.), and a distance measure : $S \times S \to \mathbb{R}$, the task is to compute a transformation $t \in T$ minimizing $\mathrm{d}(t(P), Q)$. Prominent examples of distance measures for point sets $A, B \subseteq \mathbb{R}^d$ are the *directed Hausdorff distance*

$$\mathrm{h}(A, B) := \max_{f:A \to B} \min_{a \in A} \|a - f(a)\|,$$

and the *bottleneck distance*

$$\mathrm{b}(A, B) := \max_{\substack{f:A \to B \\ f \text{ injective}}} \min_{a \in A} \|a - f(a)\|$$

(for our purposes $\|a - b\|$ always denotes the Euclidean distance between a and b, i.e., the Euclidean length of the vector $a - b$).

Geometric shape matching is often used to solve *registration problems*. In such problems the goal is to align two geometric spaces (e.g., the coordinate system of an operation theatre and the coordinate system of a $3D$-model of the patient acquired during a pre-operative MRI scan) in order to provide a mapping from one space into the other one (e.g., in order to perform computer-aided navigation during a surgery). To compute such a registration, the same structure (e.g., a prominent anatomical feature) is measured in the two spaces and a geometric shape matching is computed to align the two resulting geometric shapes. The corresponding transformation is then used as a mapping from one space into the other one. Especially in applied scenarios (e.g., soft-tissue registrations for computer-aided surgery) where local distortions and complex (e.g., non-affine) deformations can occur, geometric shape matchings as described above are too restrictive: A single transformation is computed to match the entire pattern (or to map the entire space). To address this issue, geometric shape matching problems have been generalized to *elastic geometric shape matching problems* (EGSM problems) in [5, 11].

In an EGSM problem, the pattern P consists of *subshapes*. Instead of a single transformation, a so-called *transformation ensemble* is computed. A transformation ensemble consists of a set of transformations which are (individually) applied to the subshapes of P in order to minimize the distance of the transformed pattern to the model. However, instead of independently solving a "classic" geometric shape matching problem for each subshape, the consistency (and "continuity") of the ensemble (and by that, also of the transformed pattern) is ensured by enforcing certain transformations (e.g., transformations that act on subshapes that are "close") to be similar (with respect to a suitable measure defined for the transformation class at hand). Formally, the (decision version of) a simple EGSM problem for point sets in the plane under translations can be stated as follows:

Problem 1 (EGSM for planar point sets under translations).
Given: $P = \{p_1, \ldots, p_n\} \subset \mathbb{R}^2$ a planar point set (the pattern)
         $Q = \{q_1, \ldots, q_m\} \subset \mathbb{R}^2$ a planar point set (the model)
         d a distance measure for point sets in $\mathbb{R}^2$
         $G = ([n], E)$ a graph with $[n] = \{i \mid 1 \leq i \leq n\}$
         $\delta \geq 0$ the decision parameter
Question: Are there n translation vectors $T = (t_1, \ldots, t_n)$ so that

$$\max\left(d\left(T(P), Q\right), \max_{\{i,j\} \in E} \|t_i - t_j\| \right) \leq \delta, \tag{1}$$

where $T(P) := \{t_i(p_i) \mid 1 \leq i \leq n\}$?

In this formulation each point of P forms an individual subshape of the pattern. In the context of EGSM, the graph G is called the *neighborhood graph* of the subshapes and the sequence T is called a *transformation ensemble*. The graph G encodes which subshapes have to be transformed by similar translations; in this case the (dis)similarity of two translation vectors is measured by the length of their difference.

Intuitively, the graph G should encode the concept of "neighborhood" of subshapes. For complex types of subshapes this notion is not easy to formalize but in the setting considered in Problem 1 (where each point forms a subshape on its own) a standard geometric neighborhood (or proximity) graph defined on P (e.g., the minimum-spanning tree, the Delaunay-triangulation, the Gabriel-graph, etc.) would be a natural choice. Unfortunately, to the best of our knowledge, the complexity status for Problem 1 is unknown for all the standard proximity graphs.

1.2   Related and Previous Work

Several approaches that deal with non-linear transformations can be found in the literature [1,3,7–9]. All of these strategies are heuristics that are based on relaxed ILP formulations [3,8], probabilistic methods [7], or ICP formulations [1,9]. None of these methods compute provably "good" solutions, i.e., solutions that are optimal (up to an approximation factor).

In [5] we considered several variants of Problem 1, where the pattern and the model are point sequences with a *fixed correspondence* between them. For these cases efficient (i.e., polynomial time) exact (based on a convex programming formulation) and approximate (combinatorial) algorithms were developed (for different types of neighborhood graphs).

In [11] we developed a polynomial time constant factor approximation algorithm for Problem 1 for complete neighborhood graphs under the directed Hausdorff distance (which complements our hardness result from Theorem 1 below).

1.3 Our Contribution

We show in Section 2 that Problem 1 is NP-complete for complete neighborhood graphs $G = K_n$ (the complete graph on n vertices) under the directed Hausdorff distance as well as under the bottleneck distance (c.f., Theorem 1 on page 582).

On the positive side, we show in Section 3 that variants of Problem 1, where only translations *in a prescribed, fixed direction* are allowed, permit efficient solutions. More specifically, we present algorithms

- for neighborhood graphs that are complete or trees under the Hausdorff distance that run in O $(nm \, (\log n + \log m))$ time (c.f., Theorem 2 on page 587 and Theorem 3 on page 588), and
- for complete neighborhood graphs under the bottleneck distance that run in O $\left(nm \left(\log n \, (n + m)^{1.495}\right)\right)$ time (c.f. Theorem 4 on page 588).

2 Hardness of EGSM under Translations

In the EGSM model, neighborhood graphs have been introduced as a tool to describe which transformations of the transformation ensemble are forced to be similar as their corresponding subshapes are (usually spatially) related. The second argument in the objective function (1) enforces this restriction by requiring that the length of the longest difference vector of any two translation vectors that are adjacent in the neighborhood graph G has to be smaller than the threshold value δ. In the following, we show that Problem 1 is NP-complete under the directed Hausdorff distance as well as under the bottleneck distance when G is the complete graph (the proof generalizes to any dimension $d \geq 2$).

In this specific setting (G being complete), we show that solving the decision problem is equivalent to finding a set of points in the Euclidean plane where every point has to be chosen from a specific *admissible region* and additionally the diameter of the point set must not exceed δ. For complete neighborhood graphs (and only for complete neighborhood graphs) one could also bound the deviation of the translations by minimizing the *radius* of their smallest enclosing disc. In [11, Theorem 17] we showed that this variant can be solved in O $\left(mn^2 \log(mn)\right)$ time and that such a solution can be transformed in linear time into a $(2/3)\,(1 + 1/3)$-approximation of the variant considered here. For arbitrary graphs however, the smallest enclosing disc is not a reasonable measure and the following NP-completeness proof shows the hardness of the general problem:

2.1 Moving from Object-Space to Transformation-Space

Consider the set $T_{p,q} = \{t \mid \|t(p) - q\| \leq \delta\}$ of translations that bring a point p within distance $\leq \delta$ ("δ-close") to a point q. In transformation-space, these translations form a disc of radius δ with center $q - p$. The set $T_{p,Q}$ of translations that bring a point p δ-close to some point in Q consequently is

$$T_{p,Q} = \bigcup_{q \in Q} T_{p,q}.$$

In the context of Problem 1 we call $T_{p,Q}$ the *admissible region* of p: Any $t \in T_{p,Q}$ ensures that distance of $t(p)$ to some point in Q is at most δ.

These observations allow us to investigate Problem 1 solely in transformation-space: A valid solution $T = (t_1, \ldots, t_n)$ satisfies

$$\forall 1 \leq i \leq n : t_i \in T(p_i, Q) \quad \wedge \quad \forall 1 \leq i < j \leq n : \|t_i - t_j\| \leq \delta.$$

In other words, every t_i has to lie within some disc of $T_{p_i,Q}$ and the resulting set $\{t_1, \ldots, t_n\}$ has to have a diameter of at most δ.

2.2   EGSM for Point Sets under Translations Is NP-hard

Theorem 1. *EGSM for point sets under translations and complete neighborhood graphs wrt. the directed Hausdorff distance or the bottleneck distance (Problem 1) is NP-hard.*

Proof. The NP-hardness of Problem 1 will be proven by a reduction from 3-SAT: Let ϕ be a formula in 3-CNF with n variables $v_1, \ldots, v_n$ and m clauses $\phi_1, \ldots, \phi_m$. From ϕ, an EGSM instance will be constructed that consists of $|P| = m + 1$ and $|Q| = 3 \cdot \binom{2n}{3} + 1$ points respectively. The parameter δ will be set to $2/3 - \epsilon$, for a small $\epsilon > 0$ that will be specified later. Finally, we set $G = K_n$.

We will define $P = \{p_0, p_1, \ldots, p_m\}$ and Q implicitly by describing $T_{p,Q}$ for all $p \in P$. Note, that for points $p_i, p_j \in P$ the set $T_{p_i,Q}$ is just a copy of $T_{p_j,Q}$ translated by $p_j - p_i$. To be more precise, we first define Q by defining p_0 and $T_{p_0,Q}$, and then we define p_i by describing how $T_{p_i,Q}$ is shifted relative to $T_{p_0,Q}$.

Constructing Clauses with Flowers. The clauses of ϕ will be encoded in so-called *flowers*. First we define the base-flower, see Figure 1(a): We (arbitrarily) place n discs $d_1, \ldots, d_n$ of radius δ with distinct centers on the unit circle with positive coordinates (all centers lie in the first quadrant). Let $\overline{d_1}, \ldots, \overline{d_n}$ be their reflections at the origin (so that d_i and $\overline{d_i}$ are antipodal with respect to the unit circle), and let $D = \{d_1, \ldots, d_n, \overline{d_1}, \ldots, \overline{d_n}\}$. Cabello et al. [4, Lemma 2] have shown that the centers of these discs can be found with bounded sized rational coordinates:

Lemma 1 (Cabello et al. [4, Lemma 2]). *For any $n > 0$, there exists n distinct points on the unit circle such that they all have rational coordinates with the numerators and denominators bounded by a polynomial in n, and the distance between any two points is at most $\sqrt{2}/6$.*

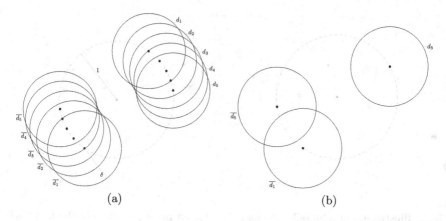

Fig. 1. Illustration of the base-flower and a clause-flower. (a) The base-flower for $n = 5$; (b) A clause-flower encoding the clause $(\overline{v_1} \vee v_3 \vee \overline{v_5})$.

We define $\binom{2n}{3}$ (the number of distinct clauses that can appear in a 3-CNF formula with n variables) *clause-flowers* by placing copies of the base-flower along the x-axis so that any two consecutive clause-flowers are 4 units apart (when we measure the distance of the centers of the translated unit circles). From each clause-flower, we remove all but three discs, in such a way that for every possible choice of three distinct discs of D, translational copies of these three discs are present in exactly one flower. The clause-flowers are now in one-to-one correspondence with all possible 3-clauses that can be formed from the literals $\{v_1, \ldots, v_n, \overline{v}_1, \ldots, \overline{v}_n\}$: Let $d_a, d_b, d_c \in D$ be the discs whose translational copies remained in the same flower. This flower corresponds to the clause $(l_a \vee l_b \vee l_c)$ where for $h \in \{a, b, c\}$ we have that $l_h = v_i$ if $d_h = d_i$ or $l_h = \overline{v}_i$ if $d_h = \overline{d}_i$ for some $1 \leq i \leq n$, see Figure 1(b). The construction of $T_{p,Q}$ is completed by placing a single disc d_x of radius δ so that its center is 4 units below the center of the leftmost clause-flower, see Figure 2.

Recall that $\delta = 2/3 - \epsilon$. The value $\epsilon > 0$ will be chosen small enough, so that for any two non-antipodal discs $d, d' \in D$ of the same flower there are two points $a \in d$ and $b \in d'$ with $\|a - b\| \leq \delta$. By construction, the closest distance of two points that lie in antipodal discs of a flower or in discs that do not belong to the same flower is $2(1 - \delta) = 2/3 + 2\epsilon > \delta$, see Figure 3 in the Appendix.

The pattern P then consists of $m + 1$ points $\{p_0, \ldots, p_m\}$. The point p_0 is chosen (independent of the actual formula ϕ) such that the center of $d_x \in T_{p_0,Q}$ lies on the origin. The point p_i for $1 \leq i \leq m$ is chosen in a way such that the flower in $T_{p_i,Q}$ that encodes the clause ϕ_i is centered in the origin.

Relation Between the Instances. The EGSM instance is constructed in such a way, that the admissible regions $T_{.,Q}$ encode all possible clauses that can appear in a 3-CNF formula. By placing the points $p_1, \ldots, p_m \in P$ as described above, the clauses ϕ_1 to ϕ_m of the formula ϕ are *selected* by placing the center

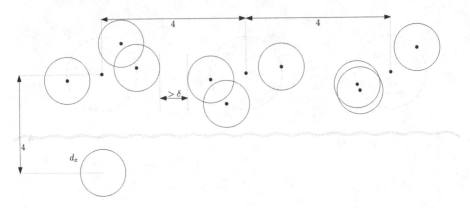

Fig. 2. Illustration of admissible regions $T_{p,Q}$ for an point $p \in P$, only the first three flowers are shown, the illustration extends to the right. The gray dotted circles indicate the translated unit discs and are not part of the construction.

of the flower that encodes the respective clause on the origin. As $T_{p_0,Q}$ has been translated so that d_x is centered in the origin, a valid solution $T = (t_0, \ldots, t_m)$ is forced to choose t_0 from the disc d_x of $T_{p_0,Q}$. To clarify this restriction on t_0, assume t_0 to be chosen from a disc that is part of some flower of $T_{p_0,Q}$: By construction, the translation $t \in T_{p_i,Q}$ that is closest to t_0 for every $1 \leq i \leq m$ is at distance $\|t_0 - t\| > \delta$.

So as $t_0 \in d_x \in T_{p_0,Q}$, all t_i for $1 \leq i \leq m$ are forced to lie in a disc that is part of the flower of $T_{p_i,Q}$ that is centered in the origin as again all other translations of $T_{p_i,Q}$ are at distance larger than δ from t_0.

If the 3-SAT instance is satisfiable, there is a solution to the EGSM instance
Let $A = (a_1, \ldots, a_n)$ with $a_i \in \{\text{TRUE, FALSE}\}$ be an assignment of the variables $v_1, \ldots, v_n$ that satisfies ϕ. Consider a circle centered in the origin with radius $1 - \delta$ that touches the afore described discs in a set X of $2n$ points. Let x_i be its touching point with disc d_i and $\overline{x_i}$ its antipodal point.
Consider the set $X_A = \{x_i \in X \mid a_i = \text{TRUE}\} \cup \{\overline{x_i} \in X \mid a_i = \text{FALSE}\}$. For $1 \leq i \leq n$ choose any point $t_i \in X_A \cap T_{p_i,Q}$. Note, that $X_A \cap T_{p_i,Q} \neq \emptyset$ as the three relevant discs of $T_{p_i,Q}$ encoded the literals of ϕ_i and at least one of these literals satisfied ϕ_i under the assignment A.

As for all $0 \leq i \leq m$: $t_i \in T_{p_i,Q}$ we have that the directed Hausdorff distance in object space is at most δ, i.e., every point $p \in P$ is at distance $\leq \delta$ to some point in Q. As for the similarity of the chosen translations we have that for $1 \leq i \leq m$: $\|t_0 - t_i\| = 1 - \delta < \delta$. By construction, we have that all non-antipodal point pairs in X_A are at distance $\leq \delta$ and X_A can not contain antipodal points.

If there is a solution for the EGSM instance, the 3-SAT instance is satisfiable
Let $T = (t_0, t_1, \ldots, t_m)$ be a solution for the EGSM instance. An assignment $A = (a_1, \ldots, a_n)$ of the variables v_1 to v_n that satisfies ϕ can be found in the

following way: For $1 \leq i \leq m$ let D_i be the set of discs of $T_{p_i,Q}$ that contain t_i (D_i may contain more than one disc, as t_i can be in the intersection of up to three discs). Pick any disc $d \in D_i$ and set

$$a_j = \text{TRUE if } d = d_j, \text{ or } a_j = \text{FALSE if } d = \overline{d_j}.$$

The disc d encoded a literal that satisfies ϕ_i, more precisely if $d = d_j$ then ϕ_i contained the literal v_j or otherwise if $d = \overline{d_j}$ then literal $\overline{v_j}$ was part of ϕ_i. Therefore choosing a_j according to d will satisfy a literal of each clause of ϕ_i for all $1 \leq i \leq m$. Note, that a valid solution T cannot contain two translations t, t' so that $t \in d_k$ and $t' \in \overline{d_k}$ for any $1 \leq k \leq n$ as by construction the smallest distance between any two points in antipodal discs is larger than δ.

It is possible, that some variables are left unassigned after this processes. As the variables assigned so far already satisfy ϕ, these remaining variables can be assigned arbitrarily to TRUE or FALSE.

This finishes the proof of the NP-hardness for the Hausdorff distance.

The Case of the Bottleneck Distance. Note, that in this reduction, all discs that are part of flowers that are centered in the origin are induced by different points of Q. This implies that the same reduction proves the NP-hardness for the bottleneck distance as well.

Essentially by following the same lines as for the NP-completeness proof above, it is easy to show that the problem remains NP-complete if every point $p \in P$ is allowed to be translated along a given direction $\vec{d_p}$:

Corollary 1. *Problem 1 remains NP-complete, even if each point $p \in P$ is restricted to be translated parallel to a a given direction $\vec{d_p}$ where the set $D_P = \{\vec{d_p} \mid p \in P\}$ is part of the input.*

3 Computing EGSM for Translations in a Fixed Direction

Given that Problem 1 is NP-hard, it is reasonable to investigate more restricted settings. In the following, each point of the pattern P can be translated along a fixed, given direction, which makes the translation-space one-dimensional.

At first this restriction to one dimension seems to be too restrictive to be helpful, but has stated in Corollary 1, the problem remains hard even for n given directions. The following positive result where a single direction is fixed (wlog. parallel to the x-axis) is therefore a first step in closing the gap between efficient strategies for and the hardness of Problem 1 for general graphs.

3.1 Admissible Transformations

A translation parallel to the x-axis will be called an x-*translation*. To simplify the presentation, we sightly abuse the notation by identifying a translation with its x-component, i.e., the x-translation with the translation vector $(t, 0)$ will be represented by the scalar t.

For points $p = (p_x, p_y) \in P$ and $q = (q_x, q_y) \in Q$ we consider the set $I_{p,q}$ of admissible transformations $I_{p,q} := \left\{ t \mid \left((p_x + t - q_x)^2 + (p_y - q_y)^2 \right)^{0.5} \leq \delta \right\}$. The set $I_{p,q}$ consists of all x-translations t that bring p at least δ-*close* to q. Note, that (for every convex distance function) $I_{p,q}$ is either empty or consists of a single interval. The set $I_{p,Q}$ of translations that bring p δ-close to some point $q \in Q$ is $I_{p,Q} := \cup_{q \in Q} I_{p,q}$. Let $I_{P,Q} := \cup_{p \in P} I_{p,Q}$ be the set of all admissible transformations for all points of P.

Lemma 2. *The set $I_{P,Q}$ (represented as a sequence of intervals sorted by their left endpoints) can be computed in* $O(m \log m)$ *time.*

Due to space limitations, the proof of Lemma 2 has been moved to the appendix.

3.2 Minimizing the Hausdorff Distance in Object-Space ...

In this section, we consider the decision problem of determining whether there is a set of x-translations for P so that the directed Hausdorff distance of the translated point set P to Q is no more than δ while at the same time requiring that all pairs of translations t_i, t_j that are connected by an edge in $G = ([n], E)$ differ by at most δ, i.e., $\|t_i - t_j\| \leq \delta$ for all $\{i, j\} \in E$.

... for Complete Neighborhood Graphs. For now, let G be the complete graph K_n, i.e., any two translations t, t' of the computed translation ensemble have to satisfy $\|t - t'\| < \delta$.

To decide whether there is a set T of translations that satisfies condition (1) (c.f., page 580), a double sweep-line algorithm will be performed. A double sweep-line algorithm can be thought of as sweeping over the scene with a vertical strip of a specific width. Events are processed when elements (here: The start- or endpoint of an admissible transformation interval) enter or leave the strip. The process is simulated by using two sweep-lines S_e and S_s representing the left and right boundary of the strip.

For the specific problem considered here, we enforce that the two sweep-lines are δ-*apart* (the strip has width δ). This implies, that if x was the event that had been processed last by sweep-line S (denoted by $x \lhd S$), the next event next(x) is determined by:

$$\text{next}(x) := \begin{cases} \min\{s \mid [s,e] \in I_{P,Q}, s \geq x\} \cup \{e \mid [s,e] \in I_{P,Q}, e \geq x - \delta\}, & \text{if } x \lhd S_s \\ \min\{s \mid [s,e] \in I_{P,Q}, s \geq x + \delta\} \cup \{e \mid [s,e] \in I_{P,Q}, e \geq x\}, & \text{if } x \lhd S_e. \end{cases}$$

Ties are broken by favoring events that appear at the right boundary S_s of the strip (among which the order is arbitrary). This means, that events that can be processed by S_s are processed first, as the interval endpoints themselves are admissible transformations.

We use the two sweep-lines to simultaneously sweep over all sets of all admissible regions $I_{p,Q}$ for all $p \in P$. Events are left and right endpoints of these admissible regions. The basic idea is to keep track of how many points of P

have non-empty admissible regions within the common sweep strip of width δ. For this purpose, we introduce counters $c_1, \ldots, c_n$, where c_i counts how many connected admissible regions of point p_i are currently intersected by the strip. Counter c_i is increased by one, whenever S_s encounters the left, and is decreased by one whenever S_e hits the right boundary of an interval of $I_{p_i,Q}$. The first event will be processed by S_s and is one of the smallest left endpoints of any interval. Note, that this algorithm also produces a witness for a *yes*-instance.

Lemma 3. *The answer to the decision problem is yes iff at some point during the sweep process all counters are non-zero.*

Theorem 2. *The EGSM problem wrt. the Hausdorff distance for complete neighborhood graphs under x-translations can be decided in* $\mathrm{O}\left(nm\left(\log n + \log m\right)\right)$ *time.*

Due to space limitations, the proofs of Lemma 3 and Theorem 2 have been moved to the appendix.

... for Neighborhood Graphs That Are Trees

In this section, we consider neighborhood graphs $G = ([n], E)$ that are trees. We start by picking an arbitrary node $r \in V$ and look at G_r, the tree G rooted at r. The n_v many children of an internal node $v \in V$ of G_r will be denoted by $c(v)_1, \ldots, c(v)_{n_v}$, the (rooted) subtree of G_r with root v by B_v.

In order to decide whether there is a set T of translations that satisfies condition (1) (c.f., page 580) we proceed iteratively. The basic idea is to propagate admissible transformations in the tree G_r from *bottom-to-top* by contracting inner nodes with their children and by appropriately merging their admissible transformations. I.e., starting with G_r, the algorithm chooses an inner node and contracts it which leads to new tree. In each iteration of the algorithm, we call the tree from which a node is selected the *current tree*.

In each step of the algorithm, a vertex v of the current tree is selected with the property that all children of v are leaves. Then, v and the children of v are contracted to a new node v' which itself becomes a leaf in the resulting tree. To compute the set $I_{v',Q}$ of admissible regions for the new leaf v' we proceed as follows: First, we *inflate* all regions $I_{c(v)_i,Q}$ by δ, i.e., for $1 \leq i \leq n_v$: $I^\delta_{c(v)_i,Q} := \left\{ [s - \delta, e + \delta] \mid [s, e] \in I_{c(v)_i,Q} \right\}$. The admissible regions for the new node v' are now defined as: $I_{v',Q} := \left(\bigcap_{i \in \{1,\ldots,n_v\}} I^\delta_{c(v)_i,Q} \right) \cap I_{v,Q}$. This process is iterated until either of the following two cases occurs:

1. For some node v we have $I(v, Q) = \emptyset$ before or after contracting v:
   The algorithm terminates returning *no*.
2. The root r is contracted and $I(r', Q) \neq \emptyset$:
   The algorithm terminates returning *yes*.

Clearly, the strategy terminates, as at some point in time the root will be contracted or case 1 is encountered earlier.

By contraction, subtrees are shrunk to nodes. The algorithm essentially detects whether either the entire tree G_r can be shrunken to r' or whether there is a subtree B_v of G_r so that condition (1) (c.f., page 580) with respect to B_v can not be satisfied. To clarify this notion, let $E_v = E[B_v]$, $P_v = \{p_i \mid \{i, \cdot\} \in E_v\}$ and $T_v = \{t_i \mid p_i \in P_v\}$. We say condition 1 can be satisfied with respect to B_v (or simply B_v can be satisfied) if there is a set of translations T_v so that

$$\max\left(\mathrm{d}\left(T_v(P_v), Q\right), \max_{\{i,j\} \in E_v} \|t_i - t_j\|\right) \le \delta.$$

If v' is the result of contracting v with its children, then the subtree B_v is shrunken to v' and we define $B_{v'} = B_v$. To prove the correctness of the algorithm, we make use of the invariant stated in Lemma 4.

Lemma 4. *After every iteration of the algorithm, the following holds: Let v be a leaf of the current tree. If $I_{v,Q} \ne \emptyset$ then B_v can be satisfied.*

Theorem 3. *The EGSM problem with respect to the Hausdorff distance under x-translations for neighborhood graphs that are trees can be decided in $\mathrm{O}\left(nm\left(n + \log m\right)\right)$ time.*

Due to space limitations, the proofs of Lemma 4 and Theorem 3 have been moved to the appendix.

3.3 Considering the Bottleneck Distance for Complete Graphs

The algorithm to decide Problem 1 for the bottleneck distance uses the same sweep-strip approach as described in Section 3.2 (for complete graphs). The main difference is that events have to be handled in a different way.

Consider the bipartite graph $G_{PQ} = (P \cup Q, E')$, where $E' \subset P \times Q$. The basic idea is to insert the edge (p_i, q_i) into E' when S_s hits the left boundary of an interval $I_{p_i, q_i, \delta}$ and to remove that edge, when S_e encounters the right boundary of this interval. Whenever an edge is deleted or removed from/into G_{PQ} we update the information of the size of the maximum matching in G_{PQ}. Clearly, the answer to the decision problem is *yes*, iff, during the course of the sweep-strip algorithm, the size of the maximum matching becomes n at some point.

Instead of recomputing the maximum matching from scratch whenever G_{PQ} is modified, we use the algorithm by Sankowski et al. [10]. Their strategy allows to maintain the information of the size of the maximum matching in $\mathrm{O}\left(|V|^{1.495}\right)$ time per edge insertion/deletion operation in/from a (not necessarily bipartite) graph.

Theorem 4. *The EGSM problem for the bottleneck distance under x-translations for complete neighborhood graphs can be decided in $\mathrm{O}\left(nm\left(\log n\left(n + m\right)^{1.495}\right)\right)$ time.*

Due to space limitations, the proof of Theorem 4 has been moved to the appendix.

4   Omitted Figures and Proofs

Proof (Proof of Lemma 2). For a fixed p, the set $I_{p,Q}$ can be computed by inserting, one-by-one for every $q \in Q$, all non-empty intervals $I_{p,q}$ (each of which can be computed in constant time) into a modified AVL-tree. Consider a standard AVL-tree T that stores start- and endpoints of intervals at internal nodes and pointers to intervals I or voids V at its leafs. Internal nodes v additionally store a pointer to the leaf that corresponds to the interval that has v as an start- or endpoint and vice versa. The structure T is initialized by setting its root to an empty void V_0 spanning the entire real line. To simplify the description of how to insert an interval $I_i = [s_i, e_i]$ into T, we introduce two pointers l and r to internal nodes of T which will be used after the insertion to perform a merge operation causing T to store the interval corresponding to $[l, r] \supset I_i$ by removing obsolete endpoints within that interval.

First, we locate the endpoints $x \in \{s_i, e_i\}$ of I_i in T by a standard AVL-tree search:

case 1 $x \in I_j = [s_j, e_j]$: The search ends in a leaf, x lies within an interval I_j
    if $x = s_i$ store s_j in l
    if $x = e_i$ store e_j in r
case 2 $x \in V_j$: The search ends in a leaf, x lies within a void V_j
    if $x = s_i$ locally replace V_j in T by a subtree with root s_i, whose left child is V_j and whose right child is a dummy node that will be replaced later, store s_i in l
    if $x = e_i$ locally replace V_j in T by a subtree with root e_i, whose right child is V_j and whose left child is a dummy node that will be replaced later, store e_i in r

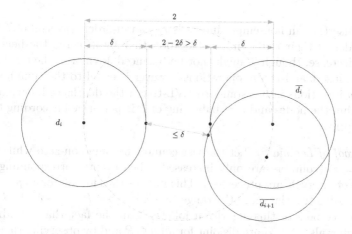

Fig. 3. Illustration showing how to choose ϵ: In this illustration, d_i and $\overline{d_{i+1}}$ are *furthest apart* among all non-antipodal discs of a flower

case 3 $x = v$: While searching, an internal node v is encountered that represents
an endpoint of an interval $I_j = [s_j, e_j]$ and that has the same value as x
if $x = s_i$ store s_j in l
if $x = e_i$ store e_j in r

Until now, the structure of T only changed in the case that the search for one
or both endpoints of I_i ends in a leaf corresponding to a void. It remains to
perform a merge operation on T to remove obsolete endpoints and potential
super-covered intermediate intervals. The only case in which a merge operation
does not have to be performed is, when I_i is completely covered by an interval
already stored in T before the insertion of I_i (which is the case if l and r point
to the same interval).

To perform a merge operation, we find the lowest common ancestor v of l
and r in T (which might be either l or r itself). The node v will be replaced by
a tree with root l whose right child is r, the left child of r becomes a new leaf
storing the interval $[l, r]$, the former left child (T_1) of l and the right child (T_4)
of r remain unchanged, see Figure 4.

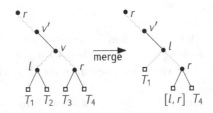

Fig. 4. Illustration of how to merge the data structure to that $[l, r]$ is stored in T

After inserting all non-empty intervals $I_{p,q}$ an in-order traversal of T reports
$I_{p,Q}$ in order of their (left) endpoints. Note, that by a merge the height of T
can only decrease, though T might not be balanced anymore. The total time to
perform m insert and merge operations however is equal to the time needed to
perform m insertions into a standard AVL-tree if the depths of inner nodes are
stored within the nodes and if re-balancing of T is performed according to these
pseudo-depths.

Proof (Proof of Lemma 3). Let the last counter become non-zero while $x \lhd S_s$,
where $x = s_i$ (counters are only increased, when events corresponding to left
endpoints of intervals are processed). This implies that for all points $p \in P$ there
is an admissible translation in the range $[x - \delta, x]$: $\forall p \in P : [x - \delta, x] \cap I_{p,Q} \neq \emptyset$.

The correctness of this algorithm follows from the fact that all admissible
regions (intervals) of $I_{p,Q}$ are disjoint for all $p \in P$ and by observing that admissible regions start to become relevant exactly when the left boundary enters the
strip and that they end being relevant when their right boundary leaves the
strip.

Proof (Proof of Theorem 2). Each set $I_{p,Q}$ consists of $O(m)$ intervals, hence there are $O(nm)$ events to be processed. Determining the next event that has to be processed takes $O(\log n)$ time as the left and right endpoints are stored in sorted order for every $I_{p,Q}$. This together with the time to compute the sets $I_{p,Q}$ (see Lemma 2) yields the stated running time.

Proof (Proof of Lemma 4). Recall, that every node $x \in V[G_r]$ represents a translation $t_x \in T$ that will be applied to point $p_x \in P$.

We proceed by induction on the number of iterations the algorithm has performed. Before any contraction has been performed, the current tree is G_r. Let v be any leaf of G_r: The subtree B_v only consists of v, hence B_v trivially can be satisfied.

For the induction step, let v be a leaf of the current tree. If v was a leaf before the contraction, then B_v can be satisfied by induction.

If v was not a leaf before the contraction, then the current tree was created by contracting a subtree B_u into v. By induction we know for all children $c(u)_i$ of u that $B_{c(u)_i}$ can be satisfied.

Now consider a translation $t \in I_{v,Q}$: We have that

- $t \in I_{v,Q}$, hence t will bring p_u δ-*close* to some point of Q,
- $t \in I^{\delta}_{c(u)_i,Q}$ for all $1 \leq i \leq n_u$, implying that $\exists [s,e] \in I_{c(u)_i,Q} : [t-\delta, t+\delta] \cap [s,e] \neq \emptyset$. This means, that there is a translation in an admissible region of $I_{c(u)_i,Q}$ at most δ-*away* from t for any $t \in I_{v,Q}$ and for any $1 \leq i \leq n_u$.

Proof (Proof of Theorem 3). Computing all admissible regions $I_{P,Q}$ takes $O(mn \log m)$ time. A single set $I_{P,Q}$ for a point $p \in P$ can consist of $O(m)$ disjoined intervals. The description complexity of the intersection of two interval sets A, B can be the description complexity of the sum of both sets $|A \cap B| \leq |A| + |B|$. As a consequence, the description complexity of the admissible regions for an (inner) node can be the sum of the description complexity of its children plus $O(m)$ (the admissible regions of the node itself). An upper bound of the complexity of an admissible region (which can be realized after collapsing the root r of G_r) is $O(mn)$ (there are n nodes each contributing $O(m)$ intervals). Assuming that the intersection of two ordered sets of segments can be intersected in linear time, the stated runtime follows, as each edge of the tree accounts for two merge operations.

Proof (Proof of Theorem 4). Each set $I_{p,Q}$ consists of $O(m)$ intervals, hence there are $O(nm)$ events to be processed. Determining the next event that has to be processed takes $O(\log n)$ time as the left and right endpoints are stored increasingly for every $I_{p,Q}$. This together with the time to compute the sets $I_{p,Q}$ (see Lemma 2) and the time for updating the size of the maximum matching in G_{PQ} (see [10]) yields the claimed running time.

References

1. Abdelmunim, H., Farag, A.A.: Elastic shape registration using an incremental free form deformation approach with the icp algorithm. In: 2011 Canadian Conference on Computer and Robot Vision (CRV), pp. 212–218, May, 2011
2. Alt, H., Guibas, L.: Discrete geometric shapes: matching, interpolation, and approximation. In: Handbook of Computational Geometry, pp. 121–153. Elsevier B.V. (2000)
3. Bazen, A.M., Gerez, S.H.: Fingerprint matching by thin-plate spline modelling of elastic deformations. Pattern Recognition 36(8), 1859–1867 (2003)
4. Cabello, S., Giannopoulos, P., Knauer, C.: On the parameterized complexity of d-dimensional point set pattern matching. In: Bodlaender, H.L., Langston, M.A. (eds.) IWPEC 2006. LNCS, vol. 4169, pp. 175–183. Springer, Heidelberg (2006)
5. Knauer, C., Kriegel, K., Stehn, F.: Non-uniform geometric matchings. In: Murgante, B., Gervasi, O., Iglesias, A., Taniar, D., Apduhan, B.O. (eds.) ICCSA 2011, Part III. LNCS, vol. 6784, pp. 44–57. Springer, Heidelberg (2011)
6. Antoine, J.B.: Maintz and Max A. Viergever. A Survey of Medical Image Registration. Medical Image Analysis 2(1), 1–36 (1998)
7. Myronenko, A., Song, X.: Point set registration: Coherent point drift. IEEE Transactions on Pattern Analysis and Machine Intelligence 32(12), 2262–2275 (2010)
8. Rohr, K., Stiehl, H.S., Sprengel, R., Buzug, T.M., Weese, J., Kuhn, M.H.: Landmark-based elastic registration using approximating thin-plate splines. IEEE Transactions on Medical Imaging 20(6), 526–534 (2001)
9. Rusinkiewicz, S., Levoy, M.: Efficient variants of the ICP algorithm. In: Proceedings of the Third International Conference on 3D Digital Imaging and Modeling, pp. 145–152 (2001)
10. Sankowski, P.: Faster dynamic matchings and vertex connectivity. In: Bansal, N., Pruhs, K., Stein, C., (eds.) SODA, pp. 118–126. SIAM (2007)
11. Stehn, F.: Geometric Hybrid Registration. PhD thesis, Freie Universität Berlin (2011)
12. Remco, C.: Veltkamp and michiel hagedoorn. principles of visual information retrieval. In: Lew, M.S. (ed.) Principles of Visual Information Retrieval Chapter State of the Art in Shape Matching. Advances in Pattern Recognition, pp. 87–119. Springer-Verlag, London (2001)
13. Venkatasubramanian, S.: Geometric Shape Matching and Drug Design. PhD thesis, Department of Computer Science, Stanford University (1999)
14. Yoshikawa, T., Koeda, M., Fujimoto, H.: Shape recognition and grasping by robotic hands with soft fingers and omnidirectional camera. In: IEEE International Conference on Robotics and Automation, ICRA 2008, pp. 299–304 (2008)

Constant Time Enumeration by Amortization

Takeaki Uno[✉]

National Institute of Informatics, Chiyoda, Japan
uno@nii.jp

Abstract. Enumeration algorithms have been one of recent hot topics in theoretical computer science. Different from other problems, enumeration has many interesting aspects, such as the computation time can be shorter than the total output size, by sophisticated ordering of output solutions. One more example is that the recursion of the enumeration algorithm is often structured well, thus we can have good amortized analysis, and interesting algorithms for reducing the amortized complexity. However, there is a lack of deep studies from these points of views; there are only few results on the fundamentals of enumeration, such as a basic design of an algorithm that is applicable to many problems. In this paper, we address new approaches on the complexity analysis, and propose a new way of amortized analysis *Push Out Amortization* for enumeration algorithms, where the computation time of an iteration is amortized by using all its descendant iterations. We clarify sufficient conditions on the enumeration algorithm so that the amortized analysis works. By the amortization, we show that many elimination orderings, matchings in a graph, connected vertex induced subgraphs in a graph, and spanning trees can be enumerated in $O(1)$ time for each solution by simple algorithms with simple proofs.

1 Introduction

Suppose that there is a simple algorithm to solve a problem, and we have two improvements on the time complexity; (a) is by developing a new algorithm with a small complexity, and (b) proves that its complexity is actually small by complexity analysis. Both types of improvements are important in theoretical computer science, but these days almost all results are on the type of (a). Developing simple algorithms in (a) is non-trivial, thus many recent algorithms and their complexity analysis are difficult to understand. Moreover, these types of algorithms often require some structures in the input, hence the problem formulations tend to be distant from the real world. On contrary, (b) type has a great advantage on these points. Even though the analysis is complicated, we can hide the difficulty by producing general statements applicable to many problems. At least, we do not have to implement the complicated proofs in a program. According to this motivation, we study on complexity analysis in this paper, that is amortized analysis for enumeration algorithms.

Amortized analysis is a paradigm of complexity analysis. In the paradigm, we charge the cost of iterations with long computation time to those with shorter

© Springer International Publishing Switzerland 2015
F. Dehne et al. (Eds.): WADS 2015, LNCS 9214, pp. 593–605, 2015.
DOI: 10.1007/978-3-319-21840-3_49

time, to make the upper bound of computation time of an iteration shorter. Compared to usual complexity analysis considering the worst case, the amortized analysis is often more powerful, for example dynamic tree, union find, and some enumeration algorithms[7,13]. In the case of dynamic tree, the cost of changing the shape of the tree is charged to the preceding changes with smaller costs, and attains $O(\log n)$ average time complexity for each change where n is the size of the tree. The time complexity is not attained by usual worst case analysis, and it seems to be hard to obtain algorithms with the same complexity by the analysis. This is similar to the union find algorithm, and the resulted time complexity is $O(n\alpha(n))$ while straightforward algorithms take $O(n^2)$ time. The concept of "charging the cost" made a paradigm shift on the design of algorithms. Some enumeration algorithms are designed so that the time complexity of an iteration is linear in the number of subproblems, to make the average computation time per child will be short[8,12].

Enumeration is now rapidly increasing its presence in theoretical computer science. One of the biggest reasons comes from its importance in application areas. An example is the pattern mining problems in data mining. The problem is to find all the patterns belonging to a class of structures, such as subsets and trees, such that the patterns satisfy some constraints in the given database, such as appearing at least k times. One more motivation is that there have not been many studies including simple problems, thus there is a great possibility. On the other hand, enumeration has several interesting aspects which we can not observe in other problems. For example, by dealing only with the difference between output solutions, we can often attain the computation time shorter than its output size, by outputting the solutions by the differences. Another example is its well-structured recursion. We can frequently have several structural results on enumeration, and it gives interesting algorithms and mathematical properties, while it is hard to characterize when a brunch and bound algorithm cuts off subproblems. Structured recursion often gives a good amortization. There is a great interest on investigating amortized analysis on enumeration algorithms.

According to this motivation and interests, this paper addresses amortized analysis of enumeration algorithms. One of our goals on this topic is to fill the gap between theory and practice. In practice, enumeration algorithms are often quite efficient and than the theoretical upper bound on the computation time. Filling the gap gives understandings for both theoretical and practical properties on data and algorithms; the properties of data accelerating algorithms, and the mechanism of the algorithms that enable us to attain smaller bounds.

We have observed that the recursive structures of enumeration algorithms satisfies a property which we call *bottom-expanded*. Iterations of enumeration algorithms generate several recursive calls. Thus, the number of iterations exponentially increases in deeper levels of the recursion. On the other hand, iterations on deeper levels often have relatively small inputs compared to upper levels. Thus, we can expect that iterations near by the root of the recursion are few and spend a long time, and iterations near by the bottom of the recursions are many and spend very short time. In practice, we can frequently observe this, especially in many kinds of pattern mining algorithms. This also implies that the amortized computation time per iteration, or even per solution, is short. This mechanism is what we call bottom-expanded. We can see this mechanism not only in practice but also classic enumeration algorithms.

This mechanism motivated us to develop a good amortized analysis. However, amortization is not easy in general, since it is hard to globally estimate the number of iterations and computation time. Thus, in many existing studies, the computation time is amortized between a parent and its children, and sometimes its grandchildren[2–5,8,12]. These local structures are easier to analyze than the global structures. Extensions of this idea to more global structures are non-trivial. For example, if we want to amortize between iterations in different subtrees of the recursion, we have to understand the relation and the correspondence between all iterations in different subtrees. This is often difficult.

In this paper, we propose a new way of carrying out amortized analysis of the time complexity of enumeration algorithms, and propose new algorithms for enumeration of matchings, elimination orderings, and connected vertex induced subgraphs. We also show that the amortized analysis can prove the existing complexity results in very simple ways, for the enumerations of spanning trees, perfect elimination orderings, and perfect sequences, while the existing algorithms often need sophisticated algorithms or data structures. We can also see that the condition in the analysis is often satisfied in practice, thus this amortized analysis explains why the enumeration algorithms are efficient in practice. These satisfy out basic motivations for this kind of studies.

Our amortization is done with all descendants of an iteration. We push out the computation time of an iteration to its children so that the assigned time is proportional to their computation time. By applying this from the root of the recursion to deeper levels, the long computation time near the root is diffused to deeper levels, that have shorter time on average. Since capturing the structure of the recursion is hard, we give a condition called *Push-out condition* such that the amortized computation time is bounded when the condition is satisfied. As the condition is given to the relation between each iteration and its children, proving the satisfiability of the condition is often not hard.

As a result, to give a bound to amortized time complexity, what we have to do is to prove that the condition holds. In this way, we propose algorithms for enumerating matchings, elimination orderings, and connected vertex induced subgraphs, and prove that the condition holds for each. These lead that these graph objects can be enumerated in constant time per solution. We also show that the condition holds for the algorithm for spanning tree enumeration, and this gives a very simple proof compared to the existing ones.

2   Preliminaries

Let $\mathcal{A}$ be an enumeration algorithm. Suppose that $\mathcal{A}$ is a recursive type algorithm, i.e., composed of a subroutine that recursively calls itself several times (or none). Thus, the recursion structure of the algorithm forms a tree. We call the subroutine, or the execution of the subroutine an *iteration*. Note that an iteration does not include the computation done in the subroutines recursively called by the iteration, thus no iteration is included in another. When the algorithm is composed of several kinds of subroutines and operations, and thus the recursion is a nest of several kind of subroutines. In such cases, we consider a series of iterations of different types as an iteration.

When an iteration X recursively calls an iteration Y, X is called the *parent* of Y, and Y is called a *child* of X. The *root iteration* is that with no parent. For non-root iteration X, its parent is unique, and is denoted by $P(X)$. The set of the children of X is denoted by $C(X)$. The parent-child relation between iterations forms a tree structure called a *recursion tree*. An iteration is called a *leaf iteration* if it has no child, and an *inner iteration* otherwise.

For iteration X, an upper bound of the execution time (the number of operations) of X is denoted by $T(X)$. Here we exclude the computation for the output process from the computation time. We remind that $T(X)$ is the time for local execution time, and thus does not included the computation time in the recursive calls generated by X. For example, when $T(X) = O(n^2)$, $T(X)$ is written as cn^2 for some constant c. T^* is the maximum $T(X)$ among all leaf iterations X. Here, T^* can be either constant, or a polynomial of the input size. If X is an inner iteration, let $\overline{T}(X) = \sum_{Y \in C(X)} T(Y)$.

In this paper, we assume that a graph is stored in a style of adjacency list. For a vertex subset U of a graph $G = (V, E)$, the *induced subgraph* of U is the graph whose vertex set is U, and whose edge set contains the edges of E connecting two vertices of U. An edge is called a *bridge* if its removal increases the number of connected components. An edge f is said to be *parallel* to e if e and f have the same endpoints, and be *series* to e if e is a bridge in $G \setminus f$ and not so in G.

For an edge e of a graph G, we denote the graph obtained by removing e from G by $G \setminus e$, and that by removing e and edges adjacent to e by $G^+(e)$. Similarly, for a vertex v of G, $G \setminus v$ is the graph obtained from G by removing v and edges incident to v. For an edge (u, v) of G, the graph *contracted* by (u, v), denoted by $G/(u, v)$, is the graph obtained by unifying the vertices u and v into one. For an edge set $F = \{e_1, \ldots, e_k\}$, G/F denotes the graph $G/e_1/e_2/\cdots/e_k$.

3   Push Out Amortization

The size of the input of each iteration for a recursive algorithm often decreases as the depth of the recursion. Thus, iterations near the root iteration take a relatively long time, and iterations near leaf iterations take a relatively short time. Motivated by this observation, we amortize the computation time by moving the computation time of each iteration to its children. We carry out this move from the top to the bottom, so that the computation time of ancestors is recursively diffused to their descendants. When we can obtain a short amortized computation time in this way, iterations with long computation times have many descendants at least proportional to their computation time; the average computation time per iteration will be long only when they have few descendants.

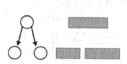

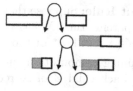

Fig. 1. An iteration, its children, and their computation time represented by rectangle lengths; seems to be inefficient if children take long time, but the descendants are many

Fig. 2. Push out rule; an iteration (center) receives computation time from its parent (while rectangle), and deliver it together with its computation time (gray rectangle) to its children, proportional to their computation time

However, it is not easy to prove that any inner iteration has sufficiently many descendants. Instead of that, we use some local conditions, related to a parent and children. Let $\alpha > 1$ and $\beta \geq 0$ be constants, and X be an iteration.

Push Out (PO) Condition: $\overline{T}(X) \geq \alpha T(X) - \beta(|C(X)| + 1)T^*.$

Fig. 1 is an example. After the assignment of the computation time of $\alpha\beta(|C(X)| + 1)T^*$ to children and the remaining to itself, the inequation $\overline{T}(X) \geq \alpha T(X)$ holds. This implies that the computation time of one level of recursion intuitively increases as the depth, unless there are not so many leaf iterations. Considering that enumeration algorithms usually spend less time in deeper levels of the recursion, we can see that this implies that each iteration has many children on average. This is in some sense not a typical condition to bound the time complexity of recursive algorithms; usually we want to decrease the total computation time in deeper levels. However, in the enumeration, the number of leaf iterations is fixed, and thereby the total computation time in the bottom level is also fixed. Thus, this condition implies that the total computation time is short.

Theorem 1. *If any inner iteration of an enumeration algorithm satisfies PO condition, the amortized computation time of an iteration is $O(T^*)$.*

Proof. To prove the lemma, we charge the computation time. We neither move the operations nor modify the algorithm, but just charge the computation time; the computation time can be considered as tokens, and we move the tokens so that each iteration has a small number of tokens. We charge the computation time from an iteration to its children, i.e., from the top of the recursion tree to the bottom. Thus, an iteration receives computation time from its parent. We charge (push out) its computation time and that received from its parent to its children. The computation time is charged to the children, in proportion of their individual computation time, using the following rule.

Push Out Rule: Suppose that iteration X receives computation time of $S(X)$ from its parent, thus X has computation time of $S(X) + T(X)$ in total. We fix $\frac{\beta}{\alpha-1}(|C(X)| + 1)T^*$ of the computation time to X, and charge (push out) the remaining computation time of $S(X) + T(X) - \frac{\beta}{\alpha-1}(|C(X)| + 1)T^*$ to its children. Each child Z of X receives computation time proportional to $T(Z)$, i.e.,

$$S(Z) = (S(X) + T(X) - \frac{\beta}{\alpha - 1}(|C(X)| + 1)T^*)\frac{T(Z)}{T(X)}.$$

See Fig. 2 as an example. According to this rule, we charge the computation time from the root iteration to leaf iterations, so that each inner iteration has $O((|C(X)| + 1)T^*)$ computation time. Since the sum of the number of children over all nodes in a tree is no greater than the number of nodes in a tree, this is equivalent to that each iteration has $O(T^*)$ time. The remaining issue is to prove the statement of the lemma by showing that each leaf iteration receives computation time of $O(T^*)$, and it is sufficient to prove the statement. To show that, we state the following claim.

Claim: if we charge computation time in the manner of the push out rule, each iteration X receives computation time of at most $T(X)/(\alpha - 1)$ from its parent, i.e., $S(X) \leq T(X)/(\alpha - 1)$

The root iteration satisfies this condition. Suppose that an iteration X satisfies it. Then, for any child Z of X, Z receives computation time of

$$(S(X) + T(X) - \frac{\beta}{\alpha - 1}(|C(X)| + 1)T^*)\frac{T(Z)}{T(X)}$$

$$\leq (T(X)/(\alpha - 1) + T(X) - \frac{\beta}{\alpha - 1}(|C(X)| + 1)T^*)\frac{T(Z)}{T(X)}$$

$$= \frac{\alpha T(X) - \beta(|C(X)| + 1)T^*}{\alpha - 1} \times \frac{T(Z)}{\overline{T}(X)}$$

$$= \frac{\alpha T(X) - \beta(|C(X)| + 1)T^*}{\overline{T}(X)} \times \frac{T(Z)}{\alpha - 1}.$$

Since PO condition is satisfied, $\overline{T}(X) \geq \alpha T(X) - \beta(|C(X)| + 1)T^*$. Thus,

$$\frac{\alpha T(X) - \beta(|C(X)| + 1)T^*}{\overline{T}(X)} \frac{T(Z)}{\alpha - 1} \leq \frac{T(Z)}{\alpha - 1}.$$

By induction, any iteration satisfies the condition in the claim.   □

Note that PO condition does not require for the iterations to have at least two children.

4   Enumeration of Elimination Ordering

Let $\mathcal{L}$ be a class of structures such as sets, graphs, and sequences. Suppose that any structure $Z \in \mathcal{L}$ consists of a set of elements called an *ground set*, that is denoted by $V(Z)$. Examples of ground sets are the vertex set of a graph, the edge set of a graph, the cells of a matrix, and the letters of a string. The empty structure $\perp$ is the unique structure that has $V(\perp) = \emptyset$, and hereafter we consider only $\mathcal{L}$ including the empty structure. For each $Z \in \mathcal{L}, Z \neq \perp$, we define the set of *removable elements* $R(Z)$, such that for each removable element $e \in R(Z)$, the removal of e from Z results in a structure $Z' \in \mathcal{L}, V(Z') = V(Z) \setminus \{e\}$. We denote the removal of e from Z by $Z \setminus e$, and we assume that no two different structures can be generated by the removal of e. By using removable elements, we define *elimination orderings*. An elimination ordering is an ordering $(z_1, \ldots, z_n)$ of elements in $V(Z)$ iteratively removed from Z until Z is $\perp$, i.e., any z_i is removable in the structure Z_i that is obtained by repeatedly removing z_1 to z_{i-1} from Z. Example of elimination ordering are removing leaves from a tree, and perfect elimination ordering of a chordal graph. A simple algorithm for enumerating elimination orderings can be described as follows.

Algorithm. EnumElimOrdering (Z, S)
1. **if** $|V(Z)| = 1$, **output** $S + z$ where $V(Z) = \{z\}$; **return**
2. **for** each element $z \in V(Z)$ **do**
       if $z \in R(Z)$, **call** EnumElimOrdering $(Z \setminus z, S + z)$

Suppose that we are given a structure Z in a class $\mathcal{L}$ and removable ground set R for ground set $V(Z)$. We suppose that for any $z \in V(Z)$, we can list all $z \in R(Z)$ in $\Theta(p(|V(Z)|)q(n))$ time, where $p(|V(Z)|)$ is a polynomial of $|V(Z)|$, and $q(n)$ is a function where n is an invariant of the input structure, such as the number of edges in the original graph. We also assume that a removal of element takes $\Theta(p(|V(Z)|)q(n))$ time.

Theorem 2. *Elimination orderings of a class $\mathcal{L}$ can be enumerated in $O(q(n))$ time for each, if $|R(Z)| \geq 2$ holds for each $Z \in \mathcal{L}$ such that $|V(Z)|$ is larger than a constant number c.*

Proof. We first bound the computation time except for the output processes, that is, step 1 of EnumElimOrdering. First, we choose two constants $\delta > c$ and $\alpha > 1$ such that $\frac{2p(i-1)}{p(i)} > \alpha$ holds for any $i > \delta$. Since p is a polynomial function, $\frac{p(i)}{p(i-1)}$ converges to 1, thus such α always exists. Let X be an iteration. When X inputs Z with $|V(Z)| \leq \delta$, the computation time is $q(n)$, except for the output process. Hence, we have $T^* = O(q(n))$. For the case $|V(Z)| \leq \delta$, the computation time of X is bounded by $q(n)$. For the case $|V(Z)| > \delta$, we have

$$\overline{T}(X) \geq 2(|V(Z)| - 1)p(|V(Z)| - 1)q(n) > \alpha|V(Z)|p(|V(Z)|)q(n),$$

since X has at least two children. Thus, X satisfies PO condition with any constant $\beta > 0$. From Theorem 1, except for the output process, the computation time is bounded by $O(q(n))$ time for each iteration whose input has at least δ elements. Since any inner iteration Y has exactly one child only if $|V(Y)| \leq c$, the number of inner iterations is bounded by the number of leaf iterations, multiplied by c. Therefore, the computation time for each elimination ordering can be bounded by $O(cq(n)) = O(q(n))$ time.

Next, let us consider the output process. Instead of explicitly outputting elimination orderings, we output each elimination ordering S by the difference from S' that is output just before S. We can output them compactly in this way. Although the difference can be large up to $|V(Z)|$, we can see that it is bounded by the number of operations done from the previous output process. Thus, the size of all output differences, except for the first one output in the usual way, is at most proportional to the total computation time. Therefore, the computation time for the output process is also bounded by $O(q(n))$ time for each.   □

The next corollary immediately follows from the theorem.

Corollary 1. *For a given set class, elimination ordering can be enumerated by EnumElimOrdering in $O(1)$ amortized time for each, if each inner iteration generates at least two recursive calls, and takes $O(p(|V(Z)|))$ time, where p is a polynomial of $|V(Z)|$.*   □

There are actually several elimination orderings to which this theorem can be applied, and they are listed below. For conciseness, we have described each by their structures and removable elements.

Example (a): perfect elimination orderings of a chordal graph[2]

For a graph, a vertex is called *simplicial* if the vertices adjacent to it form a clique. An elimination orderings of simplicial vertex is called *perfect elimination ordering*[11], and a graph is *chordal* if it has a perfect elimination ordering. We define $\mathcal{L}$ by the set of chordal graphs, $V(Z)$ by the vertex set of $Z \in \mathcal{L}$, and $R(Z)$ by the set of its simplicial vertices.

It is known that any chordal graph Z admits a clique tree whose vertices are maximal cliques of Z. If Z is a clique, all vertices in Z are simplicial. If not, it is known that there are at least two cliques that has a vertex that is not included in the other maximal cliques. Note that these cliques are leaf cliques of a clique tree, where the vertices of a clique tree are maximal cliques of Z, each edge connects overlapping cliques, and the maximal cliques including any vertex forms a subtree of the clique tree. The vertex is simplicial, hence $|R(Z)| \geq 2$ always holds. Since we can check whether a vertex is simplicial or not in $(|V(X)|^2)$ time, we can enumerate all perfect elimination orderings in $O(1)$ time for each. Note that although the algorithm in [2] already attained the same time complexity, our analysis yields much simpler algorithm and proof. **Example (b): perfect**

sequence[9]

$\mathcal{L}$ is the class of chordal graphs Z, and $V(Z)$ is the set of maximal cliques in Z. A maximal clique is removable if it is a leaf of some clique trees of Z, and the removal of a maximal clique z from Z is the removal of all vertices of z that do not belong to another maximal clique. The removal of the vertices results in the graph that includes remaining maximal cliques, and no new maximal clique appears in the graph. Note that a clique tree has at least two leaves if it has more than one vertex, thus $|R(Z)| \geq 2$. An elimination ordering is called a *perfect sequence*. Since all removable maximal cliques can be found in polynomial time in the number of maximal cliques[9], all perfect sequences are enumerated in $O(1)$ time for each.

The elimination orderings induced by following removable elements can be also enumerated in $O(1)$ time for each, such as non-cut vertices of connected graph, points on surface of convex hull of a point set in plane, leaves of a tree, and vertices of degrees less than seven of a simple planar graph.

5   Enumeration of Matchings

A *matching* of a graph is an edge subset of a graph $G = (V, E)$ such that no two edges are adjacent. The matchings are enumerated by the following algorithm.

Algorithm. EnumMatching $(G = (V, E), M)$
1: choose an edge e from E; **if** $E = \emptyset$ **then output** M; **return**
2: **call** EnumMatching $(G \setminus e, M)$ and EnumMatching $(G^+(e), M \cup \{e\})$

The time complexity of an iteration of EnumMatching is $O(|V|)$. Since each inner iteration generates two children, the computation time for each matching is $O(|V|)$, and no better algorithm has been proposed in the literature. A leaf iteration takes $O(1)$ time, thus $T^* = O(1)$. However, PO condition may not hold for some iterations. This cannot be better than $O(|V|)$ in straightforward ways.

PO condition does not hold when many edges are adjacent to e, since $G^+(e)$ has few edges, thus the subproblem of $G^+(e)$ takes short time. To avoid this situation, we modify the way of recursion as follows so that in such cases the iteration has many children. Let $u_1, \ldots, u_k$ be the vertices adjacent to v, and $e_i = (v, u_i)$. We partition the matchings to be enumerated into

- matchings including e_1
- matchings including e_2
- ...
- matchings including e_k
- matchings including no edge incident to v.

We see that any matching belongs to exactly one of these groups. To recur, we derive $G^+(e_1), \ldots, G^+(e_k)$ and $G \setminus v$. $G \setminus v$ and $G^+(e_1)$ can be derived in $O(|E|)$ time. To shorten the computation time for $G^+(e_i)$ for $i \geq 2$, we construct $G^+(e_i)$ from $G^+(e_{i-1})$. We add all edges of G incident to u_{i-1} to $G^+(e_{i-1})$, and remove all edges adjacent to u_i, and obtain $G^+(e_i)$. This can be done in $O(d(u_{i-1}) + d(u_i))$ time. To construct $G^+(e_i)$ for all $i = 2, \ldots, k$, we need

$$O(\ (d(u_1) + d(u_2)) + (d(u_2) + d(u_3)) + \cdots + (d(u_{k-1}) + d(u_k))\ ) = O(|E|)$$

time. Thus, the computation time of an iteration is bounded by $c|E|$ with a constant c. The algorithm is described as follows.

Algorithm. EnumMatching2 $(G = (V, E), M)$
1: $v :=$ a vertex of the maximum degree; if $E = \emptyset$ then **output** M; **return**
2: **call** EnumMatching2 $(G \setminus v, M)$
3: **for** each edge e incident to v, **call** EnumMatching2 $(G^+(e), M \cup \{e\})$

Theorem 3. *All matchings in a graph can be enumerated in $O(1)$ time for each, with $O(|E| + |V|)$ space.*

Proof. The amortized computation time for outputting process is bounded by $O(1)$ for each by using difference as elimination ordering. Let us consider an inner iteration X. In the iteration X, if $d(v) \geq |E|/4$, we generate at least $|E|/4$ recursive calls, thus we have $|C(X)| = \Omega(|E|)$ and PO condition is satisfied by choosing sufficiently large β. If $d(v) < |E|/4$, the subproblems of $G \setminus v$ take at least $\Theta(3c|E|/4)$ time, and the subproblems of $G^+(e_1)$ take at least $c|E|/2$ time. Hence, by setting $\alpha = 1.25$, we have

$$\overline{T}(X) \geq 3c|E|/4 + c|E|/2 = 5c|E|/4 \geq \alpha T(X) - \beta|C(X)|T^*$$

thereby PO condition holds. Remind that each inner iteration generates two or more recursive calls, the number of iterations does not exceed the twice the number of matchings. Since any inner iteration satisfies PO condition and $T^* = O(1)$, the statement holds. We remind that we assumed that there is no isolated vertex in the input graph, and thus the number of matchings in the graph is greater than the number of vertices, and the number of edges.   □

6   Enumeration of Connected Vertex Induced Subgraphs

We consider enumeration of all vertex sets of the given graph $G = (V, E)$ inducing connected subgraphs (connected induced subgraphs in short). In literature, an algorithm is proposed that runs in $O(|V|)$ time for each[1]. For the enumeration, it is sufficient to enumerate all connected induced subgraphs including the given vertex r. For a vertex v adjacent to r, the connected induced subgraphs including r are partitioned into those including v and those not including v. The former subgraphs are connected induced subgraphs in $G/(r, v)$ and the latter subgraphs are those in $G \setminus v$. We have the following algorithm according to this partition, and we prove that this algorithm satisfies PO condition.

Algorithm. EnumConnect $(G = (V, E), S, r)$
1: choose a vertex v adjacent to r; if $d(r) = 0$ then **output** S; **return**
2: **call** EnumConnect $(G/(r, v), S \cup \{v\}, r)$, and EnumConnect $(G \setminus v, S, r)$

Theorem 4. *All connected vertex induced subgraphs in a graph can be enumerated in $O(1)$ time for each, with $O(|E| + |V|)$ space.*

Proof. The correctness and the bound for memory usage are clear. Since each inner iteration generates exactly two recursive calls, the number of iterations is linearly bounded by the number of connected induced subgraphs, and $T^* = O(1)$.

Same as Theorem 3, the amortized time for outputting process is $O(1)$ for each. An inner iteration X of the algorithm takes $O(d(r)+d(v))$ time. We assume that $T(X) = c(3d(r) + d(v))$ for a constant c, and leaf iteration takes $3c$ time, since $T^* = O(1)$. The constant factor of three is a key to PO condition.

The degree of r is at least $(d(r) + d(v))/2 - 1$ in $G/(r,v)$, and $d(r) - 1$ in $G \setminus v$. Note that $d(r)$ and $d(v)$ are degrees of r and v in $\overline{G}$. From this, we can see that the child iteration of $G/(r,v)$ takes at least $3c((d(r) + d(v))/2 - 1)$ time, and that of $G \setminus v$ takes at least $3c(d(r) - 1)$ time. Their sum is at least

$$3c((d(r) + d(v))/2 - 1) + 3c(d(r) - 1) = \frac{3}{2}c(3d(r) + d(v)) - 6c = \frac{3}{2}T(X) - 6c.$$

Setting $\beta = 6$, we can see that X satisfies PO condition. Thanks to Theorem 1, the computation time for each connected induced subgraph is $O(1)$.   □

7   Spanning Trees

A subtree T of a graph $G = (V, E)$ is called a *spanning tree* if any vertex of G is incident to at least one edge of T. Any spanning tree has $|V| - 1$ edges. There have already been several studies on this problem[8, 12, 14], and [14] is the simplest and uses an amortized analysis similar to us. Without loss of generality, we assume that the input graph does not have any bridge.

Let e_1 be an edge of G. If several edges $e_2, \ldots, e_k$ are parallel to e_1, let $F = \{e_1, \ldots, e_k\}$ and $F_i = F \setminus \{e_i\}$. At most one edge from F can be included in a spanning tree, thus we enumerate spanning trees in $(G \setminus F_1)/e_1, \ldots, (G \setminus F_k)/e_k$. We further enumerate spanning trees in $G \setminus F$ if it is connected. Any spanning tree is enumerated in exactly one of these. When e_1 has no parallel edges, e_1 can have series edges. If there are several edges $e_2, \ldots, e_k$ series to e_1, again let $F = \{e_1, \ldots, e_k\}$ and $F_i = F \setminus \{e_i\}$. We also see that any spanning tree includes at least $k - 1$ edges of F, thus we enumerate spanning trees in $(G/F_1) \setminus e_1, \ldots, (G/F_k) \setminus e_k$. We further enumerate spanning trees in G/F if F is not the edges of a cycle. Also in this case, any spanning tree is enumerated once among these. By using these subdivisions, we construct the following algorithm.

Algorithm. EnumSpanningTree $(G = (V, E), T)$
1: choose an edge e_1 from E; **if** $E = \emptyset$ **then** output T; **return**
2: $F^p := \{e_1\} \cup \{e | e$ is parallel to $e_1\}$; $F^s := \{e_1\} \cup \{e | e$ is series to $e_1\} \setminus F^p$
3: **for** each $e_i \in F^p$, **call** EnumSpanningTree $((G \setminus (F^p \setminus \{e_i\})/e_i, T \cup \{e_i\})$
4: **for** each $e_i \in F^s$, **call** EnumSpanningTree $((G/(F^s \setminus \{e_i\}) \setminus e_i, T \cup (F^s \setminus \{e_i\}))$

We observe that these k subgraphs are actually isomorphic in both cases except for the edge label e_i, thus constructing these graphs takes $O(|V| + |E|)$ time.

Theorem 5. *All spanning trees in a graph can be enumerated in $O(1)$ time for each, with $O(|E| + |V|)$ space.*

Proof. The space complexity of the algorithm is $O(|E| + |V|)$ and an iteration takes $\Theta(|V| + |E|)$ time since all edges parallel/series to an edge can be found by two connected component decomposition in $O(|V| + |E|)$ time. If no edge is parallel or series to e_1, we generate two subproblems of $|E| - 1$ edges, thus PO condition holds. If k edges are parallel or series to e_1, we have at least $k + 1 \geq 2$ subproblems of $|E| - (k + 1)$ edges. When $k + 1 \geq |E|/4$, $T(X) - \beta(|C(X)| + 1)T^* = 0$ holds for some $\beta > 0$, and PO condition holds. When $k + 1 < |E|/4$, $(k + 1)(|E| - (k + 1)) \geq 1.5|E|$ holds, PO condition holds for $\alpha = 1.5$ and some $\beta > 0$. Since each iteration generates at least two recursive calls or outputs a solution, the number of iterations is at most twice the number of solutions, therefore the statement holds.   □

8   Conclusion

We introduced a new way of looking at amortizing the computation time of enumeration algorithms, by local conditions of recursion trees. We clarified the conditions that are sufficient to give non-trivial upper bounds for the average computation time of iterations that only depended on the relation between the computation time of a parent iteration and that of its child iterations. We showed that many algorithms for elimination orderings have good properties so that the conditions are satisfied, and thus enumerated in constant time for each. Several other enumeration algorithms for matchings, connected vertex induced subgraphs, and spanning trees were also described, whose time complexities are $O(1)$ for each solution.

There are many problems for those enumeration algorithms that do not satisfy the conditions. An interesting future work is to develop new algorithms for these problems, that satisfy the conditions. Another direction is to study other conditions for bounding amortized computation time. Further studies on amortized analysis will possibly fill the gaps between theory and practice, and clarify the mechanisms of enumeration algorithms.

Acknowledgments. Part of this research is supported by the Funding Program for World-Leading Innovative R&D on Science and Technology, Japan, and Grant-in-Aid for Scientific Research (KAKENHI), Japan.

References

1. Avis, D., Fukuda, K.: Reverse Search for Enumeration. Discrete Applied Mathematics **65**, 21–46 (1996)
2. Chandran, L.S., Ibarra, L., Ruskey, F., Sawada, J.: Generating and Characterizing the Perfect Elimination Orderings of a Chordal Graph. Theoretical Computer Science **307**, 303–317 (2003)
3. Eppstein, D.: Finding the k smallest spanning trees. In: Gilbert, J.R., Karlsson, R. (eds.) SWAT 90. LNCS, vol. 447, pp. 38–47. Springer, Heildelberg (1990)
4. Eppstein, D.: Finding the k shortest paths. In: FOCS 1994, pp. 154–165 (1994)

5. Ferreira, R., Grossi, R., Rizzi, R.: Output-sensitive listing of bounded-size trees in undirected graphs. In: Demetrescu, C., Halldórsson, M.M. (eds.) ESA 2011. LNCS, vol. 6942, pp. 275–286. Springer, Heidelberg (2011)
6. Frequent Itemset Mining Dataset Repository. http://fimi.cs.helsinki.fi/data/
7. Gabow, H.N., Tarjan, R.E.: A Linear-time algorithm for a special case of disjoint set union, In: STOC 1983, pp. 246–251 (1983)
8. Kapoor, H.N., Ramesh, H.: Algorithms for Enumerating all Spanning Trees of Undirected and Weighted Graphs. SIAM J. on Computing 24, 247–265 (1995)
9. Matsui, Y., Uehara, R., Uno, T.: Enumeration of the Perfect Sequences of a Chordal Graph. Theoretical Computer Science 411, 3635–3641 (2010)
10. Matsui, Y., Matsui, T.: Enumeration algorithm of the edge colorings in bipartite graphs. In: Deza, M., Euler, R., Manoussakis, L. (eds.) Combinatorics and Computer Science. LNCS, vol. 1120, pp. 18–26. Spinger, Heildelberg (1995)
11. Rose, D.J., Tarjan, R.E., Lueker, G.S.: Algorithmic Aspects of Vertex Elimination on Graphs. SIAM J. on Computing 5, 266–283 (1976)
12. Shioura, A., Tamura, A., Uno, T.: An Optimal Algorithm for Scanning All Spanning Trees of Undirected Graphs. SIAM Journal on Computing 26, 678–692 (1997)
13. Sleator, D.D., Tarjan, R.E.: A data structure for dynamic trees. In: STOC 1981, pp. 114–122 (1981)
14. Uno, T.: A new approach for speeding up enumeration algorithms and its application for matroid bases. In: Asano, T., Imai, H., Lee, D.T., Nakano, S., Tokuyama, T. (eds.) COCOON 1999. LNCS, vol. 1627, p. 349. Springer, Heidelberg (1999)

Computing the Center of Uncertain Points on Tree Networks

Haitao Wang and Jingru Zhang[(✉)]

Department of Computer Science, Utah State University, Logan, UT 84322, USA
haitao.wang@usu.edu, jingruzhang@aggiemail.usu.edu

Abstract. Uncertain data has been very common in many applications. In this paper, we consider the one-center problem for uncertain data on tree networks. In this problem, we are given a tree T and n (weighted) uncertain points each of which has m possible locations on T associated with probabilities. The goal is to find a point x^* on T such that the maximum (weighted) expected distance from x^* to all uncertain points is minimized. To the best of our knowledge, this problem has not been studied before. We propose a refined prune-and-search technique that solves the problem in linear time.

1 Introduction

In the real world, data is often associated with uncertainty because of measurement inaccuracy, sampling discrepancy, outdated data sources, resource limitation, etc. This is especially true due to the wide deployment of sensor monitoring infrastructure and increasing prevalence of technologies, such as data integration and cleaning. Hence, problems with uncertain data have been studied extensively [1–3,10,11,18,19]. In this paper, we consider the one-center problem for uncertain data on trees, where the existence (presence) of each uncertain point is described probabilistically, defined as follows.

We borrow some terminology on trees from the literature (e.g., [12,14]). Let T be a tree. Each edge $e = (u, v)$ of T has a positive length $l(e)$. We consider e as a line segment of length $l(e)$ so that we can talk about "points" on e. Formally, a point $p = (u, v, t)$ is characterized by being located at a distance of $t \le l(e)$ from the vertex u. The *distance* of any two points p and q on T, denoted by $d(p, q)$, is defined as the length of the simple path from p to q on T.

Let $\mathcal{P} = \{P_1, P_2, \ldots, P_n\}$ be a set of n uncertain points on T. Each uncertain point P_i has m possible locations on T, denoted by $\{p_{i1}, p_{i2}, \cdots, p_{im}\}$, and each location p_{ij} is associated with a probability $f_{ij} \ge 0$ that is the probability of P_i being at p_{ij} (which is independent of other locations), with $\sum_{j=1}^{m} f_{ij} = 1$; e.g., see Fig. 1. Further, each uncertain point P_i has a weight $w_i > 0$.

Consider any point x on T. For any $P_i \in \mathcal{P}$, the *(weighted) expected distance* from x to P_i, denoted by $Ed(x, P_i)$, is defined as $w_i \cdot \sum_{j=1}^{m} \{f_{ij} \cdot d(x, p_{ij})\}$.

In the following, for simplicity, we use "expected distance" to refer to "weighted expected distance". We define $R(x)$ as the maximum expected distance from x to all uncertain points of $\mathcal{P}$, i.e., $R(x) = \max_{1 \le i \le n} Ed(x, P_i)$.

This research was supported in part by NSF under Grant CCF-1317143.

F. Dehne et al. (Eds.): WADS 2015, LNCS 9214, pp. 606–618, 2015.
DOI: 10.1007/978-3-319-21840-3_50

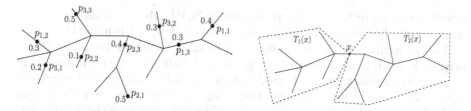

Fig. 1. Illustrating three uncertain points P_1, P_2, P_3, each with three possible locations (their probabilities are also shown)

Fig. 2. The point x has two split subtrees $T_1(x)$ and $T_2(x)$

The *center* of T with respect to $\mathcal{P}$ is defined to be a point x^* that minimizes the value $R(x)$ among all points $x \in T$. Our goal is to compute x^*.

For any edge e of T, we assume the locations of the uncertain points of $\mathcal{P}$ on e are already given sorted on e. This means that if we traverse the edge e from one end to the other, then we can encounter those locations in order.

If T is a path network, the problem has been studied in [21], where a linear time is given for computing the center. However, if T is a tree, to the best of our knowledge, the problem has not been studied before. In this paper, we give an $O(|T| + mn)$ time algorithm for the problem, where $|T|$ is the number of vertices of T. Note that since $|T| + mn$ is essentially the input size, the time complexity of our algorithm is linear, and thus our algorithm is optimal.

1.1 Related Work

Two uncertain models have been commonly considered: the *existential* model [3,10,11,18,19] and the *locational* model [1,2,19]. In the existential model, an uncertain point has a specific location but its existence is uncertain. In the locational model, an uncertain point always exists but its location is uncertain. Our one-center problem belongs to the locational model. In fact, the same problem under existential model is essentially the weighted one-center problem for deterministic data, which can be solved in linear time [14].

As mentioned before, if T is a path network, the uncertain one-center problem has been solved in linear time [21]. Algorithms for the more general uncertain k-center problems on path networks have also been given in [21].

The one-center and the more general k-center problems for the deterministic case where all data are certain have been studied extensively, as discussed below. Megiddo [14] solved the (weighted) one-center problem on trees in linear time. For the more general k-center problem on trees, Megiddo and Tamir [16] presented an $O(n \log^2 n \log \log n)$ time algorithm for the weighted case, where n is the number of vertices of the tree, and later the running time of the algorithm was reduced to $O(n \log^2 n)$ by Cole [6]. The unweighted case was solved in linear time by Frederickson [8]. If all points are on the two-dimensional plane, the unweighted one-center problem becomes the minimum enclosing circle problem, which is solvable in linear time [15]; the weighted one-center problem can be solved in $O(n \log n)$ time by the techniques in [6,17], where n is the number of input points. The general k-center problem in the plane is NP-hard [15].

Facility location or related problems under other uncertain models have also been considered. Foul [7] studied the problem of finding the center in the plane to minimize the maximum expected distance from the center to all uncertain points, where each uncertain point has a uniform distribution in a given rectangle. Jørgenson et al. [9] considered the problem of computing the distribution of the radius of the smallest enclosing ball for a set of indecisive points each of which has multiple locations associated with probabilities in the plane. Löffler and van Kreveld [13] studied the problem of finding the smallest enclosing circle and other related problems for imprecise points each of which is known to be contained in a planar region (e.g., a circle or a square). Berg. et al. [5] proposed an approximation algorithm to dynamically maintain Euclidean 2-centers for a set of moving points in the plane (the moving points are considered uncertain). See also the minmax regret problems, e.g., [4,20].

1.2 Our Techniques

Note that the uncertain points of $\mathcal{P}$ may have locations in the interior of some edges of T. A *vertex-constrained case* happens when all locations of $\mathcal{P}$ are at vertices of T and each vertex of T contains at least one location of $\mathcal{P}$. We show that the general case can be reduced to the vertex-constrained case in $O(|T| + mn)$ time. In the following, we focus our discussion on the vertex-constrained case (i.e., we assume our problem on T and $\mathcal{P}$ is a vertex-constrained case); even in this case, the center x^* may still not be at a vertex of T.

To solve our problem, one immediate option is to see whether Meggido's prune-and-search techniques [14] for solving the deterministic one-center problem on trees can be applied. However, as discussed below, there are some "enormous" difficulties to apply Meggido's techniques directly. To overcome these difficulties, we propose new techniques, which can be viewed as a refinement of Meggido's techniques and which we call the *refined prune-and-search*.

Meggido's algorithm [14] is used to find the center x^* for a tree T of n vertices, where each vertex v has a weight w_v. Meggido's algorithm first computes the centroid c of T (each of c's subtree has at most $n/2$ vertices), and then based on the weighted distances from all vertices to c, one can determine which subtree of c contains the center x^*. Suppose T' is a subtree containing x^*. The number of vertices outside T (i.e., those in $T \backslash T'$) is at least $n/2$. Consider any two vertices u and v in $T \backslash T'$. In general, by solving the equation $w_u(d(u,c)+t) = w_v(d(v,c)+t)$, one can obtain a value t_{uv} such that for every point x in T' at a distance t from c, $w_u(d(u,x)) \geq w_v(d(v,x))$ if and only if $0 \leq t \leq t_{uv}$. Based on this observation, the vertices in $T \setminus T'$ are arbitrarily arranged in roughly at least $n/4$ disjoint pairs, and for each pair u and v, the value t_{uv} is computed. Let t^* be the median of these t_{uv} values. Depending on whether x^* is within distance t^* from c, at least $n/8$ vertices of T can be pruned.

In our problem, the tree T has $O(mn)$ vertices, and we do the same thing and first find the centroid c of T. Although now we have uncertain points, by observations, we can still efficiently determine the subtree T' of c that contains the center x^*. However, we cannot proceed as above in Megiddo's algorithm. The reason is that for each uncertain point P_i, it may have locations in both T' and $T \setminus T'$, which prevents us from having an equation for two uncertain points and further prevents us from pruning uncertain points as in Megiddo's algorithm. Indeed, this is one of the major difficulties for us to apply Megiddo's algorithmic scheme. To overcome the difficulty, we continue to find the centroid c' of T' and determine which subtree of T' (rooted at c') containing x^*. One key idea is that we repeat this for $\log m + 1$ times, after which we obtain a subtree T'' with at most $nm/(2^{\log m+1}) = n/2$ vertices. One observation is that there are at most $n/2$ uncertain points that have locations in T'', and thus, at least $n/2$ uncertain points have all locations outside T''. At this moment, we show that if T'' is connected with $T \setminus T''$ by only one vertex, then we can apply Megiddo's pruning scheme. However, another major difficulty is that T'' may be connected with $T \setminus T''$ by more than one vertex (indeed, there may be as many as $\log m + 1$ such vertices), in which case we introduce new pruning techniques to further reduce T'' to a smaller subtree T''' such that $x^* \in T'''$ and T''' is connected with $T \setminus T'''$ by either one or two vertices. For either case, we develop algorithms for the pruning. All above procedures are carefully implemented so that they together take $O(mn)$ time and eventually prune at least $n/8$ uncertain points. The total time for computing the center x^* is thus $O(mn)$.

Note that although we have assumed $\sum_{j=1}^{m} f_{ij} = 1$ for each $P_i \in \mathcal{P}$, our algorithm also works if $\sum_{j=1}^{m} f_{ij} \neq 1$. But for ease of exposition, our following discussion assumes $\sum_{j=1}^{m} f_{ij} = 1$. Due to the space limit, many details are omitted but can be found in the full version of the paper.

2 Preliminaries

In the following paper, unless otherwise stated, we assume our problem is the vertex-constrained case, i.e., all locations of $\mathcal{P}$ are at vertices of T and each vertex of T has at least one location. Later in Theorem 1, we will show that the general problem can be reduced to this case in linear time. For ease of exposition, we further assume every vertex of T has only one location of $\mathcal{P}$, and thus $|T| = mn$.

In this section, we discuss a few observations, which are mainly for determining which subtree of x contains the center x^* for any given point x on T.

For any two points p and q on T, denote by $\pi(p, q)$ the simple path on T from p to q. For any subtree T' of T and any uncertain point P_i, we call the sum of the probabilities of the locations of P_i in T' the *probability sum* of P_i in T'.

Consider any point x on T. Removing x from T will produce several subtrees of T, and we call them the *split subtrees* of x in T. More specifically, if x is in the interior of an edge, then there are two split subtrees (e.g., see Fig. 2); otherwise the number of subtrees is equal to the degree of x. We consider x as a vertex in each of these subtrees. However, we assign x to be contained in only one (and an arbitrary one) split subtree, but consider x as an "open vertex" in each of other subtrees. In this way, every point of T is in one and only one split subtree of x.

Let π be any simple path on T and x be any point on π. Consider any uncertain point P_i. For any location p_{ij} of P_i, the distance $d(x, p_{ij})$ is a convex (and piecewise linear) function as x changes on π [14]. Recall that the expected distance $Ed(x, P_i) = w_i \cdot \sum_{j=1}^{m} f_{ij} \cdot d(x, p_{ij})$. Since the sum of convex functions is also convex, $Ed(x, P_i)$ is convex (and piecewise linear) on π. Therefore, in general, as x moves from one end of π to the other end, the value $Ed(x, P_i)$ first monotonically decreases and then monotonically increases. Further, recall that $R(x) = \max_{1 \leq i \leq n} Ed(x, P_i)$. Since the max of convex functions is also convex, $R(x)$ is convex (and piecewise linear) on π.

For each uncertain P_i, let p_i^* be a point $x \in T$ that minimizes $Ed(x, P_i)$. In fact, if we consider $w_i \cdot f_{ij}$ as the weight of p_{ij}, p_i^* is the *weighted median* of the points p_{ij} for all $j = 1, 2, \ldots, m$. Hence, we call p_i^* the *median* of P_i. Note that p_i^* may not be unique (in which case we use p_i^* to denote an arbitrary median of P_i). This case happens when there is an edge dividing T into two subtrees such that the probability sum of P_i in either subtree is exactly 0.5. Indeed, the above "degenerate case" also possibly makes the center x^* of T not unique, in which case we use x^* to refer to an arbitrary center of T.

The following lemma can be obtained readily from Kariv and Hakimi [12].

Lemma 1. *Consider any point x on T and any uncertain point P_i of $\mathcal{P}$.*

1. *If x has a split subtree whose probability sum of P_i is greater than 0.5, then p_i^* must be in that split subtree.*
2. *The point x is p_i^* if the probability sum of P_i in each of x's split subtree is less than 0.5.*
3. *The point x is p_i^* if x has a split subtree in which the probability sum of P_i is equal to 0.5.*

Consider any point x on T. If $R(x) = Ed(x, P_i)$ for some uncertain point P_i, then P_i is called a *dominating point* of x. Note that x may have multiple dominating points. We have the following results.

Lemma 2. *If x has a dominating point P_i with median p_i^* at x or x has two dominating points P_i and P_j whose medians p_i^* and p_j^* are in two different split subtrees of x, then x is x^*; otherwise, x^* is in the same split subtree of x as p_i^*.*

Lemma 3. *Given any point x on T, we can determine whether x is x^*, and if not, determine which split subtree of x contains x^* in $O(|T|)$ time.*

3   The Refined Prune-and-Search

In this section, we present our algorithm for computing x^*. As discussed in Section 1.2, each round of our algorithm will prune at least $\frac{n}{8}$ uncertain points of $\mathcal{P}$ in $O(mn)$ time. After $O(\log n)$ rounds, only a constant number of uncertain points remain, and then we can compute x^* in additional $O(m)$ time.

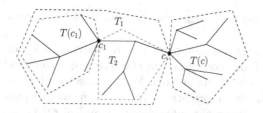

Fig. 3. Illustrating the subtrees $T(c)$, T_1, T_2, and $T(c_1)$, where c is in T_2

3.1 The Initial Pruning

A vertex c of T is called a *centroid* if every split subtree of c has no more than $|T|/2$ vertices. We first compute the centroid c of T in $O(|T|)$ time [12,14].

By Lemma 3, we determine whether c is x^*, and if not, determine which split subtree of c contains x^*. If c is x^*, we are done with the algorithm. Otherwise, let T_1 be the split subtree of c that contains x^*. To avoid repeatedly traversing $T \setminus T_1$ in future, we associate with c two *information arrays* $D_c[1 \cdots n]$ and $F_c[1 \cdots n]$, defined as follows. Note that $|T_1| \leq |T|/2+1$ (we assume $|T_1| \leq |T|/2$ for simplicity of the time analysis). Let $T(c) = (T \setminus T_1) \cup \{c\}$. For each $1 \leq i \leq n$, we define $F_c[i]$ to be the probability sum of P_i in $T(c)$, i.e., $F_c[i] = \sum_{p_{ij} \in P_i \cap T_1} f_{ij}$, and we define $D_c[i]$ to be the expected distance from c to the locations of P_i in T_1, i.e., $D_c[i] = w_i \cdot \sum_{p_{ij} \in P_i \cap T_1} f_{ij} \cdot d(c, p_{ij})$. We can compute the two information arrays in $O(mn)$ time by traversing $T(c)$.

Our algorithm will continue working on T_1. For any point $p \in T_1$ and $q \in T(c)$, the path $\pi(p,q)$ contains c. We call c a *connector* of T_1 since it connects T_1 with $T(c)$. We call $T(c)$ the *connector subtree* of c with respect to T_1.

As discussed in Section 1.2, since each uncertain point may have locations in both T_1 and $T \setminus T_1$, we cannot proceed as Megiddo's algorithm [14]. Instead, we continue to find the centroid of T_1, denoted by c_1, which can be done in $O(|T_1|)$ time. Similarly, c_1 has many split subtrees in T_1, and we want to determine whether c_1 is the center x^*, and if not, which split subtree of c_1 contains x^*. This can be done in $O(mn)$ time by Lemma 3. However, we can do faster in $O(|T_1|+n)$ time by using the two information arrays associated with the connector c without traversing the subtree $T(c)$ again. The algorithm is omitted.

If c_1 is x^*, we are done. Otherwise let T_2 denote the split subtree of c_1 in T_1 that contains x^*. Note that c may or may not be in T_2. Define $T(c_1)$ to be the subtree of T induced by c_1 and the vertices $v \in T$ such that the simple path from v to any vertex of T_2 contains c_1. In fact, $T(c_1) = (T_1 \setminus T_2) \cup \{c_1\}$ if c is in T_2 (e.g., see Fig. 3) and $T(c_1) = T(c) \cup (T_1 \setminus T_2) \cup \{c_1\}$ otherwise.

Similarly, we associate c_1 with two information arrays $F_{c_1}[1 \cdots n]$ and $D_{c_1}[1 \cdots n]$ with respect to $T(c_1)$. Specifically, for each $1 \leq i \leq n$, $F_{c_1}[i] = \sum_{p_{ij} \in P_i \cap T(c_1)} f_{ij}$ and $D_{c_1}[i] = w_i \cdot \sum_{p_{ij} \in P_i \cap T(c_1)} f_{ij} d(c_1, p_{ij})$. We can compute the above two arrays in $O(|T_1| + n)$ time. We call c_1 a *connector* of T_2 and call $T(c_1)$ the *connector subtree* of c_1. If c is in T_2, c is also a connector of T_2. Hence,

T_2 may have at most two connectors. Note that each connector of T_2 must be a leaf of T_2.

Next we continue the above procedure recursively on T_2. In general, suppose we have performed the above procedure for h recursive steps and obtain a subtree T_h, which has at most h connectors. Each connector of T_h is a leaf of T_h and is associated with two information arrays. Then, we compute the centroid c_h of T_h in $|T_h|$ time. Using the information arrays, in $O(|T_h| + nh)$ time we can determine whether c_h is x^*, and if not, which split subtree of c_h in T_h contains x^* (i.e., the algorithm spends $O(n)$ time on each connector and $O(1)$ time on each of other vertices of T_h). If c_h is x^*, we are done. Otherwise, let T_{h+1} be the split subtree of c_h containing x^*. The vertex c_h is a connector of T_{h+1}, and the connectors of T_h that are in T_{h+1} are also connectors of T_{h+1}. Hence, T_{h+1} has at most $h + 1$ connectors, each of which is a leaf of T_{h+1}. We define the *connector subtree* $T(c_h)$ of c_h similarly (i.e., $T(c_h)$ is the subtree of T induced by c_h and the vertices $v \in T$ such that the simple path from v to any vertex of T_{h+1} contains c_h). We associate c_h with two information arrays $F_{c_h}[1 \cdots n]$ and $D_{c_h}[1 \cdots n]$ with respect to $T(c_h)$, which can be computed in $O(|T_h| + nh)$ time.

We perform the above procedure for $h = 1 + \log m$ recursive steps, after which we obtain a tree T_h. By the definition of centroids, $|T_h| \leq |T|/2^h = (mn)/2^h = n/2$. Therefore, if we let $T_0 = T$, the running time of all above recursive steps is $O(\sum_{k=1}^{h}(|T_{k-1}| + n(k-1)))$, which is $O(mn + nh^2) = O(mn + n\log^2 m) = O(mn)$ since $|T| = mn$ and $|T_k| \leq |T_{k-1}|/2$ for each $1 \leq k \leq h$.

We refer to the above algorithm as *the initial pruning step*.

Since $|T_h| \leq n/2$, there are at most $n/2$ uncertain points of $\mathcal{P}$ that have locations in T_h. In other words, we have the following observation.

Observation 1. *There are at least $n/2$ uncertain points $P_i \in \mathcal{P}$ such that P_i does not have any location in T_h.*

Let C denote the number of connectors in T_h. Depending on the value of C, our algorithm will proceed accordingly for three cases: $C = 1$, $C = 2$, and $C > 2$. We first present our algorithm for the case $C = 1$. Let $\mathcal{P}'$ denote the set of uncertain points $P_i \in \mathcal{P}$ such that P_i does not have any location in T_h. We can easily find $\mathcal{P}'$ in $O(nm)$ time by traversing T_h. By Observation 1, $|\mathcal{P}'| \geq n/2$.

3.2   The Case $C = 1$

In this case, the subtree T_h has only one connector, denoted by $\hat{c}$. Although the implementation details are quite different, we can still apply Meggido's pruning scheme [14] due to the following *key property*: if an uncertain point P_i does not have any location in T_h, then all its locations must be in the connector subtree $T(\hat{c})$. Recall that $\hat{c}$ has two information arrays $D_{\hat{c}}[1 \cdots n]$ and $F_{\hat{c}}[1 \cdots n]$.

For each $P_i \in \mathcal{P}'$, since all locations of P_i are in $T(\hat{c})$, it holds that $Ed(x, P_i) = Ed(\hat{c}, P_i) + w_i \cdot t$, where x is a point in T_h at a distance t from $\hat{c}$. Note that $Ed(\hat{c}, P_i)$ is essentially $D_{\hat{c}}[i]$, and thus it is already known.

Consider any pair P_i and P_j of uncertain points in $\mathcal{P}'$. Without of loss generality, assume $Ed(\hat{c}, P_i) \geq Ed(\hat{c}, P_j)$. If $w_i < w_j$, then by solving the equation

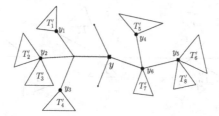

Fig. 4. Illustrating an example for the center-detecting problem: $Y = \{y_1, y_2, \ldots, y_6\}$ and $T(Y) = \{T'_1, \ldots, T'_8\}$ shown with triangles. Note that although T'_2 and T'_3 share a common point y_2, since y_2 is considered as an open vertex in each of them, T'_2 and T'_3 are disjoint. This is also the case for T'_6 and T'_8.

$Ed(\hat{c}, P_i) + w_i \cdot t = Ed(\hat{c}, P_j) + w_j \cdot t$, we can obtain a value t_{ij} such that for every x in T_h at a distance t from $\hat{c}$, $Ed(x, P_i) \geq Ed(x, P_j)$ if and only if $0 \leq t \leq t_{ij}$. If $w_i \geq w_j$, $Ed(x, P_i) \geq Ed(x, P_j)$ holds for any x in T_h, and thus P_j can be pruned immediately (since it is "dominated" by P_i).

Based on the above discussions, we arbitrarily arrange the uncertain points of $\mathcal{P}'$ into a set Σ of $|\mathcal{P}'|/2$ disjoint pairs, and for each pair (P_i, P_j), we compute the value t_{ij}. Let t^* denote the median of these t_{ij} values. Suppose we have already known whether x^* is within the distance t^* from $\hat{c}$ on T_h (we will discuss this step later); in either case, we can prune exactly one uncertain point from each pair of Σ. Since $|\mathcal{P}'| \geq n/2$ and $|\Sigma| \geq n/4$, the total number of pruned uncertain points is at least $n/8$. At this point, we have reduced our problem to the same problem on a tree T^+ of at most $7n/8$ uncertain points, defined as follows. First, let $T^+ = T_h$. Then, consider any pair (P_i, P_j) of Σ. Without of loss of generality, assume P_i is not pruned. For each location p_{ij} of P_i, we create a vertex for T^+ connecting to $\hat{c}$ directly by an edge of length $d(p_{ij}, \hat{c})$. Also let w_i still be the weight of P_i. In this way, the tree T^+ has at most $7n/8$ uncertain points and at most $7nm/8$ vertices. Based on our above pruning procedure, the center of T^+ is also the center of the original tree T. Note that the above way of constructing T^+ can be easily done in $O(mn)$ time.

It remains to determine whether x^* is within distance t^* from $\hat{c}$ on T_h. As in [14], by traversing T_h from $\hat{c}$, in $|T_h|$ time we can find all points $q \in T_h$ such that $d(\hat{c}, q) = t^*$, and we use $Q(\hat{c}, T_h)$ to denote the set of these points. Note that each point $q \in Q(\hat{c}, T_h)$ has a split subtree that contains $\hat{c}$, and we assign q to be contained in that split subtree; for any point x in other split subtrees of q, it holds that $d(\hat{c}, x) > t^*$. Hence, the set of all points x of T_h with $d(\hat{c}, x) > t^*$ can be represented as the union of split subtrees of the points in $Q(\hat{c}, T_h)$ that do not contain $\hat{c}$, and we use $\mathcal{T}(\hat{c}, T_h)$ to denote the above set of split subtrees. Our goal is to determine whether x^* is in any subtree of $\mathcal{T}(\hat{c}, T_h)$. To this end, we solve a more general problem, called a *center-detecting problem*, defined as follows. Another reason for solving this more general problem is that our algorithms for the other two cases $C = 2$ and $C > 2$ will need it.

Consider the input tree T. Let y be any vertex of T. Consider any point $x \in T$ with $x \neq y$. One split subtree of x, denoted by $\tau(x)$, contains y, and we assign

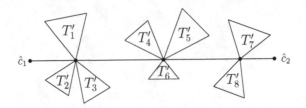

Fig. 5. Illustrating the tree T_h for the case $C = 2$, where $\mathcal{T}(V) = \{T_1', \ldots T_8'\}$ are shown with triangles

x to be in $\tau(x)$. We call other split subtrees of x than $\tau(x)$ the y-*exclusive* split subtrees of x. Note that since the y-exclusive split subtrees of x each contain x as an "open vertex", they are pairwise disjoint. Let Y be a set of points on T with $y \notin Y$ and $\mathcal{T}(Y)$ be any subset of the set of all y-exclusive subtrees of all points of Y with the following *disjoint property*: any two subtrees of $\mathcal{T}(Y)$ are disjoint (e.g., see Fig. 4). The *center-detecting problem* on $(y, Y, \mathcal{T}(Y))$ is to determine whether the center x^* is located in one of the subtrees of $\mathcal{T}(Y)$. The algorithm for Lemma 4 is omitted.

Lemma 4. *The center-detecting problem on T can be solved in $O(|T|)$ time.*

By Lemma 4, in $O(|T|)$ time, we can solve the above problem of determining whether x^* is in the subtrees of $\mathcal{T}(\hat{c}, T_h)$, as follows. Recall that $T = T(\hat{c}) \cup T_h$ and $\hat{c}$ is a leaf of T_h. By the definition of $Q(\hat{c}, T_h)$, $\mathcal{T}(\hat{c}, T_h)$ is actually the set of the $\hat{c}$-exclusive subtrees of all points of $Q(\hat{c}, T_h)$ in T and any two subtrees in $\mathcal{T}(\hat{c}, T_h)$ are disjoint. Hence, although our problem is on the subtree T_h, we can actually work on T. Thus, to determine whether x^* is in the subtrees of $\mathcal{T}(\hat{c}, T_h)$ is to solve the center-detecting problem on $(\hat{c}, Q(\hat{c}, T_h), \mathcal{T}(\hat{c}, T_h))$ and T.

3.3   The Case $C = 2$

In this case, the subtree T_h has two connectors, denoted by $\hat{c}_1$ and $\hat{c}_2$. Recall that each $\hat{c}_k$ is associated with two arrays $D_{\hat{c}_k}$ and $F_{\hat{c}_k}$ for $k = 1, 2$. The techniques for the previous case $C = 1$ do not work here. The reason is that although every uncertain point in $\mathcal{P}'$ does not have any location in T_h, it may have locations in both connector subtrees $T(\hat{c}_1)$ and $T(\hat{c}_2)$. We use a different approach.

Consider the path $\pi(\hat{c}_1, \hat{c}_2)$. Recall that $\hat{c}_1$ and $\hat{c}_2$ are leaves of T_h because they are connectors. Let V denote the set of vertices of $\pi(\hat{c}_1, \hat{c}_2)$ except $\hat{c}_1$ and $\hat{c}_2$. Consider any vertex v of V. Let $\mathcal{T}(v)$ be the set of the split subtrees of v that do not contain either $\hat{c}_1$ or $\hat{c}_2$. We assume that v is not contained in any subtree of $\mathcal{T}(v)$. Let $\mathcal{T}(V) = \cup_{v \in V} \mathcal{T}(v)$ (e.g., see Fig. 5). Note that $T_h = \mathcal{T}(V) \cup \pi(\hat{c}_1, \hat{c}_2)$.

First, we want to determine whether the center x^* is in one of the subtrees of $\mathcal{T}(V)$. To this end, notice that each subtree of $\mathcal{T}(V)$ is a $\hat{c}_1$-exclusive split subtree of some point in V, and any two subtrees of $\mathcal{T}(V)$ are disjoint. Further, $\hat{c}_1$ is a leaf of T_h. Hence, the problem is an instance of the center-detecting problem on $(\hat{c}_1, V, \mathcal{T}(V))$, which can be solved in $O(mn)$ time by Lemma 4.

If x^* is contained in a subtree τ of $\mathcal{T}(V)$, and assume τ is the split subtree of $v \in V$. Then, the problem essentially becomes the first case where $C = 1$. Indeed, we can consider v as the "connector" of τ. For each uncertain point in $\mathcal{P}'$, since it does not have any location in T_h, it does not have any location in τ. Because τ has only one connector, by using the same techniques as for the case $C = 1$, we can prune at least $n/8$ uncertain points.

If x^* is not contained in any subtree of $\mathcal{T}(V)$, then x^* must be in the path $\pi(\hat{c}_1, \hat{c}_2)$. Consider any uncertain point $P_i \in \mathcal{P}'$. Since P_i does not have any location in T_h, P_i does not have any location in $\pi(\hat{c}_1, \hat{c}_2)$. Hence, if x is a point on $\pi(\hat{c}_1, \hat{c}_2)$ at a distance t from $\hat{c}_1$, then it is not difficult to see that $Ed(x, P_i) = Ed(\hat{c}_1, P_i) + w_i \cdot (F_{\hat{c}_1}[i] - F_{\hat{c}_2}[i]) \cdot t$ (recall that $F_{\hat{c}_k}[i]$ is the probability sum of P_i in $T(\hat{c}_k)$ for $k = 1, 2$). Note that $F_{\hat{c}_1}[i] - F_{\hat{c}_2}[i]$ is constant as long as x is in $\pi(\hat{c}_1, \hat{c}_2)$ since P_i does not have any location in $\pi(\hat{c}_1, \hat{c}_2)$. Hence, as x moves from $\hat{c}_1$ to $\hat{c}_2$ along $\pi(\hat{c}_1, \hat{c}_2)$, the value $Ed(x, P_i)$ changes linearly.

Based on the above observation, we do the pruning as follows. We arbitrarily arrange the points of $\mathcal{P}'$ into $|\mathcal{P}'|/2$ pairs. In general, for each such pair (P_i, P_j), by solving the equation $Ed(\hat{c}_1, P_i) + w_i \cdot (F_{\hat{c}_1}[i] - F_{\hat{c}_2}[i]) \cdot t = Ed(\hat{c}_1, P_j) + w_j \cdot (F_{\hat{c}_1}[j] - F_{\hat{c}_2}[j]) \cdot t$, we can determine a value t_{ij} such that for a point x in $\pi(\hat{c}_1, \hat{c}_2)$ at a distance t from $\hat{c}_1$, $Ed(x, P_i) \geq Ed(x, P_j)$ if and only if $0 \leq t \leq t_{ij}$. In this way, we can obtain $|\mathcal{P}'|/2$ such values t_{ij}, and let t^* be the median of them. Let q^* be the point on $\pi(\hat{c}_1, \hat{c}_2)$ at distance t^* from $\hat{c}_1$. Again, by Lemma 3, we can determine in $O(mn)$ time whether q^* is x^*, and if not, which split subtree of q^* contains x^* (and thus determine whether x^* is within the distance t^* from $\hat{c}_1$). In either case, we can prune an uncertain point from each of the above pairs of $\mathcal{P}'$, and thus prune a total of at least $n/8$ uncertain points due to $|\mathcal{P}'| \geq n/2$.

3.4   The Case $C > 2$

In this case, T_h has more than two connectors. Indeed, this is the most general case. Clearly, the techniques for the case $C = 2$, which reply on a path $\pi(\hat{c}_1, \hat{c}_2)$, are not applicable any more. We use a new approach by "shrinking" T_h until the problem is reduced to one of the previous two cases.

A vertex z of T_h is called a *connector-centroid* if each split subtree of z has no more than $C/2$ connectors (z may not be unique). The main idea of our algorithm is similar to the scheme of the initial pruning in Section 3.1. We first find a connector-centroid z of T_h and then remove the split subtrees of z that do not contain the center x^*. We work on the remaining split subtree of z recursively until there are at most two connectors left, at which moment we have reduced the problem to the case of either $C = 2$ or $C = 1$. The details are given below.

We first find a connector-centroid z of T_h. This can be done in $O(|T_h|)$ time by a traversal of T_h, always moving in the direction in which the number of connectors, in the subtree entered into, is being maximized. As the algorithm in Section 3.1, by traversing T_h and using the information arrays associated with the connectors, we can determine in $O(Cn+|T_h|)$ time whether z is the center x^*, and if not, which split subtree of z contains x^*. If z is x^*, we are done. Otherwise, let $T_h^1(z)$ denote the split subtree of z in T_h that contains x^*. Further, we consider

z as a "connector" of $T_h^1(z)$. By the definition of the connector-centroid, $T_h^1(z)$ has at most $C/2+1$ connectors. As in Section 3.1, we define the *connector subtree* $T(z)$ for z as the subtree of T induced by z and the vertices $v \in T$ such that the simple path from v to any vertex of $T_h^1(z)$ contains z. Similarly, we associate two information arrays $F_z[1 \cdots n]$ and $D_z[1 \cdots n]$ with z (i.e., for each $1 \leq i \leq n$, $F_z[i] = \sum_{p_{ij} \in P_i \cap T(z)} f_{ij}$ and $D_z[i] = w_i \cdot \sum_{p_{ij} \in P_i \cap T(z)} f_{ij} d(z, p_{ij})$). As in Section 3.1, the two arrays can be computed in $O(Cn + |T_h|)$ time by traversing T_h.

We continue the above algorithm recursively on $T_h^1(z)$ until after l steps we obtain a subtree $T_h^l(z)$ of at most two connectors, at which moment we have reduced the problem to one of the previous two cases. Clearly, $l = O(\log C)$. For the running time, suppose for each $1 \leq k \leq l$, we refer to the *k-th step* as for determining the subtree $T_h^k(z)$. As discussed above, the first step takes $O(nC + |T_h|)$ time, and similarly, the second step can be done in $O(nC/2 + |T_h|)$ time because there are only at most $C/2+1$ connectors in $T_h^1(z)$. In general, the k-th step can be done in $O(nC/2^{k-1} + |T_h|)$ time for each $1 \leq k \leq l$. Recall that T_h was obtained in the initial pruning step in Section 3.1 and $|T_h| \leq n/2$. Since $l = O(\log C)$ and $C = O(\log m)$, the total running time for obtaining the subtree $T_h^l(z)$ (and thus reducing the problem to the previous two cases) is bounded by $\sum_{k=1}^{l}(nC/2^{k-1} + n/2) = O(nC + nl) = O(n \log m)$. Hence, for the case $C > 2$, within $O(mn)$ time we can also prune at least $n/8$ uncertain points.

The above gives an $O(mn)$ time algorithm that computes a tree T^+ of at most $7n/8$ uncertain points and at most $7mn/8$ vertices, such that the center of T^+ is x^*. We continue the same procedure recursively on T^+ until we obtain a tree T^* with only a constant number of uncertain points (hence T^* has $O(m)$ vertices). The total time is $O(mn)$. Finally we compute x^* on T^* in $O(m)$ time. The algorithm is similar to the scheme of the initial pruning step and we omit it. We conclude that the center x^* on T can be found in $O(|T|) = O(mn)$ time.

Recall that the above only considered the vertex-constrained case. For the general case, the following theorem solves it in $O(mn + |T|)$ time.

Theorem 1. *The center x^* of T and $\mathcal{P}$ can be found in $O(mn + |T|)$ time.*

Proof. Recall that $\mathcal{P}$ consists of n uncertain points, each of which has m locations on T. We reduce the problem to a problem instance of the vertex-constrained case and then apply our algorithm for the vertex-constrained case. More specifically, we modify the tree T to obtain another tree T' of size $O(mn)$. We also compute another set $\mathcal{P}'$ of n uncertain points on T', which correspond to the uncertain points of $\mathcal{P}$ with the same weights, but each uncertain point of $\mathcal{P}'$ has at most $2m$ locations on T' (some locations have zero probabilities). Further, each location of $\mathcal{P}'$ is at a vertex of T' and each vertex of T' holds at least one location of $\mathcal{P}'$. We can construct T' and $\mathcal{P}'$ in $O(mn + |T|)$ time. Finally, given the center x' of T' and $\mathcal{P}'$, we can find x^* on T in $O(mn + |T|)$ time. The details of the above problem reduction are omitted. □

References

1. Agarwal, P., Cheng, S.W., Tao, Y., Yi, K.: Indexing uncertain data. In: Proc. of the 28th Symposium on Principles of Database Systems (PODS), pp. 137–146 (2009)
2. Agarwal, P., Efrat, A., Sankararaman, S., Zhang, W.: Nearest-neighbor searching under uncertainty. In: Proc. of the 31st Symposium on Principles of Database Systems (PODS), pp. 225–236 (2012)
3. Agarwal, P.K., Har-Peled, S., Suri, S., Yıldız, H., Zhang, W.: Convex hulls under uncertainty. In: Schulz, A.S., Wagner, D. (eds.) ESA 2014. LNCS, vol. 8737, pp. 37–48. Springer, Heidelberg (2014)
4. Averbakh, I., Bereg, S.: Facility location problems with uncertainty on the plane. Discrete Optimization 2, 3–34 (2005)
5. de Berg, M., Roeloffzen, M., Speckmann, B.: Kinetic 2-centers in the black-box model. In: Proc. of the 29th Annual Symposium on Computational Geometry (SoCG), pp. 145–154 (2013)
6. Cole, R.: Slowing down sorting networks to obtain faster sorting algorithms. Journal of the ACM 34(1), 200–208 (1987)
7. Foul, A.: A 1-center problem on the plane with uniformly distributed demand points. Operations Research Letters 34(3), 264–268 (2006)
8. Frederickson, G.N.: Parametric search and locating supply centers in trees. In: Dehne, F., Sack, J.-R., Santoro, N. (eds.) WADS 1991. LNCS, vol. 519, pp. 299–319. Springer, Heidelberg (1991)
9. Jørgensen, A., Löffler, M., Phillips, J.M.: Geometric computations on indecisive points. In: Dehne, F., Iacono, J., Sack, J.-R. (eds.) WADS 2011. LNCS, vol. 6844, pp. 536–547. Springer, Heidelberg (2011)
10. Kamousi, P., Chan, T.M., Suri, S.: Closest pair and the post office problem for stochastic points. In: Dehne, F., Iacono, J., Sack, J.-R. (eds.) WADS 2011. LNCS, vol. 6844, pp. 548–559. Springer, Heidelberg (2011)
11. Kamousi, P., Chan, T., Suri, S.: Stochastic minimum spanning trees in euclidean spaces. In: Proc. of the 27th Annual Symposium on Computational Geometry (SoCG), pp. 65–74 (2011)
12. Kariv, O., Hakimi, S.: An algorithmic approach to network location problems. II: The p-medians. SIAM Journal on Applied Mathematics 37(3), 539–560 (1979)
13. Löffler, M., van Kreveld, M.: Largest bounding box, smallest diameter, and related problems on imprecise points. Computational Geometry: Theory and Applications 43(4), 419–433 (2010)
14. Megiddo, N.: Linear-time algorithms for linear programming in R^3 and related problems. SIAM Journal on Computing 12(4), 759–776 (1983)
15. Megiddo, N., Supowit, K.: On the complexity of some common geometric location problems. SIAM Journal on Comuting 13, 182–196 (1984)
16. Megiddo, N., Tamir, A.: New results on the complexity of p-centre problems. SIAM Journal on Computing 12(4), 751–758 (1983)
17. Megiddo, N., Zemel, E.: An $O(n \log n)$ randomizing algorithm for the weighted Euclidean 1-center problem. Journal of Algorithms 7, 358–368 (1986)
18. Suri, S., Verbeek, K.: On the Most Likely Voronoi Diagram and Nearest Neighbor Searching. In: Ahn, H.-K., Shin, C.-S. (eds.) ISAAC 2014. LNCS, vol. 8889, pp. 338–350. Springer, Heidelberg (2014)

19. Suri, S., Verbeek, K., Yıldız, H.: On the Most Likely Convex Hull of Uncertain Points. In: Bodlaender, H.L., Italiano, G.F. (eds.) ESA 2013. LNCS, vol. 8125, pp. 791–802. Springer, Heidelberg (2013)
20. Wang, H.: Minmax regret 1-facility location on uncertain path networks. European Journal of Operational Research **239**, 636–643 (2014)
21. Wang, H., Zhang, J.: One-dimensional k-center on uncertain data. In: Cai, Z., Zelikovsky, A., Bourgeois, A. (eds.) COCOON 2014. LNCS, vol. 8591, pp. 104–115. Springer, Heidelberg (2014)

Swapping Colored Tokens on Graphs

Katsuhisa Yamanaka[1]([✉]), Takashi Horiyama[2], David Kirkpatrick[3],
Yota Otachi[4], Toshiki Saitoh[5], Ryuhei Uehara[4], and Yushi Uno[6]

[1] Iwate University, Morioka, Japan
yamanaka@cis.iwate-u.ac.jp
[2] Saitama University, Saitama, Japan
horiyama@al.ics.saitama-u.ac.jp
[3] University of British Columbia, Vancouver, Canada
kirk@cs.ubc.ca
[4] Japan Advanced Institute of Science and Technology, Nomi, Japan
{otachi,uehara}@jaist.ac.jp
[5] Kobe University, Kobe, Japan
saitoh@eedept.kobe-u.ac.jp
[6] Osaka Prefecture University, Sakai, Japan
uno@mi.s.osakafu-u.ac.jp

Abstract. We investigate the computational complexity of the following problem. We are given a graph in which each vertex has the current and target colors. Each pair of adjacent vertices can swap their current colors. Our goal is to perform the minimum number of swaps so that the current and target colors agree at each vertex. When the colors are chosen from $\{1, 2, \ldots, c\}$, we call this problem c-COLORED TOKEN SWAPPING since the current color of a vertex can be seen as a colored token placed on the vertex. We show that c-COLORED TOKEN SWAPPING is NP-complete for every constant $c \geq 3$ even if input graphs are restricted to connected planar bipartite graphs of maximum degree 3. We then show that 2-COLORED TOKEN SWAPPING can be solved in polynomial time for general graphs and in linear time for trees.

1 Introduction

Sorting problems are fundamental and important in computer science. In this paper, we consider a problem of sorting on graphs. Let $G = (V, E)$ be an undirected unweighted graph with vertex set V and edge set E. Suppose that each vertex in G has a color in $C = \{1, 2, \ldots, c\}$. A token is placed on each vertex in G, and each token also has a color in C. Then, we wish to transform the current token-placement into the one such that a token of color i is placed on a vertex of color i for all vertices by swapping tokens on adjacent vertices in G. See Fig.1 for an example. If there exists a color i such that the number of vertices of color i is not equal to the number of tokens of color i in the current token-placement, then we cannot transform the current token-placement into the target one. Thus, without loss of generality, we assume that the number of vertices of color i for each $i = 1, 2, \ldots, c$ is equal to the number of tokens of the same color. As we

© Springer International Publishing Switzerland 2015
F. Dehne et al. (Eds.): WADS 2015, LNCS 9214, pp. 619–628, 2015.
DOI: 10.1007/978-3-319-21840-3_51

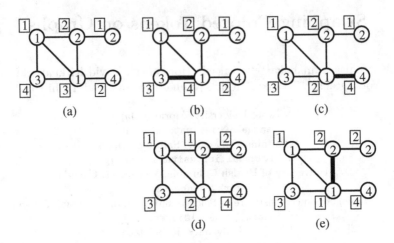

Fig. 1. An example of 4-COLORED TOKEN SWAPPING. Colors of vertices are written inside circles and tokens are drawn as rectangles with their colors. We swap the two tokens along each thick edge. (a) An initial token-placement. (b)–(d) Intermediate token-placements. (e) The target token-placement.

will see in the next section, any token-placement can be transformed into the target one by $O(n^2)$ token-swappings, where n is the number of vertices in G. We thus consider the problem of minimizing the number of token-swappings to obtain the target token-placement.

If vertices have distinct colors and tokens also have distinct colors, then the problem is called TOKEN SWAPPING [11]. This has been investigated for several graph classes. TOKEN SWAPPING can be solved in polynomial time for paths [7, 8], cycles [7], stars [10], complete graphs [1,7], and complete bipartite graphs [11]. Heath and Vergara [6] gave a polynomial-time 2-approximation algorithm for squares of paths, where the square of a path is the graph obtained from the path by adding a new edge between two vertices with distance exactly two in the path. For squares of paths, some upper bounds of the minimum number of token-swappings are known [3,4,6]. Yamanaka et al. [11] gave a polynomial-time 2-approximation algorithm for trees. TOKEN SWAPPING is solved for only restricted graph classes. However no hardness result is known, even if input graphs are general graphs, to the best of our knowledge.

The c-COLORED TOKEN SWAPPING problem is a generalization of TOKEN SWAPPING. We investigate c-COLORED TOKEN SWAPPING and clarify its computational complexity in the sense that we found the boundary of easy and hard cases with respect to the number of colors. For $c = 2$, the problem can be solved in polynomial time for general graphs and in linear time for trees. However, the problem for $c = 3$ is hard even if input graphs are quite restricted. We show that the problem is NP-complete for connected planar bipartite graphs with maximum degree 3.

2   Preliminaries

In this paper, we assume without loss of generality that graphs are simple and connected. Let $G = (V, E)$ be an undirected unweighted graph with vertex set V and edge set E. We sometimes denote by $V(G)$ and $E(G)$ the vertex set and the edge set of G, respectively. We always denote $|V|$ by n. For a vertex v in G, let $N(v)$ be the set of all neighbors of v. Each vertex of a graph G has a color in $C = \{1, 2, \ldots, c\}$. We denote by $c(v)$ the color of a vertex $v \in V$. A token is placed on each vertex in G, and each token also has a color in C. For a vertex v, we denote by $f(v)$ the color of the token placed on v. Then, we call the function $f : V \rightarrow C$ a *token-placement* of G. Two token-placements f and f' of G are said to be *adjacent* if the following two conditions (a) and (b) hold:

(a) there exists exactly one edge $(u, v) \in E$ such that $f'(u) = f(v)$ and $f'(v) = f(u)$; and

(b) $f'(w) = f(w)$ for all vertices $w \in V \setminus \{u, v\}$.

In other words, the token-placement f' is obtained from f by *swapping* the tokens on the two adjacent vertices u and v. Note that swapping two tokens of the same color gives the same token-placement. Thus, to eliminate redundancy, we assume that tokens of the same color are never swapped. For two token-placements f and f' of G, a sequence $\mathcal{S} = \langle f_0, f_1, \ldots, f_h \rangle$ of token-placements is a *swapping sequence* between f and f' if the following three conditions (1)–(3) hold:

(1) $f_0 = f$ and $f_h = f'$;

(2) f_k is a token-placement of G for each $k = 0, 1, \ldots, h$; and

(3) f_{k-1} and f_k are adjacent for every $k = 1, 2, \ldots, h$.

The *length* of a swapping sequence $\mathcal{S}$, denoted by $\text{len}(\mathcal{S})$, is defined to be the number of token-placements in $\mathcal{S}$ minus one, that is, $\text{len}(\mathcal{S})$ indicates the number of token swappings in $\mathcal{S}$. For two token-placements f and f' of G, we denote by $\text{OPT}(f, f')$ the minimum length of a swapping sequence between f and f'. As we will prove in Lemma 1, there always exists a swapping sequence between any two token-placements f and f' if the number of vertices of color i for each $i = 1, 2, \ldots, c$ is equal to the number of tokens of the same color. For the two token-placement f and f', $\text{OPT}(f, f')$ is well-defined.

Given two token-placements f and f' of a graph G and a nonnegative integer ℓ, the c-COLORED TOKEN SWAPPING problem is to determine whether or not $\text{OPT}(f, f') \leq \ell$ holds. From now on, we always denote by f and f' the *initial* and *target* token-placements of G, respectively, and we may assume without loss of generality that f' is a token-placement of G such that $f'(v) = c(v)$ for all vertices $v \in V$.

We show that the length of any swapping sequence need never exceed n^2. This claim is derived by slightly modifying the proof of Theorem 1 in [11].

Lemma 1. *For any pair of token-placements f and f' of a graph G, $OPT(f, f') \leq n^2$.*

Proof. Let T be any spanning tree of a graph G. Choose an arbitrary leaf v of T. Then, we move a nearest token of color $c(v)$ in T from the current position

u to its target position v. Note that there is no token of color $c(v)$ placed on a vertex of the path in T from u to v except u. Let $(p_1, p_2, \ldots, p_q)$ be a unique path in T from $p_1 = u$ to $p_q = v$. Then, we swap the tokens on p_k and p_{k+1} for each $k = 1, 2, \ldots, q-1$ in this order, and obtain the token-placement f of G such that $f(v) = c(v)$. We then delete the vertex v from G and T, and repeat the process until we obtain f'.

Each vertex obtains a token of the same color via a swapping sub-sequence of length in n. Therefore, the swapping sequence S above between f and f' satisfies $\text{len}(S) \leq n^2$. Since $\text{OPT}(f, f') \leq \text{len}(S)$, we have $\text{OPT}(f, f') \leq n^2$. $\square$

From Lemma 1, any token-placement for an input graph can be transformed into the target one by $O(n^2)$ token-swappings, and a swapping sequence of length $O(n^2)$ can be computed in polynomial time.

3   Hardness Results

In this section, we show that c-COLORED TOKEN SWAPPING problem is NP-complete for any constant $c \geq 3$ by constructing a polynomial-time reduction from PLANAR 3DM [2]. To define PLANAR 3DM, we first introduce the following well-known NP-complete problem.

Problem: 3-DIMENSIONAL MATCHING (3DM) [5, SP1]
Instance: Set $T \subseteq X \times Y \times Z$, where X, Y, and Z are disjoint sets having the same number m of elements.
Question: Does T contain a matching, i.e., a subset $T' \subseteq T$ such that $|T'| = m$ and it contains all elements of X, Y, and Z?

PLANAR 3DM is a restricted version of 3DM in which the following bipartite graph G is planar. The graph G has the vertex set $V(G) = T \cup X \cup Y \cup Z$ with a bipartition $(T, X \cup Y \cup Z)$. Two vertices $t \in T$ and $w \in X \cup Y \cup Z$ are adjacent in G if and only if $w \in t$. PLANAR 3DM is NP-complete even if G is a connected graph of maximum degree 3 [2].

Theorem 1. 3-COLORED TOKEN SWAPPING *is NP-complete even for connected planar bipartite graphs of maximum degree 3.*

Proof. By Lemma 1, there is a polynomial-length swapping sequence for any initial token-placement, and thus 3-COLORED TOKEN SWAPPING is in NP.

Now we present a reduction from PLANAR 3DM. Let $(X, Y, Z; T)$ be an instance of PLANAR 3DM and $m = |X| = |Y| = |Z|$. As mentioned above, we construct a bipartite graph $G = (T, X \cup Y \cup Z; E)$ from $(X, Y, Z; T)$. We set $c(x) = 1$ and $f(x) = 2$ for every $x \in X$, set $c(y) = 2$ and $f(y) = 3$ for every $y \in Y$, set $c(z) = 3$ and $f(z) = 1$ for every $z \in Z$, and set $c(t) = 1$ and $f(t) = 1$ for every $t \in T$. See Fig.2. From the assumptions, G is a planar bipartite graph of maximum degree 3. The reduction can be done in polynomial time. We prove that the instance $(X, Y, Z; T)$ is a yes-instance if and only if $\text{OPT}(f, f') \leq 3m$.

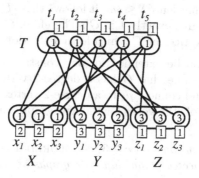

Fig. 2. The graph constructed from an instance $(X = \{x_1, x_2, x_3\}$, $Y = \{y_1, y_2, y_3\}$, $Z = \{z_1, z_2, z_3\}$, $T = \{t_1 = (x_1, y_1, z_3), t_2 = (x_3, y_2, z_1), t_3 = (x_1, y_1, z_2), t_4 = (x_3, y_3, z_2), t_5 = (x_2, y_2, z_1)\})$

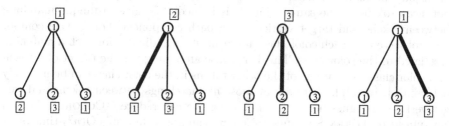

Fig. 3. A swapping sequence to resolve the token-placement of a triple

To show the only-if part, assume that there exists a subset T' of T such that $|T'| = m$ and T' contains all elements of X, Y, and Z. Since the elements of T' are pairwise disjoint, we can cover the subgraph of G induced by $T' \cup X \cup Y \cup Z$ with m disjoint stars of four vertices, where each star is induced by an element t of T' and its three elements. To locally move the tokens on the target place in such a star, we need only three swappings. See Fig.3. This implies that a swapping sequence of length $3m$ exists.

To show the if part, assume that there is a swapping sequence S from f to f' with at most $3m$ token-swappings. Let $T' \subseteq T$ be the set of vertices such that the tokens on them are moved in S. Let G' be the subgraph of G induced by $T' \cup X \cup Y \cup Z$. Let $w \in X \cup Y \cup Z$. Since $c(w) \neq f(w)$ and $N(w) \subseteq T$, the sequence S swaps the tokens on w and on a neighbor $t \in T'$ of w at least once. This implies that w has degree at least 1 in G'. Since each $t \in T'$ has degree at most 3 in G', we can conclude that $|T'| \geq \frac{1}{3}|X \cup Y \cup Z| = m$. In S, the token placed on a vertex in $X \cup Y$ in the initial token-placement is moved at least twice, while the token placed on a vertex in $Z \cup T'$ is moved at least once. As a token-swapping moves two tokens at the same time,

$$\text{len}(S) \geq \frac{1}{2}(2|X| + 2|Y| + |Z| + |T'|) \geq 3m.$$

From the assumption that $\text{len}(\mathcal{S}) \leq 3m$, it follows that $|T'| = m$, and hence each $w \in X \cup Y \cup Z$ has degree exactly 1 in G'. Therefore, G' consists of m disjoint stars centered at the vertices of T' which form a solution of PLANAR 3DM.   □

The proof above can be extended for any constant number of colors. It is known that we can assume that G has a degree-2 vertex [2]. We add a path $(p_4, p_5, \ldots, p_c)$ to G, and connect p_4 to a degree-2 vertex in G. We set $c(p_i) = i$ and $f(p_i) = i$. The proof still works for the new graph, and hence we obtain the following corollary.

Corollary 1. *For every constant $c \geq 3$, c-COLORED TOKEN SWAPPING is NP-complete even for connected planar bipartite graphs of maximum degree 3.*

Note that the degree bound in the corollary above is tight. If a graph has maximum degree 2, then we can solve c-COLORED TOKEN SWAPPING in polynomial time for every constant c as follows. A graph of maximum degree 2 consists of disjoint paths and cycles. Observe that a shortest swapping sequence does not swap tokens of the same color. This immediately gives a unique matching between tokens and target vertices for a path component. For a cycle component, observe that each color class has at most n candidates for such a matching restricted to the color class. This is because after we guess the target of a token in a color class, the targets of the other tokens in the color class can be uniquely determined. In total, there are at most n^c matchings between tokens and target vertices. By guessing such a matching, we can reduce c-COLORED TOKEN SWAPPING to TOKEN SWAPPING. Now we can apply Jerrum's $O(n^2)$-time algorithms for solving TOKEN SWAPPING on paths and cycles [7]. Therefore, we can solve c-COLORED TOKEN SWAPPING in $O(n^{c+2})$ time for graphs of maximum degree 2.

Theorem 2. *For every constant $c \geq 1$, c-COLORED TOKEN SWAPPING is solvable in polynomial time for graphs of maximum degree 2.*

4   Polynomial-Time Algorithms

In this section, we give some positive results. We first show that 2-COLORED TOKEN SWAPPING for general graphs can be solved in polynomial time. We next show that 2-COLORED TOKEN SWAPPING problem for trees can be solved in linear time without constructing a swapping sequence.

4.1   General Graphs

Let $C = \{1, 2\}$ be the color set. Let $G = (V, E)$ be a graph, and let f and f' be an initial token-placement and the target token-placement. We construct a weighted complete bipartite graph $G_B = (X, Y, E_B, w)$, as follows. The vertex sets X, Y and the edge set E_B are defined as follows:

$$X = \{x_v \mid v \in V \text{ and } f(v) = 1\}$$
$$Y = \{y_v \mid v \in V \text{ and } c(v) = 1\}$$
$$E_B = \{(x, y) \mid x \in X \text{ and } y \in Y\}.$$

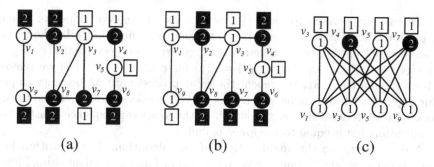

Fig. 4. (a) An initial token-placement. (b) The target token-placement. (c) The weighted complete bipartite graph constructed from (a) and (b) (the weight of each edge is omitted).

Intuitively, X is the copies of vertices in V having tokens of color 1, and Y is the copies of vertices in V of color 1. The weight function w is a mapping from E_B to positive integers. For $x \in X$ and $y \in Y$, the weight $w(e)$ of the edge $e = (x, y)$ is defined as the length of a shortest path from x to y in G. Fig.4 gives an example of an initial token-placement, the target token-placement, and the associated weighted complete bipartite graph.

We bound $\mathrm{OPT}(f, f')$ from below, as follows. Let $\mathcal{S}$ be a swapping sequence between f and f'. The swapping sequence gives a perfect matching of G_B, as follows. For each token of color 1, we choose an edge (x, y) of G_B if the token is placed on $x \in X$ in f and on $y \in Y$ in f'. The obtained set is a perfect matching of G_B. A token corresponding to an edge e in the matching needs $w(e)$ token-swappings, and two tokens of color 1 are never swapped in $\mathcal{S}$. Therefore, for a minimum weight matching M of G_B, we have the following lower bound:

$$\mathrm{OPT}(f, f') \geq \sum_{e \in M} w(e).$$

Now we describe our algorithm. First we find a minimum weight perfect matching M of G_B. We choose an edge e in M. Let $P_e = \langle p_1, p_2, \ldots, p_q \rangle$ of G be a shortest path corresponding to e. We have the following lemma.

Lemma 2. *Suppose that the two tokens on endpoints of P_e have different colors. The two tokens can be swapped by $w(e)$ token-swappings such that the color of the token on each internal vertex does not change.*

Proof. Without loss of generality, we assume that $f(p_1) = 2$ and $f(p_q) = 1$ hold. We first choose the minimum i such that $f(p_i) = 1$ holds. We next move the token on p_i to p_1 by $i - 1$ token-swappings. We repeat the same process to the subpath $\langle p_i, p_{i+1}, \ldots, p_q \rangle$. Finally, we obtain the desired token-placement. Recall that there are only two colors on graphs, and so the above "color shift" operation works. Since each edge of P_e is used by one token-swapping, the total number of token-swapping is $w(e) = q - 1$. □

This lemma permits to move the two tokens on the two endpoints p_1 and p_q of P_e to their target positions in $w(e)$ token-swappings. Let g be the token-placement obtained after the token-swappings. We can observe that $f(v) = g(v)$ for every $v \in V \setminus \{p_1, p_q\}$ and $g(v) = c(v)$ for $v \in \{p_1, p_q\}$. Then we remove e from the matching M. We repeat the same process until M becomes empty. Our algorithm always exchanges tokens on two vertices using a shortest path between the vertices. Hence, the length of the swapping sequence constructed by our algorithm is equal to the lower bound.

Now we estimate the running time of our algorithm. The algorithm first constructs the weighted complete bipartite graph. This can be done using Floyd-Warshall algorithm in $O(n^3)$ time. Then, our algorithm constructs a minimum weight perfect matching. This can be done in $O(n^3)$ time [9, p.252]. Finally, for each of the $O(n)$ paths in the matching, our algorithm moves the tokens on the endpoints of the path in linear time. We have the following theorem.

Theorem 3. 2-COLORED TOKEN SWAPPING *is solvable in* $O(n^3)$ *time. Furthermore, a swapping sequence of the minimum length can be constructed in the same running time.*

4.2   Trees

In this subsection, we show that 2-COLORED TOKEN SWAPPING for trees can be solved in linear time without constructing a swapping sequence.

Let T be an input tree, and let f and f' be an initial token-placement and the target token-placement of T. Let $e = (x, y)$ be an edge of T. Removal of e disconnects T into the two subtrees. We denote by $T(x)$ the subtree containing x and denote by $T(y)$ the subtree containing y.

Now we define the value $\mathrm{diff}(e)$ for each edge e of T. Intuitively, $\mathrm{diff}(e)$ is the number of tokens of color 1 which we wish to move from $T(x)$ to $T(y)$ along e. More formally, we give the definition of $\mathrm{diff}(e)$, as follows. Let n_t^1 be the number of tokens of color 1 in $T(x)$, and let n_v^1 be the number of vertices of color 1 in $T(x)$. Then, we define $\mathrm{diff}(e) = |n_t^1 - n_v^1|$. (Note that, even if we count tokens and vertices of color 1 in $T(y)$ instead of $T(x)$, the value $\mathrm{diff}(e)$ takes the same value.) See Fig.5 for an example. For each edge e of T, we need to move at least $\mathrm{diff}(e)$ tokens of color 1 from a subtree to the other one along e. Therefore, $\mathrm{OPT}(f, f')$ is lower bounded by the sum $D = \sum_{e \in E(T)} \mathrm{diff}(e)$.

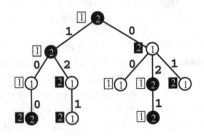

Fig. 5. An example of $\mathrm{diff}(e)$. For each edge e, the value $\mathrm{diff}(e)$ is written beside e.

To give an upper bound of $\text{OPT}(f, f')$, we next show that there exists a swapping sequence of length D. The following lemma is key to construct the swapping sequence.

Lemma 3. *If $D \neq 0$, then there exists an edge e such that the token-swapping on e decreases D by one.*

Proof. We first give an orientation of edges of T. For each edge $e = (x, y)$, we orient e from x to y if the number of tokens of color 1 in $T(x)$ is greater than the number of vertices of color 1 in $T(x)$. Intuitively, the direction of an edge means that we need to move one or more tokens of color 1 from $T(x)$ to $T(y)$. If the two numbers are equal, we remove e from T. Let T' be the obtained directed tree. For an edge $e = (x, y)$ oriented from x to y, if x has a token of color 1 and y has a token of color 2, swapping the two token decreases D by one. We call such an edge a *desired* edge. We now show that there exists a desired edge in T'. Observe that if no vertex u with $f(u) = 1$ is incident to a directed edge in T', then indeed T' has no edge and $D = 0$. Let u be a vertex with $f(u) = 1$ that has at least one incident edge in T'. If u has no out-going edge, then the number of the color-1 tokens in T exceeds the number of the color-1 vertices in T. Thus we can choose an edge (u, v) oriented from u to v. If $f(v) = 2$ holds, the edge is desired. Now we assume that $f(v) = 1$ holds. We apply the same process for v, then an edge (v, w) oriented from v to w can be found. Since trees have no cycle, by repeating the process, we always find a desired edge. $\square$

From Lemma 3, we can find a desired edge, and we swap the two tokens on the endpoints of the edge. Since a token-swapping on a desired edge decreases D by one, by repeatedly swapping on desired edges, we obtain the swapping sequence of length D. Note that $D = 0$ if and only if $c(v) = f(v)$ for every $v \in V(T)$. Hence, we have $\text{OPT}(f, f') \leq D$.

Therefore $\text{OPT}(f, f') = D$ holds, and so we can solve 2-COLORED TOKEN SWAPPING by calculating D. The value $\text{diff}(e)$ for every edge e, and thus the value D, can be calculated in a bottom-up manner in linear time in total. We have the following theorem.

Theorem 4. 2-COLORED TOKEN SWAPPING *is solvable in linear time for trees.*

5   Conclusions

We have investigated computational complexity of c-COLORED TOKEN SWAPPING. We first showed the NP-completeness for 3-COLORED TOKEN SWAPPING by a reduction from PLANAR 3DM, even for connected planar bipartite graphs of maximum degree 3. We next showed that 2-COLORED TOKEN SWAPPING can be solved in $O(n^3)$ time for general graphs and in linear time for trees.

We showed that c-COLORED TOKEN SWAPPING for every constant c can be solved in polynomial time for graphs of maximum degree 2 (disjoint paths and cycles). If c is not a constant, can we solve c-COLORED TOKEN SWAPPING for such graphs in polynomial time? For TOKEN SWAPPING on cycles, Jerrum [7] proposed an $O(n^2)$-time algorithm. As mentioned in [7], the proof of the correctness of the algorithm needs complex discussions.

Acknowledgments. This work is partially supported by MEXT/JSPS KAKENHI, including the ELC project (Grant Numbers 15H00853, 24106004, 24106007, 24700130, 25330001, 25730003, and 26330009), and the National Science and Engineering Research Council of Canada.

References

1. Cayley, A.: Note on the theory of permutations. Philosophical Magazine **34**, 527–529 (1849)
2. Dyer, M.E., Frieze, A.M.: Planar 3DM is NP-complete. Journal of Algorithms **7**, 174–184 (1986)
3. Feng, X., Meng, Z., Sudborough, I.H.: Improved upper bound for sorting by short swaps. In: Proc. IEEE Symposium on Parallel Architectures, Algorithms and Networks, pp. 98–103 (2004)
4. Feng, X., Sudborough, I.H., Lu, E.: A fast algorithm for sorting by short swap. In: Proc. IASTED International Conference Computational and Systems Biology, pp. 62–67 (2006)
5. Garey, M.R., Johnson, D.S.: Computers and Intractability: A Guide to the Theory of NP-Completeness. Freeman (1979)
6. Heath, L.S., Vergara, J.P.C.: Sorting by short swaps. Journal of Computational Biology **10**(5), 775–789 (2003)
7. Jerrum, M.R.: The complexity of finding minimum-length generator sequence. Theoretical Computer Science **36**, 265–289 (1985)
8. Knuth, D.E.: The art of computer programming, 2nd edn., vol. 3. Addison-Wesley (1998)
9. Korte, B., Vygen, J.: Combinatorial Optimization: Theory and Algorithms, 2nd edn. Springer (2005)
10. Pak, I.: Reduced decompositions of permutations in terms of star transpositions, generalized catalan numbers and k-ary trees. Discrete Mathematics **204**, 329–335 (1999)
11. Yamanaka, K., Demaine, E.D., Ito, T., Kawahara, J., Kiyomi, M., Okamoto, Y., Saitoh, T., Suzuki, A., Uchizawa, K., Uno, T.: Swapping labeled tokens on graphs. Theoretical Computer Science (2015) (accepted)

Positive Semidefinite Zero Forcing: Complexity and Lower Bounds

Boting Yang[(✉)]

Department of Computer Science, University of Regina, Regina, Canada
`boting.yang@uregina.ca`

Abstract. The positive semidefinite zero forcing number of a graph is a parameter that is important in the study of minimum rank problems. In this paper, we focus on the algorithmic aspects of computing this parameter. We prove that it is NP-complete to find the positive semidefinite zero forcing number of a given graph, and this problem remains NP-complete even for graphs with maximum vertex degree 7. We present a linear time algorithm for computing the positive semidefinite zero forcing number of generalized series-parallel graphs. We introduce the constrained tree cover number and apply it to improve lower bounds for positive semidefinite zero forcing. We also give formulas for the constrained tree cover number and the tree cover number on graphs with special structures.

1 Introduction

The zero forcing number was introduced in [1]. The interest in this parameter is to apply zero forcing as upper bounds on the maximum nullities (or, equivalently, the minimum ranks) of certain symmetric matrices associated with graphs. Independently, this parameter has been studied in physics, referring to it as the graph infection number [7], and it has also been studied in the field of graph searching within the context of fast-mixed search number [20].

The most widely-studied variant of the zero forcing, called the positive semidefinite zero forcing (PSZF for short), was introduced in [3] (see also [9–11] for recent results). Similar to the zero forcing number, a primary reason to study the PSZF number of a graph is its relationship to the maximum positive semidefinite nullity of certain positive semidefinite symmetric matrices associated with the graph.

Let G be a finite graph with no loops or multiple edges. In the zero forcing process or the positive semidefinite zero forcing process, we begin by specifying a set of vertices of G that are colored black initially, while all other vertices are colored white. Then, using a color change rule to these vertices, we progressively change the color of white vertices in the graph to black. Our objective is to color all vertices black by repeated application of the color change rule. In general,

---

Research supported in part by an NSERC Discovery Research Grant, Application No.: RGPIN-2013-261290.

F. Dehne et al. (Eds.): WADS 2015, LNCS 9214, pp. 629–639, 2015.
DOI: 10.1007/978-3-319-21840-3_52

we want to determine the smallest set of vertices needed to be black initially, to eventually change all of the vertices in the graph to black.

The zero forcing (resp. positive semidefinite zero forcing) rule results in a partition of the vertices of G into sets, so that each such set induces a path (resp. tree) in G.

As mentioned above, one of the original motivations for studying these parameters is that they provide upper bounds on the maximum nullities of symmetric or positive semidefinite matrices associated with graphs (see [2,3]). Let $\mathcal{S}_n(\mathbb{R})$ denote the set of $n \times n$ real symmetric matrices. The *graph of* $A = [a_{ij}] \in \mathcal{S}_n(\mathbb{R})$, denoted by $\mathcal{G}(A)$, is the graph with vertex set $\{1, 2, \ldots, n\}$ and edge set $\{\{i, j\} : a_{ij} \neq 0, 1 \leq i < j \leq n\}$. For a given graph G, the set of symmetric matrices described by G is defined to be $\mathcal{S}(G) = \{A \in \mathcal{S}_n(\mathbb{R}) : \mathcal{G}(A) = G\}$. Let $\mathcal{S}_+(G)$ denote the subset of positive semidefinite matrices in $\mathcal{S}(G)$. For a matrix B, we use null(B) to denote its nullity. The *maximum positive semidefinite nullity* of G is defined to be $\mathrm{M}_+(G) = \max\{\mathrm{null}(B) : B \in \mathcal{S}_+(G)\}$.

Another motivation for investigating the zero forcing number and the positive semidefinite zero forcing number are that they provide parameters to measure some graph structures, just like pathwidth and treewidth [16,17]. At its most basic level, edge search and node search models represent two significant graph search problems [13,15], where the node search number of a graph G is equal to the pathwidth of G plus one and the edge search number is at least the pathwidth and is at most the pathwidth plus two. Bienstock and Seymour [5] introduced the mixed search problem that combines the edge search and the node search problems. The mixed search number of G is at least the pathwidth of G and is at most the pathwidth plus one [13,21]. Dyer et al. [8] introduced the fast search problem, and Yang [20] introduced the fast-mixed search model that is a combination of the fast search and mixed search models [20]. Note that the fast-mixed search number of G is equal to the zero forcing number of G [11]. Since the fast-mixed searching is a variant of the mixed searching, we know that the zero forcing number can be considered as a variant of pathwidth (or the mixed search number). From the relation of zero forcing to positive semidefinite zero forcing and the relation of pathwidth to treewidth, the positive semidefinite zero forcing number can be considered as a variant of treewidth, which is one of the most important parameters in structural graph theory and it plays an important role in graph algorithms.

2   Preliminaries

Throughout this paper, we only consider finite graphs with no loops or multiple edges. We use $G = (V, E)$ to denote a graph with vertex set V and edge set E, and we also use $V(G)$ and $E(G)$ to denote the vertex set and edge set of G respectively. For $V' \subseteq V$, the vertex set $\{u : uv \in E, u \in V \setminus V'$ and $v \in V'\}$ is the *neighborhood* of V', denoted as $N_G(V')$. We use $G[V']$ to denote the subgraph induced by V', which consists of all vertices of V' and all of the edges that connect vertices of V' in G. We use $G - V'$ to denote the induced subgraph $G[V \setminus V']$. Definitions omitted here can be found in [19].

The *positive semidefinite zero forcing*, or simply *PSZF*, is based on a color change rule. Let G be a graph in which every vertex is initially colored either black or white. Let B be the set of black vertices and $W_1, \ldots, W_k$ be the sets of white vertices in each of the connected components of $G - B$ (note that it is possible that $k = 1$). The *PSZF color change rule* is: If v is a black vertex in B and w is the only white neighbor of v in the graph $G[W_i \cup B]$, than change the color of w to black; in this case we say "v forces w" and write $v \to w$. Given an initial coloring of G, in which a set of the vertices is black and all other vertices are white, if all white vertices are forced to black after repeatedly applying the PSZF color change rule, then the set of initial black vertices is called a *PSZF set*; it is called a *minimum PSZF set* if the initial set of black vertices is a PSZF set of the smallest possible size. The *PSZF number* of a graph G, denoted by $Z_+(G)$, is the size of the smallest PSZF set of G. The procedure of coloring a graph using the PSZF color change rule is called a *PSZF process*. A PSZF process is called *optimal* if the initial set of black vertices is a minimum PSZF set.

The zero forcing number, denoted $Z(G)$, has a different color change rule. Let B be the set of black vertices in G and all other vertices are white. The *zero forcing color change rule* is: If v is a black vertex in B and w is the only white neighbor of v in $G - B$, then change the color of w to black.

If B is a PSZF set of a graph G, then we can produce a set of rooted trees that give us the order in which forces are performed in the PSZF process. When we apply the PSZF color change rule once, a black vertex in B can force multiple white vertices from different connected components W_i to black at the same time. Then we update the black vertex set and apply the PSZF color change rule again. We repeat this process until all vertices are black. So, for each initial black vertex v, the forces determine an induced rooted-tree T, referred to as a *forcing tree*. The root of the forcing tree T is the vertex v and a vertex w is on T if and only if there is a sequence of vertices $(v_1, \ldots, v_k)$, where $v = v_1$ and $v_k = w$, on T such that $v_i \to v_{i+1}$, for $i = 1, \ldots, k-1$. In this case we use $v \Rightarrow T$ to denote that the vertex v produces the forcing tree T in the PSZF process. If B is a PSZF set of G, then the set of rooted-trees $\{T : v \in B, v \Rightarrow T\}$ is called a *PSZF tree cover* of G; furthermore, if B is a minimum PSZF set of G, then this set is called a *minimum PSZF tree cover* of G. Observe that every vertex not in the PSZF set B is forced by exactly one vertex. Thus $\{T : v \in B, v \Rightarrow T\}$ is a set of vertex-disjoint induced rooted-trees that partition $V(G)$. In a PSZF tree cover, we will use terms "rooted-tree" and "forcing tree" interchangeably if there is no ambiguity.

Given a PSZF tree cover F of G, a *forcing chain* is a path $v_1 \ldots v_k$ on a forcing tree in F satisfying $v_i \to v_{i+1}$, $1 \le i \le k - 1$. Such a forcing chain is denoted by $v_1 \to \cdots \to v_k$.

More generally, a *tree cover* of a graph G is a family of vertex-disjoint induced trees in G that cover all vertices of G [4]. The minimum number of such trees is called the *tree cover number* of G and is denoted by tc(G). A *minimum tree cover* of G is a tree cover of G whose cardinality equals to tc(G). Barioli et al.

[4] showed that for any outerplanar graph G, $M_+(G) = \mathrm{tc}(G)$; and Ekstrand et al. [9] showed that for any graph G, $\mathrm{tc}(G) \leq Z_+(G)$.

We will consider series-parallel graphs and generalized series-parallel graphs. The class of *two-terminal series-parallel graphs* are defined inductively as follows:

1. The graph with the single edge st is a two-terminal series-parallel graph, where the distinguished vertices s and t are *terminals*.
2. If G_1 is a two-terminal series-parallel graph with terminals s_1 and t_1, and G_2 is a two-terminal series-parallel graph with terminals s_2 and t_2, then
   (a) Create a new two-terminal series-parallel graph G by identifying t_1 with s_2. Define s_1 and t_2 as two terminals of G. This is known as the *series composition* of G_1 and G_2, denoted by $G_1 \oplus G_2$.
   (b) Create a new two-terminal series-parallel graph G by identifying $s = s_1 = s_2$ and $t = t_1 = t_2$. Define s and t as two terminals of G. This is known as the *parallel composition* of G_1 and G_2, denoted by $G_1\|G_2$.

A graph G is *series-parallel* if there is a pair of vertices s and t on G such that G is a two-terminal series-parallel graph with terminals $\{s, t\}$.

A connected graph is called *biconnected* if the graph remains connected after the deletion of any vertex. A biconnected component of a graph is a maximal biconnected subgraph. Note that an edge could be a biconnected component. A vertex whose removal produces a graph with more connected components is called a *cut vertex*.

A connected graph is called a *generalized series-parallel graph* if all its biconnected components are two-terminal series-parallel graphs such that any two such series-parallel graphs can only share a vertex that is a terminal of both of them.

3   The Complexity of PSZF

In [20], Yang proved that finding the fast-mixed search number of a given graph is NP-complete. Since the fast-mixed search number of a graph is equal to its zero forcing number, this NP-completeness result is also held for the zero forcing problem [11]. In this section we investigate the complexity of finding PSZF numbers. The decision version of this problem is as follows.

POSITIVE ZERO FORCING
Instance: A graph G and a positive integer k.
Question: Is $Z_+(G) \leq k$?

We can prove the NP-completeness of this problem by a reduction from the following problem [14] (The proof is omitted due to space constraints).

POSITIVE NOT-ALL-EQUAL 3SAT
Instance: A boolean formula ϕ in 3-CNF such that no clause contains a negated literal.
Question: Is there a satisfying assignment for ϕ such that each clause has at least one true literal and at least one false literal?

Theorem 1. POSITIVE ZERO FORCING *is NP-complete. The problem remains NP-complete even for graphs with maximum vertex degree* 7.

4 Series-Parallel Graphs

In this section we will establish properties of PSZF numbers on series-parallel graphs. As mentioned in Section 2, we only consider graphs with no multiple edges or loops.

Definition 1. (Terminal-unlinked, -semi-linked, and -linked.) A series-parallel graph G with terminals $\{s, t\}$ is called *terminal-unlinked* if there is a minimum PSZF tree cover of G which contains two different rooted-trees T_1 and T_2 such that $s \Rightarrow T_1$, $t \Rightarrow T_2$ and $N_G(V(T_1)) \cap V(T_2) = \emptyset$; G is called *terminal-semi-linked* if G is not terminal-unlinked and there is a minimum PSZF tree cover of G which contains two different rooted-trees T_1 and T_2 such that $s \Rightarrow T_1$, and $t \Rightarrow T_2$; and G is called *terminal-linked* if every minimum PSZF tree cover of G contains a rooted-tree T such that $s \Rightarrow T$ and $t \in V(T)$ or $t \Rightarrow T$ and $s \in V(T)$.

Lemma 1. *Let G be a terminal-linked series-parallel graph with terminals $\{s, t\}$. Let F^* be a minimum PSZF tree cover of G that contains a rooted-tree T such that $t \Rightarrow T$ and $s \in V(T)$. If we change the root of T from t to s, then the new F^* is still a minimum PSZF tree cover of G that contains the modified rooted-tree T such that $s \Rightarrow T$ and $t \in V(T)$.*

Theorem 2. *Let G_1 and G_2 be two series-parallel graphs with terminals $\{s_1, t_1\}$ and $\{s_2, t_2\}$, respectively, and let $G = G_1 \| G_2$ with terminals $s(= s_1 = s_2)$ and $t(= t_1 = t_2)$.*

  (i) *If both G_1 and G_2 are terminal-unlinked, then G is terminal-unlinked and $Z_+(G) = Z_+(G_1) + Z_+(G_2) - 2$.*
  (ii) *If one of G_1 and G_2 is terminal-linked and the other is terminal-semi-linked, then G is terminal-semi-linked and $Z_+(G) = Z_+(G_1) + Z_+(G_2) - 1$.*
  (iii) *If one of G_1 and G_2 is terminal-linked and the other is terminal-unlinked, then G is terminal-linked and $Z_+(G) = Z_+(G_1) + Z_+(G_2) - 2$.*
  (iv) *If both G_1 and G_2 are terminal-linked, then G is terminal-semi-linked and $Z_+(G) = Z_+(G_1) + Z_+(G_2)$.*
  (v) *If one of G_1 and G_2 is terminal-unlinked and the other is terminal-semi-linked, then G is terminal-semi-linked and $Z_+(G) = Z_+(G_1) + Z_+(G_2) - 2$.*
  (vi) *If both G_1 and G_2 are terminal-semi-linked, then G is terminal-semi-linked and $Z_+(G) = Z_+(G_1) + Z_+(G_2) - 2$.*

When a group of series-parallel graphs share two terminals, we have the following results.

Corollary 1. *Let G_i, $1 \leq i \leq k$, be series-parallel graphs with terminals $\{s_i, t_i\}$, and let $G = G_1 \| \cdots \| G_k$ with terminals $s(= s_1 = \cdots = s_k)$ and $t(= t_1 = \cdots = t_k)$.*

(i) *If all G_i, $1 \leq i \leq k$, are terminal-unlinked, then G is terminal-unlinked and $Z_+(G) = \sum_{i=1}^{k} Z_+(G_i) - 2k + 2$.*

(ii) *If only one of $G_1, \ldots, G_k$ is terminal-linked and at least another one is terminal-semi-linked, then G is terminal-semi-linked and $Z_+(G) = \sum_{i=1}^{k} Z_+(G_i) - 2k + 3$.*

(iii) *If only one of $G_1, \ldots, G_k$ is terminal-linked and all others are terminal-unlinked, then G is terminal-linked and $Z_+(G) = \sum_{i=1}^{k} Z_+(G_i) - 2k + 2$.*

(iv) *If k_1 ($k_1 \geq 2$) of $G_1, \ldots, G_k$ are terminal-linked, then G is terminal-semi-linked and $Z_+(G) = \sum_{i=1}^{k} Z_+(G_i) - k_1 - 2(k - k_1) + 2$.*

(v) *If none of $G_1, \ldots, G_k$ is terminal-linked and at least one of them is terminal-semi-linked, then G is terminal-semi-linked and $Z_+(G) = \sum_{i=1}^{k} Z_+(G_i) - 2k + 2$.*

Theorem 3. *Let G_1 and G_2 be two series-parallel graphs with terminals $\{s_1, t_1\}$ and $\{s_2, t_2\}$, respectively, and let $G = G_1 \bigoplus G_2$ with terminals $s(= s_1)$ and $t(= t_2)$. Then $Z_+(G) = Z_+(G_1) + Z_+(G_2) - 1$; and furthermore,*

(i) *if both G_1 and G_2 are terminal-linked, then G is terminal-linked;*

(ii) *if one of G_1 and G_2 is terminal-linked and the other is terminal-semi-linked, then G is terminal-semi-linked;*

(iii) *if both G_1 and G_2 are terminal-semi-linked, then G is terminal-unlinked;*

(iv) *if at least one of G_1 and G_2 is terminal-unlinked, then G is terminal-unlinked.*

5   Algorithm

In this section, we give a linear time algorithm for finding PSZF numbers of series-parallel graphs.

A series-parallel graph can be decomposed into a set of primitive graphs (e.g., single edges) by series compositions and parallel compositions. The decomposition structure of a series-parallel graph G can be represented by a decomposition tree, denoted as $D_T(G)$, which is a rooted binary tree having the following properties:

1. Each leaf corresponds to an edge of G.
2. Each internal node is called an S-node or a P-node.
3. Each S-node corresponds to a series-parallel graph that is obtained from the two series-parallel graphs corresponding to its two children by a series composition.
4. Each P-node corresponds to a series-parallel graph that is obtained from the two series-parallel graphs corresponding to its two children by a parallel composition.

In our algorithm, we first construct a decomposition tree $D_T(G)$. Let r be the root of $D_T(G)$. Each node of $D_T(G)$ except r is a descendant of r. For any two nodes v_1 and v_2 in $D_T(G)$, if there is a directed path from r to v_2 containing

v_1, then we say that v_2 is a *descendant* of v_1; specifically, if v_2 is adjacent to v_1, we say v_2 is a *child* of v_1. For a node v of $D_T(G)$, let $D_T[v]$ be the subtree of $D_T(G)$ induced by v and all its descendants. We will use G^v to denote the series-parallel graph whose decomposition tree is $D_T[v]$. So G^v is a subgraph of G. Note that $D_T[r] = D_T(G)$ and $G^r = G$.

After we construct a decomposition tree $D_T(G)$, we assign labels to all nodes of $D_T(G)$. The *label* of a node v records the essential structural information of the subgraph G^v. The label of a node v consists of two components (type(G^v), $Z_+(G^v)$)), where the first component indicates the type of G^v, which is one of the three types: terminal-unlinked, terminal-semi-linked or terminal-linked.

Algorithm. ZPLUS-SP

Input: a series-parallel graph G.

Output: $Z_+(G)$.

1. Construct a decomposition tree $D_T(G)$.
2. Use the post-order traversal to visit each node v of $D_T(G)$ and label it in one of the following three cases:
   (a) If v is a leaf of $D_T(G)$, label it as (terminal-linked, 1).
   (b) If v is a P-node, then compute the label of v using theorem 2.
   (c) If v is an S-node, then compute the label of v using theorem 3.
3. Return the label of the root of $D_T(G)$.

Using mathematical induction, we can prove the correctness of the above algorithm, that is, the second component of the label of the root is equal to $Z_+(G)$. We now consider the running time.

Theorem 4. *For any series-parallel graph G with n vertices and m edges, $Z_+(G)$ can be computed in $O(n + m)$ time.*

Note that Ekstrand et al. [10] showed that $Z_+(G) = \text{tc}(G) = M_+(G)$ for any graph that is a partial 2-tree. Since every series-parallel graph is a partial 2-tree, we have the following.

Corollary 2. *For any series-parallel graph G with n vertices and m edges, $\text{tc}(G)$ and $M_+(G)$ can be computed in $O(n + m)$ time.*

6   Lower Bounds on $Z_+(G)$

In [9], Ekstrand et al. showed that for any graph G, $Z_+(G) \geq \text{tc}(G)$. In this section, we will introduce the constrained tree cover number which can be applied to improve lower bounds for $Z_+(G)$. We will use the set of all cut vertices to determine the PSZF number of cacti, simply-linked "clique" graphs and generalized series-parallel graphs.

Similar to Corollary 3.9 in [20], we can prove the following lemma that describes a structural property between two forcing chains.

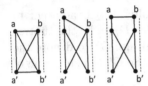

Fig. 1. The forbidden graphs induced by the vertices of the two forcing chains $a' \to \cdots \to a$ and $b' \to \cdots \to b$, where any edge marked by a dashed line can be replaced by a path of length at least one

Lemma 2. *Let G be a graph and F be a PSZF tree cover of G. For any two forcing chains $a' \to \cdots \to a$ and $b' \to \cdots \to b$ in F, the subgraph of G induced by the vertices of the two chains cannot contain any graph illustrated in Figure 1 as a subgraph.*

We now introduce the constrained tree cover which is an extension of the tree cover.

Definition 2. (Constrained tree cover.) *A constrained tree cover* of a graph G is a set of trees $\{T_1, \ldots, T_k\}$ in G satisfying the following conditions: (1) each T_i, $1 \leq i \leq k$, is an induced subgraph of G, (2) $\cup_{i=1}^{k} V(T_i) = V(G)$, (3) $V(T_i) \cap V(T_j) = \emptyset$, for any $1 \leq i < j \leq k$, and (4) for any two paths P_1 in T_i and P_2 in T_j, $1 \leq i < j \leq k$, the induced subgraph $G[V(P_1) \cup V(P_2)]$ is K_4-minor-free. A *minimum constrained tree cover* of G is a constrained tree cover of G that has the smallest possible size. The *constrained tree cover number* of G, denoted by $\mathrm{ctc}(G)$, is the size of a minimum constrained tree cover of G.

Lemma 3. *For any graph G, $Z_+(G) \geq \mathrm{ctc}(G) \geq \mathrm{tc}(G)$.*

Although we only need formulas for $\mathrm{ctc}(G)$ in the following theorems and corollaries, we will also give formulas for $\mathrm{tc}(G)$ because the minimum tree cover is interesting in its own right. Note that our formula for $\mathrm{tc}(G)$ improves Proposition 2.5 in [9].

Theorem 5. *Let G be a connected graph, X be the set of all cut vertices of G, and $G_1, \ldots, G_k$ be all biconnected components of G. Then $\mathrm{ctc}(G) = \sum_{i=1}^{k} \mathrm{ctc}(G_i) - \sum_{v \in X} \delta(v) + |X|$, and $\mathrm{tc}(G) = \sum_{i=1}^{k} \mathrm{tc}(G_i) - \sum_{v \in X} \delta(v) + |X|$, where $\delta(v) = |\{G_i : G_i \text{ contains } v, 1 \leq i \leq k\}|$.*

Corollary 3. *Let G be a connected graph, and $G_i = (V_i, E_i)$, $1 \leq i \leq k$, be all biconnected components of G. Suppose that F^* is a minimum (constrained) tree cover of G and $F^*[V_i]$ is a (constrained) tree cover of G_i induced by V_i. Then each $F^*[V_i]$, $1 \leq i \leq k$, is a minimum (constrained) tree cover of G_i.*

While Theorem 5 characterizes $\mathrm{tc}(G)$ and $\mathrm{ctc}(G)$ using the number of cut vertices, the next theorem characterizes $\mathrm{tc}(G)$ and $\mathrm{ctc}(G)$ using the number of biconnected components.

Theorem 6. *Let G be a connected graph, and $G_1, \ldots, G_k$ be all biconnected components of G. Then $\mathrm{ctc}(G) = \sum_{i=1}^{k} \mathrm{ctc}(G_i) - k + 1$, and $\mathrm{tc}(G) = \sum_{i=1}^{k} \mathrm{tc}(G_i) - k + 1$.*

Ekstrand et al. [10] showed that $Z_+(G) = \text{tc}(G) = M_+(G)$ for any graph that is a partial 2-tree. Since each cactus is a partial 2-tree, we have $Z_+(G) = \text{tc}(G) = M_+(G)$ for any cactus G. The following corollary enhances this result by establishing a relation between $Z_+(G)$ and the number of simple cycles in G.

Corollary 4. *For a cactus G with ℓ simple cycles, $Z_+(G) = \text{ctc}(G) = \text{tc}(G) = M_+(G) = \ell + 1$.*

For a complete graph K_n, $n \geq 4$, it is easy to see that $Z_+(K_n) = \text{ctc}(K_n) = n - 1$ while $\text{tc}(K_n) = \lceil n/2 \rceil$. Since the gap between $Z_+(K_n)$ and $\text{tc}(K_n)$ can be arbitrarily large, we know if a graph contains large cliques, then $\text{ctc}(G)$ usually gives us a better lower bound than $\text{tc}(G)$. This is the reason that we can use $\text{ctc}(G)$ as a lower bound for $Z_+(G)$ to show Corollaries 5 and 8, but it would be hard to prove them if we use $\text{tc}(K_n)$ as a lower bound.

Corollary 5. *Let $G = (V, E)$ be a connected graph and $G_1, \ldots, G_k$ be all biconnected components of G. If each G_i, $1 \leq i \leq k$, is a clique, then $Z_+(G) = \text{ctc}(G) = M_+(G) = |V| - k$.*

From Theorem 6, we have the following result for generalized series-parallel graphs.

Corollary 6. *Let G be a generalized series-parallel graph and $G_1, \ldots, G_k$ be all biconnected components of G. Then $Z_+(G) = \text{ctc}(G) = \text{tc}(G) = \sum_{i=1}^{k} Z_+(G_i) - k + 1$.*

Corollary 7. *For any generalized series-parallel graph G with n vertices and m edges, $Z_+(G)$ can be computed in $O(n + m)$ time.*

In order to use the constrained tree cover as lower bounds on $Z_+(G)$ for more graphs, we weaken the condition of Theorem 5 to obtain a lower bound for $\text{ctc}(G)$ and $\text{tc}(G)$.

Lemma 4. *Let $G = (V, E)$ be a connected graph and $\mathcal{G} = \{G_1, \ldots, G_k\}$ be a set of connected subgraphs of G such that these subgraphs include all vertices and edges of G and any two distinct subgraphs in $\mathcal{G}$ can share at most one vertex. Let $\chi(V) = \{v \in V : v$ is contained in at least two subgraphs of $\mathcal{G}\}$. Then $\text{ctc}(G) \geq \sum_{i=1}^{k} \text{ctc}(G_i) - \sum_{v \in \chi(V)} \delta(v) + |\chi(V)|$, and $\text{tc}(G) \geq \sum_{i=1}^{k} \text{tc}(G_i) - \sum_{v \in \chi(V)} \delta(v) + |\chi(V)|$, where $\delta(v) = |\{G_i : G_i$ contains $v, 1 \leq i \leq k\}|$.*

Corollary 8. *Let $G = (V, E)$ be a connected graph and $G_i = (V_i, E_i)$, $1 \leq i \leq k$, be maximal cliques of G satisfying the following conditions: (1) $\cup_{i=1}^{k} V_i = V$ and $\cup_{i=1}^{k} E_i = E$, (2) for any $1 \leq i < j \leq k$, $|V_i \cap V_j| \leq 1$, and (3) each G_i, $1 \leq i \leq k$, contains a vertex v_i that is not in any G_j, $j \neq i$. Then $Z_+(G) = \text{ctc}(G) = M_+(G) = |V| - k$.*

Corollary 9. *Let* $G = (V, E)$ *be a connected graph and* $\mathcal{G} = \{G_1, \ldots, G_k\}$ *be a set of series-parallel graphs of* G *satisfying the following conditions: (1)* $\cup_{i=1}^{k} V(G_i) = V$ *and* $\cup_{i=1}^{k} E(G_i) = E$, *(2) for any* G_i *and* G_j, $1 \leq i < j \leq k$, *they can share at most one vertex and this vertex must be their common terminal, and (3) each* G_i *(1 $\leq i \leq k$) is terminal-unlinked or terminal-semi-linked. Let* $Y = \{v \in V : v \text{ is a terminal of some } G_i\}$. *Then* $Z_+(G) = \mathrm{ctc}(G) = \mathrm{tc}(G) = \sum_{i=1}^{k} Z_+(G_i) - 2k + |Y|$.

Corollary 10. *Let* H *be a connected graph. Let* G_H *be a graph obtained from* H *by replacing every edge* e *of* H *by a terminal-semi-linked series-parallel graph* G_e *with* $Z_+(G_e) = 2$ *such that the two endpoints of* e *correspond to the two terminals of* G_e. *Then* $Z_+(G_H) = \mathrm{ctc}(G_H) = \mathrm{tc}(G_H) = |V(H)|$.

References

1. AIM Minimum Rank-Special Graphs Work Group: Zero forcing sets and the minimum rank of graphs. Linear Algebra Appl. **428**(7), 1628–1648 (2008)
2. Barioli, F., Barrett, W., Fallat, S., Hall, H.T., Hogben, L., Shader, B., van den Driessche, P., van der Holst, H.: Parameters related to tree-width, zero forcing, and maximum nullity of a graph. J. Graph Theory **72**, 146–177 (2013)
3. Barioli, F., Barrett, W., Fallat, S., Hall, H.T., Hogben, L., Shader, B., van den Driessche, P., van der Holst, H.: Zero forcing parameters and minimum rank problems. Linear Algebra Appl. **433**(2), 401–411 (2010)
4. Barioli, F., Fallat, S., Mitchell, L., Narayan, S.: Minimum semidefinite rank of outerplanar graphs and the tree cover number Electron. J. Linear Algebra **22**, 10–21 (2011)
5. Bienstock, D., Seymour, P.: Monotonicity in graph searching. J. Algorithms **12**, 239–245 (1991)
6. Booth, M., Hackney, P., Harris, B., Johnson, C.R., Lay, M., Mitchell, L.H., Narayan, S.K., Pascoe, A., Steinmetz, K., Sutton, B.D., Wang, W.: On the minimum rank among positive semidefinite matrices with a given graph. SIAM J. Matrix Anal. Appl. **30**, 731–740 (2008)
7. Burgarth, D., Giovannetti, V.: Full control by locally induced relaxation. Phys. Rev. Lett. **99**(10), 100–501 (2007)
8. Dyer, D., Yang, B., Yaşar, Ö.: On the fast searching problem. In: Fleischer, R., Xu, J. (eds.) AAIM 2008. LNCS, vol. 5034, pp. 143–154. Springer, Heidelberg (2008)
9. Ekstrand, J., Erickson, C., Hall, H.T., Hay, D., Hogben, L., Johnson, R., Kingsley, N., Osborne, S., Peters, T., Roat, J., Ross, A., Row, D., Warnberg, N., Young, M.: Positive semidefinite zero forcing. Linear Algebra Appl. **439**, 1862–1874 (2013)
10. Ekstrand, J., Erickson, C., Hay, D., Hogben, L., Roat, J.: Note on positive semidefinite maximum nullity and positive semidefinite zero forcing number of partial 2-trees. Electron. J. Linear Algebra **23**, 79–87 (2012)
11. Fallat, S., Meagher, K., Yang, B.: On the complexity of the positive semidefinite zero forcing number. Linear Algebra Appl. (accepted March 2015), doi:10.1016/j.laa.2015.03.011
12. Hopcroft, J., Tarjan, R.: Efficient algorithms for graph manipulation. Communications of the ACM **16**(6), 372–378 (1973)

13. Kirousis, L., Papadimitriou, C.: Searching and pebbling. Theoret. Comput. Sci. **47**, 205–218 (1986)
14. Kratochvíl, J., Tuza, Z.: On the complexity of bicoloring clique hypergraphs of graphs. J. Algorithms **45**, 40–54 (2002)
15. Megiddo, N., Hakimi, S., Garey, M., Johnson, D., Papadimitriou, C.: The complexity of searching a graph. J. ACM **35**, 18–44 (1988)
16. Robertson, N., Seymour, P.: Graph minors I: Excluding a forest. J. of Combinatorial Theory, Series B **35**, 39–61 (1983)
17. Robertson, N., Seymour, N.: Graph minors III: Planar tree-width. J. of Combinatorial Theory, Series B **36**, 49–64 (1984)
18. Valdes, J., Tarjan, R.E., Lawler, E.L.: The Recognition of Series Parallel Digraphs. SIAM J. Comput. **11**, 289–313 (1982), and Proc. 11th ACM Symp. Theory of Computing, vol. 1-12 (1979)
19. West, D.B.: Introduction to Graph Theory, 2nd edn. Prentice Hall (2001)
20. Yang, B.: Fast-mixed searching and related problems on graphs. Theoret. Comput. Sci. **507**(7), 100–113 (2013)
21. Yang, B.: Strong-mixed searching and pathwidth. J. of Combinatorial Optimization **13**, 47–59 (2007)

Communication and Dynamic Networks

Bernard Chazelle

Department of Computer Science,
Princeton University,
Princeton, NJ 08540, USA

Abstract. This talk will discuss an algorithmic approach to the study of dynamical systems on time-varying graphs. Diffusive influence systems have been used to model all sorts of dynamics, from political polarization to firefly, power grid, and heart pacemaker cell synchronization. We review a suite of new techniques for analyzing such systems, including the s-energy, network sequence parsing, multiagent renormalization, tensor lifts, and message-passing methods for resolving entropy-dissipation tension around critical points. Our analytical framework allows us to formulate new criteria for ensuring the asymptotic periodicity of diffusive influence systems. The main novelty of our approach is to make algorithms the central ingredient in the investigation of multiagent dynamics.

This work was supported in part by NSF grants CCF-0963825 and CCF-1420112.

F. Dehne et al. (Eds.): WADS 2015, LNCS 9214, p. 641, 2015.
DOI: 10.1007/978-3-319-21840-3

Inferring People's Social Behavior by Exploiting Their Spatiotemporal Location Data

Cyrus Shahabi

University of Southern California,
Department of Computer Science,
Los Angeles, CA 90089-0781
Shahabi@usc.edu

For decades, social scientists have been studying people's social behaviors by utilizing sparse datasets obtained by observations and surveys. These studies received a major boost in the past decade due to the availability of web data (e.g., social networks, blogs and review web sites). However, due to the nature of the utilized dataset, these studies were confined to behaviors that were observed mostly in the virtual world. Differing from all the earlier work, here, we aim to study social behaviors by observing people's behaviors in the real world. This is now possible due to the availability of large high-resolution spatiotemporal location data collected by GPS-enabled mobile devices through mobile apps (Google's Map/Navigation/Search/Chrome, Facebook, Foursquare, WhatsApp, Twitter) or through online services, such as geo-tagged contents (tweets from Twitter, pictures from Instagram, Flickr or Google+ Photo), etc.

In particular, we focus on *inferring* and quantifying two specific social measures: 1) *pairwise* strength -- the strength of social connections between a pair of users, and 2) *pairwise* influence - the amount of influence that an individual exerts on another, by utilizing the available high-fidelity location data representing people's movements.

To compute pairwise strength, we study an Entropy-Based model (EBM) [1] to infer social strength (the closeness of friendships) between two people based on the knowledge that two people were at the same places and at the same time, called *co-occurrences*. The model considers the impact on social strength by different factors, including the frequency and the diversity of co-occurrences, the popularity of locations and coincidences.

To compute pairwise influence, we try to identify when an individual visits a location (e.g., a restaurant) due to the influence of another individual who visited that same location in the past. We define this behavior as *followship*. Hence, followship is an indication of pair-wise influence between people in the real world. Subsequently, we introduce and study *spatial influence* - a concept of inferring pairwise influence from spatiotemporal data by quantifying the followship influence that an individual exerts on another in the real world. Quantifying spatial influence has many challenges. First, we need to distinguish actual followship from other successive visits that are *not* due to influence, which we call *coincidences*. Second, even if we can identify

© Springer International Publishing Switzerland 2015
F. Dehne et al. (Eds.): WADS 2015, LNCS 9214, pp. 642–643, 2015.
DOI: 10.1007/978-3-319-21840-3

successive visits as followship, how should we quantify followship? What are the factors that impact it? Should it be a function of location, the participants and/or the time delay (the time interval) between visits? Third, how should we measure the individual contribution of each factor and then combine them in a meaningful manner? Among the above-mentioned issues, the impact of locations and the issues related to coincidences are critical to spatial influence particularly due to the inclusion of users' locations.

Reference

1. Pham, H., Shahabi, C., Liu, Y.: EBM: an entropy-based model to infer social strength from spatiotemporal data. In: ACM SIGMOD, pp. 265–276. ACM (2013)

Author Index

Printed in the United States
By Bookmasters

Printed in the United States
By Bookmasters